Advances in Functional Rubber and Elastomer Composites II

Advances in Functional Rubber and Elastomer Composites II

Guest Editors

Md Najib Alam
Vineet Kumar

MDPI

Basel • Beijing • Wuhan • Barcelona • Belgrade • Novi Sad • Cluj • Manchester

Guest Editors

Md Najib Alam
School of Mechanical Engineering
Yeungnam University
Gyeongsan
Republic of Korea

Vineet Kumar
School of Mechanical Engineering
Yeungnam University
Gyeongbuk
Republic of Korea

Editorial Office
MDPI AG
Grosspeteranlage 5
4052 Basel, Switzerland

This is a reprint of the Special Issue, published open access by the journal *Polymers* (ISSN 2073-4360), freely accessible at: https://www.mdpi.com/journal/polymers/special_issues/WV2LEFJKEZ.

For citation purposes, cite each article independently as indicated on the article page online and as indicated below:

Lastname, A.A.; Lastname, B.B. Article Title. *Journal Name* **Year**, *Volume Number*, Page Range.

ISBN 978-3-7258-5327-4 (Hbk)
ISBN 978-3-7258-5328-1 (PDF)
https://doi.org/10.3390/books978-3-7258-5328-1

Contents

About the Editors

Md Najib Alam

Md Najib Alam obtained his B.Sc. and M.Sc. degrees from the University of Kalyani, India, in 2006 and 2008, respectively. He completed his Ph.D. at the same university in 2014, where he addressed numerous challenges related to rubber vulcanization chemistry. Following his Ph.D., Dr. Alam pursued postdoctoral research at Chulalongkorn University, Thailand, from 2015 to 2019. From 2019 to 2022, he served as a research fellow at the School of Mechanical Engineering, Yeungnam University, Republic of Korea. In 2022, he transitioned to the role of assistant professor at the same institution. Throughout his research career, Dr. Alam has published over 50 original research articles in internationally recognized peer reviewed journals, including the *Journal of Industrial and Engineering Chemistry*, *Composites Science and Technology*, and *Composites Part B: Engineering*. Currently, he is serving as a guest editor for various Special Issues in MDPI *Polymers*. His research focuses on rubber vulcanization, novel rubber composites, stretchable sensors, and smart materials.

Vineet Kumar

Vineet Kumar has been an Assistant Professor since 2016 at the Department of Mechanical Engineering, Yeungnam University, Gyeongsan-si, Republic of Korea. He completed his Ph.D. in the Department of Materials Science at the University of Milan-Bicocca, Milan, Italy, in 2014. He finished his Master of Science in Environment Management from the Forest Research Institute, Dehradun, India, in 2008. He has co-authored more than 75 peer-reviewed articles in reputed SCI journals. He has also served as a Topic Editor for *Polymers*—MDPI since 2021. Moreover, he edited as a Guest Editor for more than six Special Issues in *Polymers*—MDPI and *Frontiers in Materials*.

Preface

Since the invention of the crosslinking process, or vulcanization, rubber materials have found widespread use in the fashion and automobile industries. More recently, rubber composites have evolved far beyond their traditional applications, finding roles in smart and intelligent systems. To support these advanced applications, new formulations, fabrication methods, characterization techniques, and conceptual proposals are being continuously developed. However, many of these advancements remain unfamiliar to young rubber researchers due to the lack of a comprehensive collection of subject-specific research papers. This reprint aims to bridge that gap by offering a curated collection of papers focused on the advanced applications of rubber composites.

With ongoing progress in materials science, rubber composites are now being used in cutting-edge fields such as medical applications with antibacterial properties, energy harvesting, smart structures, and other intelligent technologies. Nevertheless, the rising demand for rubber products presents both economic and environmental challenges. Extensive research has therefore been carried out to address these issues and offer sustainable solutions.

This reprint provides readers with valuable insights into the diverse techniques used to fabricate rubber composites and their emerging applications. It presents up-to-date concepts and a state-of-the-art overview of the interactions between rubber and other functional components. Furthermore, the reprint explores potential alternative fillers that could offer both environmental and economic benefits. Each chapter presents a thorough background of the innovations discussed, followed by detailed results and analyses, concluding with key findings.

We are confident that readers will gain a wealth of innovative ideas and a deeper understanding of the latest developments and trends in rubber composite research.

Md Najib Alam and Vineet Kumar
Guest Editors

polymers

Editorial

Advances in Functional Rubber and Elastomer Composites II

Vineet Kumar and Md Najib Alam *

School of Mechanical Engineering, Yeungnam University, 280, Daehak-ro, Gyeongsan 38541, Republic of Korea; vineetfri@gmail.com
* Correspondence: mdnajib.alam3@gmail.com

1. Introduction

Recently, rubber and elastomer materials have expanded their applications from traditional mechanical uses to advanced mechanical, electrical, and sensor applications [1]. To achieve these goals, functionalization of both the rubber matrix and the fillers is often necessary. For example, rubber is a good dielectric material, and incorporating electrically conductive fillers significantly enhances its dielectric properties, making it useful for nanogenerators, actuators, and sensor applications [2–5]. Similarly, some advanced elastomer composites, such as magnetorheological elastomers, have smart mechanical applications because their properties can be tuned by an external magnetic field [6,7]. However, in many rubber composite applications, the fillers used are sometimes synthetic and non-renewable [8–10]. From environmental and health perspectives, rubber scientists are striving to address these issues by developing eco-friendly alternatives [8–10].

The previous related Special Issue, "Advances in Functional Rubber and Elastomer Composites," discussed developments in rubber and polymer composites with enhanced mechanical, thermal, dielectric, and functional properties [11–28]. This Special Issue focuses on further advancements in mechanical, magnetomechanical, electrical, energy harvesting, and sensing properties, critically examining the roles of rubber and functional filler materials in these composites.

2. Overview of Published Articles

Dai et al. [29] proposed a novel alkaline protease/calcium chloride coagulation method that enhances natural rubber's mechanical properties while reducing environmental impact. Compared to acid coagulation, this new method improves strength, abrasion resistance, and thermal stability, while minimizing soil acidification—enabling eco-friendly, high-performance rubber without chemical modification. Torres et al. [30] created sustainable composites with enhanced mechanical and thermal properties by combining bovine leather waste with natural rubber (TSR-20). The resulting material shows strong rubber–leather interaction, high density, and PVC-like hardness, offering eco-friendly applications in footwear and fashion, while promoting waste valorization and resource recycling. Alam et al. [31] compared various properties of CNT-reinforced natural rubber (NR) and nitrile butadiene rubber (NBR) composites. NR shows greater mechanical improvement due to stronger CNT bonding, while NBR exhibits superior piezoresistive sensitivity and electromechanical actuation. NR is suitable for dynamic applications; NBR excels in stretchable sensors and actuators due to its polar functionality. Luengchavanon et al. [32] synthesized Ag_{mgt}/Zn_{mgt} nanocomposites using mangosteen peel extract, which were applied to rubber gloves and showed strong antibacterial activity, including against MRSA, with MICs of 40–320 μg/mL. Despite effective bacterial inhibition in suspensions, glove coatings required at least 30 min of contact time for antimicrobial efficacy. Yu et al. [33] studied three 5%

 https://doi.org/10.3390/polym17162247

fiber-reinforced TPU composites for water-lubricated bearings. Basalt fiber showed the best tribological performance, reducing friction by 70.57% and wear by 98.69% compared to pure TPU. UHMWPE and bamboo fibers performed well under high loads, enhancing surface smoothness and wear resistance. Ding et al. [34] examined how different coagulation methods affect the hyperelastic behavior of natural rubber using tensile tests and an improved Yeoh model. Results reveal distinct mechanical properties across methods, highlighting the importance of coagulation in rubber performance and offering insights for optimizing natural rubber processing. Li et al. [35] presented a simplified GRU-attention-based constitutive model for high-damping rubber (HDR), implemented as HDRGA material in OpenSees. Validation using structural models shows the HDRGA material accurately simulates HDR's seismic behavior, offering a reliable tool for earthquake-resistant design in civil engineering. Beknazarov et al. [36] evaluated shungite ore and its concentrate as partial replacements for carbon black in nitrile butadiene rubber. Flotation increased carbon content and improved tensile strength, reduced viscosity, and extended curing times without affecting oil resistance, demonstrating shungite's potential as an alternative rubber filler. Yangthong et al. [37] studied epoxidized natural rubber (ENR) composites filled with alumina, silica, and hybrid fillers, showing enhanced thermal and mechanical properties. The ENR/hybrid (25/25 phr) filler achieved superior conductivity (2.23 W/mK) and strength. This cost-effective alternative to silicone rubber is promising for electronics, batteries, and energy-related applications. Ge et al. [38] studied a self-oscillating liquid crystal elastomer (LCE) fiber-slide system driven by a self-flickering light source under constant voltage. The design enables synchronized energy input, enhancing efficiency and simplicity. It offers potential applications in dynamic circuits, sensing systems, and optical devices due to its autonomous and periodic motion. Inphonlek et al. [39] investigated enhanced deproteinized natural rubber by grafting acrylic acid and acrylamide, enabling coordination with silver and titanium dioxide. The resulting MDPNR/Ag-TiO_2 composites showed improved dielectric properties, thermal stability, stiffness, and antibacterial activity, with well-dispersed silver and increased dielectric constant compared to unmodified rubber. Akkenzheyeva et al. [40] fabricated the rubber–polymer composite ELTC, derived from devulcanized tires and plastics, which was used to modify asphalt binders. ELTC improved binder homogeneity, thermal stability, softening temperature, and resistance to summer plastic deformation. Its even distribution and good component compatibility enhanced the performance of the modified asphalt binders. Wang et al. [41] examined how free and bonded proteins affect natural rubber's vulcanized network and mechanical properties. Reduced protein content decreased entanglement, crosslink density, and strength. Bonded proteins influenced vulcanization, while free proteins affected crosslink density. Both protein types contribute to stronger, more robust vulcanized rubber. Glavan et al. [42] investigated the dynamic strain response of soft magnetoactive elastomers (MAEs) under alternating magnetic fields. MAEs exhibit significantly higher piezomagnetic strain coefficients than conventional magnetostrictive composites, with notable phase lag. The results highlight nonlinear viscoelasticity as a key factor influencing MAE actuator performance in soft robotics. Bao et al. [43] synthesized siloxane polyurea copolymer elastomers using low and high Mn monomers to study structure–property relationships. Film analyses revealed that tensile properties, transparency, Tg, and elasticity depend on composition. Hydrogen bonding strength and recovery, assessed via IR and tensile tests, correlate with hard segment Mn and influence macroscopic performance. Alam et al. [44] developed cost-effective, multifunctional styrene–butadiene rubber composites using iron, aluminum, and hybrid metal fillers. Iron enhances electrical and magnetic properties but reduces crosslink density, while aluminum improves modulus. At 20 vol%, iron-filled composites show strong magnetic anisotropy and conductivity, suggesting potential for smart applications with future reinforcement improvements. Liu et al. [45] synthesized three reactive POSS modifiers to toughen epoxy resins without compromising key properties. The resulting hybrid resins showed good compatibility, low viscosity, low curing temperature, and significantly improved toughness. Notably, OGCPS-EP-0.6-C achieved a 58.75% increase in KIC with minimal changes in glass transition temperature and flexural strength. Abdollahi et al. [46] developed a 3D finite element model using a microscale representative volume element (RVE) to predict the shear behavior of magnetorheological elastomers (MREs) under magnetic fields. Incorporating different silicone matrices and carbonyl iron particles, the model accurately captures field-dependent responses, offering insights for adaptive vibration absorber design. Magaletti et al. [47] replaced silica with SP-functionalized carbon black (CB/SP) in elastomer composites for tire applications. Using TESPT as a coupling agent, CB/SP-based composites show lower hysteresis, reduced Payne effect, and higher dynamic rigidity. While slightly lower in ultimate properties, they offer environmental benefits and reduced weight compared to silica-based systems. Alam et al. [48] showed that adding 1–2 phr MgO to ZnO-based sulfur vulcanization systems significantly accelerates curing kinetics. MgO reduces vulcanization time and allows curing at lower temperatures (down to 100 °C), maintaining or improving mechanical properties. This approach offers a more energy-efficient, cost-effective, and eco-friendly rubber vulcanization process. Leyva-Porras et al. [49] used swelling and mechanical tests with the Flory–Rehner equation to estimate the Flory–Huggins interaction parameter ($\chi 12^*$) in crosslinked elastomer blends of BR, SBR, and NBR. Results indicate preferential crosslinking within phases, blend immiscibility, and slightly higher $\chi 12^*$ values than literature reports. Kitsawat et al. [50] developed green natural rubber composites reinforced with up to 60 phr synthetic graphite platelets using alginate, showing significantly enhanced mechanical strength, chemical resistance, hydrophilicity, and electrical conductivity. The composites maintained biodegradability (21–30% after 90 days) and demonstrated potential for electronic substrate applications without agglomeration issues. Kodal et al. [51] demonstrated that adding multi-walled carbon nanotubes (MWCNTs) to carbon black-reinforced NR/SBR tire compounds improves filler dispersion, vulcanization, mechanical properties, and thermal conductivity. Enhanced heat dissipation reduces heat build-up, with a percolation model explaining the synergistic effects of hybrid carbon fillers in the composites. Rebane et al. [52] compared hydroxyl-bearing blowing agents in silica- and mica-filled silicone foams, achieving low-density, uniform porosity, and enhanced mechanical properties. Adjusting diol chain length and combining agents like glycerol with water reduced foam density by up to 30% while improving tensile strength, with fast curing and suitable viscosity for industrial processing. Zhao et al. [53], in a review, explored wet mixing technology for dispersing fillers like graphene, carbon nanotubes, silica, and carbon black in natural rubber (NR) composites. The review highlights improved filler dispersion and reinforcement mechanisms, discusses preparation techniques, and identifies current limitations and challenges to guide future advancements in wet mixing methods.

3. Summary and Future Outlook

This collection of recent research highlights major advances in sustainable and high-performance rubber and elastomer composites. A novel alkaline protease and calcium chloride coagulation method has been developed to improve natural rubber's mechanical properties while reducing environmental impact. Sustainable composites using bovine leather waste in natural rubber have shown enhanced mechanical and thermal properties, making them suitable for eco-friendly applications. Comparisons between CNT-reinforced natural rubber and nitrile butadiene rubber reveal that natural rubber offers better mechanical reinforcement, while nitrile rubber demonstrates superior electromechanical performance for sensing and actuation.

Antibacterial nanocomposites synthesized using mangosteen peel extract have shown promising applications in rubber gloves. Fiber-reinforced thermoplastic polyurethane composites, especially those with basalt fiber, exhibit significant improvements in friction and wear resistance for water-lubricated bearings. Studies on coagulation methods indicate their significant impact on the hyperelastic behavior of natural rubber. A new GRU-attention-based model accurately simulates the seismic response of high-damping rubber, aiding structural design.

Further innovations include using shungite ore as a filler alternative, hybrid fillers in epoxidized rubber for improved thermal conductivity, and self-oscillating liquid crystal elastomer systems for energy-efficient motion. Modified deproteinized rubber with metal oxides enhances dielectric and antibacterial properties. Recycled rubber–plastic composites improve asphalt performance. Protein content in rubber has been shown to influence crosslinking and mechanical strength.

Magnetoactive elastomers exhibit enhanced piezomagnetic responses under magnetic fields. Siloxane polyurea copolymers demonstrate structure–property relationships depending on hydrogen bonding. Hybrid metal fillers in styrene–butadiene rubber enable multifunctional performance, and POSS modifiers toughen epoxy resins without sacrificing key characteristics. Additional advancements include predictive models for magnetorheological elastomers, functional carbon black for tire applications, energy-efficient vulcanization using MgO, and biodegradable natural rubber composites with synthetic graphite. Wet mixing technology continues to evolve, offering improved filler dispersion and reinforcing effects in rubber composites.

Despite significant advancements, challenges still remain in achieving the required mechanical, electrical, and thermal properties for rubber composites to meet industrial application standards. Therefore, addressing these issues will be essential in future research to fully realize their practical potential.

Author Contributions: Conceptualization: V.K. and M.N.A.; Validation: V.K. and M.N.A.; writing—original draft preparation: M.N.A.; writing—review and editing: V.K. and M.N.A.; visualization: M.N.A. and V.K.; Supervision: M.N.A.; project administration: M.N.A. All authors have read and agreed to the published version of the manuscript.

Acknowledgments: The authors thank all the contributors and reviewers for their valuable contributions and support from section editors of this Special Issue.

Conflicts of Interest: The authors declare no conflicts of interest.

References

1. Alarifi, I.M. A Comprehensive Review on Advancements of Elastomers for Engineering Applications. *Adv. Ind. Eng. Polym. Res.* **2023**, *6*, 451–464. [CrossRef]
2. Gurjar, K.V.S.; Sadangi, A.S.; Kumar, A.; Ahmad, D.; Patra, K.; Collins, I.; Hossain, M.; Ajaj, R.M.; Zweiri, Y. Dielectric Elastomer Generators: Recent Advances in Materials, Electronic Circuits, and Prototype Developments. *Adv. Energy Sustain. Res.* **2025**, *6*, 2400221. [CrossRef]
3. Thongthapthai, W.; Harnchana, V.; Chanthad, C.; Amornkitbamrung, V.; Chindaprasirt, P. The Fabrication of Calcium Silicate-Natural Rubber Composite for Mechanical Energy Harvesting. *Surf. Interfaces* **2021**, *25*, 101180. [CrossRef]
4. Ghevondyan, M.; Davtyan, M.; Aghayan, M. Dielectric Elastomer Actuators: Medical Applications Review. *Discov. Mater.* **2025**, *5*, 43. [CrossRef]
5. Qu, M.; Zhu, M.; Liu, Q.; Li, J.; Gao, Y.; Zhang, J.; Cao, M.; Wei, X.; He, J. Building a Multi-Performance Wearable Rubber-Based Strain Sensor: For Human Motion Capture, Optical Heating and Underwater Sensing. *Carbon* **2025**, *238*, 120274. [CrossRef]
6. Bastola, A.K.; Hossain, M. A Review on Magneto-Mechanical Characterizations of Magnetorheological Elastomers. *Compos. Part B Eng.* **2020**, *200*, 108348. [CrossRef]

7. Kang, S.S.; Choi, K.; Nam, J.D.; Choi, H.J. Magnetorheological Elastomers: Fabrication, Characteristics, and Applications. *Materials* **2020**, *13*, 4597. [CrossRef]
8. Omar, M.F.; Ali, F.; Jami, M.S.; Azmi, A.S.; Ahmad, F.; Marzuki, M.Z.; Muniyandi, S.K.; Zainudin, Z.; Kim, M.P. A Comprehensive Review of Natural Rubber Composites: Properties, Compounding Aspects, and Renewable Practices with Natural Fibre Reinforcement. *J. Renew. Mater.* **2025**, *13*, 497–538. [CrossRef]
9. Greenough, S.; Kurian, J.K.; Gariépy, Y.; Prasher, S.; Khripin, C.Y.; Mehlem, J.J.; Dumont, M.J. Protein-Based Bio-Chars as Potential Renewable Fillers in Styrene-Butadiene Rubber Composites. *J. Anal. Appl. Pyrolysis* **2022**, *167*, 105697. [CrossRef]
10. Bardha, A.; Prasher, S.; Dumont, M.J. Waste Biomass-Derived Rubber Composite Additives: Review of Current Research and Future Investigations into Biowaste Tire Formulation. *Biomass Bioenergy* **2024**, *183*, 107149. [CrossRef]
11. Alam, M.N. Advances in Functional Rubber and Elastomer Composites. *Polymers* **2024**, *16*, 1726. [CrossRef]
12. Fasolt, B.; Albuquerque, F.B.; Hubertus, J.; Schultes, G.; Shea, H.; Seelecke, S. Electrode Impact on the Electrical Breakdown of Dielectric Elastomer Thin Films. *Polymers* **2023**, *15*, 4071. [CrossRef]
13. Liu, M.; Imiete, I.E.; Staropoli, M.; Steiner, P.; Duez, B.; Lenoble, D.; Scolan, E.; Thomann, J.S. Hydrophobized MFC as Reinforcing Additive in Industrial Silica/SBR Tire Tread Compound. *Polymers* **2023**, *15*, 3937. [CrossRef]
14. Wang, Z.; Lin, Y.; Li, Z.; Yang, Y.; Lin, J.; He, S. Effect of Fluorosilicone Rubber on Mechanical Properties, Dielectric Breakdown Strength and Hydrophobicity of Methyl Vinyl Silicone Rubber. *Polymers* **2023**, *15*, 3448. [CrossRef]
15. Magaletti, F.; Margani, F.; Monti, A.; Dezyani, R.; Prioglio, G.; Giese, U.; Barbera, V.; Galimberti, M.S. Adducts of Carbon Black with a Biosourced Janus Molecule for Elastomeric Composites with Lower Dissipation of Energy. *Polymers* **2023**, *15*, 3120. [CrossRef]
16. Song, S.; Choi, H.; Jeong, J.; Kim, S.; Kwon, M.; Kim, M.; Kim, D.; Jeon, H.; Paik, H.J.; Chung, S.; et al. Optimized End Functionality of Silane-Terminated Liquid Butadiene Rubber for Silica-Filled Rubber Compounds. *Polymers* **2023**, *15*, 2583. [CrossRef]
17. Slobodinyuk, D.; Slobodinyuk, A.; Strelnikov, V.; Kiselkov, D. Simple and Efficient Synthesis of Oligoetherdiamines: Hardeners of Epoxyurethane Oligomers for Obtaining Coatings with Shape Memory Effect. *Polymers* **2023**, *15*, 2450. [CrossRef]
18. Bakhsh, A.A. Optimization of Polyolefin-Bonded Hydroxyapatite Graphite for Sustainable Industrial Applications. *Polymers* **2023**, *15*, 1505. [CrossRef]
19. Razzaq, M.Y.; Gonzalez-Gutierrez, J.; Farhan, M.; Das, R.; Ruch, D.; Westermann, S.; Schmidt, D.F. 4D Printing of Electroactive Triple-Shape Composites. *Polymers* **2023**, *15*, 832. [CrossRef]
20. Al-Mhyawi, S.R.; Abdel-Tawab, N.A.H.; El Nashar, R.M. Synthesis and Characterization of Orange Peel Modified Hydrogels as Efficient Adsorbents for Methylene Blue (MB). *Polymers* **2023**, *15*, 277. [CrossRef]
21. Jung, J.K.; Lee, J.H.; Jeon, S.K.; Baek, U.B.; Lee, S.H.; Lee, C.H.; Moon, W.J. H_2 Uptake and Diffusion Characteristics in Sulfur-Crosslinked Ethylene Propylene Diene Monomer Polymer Composites with Carbon Black and Silica Fillers after High-Pressure Hydrogen Exposure Reaching 90 MPa. *Polymers* **2022**, *15*, 162. [CrossRef]
22. Do, Q.V.; Kida, T.; Yamaguchi, M.; Washizu, K.; Nagase, T.; Tada, T. Anomalous Strain Recovery after Stress Removal of Graded Rubber. *Polymers* **2022**, *14*, 5477. [CrossRef]
23. Alam, M.N.; Kumar, V.; Park, S.S. Advances in Rubber Compounds Using ZnO and MgO as Co-cure Activators. *Polymers* **2022**, *14*, 5289. [CrossRef]
24. Kumar, V.; Alam, M.N.; Park, S.S.; Lee, D.J. New Insight into Rubber Composites Based on Graphene Nanoplatelets, Electrolyte Iron Particles, and Their Hybrid for Stretchable Magnetic Materials. *Polymers* **2022**, *14*, 4826. [CrossRef]
25. Kumar, V.; Azam, S.; Alam, M.N.; Hong, W.B.; Park, S.S. Novel Rubber Composites Based on Copper Particles, Multi-Wall Carbon Nanotubes and Their Hybrid for Stretchable Devices. *Polymers* **2022**, *14*, 3744. [CrossRef]
26. Jung, Y.S.; Lee, S.; Park, J.; Shin, E.J. One-Shot Synthesis of Thermoplastic Polyurethane Based on Bio-Polyol (Polytrimethylene Ether Glycol) and Characterization of Micro-Phase Separation. *Polymers* **2022**, *14*, 4269. [CrossRef]
27. Yang, C.; Ma, W.; Zhang, Z.; Zhong, J. Low-Velocity Impact Behavior of Sandwich Plates with FG-CNTRC Face Sheets and Negative Poisson's Ratio Auxetic Honeycombs Core. *Polymers* **2022**, *14*, 2938. [CrossRef]
28. Alhashmi Alamer, F.; Almalki, G.A. Fabrication of Conductive Fabrics Based on SWCNTs, MWCNTs and Graphene and Their Applications: A Review. *Polymers* **2022**, *14*, 5376. [CrossRef]
29. Dai, T.; Li, Y.; Huang, H.; Ding, L.; Li, J.; Geng, H.; Song, Y.; Zhao, T.; Zhao, L.; Gui, H. A Study on the Preparation of Environmentally Friendly High-Performance Natural Rubber Using the Interaction Mechanism of Alkaline Protease and Calcium Ions. *Polymers* **2025**, *17*, 490. [CrossRef]
30. Barrera Torres, G.; Gutierrez Aguilar, C.M.; R. Lozada, E.; Tabares Montoya, M.J.; Ángel Álvarez, B.E.; Sánchez, J.C.; Jaramillo Carvalho, J.A.; Santos, R.J. Application of Post-Industrial Leather Waste for the Development of Sustainable Rubber Composites. *Polymers* **2025**, *17*, 190. [CrossRef]
31. Alam, M.N.; Azam, S.; Yun, J.; Park, S.S. Critical Role of Rubber Functionalities on the Mechanical and Electrical Responses of Carbon Nanotube-Based Electroactive Rubber Composites. *Polymers* **2025**, *17*, 127. [CrossRef]

32. Luengchavanon, M.; Ananacharoenwong, E.; Marthosa, S.; Pengsakul, T.; Szekely, J. Application of Antimicrobial Rubber-Coated Cotton Gloves for Mangosteen-Peel-Extract-Mediated Biosynthesis of Ag–ZnO Nanocomposites. *Polymers* **2025**, *17*, 32. [CrossRef]
33. Yu, Y.; Jiang, P.; Yu, W.; Guo, Z. Tribological Performance of Short Fibers Reinforced Thermoplastic Polyurethane Composite Materials Under Water-Lubricated Condition. *Polymers* **2025**, *17*, 30. [CrossRef]
34. Ding, L.; Huang, H.; Wang, Y.; Li, J.; Gui, H.; Chen, Y. The Influence of Fresh Latex Coagulation on the Parameter Characteristics of the Yeoh Hyperelastic Constitutive Model for Natural Rubber. *Polymers* **2024**, *16*, 3601. [CrossRef]
35. Li, F.; Peng, T. An OpenSees Surrogate Constitutive Model for High-Damping Rubber Based on Machine Learning. *Polymers* **2024**, *16*, 3424. [CrossRef]
36. Beknazarov, K.; Tokpayev, R.; Nakyp, A.; Karaseva, Y.; Cherezova, E.; El Fray, M.; Volfson, S.; Nauryzbayev, M. Influence of Kazakhstan's Shungites on the Physical–Mechanical Properties of Nitrile Butadiene Rubber Composites. *Polymers* **2024**, *16*, 3370. [CrossRef]
37. Yangthong, H.; Nun-Anan, P.; Krainoi, A.; Chaisrikhwun, B.; Karrila, S.; Limhengha, S. Hybrid Alumina–Silica Filler for Thermally Conductive Epoxidized Natural Rubber. *Polymers* **2024**, *16*, 3362. [CrossRef]
38. Ge, D.; Hong, Q.; Liu, X.; Liang, H. Self-Oscillation of Liquid Crystal Elastomer Fiber-Slide System Driven by Self-Flickering Light Source. *Polymers* **2024**, *16*, 3298. [CrossRef]
39. Inphonlek, S.; Kotchapradit, S.; Marungsri, B.; Ruksakulpiwat, Y.; Ruksakulpiwat, C. Enhanced Dielectric Properties and Antibacterial Activity of Natural Rubber by Modification with Poly(Acrylic Acid-Co-Acrylamide) Incorporating Silver Nanoparticles and Titanium Dioxide. *Polymers* **2024**, *16*, 3218. [CrossRef]
40. Akkenzheyeva, A.; Haritonovs, V.; Bussurmanova, A.; Merijs-Meri, R.; Imanbayev, Y.; Serikbayeva, A.; Sydykov, S.; Ayapbergenov, Y.; Jankauskas, M.; Trumpels, A.; et al. The Use of Rubber-Polymer Composites in Bitumen Modification for the Disposal of Rubber and Polymer Waste. *Polymers* **2024**, *16*, 3177. [CrossRef]
41. Wang, Y.; Su, S.; Liu, H.; Wang, R.; Liao, L.; Peng, Z.; Li, J.; Wu, H.; He, D. Effect of Proteins on the Vulcanized Natural Rubber Crosslinking Network Structure and Mechanical Properties. *Polymers* **2024**, *16*, 2957. [CrossRef]
42. Glavan, G.; Belyaeva, I.A.; Shamonin, M. On the Piezomagnetism of Magnetoactive Elastomeric Cylinders in Uniform Magnetic Fields: Height Modulation in the Vicinity of an Operating Point by Time-Harmonic Fields. *Polymers* **2024**, *16*, 2706. [CrossRef] [PubMed]
43. Bao, M.; Liu, T.; Tao, Y.; Ni, X. The Hydrogen Bonding in the Hard Domains of the Siloxane Polyurea Copolymer Elastomers. *Polymers* **2024**, *16*, 2438. [CrossRef]
44. Alam, M.N.; Kumar, V.; Jeong, S.-U.; Park, S.-S. The Effect of Rubber–Metal Interactions on the Mechanical, Magneto–Mechanical, and Electrical Properties of Iron, Aluminum, and Hybrid Filler-Based Styrene–Butadiene Rubber Composites. *Polymers* **2024**, *16*, 2424. [CrossRef]
45. Liu, W.; Wang, C.; Feng, Y.; Chen, Y.; Wan, L.; Huang, F.; Liu, Z.; Qian, J.; Liu, W. Novel Reactive Polyhedral Oligomeric Silsesquioxane-Reinforced and Toughened Epoxy Resins for Advanced Composites. *Polymers* **2024**, *16*, 1877. [CrossRef]
46. Abdollahi, I.; Sedaghati, R. Investigation of Macroscopic Mechanical Behavior of Magnetorheological Elastomers under Shear Deformation Using Microscale Representative Volume Element Approach. *Polymers* **2024**, *16*, 1374. [CrossRef]
47. Magaletti, F.; Galbusera, M.; Gentile, D.; Giese, U.; Barbera, V.; Galimberti, M. Carbon Black Functionalized with Serinol Pyrrole to Replace Silica in Elastomeric Composites. *Polymers* **2024**, *16*, 1214. [CrossRef]
48. Alam, M.N.; Kumar, V.; Jeong, S.U.; Park, S.S. Enhancing Rubber Vulcanization Cure Kinetics: Lowering Vulcanization Temperature by Addition of MgO as Co-Cure Activator in ZnO-Based Cure Activator Systems. *Polymers* **2024**, *16*, 876. [CrossRef]
49. Leyva-Porras, C.; Estrada-Moreno, I.A.; Piñón-Balderrama, C.I.; Flores-Gallardo, S.G.; Márquez-Lucero, A. Thermodynamic Parameters of Crosslinked Elastomers (BR, SBR and NBR) and Their Blends. *Polymers* **2024**, *16*, 351. [CrossRef]
50. Kitsawat, V.; Siri, S.; Phisalaphong, M. Electrically Conductive Natural Rubber Composite Films Reinforced with Graphite Platelets. *Polymers* **2024**, *16*, 288. [CrossRef]
51. Kodal, M.; Yazıcı Çakır, N.; Yıldırım, R.; Karakaya, N.; Özkoç, G. Improved Heat Dissipation of NR/SBR-Based Tire Tread Compounds via Hybrid Fillers of Multi-Walled Carbon Nanotube and Carbon Black. *Polymers* **2023**, *15*, 4503. [CrossRef] [PubMed]
52. Rebane, I.; Levin, K.J.; Mäeorg, U.; Johanson, U.; Piirimägi, P.; Tätte, T.; Tamm, T. Enhanced Low-Density Silicone Foams Blown by Water–Hydroxyl Blends. *Polymers* **2023**, *15*, 4425. [CrossRef] [PubMed]
53. Zhao, Q.; Niu, F.; Liu, J.; Yin, H. Research Progress of Natural Rubber Wet Mixing Technology. *Polymers* **2024**, *16*, 1899. [CrossRef] [PubMed]

polymers

MDPI

Article

A Study on the Preparation of Environmentally Friendly High-Performance Natural Rubber Using the Interaction Mechanism of Alkaline Protease and Calcium Ions

Tuo Dai [1,†], Yun Li [2,†], Honghai Huang [1], Li Ding [1], Jianwei Li [1], Haoran Geng [1], Yazhong Song [1], Tao Zhao [1], Liguang Zhao [1] and Hongxing Gui [1,*]

1 Hainan Natural Rubber Technology Innovation Center, Rubber Research Institute, Chinese Academy of Tropical Agricultural Science, Haikou 571101, China; catasdaituo@163.com (T.D.)
2 Mengla Tianye Rubber Sales Co., Ltd., Xishuangbanna 666100, China
* Correspondence: guihongxing@263.net; Tel.: +86-15289761203
† These authors contributed equally to this work and should be considered co-first authors.

Abstract: Natural rubber (NR) is a material with a wide range of industrial and commercial applications, including agriculture, defense, transportation, and domestic use. However, the mechanical properties of natural rubber treated by traditional acid coagulation are limited, which restricts its application in high-end products. Furthermore, the wastewater generated also causes soil acidification. Consequently, there is a necessity to investigate new coagulation methods to enhance the comprehensive performance of natural rubber and reduce environmental pollution. In this work, a novel method for the preparation of environmentally friendly high-performance natural rubber by alkaline protease/calcium chloride coagulation of natural rubber (AC-NR) is reported. The research demonstrates that the products of protein cleavage by alkaline protease together with calcium ions can greatly enhance the cross-linking between rubber particles, form the network structure of natural rubber well. Furthermore, increasing the pH at the isoelectric point of the discharged wastewater reduces the impact on soil acidification. In comparison with those from conventional acid coagulation of natural rubber (A-NR), the tensile strength of AC-NR was increased by 7.9 MPa, the tear strength was increased by 5.3 kN/m, the final temperature rise was lowered by 6.5 °C, and abrasion performance was improved. This study demonstrates that by utilizing the collaborative impact of alkaline protease and calcium chloride on the rubber molecular chain during the coagulation process of natural rubber, environmentally friendly high-performance natural rubber with excellent mechanical properties and reduced environmental pollution can be prepared without the necessity for chemical modification or cumbersome processes, which is conducive to the green development and high-quality pursuit of NR materials.

Keywords: natural rubber (NR); coagulation; alkaline protease; high-performance; environmentally friendly

1. Introduction

Natural rubber (NR) is a natural polymer compound with polyisoprene as the main component, 91–94% of which is rubber hydrocarbon (polyisoprene), and the remainder is non-rubber substances such as proteins, fatty acids, ash, and sugars [1]. As a crucial industrial raw material, natural rubber possesses advantageous properties, including elasticity, plasticity, tensile strength, and resistance to abrasion, making it a widely utilized

 https://doi.org/10.3390/polym17040490

material in various industrial, agricultural, national defense, transportation, and daily applications [2]. However, due to inherent limitations of the conventional coagulation process and the presence of quality issues in domestic natural rubber, China is compelled to rely on imports for its high-end natural rubber requirements. This dependence poses a major threat to national security and is a 'bottleneck' that hinders national development. Furthermore, the wastewater produced by the conventional coagulation process of natural rubber possesses a low pH at the isoelectric point, thereby inducing environmental concerns such as soil acidification [3].

The conventional coagulation process of natural rubber involves acid coagulation, wherein the principle is that H^+ ions cause the pH at the isoelectric point to decline to a specific critical value. This results in the rubber particles attaining the isoelectric state, thereby eliminating their surface charge and leading to the dissolution of the charge repulsion between the rubber particles. Consequently, the rubber particles adhere to each other and coalesce [4]. However, extensive research has demonstrated that the acid coagulation process results in a significant reduction in the tensile strength of standard rubber. Furthermore, the discharge of rubber wastewater is characterized by high acidity, elevated levels of chemical oxygen consumption (COD), and the presence of other pollutants, thereby causing substantial environmental pollution [5].

The process of natural rubber microbial coagulation involves the utilization of wastewater from rubber gardens and other environments as a medium for the cultivation of microbial strains that possess the capacity to produce acid. This process entails the addition of sugar substances to the original substrate, thereby facilitating continuous decomposition. The cultivation of these microbial strains results in the acidification of the substrate, leading to the production of an acidic liquid that possesses a low isoelectric point. This, in turn, significantly reduces the duration of natural rubber coagulation [6].The duration of natural rubber coagulation has been found to be a crucial factor in the process. The selection of microbial strains that demonstrate effective performance in natural rubber production is less damaging, while also reducing production costs [7]. However, it should be noted that the repeated treatment of fresh latex with alkali (ammonia preservation) and acid (low microbial liquid pH at the isoelectric point) will result in the destruction of the original antioxidant properties of natural rubber, its cross-linking structure, and ageing properties [8].

The coagulation mechanism of natural rubber involves the elimination of the repulsive force between the rubber particles, thereby inducing coagulation. However, there are divergent perspectives on the coagulation mechanism of natural rubber. One perspective attributes the coagulation of natural rubber to the dissociation of salt electrolytes in water, which produces cations. This process leads to the compression of the double layer, resulting in a thinning of the hydration film surrounding the natural rubber particles. The electric potential is reduced to zero, causing the natural rubber particles to lose their repulsive force and agglomerate, leading to coagulation. Alternatively, the presence of two-valent or three-valent metal ions, along with rubber particles derived from fatty acids, has been identified as a crucial factor in the formation of insoluble soaps within the protective layer. This, in turn, has been shown to compromise the stability of the rubber particles, thereby initiating the process of latex coagulation [9]. The pH at the isoelectric point of the natural rubber salt coagulation curve is biased in the weakly acidic region, which reduces the corrosion of instrumentation and environmental pollution caused by acids. However, excessive variable metal ions (e.g., Cu^{2+}, Mn^{2+}, etc.) accelerate the aging of natural rubber and vulcanized rubber [10], and calcium ions do not affect the aging properties of natural rubber and vulcanized rubber. Studies on the mechanism of action of calcium ions and alkaline protease have shown that alkaline protease has two Ca^{2+} binding sites, and that there is a strong binding force between Ca^{2+} and alkaline protease. Calcium ions and alkaline protease have been shown to greatly increase the activity of alkaline protease after binding calcium ions to alkaline protease [11].

In this study, a novel technique for natural rubber coagulation using a combination of alkaline protease and calcium ions is proposed, and the mechanism of action of calcium ions to promote alkaline protease activity is demonstrated in Figure 1. The utilization of alkaline protease serves to negate the impact of the low pH at the isoelectric point of acid and microbial coagulation systems on the overall characteristics of natural rubber, as well as the effects of wastewater on soil acidification. The incorporation of calcium ions has been shown to expedite the coagulation process, thereby enhancing the productivity of rubber-processing facilities and optimizing the overall performance of natural rubber. The study also involves a comparative analysis of traditional acid coagulation and microbial coagulation with the novel alkaline protease–salt coagulation technique. This analysis aims to compare the effects of changes in pH at the isoelectric point during coagulation processes on the structure of the rubber network and various performance indicators. The new coagulation technology formed in this project has the potential to enhance the quality of natural rubber, reduce the production cost and wastewater treatment cost of rubber-processing enterprises, and play an important role in promoting the development of high-performance natural rubber industry in China.

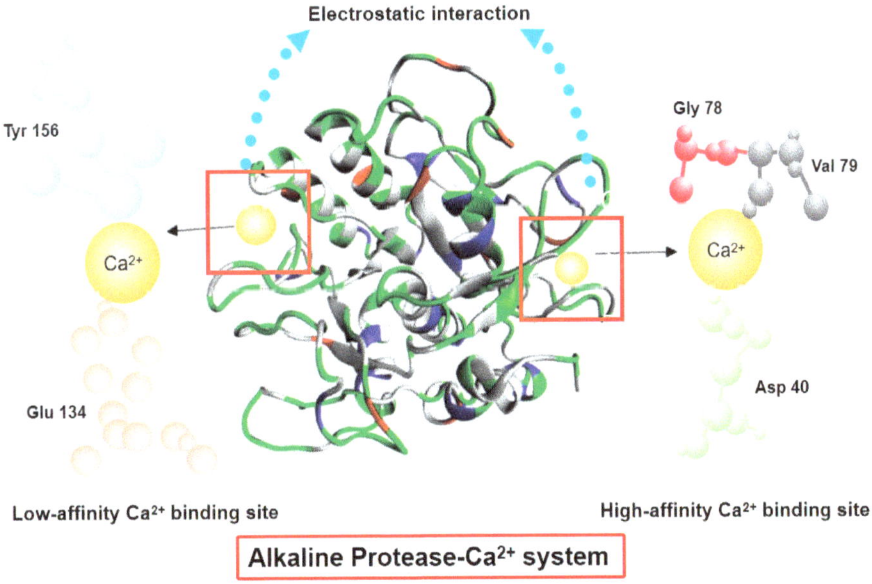

Figure 1. The interaction mechanism between alkaline protease and Ca^{2+}.

2. Experimental

2.1. Materials

Natural latex from Mengla County Tianye Rubber Sales Ltd., Xishuangbanna, China; Alkaline protease from Nanning Pangbo Bioengineering Ltd., Nanning, China. Acetic acid (98%), Calcium chloride (95%), Zinc oxide (ZnO), Stearic acid (SA), Quebrith, accelerator N-tert-butylbenzothiazole-2-sulfenamide, Concentrated sulfuric acid, Sodium hydroxide, Toluene all from Xilong Reagent Ltd., Shantou, China (Table 1).

Table 1. Types and sources of main raw materials.

Materials	Specification	Manufacture Factory
Nature rubber	LR	Mengla County Tianye Rubber Sales Ltd., Xishuangbanna, China
Alkaline protease	AR	Nanning Pangbo Bioengineering Ltd., Nanning, China
Calcium chloride	AR	Xilong Reagent Ltd., Shantou, China
Acetic acid	AR	Xilong Reagent Ltd., Shantou, China
Zinc oxide (ZnO)	AR	Xilong Reagent Ltd., Shantou, China
Stearic acid (SA)	AR	Xilong Reagent Ltd., Shantou, China
Sulfur(S)	AR	Xilong Reagent Ltd., Shantou, China
TBBS	AR	Xilong Reagent Ltd., Shantou, China
Accelerator M	AR	Xilong Reagent Ltd., Shantou, China
Carbon black	AR	Xilong Reagent Ltd., Shantou, China

2.2. Instruments and Equipment

The following items were available from Yuanfeng Inspection Equipment Ltd., Yangzhou, China: an open-type rubber-refining machine, a closed-type rubber-refining machine, a plate-vulcanizing machine, and a Fourier infrared spectroscopy tester. The following items were available from NETZSCH Instruments Ltd., Weimar, Germany: a thermogravimetric analyzer and a differential scanning calorimeter. The following item was available from Shimadzu Ltd., Tokyo, Japan: an electric blast drying box. The following item was available from MonTech Trade Ltd., Buchen, Germany: D-RPA-3000 rubber-processing analyzer.

2.3. Preparation of Natural Rubber Samples

Natural rubber latex was collected in rubber forests, and the dry rubber content of natural rubber latex (approximately 33–39%) was measured and uniformly diluted to 23%. The initial pH at the isoelectric point of the natural rubber latex was measured and adjusted to neutral, and the natural rubber latex was subsequently coagulated by different coagulation processes. Acetic acid-NR (hereafter referred to as A-NR) was obtained by adding appropriate amounts of acetic acid to the natural rubber latex and mixing uniformly; microorganism-NR (hereafter referred to as M-NR) was obtained by adding appropriate amounts of microbial bacterial liquid to the natural rubber latex. Alkaline protease/calcium chloride-NR (henceforth referred to as AC-NR) was obtained by adding 0.07% alkaline protease and 0.1% calcium chloride to natural rubber latex with homogeneous mixing. The process of natural rubber sample preparation is shown in Figure 2. After the latex had completely solidified, the rubbers were placed in a coagulation tank for 1 day. The wastewater in the three coagulation tanks was extracted to measure the pH at the isoelectric point, and then solidified. The solidified rubber samples were pressed five times, crushed, dried at 110 °C, and extruded to obtain raw rubber blocks, which were mixed. These samples were then subjected to various compounding agents (formulations as shown in Table 2) in the open roll mill, in accordance with the prevailing operating standards [9]. The vulcanization process was then initiated in the vulcanizing machine at 140 °C, yielding vulcanized rubber samples of pure rubber/carbon black formulation.

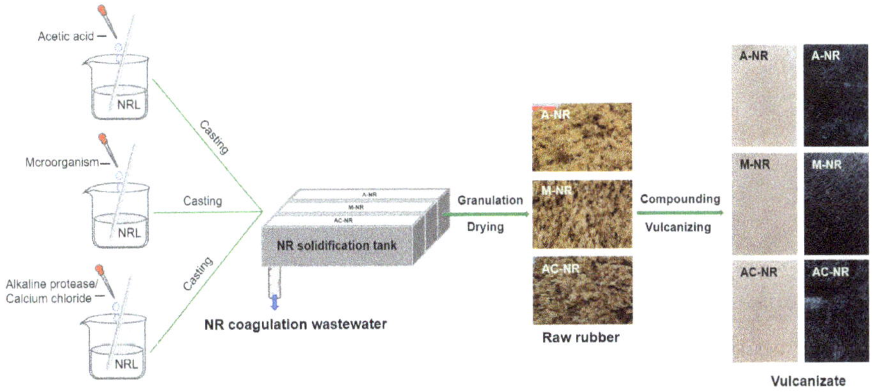

Figure 2. Schematic diagram of natural rubber samples prepared by three coagulation methods.

Table 2. Vulcanization formulas of NR.

Samples	NR	ZnO	Sulfur	Stearic Acid	Accelerator M	TBBS	Carbon Black
Formula of pure rubber compound, g	100	6	3.5	0.5	0.5	—	—
Formula of carbon black rubber compound, g	100	5	2.25	2	—	0.7	35

2.4. Characterizations

Gel content: About 0.1 g (noted as m_0) of natural raw rubber was weighed and soaked for 1 week in tetrahydrofuran as a solvent, then centrifuged by a GL-21M refrigerated centrifuge to get the precipitate. A weighing dish was taken and weighed (m_1), into which the precipitate was then poured and dried at 100 °C until constant weight, and then the total weight (m_2) was measured. The gel content (A) was calculated from Equation (1):

$$A = \frac{(m_2 - m_1)}{m_0} \times 100\% \quad (1)$$

Molecular weight size and its distribution: 15 mg of raw rubber was sheared and statically dispersed, then gradually dissolved in a room-temperature tetrahydrofuran (THF) solution, sealed from light, dissolved for 72 h until the sample was completely dissolved, and the sample solution was filtered using LC-100 high-performance gel permeation chromatography (GPC) to determine the relative molecular weight size of the natural rubber samples and its distribution. The test preparation conditions were as follows: polystyrene was the test standard sample, the mobile phase was a THF solution, the concentration of rubber filtrate was 3 mg/mL, and the flow rate was 1 mL/min. The test preparation conditions were as follows: polystyrene was used as the test standard, the mobile phase was a THF solution, the concentration of rubber filtrate was 3 mg/mL, the flow rate was 1 mL/min, the temperature of the test sample was stabilized at 30 °C, and the injection volume was 10 μL.

Vulcanization characteristics: 5 g of rubber mixing samples were taken to determine the vulcanization characteristics of natural rubber using a CZ-3001D rotorless rheometer, 143 °C × 60 min.

Infrared spectroscopy: Preparation of thin samples of natural rubber, a Nicolet iS5 Fourier infrared spectrometer was used to perform attenuated total reflectance (ATR) infrared spectroscopy scanning on thin samples. The necessary conditions for the test were: wave number scanning range of 600–4000 cm^{-1}, spectral image resolution of 4 cm^{-1}, and the number of scanning completion of 16.

Physicochemical indicators: Mooney viscosity test was performed according to the GB/T1232.1-2000 standard by taking homogenized raw rubber samples of about 8 g, sourced from Abundant Testing Equipment Co., Ltd. (Delhi, India) and produced by a YF-8005 type Mooney viscosity meter. The initial value of plasticity (P_0) was tested using a KS-II rapid plasticity meter at 100 °C, preheated for 1min and tested for 4 min. P_{30} was tested using an aging test chamber for aging the samples (140 °C, 30 min) and tested using a KS-II rapid plasticity meter. The plasticity retention rate (PRI) was calculated according to Equation (2):

$$\mathrm{PRI} = \frac{P_{30}}{P_0} \times 100\% \quad (2)$$

Physical and mechanical properties: Natural rubber samples were vulcanised and prepared in the shape of a dumbbell and hardness and thickness were measured to test the tensile strength of natural rubber. Natural rubber samples were prepared in right-angle shape after vulcanisation and the thickness was measured to test the tear strength of natural rubber. Tensile strength and tear strength were measured on a PT-307 universal material tensile machine produced by Puzet Testing Equipment Ltd., Quanzhou, China. with tensile speed of (500 ± 50) mm/min (Figure 3A,B).

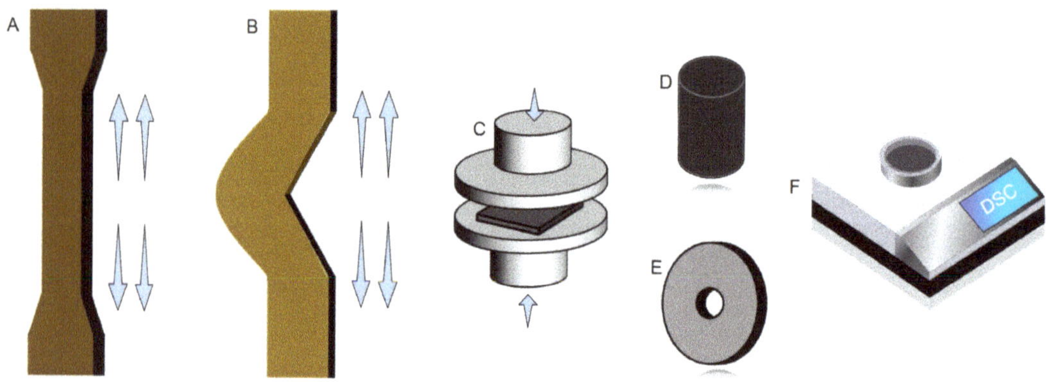

Figure 3. (**A**) Schematic of tensile strength test. (**B**) Schematic of tear strength test. (**C**) Schematic of RPA test of dynamic mechanical properties. (**D**) Schematic of DIN abrasion sample. (**E**) Schematic of Akron abrasion sample. (**F**) Schematic of the DSC instrument.

RPA test dynamic mechanical properties: D-RPA-3000 rubber-processing analyzer test was used to evaluate the rheological properties of the natural raw rubber samples. Regarding the Payne effect ($\Delta G'$), the rubber was mixed at 100 °C, with 0.5 Hz strain scanning 0.28–1000%, and then warmed up to 143 °C to complete the vulcanization (time 1.2 × t_{90}min), and finally the Payne effect was calculated according to the following formula: $\Delta G' = G'$0.28–100%. Finally, the temperature was lowered to 100 °C to complete the strain scan test on the vulcanized rubber (range 0.28–100%), and the Payne effect was calculated according to the following formula: $\Delta G' = G'0.56 - G'100\%$ (Figure 3C).

Compression heat generation: The height, diameter, and hardness of the prepared cylindrical vulcanized rubber samples were measured, and a compression fatigue temperature rise testing machine RH-2000N was used to carry out the heat generation test of the samples. The sample was preheated for 30 min, the temperature was generally constant at 55 ± 1 °C, the stroke was 4.45 mm, the load was 1.0 MPa, the test frequency was set at 20 Hz. The sample was preheated for 30 min and then tested for 25 min, and the final temperature rise of the sample was recorded.

DIN abrasion: Cylindrical samples of natural rubber vulcanised with carbon black were prepared first (Figure 3D). The test method involved the utilization of a DIN abrasion tester, wherein the specimen was pressed against a rotating drum coated with sandpaper, with a contact pressure of 10 N. The drum was set to rotate at a speed of 40 r/min, and the specimen was moved horizontally over the drum. The wear volume of the rubber specimen was then measured after a stroke of 40 m. It is imperative to note that the test was conducted in a circular motion, with the specimen being moved horizontally over the drum.

Akron abrasion: A long sample of carbon black vulcanised natural rubber was first prepared, the sample was fixed on a rubber wheel axle and the motor was started (Figure 3E). After pre-grinding for 15 min (500 revolutions), the sample was removed, the crumbs were brushed, weighed, and recorded as m_1. The test was carried out with a pre-milled sample. After driving 1.61 km (3418 revolutions) [12], the motor was turned off, the sample was removed, brushed with the crumbs, and weighed within 1 h, recorded as m_2. The density ρ of the sample was measured using a rubber gravimeter. The wear volume (V) of the specimen was calculated as follows:

$$V = \frac{m_1 - m_2}{\rho} \quad (3)$$

Thermogravimetry analysis: The thermal degradation of raw rubber samples was studied in depth using a thermogravimetric analyzer (TGA 101). First, 5–10 mg of raw rubber samples were cut into pieces and placed in an alumina crucible in the TG sample chamber. The experimental conditions consisted of a nitrogen atmosphere, with a constant nitrogen flow rate of 20 mL/min, and a temperature increase rate that varied at 10 °C per minute. The temperature interval ranged from 30 to 550 °C.

Differential scanning calorimetry: The glass transition temperature (Tg) of the raw rubber samples was tested using a JB-DSC-500 differential scanning calorimetry (DSC) analyzer, and 5–10 mg samples were cut into pieces and put into an aluminum crucible, placed in the sample chamber of DSC (Figure 3F), and the process of going from the low temperature of −80 °C to the high temperature of 100 °C was repeated twice to eliminate the thermal effect. The test conditions were as follows: nitrogen atmosphere, temperature increases and decreases at a rate of 10 °C per minute, and temperature range of −80 to 100 °C, and nitrogen and liquid nitrogen flow rate of 20 mL/min.

3. Results and Discussion

3.1. Changes in pH at the Isoelectric Point of Natural Rubber During Coagulation with Different Coagulation Methods

Natural rubber latex, composed of rubber particles of varying sizes, exhibited polydispersity, thereby resulting in inconsistent stability amongst the particles. The most unstable particles, approaching the isoelectric point of cohesion, initially exhibited a tendency to congregate in proximity to the isoelectric point, gradually disseminating to all rubber particles over time. In the event that the isoelectric point of latex is not fully attained, but instead only approaches the pH level, the coagulation process of natural rubber is deemed incomplete. This phenomenon is referred to as "natural rubber seepage white water". Consequently, the coagulation process is not instantaneous, but rather undergoes a series of changes during its progression [9]. The study of the pH at the isoelectric point change in this process is of great significance to investigate the changes in the internal composition and network structure of natural rubber in different coagulation methods, and the effect of wastewater on the degree of soil acidification.

Figure 4A illustrates the alterations in pH at the isoelectric point of A-NR, M-NR, and AC-NR throughout the coagulation process. The initial and overall pH at the isoelectric point of A-NR and M-NR were lower than that of AC-NR due to the lower pH at the isoelectric point of the added coagulant. The overall trend of pH at the isoelectric point of the three types of coagulation modes (i.e., A-NR, M-NR, and AC-NR) was shown to be the same, with a decrease first and then an increase. The time period from 0 to 60 min was shown to be the time period for alkaline protease of bacillus a class of serine protease. The hydrolysis reaction for a protein by serine proteases is carried out by combining active sites with serine proteases. The hydrolysis reaction for a protein by serine proteases involves the combination of active sites with substrates, resulting in the production of a histidine containing the imidazole group and a serine residue with a β-OH as a functional group and other basic amino acids [11]. These amino acids have a pH > 7 at the isoelectric point [13], leading to a modest increase in the pH at the isoelectric point of AC-NR. A-NR and M-NR are devoid of alkaline protease, the enzyme responsible for the breakdown of proteins to produce alkaline amino acids. Instead, these microorganisms are capable of producing acidic substances through a different mechanism. Consequently, the pH at the isoelectric point of A-NR and M-NR was decreasing since the onset of the reaction. Furthermore, a substantial number of exotic microorganisms are present in M-NR, which serve to augment the "acidification" reaction of the original substrate [14]. This further exacerbates the decline in pH at the isoelectric point.

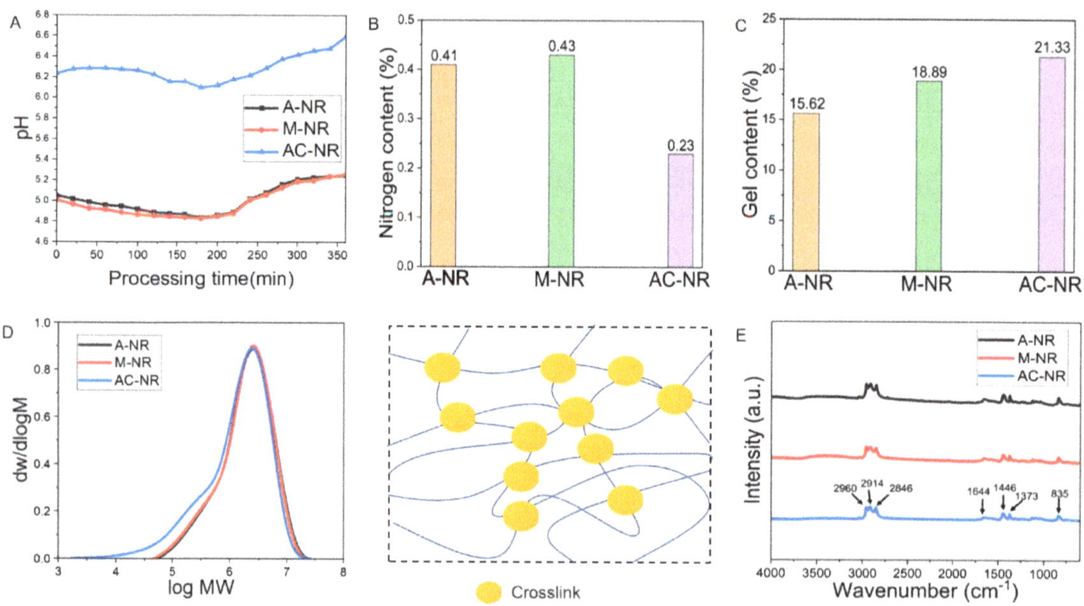

Figure 4. (**A**) Trends in pH at the isoelectric point during the coagulation of NR. (**B**) Nitrogen content of NR. (**C**) Gel content of NR. (**D**) Molecular weight distribution of NR. (**E**) Infrared spectra of NR.

In the 60–180 min time period, the coagulation of natural rubber was observed. The process of microbial reproduction in natural rubber leads to the decomposition of its organic matter, resulting in the production of a substantial quantity of acidic substances, namely A-NR, M-NR, and AC-NR. This process concurrently results in a decrease in the pH at the isoelectric point of the material. After 180 min, the three coagulation methods of NR were found to be essentially solidified. The coagulation of natural rubber is characterized by the mechanism of "agglomerate maturation". At this stage, H^+ participates in the construction of the network between natural rubber hydrocarbons and is consumed. This process serves as a conduit, connecting the rubber particles [9]. As H^+ is utilized and reduced within the natural rubber coagulation system, the pH at the isoelectric point in the residual liquid is observed to undergo an increase. Consequently, A-NR, M-NR, and AC-NR attained a higher final pH at the isoelectric point compared to that at the initial stage of coagulation. The alkaline protease enzyme dissolves the bound proteins on the surface of the rubber particles, and the decomposition products of the proteins and the dimethylallyl groups on the end groups of the molecular chains are connected by van der Waals forces, hydrogen bonding, or ionic bonding. This process leads to the formation of omega-terminal groups and the establishment of a cross-linking network structure [15,16]. Calcium ions act as ionic bridges between rubber particles and H^+ ions during the coagulation reaction [9]. The interplay between these two effects results in the entanglements of additional rubber molecular chains, thereby enhancing the network structure of natural rubber and improving its quality.

The prolongation of the "agglomerate ripening" time of natural rubber results in a more compact internal network structure, with continuous consumption of H^+ and a pH at the isoelectric point of the liquid phase that is AC-NR > A-NR ≈ M-NR. The wastewater from the coagulation of AC-NR is weakly acidic, which has less impact on soil acidification and rubber tree growth than the strongly acidic wastewater from A-NR and M-NR [17]. In conclusion, it can be posited that AC-NR possesses superior quality and also mitigates the effect of wastewater from natural rubber processing on soil acidification.

3.2. *Effect of Different Coagulation Methods on Gel Content of Natural Rubber*

Gel content is employed to denote the macromolecules present within natural rubber that are insoluble in organic solvents, such as toluene. These macromolecules have the capacity to respond to the extent of protein and rubber molecule cross-linking with each other, thereby resulting in the formation of a network structure [18]. As demonstrated in Figure 4B,C, which depicts the protein content and gel content respectively, under three distinct coagulation modes, it is evident that AC-NR exhibited reduced protein content, accompanied by augmented gel content. This phenomenon can be attributed to the process of alkaline protease digestion of proteins, which results in the production of amino acids. The opening of the epoxy group on the rubber molecular chain, consequent to this process, serves to promote the cross-linking of molecular chains; at the same time, calcium ions play a role in the rubber particles between the "bonding bridge". These two factors influence the micro-cross-linking network structure in the AC-NR coagulation system more than in the A-NR and M-NR coagulation systems. The nature of the A-NR and M-NR coagulation systems is such that the addition of strong acid reagents leads to the isoelectric point of natural rubber within a short period of time, causing a strong change in the stability of the coagulation and also damaging the hydrogen bond formation of the rubber and the coagulation of the rubber. In addition, it has been demonstrated that the process of A-NR and M-NR coagulation results in the destruction of the pivot point of hydrogen bond formation [19]. This phenomenon resulted in a reduced degree of cross-linking of the terminal groups of the rubber molecular chains in A-NR and M-NR, and the gel content of both was lower than that of AC-NR.

3.3. *Effects of Different Coagulation Methods on Molecular Weight Size and Distribution of Natural Rubber*

As illustrated in Table 3, the number average molecular weight (Mn), the weight average molecular weight (Mw), and the distribution coefficients of A-NR, M-NR, and AC-NR are reported. The molecular weight size and its distribution represent the basic structural parameters of natural rubber polymers polymerized from repeated rubber hydrocarbon units. The number of macromolecular structures contained in natural rubber has an impact on the molecular weight size [20], so the most appropriate method for determining the molecular weight size is to use Mw (weight average molecular weight). It is evident that the number average molecular weight (Mn) of the three coagulation modes of natural rubber did not vary significantly, with AC-NR exhibiting the largest weight average molecular weight of 24.15×10^5 daltons. The molecular weight distribution in Figure 4D reveals a substantially larger area for AC-NR in comparison to that of A-NR and M-NR. The results of molecular weight size and gel content also correspond to each other, which can be attributed to the presence of rubber molecules with aldehyde-containing groups. These undergo condensation reactions with other groups to produce cross-linking between rubber molecules [21], and these condensation reactions increase the molecular weight of rubber and gel content [22]. The low pH at the isoelectric point of A-NR and M-NR has been shown to affect the formation of physical cross-linking bonds in natural rubber, leading to an increase in the proportion of low molecular weight compounds and a decrease in the molecular weight of natural rubber.

Table 3. Molecular weight size and distribution of NR.

Samples	Number Average Molecular Weights/Mn	Weight Average Molecular Weight/Mw	Molecular Weight Distribution Width Coefficient/Mw/Mn
A-NR	3.01×10^5	11.05×10^5	3.67
M-NR	3.45×10^5	13.35×10^5	3.87
AC-NR	3.51×10^5	24.15×10^5	6.88

The molecular weight distribution curves of the same fresh latex raw materials were found to be largely similar, a bimodal distribution with one shoulder and one peak in both the low and high molecular weight regions. AC-NR was co-treated with alkaline protease and calcium ions, which resulted in an augmentation of the "phospholipid-α-terminal group" network structure, accompanied by an enhancement in the heavy average molecular weight of natural rubber (Mw) and a broadening of the distribution [23].

3.4. *Influence of Different Coagulation Methods on Vulcanization Characteristics of Natural Rubber*

As illustrated in Table 4, the torque values and vulcanization rates of A-NR, M-NR, and AC-NR during vulcanization are presented. It is evident that the difference between the highest torque (M_H) and the lowest torque (M_L) of AC-NR was the largest at 6.88 dN·m. The magnitude of M_H-M_L has been shown to reflect the number of cross-linking bonds formed in natural rubber during vulcanization [24], indicating that alkaline protease and calcium ions facilitate the formation of more cross-linking bonds and greater cross-linking density in natural rubber. Scorch time is defined as the time required for torque to reach 10%, indicating the duration of the heat accumulation effect of natural rubber during the mixing of the filler. The remaining scorch time is the time that the rubber can flow before setting, and the scorch time (t_{10}) and the optimum curing time (t_{90}) of the AC-NR were shorter than those of the A-NR and the M-NR. This was due to the fact that the alkaline protease decomposed proteins and produced a large number of peptide bonds, and the amide group of the peptide bonds formed a ligand complex with the zinc ions in zinc oxide, increasing the solubility of zinc oxide in the vulcanization reaction matrix, thus playing a role in promoting vulcanization. Within the vulcanization reaction system, an increase in the concentration of zinc oxide is observed, and the primary formation between the promoter and sulfur is a brief cross-linking bond, which increases the active site of the cross-linking reaction [25]. It can be discerned that alkaline protease and calcium ions have the capacity to regulate the number of cross-linking bonds and scorch time of natural rubber during the vulcanization process, thereby affecting the comprehensive performance of natural rubber.

Table 4. Vulcanization characteristics of NR.

Samples	M_H/dN·m	M_L/dN·m	M_H-M_L/dN·m	t_{10}/min	t_{90}/min
A-NR	0.50	6.21	5.71	2:01	17:57
M-NR	0.52	6.58	6.06	1:56	18:25
AC-NR	0.79	7.08	6.29	1:31	12:25

3.5. Comparison of Infrared Spectra of Natural Rubber Under Different Coagulation Methods

As illustrated in Figure 4E, the infrared spectra of AC-NR, M-NR, and A-NR exhibited characteristic peaks that demonstrated comparable trends, suggesting that the incorporation of alkaline protease and calcium ions into natural rubber did not induce substantial alterations to the main chain or branched structures. This observation signifies that natural rubber maintains its inherent benefits over synthetic rubber. There, 2960 cm^{-1} corresponded to the antisymmetric stretching vibration peak of -CH_3, 2960 cm^{-1} corresponded to the antisymmetric stretching vibration peak of -CH_3, 2914 cm^{-1} corresponded to the antisymmetric stretching vibration peak of -CH_2-, and 2846 cm^{-1} corresponded to the symmetric stretching vibration peak of -CH_2-. The amide I band absorption peak at 1644 cm^{-1} [26] is indicative of protein presence in the latex, with the area under the amide I vibration absorption band corresponding to the protein content. This suggests that the protein content of AC-NR was low. The peaks at 1446 cm^{-1} and 1373 cm^{-1} corresponded to methyl δC-H symmetric bending vibrations, while the peak at 835 cm^{-1} was attributed to C-H deformation vibrations external to the plane of the natural rubber cis-structure. It is evident that natural rubber maintained its distinct advantageous properties and enhanced its overall performance when subjected to the addition of alkaline protease and calcium ions.

3.6. Effects of Different Coagulation Methods on Physical and Chemical Indexes of Natural Rubber

As demonstrated in Figure 5A,B, the initial plasticity value P_0 and Mooney viscosity of AC-NR were greater than those of A-NR and M-NR. The initial plasticity value P_0 is indicative of the plasticity of natural rubber, which has a larger correlation with gel content, molecular weight, and mechanical properties of natural rubber [27], which suggests that AC-NR samples have more network structures formed by intertwining the molecular chains of the rubber molecules in the coagulation process, but it will slightly reduce its plasticity. Mooney viscosity is a significant indicator of the performance of natural rubber processing, primarily employed to reflect the processing difficulty of natural rubber. An elevated Mooney viscosity value indicates poor natural rubber plasticity, which in turn affects the subsequent calendering and extrusion operations. This can result in difficulties in achieving uniformity in the rubber mixed filler, and in the subsequent extrusion process. Conversely, if the Mooney viscosity value is too low, natural rubber can adhere to the rollers, which can have a detrimental effect on the calendering and extrusion operations [28]. As AC-NR contains alkaline protease, it is reasonable to hypothesize that a greater number of binding proteins will be consumed in the rubber molecular chain. Furthermore, it is predicted that an increased number of proteins will be forced to undergo cross-linking to build a network structure, which may impede the flow between rubber molecules. The hypothesis is supported by the higher Mooney viscosity of AC-NR. The plasticity retention index (PRI) in Figure 5C represents the oxygen ageing resistance of natural rubber, which is positively related to the oxygen ageing resistance of natural rubber [29]. Protein is a natural antioxidant [30], and AC-NR had the lowest protein content but the highest plasticity retention PRI, which may have been due to the addition of alkaline protease and calcium ions to AC-NR. The variable metal particles in natural rubber cause electron transfer when heated, which promotes the generation of free radicals and thus the decomposition of rubber hydrocarbons [31]. Amino acids, produced by the breakdown of proteins by alkaline proteases, have been shown to bind readily to metal particles, reducing the abundance of metal ions and thereby inhibiting the free radical chain reaction, thus slowing down the aging of rubber [32].

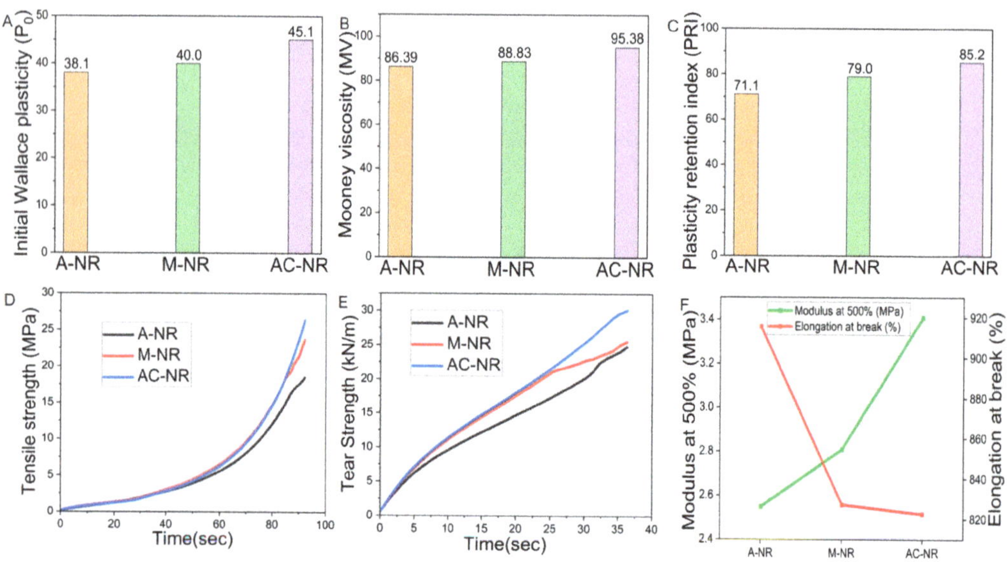

Figure 5. (**A**) Initial plastic value P_0 of NR. (**B**) Mooney viscosity of NR. (**C**) Plasticity retention index (PRI) of NR. (**D**) Tensile strength of NR. (**E**) Tear strength of NR. (**F**) Comparison of trends in elongation at break and 500% modulus of NR.

3.7. Influence of Different Coagulation Methods on Physical and Mechanical Properties of Natural Rubber

The physical and mechanical properties of natural rubber are significant metrics for evaluating its quality, particularly in relation to its application in various rubber product areas. These properties must be met to ensure compliance with production and utilization requirements, both in the formulation of products and the design of processes. As demonstrated in Figure 5D, which illustrates the tensile strength of the samples, AC-NR exhibited higher tensile strength in comparison to that of A-NR and M-NR. This indicates that AC-NR possesses a superior ability to withstand maximum tensile stress during the fracture process when compared to that of A-NR and M-NR. Furthermore, this finding suggests that the destructive force of AC-NR is greater than that of AC-NR under the action of a certain external force [33]. This suggests that AC-NR has the capacity to enhance polar substituents and crystallinity within the rubber molecular chain in the presence of alkaline protease and metal ions, thereby augmenting the intermolecular forces between the rubber molecules [34]. This phenomenon is absent in the acid-solidified A-NR system. The Figure 5E represents the tear strength, and the findings demonstrate that AC-NR exhibited superior tear strength in comparison to that of A-NR and M-NR. This observation is consistent with the comparative results of MH-ML. The reason for this is that there is a strong correlation between tear strength and cross-linking density. In a specific cross-linking density range, the tear strength increases with the increase of cross-linking density [35]. The results demonstrate that AC-NR exhibited the capacity to resist continuous rupture at the crack under unit thickness.

As demonstrated in Figure 5F, the modulus at 500% and the elongation at break of A-NR, M-NR, and AC-NR exhibited a close correlation with the network-like structure of rubber hydrocarbon linear polymers, which were formed by cross-linking following the vulcanization of natural rubber [36]. Furthermore, an inverse proportional relationship was observed between these two parameters. The modulus at 500% of AC-NR and M-NR were 3.41 MPa and 3.21 MPa, respectively, which were higher than that of A-NR [37]. Figure 6 illustrates the potential mechanisms through which alkaline proteases enhance the overall properties of natural rubber. With the increase of cross-link density, constant tensile stress increased, which indicates that AC-NR vulcanized rubber built more network structures to resist external deformation, but there was a partial loss of resilience and increased permanent deformation of AC-NR vulcanized rubber, and the elongation at break was at the normal level, although slightly lower.

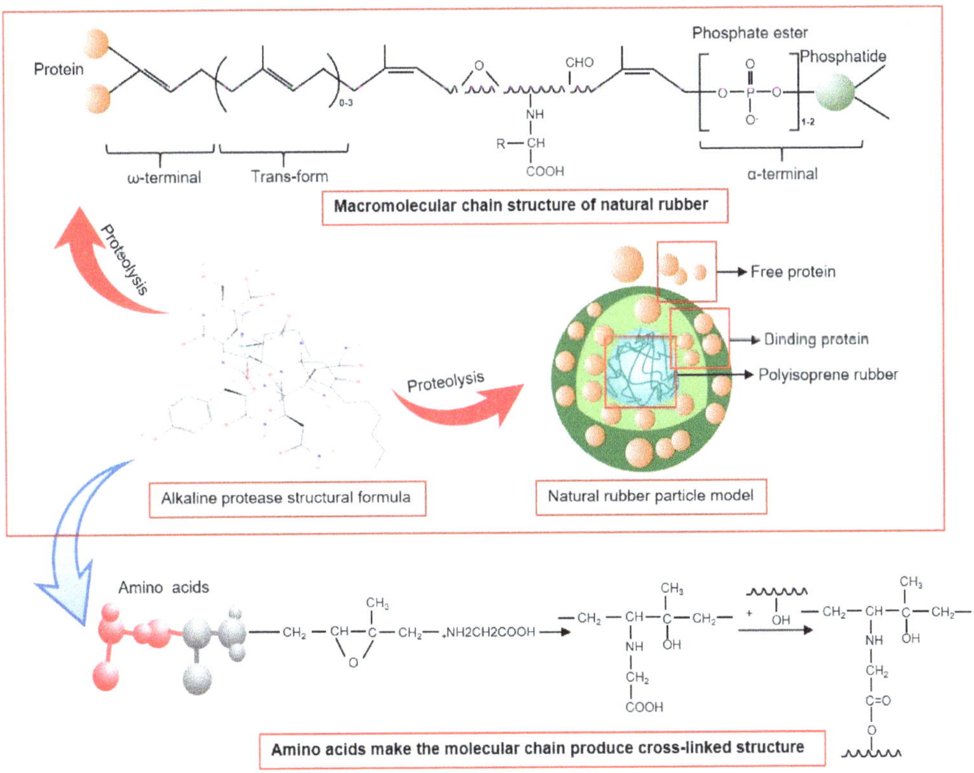

Figure 6. Alkaline protease enhances overall performance of NR.

3.8. Effects of Different Coagulation Methods on the Processing Properties of Natural Rubber

As natural rubber is subject to dynamic influence during processing and use, the testing of dynamic mechanical properties can provide an objective assessment of the comprehensive performance of natural rubber, thereby characterizing the viscoelastic trend throughout the process before and after vulcanization of the rubber material [38]. The trends of energy storage modulus G′ for A-NR, M-NR, and AC-NR are represented in Figure 7A, and it is evident that the trends were essentially similar within the same strain angle range. AC-NR exhibited the highest energy storage modulus G′, indicating its greater gel content, extensive molecular chain branching and cross-linking, and pronounced rigidity. As demonstrated in Figure 7B, AC-NR displayed a pronounced Payne effect (ΔG′), which was particularly evident under cyclic loading conditions with a large strain amplitude, as evidenced by the dependence of the viscoelastic energy storage modulus on the amplitude of the applied strain. When the strain amplitude was greater than 0.1%, the increased friction generated endogenous heat due to the intermolecular movement of natural rubber inside the fracture. The molecular chain of the rubber, composed of a cross-linked network, was easily destroyed, resulting in a slow decrease in the AC-NR storage modulus with an increasing amplitude. At a strain amplitude of 20% or greater, the storage modulus of natural rubber decreased rapidly and tended to a lower limit value, in the region where the decrease in the storage modulus G′ was accompanied by an increase in the loss modulus G″ [39], and this tendency was reflected in the trend of the value of the dielectric loss factor tanδ (Figure 7C). The dielectric loss factor tanδ of AC-NR was minimal, indicating that the alkaline protease enzyme caused cross-linking between the rubber molecular chains to form a network of entanglements. This has been shown to increase the rolling resistance of natural rubber during processing, resulting in a decrease in the processing performance of the material.

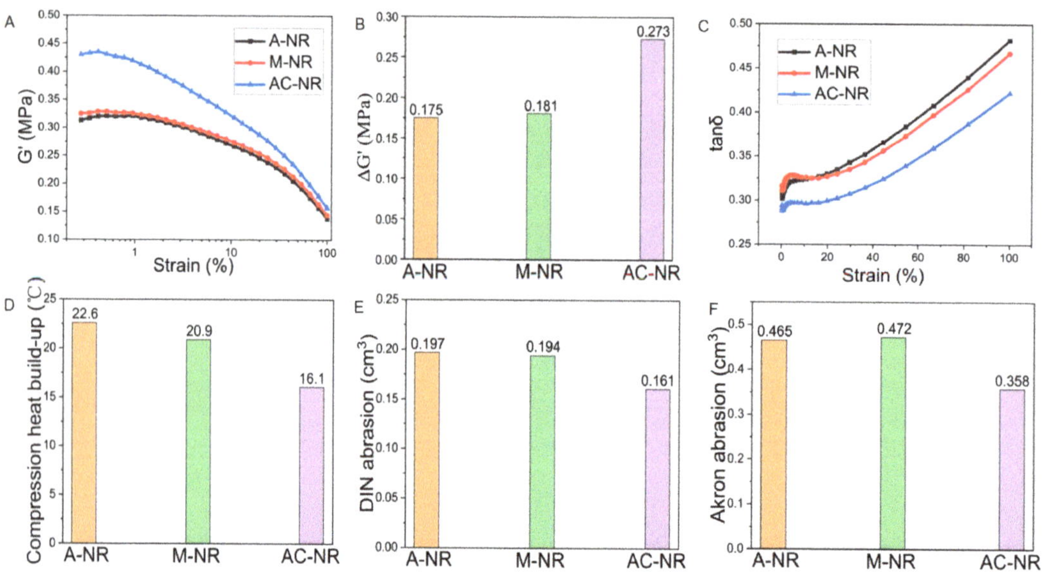

Figure 7. (**A**) Elastic modulus G′ of NR. (**B**) Payne effect of NR. (**C**) Dielectric loss factor tanδ of NR. (**D**) Final temperature rise of NR. (**E**) DIN abrasion of NR. (**F**) Akron abrasion of NR.

3.9. Influence of Different Coagulation Methods on Thermal Performance of Natural Rubber

The final temperature rise is defined as the phenomenon of delayed dissipation of heat that occurs when the molecular chains of rubber rub against each other under high-frequency loaded compression vibration of a vulcanized rubber cylinder sample [40]. Vulcanized rubber samples generate and accumulate heat and radiate it throughout the compression resilience test, and the lower the final temperature rise, the better the heat generation performance [41]. As demonstrated in Figure 7D, the final temperature rises of A-NR, M-NR, and AC-NR were 22.6 °C, 20.9 °C, and 16.1 °C, respectively, with AC-NR exhibiting the optimal thermogenic performance. This phenomenon can be attributed to the reduced protein content of AC-NR, where the presence of rubber particles with low protein content facilitates the reduction of deformation through particle rotation, thereby diminishing endogenous heat generation [42]. The findings demonstrate the efficacy of the alkaline protease–salt coagulation system in mitigating the final temperature rise of natural rubber, thereby enhancing its heat generation performance to a certain extent.

3.10. Influence of Different Coagulation Methods on the Abrasion Performance of Natural Rubber

DIN abrasion tests the static loss of natural rubber, and Akron abrasion tests the dynamic loss of natural rubber (see Figure 7E,F). These figures show that AC-NR had the best abrasion performance, with the smallest DIN abrasion loss volume and the smallest Akron abrasion loss volume of 0.161 cm^3 and 0.358 cm^3, respectively. This finding suggests that alkaline protease and calcium ions could enhance the structure of natural rubber molecular chain entanglement and cross-linking network construction, thereby improving resistance to physical friction damage. Furthermore, the final temperature rise of AC-NR was also the lowest, indicating that the improvement of natural rubber's thermogenic properties could reduce the damage caused by frictional heat to the network structure. DIN abrasion is a method of evaluating the abrasion resistance of natural rubber used in the resistance of natural rubber to abrasion in the manufacturing of softer rubber flooring and hose compounds, while Akron abrasion is utilized to assess the abrasion resistance of harder shoe soles and tire compounds. These findings suggest that AC-NR has a wide range of applications in the field of both soft and hard rubber products, thereby underscoring its versatility and significance in various industrial contexts.

3.11. Thermogravimetric Change Analysis of Natural Rubber Under Different Coagulation Methods

The ability of natural rubber structures to withstand impacts at high temperatures, and the thermal stability relationship between mass and temperature were evaluated by thermogravimetric analysis [43]. As demonstrated in Figure 8A, natural rubber undergwent a thermal degradation reaction, which is a smooth curve under a nitrogen atmosphere, and is a one-step reaction with one peak. The thermogravimetric (TG) curves were processed using software employing the double tangent method to obtain three heat loss characteristic temperatures: T_O (onset of degradation temperature), T_P (maximum degradation rate temperature), and T_f (termination of degradation temperature). The natural rubber temperature curves were calculated for percentage of mass consumption at T_P and T_f to the two degradation rates, degradation rate at peak (C_P) and final degradation rate (Cf), were then obtained. As presented in Table 5, the highest T_O, T_P, and T_f were found for AC-NR, with 329 °C, 401 °C, and 510 °C, respectively, which indicates the better thermal stability of AC-NR. The residual mass of AC-NR was the largest among the three, which also verifies that AC-NR had good thermal stability. The temperature at which the maximum rate of mass change occurred (i.e., the peak of the DTG curve in Figure 8B) is indicative of the thermal stability of AC-NR, with the peak temperature of AC-NR occurring on the rightmost side of the curve. A comparison of the degradation rates at T_P and T_f temperatures reveals that the characteristic temperature thermal degradation rates C_P and C_f of AC-NR were the lowest, indicating that the addition of alkaline protease and calcium ions could improve the thermal stability of natural rubber and better fight against its own decomposition at high temperatures.

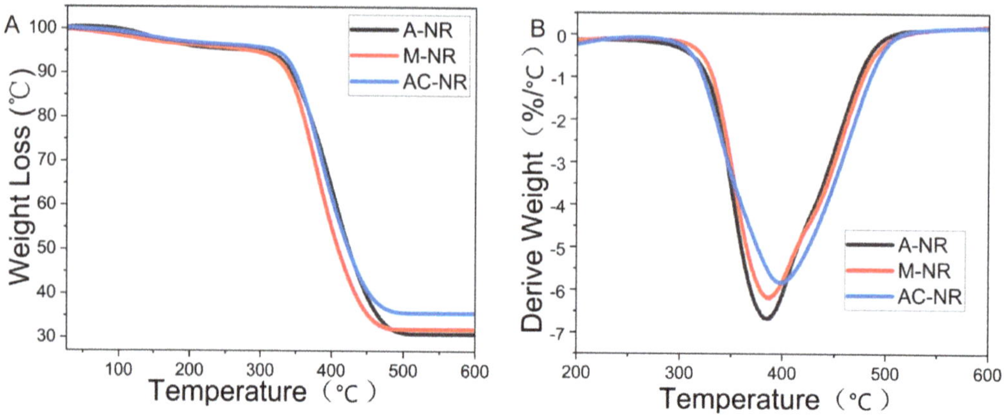

Figure 8. (**A**) Thermogravimetric curve of NR. (**B**) DTG curve of NR.

Table 5. Characteristic temperature and degradation rate of NR.

Samples	T_O/°C	T_P/°C	T_f/°C	C_P/%	C_f/%
A-NR	344	365	381	47	80
M-NR	347	369	390	45	79
AC-NR	350	373	396	40	76

3.12. Influence of Different Coagulation Methods on Glass Transition Temperature of Natural Rubber

When natural rubber is at the glass transition temperature (Tg), the specific heat capacity, pressure and expansion modulus coefficients, and viscosity all undergo a sudden change [44]. As shown in Table 6, the glass transition temperature (Tg) of AC-NR was −59.62 °C, which was the highest among the three. In the context of biomolecular structure, at the point of complete glass transition, the molecular chain of natural rubber is unable to move freely, but one segment can begin to move rapidly, due to the smooth nature of the rubber molecular chain segments, which exhibit a high-speed rebound. As the temperature continues to rise, the interconnected molecular chains of natural rubber are prompted to move rapidly in the direction of non-rebounding, resulting in the production of viscous fluidity on the surface [45]. The temperature change of AC-NR is distinct from that of A-NR and M-NR due to the decomposition of proteins by alkaline protease and the reaction of other rubber molecules with each other, thereby enhancing the cross-linking reaction between the rubber, and as the force between the molecular chains of proteins gradually increases, so too does the energy required for movement. Consequently, the glass transition temperature Tg of natural rubber is slightly accelerated, moving slightly towards higher temperatures.

Table 6. Glass transition temperature of NR.

Samples	Glass Transition Temperature/°C
A-NR	−60.58
M-NR	−60.15
AC-NR	−59.77

4. Conclusions

In this study, natural rubber samples with three different coagulation systems (i.e., acid, microbial, and alkaline protease/calcium chloride) were prepared in order to compare the differences in the comprehensive performance indexes of natural rubber under the new coagulation system and traditional coagulation systems. The investigation encompassed a comprehensive set of parameters, including the pH at the isoelectric point change during the coagulation of fresh rubber latex; protein content, gel content, molecular weight size and distribution, cross-linking density and scorch time, physicochemical indexes, mechanical properties, processing properties, thermal properties, abrasion properties, and other performance indexes of raw rubber. The results demonstrated that the utilization of alkaline protease/calcium chloride coagulation of natural rubber, in comparison with conventional acid coagulation, was capable of reducing the impact on soil acidification, enhancing molecular chain entanglement, gel content, and molecular weight of natural rubber, increasing tensile strength by 7.9 MPa, increasing tear strength by 5.3 kN/m, reducing the final temperature rise by 6.5 °C, and improving abrasion performance. In summary, the experimental results provide theoretical support for the preparation of environmentally friendly, high-performance natural rubber. The addition of alkaline protease and calcium chloride to the coagulation process of natural rubber has been shown to yield superior overall performance, while concomitantly reducing production costs and wastewater treatment expenses for rubber-processing enterprises. Consequently, this approach can be widely applied in the field of natural rubber primary processing.

Author Contributions: Conceptualization, T.D. and H.G. (Hongxing Gui); methodology, T.D., H.H. and L.D.; validation, T.D. and H.G. (Hongxing Gui); formal analysis, T.D. and T.Z.; investigation, L.D.; resources, H.G. (Hongxing Gui); data curation, T.D., Y.L. and J.L.; writing—original draft preparation, T.D.; writing—review and editing, T.D. and H.G. (Haoran Geng); project administration, T.D., Y.S. and L.Z. All authors have read and agreed to the published version of the manuscript.

Funding: This research was supported by the National Key R&D Program of China (2024YFD2300903; 2022YFD2301201), the Central Public-interest Scientific Institution Basal Research Fund for the Chinese Academy of Tropical Agricultural Sciences (No. 1630022020029), the Yunnan Provincial Major Science and Technology Special Programme (202302AB080059), and the Earmarked Fund for China Agriculture Research System (CARS-33).

Institutional Review Board Statement: Not applicable.

Data Availability Statement: The data presented in this study are available on request from the corresponding author.

Acknowledgments: We would like to thank the Ministry of Agriculture and Rural Affairs of China and the Department of Science and Technology of Hainan Province for their support.

Conflicts of Interest: Author Yun Li was employed by the company Mengla Tianye Rubber Sales Co., Ltd. The remaining authors declare that the research was conducted in the absence of any commercial or financial relationships that could be construed as a potential conflict of interest.

References

1. Ehabe, E.; Le Roux, Y.; Ngolemasango, F.; Bonfils, F.; Nkeng, G.; Nkouonkam, B.; Sainte-Beuve, J.; Gobina, M.S. Effect of maturation on the bulk viscosity and molecular chain length of cuplump natural rubber. *J. Appl. Polym. Sci.* **2002**, *86*, 703–708. [CrossRef]
2. Li, S.D.; Yu, H.P.; Peng, Z.; Li, P.S. Study on variation of structure and properties of natural rubber during accelerated storage. *J. Appl. Polym. Sci.* **1998**, *70*, 1779–1783. [CrossRef]
3. Lai, Y.; Jiang, Y.; Pan, J. Current situation and problems in the treatment of wastewater and exhaust gas of natural rubber processing in China. *Guangdong Chem. Ind.* **2018**, *13*, 179–180+198.
4. Liu, Y.; Liu, H.; Cheng, Y. Research progress of natural latex coagulation methods. *Guangdong Chem. Ind.* **2016**, *43*, 104–105.

5. Ding, L.; Chen, M.; Liu, P. Research progress of natural latex coagulation process. *Trop. Agric. Sci.* **2007**, *2*, 67–71.
6. Li, X.; Liao, S. Properties of microbiologically solidified natural rubber. *J. Mater. Sci. Eng.* **2012**, *30*, 779–782.
7. John, C.K.; Pillai, N.M. Improvements to assisted biological coagulation of Hevea latex. *J. Rubber Res. Inst. Malaya* **1971**, *23*, 138–146.
8. Deng, D.H.; Deng, P.Y. Discussion on several problems of current domestic standard rubber production. *Trop. Agric. Eng.* **1999**, *2*, 1–6.
9. He, Y. *Natural Rubber Processing Science*; Hainan Publishing House: Haikou, China, 2007; pp. 33–37.
10. Yuan, T. Research on the Effect of Metal Ions on the Aging Properties of Natural Rubber. Master's Thesis, Hainan University, Haikou, China, 2017.
11. Zhang, J.; Wang, J.; Zhao, Y.; Li, J.; Liu, Y. Study on the interaction between calcium ions and alkaline protease of bacillus. *Int. J. Biol. Macromol.* **2018**, *124*, 121–130. [CrossRef] [PubMed]
12. Sun, J.T.; Wang, W.; Zhang, P.; Zhao, S.G. Effect of synergistic effect of hardness and resilience on the wear properties of SBR vulcanised rubber Akron. *J. Macromol. Sci. B* **2012**, *51*, 1658–1667. [CrossRef]
13. Li, B.G. An investigation of problems involving amino acids → proteins. *Teach. Biol.* **2005**, *30*, 71–72.
14. John, C.K.; Pillai, N.M. Improvement of Hevea latex-assisted biocoagulation. *J. Malay. Rubber Inst.* **1971**, *23*, 138–146.
15. Mekkriengkrai, D.; Ute, K.; Swiezewska, E.; Chojnacki, T.; Tanaka, Y.; Sakdapipanich, J.T. Structural characterization of rubber from jackfruit and euphorbia as a model of natural rubber. *Biomacromolecules* **2004**, *5*, 2013–2019. [CrossRef] [PubMed]
16. Mekkriengkrai, D.; Sakdapipanich, J.T.; Tanaka, Y. Structural characterisation of terminal groups in natural rubber: Origin of nitrogen-containing groups. *Rubber Chem. Technol.* **2006**, *79*, 366–379. [CrossRef]
17. Lei, Y.; Lin, Q.; Hua, Y. Correlation analysis between soil pH and nutrient content of rubber plantation in Yangjiang Farm. *Trop. Agric. Sci.* **2013**, *33*, 30–32+37.
18. Suchiva, K.; Kowitteerawut, T.; Srichantamit, L. Structure properties of purified natural rubber. *J. Appl. Polym. Sci.* **2015**, *78*, 1495–1504. [CrossRef]
19. Li, H.; Fang, Q.; Wang, N.; Zhang, X.; Wang, Z. Relationship between aging and dynamic properties of NR with different vulcanization systems. *Rubber Ind.* **2010**, *57*, 659–663.
20. Angulo-Sanchez, J.L.; Caballero-Mata, P. Long chain branching in natural Hevea rubber-determination by gel permeation chromatography. *Rubber Chem. Technol.* **1981**, *54*, 34–41. [CrossRef]
21. Sekhar, B.C. Inhibition of hardening in Natural Rubber. *Rubber Chem. Technol.* **1962**, *35*, 889–895. [CrossRef]
22. Li, S.D.; Yu, H.P. Study on Thermal Degradation of Sol and Gel of Natural Rubber. *J. Appl. Polym. Sci.* **2000**, *75*, 1339. [CrossRef]
23. Tanaka, Y.; Tarachiwin, L. Recent advances in structural characterization of natural rubber. *Rubber Chem. Technol.* **2009**, *82*, 283–314. [CrossRef]
24. Li, Z.J. *Analysis and Testing of Natural Rubber*; China Agricultural University Press: Beijing, China, 2007; pp. 67–200.
25. Wei, Y. Analysis and Characterisation of the Molecular Structure of Natural Rubber. Ph.D. Thesis, Hainan University, Haikou, China, 2020. [CrossRef]
26. Zhu, W. Identification of different types of polyamides by infrared spectroscopy. *Plastics* **2009**, *38*, 114–117.
27. Zhang, H.; Jiang, B.; Zhao, F.; Peng, Z.; Liao, S. Analysis of molecular cross-linking network during storage of natural rubber. *J. Trop. Crops* **2017**, *38*, 939–944.
28. Liu, Q.; Li, H. Influence of processing technology on Mooney Viscosity and Mooney scorch time of rubber compound. *Rubber Sci. Technol.* **2020**, *18*, 641–643.
29. Bateman, L.; Sekhar, B.C. Significance of PRI in raw and vulcanised natural rubber. *Rubber Chem. Technol.* **1966**, *19*, 133.
30. Zhao, T.; Lu, Y.; Li, C. Study on the effect of proteins in natural latex on its thermal aging properties. *Chem. Eng.* **2018**, *22*, 1–3.
31. Zeng, Z.; Huang, M.; Chen, M. Aging resistances of natural rubber Vulcanizates from different clones of rubber trees. *Mater. Rep.* **2009**, *23*, 478–480.
32. Qu, Q.; Wang, X.; Liu, S.; Cui, J.; Xin, Z.; Wang, H.; Ding, S. Effect of alkaline protease content on the structure and properties of natural rubber. *Polym. Adv. Technol.* **2024**, *35*, e6549. [CrossRef]
33. Nasir, M.; Teh, G.K. The effects of various types of crosslinks on the physical properties of natural xubbex. *Eur. Polym. J.* **1988**, *24*, 733–736. [CrossRef]
34. Zhao, B. Flow Field-Induced Crystallisation of Polymers. Master's Thesis, University of Science and Technology of China, Hefei, China, 2010.
35. Wang, B.X.; Feng, X.M.; Liu, L. Effect of carbon black on crosslink density and mechanical properties of NBR composites. *Elastomer Body* **2013**, *23*, 16–20.
36. Yao, X.; Zhao, P.; Zhao, S. Research progress on the network structure of rubber vulcanization. *J. Qingdao Univ. Sci. Technol. Nat. Sci. Ed.* **2003**, *2*, 54–57+71.
37. Hu, W.; Lu, Q.; Zhou, W. Analysis of factors affecting the constant tensile stress of SBR-1500E. *Contemp. Chem. Ind.* **2015**, *44*, 2257–2259.

38. Wang, G. Application of RPA2000 rubber processing analyser in rubber research. *Spec. Rubber Prod.* **2001**, *1*, 56–62.
39. Xu, Z.; Song, Y.; Zheng, Q. Payne effect of carbon black filled natural rubber compounds and their carbon black gels. *Polymer* **2019**, *185*, 121953. [CrossRef]
40. Ni, L.; Cheng, L.; Zhou, M.; Tao, L.; Wang, J. Study on the compression heat generation performance of rubber materials for elastic elements. *Spec. Rubber Prod.* **2017**, *38*, 32–34.
41. Anlin. Compression Heat Generation Thermodynamic Coupling Finite Element Analysis of Carbon Black Filled Rubber. Master's Thesis, Qingdao University of Science and Technology, Qingdao, China, 2020.
42. Zhang, W. Research on the Low Proteinisation and Properties of Natural Rubber. Master's Thesis, Tianjin University, Tianjin, China, 2012.
43. Chen, M.; Yu, H.; Qian, J. Drying characteristics of epoxidised natural rubber by thermogravimetric method. *Trop. Agric. Sci.* **2011**, *31*, 70–72.
44. Li, F. Simulation of Fatigue Failure and Hysteresis Temperature Rise of Natural Rubber Nanocomposites with Thermodynamic Coupling. Ph.D. Thesis, Beijing University of Chemical Technology, Beijing, China, 2017.
45. Liu, J.; Chang, C.; Yu, Q. Comprehensive application of polymer physics knowledge points from rubber materials. *Polym. Bull.* **2017**, *1*, 83–86.

polymers

MDPI

Article

Critical Role of Rubber Functionalities on the Mechanical and Electrical Responses of Carbon Nanotube-Based Electroactive Rubber Composites

Md Najib Alam, Siraj Azam, Jongwan Yun and Sang-Shin Park *

School of Mechanical Engineering, Yeungnam University, 280, Daehak-ro, Gyeongsan 38541, Republic of Korea; mdnajib.alam3@gmail.com (M.N.A.); sirajazam@gmail.com (S.A.)
* Correspondence: pss@ynu.ac.kr

Abstract: Carbon nanomaterials, particularly carbon nanotubes (CNTs), are widely used as reinforcing fillers in rubber composites for advanced mechanical and electrical applications. However, the influence of rubber functionality and its interactions with CNTs remains underexplored. This study investigates electroactive elastomeric composites fabricated with CNTs in two common diene rubbers: natural rubber (NR) and acrylonitrile-butadiene rubber (NBR), each with distinct functionalities. For NR-based composites containing 2 vol% CNTs, mechanical properties, such as elastic modulus (2.24 MPa), tensile strength (12.48 MPa), and fracture toughness (26.92 MJ/m^3), show significant improvements of 125%, 215%, and 164%, respectively, compared to unfilled rubber. Similarly, for NBR-based composites, the elastic modulus (5.46 MPa), tensile strength (13.47 MPa), and fracture toughness (82.89 MJ/m^3) increase by 94%, 22%, and 65%, respectively, over the unfilled system. Although NBR-based composites exhibit higher mechanical properties, NR systems show more significant improvements, suggesting stronger chemical bonding between NR chains and CNTs, as evidenced by dynamic mechanical, X-ray diffraction, thermogravimetric, and thermodynamic analyses. The NBR-based composite at 1 vol% CNT content exhibits 261% higher piezoresistive strain sensitivity (GF = 65 at 0% $\leq \Delta\varepsilon \leq$ 200%) compared to the NR-based composite (GF = 18 at 0% $\leq \Delta\varepsilon \leq$ 200%). The highest gauge factor of 39,125 (1000% $\leq \Delta\varepsilon \leq$ 1220) was achieved in NBR-based composites with 1 vol% CNT content. However, 1.5 vol% CNT content in NBR provides better strain sensitivity and linearity than other composites. Additionally, NBR demonstrates superior electromechanical actuation properties, with 1317% higher actuation displacement and 276% higher electromechanical pressure compared to NR at an applied electric field of 12 kV. Due to the stronger chemical bonding between the rubber and CNT, NR-based composites are more suitable for dynamic mechanical applications. In contrast, NBR-based CNT composites are ideal for stretchable electromechanical sensors and actuators, owing to the high dielectric constant and polarizable functional groups in NBR.

Keywords: functional rubber; dielectric elastomers; mechanical properties; thermodynamics; filler–polymer interactions; electromechanical sensors and actuators

1. Introduction

Dielectric elastomer-based nanocomposites have recently garnered significant attention in the fields of science, engineering, and technology [1,2]. Modern engineering devices require high stretchability, durability, conductivity, and robustness. Rubber polymers

https://doi.org/10.3390/polym17020127

exhibit these features due to their low glass transition temperatures and excellent dynamic properties. However, despite these advantages, rubbers often present drawbacks, such as poor thermal, electrical, and magnetic properties, which limit their application as electroactive or magnetoactive materials [3,4]. To enhance these properties, rubber can be blended with particulate fillers that provide special functionalities [5,6]. For instance, blending rubber with magnetic fillers can introduce magnetic properties into the composite [7,8]. Similarly, incorporating thermal or electrical conducting fillers can significantly improve its thermal and electrical properties [9]. Additionally, the fabrication of rubber composites is cost-effective, straightforward, and reliable, making it a viable option for industrial applications.

Different types of reinforcing and non-reinforcing fillers can be used to enhance the desired properties of rubber composites. Filler morphology, surface characteristics, and filler dispersion are critical factors that govern the ultimate properties of these composites [10]. Inorganic fillers, such as silica, clay, and minerals, are well-known as effective reinforcing fillers for rubbers [11]. However, to achieve improved properties, these fillers often require surface treatments [12], which can be expensive and time-consuming. For over a century, carbon materials, such as carbon blacks, have been utilized as reinforcing fillers in tires [13]. Carbon materials exhibit strong π–π interactions, in addition to van der Waals forces with diene rubbers, making them suitable for reinforcement without requiring surface treatments. Furthermore, carbon materials significantly enhance heat buildup, abrasion, and aging resistance properties [14,15]. Consequently, carbon materials continue to be regarded as effective reinforcing fillers for commercial diene rubbers. However, due to their low aspect ratio and surface area, carbon blacks require larger quantities to achieve effective reinforcement. Among various carbon materials, graphitic nanocarbons, such as nanographite, graphene, and carbon nanotubes (CNTs), significantly improve the thermal and electrical properties of rubbers, making them ideal for advanced mechanical and electrical applications [16]. For example, multiwalled carbon nanotubes (CNTs) exhibit electrical conductivities of approximately 20–2 $\times$ 10^7 S/m [17] and thermal conductivity values around 3000 W/mK [18], which are suitable for preparing rubber composites with exceptional thermal and electrical conductivities. Recently, CNT-based elastomer nanocomposites have found numerous applications in flexible devices for strain sensing [19], robotics [20], energy harvesting [21], electromagnetic interference shielding [22], and other fields [23]. The properties of these composites depend on various factors, including filler morphology, dispersion, and filler–polymer interactions. For many advanced applications, mechanical stability is crucial while maintaining uniform conductivity in the nanocomposites [24–26]. Elastomers, such as natural rubber [27], styrene-butadiene-styrene rubber [28], carboxylated acrylonitrile-butadiene rubber [29], and silicone rubber [30–33], have been used to fabricate CNT-based nanocomposites, and their properties have been extensively investigated. Most studies have focused on properties related to filler characteristics, morphology, and distribution within the rubber matrix. Some researchers have developed strain-sensing composites based on natural rubber and acrylonitrile-butadiene rubber, incorporating conducting carbon materials [34–36]. From the reported gauge factor values, it has been observed that nitrile rubber-based composites exhibit better sensitivities compared to natural rubber-based composites [34–36]. Since strain sensitivities were studied separately, there has been limited exploration of how sensitivities vary with different functional rubbers. Several types of strain-sensing composites can be fabricated with conducting fillers embedded in dielectric polymer materials [37]. Among these, resistive and capacitive types are the most widely studied [37]. It has been demonstrated that embedding conducting materials in insulating materials leads to enhanced dielectric properties due to the formation of micro capacitors [38–41]. Since different rubbers have varying dielectric constant values, the capacitance of rubber composites can differ, which in turn, affects their strain sensitivities. In addition to electrical applications, acrylonitrile-butadiene rubber is widely used in membrane and oil field technologies due to its excellent oil resistance properties [42,43].

Although electroactive rubber composites have been developed and continue to be improved for enhanced sensitivity and mechanical robustness, it is well known that different rubbers exhibit varying affinities for filler materials, leading to differences in filler dispersion. Filler dispersion and, ultimately, filler–polymer interactions can result in variations in mechanical properties. Moreover, while rubber-based composites have been shown to be electroactive, the effects of functional groups on their performance have yet to be fully explored. Due to the dielectric variation and polarizability of the functional groups in rubber, these materials may exhibit different behaviors in the presence of an electric field. Some rubbers, such as acrylonitrile-butadiene rubber, contain polarizable nitrile functional groups that can rearrange their dipoles in the direction of the induced electric field, resulting in rubber chain contraction. These characteristics are particularly important for achieving variable stiffness in artificial muscles or soft robotics [44,45].

In this research, we aim to demonstrate that functional rubbers with different functional groups play a critical role in the properties of electroactive materials. To achieve this, we selected two different diene rubbers: natural rubber (cis-polyisoprene) and acrylonitrile-butadiene rubber (a copolymer of butadiene and acrylonitrile), each containing methyl and nitrile side chain functional groups, respectively. Due to differences in electronegativity and bonding environments, these rubbers exhibit distinct dipole moments. To create stretchable electroactive rubber composites, varying amounts of multiwalled carbon nanotubes (CNTs) were incorporated through solution blending. Diene rubbers are known for their high stretchability and outstanding fracture toughness, which make them ideal for producing stretchable thin films for various electronic applications. The physical, mechanical, and electroactive properties of the composites were thoroughly investigated and compared, with a focus on how the rubber functionalities influence these properties. The mechanism of filler–polymer interactions and their relationship with the observed properties is critically discussed. Finally, the composites were evaluated for electromechanical activities, such as strain sensing and actuation performance.

2. Materials and Methods

2.1. Materials

Two different types of rubber were used: natural rubber (NR, STR 5L, >95% cis-polyisoprene content) and acrylonitrile-butadiene rubber (NBR, KNB 35H, ~34% acrylonitrile content), both purchased from Kumho Petrochemicals, Seoul, Republic of Korea. Various rubber-grade curing additives, including zinc oxide, stearic acid, sulfur, accelerator tetramethyl thiuram disulfide (TMTD), and accelerator N-tert-butyl-benzothiazole sulfonamide (TBBS), were utilized to prepare the masterbatch rubber. The primary role of curing additives is to form crosslinks between the rubber chains, resulting in significant improvements in the elasticity and mechanical stability of the rubber compound. Masterbatches from different rubbers were prepared by mixing on a two-roll mill for 30 min using similar amounts of curing additives, expressed as parts per hundred rubber (phr). The compositions in the masterbatches were as follows: 100 g of rubber, 5 phr zinc oxide, 2 phr stearic acid, 1 phr TMTD, 1.75 phr TBBS, and 1.5 phr sulfur. The masterbatches were prepared using the same sequences and methods as described in previous studies [46,47]. Toluene (GR grade) was purchased from Daejung Chemicals & Metals, Siheung-si, Republic of Korea. The multiwalled carbon nanotube filler (CNT, CM-100, density = 1.74 g/cm^3) was obtained from Hanwha Nanotech Corporation Ltd., Seoul, Republic of Korea.

Raman spectra, X-ray diffraction patterns, scanning electron microscope (SEM), and transmission electron microscope (TEM) images are shown in Figure 1a–d. The characteristic D band and G band peaks at ~1345 cm^{-1} and 1575 cm^{-1}, respectively, and their relative intensity ratio in the Raman spectroscopy (Figure 1a) indicate that the CNT contains sp^2 and sp^3 carbon defects, which may exhibit chemical reactivity [48]. The X-ray diffraction pattern of the CNT corresponds well to the hexagonal graphite structure (JCPDS card no. 00-008-0415) with various planes, as shown in Figure 1b. However, some major peaks also match with the orthorhombic trans-polyacetylene crystal system (JCPDS card no. 00-047-2029), suggesting that the CNT may contain other carbonaceous impurities. The SEM image (Figure 1c) shows that the diameter of the CNT is less than 25 nm, and it exhibits a very high aspect ratio. The TEM image (Figure 1d) clearly shows that the CNT is multiwalled in structure.

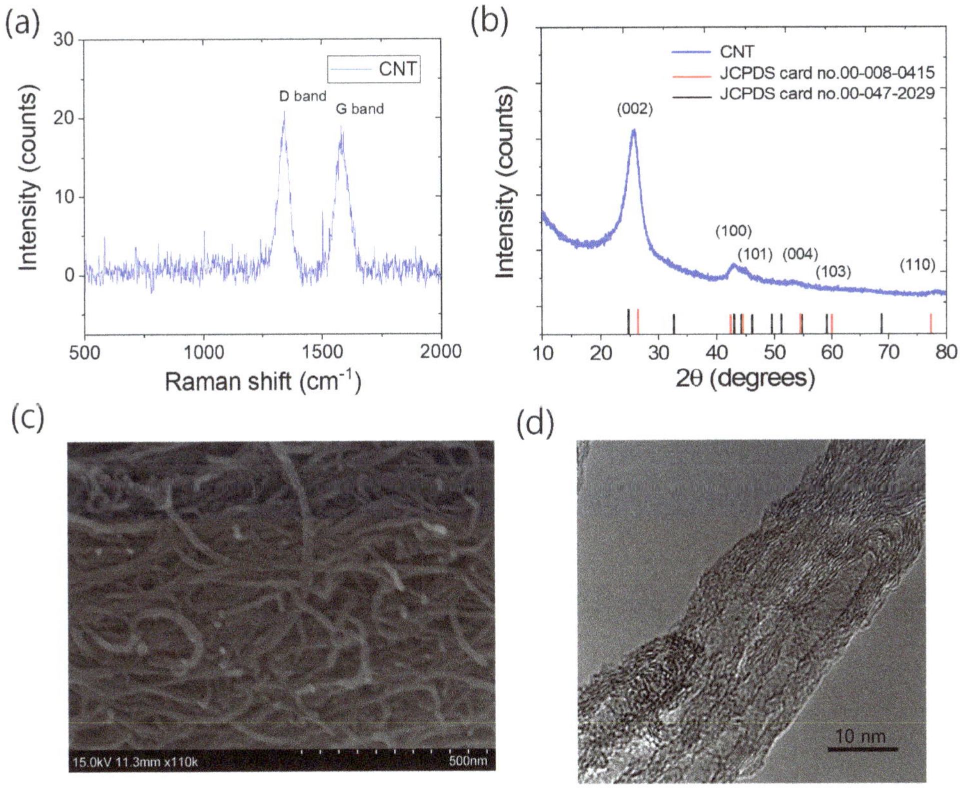

Figure 1. Characteristics of used carbon nanotubes. (**a**) Raman spectra, (**b**) XRD, (**c**) SEM image, and (**d**) TEM image.

2.2. Fabrication of CNT-Based Rubber Nanocomposites

To maintain uniform electrical conductivity in rubber nanocomposites, the solvent blending technique was employed instead of dry mixing [49]. In solvent mixing, the rubber chains are sufficiently free from intermolecular interactions, allowing the deposition of rubber molecules onto the filler surfaces. This process enhances the wettability of the filler surfaces by the rubber chains, thereby improving filler dispersion as well [34,35]. However, the environmental pollution associated with the use of toluene as a solvent can be mitigated through the proper handling of the solvent [46,47].

The method for fabricating CNT-based rubber composites is described elsewhere [33]. Initially, 25 g of masterbatch rubber was placed in a jar, and 100 mL of toluene was added and allowed to soak for approximately 24 h. After soaking, the rubber was homogenized by mechanical stirring to create a slurry. In a separate container, the required amount of CNT (0 to 2.5 vol%) was combined with 100 mL of toluene and placed in an ultrasonication bath for 30 min. The entire CNT solution was then added to the rubber slurry and mixed thoroughly with a mechanical stirrer for 10 min. The CNT-mixed rubber slurry was transferred to a flat tray and dried at 80 °C for approximately 24 h until it reached a constant weight. The dried compounded rubber was then placed in a mold and cured under compression at 150 °C for 15 min to ensure complete curing [49]. The cured samples were stored in a refrigerator to minimize the effects of environmental aging. Prior to the next characterization, the samples were brought to room temperature and allowed to rest for approximately 24 h. The different CNT concentrations in the masterbatch rubber are provided in Table 1.

Table 1. Mix compositions of different rubber composites.

Formulation	Masterbatch (g)	Amount of CNT in vol%
NR-unfilled	100	-
NR/(0.5 vol%)CNT	100	0.5 *
NR/(1.0 vol%)CNT	100	1
NR/(1.5 vol%)CNT	100	1.5
NR/(2.0 vol%)CNT	100	2
NR/(2.5 vol%)CNT	100	2.5
NBR-unfilled	100	-
NBR/(0.5 vol%)CNT	100	0.5
NBR/(1.0 vol%)CNT	100	1
NBR/(1.5 vol%)CNT	100	1.5
NBR/(2.0 vol%)CNT	100	2
NBR/(2.5 vol%)CNT	100	2.5

* To calculate the amounts of CNT in volume %, the densities of the unfilled rubbers were considered: 0.92 g/cm^3 for NR and 0.95 g/cm^3 for NBR.

2.3. Compressive and Tensile Mechanical Analyses

For compressive mechanical studies, cylindrical samples (height = 10 mm, diameter = 20 mm) were prepared. For tensile mechanical studies, dumbbell-shaped samples (ISO 37, type 2, gauge length = 25 mm) were punched from 1 mm thick vulcanized sheets. Static mechanical testing was conducted using a universal testing machine (UTM) equipped with a 1 kN load cell. The motor speeds of the UTM were set at 2 mm/min for compressive testing and 300 mm/min for tensile testing. For each specific value, the average of four test specimens is reported.

2.4. Dynamic Mechanical Studies

Dynamic mechanical thermal analyses were conducted using a dynamic mechanical analyzer (DMA Q800, New Castle, DE, USA) in tension mode, with a temperature range of −100 to +100 °C and a heating rate of 2 °C/min, in a nitrogen atmosphere. Rectangular test specimens with uniform dimensions (10 mm × 4 mm × 1 mm) were punched from the rubber sheets and used for the DMA analyses.

2.5. X-Ray Diffraction Studies of Rubber Composites

XRD analyses of the rubber composites were conducted using 2 mm thick rubber sheets in an X-ray diffractometer (PANalytical XpertPro, Malvern, Worcestershire, UK) with CuKα (0.154 nm) radiation. The scans were performed by varying Bragg's angles (2θ) from 10° to 30° at a scanning rate of approximately 4° per minute.

2.6. Scanning Electron Microscopic Studies

The morphologies of the tensile fractured surfaces were investigated using a scanning electron microscope (FE-SEM, S-4800, Hitachi, Tokyo, Japan). Prior to SEM analysis, the surfaces were coated with platinum using a sputtering machine.

2.7. Thermogravimetric Analyses

Thermogravimetric analyses of the rubber compounds were conducted using a TA instrument (SDT Q600, New Castle, DE, USA) with a temperature range of 20–800 °C in a nitrogen atmosphere.

2.8. Swelling and Thermodynamic Studies

To study the solvent swelling characteristics of the different rubber composites, known amounts of vulcanized rubber were immersed in toluene for 7 days to reach equilibrium at room temperature. Afterward, the samples were removed, and the weight of the swollen sample was measured. The weight of the swollen solvent was determined by subtracting the initial weight from the weight of the swollen sample. Finally, the swelling index was calculated based on the weight of the swollen solvent relative to 100 g of rubber compound.

The swelling of the solvent by rubber is a thermodynamic process where no internal energy change is assumed [50]. Hence, the Gibbs equation for thermodynamic equilibrium can be expressed as shown in Equation (1):

$$\Delta G = -T\Delta S \tag{1}$$

where ΔG represents the Gibbs free energy, and ΔS represents the conformational entropy change due to solvent swelling. Since mixing filler with rubber is also related to an entropic change, the change in entropy from unfilled to filled rubber may be attributed to filler–polymer interactions. ΔG can be measured according to the Flory–Huggins equation [51], as shown in Equation (2):

$$\Delta G = RT\left[\ln(1 - V_r) + V_r + \chi V_r^2\right] \tag{2}$$

where R is the gas constant, T is the temperature in the absolute scale, V_r is the volume fraction of rubber in the swollen specimen, and χ is the interaction parameter in the rubber–solvent system. The V_r values were obtained from equilibrium swelling data using Equation (3):

$$V_r = \{(w_r/d_r)\}/\{(w_r/d_r) + (w_s/d_s)\} \tag{3}$$

where w_r and w_s are the weights of rubber and solvent, respectively, and d_r and d_s are the densities of rubber and solvent, respectively. The density of the solvent (toluene) was 0.87 g/cm^3, and the χ values for the NR–toluene and NBR–toluene systems were taken as 0.378 and 0.425, respectively, from existing literature [52,53].

2.9. Strain Sensing and Electromechanical Actuation Studies

The piezoresistive strain sensing behavior of the rubber composites was studied using a Keithley 2634B source meter under uniaxial tensile strain, employing dumbbell-shaped specimens. Uniform strain was applied using a tensile testing machine, and the corresponding resistances were recorded. The relative resistance was calculated using the formula presented in Equation (4):

$$\text{Relativeresistance}(\%) = (R - R_0) \times 100/R_0 \tag{4}$$

where R is the resistance at a certain strain, and R_0 is the initial resistance.

The capacitance of the composites was measured using cylindrical samples (d = 20 mm, h = 10 mm), with two electrodes placed on opposite sides.

The dielectric actuation performances of different unfilled rubbers were studied using 0.1 mm thick electrodes, based on 2 phr CNT and room temperature vulcanized silicone rubber, which were painted on both sides of 1 mm thick, 6 cm × 6 cm square-shaped test specimens, as described elsewhere [54,55]. The actuation displacement under DC voltages was measured using a laser sensor (Opto NCDT 1302, Micro-Epsilon Messtechnik, Ortenburg, Germany). The schematics of the actuator sample and the actuator setup are provided in Figure 2a,b.

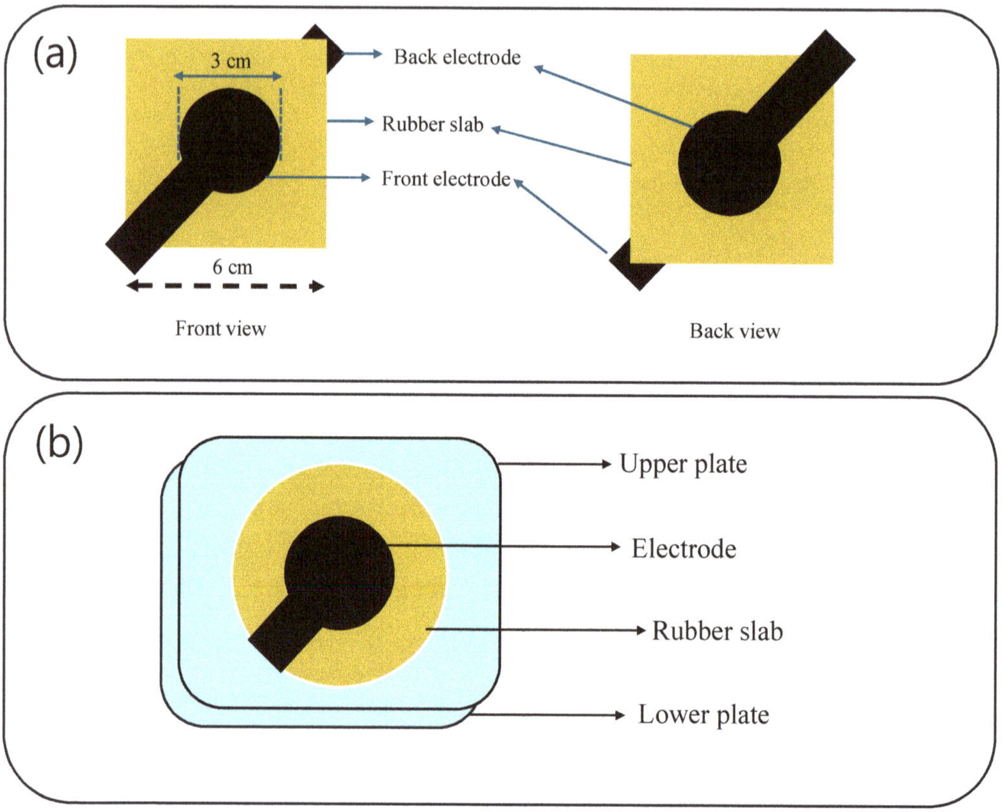

Figure 2. (**a**) Schematic of actuator samples and (**b**) schematic of actuator.

The equivalent electromechanical pressure (P_{eq}), which is twice that of the electrostatic pressure, was determined according to the formula presented in Equation (5):

$$P_{eq} = \varepsilon_0 \varepsilon_r V^2 / z^2 \quad (5)$$

where ε_0 is the vacuum permittivity (8.85×10^{-12} F/m), ε_r is the dielectric constant of the rubber, V is the applied voltage, and z is the thickness of the rubber slab during deformation. The value of z was calculated by considering the change in the diameter of the electrode, assuming no net volume change after actuation. The ε_r values of unfilled NR and NBR were taken as 2.70 and 8.28, respectively, at room temperature and 1 kHz frequency, as reported in the literature [56,57].

3. Results and Discussion

3.1. Mechanical and Physical Properties

The compressive mechanical properties of the rubber compounds are presented in Figure 3a–c. Figure 3a,b shows the compressive stress–strain relationships for NR- and NBR-based rubber composites. From these figures, it is evident that the stress value increases with increasing strain and CNT content. However, no significant increase in the ultimate stress values at 35% compressive strain is observed beyond 1.5 vol% CNT content. The substantial increase in ultimate stress values with respect to the filler content up to 1.5 vol% CNT may be attributed to the development of filler–rubber interfacial interactions, which become saturated at this level [58]. The slight increase in ultimate stress values beyond this critical filler content suggests that no significant additional filler–rubber networks are developed, and filler aggregation may occur beyond 1.5 vol% CNT content. It is believed that, above a certain threshold, the filler particles are weakly adhered to each other via van der Waals forces, resulting in minimal additional stress in the rubber composites. This hypothesis is further supported by the elastic modulus values presented in Figure 3c. From this figure, it is clear that the elastic modulus increases more rapidly up to 1.5 vol% CNT loading in the rubber composites. Beyond this threshold, there is only a slight increase in the elastic modulus due to the rigidity of the CNT, which is much higher compared to the rubber matrix.

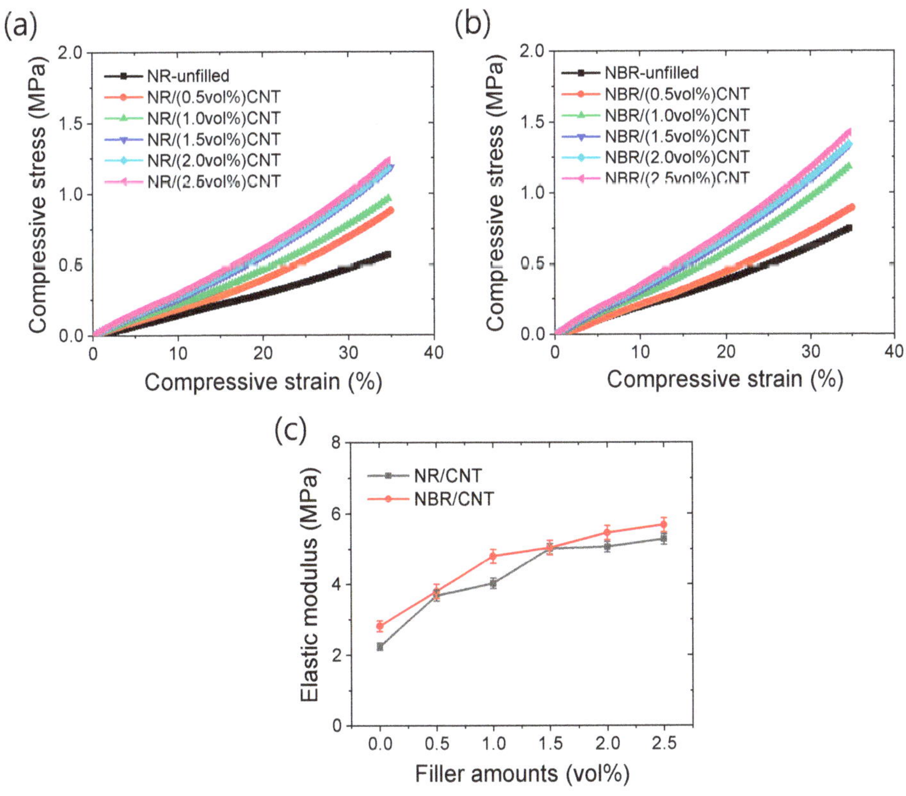

Figure 3. Compressive mechanical properties. (**a**) Stress–strain curves of NR-based nanocomposites, (**b**) stress–strain of NBR-based nanocomposites, and (**c**) elastic modulus.

The tensile mechanical properties of the rubber compounds are presented in Figure 4a–f. Figure 4a and 4b show the tensile stress–strain curves for NR- and NBR-based rubber compounds, respectively. The stress values of the unfilled rubber are significantly enhanced by the addition of reinforcing CNT in the rubber matrix. The stress–strain curves also reveal that there are no significant enhancements in the stress–strain slopes after the addition of 2 vol% CNT. This suggests that rubber–filler interfacial interactions are optimized at this filler concentration. A slight enhancement in the stress–strain slope beyond the optimum filler amount may be attributed to filler–filler interactions, which could result from anisotropic filler distribution [58]. Interestingly, when comparing the stress–strain slope values of different rubber systems, the slope values are higher for NBR systems than for NR systems at lower strain amplitudes. However, the slope value for NBR systems decays more noticeably than that for NR systems, as strain increases. This reduction in the stress–strain slope at a higher strain may be due to the breakdown of filler networks [59].

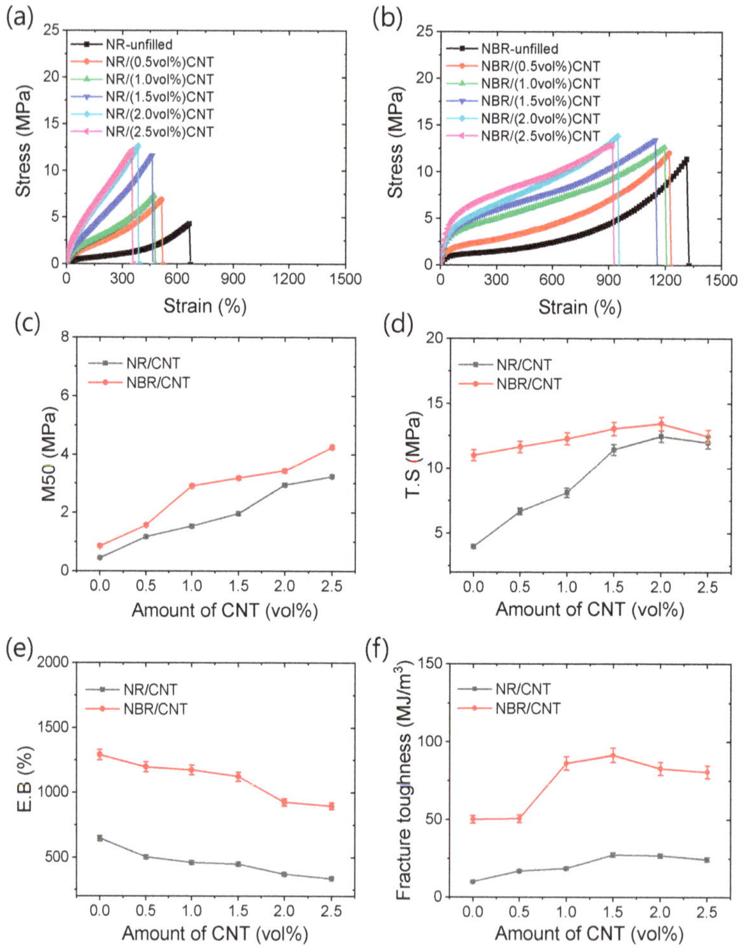

Figure 4. Tensile mechanical properties of rubber nanocomposites. (**a**) Stress–strain of NR-based composites, (**b**) stress–strain of NBR-based composites, (**c**) M50, (**d**) T.S, (**e**) E.B, and (**f**) fracture toughness.

From Figure 4c, it can be observed that the moduli at 50% strain (M50) are higher for NBR-based composites compared to NR-based composites at similar filler contents. However, at strains exceeding 300%, the moduli of NBR-based composites are lower than those of NR-based composites. These results suggest that the addition of CNTs to NBR rubber primarily enhances the viscous modulus rather than the elastic modulus, likely due to weak physical bonding, which degrades more easily with increased strain [59]. This hypothesis is further supported by the dynamic mechanical studies presented in a later section. Since physical bonding, such as that through van der Waals forces, diminishes more rapidly with increasing deformation, it can be inferred that the CNT filler is primarily bound by a higher number of physical bonds. As a result, the modulus is higher at lower deformation but degrades more quickly with increasing strain. In contrast, the NR matrix is likely to form a higher number of chemical bonds with the filler, as discussed later in the dynamic mechanical and thermal properties sections. These chemical bonds are more stable and do not degrade as easily as physical bonds, allowing the NR-based composites to retain their strength at higher strains. Additionally, the stress–strain slope may degrade up to a certain strain and then increase again due to the strain-induced crystallization of the rubber chains.

The specific tensile mechanical properties, including tensile strength (T.S.), elongation at break (E.B.), and fracture toughness, for the composite materials are shown in Figure 4d, Figure 4e, and Figure 4f, respectively. From Figure 4d, it is evident that NBR-based composites exhibit higher tensile strength than NR-based composites. Additionally, the NBR-unfilled compound demonstrates higher tensile strength than the NR-unfilled compound. However, the tensile strength values increase more rapidly with the addition of filler in the NR systems. In both rubber systems, the tensile strength reaches a maximum at 2 vol% CNT content, indicating that uniform filler dispersion is optimized at this level. The elongation properties are presented in Figure 4e, where it is observed that the addition of filler reduces the polymer chain stretchability. These results suggest that the filler introduces additional forces, such as chemical or physical bonding, which restrict the movement of the rubber chains. More importantly, from Figure 4f, it can be seen that fracture toughness increases rapidly with the addition of filler. The maximum toughness values are achieved at 1.5 vol% CNT content, and beyond this amount, the fracture toughness slightly decreases. It is clear that the tensile strengths of NR- and NBR-based composites at 2 vol% CNT content reach maximum values of 12.48 MPa and 13.47 MPa, which are approximately 216% and 22% higher, respectively, than those of the neat rubbers. Similarly, the fracture toughness values of NR- and NBR-based nanocomposites at 1.5 vol% CNT content reach peak values of 27.46 MJ/m^3 and 91.43 MJ/m^3, which are 169% and ~82% higher, respectively, than those of the neat rubbers. These results suggest that both types of rubber exhibit a similar trend in mechanical properties with respect to the filler amounts, with the optimal values for tensile mechanical properties obtained at 1.5 to 2 vol% CNT content. Beyond this point, the filler particles tend to aggregate significantly, leading to a reduction in mechanical properties.

To better understand the interfacial interactions between rubber and CNT, the dynamic mechanical analyses of unfilled rubber and 2 vol% CNT-containing rubber composites were conducted. The different dynamic mechanical properties are presented in Figure 5a–c. From Figure 5a, it can be seen that the loss modulus reaches its maximum value at the glass transition temperature (T_g). As the temperature increases beyond T_g, the loss modulus decreases. The loss modulus after T_g is higher in the 2 vol% CNT-filled rubber composites compared to the unfilled rubber systems. This may be due to the enhanced viscosity resulting from rubber–filler interfacial interactions. Higher loss modulus values in neat NBR compared to neat NR in the rubbery state suggest that stronger polymer–polymer interactions exist in NBR. To confirm a more homogeneous filler dispersion, the solvent blending technique was employed in this study, which involves solvent-diluted polymer chains. The addition of CNT filler significantly improves the loss modulus values, likely due to the development of filler–rubber interactions. Similarly, the storage modulus in Figure 5b increases with the addition of the rigid filler. The loss factor (tan δ) plots for different rubber compounds are shown in Figure 5c. It was found that the height of the tan δ peak at T_g is reduced with the addition of CNT. However, a slight increase in T_g suggests that additional chemical bonds may form between the active sites of CNT and the NR matrix, or that CNT may enhance the number of sulfur crosslinks, as NR has a higher number of crosslinking sites than NBR. In the case of NBR, a slight decrease in T_g indicates that CNT may interact with the rubber through van der Waals-type physical forces, which enhances the modulus values. For strong filler–polymer interactions, the height of the tan δ peak always decreases but does not always result in an increase in T_g compared to the neat polymer [60]. From the loss factors at higher temperatures, it can be inferred that NBR-based composites exhibit higher heat buildup properties than NR-based composites. Therefore, for dynamic applications, such as in tires, CNT-based NR composites may be more preferable.

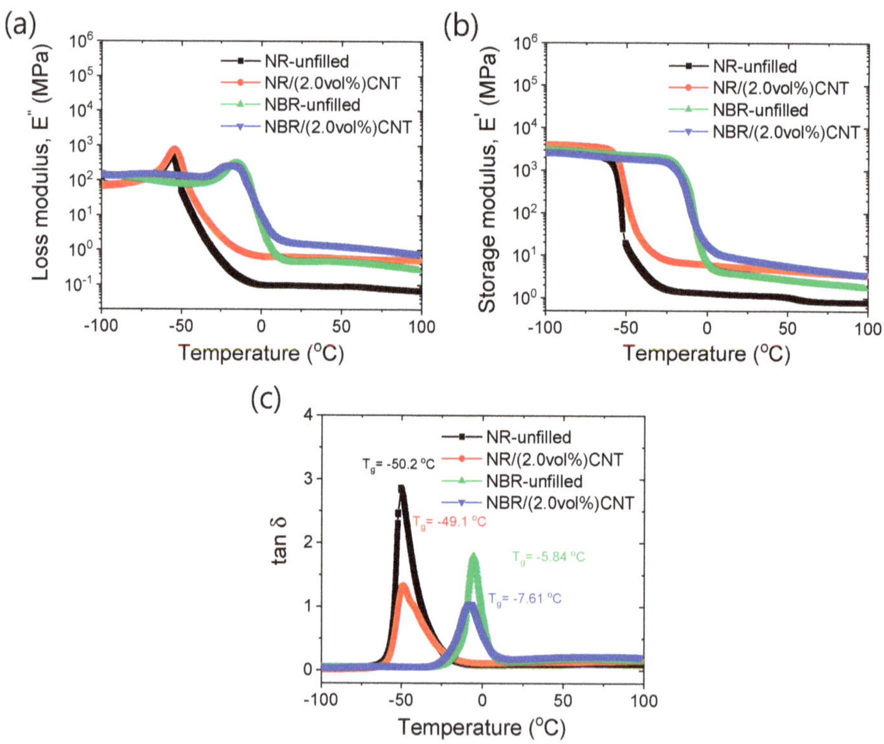

Figure 5. Dynamic mechanical properties. (**a**) Loss modulus, (**b**) storage modulus, and (**c**) tan δ.

To investigate the state of filler distribution in the rubber matrix, XRD and SEM studies were performed. The XRD patterns of different rubber composites are provided in Figure 6a,b. The broad XRD peak starting from 15° to 30° can be attributed to the amorphous rubber phase [61]. From Figure 6a,b, it is evident that, as the amount of CNT increases, a stronger 002 plane peak at 26.426° for CNT is observed. The XRD curves show that below 2 vol%, there is no intense 002 plane peak. However, with increasing filler amounts starting from 2 vol%, more intense 002 plane peaks are obtained for both rubber composites. The intense 002 plane peak may be attributed to the aggregation of CNT particles in the rubber matrix [61]. The smaller increments in mechanical properties and thermodynamic parameters (discussed in a later section) at filler contents higher than 1.5 vol% suggest filler aggregation in the rubber matrix. Furthermore, the higher intensity of CNT peaks in NR systems compared to NBR systems after the optimum filler amount may indicate greater filler particle aggregation. This could be due to the lower physical interactions of NR chains with CNTs, which promote the more pronounced aggregation of the filler particles.

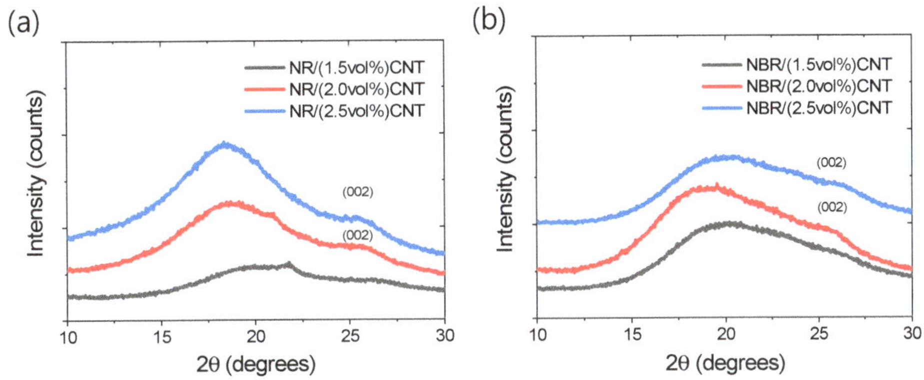

Figure 6. XRD patterns of rubber composites. (**a**) NR-based and (**b**) NBR-based.

The SEM images of unfilled and filled rubber composites are provided in Figure 7a–d. Comparing Figure 7a with 7b and Figure 7c with 7d, it is evident that both rubbers exhibit excellent filler dispersion, with the rubber additives well distributed in the 2 vol% CNT-containing composites. The rough structures of the fractured surfaces in the SEM images (Figure 7b,d) clearly indicate that the composites were toughened after the addition of reinforcing CNT to the neat rubber (Figure 7a,c).

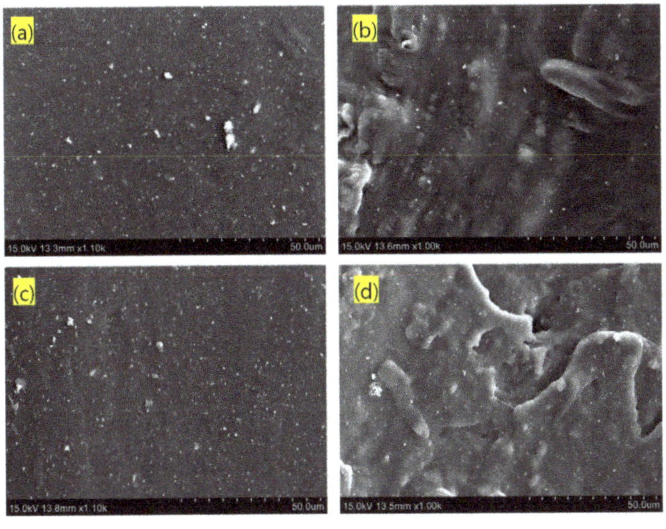

Figure 7. SEM images of tensile fractured surfaces. (**a**) NR-unfilled, (**b**) NR/(2.0 vol%)CNT, (**c**) NBR-unfilled, and (**d**) NBR/(2.0 vol%)CNT.

3.2. Thermal Degradation Behaviors

The thermal properties of selected rubber compounds were studied, and the results are provided in Figure 8a–c. Detailed discussions of different fragments of thermally degraded NR and NBR can be found elsewhere [62–65]. From the TGA curves in Figure 8a,b, it is evident that CNT at 2 vol% significantly enhances the initial degradation temperature (~3% weight loss) of NR but has a negligible effect on NBR. However, the initial degradation temperature of NBR-based compounds is much higher compared to NR-based compounds. Additionally, the char residues of NBR-based compounds are comparatively higher than those of NR-based compounds. These results indicate that NBR inherently has higher thermal stability but exhibits less chemical interaction with CNT. On the other hand, NR has lower thermal stability, but its interaction with CNT, likely through stronger chemical bonding, improves the thermal stability by increasing the initial degradation temperature and char residue. It is evident that 2 vol% CNT can enhance the initial degradation temperature by approximately 15 °C in the NR/(2.0 vol%) CNT composite. The derivative thermogravimetry (DTG) curves in Figure 8c suggest that NBR-based compounds have higher thermal stability and a lower degradation rate compared to the NR-based compound [62–65].

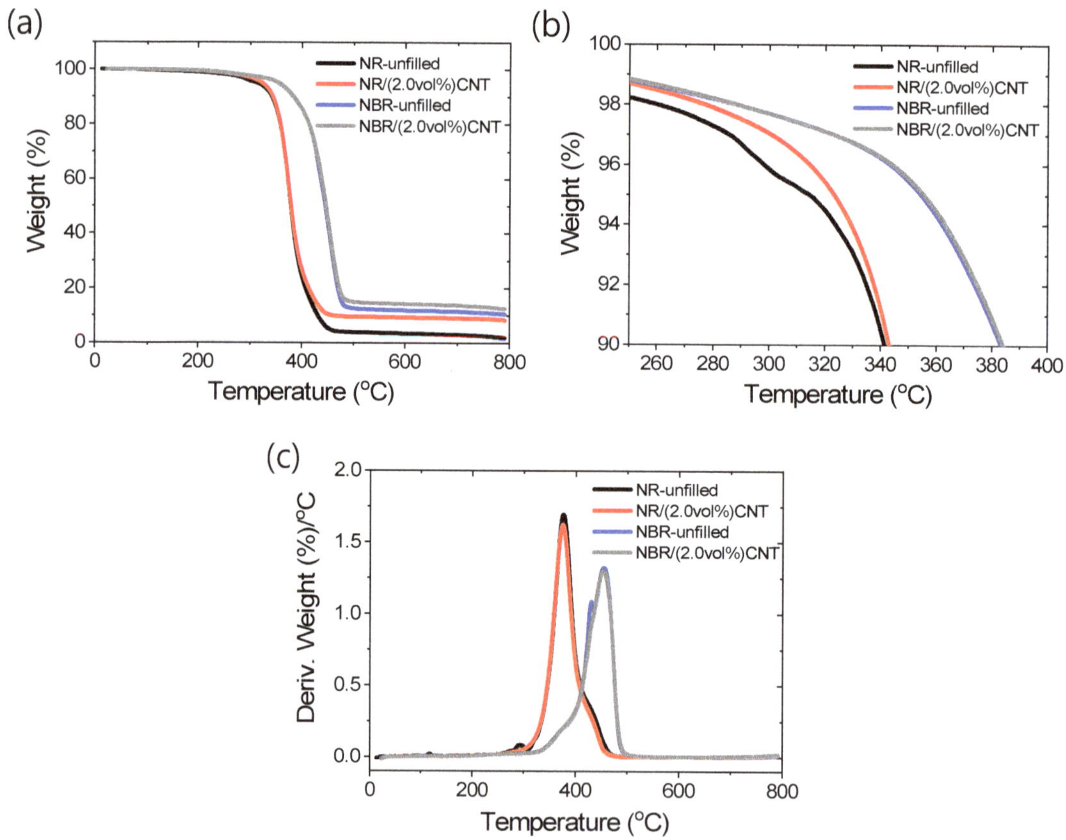

Figure 8. Thermal degradation behavior of rubber compounds. (**a**,**b**) TGA curves and (**c**) DTA curves.

The DMA analyses reveal that NR exhibits stronger chemical interactions with CNTs, forming chemical bonds between the rubber and filler, which results in a shift of T_g to a higher temperature. In contrast, the T_g of CNT-containing NBR composites shifts to a lower temperature, while the height of the tan δ peak decreases, suggesting that physical-type interactions exist between NBR and the filler. This finding is further supported by the TGA analyses, which confirm that NR chains are grafted with stronger chemical bonds to CNT particles, enhancing the thermal stability of the composite compared to the physically bonded CNT in NBR.

Regarding the chemical structures, NR consists of polyisoprene units, while NBR contains butadiene and acrylonitrile units. The methyl group in NR is highly reactive [66], which may form chemical bonds with the sp3 or reactive carbon atoms on the CNT surface. Since the vulcanization reaction is primarily a radial-type reaction, curing additives generate radicals [66]. These radicals can interact chemically with the active sites of the rubber chains and CNT particles, forming chemical bridges between them. On the other hand, NBR contains nitrile groups that are chemically less reactive for crosslinking, and the backbone methylene protons cannot form crosslinks with the CNT particles, likely due to steric hindrance. However, due to its permanent dipole, NBR has more physical interactions, such as dipole–dipole and n–π interactions, compared to NR-based composites.

3.3. Swelling and Thermodynamic Properties

The swelling study is an essential and straightforward method for understanding the polarity of rubber and its intermolecular interactions [67]. The different swelling and thermodynamic properties are presented in Figure 9a–d. From Figure 9a, it is evident that neat NBR has a lower swelling index value than neat NR. This can be attributed to the polar nitrile group in NBR, which leads to higher dipole–dipole interactions, as well as other van der Waals interactions. Consequently, NBR exhibits stronger polymer–polymer interactions, resulting in lower swelling in toluene compared to NR. With the addition of CNT, NR shows a more pronounced reduction in its swelling properties compared to NBR. Composites with 1.5 to 2 vol% CNT content exhibit the lowest swelling index values. However, beyond 1.5 vol% CNT, the swelling index does not significantly decrease and may even increase with further CNT addition. Therefore, to achieve a reduced swelling index, the CNT content should not exceed 2 vol% in the rubber.

The types of interactions present in the rubber compounds are further clarified in Figure 9b–d. From Figure 9b,c, it is clear that the ΔG values are more negative, and the ΔS values are significantly higher for NBR-based composites compared to NR-based composites. From a thermodynamic perspective, this indicates that NBR-based composites have stronger total interactions, including polymer–polymer and filler–polymer interactions, compared to NR-based composites. The higher binding strength in NBR-based composites is also reflected in their higher fracture toughness values and thermal stabilities compared to NR-based composites. The contribution of filler to the total strength is determined by subtracting the ΔS value of unfilled rubber from that of filled rubber. Figure 9d represents the specific contribution of filler to the ΔS value by ΔΔS. It is evident from this figure that ΔΔS values are lower for NBR-based rubber composites compared to NR-based composites at filler contents below 1 vol%. However, the trend reverses at higher filler content. This could be due to higher polymer–polymer interactions controlling filler dispersion at lower filler concentrations, while at higher filler contents, the increased filler–polymer interactions contribute more to reinforcement. At higher levels of filler dispersion, the n–π stacking interactions between the nitrogen atom from the nitrile group in NBR and the π-electrons of CNT play a significant role in contributing to filler reinforcement [68]. From this figure, it is evident that the contribution of filler–polymer interactions sharply decreases with more than 2 vol% filler content. Therefore, the CNT filler content in the rubber should not exceed 2 vol%.

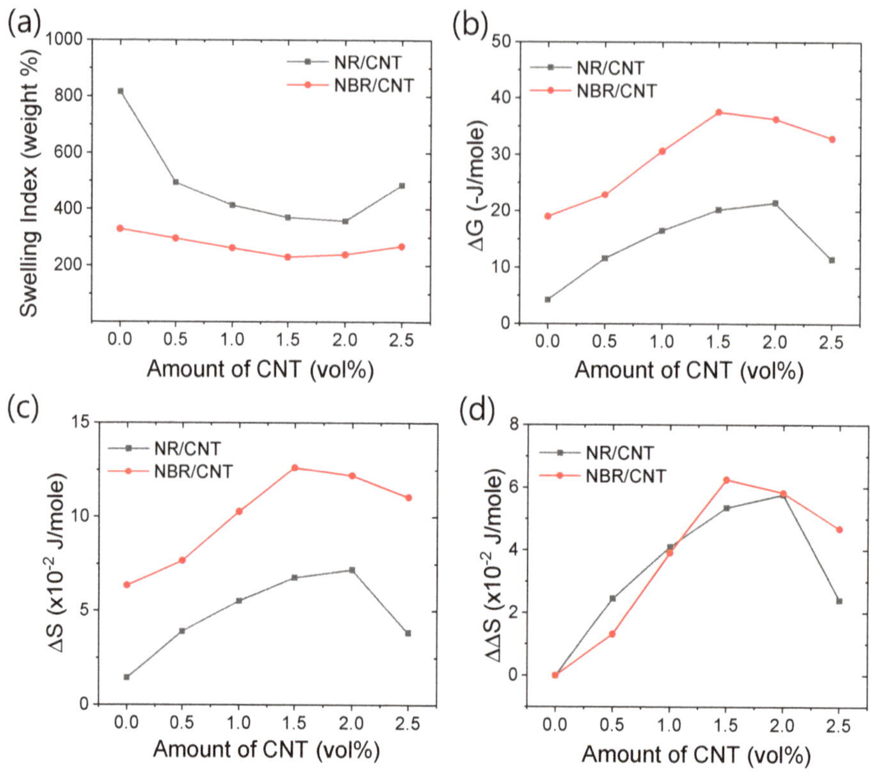

Figure 9. Characteristics of swelling and thermodynamic properties. (**a**) Swelling index, (**b**) ΔG, (**c**) ΔS, and (**d**) ΔΔS.

3.4. Electrical Responses of Rubber Nanocomposites

Rubber composites become electroactive upon the introduction of electrical conductivity. To achieve electrical conductivity in the rubber composite, the conducting filler content must reach the percolation threshold. In this study, significant electrical conductivity was observed in both NR- and NBR-based composites after the addition of 1 vol% CNT. Based on the mechanical properties of the rubber composites, it can be concluded that the filler content should be kept below 2 vol% to obtain optimal mechanical performance. Therefore, composites containing 1 to 2 vol% CNT were further investigated for their strain-sensing properties.

The strain-sensing behaviors of the rubber composites are shown in Figure 10a–d, with gauge factor (GF) values calculated for the strain range ($\Delta\varepsilon$) of 200%, excluding the end ranges. From these figures, it is clear that the relative resistance change increases with increasing strain amplitude. These figures also demonstrate that the linearity in the relative resistance changes with strain decreases as the strain increases. However, below 100% strain, the strain sensitivity curves maintain better linearity (Figure 10b,d). Additionally, it is evident that the linearity of strain sensitivity can be enhanced at the expense of reduced sensitivity with higher filler amounts. Figure 10e shows that, for similar strain ranges ($0\% \leq \Delta\varepsilon \leq 200\%$), NBR-based composites exhibit higher GFs compared to NR-based rubber composites. From this figure, it is clear that the GF values of NBR-based composites are significantly higher than those of NR-based composites at similar filler amounts. The NBR/(1 vol%) CNT composite shows a GF of 65 at the $0\% \leq \Delta\varepsilon \leq 200\%$ strain range, which is 260% higher than the GF of NR/(1 vol%) CNT. Higher GFs are obtained at lower filler amounts, because the highest resistance change occurs within the filler percolation region. As a result, higher GFs can be achieved at this level. Above the filler percolation threshold, sensitivity or GF may decrease, but linearity can be enhanced [33]. The gauge factor can reach as high as 39,125 at maximum deformation for the 1 vol% CNT-containing NBR composite. However, larger deformations or higher working strains should be avoided due to the potential for high hysteresis loss [32,33].

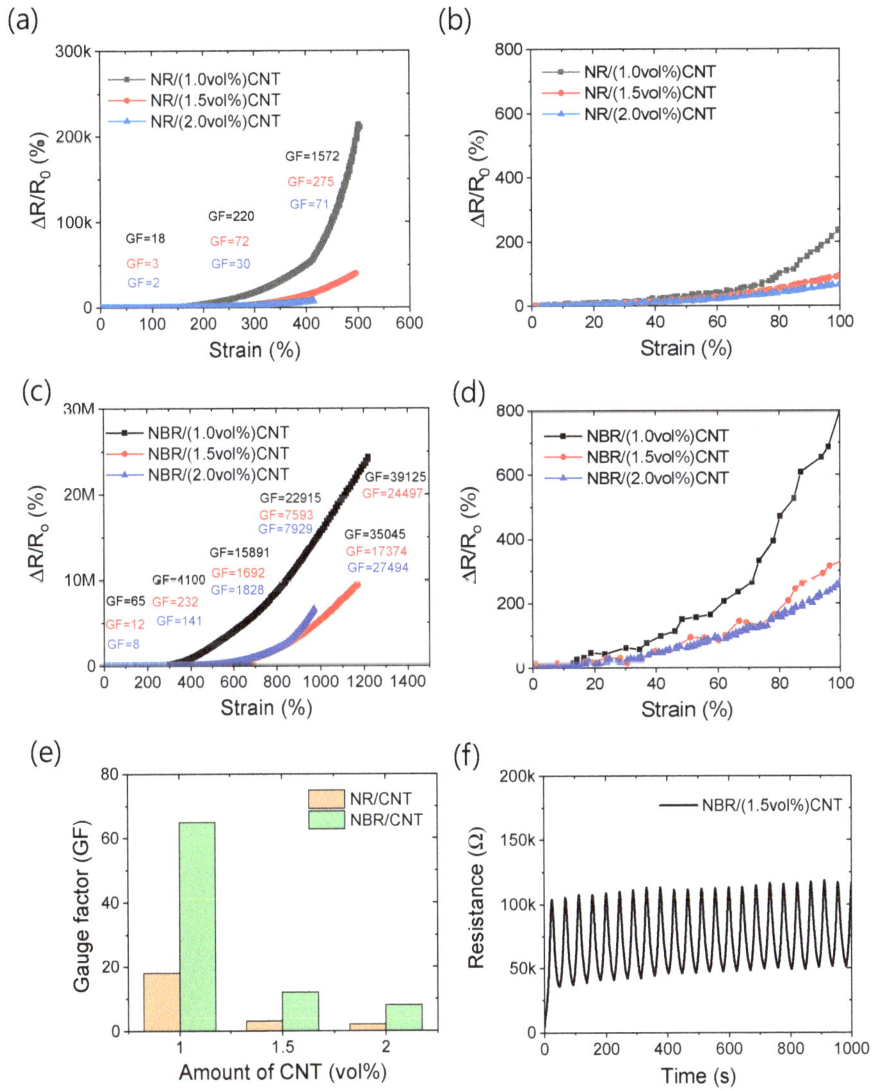

Figure 10. Strain sensing properties. (**a**) Strain sensitivity curves of NR-based composites with different gauge factors with interval of 200% elongation, (**b**) strain sensitivity curves of NR-based composites with 100% elongation, (**c**) strain sensitivity curves of NBR-based composites with different gauge factors with interval of 200% elongation, (**d**) strain sensitivity curves of NBR-based composites with 100% elongation, (**e**) gauge factors within 0–200% strain, and (**f**) real-time strain sensing at 200% cyclic strain.

Figure 10f represents typical strain sensitivity at 200% cyclic strain for NBR composites containing 1.5 vol% CNT. Although the sensitivity curves are not perfectly linear, the figure demonstrates that the sensitivity peaks exhibit good repeatability, which can be useful for measuring strain sensitivity more accurately [37,69,70]. Compared to NR-based composites, NBR-based composites show considerably better strain sensitivity.

It is well-established that the strain sensitivity of polymer composites depends on several factors, with the characteristics of the polymer materials, fillers, and fabrication techniques being the most critical. For reliable strain sensors, the strain sensitivity and detection limit are two of the most important factors. For clarity, a comparison table (Table 2) is provided, showing that strain-sensing composites were made by in situ filler dispersion in the polymer matrix. From this table, it is evident that both NR- and NBR-based CNT composites offer better gauge factors and superior working strain ranges compared to previously established strain-sensing composites [32,33,71,72].

Table 2. Sensitivity, working strain, and prepared methods comparison of CNT-based composite type strain sensors.

Composites	Fillers	Gauge Factor	Working Strain	Preparation Method	References
Epoxy resin/CNT	CNT	6	0.6–14%	Solvent casting with sonication and drying	[71]
PMVS/CNT/CB	CNT/CB	10	0–60%	Dry mixing in two-roll mill and vulcanized at 170 °C	[72]
RTV-SR/CNT/MoS_2	CNT/MoS_2	25.95	100–155%	Solution blending and vulcanization at ~25 °C (room temperature)	[32]
SBR/CNT	CNT	58.79	100–150%	Solvent blending and vulcanization at 150 °C	[33]
NR/CNT	CNT	1572	400–505	Solvent blending and vulcanization at 150 °C	Present work
NBR/CNT	CNT	39,125	1000–1220%	Solvent blending and vulcanization at 15 0°C	Present work

(PMVS: polymethylvinylsiloxane; RTV-SR: room temperature vulcanized silicone rubber; SBR: styrene butadiene rubber).

To investigate the origin of the observed behaviors, we measured the capacitance of the rubber composites, as presented in Table 3. From this table, it is evident that the capacitance of the composites significantly increases with the addition of an optimal amount of CNT. However, beyond this optimal CNT concentration, the capacitance may decrease or even disappear due to reduced resistance and higher electrical conductivity, which prevents charge accumulation. The table also shows that, at similar filler concentrations (typically 0.5 vol%), the capacitance of the NBR-based composite is more than 10,000 times higher than that of the NR-based composite. This substantial difference in capacitance can be attributed to the significantly different dielectric constant values of the two rubbers.

Table 3. Volume capacitance of different rubber composites.

Formulation	Capacitance (F/cm^3)
NR-unfilled	0.0095×10^{-9}
NR/(0.5 vol%)CNT	0.1338×10^{-9}
NR/(1.0 vol%)CNT	15.9235×10^{-9}
NR/(1.5 vol%)CNT	9.6306×10^{-6}
NR/(2.0 vol%)CNT	6.5159×10^{-6}
NR/(2.5 vol%)CNT	-
NBR-unfilled	0.0095×10^{-9}
NBR/(0.5 vol%)CNT	1.3949×10^{-6}
NBR/(1.0 vol%)CNT	-
NBR/(1.5 vol%)CNT	-
NBR/(2.0 vol%)CNT	-
NBR/(2.5 vol%)CNT	-

It can be inferred that, as the strain increases, the conductivity of the composite decreases, indicating strain sensitivity. Several mechanisms contribute to this strain sensitivity [37]. Among them, the disconnection of filler particles is the primary reason for the change in conductivity with increasing strain. It is well known that, instead of direct contact between filler particles, electrons can be transported via the tunneling effect over a critical distance between two filler particles [73]. This critical distance is covered by the rubber matrix. Due to the higher dielectric constant of NBR, electron transport is less permeable compared to NR. As a result, even slight deformation can significantly alter the conductivity of the composites, leading to a higher resistance change in NBR-based composites than in NR-based composites at similar filler contents.

Since tunneling resistance is more sensitive to strain-induced contact resistance, conductivity changes more rapidly at higher strains, making the composites more sensitive and leading to an increase in the gauge factor (GF). At similar strain levels, the tunneling resistance of NBR-based composites is believed to change more significantly due to the higher dielectric constant of NBR compared to NR, resulting in higher GF values for NBR-based composites at similar filler amounts. The higher strain sensitivities at lower filler amounts are attributed to the greater change in tunneling resistance. With a lower filler content, maximum conductivity arises from the tunneling effect. However, a lower filler content shows less linearity in the sensing curves, suggesting a nonlinear change in conductivity with strain. Therefore, to obtain a reliable strain-sensitive composite, both the rubber type and filler amount should be carefully considered.

The actuation performance of dielectric rubber depends on several factors, including mechanical properties, but it is primarily influenced by the applied voltage and the dielectric constant of the rubber [74–76]. Among these, the dielectric constant is the most important factor for generating higher mechanical properties. The actuator performances of different unfilled rubber systems are shown in Figure 11a,b.

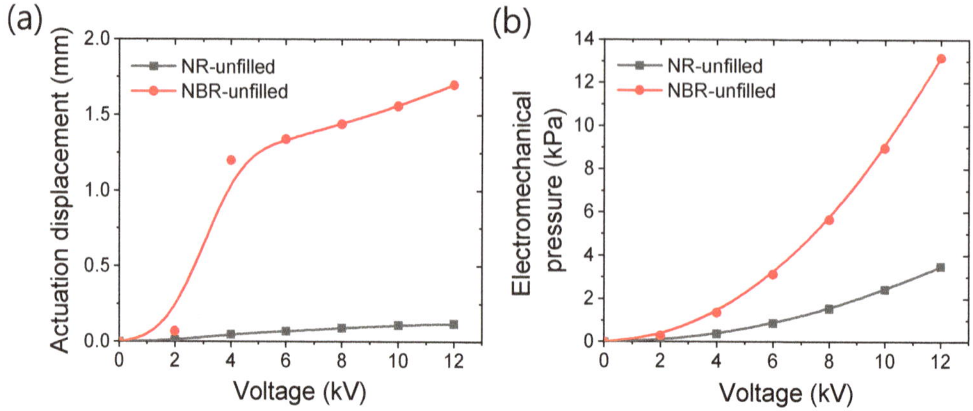

Figure 11. Variation in electromechanical properties in dielectric rubbers. (**a**) In-plane actuation displacement with respect to applied voltage, and (**b**) electromechanical pressure with respect to applied voltage.

From Figure 11a, it is evident that the in-plane electrode displacement values are higher in unfilled NBR compared to NR. This figure also shows that, after reaching the percolation voltage, a significant displacement can be observed at higher voltages. This result indicates that NBR rubber is more sensitive to the electric field compared to NR. In the presence of an electric field, an induced dipole moment is generated. These dipoles tend to orient in the direction of the applied electric field, causing actuation displacement. From this perspective, NBR, being a polar rubber, has better orientation and attraction between the dipoles, resulting in larger actuation compared to the nonpolar NR. Suresh et al. found that the chemical microstructure plays a vital role in improving the dielectric constant, and NBR has a higher dielectric constant than other rubbers [75].

From Figure 11a, it can be seen that a threshold voltage is needed to achieve higher actuation displacement. This might indicate that the nitrile groups in NBR are anisotropically distributed, with stronger dipole–dipole interactions, enhancing the inherent mechanical properties, as observed in the previous sections. Thus, a threshold voltage is required to overcome these interactions and align the dipoles in the direction of the applied electric field. The calculated electromechanical pressure generated in the actuators is provided in Figure 11b. From this figure, it is clear that the NBR-based actuator produces higher electromechanical pressure at similar applied voltages. For example, at a 12 kV applied voltage, the NBR-based actuator produces approximately 1317% higher displacement and 276% higher pressure compared to the NR-based actuator. The higher electromechanical pressure in the NBR-based actuator indicates that, for dielectric actuators, the rubber should be more polar with a higher dielectric constant.

Similar to unfilled dielectric elastomers, CNT-based elastomer composites may also have electromechanical actuation properties, but they were not studied in this work. It is believed that, due to their higher electrical conductivity, CNT-based composites can easily reach the breakdown voltage, thereby losing their actuation performance.

Based on the above results, a mechanism of behavior for different rubbers in an electric field can be presented, as shown in Figure 12. NBR contains numerous nitrile groups that introduce polarity to the rubber, and these polar groups are in an isotropic state in the absence of an electric field. In the presence of an electric field, these groups tend to orient. Since these groups exist as side chains, they are flexible and can easily orient in the direction of the applied electric field. At a threshold voltage, the flexible nitrile groups almost completely orient, and thereafter, they come closer, causing actuation in the rubber slab. On the other hand, NR is a nonpolar rubber with a low dielectric constant compared to NBR. Hence, the induced dipoles are very weak in the presence of applied electric fields, resulting in lower actuation properties.

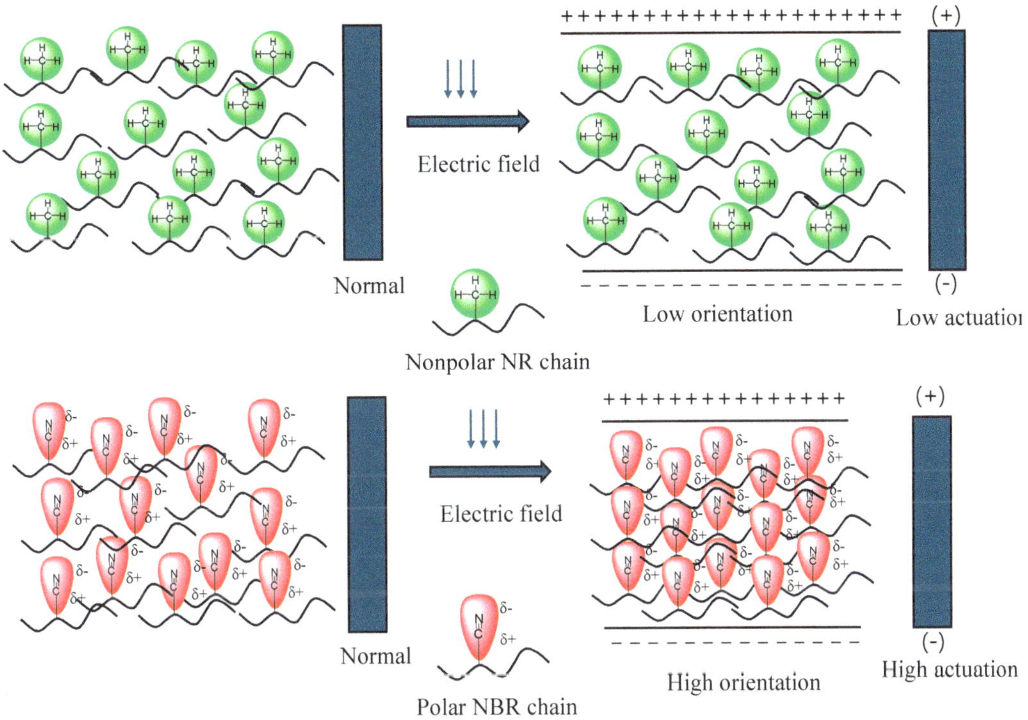

Figure 12. Possible mechanism of electromechanical actuation in different functional rubbers.

4. Conclusions

The aim of this article was to investigate the role of functionality in diene rubbers concerning mechanical reinforcement and electromechanical activities in carbon nanotube-based electroactive rubber composites. Static and dynamic mechanical studies suggest that CNT strongly reinforces the rubbers at an optimum amount of approximately 2 vol% in both NR- and NBR-based composites. From the elastic modulus, tensile strength, and fracture toughness values, it is evident that CNT has higher reinforcing efficiency toward NR, with more significant changes in these properties compared to NBR. At 2 vol% CNT content, the glass transition temperature value of the NR-based composite shifted to a higher temperature, and the thermal stability of the composite improved, with about a 15 °C enhancement in the initial degradation temperature compared to the NBR-based composite. X-ray diffraction and SEM analyses of the rubber composites indicate that CNT particles form aggregates above 1.5 vol% CNT content, with this tendency being more pronounced in NR-based composites.

Thermodynamic studies also reveal that NR-based composites show greater changes in thermodynamic parameters at lower concentrations of CNT, but the changes decrease with increasing CNT content compared to NBR-based composites. From the higher changes in mechanical properties, increased glass transition temperature, improved thermal stability, and enhanced thermodynamic parameters with respect to unfilled rubber systems, it can be concluded that CNT forms stronger chemical bonds with NR rather than physical bonds with NBR.

The significant improvements in mechanical and dynamic mechanical properties with small amounts of CNT suggest that the composites can be useful for mechanical applications. Since CNT filler particles are attached to NR chains via stronger chemical bonds, NR-based composites may be more suitable for dynamic mechanical applications with lower heat build-up properties. Although a certain amount of chemical bonding between rubber and filler may form in NR-based composites, the better elasticity in NBR-based composites could be due to its higher overall bonding capacity, including both physical and chemical interactions.

Strain-sensing activities of the rubber composites suggest that NBR-based composites exhibit higher strain-sensing properties, with higher gauge factor or sensitivity values and a wider detection limit compared to NR-based electroactive composites at similar CNT contents. The best strain-sensitive composite, with the highest gauge factor values, can be obtained with 1 vol% CNT in the NBR-based composite. However, to achieve better linearity in the strain sensitivity with strain amplitude, 1.5 vol% CNT is necessary in the NBR-based strain-sensing composite. The strain sensor made from 1.5 vol% CNT and NBR rubber maintains its efficiency without significant loss up to 200% cyclic stretching and releasing.

The higher strain sensitivities in NBR-based composites compared to NR-based composites are likely due to the stronger electronic tunneling effect in the former. Due to the presence of polar functional groups and a higher dielectric constant, NBR demonstrates high strain sensitivity, significant actuation, and strong electromechanical pressure—260%, 1317%, and 276% higher, respectively, compared to NR. Hence, for fabricating high-performance electroactive rubber composites, selecting polar NBR with a high dielectric constant value may be more advantageous than using nonpolar NR. Moreover, as electrical energy is converted to mechanical energy as a result of actuation, these rubber composites, either with this filler or hybridized with other fillers, could be useful for converting mechanical energy into electrical power. Therefore, high-performance NBR-based composites have potential applications for producing clean energy from renewable resources in the future.

Author Contributions: Conceptualization, M.N.A.; methodology, M.N.A.; validation, M.N.A., S.A, J.Y. and S.-S.P.; formal analysis, M.N.A., S.A. and J.Y.; investigation, M.N.A.; data curation, M.N.A.; writing—original draft preparation, M.N.A.; writing—review and editing, M.N.A., S.A., J.Y. and S.-S.P.; visualization, M.N.A. and S.-S.P.; supervision, S.-S.P.; project administration, S.-S.P.; funding acquisition, S.-S.P. All authors have read and agreed to the published version of the manuscript.

Funding: This research received no external funding.

Institutional Review Board Statement: Not applicable.

Data Availability Statement: Data will be available based on request to the corresponding author.

Conflicts of Interest: The authors declare no conflicts of interest.

References

1. Thomas, S.; Stephen, R. (Eds.) *Rubber Nanocomposites: Preparation, Properties, and Applications*; John Wiley & Sons (Asia) Pte Ltd.: Singapore, 2010.
2. Wibowo, A.F.; Han, J.W.; Kim, J.H.; Prameswati, A.; Entifar, S.A.N.; Park, J.; Lee, J.; Kim, S.; Lim, D.C.; Moon, M.W.; et al. Highly Stretchable and Robust Transparent Conductive Polymer Composites for Multifunctional Healthcare Monitoring. *Sci. Technol. Adv. Mater.* **2022**, *23*, 332–340. [CrossRef]
3. Lorenz, H.; Fritzsche, J.; Das, A.; Stöckelhuber, K.W.; Jurk, R.; Heinrich, G.; Klüppel, M. Advanced Elastomer Nanocomposites Based on CNT-Hybrid Filler Systems. *Compos. Sci. Technol.* **2009**, *69*, 2135–2143. [CrossRef]

4. Kitamura, M.; Hata, Y.; Yasuoka, H.; Kurotsu, T.; Asano, A. Strain-Induced ^{13}C Chemical Shift Change of Natural Rubber. *Polym. J.* **2012**, *44*, 778–785. [CrossRef]
5. Edwards, D.C. Polymer-Filler Interactions in Rubber Reinforcement. *J. Mater. Sci.* **1990**, *25*, 4175–4185. [CrossRef]
6. Abdul Salim, Z.A.S.; Hassan, A.; Ismail, H. A Review on Hybrid Fillers in Rubber Composites. *Polym.-Plast. Technol. Eng.* **2018**, *57*, 523–539. [CrossRef]
7. Tikhonova, S.A.; Evdokimov, P.V.; Filippov, Y.Y.; Safronova, T.V.; Garshev, A.V.; Shcherbakov, I.M.; Dubrov, V.E.; Put-lyaev, V.I. Electro- and Magnetoactive Materials in Medicine: A Review of Existing and Potential Areas of Application. *Inorg. Mater.* **2020**, *56*, 1319–1337. [CrossRef]
8. Alam, M.N.; Kumar, V.; Jo, C.R.; Ryu, S.R.; Lee, D.J.; Park, S.S. Mechanical and Magneto-Mechanical Properties of Styrene-Butadiene-Rubber-Based Magnetorheological Elastomers Conferred by Novel Filler-Polymer Interactions. *Compos. Sci. Technol.* **2022**, *229*, 109669. [CrossRef]
9. Damampai, K.; Pichaiyut, S.; Dasa, A.; Nakason, C. Internal Polymerization of Epoxy Group of Epoxidized Natural Rubber by Ferric Chloride Filled with Carbon Nanotubes: Mechanical, Morphological, Thermal and Electrical Properties of Rubber Vulcanizates. *Express Polym. Lett.* **2022**, *16*, 812–826. [CrossRef]
10. Utrera-Barrios, S.; Perera, R.; León, N.; Santana, M.H.; Martínez, N. Reinforcement of Natural Rubber Using a Novel Combination of Conventional and In Situ Generated Fillers. *Compos. Part C Open Access* **2021**, *5*, 100133. [CrossRef]
11. Shui, Y.; Huang, L.; Wei, C.; Sun, G.; Chen, J.; Lu, A.; Sun, L.; Liu, D. How the Silica Determines Properties of Filled Silicone Rubber by the Formation of Filler Networking and Bound Rubber. *Compos. Sci. Technol.* **2021**, *215*, 109024. [CrossRef]
12. Roy, K.; Debnath, S.C.; Potiyaraj, P. A Critical Review on the Utilization of Various Reinforcement Modifiers in Filled Rubber Composites. *J. Elastomers Plast.* **2020**, *52*, 167–193. [CrossRef]
13. Robertson, C.G.; Hardman, N.J. Nature of Carbon Black Reinforcement of Rubber: Perspective on the Original Polymer Nanocomposite. *Polymers* **2021**, *13*, 538. [CrossRef] [PubMed]
14. Hong, C.K.; Kim, H.; Ryu, C.; Nah, C.; Huh, Y.I.; Kaang, S. Effects of Particle Size and Structure of Carbon Blacks on the Abrasion of Filled Elastomer Compounds. *J. Mater. Sci.* **2007**, *42*, 8391–8399. [CrossRef]
15. Alam, M.N.; Kumar, V.; Potiyaraj, P.; Lee, D.J.; Choi, J. Mutual Dispersion of Graphite-Silica Binary Fillers and Its Effects on Curing, Mechanical, and Aging Properties of Natural Rubber Composites. *Polym. Bull.* **2022**, *79*, 2707–2724. [CrossRef]
16. Sadasivuni, K.K.; Ponnamma, D.; Thomas, S.; Grohens, Y. Evolution from Graphite to Graphene Elastomer Composites. *Prog. Polym. Sci.* **2014**, *39*, 749–780. [CrossRef]
17. Ebbesen, T.W.; Lezec, H.J.; Hiura, H.; Bennett, J.W.; Ghaemi, H.F.; Thio, T. Electrical Conductivity of Individual Carbon Nanotubes. *Nature* **1996**, *382*, 54–56. [CrossRef]
18. Kim, P.; Shi, L.; Majumdar, A.; McEuen, P.L. Thermal Transport Measurements of Individual Multiwalled Nanotubes. *Phys. Rev. Lett.* **2001**, *87*, 215502. [CrossRef] [PubMed]
19. Yang, H.; Yao, X.; Yuan, L.; Gong, L.; Liu, Y. Strain-Sensitive Electrical Conductivity of Carbon Nanotube-Graphene-Filled Rubber Composites under Cyclic Loading. *Nanoscale* **2019**, *11*, 578–586. [CrossRef]
20. O'Neill, J.; Lu, J.; Dockter, R.; Kowalewski, T. Stretchable, Flexible, Scalable Smart Skin Sensors for Robotic Position and Force Estimation. *Sensors* **2018**, *18*, 953. [CrossRef] [PubMed]
21. Kim, S.H.; Haines, C.S.; Li, N.; Kim, K.J.; Mun, T.J.; Choi, C.; Di, J.; Oh, Y.J.; Oviedo, J.P.; Bykova, J.; et al. Harvesting Electrical Energy from Carbon Nanotube Yarn Twist. *Science* **2017**, *357*, 773–778. [CrossRef]
22. Zhan, Y.; Oliviero, M.; Wang, J.; Sorrentino, A.; Buonocore, G.G.; Sorrentino, L.; Lavorgna, M.; Xia, H.; Iannace, S. Enhancing the EMI Shielding of Natural Rubber-Based Supercritical CO_2 Foams by Exploiting Their Porous Morphology and CNT Segregated Networks. *Nanoscale* **2019**, *11*, 1011–1020. [CrossRef] [PubMed]
23. Qian, H.; Greenhalgh, E.S.; Shaffer, M.S.; Bismarck, A. Carbon Nanotube-Based Hierarchical Composites: A Review. *J. Mater. Chem.* **2010**, *20*, 4751–4762. [CrossRef]
24. Lee, S.J.; Yoon, S.J.; Jeon, I.Y. Graphene/Polymer Nanocomposites: Preparation, Mechanical Properties, and Application. *Polymers* **2022**, *14*, 4733. [CrossRef] [PubMed]
25. Donchak, V.; Stetsyshyn, Y.; Bratychak, M.; Broza, G.; Harhay, K.; Stepina, N.; Kostenko, M.; Voronov, S. Nanoarchitectonics at Surfaces Using Multifunctional Initiators of Surface-Initiated Radical Polymerization for Fabrication of Nanocomposites. *Appl. Surf. Sci. Adv.* **2021**, *5*, 100104. [CrossRef]
26. Xu, Y.; Xu, Z.D.; Guo, Y.Q.; Jia, H.; Huang, X.; Wen, Y. Mathematical Modeling and Test Verification of Viscoelastic Materials Considering Microstructures and Ambient Temperature Influence. *Mech. Adv. Mater. Struct.* **2022**, *29*, 7063–7074. [CrossRef]
27. Medupin, R.O.; Abubakre, O.K.; Abdulkareem, A.S.; Muriana, R.A.; Abdulrahman, A.S. Carbon Nanotube Reinforced Natural Rubber Nanocomposite for Anthropomorphic Prosthetic Foot Purpose. *Sci. Rep.* **2019**, *9*, 20146. [CrossRef]
28. Costa, P.; Silva, J.; Ansón-Casaos, A.; Martínez, M.T.; Abad, M.J.; Viana, J.; Lanceros-Méndez, S. Effect of Carbon Nanotube Type and Functionalization on the Electrical, Thermal, Mechanical and Electromechanical Properties of Carbon Nanotube/Styrene-Butadiene-Styrene Composites for Large Strain Sensor Applications. *Compos. Part B Eng.* **2014**, *61*, 136–146. [CrossRef]

29. Nair, K.P.; Thomas, P.; Joseph, R. Latex Stage Blending of Multiwalled Carbon Nanotube in Carboxylated Acrylonitrile Butadiene Rubber: Mechanical and Electrical Properties. *Mater. Des.* **2012**, *41*, 23–30. [CrossRef]
30. Song, P.; Song, J.; Zhang, Y. Stretchable Conductor Based on Carbon Nanotube/Carbon Black Silicone Rubber Nanocomposites with Highly Mechanical, Electrical Properties and Strain Sensitivity. *Compos. Part B Eng.* **2020**, *191*, 107979. [CrossRef]
31. Georgopoulou, A.; Clemens, F. Piezoresistive Elastomer-Based Composite Strain Sensors and Their Applications. *ACS Appl. Electron. Mater.* **2020**, *2*, 1826–1842. [CrossRef]
32. Alam, M.N.; Kumar, V.; Lee, D.J.; Choi, J. Synergistically Toughened Silicone Rubber Nanocomposites Using Carbon Nanotubes and Molybdenum Disulfide for Stretchable Strain Sensors. *Compos. Part B Eng.* **2023**, *259*, 110759. [CrossRef]
33. Alam, M.N.; Kumar, V.; Lee, D.J.; Park, S.S. Styrene-Butadiene Rubber-Based Nanocomposites Toughened by Carbon Nanotubes for Wide and Linear Electromechanical Sensing Applications. *Polym. Compos.* **2024**, *45*, 2485–2499. [CrossRef]
34. Liu, H.; Gao, H.; Hu, G. Highly Sensitive Natural Rubber/Pristine Graphene Strain Sensor Prepared by a Simple Method. *Compos. Part B Eng.* **2019**, *171*, 138–145. [CrossRef]
35. Qu, M.; Qin, Y.; Sun, Y.; Xu, H.; Schubert, D.W.; Zheng, K.; Xu, W.; Nilsson, F. Biocompatible, Flexible Strain Sensor Fabricated with Polydopamine-Coated Nanocomposites of Nitrile Rubber and Carbon Black. *ACS Appl. Mater. Interfaces* **2020**, *12*, 42140–42152. [CrossRef]
36. He, S.; Wu, J.; Liu, S.; Wei, Q.; Liu, S.; Su, B.; Wang, Y. A Fully Integrated Multifunctional Flexible Sensor Based on Nitrile Rubber/Carbon Nanotubes/Graphene Composites for Smart Tire. *Chem. Eng. J.* **2024**, *486*, 150104. [CrossRef]
37. Souri, H.; Banerjee, H.; Jusufi, A.; Radacsi, N.; Stokes, A.A.; Park, I.; Sitti, M.; Amjadi, M. Wearable and Stretchable Strain Sensors: Materials, Sensing Mechanisms, and Applications. *Adv. Intell. Syst.* **2020**, *2*, 2000039. [CrossRef]
38. Dang, Z.M.; Wu, J.P.; Xu, H.P.; Yao, S.H.; Jiang, M.J.; Bai, J. Dielectric Properties of Upright Carbon Fiber Filled Poly (Vinylidene Fluoride) Composite with Low Percolation Threshold and Weak Temperature Dependence. *Appl. Phys. Lett.* **2007**, *91*, 073507. [CrossRef]
39. Panda, M.; Srinivas, V.; Thakur, A.K. On the Question of Percolation Threshold in Polyvinylidene Fluoride/Nanocrystalline Nickel Composites. *Appl. Phys. Lett.* **2008**, *92*, 133506. [CrossRef]
40. Huang, X.; Jiang, P.; Xie, L. Ferroelectric Polymer/Silver Nanocomposites with High Dielectric Constant and High Thermal Conductivity. *Appl. Phys. Lett.* **2009**, *95*, 242905. [CrossRef]
41. He, F.; Lau, S.; Chan, H.L.; Fan, J. High Dielectric Permittivity and Low Percolation Threshold in Nanocomposites Based on Poly (Vinylidene Fluoride) and Exfoliated Graphite Nanoplates. *Adv. Mater.* **2009**, *21*, 710–715. [CrossRef]
42. Ghamarpoor, R.; Jamshidi, M. Preparation of Superhydrophobic/Superoleophilic Nitrile Rubber (NBR) Nanocomposites Containing Silanized Nano Silica for Efficient Oil/Water Separation. *Sep. Purif. Technol.* **2022**, *291*, 120854. [CrossRef]
43. Liu, G.; Wang, H.; Ren, T.; Chen, Y.; Liu, S. Systematic Investigation of the Degradation Properties of Nitrile-Butadiene Rubber/Polyamide Elastomer/Single-Walled Carbon Nanotube Composites in Thermo-Oxidative and Hot Oil Environments. *Polymers* **2024**, *16*, 226. [CrossRef] [PubMed]
44. Shi, M.; Yeatman, E.M. A Comparative Review of Artificial Muscles for Microsystems Applications. *Microsyst. Nanoeng.* **2021**, *7*, 95. [CrossRef]
45. Kalita, B.; Leonessa, A.; Dwivedy, S.K. A Review on the Development of Pneumatic Artificial Muscle Actuators: Force Model and Application. *Actuators* **2022**, *11*, 288. [CrossRef]
46. Alam, M.N.; Kumar, V.; Park, S.S. Advances in Rubber Compounds Using ZnO and MgO as Co-Cure Activators. *Polymers* **2022**, *14*, 5289. [CrossRef] [PubMed]
47. Alam, M.N.; Choi, J. Highly Reinforced Magneto-Sensitive Natural-Rubber Nanocomposite Using Iron Oxide/Multilayer Graphene as Hybrid Filler. *Compos. Commun.* **2022**, *32*, 101169. [CrossRef]
48. Pumera, M.; Ambrosi, A.; Chng, E.L.K. Impurities in Graphenes and Carbon Nanotubes and Their Influence on the Redox Properties. *Chem. Sci.* **2012**, *3*, 3347–3355. [CrossRef]
49. Alam, M.N.; Kumar, V.; Ryu, S.R.; Ko, T.J.; Lee, D.J.; Choi, J. Synergistic Magnetorheological NR-NBR Elastomer Blend with Electrolytic Iron Particles. *Rubber Chem. Technol.* **2021**, *94*, 642–656. [CrossRef]
50. El-Sabbagh, S.A.; Yehia, A.A. Detection of Crosslink Density by Different Methods for Natural Rubber Blended with SBR and NBR. *Egypt. J. Solids* **2007**, *30*, 157–173.
51. Usuki, A.; Kojima, Y.; Kawasumi, M.; Okada, A.; Fukushima, Y.; Kurauchi, T.; Kamigaito, O. Synthesis of Nylon 6-Clay Hybrid. *J. Mater. Res.* **1993**, *8*, 1179–1184. [CrossRef]
52. Bristow, G.M.; Watson, W.F. Viscosity-Equilibrium Swelling Correlations for Natural Rubber. *Trans. Faraday Soc.* **1958**, *54*, 1567–1573. [CrossRef]
53. Campos, G.N.; Coimbra, A.C.R.; Silva, A.A.D.; Rocha, E.B.D.D.; Linhares, F.N.; Furtado, C.R.G.; Sousa, A.M.F.D. Cross-Link Density Measurement of Nitrile Rubber Vulcanizates Using Dynamic Shear Test. *Polímeros* **2022**, *32*, e2022011. [CrossRef]
54. Kumar, V.; Lee, G.; Choi, J.; Lee, D.J. Studies on Composites Based on HTV and RTV Silicone Rubber and Carbon Nanotubes for Sensors and Actuators. *Polymers* **2020**, *190*, 122221. [CrossRef]

55. Kumar, V.; Alam, M.N.; Manikkavel, A.; Park, S.S. Efficacy of Graphitic Allotrope's Surface Area in Silicone Rubber Based Composites for Reverse Piezoelectric Actuation: Nano Effect. *Sens. Actuators A Phys.* **2022**, *336*, 113411. [CrossRef]
56. Al-Hartomy, O.; Al-Ghamdi, A.; Dishovsky, N.; Shtarkova, R.; Iliev, V.; Mutlay, I.; El-Tantawy, F. Dielectric and Microwave Properties of Natural Rubber Based Nanocomposites Containing Graphene. *Mater. Sci. Appl.* **2012**, *3*, 453–459. [CrossRef]
57. Morsi, R.M.M.; Ismail, M.; Yehia, A.A. Conductivity Studies of Acrylonitrile Butadiene Rubber Composites Loaded with Different Types of Carbon Blacks. *J. Text. Color. Polym. Sci.* **2010**, *14*, 113–126. [CrossRef]
58. Fröhlich, J.; Niedermeier, W.; Luginsland, H.D. The Effect of Filler-Filler and Filler-Elastomer Interaction on Rubber Reinforcement. *Compos. Part A Appl. Sci. Manuf.* **2005**, *36*, 449–460. [CrossRef]
59. Bokobza, L. Elastomer Nanocomposites: Effect of Filler-Matrix and Filler-Filler Interactions. *Polymers* **2023**, *15*, 2900. [CrossRef]
60. Bashir, M.A. Use of Dynamic Mechanical Analysis (DMA) for Characterizing Interfacial Interactions in Filled Polymers. *Solids* **2021**, *2*, 108–120. [CrossRef]
61. Farida, E.; Bukit, N.; Ginting, E.M.; Bukit, B.F. The Effect of Carbon Black Composition in Natural Rubber Compound. *Case Stud. Therm. Eng.* **2019**, *16*, 100566. [CrossRef]
62. Brandrup, J.; Immergut, E.H.; McDowell, W. *Polymer Handbook*, 2nd ed.; Wiley: New York, NY, USA, 1975.
63. Mathew, A.P.; Packirisamy, S.; Thomas, S. Studies on the Thermal Stability of Natural Rubber/Polystyrene Interpenetrating Polymer Networks: Thermogravimetric Analysis. *Polym. Degrad. Stab.* **2001**, *72*, 423–439. [CrossRef]
64. Hiltz, J.A. Pyrolysis Gas Chromatography/Mass Spectrometry Identification of Poly(Butadiene-Acrylonitrile) Rubbers. *J. Anal. Appl. Pyrolysis* **2000**, *55*, 135–150. [CrossRef]
65. Pappa, A.; Mikedi, K.; Agapiou, A.; Karma, S.; Pallis, G.C.; Statheropoulos, M.; Burke, M. TG–MS Analysis of Nitrile Butadiene Rubber Blends (NBR/PVC). *J. Anal. Appl. Pyrolysis* **2011**, *92*, 106–110. [CrossRef]
66. Akiba, M.; Hashim, A.S. Vulcanization and Crosslinking in Elastomers. *Prog. Polym. Sci.* **1997**, *22*, 475–521. [CrossRef]
67. Metze, F.K.; Sant, S.; Meng, Z.; Klok, H.A.; Kaur, K. Swelling-Activated, Soft Mechanochemistry in Polymer Materials. *Langmuir* **2023**, *39*, 3546–3557. [CrossRef] [PubMed]
68. Li, S.; Gou, R.; Zhang, C. n-π Stacking in Energetic Crystals. *Cryst. Growth Des.* **2022**, *22*, 1991–2000. [CrossRef]
69. Wang, X.; Li, J.; Song, H.; Huang, H.; Gou, J. Highly Stretchable and Wearable Strain Sensor Based on Printable Carbon Nanotube Layers/Polydimethylsiloxane Composites with Adjustable Sensitivity. *ACS Appl. Mater. Interfaces* **2018**, *10*, 7371–7380. [CrossRef]
70. Li, S.; Liu, G.; Li, R.; Li, Q.; Zhao, Y.; Huang, M.; Zhang, M.; Yin, S.; Zhou, Y.; Tang, H.; et al. Contact-Resistance-Free Stretchable Strain Sensors with High Repeatability and Linearity. *ACS Nano* **2021**, *16*, 541–553. [CrossRef]
71. Nag-Chowdhury, S.; Bellegou, H.; Pillin, I.; Castro, M.; Longrais, P.; Feller, J.F. Non-Intrusive Health Monitoring of Infused Composites with Embedded Carbon Quantum Piezo-Resistive Sensors. *Compos. Sci. Technol.* **2016**, *123*, 286–294. [CrossRef]
72. Ning, N.; Wang, S.; Zhang, L.; Lu, Y.; Tian, M.; Chan, T.W. Synchronously Tailoring Strain Sensitivity and Electrical Stability of Silicone Elastomer Composites by the Synergistic Effect of a Dual Conductive Network. *Polymers* **2016**, *8*, 100. [CrossRef]
73. Bao, W.S.; Meguid, S.A.; Zhu, Z.H.; Weng, G.J. Tunneling Resistance and Its Effect on the Electrical Conductivity of Carbon Nanotube Nanocomposites. *J. Appl. Phys.* **2012**, *111*, 093726. [CrossRef]
74. Yin, L.J.; Zhao, Y.; Zhu, J.; Yang, M.; Zhao, H.; Pei, J.Y.; Zhong, S.L.; Dang, Z.M. Soft, Tough, and Fast Polyacrylate Dielectric Elastomer for Non-Magnetic Motor. *Nat. Commun.* **2021**, *12*, 4517. [CrossRef]
75. Suresh, J.N.; Arief, I.; Naskar, K.; Heinrich, G.; Tahir, M.; Wießner, S.; Das, A. The Role of Chemical Microstructures and Compositions on the Actuation Performance of Dielectric Elastomers: A Materials Research Perspective. *Nano Select* **2023**, *4*, 289–315. [CrossRef]
76. Suresh, J.N.; Liebscher, H.; Komber, H.; Tahir, M.; Gerlach, G.; Wießner, S. Network Formation, Properties, and Actuation Performance of Functionalized Liquid Isoprene Rubber. *ACS Omega* **2024**, *9*, 4754–4761. [CrossRef]

polymers

MDPI

Article

The Influence of Fresh Latex Coagulation on the Parameter Characteristics of the Yeoh Hyperelastic Constitutive Model for Natural Rubber

Li Ding [1,2,3], Honghai Huang [2,3], Yuekun Wang [2,3], Jianwei Li [2,3], Hongxing Gui [2,3,*] and Yongping Chen [1,*]

1 School of Materials Science and Engineering, Hainan University, Haikou 570228, China; gdzjding@163.com
2 Rubber Research Institute, Chinese Academy of Tropical Agricultural Science, Haikou 570101, China; honghaihuang2009@163.com (H.H.); xwangyk@126.com (Y.W.); 18272733025@163.com (J.L.)
3 Hainan Natural Rubber Technology Innovation Center, Haikou 571101, China
* Correspondence: guihongxing@263.net (H.G.); chenyp@hainanu.edu.cn (Y.C.); Tel.: +86-15289761203 (H.G.); +86-13697520538 (Y.C.)

Abstract: The coagulation of fresh latex is one of the critical processes that impacts rubber quality during natural rubber processing. Constitutive relationships are the basis for the study of the mechanical properties of rubber materials and serve as a prerequisite for material simulation studies. However, studies on the effect of different coagulation methods on natural rubber constitutive relationships have yet to be carried out, and the current models used for natural rubber constitutive relationships need to be improved. In order to investigate the effects of different coagulation methods on the hyperelastic properties of natural rubber, the impact of natural coagulation, enzyme coagulation, acid coagulation, microbial coagulation, and enzyme-assisted microbial coagulation on the hyperelastic constitutive relationship of natural rubber were analyzed in detail based on tensile experiments and the Yeoh model. The results show that after introducing a strain rate-related factor, the Yeoh model can describe well the mechanical behavior of natural rubber carbon black composites in different deformation regions, and the rubber, studied with varying coagulation methods, exhibits different mechanical properties in different deformation regions. This study provides new evidence for the study of high-performance natural rubber and serves as a reference for process selection in the primary processing of natural rubber.

Keywords: natural rubber (NR); coagulation; constitutive model; Yeoh model; hyperelastic

1. Introduction

Natural rubber is a primary processed product obtained from the fresh latex of the Brazilian rubber tree through processes such as coagulation, dehydration (sheet and crepe processing), and drying. The structural composition of natural rubber and other bioactive components endow it with superior resilience, insulation, waterproofness [1], and flexibility compared to synthetic rubber. As a result, natural rubber is extensively applied in the tire manufacturing industry and remains irreplaceable in some critical industrial sectors and cutting-edge equipment. With the continuous advancement of science and technology and the expansion of various fields, higher demands have been placed on the comprehensive performance of natural rubber.

The coagulation of fresh latex is a critical step in the initial processing of natural rubber that significantly affects rubber quality [2], primarily because non-rubber substances such as proteins and phospholipids are key components of the natural rubber network structure and are closely related to the vulcanization characteristics and mechanical properties of natural rubber. The types and content of these non-rubber components can change variably under different coagulation processes, thus causing variations in the quality and application performance of natural rubber [3]. The common coagulation processes and their characteristics are as follows: Acid coagulation is the most widely used method but can corrode equipment and affect product performance [4]. Natural coagulation requires no operation but is time-consuming and may result in incomplete coagulation. It is the method used for obtaining cup lump raw materials [5,6]. Microbial coagulation produces dry rubber with better physical–mechanical properties and vulcanization characteristics [7], but wet coagulated blocks can develop an odor during storage, similar to naturally coagulated blocks [8,9]. Enzymatic coagulation has a rapid coagulation rate and high product physical–mechanical properties but leads to fast vulcanization and the risk of scorching the rubber compound. Enzyme-assisted microbial coagulation combines the advantages of microbial and enzymatic coagulation, improving vulcanization processing characteristics while obtaining higher mechanical properties [10].

Choosing the appropriate coagulation process is an important method to improve the performance of rubber products. Currently, evaluations of fresh latex coagulation processes mainly focus on the effects of latex coagulation and specific performance indicators of raw rubber and vulcanized rubber [11–13]. Ng et al. [2] described the subsequent effects of different coagulation methods on natural rubber from the perspective of processing characteristics, vulcanization characteristics, mechanical properties, and the aging retention performance of vulcanized rubber. However, to establish a connection between latex coagulation technology and the rubber industry, it is necessary to research the impact of different coagulation methods on the application performance of end products. Forming and structural simulations are common methods for predicting the performance of rubber composites and the application performance of end products due to their convenience, speed, and intuitiveness [14–18]. Engineers often use software such as Ansys and Abaqus for simulations [19–22], and the constitutive model and parameters play an important role in the reasonableness and precision of the calculations.

There are few references, mainly in Chinese, on using constitutive models to study the hyperelastic properties of natural rubber. The Mooney–Rivlin [23–25] and Yeoh [26,27] models are commonly used to investigate the constitutive relationships of natural rubber. Zhou et al. [28] found that the Mooney–Rivlin model is suitable for small to medium deformations, generally for strains of about 100% for tension and 30% for compression, and gives relatively accurate results for rubber without carbon black. However, it fails to simulate the behavior of carbon black-filled rubber composites precisely. Zhao et al. also found that the Mooney–Rivlin model cannot accurately reflect the stress–strain relationship of hyperelastic material under large deformation, and the fitting degree of uniaxial tensile data of two diaphragm materials is relatively low [29]. The Yeoh model has the advantage of simulating mechanical behavior under other deformation conditions using simple uniaxial tensile test data [30]. Wang found that the Yeoh model is better suited to simulating the significant deformation behavior of carbon black-filled natural rubber [31]. However, the Yeoh model cannot accurately simulate the mechanical behavior of small deformation regions. Huang et al. compared the displacement and stress maps of the two models with the Ansys finite element analysis software. They found that the Mooney–Rivlin model is suitable for simulating medium and small deformation behaviors, and the Yeoh model is more suitable for simulating the significant deformation behavior of carbon black-filled natural rubber [32]. Carbon black is the most commonly used filler in the natural rubber industry. The Yeoh model cannot accurately simulate the small deformation region primarily because the initial tensile speed must rapidly increase from zero to the set speed within a short period, resulting in a non-constant strain rate in the small strain area, and thus, inaccurate simulation results [28].

In order to investigate the effects of different coagulation methods on the hyperelastic properties of natural rubber, this study applies the Yeoh model in conjunction with a tensile testing machine to explore the constitutive relationships of carbon black-filled natural rubber (NR) samples obtained through various coagulation methods. It examines the variations in Yeoh model parameters and the mechanical properties of natural rubber under different coagulation techniques. Furthermore, a strain-rate-related factor is introduced to refine the Yeoh model, addressing its limitation in accurately describing the mechanical behavior of natural rubber carbon black composites in the low-strain region. This article lays the foundation for further research on material forming and structural simulation.

2. Materials and Methods

2.1. Raw Materials and Reagents

Fresh latex was collected from the Chinese Academy of Tropical Agricultural Sciences experimental field in Danzhou, Hainan, China. The microbial coagulation liquid was prepared and preserved by the Rubber Research Institute of the Chinese Academy of Tropical Agricultural Sciences. The alkaline protease was sourced from Shengwanjia New Materials Co., Ltd., Ji'nan, China. The other chemical reagents were commercially available.

2.2. Instruments and Equipment

The instruments and equipment were sourced from the following locations: the electronic tensile machine: AI-7000-SGD1, Gotech Testing Machines Inc., Qingdao, China. the open mill: XK-300B, Shanghai Kechuang Rubber and Plastic Machinery Co., Ltd., Shanghai, China. the rheometer: MDR-2000E, Liyuan Electronic Chemical Equipment Co., Ltd., Dongguan, China. and the flat vulcanizing machine: JH-PB-600T, Shanghai Kuntian Laboratory Instrument Co., Ltd., Shanghai, China.

2.3. Sample Preparation

Homogeneous ammonia-free fresh latex with a dry rubber content of 23% was taken and sampled using different coagulation methods, which are shown in Table 1.

Table 1. Sample preparation.

Samples Number	Coagulation Methods	Type of Coagulant	Dosage of Coagulant	Note
J-1-2	Natural coagulation	—	—	
J-2-2	Enzymatic coagulation	Alkaline protease	Weight of dry rubber × 0.09%	
J-3-2	Acid coagulation	Methanoic acid	Weight of dry rubber × 0.35%	
J-4-2	Microbial coagulation	Microbial coagulation liquid	Weight of latex × 10%	
J-5-2	Enzyme-assisted microbial coagulation	a. Alkaline protease	Weight of dry rubber × 0.05%	"a" was added a first and mixed well.
		b. Microbial coagulation liquid	Weight of latex × 10%	After 20 min, "b" was added

(1) Natural coagulation: No additives were used, allowing the latex to coagulate by standing still.
(2) Enzymatic coagulation: A protease solution was added to the latex, mixed well, and then left to coagulate by standing still. The amount of protease used was 0.09% of the latex's dry rubber content.
(3) Acid coagulation: A formic acid solution was added to the latex, mixed well, and then left to coagulate by standing still. The amount of formic acid used was 0.35% of the latex's dry rubber content.
(4) Microbial coagulation: Microbial coagulation liquid was added to the latex, mixed well, and left to coagulate by standing still. The microbial coagulation liquid used was 10% of the latex weight.
(5) Enzyme-assisted microbial coagulation: The protease solution was first added to the latex, mixed well, and left to stand for 20 min. Then, the microbial coagulation liquid was added, mixed well, and left to coagulate by standing still. The amount of protease used was 0.05% of the latex's dry rubber content, and the amount of microbial coagulation liquid used was 10% of the latex weight.

After the above latex coagulated and matured for 2 days, it was pressed into sheets, hung to dry in a ventilated area for 3 days, and dried at 80 °C to produce raw rubber samples prepared by different coagulation methods.

2.4. Preparation of NR Vulcanizates

Compounded rubber was prepared as shown in Table 2. Firstly, the rubber sample was loaded into an open mill and plasticized for 4 min. Then, the vulcanization ingredient (ZnO, Stearic acid, sulfur, TBBS, Oil furnace Black) was added to the sample and mixed for another 7 min at 67 °C. Finally, the curing properties of the rubber compounds were characterized with a rheometer, and the vulcanization of the NR sheets was completed at 143 °C for the optimum cure time (tc90) by a flat vulcanizing machine.

Table 2. Formulation of NR compounds.

Chemical	Quantity (phr)
Natural rubber	100.00
Stearic acid	2.00
N-tert-butyl-2-benzothiazole sulfenamide (TBBS)	0.70
Zinc oxide	5.00
Sulfur	2.25
Oil furnace Black	35.00

2.5. Sample Testing

The dumbbell-shaped vulcanizates were prepared as shown in Figure 1a.

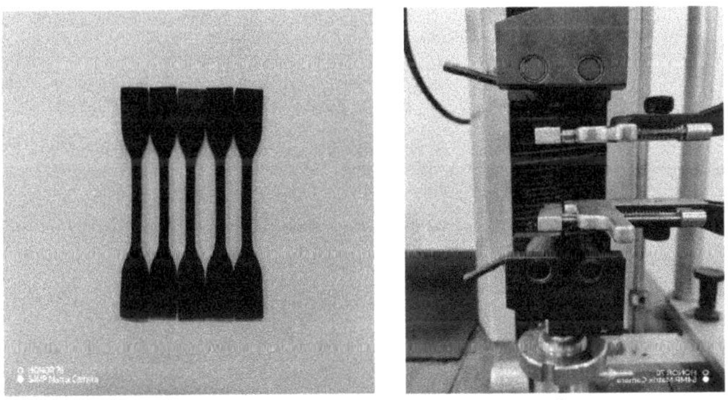

(a) (b)

Figure 1. Tensile samples and instrument: (**a**): samples; (**b**): instrument.

(1) Uniaxial tensile test: The electronic tensile machine obtained the tensile properties and stress–strain curves, shown in Figure 1b.
(2) Cyclic tensile test: A tensile strain of 400% was selected for five load–unload cycles at a stretch rate of 500 mm/min by the electronic tensile machine.

2.6. Yeoh Model Under Uniaxial Tensile Conditions for Rubber [31]

The Yeoh model is a commonly used molecular statistical constitutive model for rubber materials, which is particularly suitable for simulating the large deformation behavior of carbon black-filled natural rubber. The ANSYS finite element analysis software also categorizes it into various parameter forms such as first, second, third, fourth, and fifth orders. Rubber, being an incompressible material, has its strain energy density function model under uniaxial tensile conditions with three parameters, as follows [31]:

$$W = \sum_{i=1}^{3} C_{i0}(I_1 - 3)^i \quad (1)$$

where W is the strain energy function, I_1 is the first invariant of the deformation tensor, and C_{i0} is a material constant determined by material experimentation, with the initial shear modulus μ = $2C_{10}$. For uniaxial tension and incompressible materials [33] the following equation applies:

$$I_1 = {\lambda_1}^2 + \frac{2}{\lambda_1} \quad (2)$$

Because [34]

$$\lambda_1 = 1 + \frac{\Delta L}{L_0} = 1 + \varepsilon \quad (3)$$

Then,

$$I_1 = (1+\varepsilon)^2 + 2(1+\varepsilon)^{-1} \quad (4)$$

where λ_1 is the elongation, L_0 is the initial length, ΔL is the deformation length, and ε represents the engineering strain. The three-parameter Yeoh constitutive relationship model for rubber is as follows [31]:

$$= 2\left[(1+\varepsilon) - \left(1+\varepsilon\right)^{-2}\right]\left\{C_{10} + 2C_{20}\left[(1+\varepsilon)^2 + 2(1+\varepsilon)^{-1} - 3\right] + 3C_{30}\left[(1+\varepsilon)^2 + 2(1+\varepsilon)^{-1} - 3\right]^2\right\} \quad (5)$$

By conducting uniaxial tensile tests on rubber samples prepared using different coagulation methods, the strain and stress data obtained from the tests were nonlinearly curve-fitted by the above equations in the software Origin 2021 to obtain the parameters C_{10}, C_{20}, and C_{30} of the Yeoh model.

2.7. Strain Rate-Related Three-Parameter Yeoh Model

The Yeoh model cannot well simulate the stress–strain curve in the small strain region, mainly because, at the initial stretch, the stretching speed must rapidly increase from zero to the specified rate within a short period, resulting in a non-constant strain rate in the small strain region. Since the Yeoh model is a function related to the principal strain obtained by fitting uniaxial tensile test data and does not contain parameters and variables related to strain rate and time, it causes the fitting curve in the small strain area (<100%) to not coincide with the experimental curve. To solve this problem, different strain rates are designed for uniaxial tensile tests in the small strain region to refit the model parameters, and a strain rate-related factor υ is introduced to modify the Yeoh model equation.

2.8. The Determination and Application of the Modified Yeoh Model Parameter μ [28]

A group of samples were stretched to 100% strain at different tensile speeds of 100 mm/min, 200 mm/min, 300 mm/min, 400 mm/min, and 500 mm/min to measure the stress–strain curve data at each tensile speed. Selecting a certain tensile speed as the reference speed, the strain rate-related parameter μ was linearly fitted based on the introduced strain rate-related factor relationship, combined with unidirectional tensile test data and parameters simulated by the Yeoh model.

The fitted strain rate-related parameter μ was substituted into the modified Yeoh model, which was used to predict the uniaxial tensile stress–strain curves of the group of rubber materials at different strain rates within the small strain range. The predictions were then compared with the actual stress–strain curves to verify the model's accuracy.

3. Results and Discussion

3.1. Tensile Property and Stress–Strain Curves of Natural Rubber Materials Under Different Coagulation Methods

The tensile property of natural rubber materials by different coagulation methods is shown in Table 3. Using the common acid coagulation process in production as a reference group, we explored the relative significance of the differences between natural coagulation, enzymatic coagulation, microbial coagulation, and enzyme-assisted microbial coagulation with acid coagulation and found that the relative differences between the different coagulation modes were highly significant, which are shown in Tables 4 and 5.

Table 3. Tensile property of natural rubber materials by different coagulation methods.

Samples Number	Tensile Strength/MPa	Elongation at Break/%	Modulus at 100%/MPa	Modulus at 300%/MPa	Modulus at 500%/MPa
J-1-2	23.72	558.10	2.11	9.18	20.23
J-2-2	26.76	572.12	2.24	9.30	21.71
J-3-2	24.02	568.32	1.92	8.48	19.67
J-4-2	26.22	573.26	2.18	9.37	20.18
J-5-2	27.06	637.60	2.10	8.44	19.01

Table 4. Variance analysis of group.

Group	Number of Observations	Summation	On Average	Variance (Statistics)
J-1-2	5	0.179503	0.035901	0.002863
J-2-2	5	0.487834	0.097567	0.003336
J-3-2	5	0	0	0
J-4-2	5	5.36658	1.073316	0.002903
J-5-2	5	5.303944	1.060789	0.005585

Table 5. Variance analysis of intergroup.

Source of Variation	SS	df	MS	F	*p*-Value	F Crit
intergroup	6.298557	4	1.574639	536.0519	4.95×10^{-20}	2.866081

The stress–strain curves for rubber obtained through five different coagulation methods exhibit a consistent trend, which is shown in Figure 2. The differences become increasingly evident in the medium-to-large strain region, i.e., beyond 200%. Regarding mechanical properties, the samples produced by enzyme-assisted microbial coagulation exhibited higher elongation at break and slightly greater tensile strength. In contrast, the tensile strength of natural and acid coagulation was relatively lower. It is also noticeable that at the same strain, the modulus of the samples from enzymatic and microbial coagulation was relatively higher, which is primarily because non rubber components facilitate the construction of the natural rubber vulcanization network [3]. Enzymatic and microbial coagulation involves more intense material transformations and a richer variety of non-rubber components, resulting in a more extensive network crosslinking and a better overall performance than rubber obtained through natural and acid coagulation [5,35]. Because enzymes can degrade non-rubber components such as proteins to produce small molecules [10], and the growth and metabolism of microorganisms can also convert these non-rubber components into small molecules [35], they have similar effects, and so the stress–strain curves of enzymatic coagulation and microbial coagulation rubber are close to each other. As for natural coagulation and the acid coagulation, the exogenous substances that promote the change in their non-rubber components are relatively few, so they also have similar stress–strain curves.

3.2. The Impact of Coagulation Method on the Yeoh Model Parameters Under the Uniaxial Tensile Conditions for Rubber

The stress–strain data from uniaxial tensile tests on rubber samples produced by different coagulation methods were fitted using the Yeoh model equation. Comparisons between the five fitted curves and the experimental stress–strain curves are shown in Figure 3, and the Yeoh model parameters C_{10}, C_{20}, and C_{30} were obtained, as shown in Table 6. These parameters directly affect the accuracy and effectiveness of the simulation results.

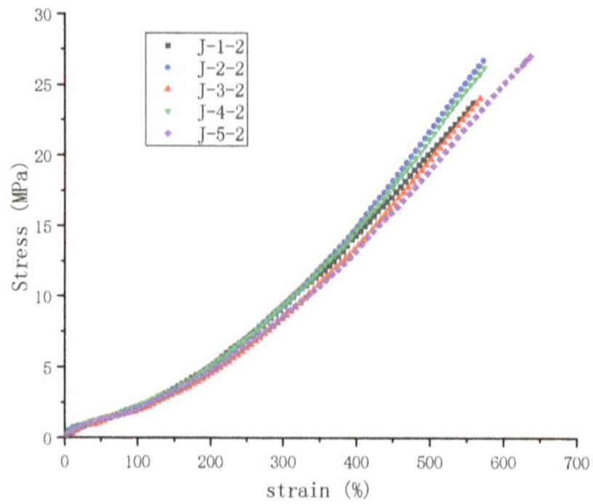

Figure 2. Stress–strain curves of natural rubber materials under different coagulation methods.

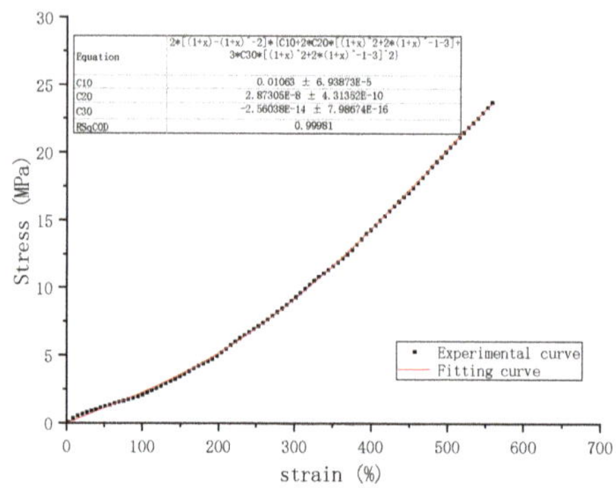

(a)

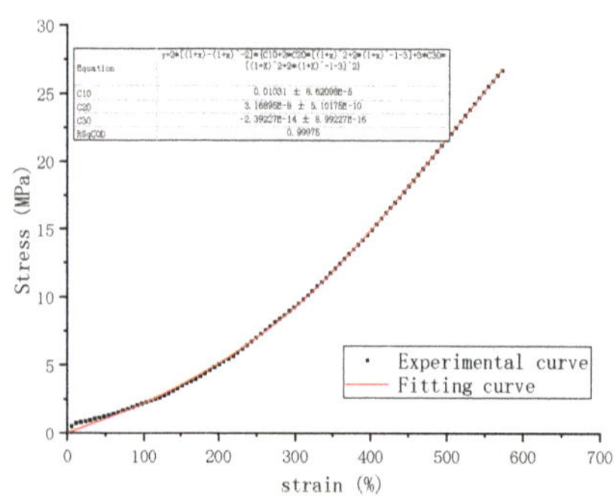

(b)

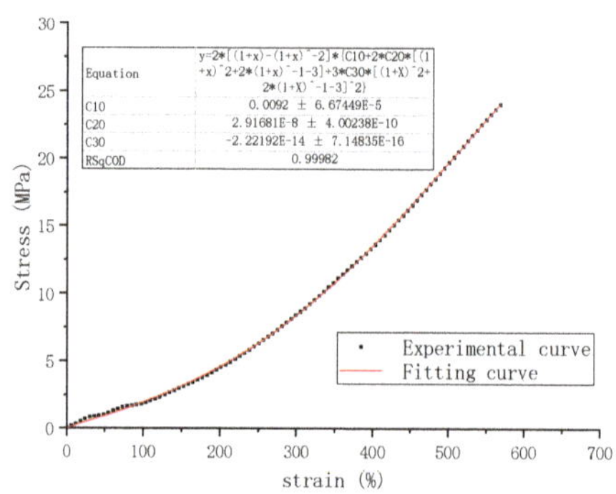

(c)

Figure 3. *Cont.*

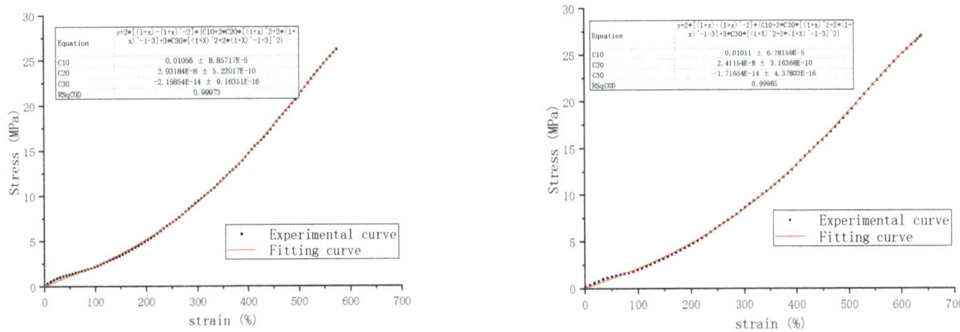

(d) (e)

Figure 3. Yeoh model's fitting curve of rubber samples: (**a**) J-1-2; (**b**) J-2-2; (**c**) J-3-2; (**d**) J-4-2; and (**e**) J-5-2.

Table 6. Constants of Yeoh model.

Samples Number	C_{10}	C_{20}	C_{30}
J-1-2	$0.01063 \pm 6.93873 \times 10^{-5}$	$2.87305 \times 10^{-8} \pm 4.31352 \times 10^{-10}$	$-2.56038 \times 10^{-14} \pm 7.98674 \times 10^{-16}$
J-2-2	$0.01031 \pm 8.62098 \times 10^{-5}$	$3.16895 \times 10^{-8} \pm 5.10175 \times 10^{-10}$	$-2.39227 \times 10^{-14} \pm 8.99227 \times 10^{-16}$
J-3-2	$0.0092 \pm 6.67449 \times 10^{-5}$	$2.91681 \times 10^{-8} \pm 4.00238 \times 10^{-10}$	$-2.22192 \times 10^{-14} \pm 7.14835 \times 10^{-16}$
J-4-2	$0.01056 \pm 8.85717 \times 10^{-5}$	$2.93184 \times 10^{-8} \pm 5.22017 \times 10^{-10}$	$-2.15854 \times 10^{-14} \pm 9.16351 \times 10^{-16}$
J-5-2	$0.01011 \pm 6.78159 \times 10^{-5}$	$2.41154 \times 10^{-8} \pm 3.16368 \times 10^{-10}$	$-1.71654 \times 10^{-14} \pm 4.37802 \times 10^{-16}$

C_{10} represents the initial shear modulus. As the second coefficient, C_{20}, is positive, the material hardens during moderate deformation, but due to the third coefficient, C_{30}, being negative, it softens under large deformations. The small strain region has a larger error, but the results align well with experimental outcomes in the moderate to large strain regions.

Since the material parameter C_{10} represents half of the shear modulus at a small strain region [36], the initial shear modulus of the rubber samples prepared by the five coagulation methods, from largest to smallest, is as follows: natural coagulation, microbial coagulation, enzymatic coagulation, enzyme-assisted microbial coagulation, and acid coagulation. The material parameter C_{20} indicates the hardening phenomenon of filled rubber under moderate deformation. The greater C_{20}, the harder the rubber material is [37]. It can be seen that in moderate deformation, the hardness of rubber samples prepared by the five coagulation methods, from hardest to softest, is as follows: enzymatic coagulation, microbial coagulation, acid coagulation, natural coagulation, and enzyme-assisted microbial coagulation. The material parameter C_{30} indicates the softening of filled rubber under large deformation. The larger C_{30}, the more easily the rubber material softens [37]. Therefore, the order of the rubber samples prepared by the five coagulation methods in terms of ease of softening under large deformation is as follows: natural coagulation, enzymatic coagulation, acid coagulation, microbial coagulation, and enzyme-assisted microbial coagulation.

3.3. Effect of Coagulation Method on Energy Dissipation Capacity of Natural Rubber Composites During Cyclic Stretching

Observations from the cyclic stretching curves at a fixed strain of 400% and the energy dissipation of the five samples show that the energy dissipation decreases gradually with an increasing number of cycles, exhibiting the typical Mullins effect [38], which is shown in Figure 4. The differences in energy dissipation are more pronounced during the first stretch but become less significant thereafter, as shown in Figure 5. For rubber materials, a portion of the dissipated energy contributes to deforming the network structure, leading to material deformation and thus providing the rubber with shock-absorbing functionality. The extent of deformation is related to the softness or hardness of the material; hence, samples with greater energy dissipation indicate a tendency for easier deformation or relatively softer characteristics. Therefore, the results of the cyclic stretching tests suggest that the coagulation methods leading to easier deformation of the rubber material are, in order, natural coagulation, enzymatic coagulation, acid coagulation, microbial coagulation, and enzyme-assisted microbial coagulation—consistent with the hardness rankings under the large strain predicted by the Yeoh model.

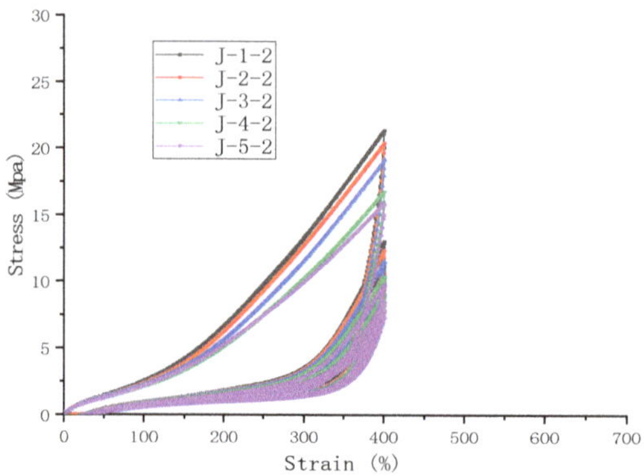

Figure 4. The effect of the coagulation method on the cyclic stretching curves.

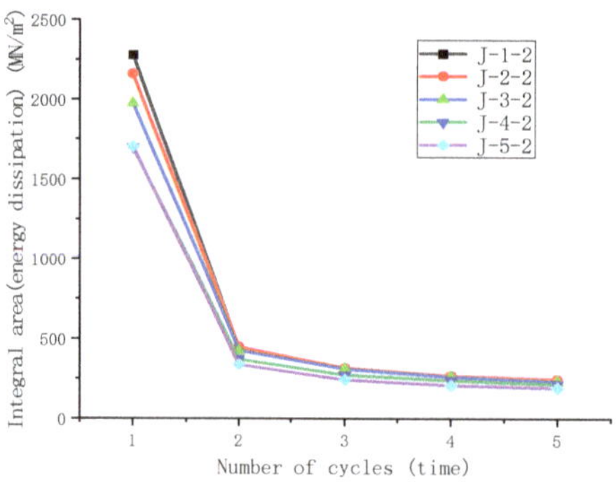

Figure 5. The effect of the coagulation method on energy dissipation.

3.4. *Optimization of the Yeoh Model Parameters*

Comparing the fitted curves of the Yeoh model with the experimental curves, the Yeoh model can describe the mechanical behavior of natural rubber carbon black composites well in the moderate and large strain regions. However, there is a significant error in the small strain region. It has been proposed that different forms of constitutive models must be adopted within different strain ranges [39], but this approach is cumbersome. To address this issue, uniaxial tensile tests with different strain rates were conducted in the small strain region to refit the model parameters, introducing a strain rate-related factor, υυ, to modify the Yeoh model equation.

Based on previous research, the strain rate-related factor, υ, can be represented by the following exponential function [28]:

$$\upsilon = 1 + \mu \ln V_{\varepsilon}^{*} \quad (6)$$

$$V_\varepsilon^* = \frac{V_\varepsilon^f}{V_{\varepsilon 0}} \tag{7}$$

where μ is the strain rate-related parameter, V_ε^* is the ratio of strain rates, V_ε^f is the actual strain rate obtained by the tensile machine at a certain stretching speed, and V_{ε_0} is a reference strain rate taken from one of the rates in a set of experiments.

After introducing the strain rate-related factor, the Yeoh model can be written as follows:

$$+\varepsilon) - \left(1+\varepsilon\right)^{-2}\Big]\Big\{C_{10} + 2C_{20}\Big[(1+\varepsilon)^2 + 2(1+\varepsilon)^{-1} - 3\Big] + 3C_{30}\Big[(1+\varepsilon)^2 + 2(1+\varepsilon)^{-1} - 3\Big]^2\Big\}(1+\mu ln V_\varepsilon^*) \tag{8}$$

This is the three-parameter Yeoh model related to the strain rate, and the modified Yeoh model can reflect the influence of the strain rate on the hyperelastic properties of the rubber material in the small strain region.

3.5. Determination of the Modified Yeoh Model Parameter μ

A rubber sample prepared by the natural coagulation method was selected and cut into several dumbbell-shaped films. The films were stretched to 100% strain at different stretching speeds, and the differences between the theoretical and actual strain rates at different speeds are shown in Table 7. The strain rate was calculated as the derivative of strain with respect to time because the scale distance of the tension machine was 25 mm, and the material was stretched to 100% with 100 mm/min, which is equivalent to four strains in 60 s, so the theoretical strain rate was 0.0667 s^{-1}, but because the stretching speed needed to be increased from 0 to 100% mm/min, the actual stretching required a more extended period, leading to the actual strain rate being lower than the theoretical strain rate.

Table 7. The differences between the theoretical and actual strain rates at different speeds.

Stretching Speeds (mm/min)	100	200	300	400	500
V_ε^f theoretical strain rate (/s)	0.0667	0.1333	0.2	0.2667	0.3333
Strain time (s)	31.27	16.11	9.05	7.03	6.02
V_ε^f actual strain rates (/s)	0.032	0.0621	0.1105	0.1422	0.1661

A stretching speed of 300 mm/min as the reference speed was selected, the actual strain rate at this speed was set as the reference strain rate V_{ε_0}, and the model parameter C_{10} at this speed was set as the reference $C_{10}^{\varepsilon 0}$; from this, the relationship between the strain rate V_ε^f, the model parameter C_{10}^f at other stretching speeds and the reference speed V_{ε_0}, $C_{10}^{\varepsilon 0}$ could be calculated as follows [29]:

$$C_{10}^f = C_{10}^{\varepsilon 0}[1 + \mu \ln\left(\frac{V_\varepsilon^f}{V_{\varepsilon 0}}\right)] \tag{9}$$

Equation (9) was transformed into:

$$\frac{C_{10}^f}{C_{10}^{\varepsilon 0}} - 1 = \mu \ln\left(\frac{V_\varepsilon^f}{V_{\varepsilon 0}}\right) \tag{10}$$

Combining the uniaxial tensile test data and the parameters simulated by the Yeoh model, the strain rate-related parameter μ could be linearly fitted using the equation above.

The actual stress–strain curve at a reference speed of 300 mm/min and the fitted curve of the Yeoh model are shown in Figure 6.

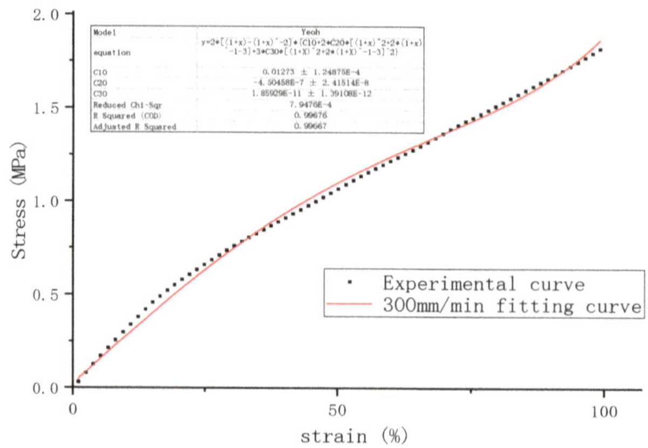

Figure 6. The actual stress–strain curve and the fitting curve at the reference speed of 300 mm/min.

The reference strain rate V_{ε_0} at the reference speed was 0.1105/s, and the simulated C10, which is $C_{10}^{\varepsilon_0}$, was 0.01273. The strain rates V_ε^f and simulated parameters C_{10}^f at other stretching speeds are shown in Table 8. According to Equation (10), the strain rate-related parameter μ can be linearly fitted to be 0.251, which is shown in Figure 7.

Table 8. The differences between the theoretical and actual strain rates at different speeds.

Stretching Speeds (mm/min)	V_ε^f Actual Strain Rates (/s)	C_{10}^f	$\ln(V_\varepsilon^f/V_{\varepsilon_0})$	$C_{10}^f/C_{10}^{\varepsilon_0}-1$
100	0.032	0.0091	−1.239279618	−0.285153181
200	0.0621	0.01119	−0.609005679	−0.120974077
300	0.1105 (V_{ε_0})	0.01273 ($C_{10}^{\varepsilon_0}$)	0	0
400	0.1422	0.01377	0.252218996	0.081696779
500	0.1661	0.0146	0.407574496	0.146897093

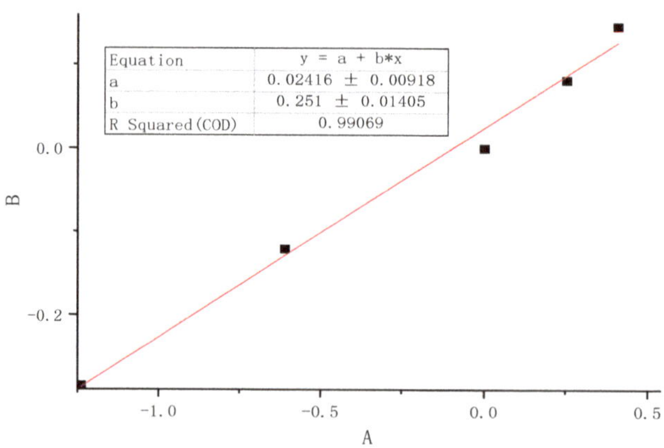

Figure 7. Linear fitting of the strain rate-related parameter μ.

For the same batch of rubber samples, the strain rate-related parameter μ was constant. For different batches of samples, the strain rate-related parameter μ would not the same and would need to be recalculated according to the same method.

3.6. Application of the Modified Yeoh Rubber Model

The strain rate-related parameter (μ = 0.251) was substituted into the Yeoh-modified model (8) and the same sample was selected in 2.5 as the experimental object. Using the stress–strain curve data of the sample at a stretching speed of 300 mm/min as a basis, the theoretical strain data for each speed were calculated by substituting the values of C_{10}, C_{20}, and C_{30} at a 300 mm/min tensile speed and "$\ln(V_\varepsilon^f/V_{\varepsilon 0})$" at this speed under a 300 mm/min tensile speed, as seen in Table 7 into Equation (8). The mechanical behavior at different strain rates was forecasted, with results depicted, which are shown in Figure 8. It can be observed that the predictions of the modified model at various stretching speeds align well with the actual stress–strain curves, especially at a stretching speed of 500 mm/min, where the stress–strain curve shows a marked improvement in the small strain region compared to the unmodified model.

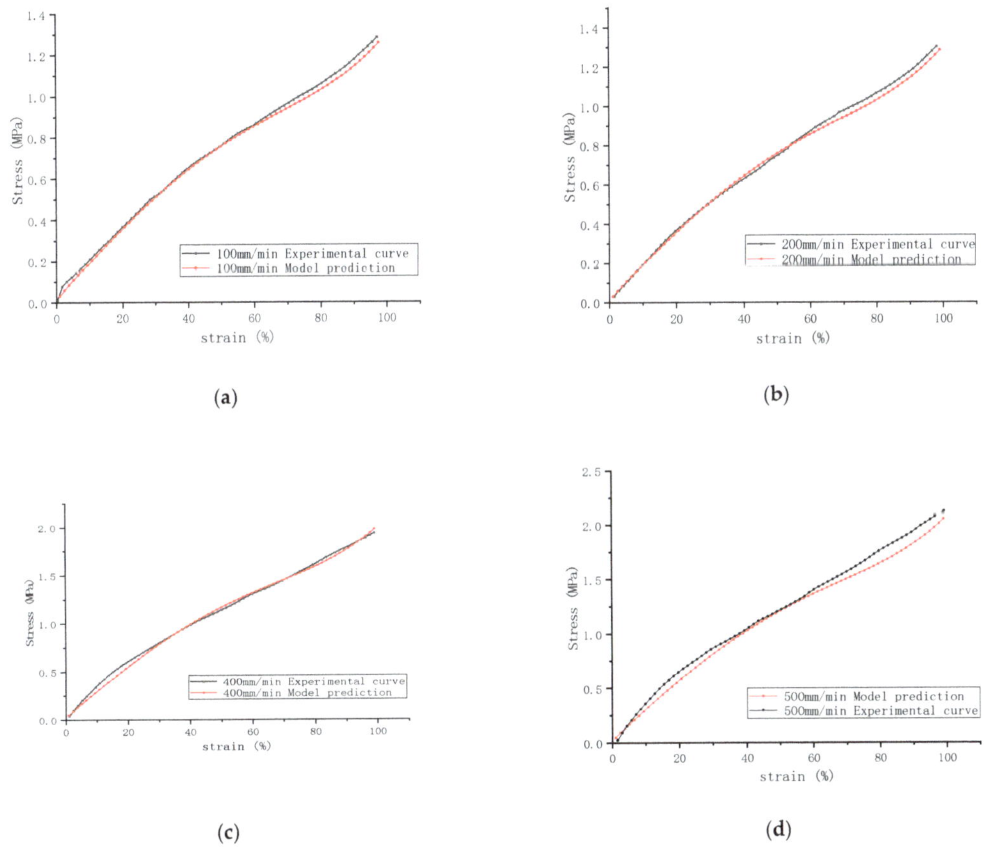

Figure 8. A comparison between the curve calculated by the modified model and the actual curve at different speeds: (**a**) 100 mm/min; (**b**) 200 mm/min; (**c**) 400 mm/min; and (**d**) 500 mm/min.

The stress–strain data at different speeds predicted by the model and obtained from the actual test were fitted to a nonlinear curve in the software Origin 2021. The fitting function was Equation (8), which can be obtained as the goodness-of-fit R^2, which is shown in Table 9. The higher the value of R^2, the better the model fits the data, and R^2 of the model prediction was 1, which indicates that it fits perfectly. In contrast, the actual test had an R^2 of more than 0.99, which is close to the model prediction, indicating that the two curves have a high degree of approximation, simultaneously indicating that the model prediction has a high degree of accuracy.

Table 9. Accuracy of model predictions.

Stretching Speeds (mm/min)	R^2 (COD)
100 (actual test)	0.99932
100 (model prediction)	1
200 (actual test)	0.99921
200 (model prediction)	1
400 (actual test)	0.99624
400 (model prediction)	1
500 (actual test)	0.99657
500 (model prediction)	1

Predicted uniaxial tensile stress–strain curves at five different stretching speeds are shown in Figure 9, indicating that the modified model accurately forecasts the stress–strain relationship trends of rubber materials at different strain rates without the need to refit parameters. It is also evident that varying stretching speeds significantly impact the material's stress–strain curves, with higher speeds resulting in steeper slopes, which further confirms the inability of the Yeoh model to simulate behaviors in the small strain range accurately.

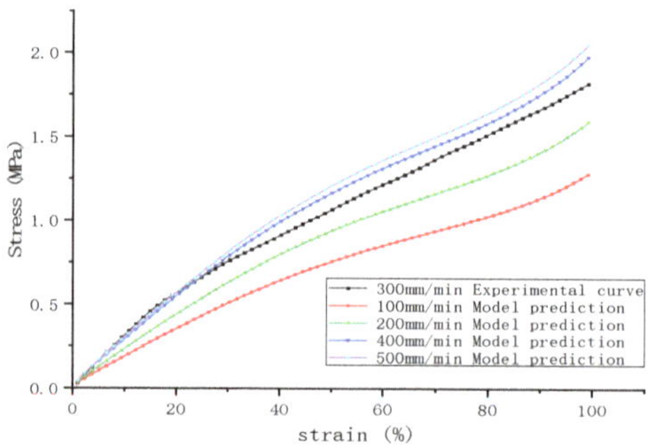

Figure 9. Predicted uniaxial tensile stress–strain curves.

The figure also reflects the influence of the strain rate on the hyperelastic properties of the materials within the small strain region. Rubber materials display similar nonlinear behaviors at different strain rates in the forecasted stretching results. The material's stiffness decreases with increasing applied strain, while at a constant strain, the stiffness increases with an increasing strain rate. Therefore, the initial stiffness is closely related to the strain rate—the higher the strain rate, the greater the material stiffness (i.e., modulus). This phenomenon is consistent with the conclusions of Sanghyeub Kim et al. [40], who found that different strain rates lead to variations in the entanglement and disentanglement of chains in uncured rubber. In vulcanized rubber, a network of chain entanglements also exists, and the contribution of this network to the modulus significantly increases under high-speed strain [41].

3.7. The Impact of the Coagulation Method on the Modified Model Parameters of Rubber in Small Strain Regions

The five rubber samples were stretched to 100% strain at different tensile speeds of 300 mm/min and 500 mm/min. The tensile speed of 300 mm/min was selected as the reference speed, and the model parameters were calculated according to the Formula (9). The changes in the model parameters (C_{10}) before and after the modification are shown in Table 10.

Table 10. The model parameters (C_{10}) before and after the modification.

Samples Number	V_ε^f Actual Strain Rates (/s)		Yeoh Model C_{10}		Yeoh Modified Model C_{10}
	300 mm/min	500 mm/min	300 mm/min	500 mm/min	500 mm/min
J-1-2	0.1105	0.1661	0.01615	0.0146	0.017802
J-2-2	0.0994	0.1661	0.01427	0.01353	0.016109
J-3-2	0.0994	0.1661	0.01263	0.01237	0.014258
J-4-2	0.0903	0.1658	0.01438	0.01455	0.016233
J-5-2	0.0994	0.1661	0.01273	0.01288	0.014672

After the modification, the model parameter C_{10} increased for all five samples, and the differences between them were amplified. However, there was no change in the order of their values, indicating that the natural coagulant is still the largest and the acid coagulant is the smallest in terms of modulus, which is in agreement with the results of the study before the model modification.

4. Conclusions

Based on tensile experiments and the Yeoh model, the effects of natural coagulation, enzyme coagulation, acid coagulation, microbial coagulation, and enzyme-assisted microbial coagulation modes on the superelastic ontological relationships of natural rubber were analyzed in detail. Given the limitations of the Yeoh model in describing mechanical behavior in the low-strain region, a strain rate-related factor was introduced, and a more accurate hyperelastic constitutive relationship of natural rubber was established. The results demonstrate that the Yeoh model effectively characterizes the mechanical behavior of natural rubber carbon black composites in the moderate and large strain regions. Additionally, the material characteristics reflected by the model parameters align with those observed in cyclic tensile tests. The performance of rubber under small, medium, and large deformations is different for different coagulation methods. The modulus of natural coagulation rubber is the highest in small deformations, the modulus of enzyme coagulation rubber is the highest in medium deformations, and the modulus of enzyme-assisted microbial coagulation rubber is the highest in large deformations, and thus the appropriate coagulation method can be selected according to the degree of deformation of the rubber required by application scenarios in the practical application. Following the incorporation of a strain rate-related factor, the modified Yeoh model accurately predicts stress–strain curves across different tensile speeds within the low-strain range.

Author Contributions: Conceptualization, L.D. and H.H.; methodology, L.D. and H.G.; software, L.D.; validation, Y.W., H.H. and L.D.; formal analysis, L.D. and J.L.; investigation, L.D. and H.H.; resources, H.G. and Y.C.; data curation, L.D., Y.W. and H.H.; writing—original draft preparation, L.D.; writing—review and editing, Y.C.; visualization, L.D.; supervision, J.L.; project administration, Y.C. and H.G.; funding acquisition, H.G. All authors have read and agreed to the published version of the manuscript.

Funding: This research was funded by the National Key R&D Program of China (No. 2022YFD2301203), the Hainan Province Science and Technology Special Fund (No. ZDYF2024XDNY284) and the Agricultural Technology Experiment Demonstration and Service Support Project of the Ministry of Agriculture and Rural Affairs (No. XJSHT-02).

Institutional Review Board Statement: Not applicable.

Data Availability Statement: Data are contained within the article.

Conflicts of Interest: The authors declare no conflicts of interest.

References

1. He, Y.P. *Natural Rubber Processing Science*; Hainan Publishing House: Haikou, China, 2007; pp. 33–37.
2. Ng, J.W.; Othman, N.; Yusof, N.H. Various coagulation techniques and their impacts towards the properties of natural rubber latex from *Hevea brasiliensis*—A comprehensive review related to tyre application. *Ind. Crops Prod.* **2022**, *181*, 114835. [CrossRef]
3. Wei, Y.C.; Liu, G.X.; Zhang, L.; Zhao, F.; Liao, S.; Luo, M.C.; Liao, S. Non-rubber components tuning mechanical properties of natural rubber from vulcanization kinetics. *Polymer* **2019**, *183*, 121911. [CrossRef]
4. Wang, G.S.; Fu, Z.N.; Fu, X. Effect of different coagulation methods on the properties of natural rubber. *Chem. Eng.* **2014**, *28*, 11–13.
5. Salomez, M.; Subileau, M.; Vallaeys, T.; Santoni, S.; Bonfils, F.; Sainte-Beuve, J.; Intapun, J.; Granet, F.; Vaysse, L.; Dubreucq, E. Microbial communities in natural rubber coagula during maturation: Impacts on technological properties of dry natural rubber. *J. Appl. Microbiol.* **2018**, *124*, 444–456. [CrossRef] [PubMed]
6. Bai, W.N.; Guan, J.; Liu, H.; Cheng, S.H.; Zhao, F.C.; Liao, S.Q. Microstructure and Lamellae Phase of Raw Natural Rubber via Spontaneous Coagulation Assisted by Sugars. *Polymers* **2021**, *13*, 4306. [CrossRef] [PubMed]
7. Liang, Y.; Huang, M.F.; Zeng, Z.Q. Molecular Structures and Mechanical Properties of Microbe Rapid Coagulation Natural Rubber. *Chin. J. Struct. Chem.* **2011**, *30*, 1810–1819. [CrossRef]
8. Chen, J.; Li, S.D.; Li, L.F.; Wang, Z.F.; Yang, L.; Zhong, J.P. Contrastive Study on Properties of Acid and Microorganisms Coagulated Natural Rubber during Accelerated Storage. *J. Rubber Res.* **2018**, *21*, 17–29. [CrossRef]
9. Li, S.D.; Chen, J.; Li, L.F.; Wang, Z.F.; Zhong, J.P.; Yang, L. Vulcanization characteristics of natural rubber coagulated by microorganisms. *Rubber Chem. Technol.* **2018**, *91*, 64–78. [CrossRef]
10. Deng, D.Y.; Zhao, T.; Ding, L.; Li, J.; Huang, H.H.; Liao, J.H.; Gui, H.X. Effect of enzyme dosage assisted microbial coagulation on properties of natural rubber. *China Elastomerics* **2023**, *33*, 1–4+39.
11. Hutapea, Y.; Eliza Maryana, Y.; Karman, J.; Hanapi, S.; Raharjo, B. The effects of coagulant types on the quality and added value of raw rubber materials. *E3S Web Conf.* **2022**, *361*, 01001. [CrossRef]
12. Jidapa, N.; Frédéric, B.; Natedao, M.; Sainte-Beuve, J.; Albert, F.; Siriluck, L.; Kittipong, R.; Francoise, G.; Laurent, V. Post-harvest maturation of *Hevea brasiliensis* latex coagula: Ranking of the key drivers of the mesostructure and physical properties of natural rubber. *J. Rubber Res.* **2022**, *25*, 5–18. [CrossRef]
13. Achmad, F.; Deviany, D. The Effect of *Averrhoa bilimbi* Extract as Natrual Coagulants on the Characteristics of rubber. *Konversi* **2022**, *11*, 44–51. [CrossRef]
14. Chen, L.; Zhu, X.; Wang, G.; Li, C. Test and Simulation Analysis of Tire Inflation Pressure Loss. *Tire Sci. Technol.* **2020**, *48*, 329–353. [CrossRef]
15. Luyen, T.T.; Mac, T.B.; Nguyen, D.T. Simulation and experimental comparison study based on predicting forming limit curve of sus304 sheet material. *Mod. Phys. Lett. B* **2023**, *37*, 257271272. [CrossRef]
16. Nguyen, D.T.; Kim, Y. Combined numerical and experimental study to predict the forming limit curve of boron steel sheets at elevated temperatures. *Proc. Inst. Mech. Eng. Part B J. Eng. Manuf.* **2020**, *234*, 189–203. [CrossRef]
17. Berger, T.; Kim, S.; Kaliske, M. A Microsphere-Based Rubber Curing Model for Tire Production Simulation. *Tire Sci. Technol.* **2023**, *51*, 82–113. [CrossRef]
18. Wan, Q.; Xiong, Z. Simulation Analysis of Stir Tool Wear Considering the Effect of Temperature Rise caused by Material Deformation on Flow Stress. *Curr. Mater. Sci.* **2023**, *16*, 225–234. [CrossRef]
19. Han, F.; Li, Z.B.; Zhang, J.Y.; Liu, Z.Y.; Yao, C.; Han, W.P. ABAQUS and ANSYS Implementations of the Peridynamics-Based Finite Element Method (PeriFEM) for Brittle Fractures. *Comput. Model. Eng. Sci.* **2023**, *136*, 2715–2740. [CrossRef]
20. Parvathidevi, A.; Naga Satish, K.C. Thermal analysis of concrete slabs with insulating materials using ANSYS. *World J. Eng.* **2024**, *21*, 156–169. [CrossRef]
21. Zeng, J.; Li, J.Y.; Dong, S.; Wang, F.H.; Wang, F.L.; Jin, L.; Dong, J. Optimization of Hot Backward Extrusion Process Parameters for Seamless Tube of Mg-8Gd-3Y Alloy by Finite Element Simulation. *J. Mater. Eng. Perform.* **2024**, *33*, 2453–2461. [CrossRef]
22. Yoon, I.C.; Kang, I.S.; Park, K.H.; Heo, J.Y.; Lee, J.R.; Jung, Y.C. Three-Dimensional Measurement and Finite-Element Simulation for Tool Wear Estimation in Cutting of Inconel 718 Superalloy. *Int. J. Precis. Eng. Manuf.* **2024**, *25*, 21–34. [CrossRef]
23. Niu, W.; Li, Y.; Ma, Y.; Zhao, G. Determination and Prediction of Time-Varying Parameters of Mooney-Rivlin Model of Rubber Material Used in Natural Rubber Bearing under Alternating of Aging and Seawater Erosion. *Materials* **2023**, *16*, 4696. [CrossRef] [PubMed]
24. Santos, R.J.; Hiranobe, C.T.; Dognani, G.; Silva, M.J.; Paim, L.L.; Cabrera, F.C.; Torres, G.B.; Job, A.E. Using the Lorenz–Park, Mooney–Rivlin, and dynamic mechanical analysis relationship on natural rubber/leather shavings composites. *J. Appl. Polym. Sci.* **2022**, *139*, 244299978. [CrossRef]
25. Zhao, G.F.; Ma, Y.H.; Li, Y.M.; Luo, J.R.; Chang, D. Development of a modiffed Mooney-Rivlin constitutive model for rubber to investigate the effects of aging and marine corrosion on seismic isolated bearings. *Earthq. Eng. Eng. Vib.* **2017**, *16*, 815–826. [CrossRef]
26. Kuang, Y.; Fan, P.; Lin, H.W. Research on Yeoh-Revised hyperelastic constitutive model considering the volume almost incompressible premise for rubber materials. *J. Elastomers Plast.* **2023**, *55*, 1123–1151. [CrossRef]
27. Mars, W.V.; Fatemi, A. Observations of the Constitutive Response and Characterization of Filled Natural Rubber Under Monotonic and Cyclic Multiaxial Stress States. *J. Eng. Mater. Technol.* **2004**, *126*, 19–28. [CrossRef]

28. Zhou, W.H. *The Study of Mechanical Constitutive Relation Rubber Materials*; Nanjing University of Aeronautics and Astronautics: Nanjing, China, 2021.
29. Zhao, Y.Q.; Lin, D.Z.; Chen, H.; Tang, J.L.; Chen, P. Performance test and constitutive model selection of diaphragms in hot diaphragm forming. *J. Beijing Univ. Aeronaut. Astronaut.* **2023**. [CrossRef]
30. Yeoh, H.O. Some Forms of the Strain Energy Function for Rubber. *Rubber Chem. Technol.* **1993**, *66*, 754–771. [CrossRef]
31. Wang, H. *Application of Hyper-Elastic Constitutive Models on Rubbers in Tyre Analysis*; Harbin Institute of Technology: Harbin, China, 2008.
32. Huang, J.L.; Xie, G.J.; Liu, Z.W. Finite element analysis of super-elastic rubber materials based on the Mooney-Rivlin and Yeoh model. *China Rubber Plast. Technol. Equip.* **2008**, *34*, 22–26. [CrossRef]
33. Belytschko, T.; Liu, W.K.; Moran, B. *Nonlinear Finite Elements for Continua and Structures*; John Wiley: New York, NY, USA, 2000; Volume 147.
34. Lu, M.W.; Luo, X.F. *Foundations of Elasticity*; Tsinghua University Press: Beijing, China, 1990.
35. Intapun, J.; Sainte-Beuve, J.; Bonfils, F. Effect of microorganisms during the initial coagulum maturation of hevea natural rubber. *J. Appl. Polym. Sci.* **2010**, *118*, 1341–1348. [CrossRef]
36. Du, K.; Zhou, H.W.; Ding, M.M.; Ye, F.; Shi, T.F. Cluster Analysis of Correlation Between Rubber Carbon Black Filling and Yeoh Model Parameters. *Chin. J. Appl. Chem.* **2021**, *38*, 675–684. [CrossRef]
37. Hu, X.L. *Micro- and Macro-Viscohyperelastic Behavior of Carbon Black Filled Rubber*; Xiangtan University: Xiangtan, China, 2013.
38. Diani, J.; Fayolle, B.; Gilormini, P. A review on the Mullins effect. *Eur. Polym. J.* **2009**, *45*, 601–612. [CrossRef]
39. Wang, L.R.; Lu, Z.H. Modeling Method of Constitutive Law of Rubber Hyperelasticity Based on Finite Element Simulations. *Rubber Chem. Technol.* **2003**, *76*, 271–285. [CrossRef]
40. Sanghyeub, K.; Thomas, B.; Michael, K. Strain Rate-Dependent Behavior of Uncured Rubber: Experimental Investigation and Constitutive Modeling. *Rubber Chem. Technol.* **2022**, *95*, 175–199. [CrossRef]
41. Huang, C.; Huang, G.S.; Li, S.Q.; Luo, M.C.; Liu, H.; Fu, X.; Qu, W.; Xie, Z.T.; Wu, J.R. Research on architecture and composition of natural network in natural rubber. *Polymer* **2018**, *154*, 90–100. [CrossRef]

polymers

MDPI

Article

Application of Post-Industrial Leather Waste for the Development of Sustainable Rubber Composites

G. Barrera Torres [1,*], Carlos M. Gutierrez Aguilar [1,*], Elizabeth R. Lozada [1], Manuel J. Tabares Montoya [2], Beatriz E. Ángel Álvarez [2], Juan C. Sánchez [3,4], Jaime A. Jaramillo Carvalho [3,4] and Renivaldo J. Santos [5]

1 Faculty of Arts and Humanities, Instituto Tecnológico Metropolitano (ITM), Medellín 050036, Colombia; eliza.rico.lozada@gmail.com
2 School of Engineering, Universidad Pontificia Bolivariana (UPB), Medellín 050031, Colombia; manuel.tabares@upb.edu.co (M.J.T.M.); beatriz.angel@upb.edu.co (B.E.Á.Á.)
3 Advanced Manufacturing Technology Center, Servicio Nacional de Aprendizaje, Medellín 050036, Colombia; jcsanchezg@sena.edu.co or juan.sanchezg@pascualbravo.edu.co (J.C.S.); jaime.jaramilloc@pascualbravo.edu.co (J.A.J.C.)
4 Mechanical Department, Institución Universitaria Pascual Bravo (IUPB), Medellín 050036, Colombia
5 Faculty of Engineering and Science, Sao Paulo State University (UNESP), Rosana 19274-000, SP, Brazil; renivaldo.santos@unesp.br
* Correspondence: giovannibarrera@itm.edu.co (G.B.T.); carlosgutierrez@itm.edu.co (C.M.G.A.)

Abstract: The substantial waste generated during the processing of hides and skins as well as at other stages of manufacturing is a recurring issue in the leather industry that this article attempts to address. To investigate the mechanical and thermal characteristics of the resultant composites, this study suggests using leather waste from the bovine leather industry, analyzes the tanning process, and assesses the viability of mixing this waste with natural rubber (TSR-20). Without the inclusion of leather waste, the resulting composites had exceptional tensile strength, surpassing 100% of rubber's strength. The effective interaction of the recycled leather particles with the natural rubber matrix was evidenced using the Lorentz–Park equation. This better performance points to a competitive relationship between rubber and leather waste. The samples' density was 10% greater than that of polybutadiene elastomers and 10% greater than that of natural leather, while the hardness was comparable to that of PVC, which is frequently utilized in the design of general-purpose soles. This suggests that waste from the leather industry can be efficiently utilized in sustainable applications, particularly in the production of leather goods and shoes, helping to valorize waste that is typically discarded. Furthermore, by encouraging the use of recycled resources in the creation of new compounds, this plan provides the rubber sector with a sustainable option. To optimize this proposal, perhaps will be necessary to identify different vulcanization systems to improve the physical mechanical properties and other uses derived from the optimizations realized. This composite can be applied in the fashion industry in order to develop new trends around the application of waste and residues for a natural design line. Through the research process, it was possible to integrate the residues into the natural rubber matrix, as evidenced in the characterization process.

Keywords: leather waste; natural rubber; rubber composites

1. Introduction

Industrial processes aim to obtain products by transforming raw materials, which often requires the use of natural resources. In recent decades, growing concerns about pollution and climate change have drawn attention to the way these resources are used [1]. The industrial sector, one of the major contributors to these problems, is becoming increasingly focused on reducing resource consumption and pollutant emissions. An example is the bovine leather manufacturing and processing sector, usually referred to as tanning. Several studies [2–5] have proposed alternatives to mitigate environmental impact in this industry; however, they remain insufficient. This problem is also observed in the Colombian tanning industry, which lacks programs such as those aligned with the 2030 Sustainable Development Agenda, responsible consumption, and climate action [6–8]. To promote sustainable production and manufacturing and the consideration of global factors in the process, the manufacturing industry could support the fulfillment of sustainable development goals by seeking to reduce the emission of waste by incorporating it into the production cycle rather than dumping it in landfills.

In the global context, the United Nations Environment Program (UNEP) introduced in the 1980s the concept of Cleaner Production as a strategy to reduce the excessive consumption of natural resources and make production processes more efficient [9]. The main objective of such initiatives was to reorient industrial processes, products, and services towards preventing environmental impacts. Examples include the treatment of contaminated water [10–12]. Furthermore, global leather supply chains constitute a socio-technical system, consisting of stakeholders, such as consumers, civil society, and scientific communities; contextual factors, such as demand, product requirements, production, transportation, and communication; and social mechanisms, such as collaboration and working conditions [13].

Around the world, various industries are implementing regulations to support sustainability, and some businesses are adopting more conscientious production methods. However, some sectors continue to rely on outdated and environmentally harmful technologies instead of taking preventative measures [14,15]. The tanning industry has faced challenges due to its reliance on traditional processes for treating animal skins. Consequently, research is exploring alternatives to the conventional chrome tanning process. Although the leather tanning industry has seen minimal changes in its core production processes, increasing demand has driven efforts to automate certain stages and incorporate new technologies to improve production efficiency [16,17].

The leather industry generates solid waste categorized into non-tanned and tanned waste. Non-tanned waste includes by-products from fleshing, cutting, trimming, and sanding operations. After tanning, semi-finished and finished leather shavings and sawdust are also generated. Leather waste from the tanning process takes over 100 years to biodegrade, making it one of the least valued and most problematic wastes in the industry. Moreover, out of 1000 kg of raw hide, nearly 850 kg becomes solid waste in tanneries, while only 150 kg is transformed into leather. In other words, although part of this material is used in the manufacturing of leather goods, most ends up in public landfills [18].

A global study revealed that China leads in leather production, followed by Italy, Vietnam, and Germany. India's share is 2.7%, worth USD 1.4 billion. The largest importers are the USA, Germany, and EU countries. Despite its economic significance, leather production negatively impacts the environment [19]. The study also highlights the presence of tanneries in 11 departments in Colombia, and the country's leather industry is considered a key production sector in the National Development Plan, with trade surpluses recorded since 2003.

This industry is particularly important in the footwear and leather goods sectors [20]. In the case of Colombia, the footwear sector experienced a 13.3% increase in exports between 2017 and 2018, with a 4.73% rise in real annual spending on footwear purchases by November 2018. This underscores the importance of adopting alternative waste management strategies in footwear production [21]. Efforts have been made to repurpose leather waste as a reinforcement phase in the production of rubber-based materials. For that reason, this study focuses on combining leather residue (LR) with natural rubber (NR), considering its excellent physical, mechanical, and chemical properties, which offer an environmentally friendly material. As a polymer, natural rubber allows for the incorporation of particles and fibers, enhancing its properties for new industrial applications [22–25].

Furthermore, environmentally friendly composites—also known as green composites, ecological composites, or biocomposites—show physical, mechanical, thermal, and chemical characteristics comparable to traditional materials but with the added benefit of environmental sustainability [26,27]. Natural rubber, as a sustainable material, is increasingly being used in new developments. In fact, over the past six years, the global production of polymers from sustainable natural sources has increased by more than 400% [28].

Given their favorable technical qualities, affordability, and environmental attributes, rubber composites reinforced with plant and animal fibers are being used in numerous industries [29]. Moreover, the mechanical properties of reinforced natural rubber composites, such as strength and modulus, elongation at break, creep resistance, increased toughness, and shear and tear strengths, are markedly superior to those of particulate-reinforced rubber [30].

As a result, a variety of products have been developed in Colombia using these composites, including conveyor belts and footwear [31]. This has been achieved through a strategic alliance with the Colombian National Training Service (abbreviated SENA in Spanish), which provides the necessary infrastructure to formulate novel proposals on the use of leather residues in the rubber industry and introduce them to the productive sector. Consequently, this study aims to evaluate the technical feasibility of producing blends of natural rubber and leather residues for use in footwear products.

The methodologies consulted for this project include steps for ensuring efficient and sustainable production in the leather tanning industry (see Figure 2). A comparison was made between the technical information found and the documentation applicable to the Colombian industry, aiming to identify the benefits of implementing reuse alternatives for leather waste. These efforts contribute to mitigating the environmental damage caused by improper disposal of leather waste while proposing viable solutions for the development of new products within the leather industry [32–35].

The production model around waste use focuses on reducing the environmental impact of production activities through comprehensive preventive measures. It emphasizes resource reduction by adopting advanced process technologies and equipment, thereby achieving sustainable development [36]. Figure 1 illustrates each stage of the tanning process, which proceeds in the following order: pre-treatment (receiving, pre-fleshing, liming, fleshing, and splitting), tanning (dewatering, trimming, dyeing, and drying), and post-treatment (conditioning, finishing, packaging, and shipping).

One of the least utilized residues throughout the production process is post-industrial leather, generated after the cutting of pieces. This residue is typically discarded in municipal dumps without any probability of use, which is why it was selected as the focus of this study.

The study identified several processes in which this waste could be reprocessed. Figure 2 illustrates some potential alternatives, where the decision was made to reduce post-industrial leather waste by adopting the Level II strategy—internal recycling of waste. This strategy involves reusing residues by mixing them with a binder, in this case, natural rubber. The implementation of this approach holds great promise, especially for small entrepreneurs, as the potential for designing new products with this composite material is highly feasible and within their technological reach.

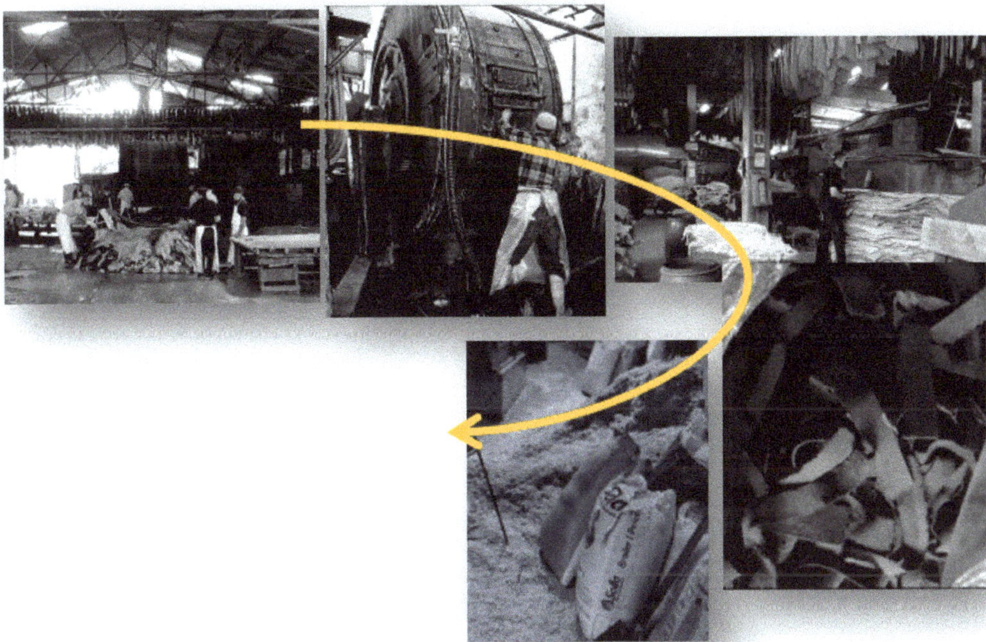

Figure 1. From left to right: reception of skins, tanning drum, drying, residues. Source: Authors' own work.

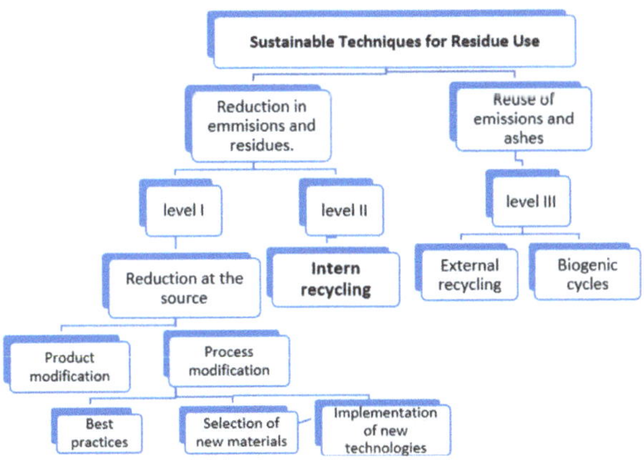

Figure 2. Typical sustainable production process [37].

Figure 2 presents three levels of sustainable production methods, which serve as mitigation alternatives to reduce the environmental impact of production processes [38,39]. These levels represent strategies for optimizing factory operations. Level I involves actions at the product or process level aimed at reducing or eliminating waste generation. This includes improvements in the production process, the substitution of raw materials with less polluting or more reusable alternatives that generate fewer by-products, or the adoption of more efficient technology. Level II focuses on internal recycling actions, where by-products and waste are reintegrated into the production process to maximize resource use. Finally, Level III encompasses recycling actions that are external to the company; i.e., waste generated within the company is repurposed as raw material for other industries. This study explored the Level II alternative, as previous research has demonstrated the potential of developing synthetic leather by combining it with natural rubber to create a useful composite for the production sector [40].

The methodology for rubber conformation used in developing the mixtures follows the process described in [21,29,41,42]. First, the mixtures of leather residues and natural rubber were analyzed and characterized. This process began by identifying the residues generated during production, followed by an analysis of their functionality when mixed with a binder such as natural rubber, aiming to project potential industrial applications. To prepare NR mixtures with leather waste, vulcanization activators were added to a roller mixer, and proportional calculations were made per hundred parts of rubber (phr). The mixture was processed at room temperature for 6 min to achieve the plasticization of the elastomer. The leather residue was then added, and the mixing process continued until the material was completely dispersed.

2. Materials and Methods

The leather residues were initially shredded using an industrial shredder and left to dry at 80 °C for 24 h. The dried material was then sieved using a Ro-Tap sieve shaker W.S. Tyler®—ro-tap Shaker, passing through a 30-mesh sieve with an aperture size of ≤600 μm, after which the sieved material was reserved for further use. Composites of natural rubber and finished leather particles were prepared with varying concentrations of leather particles at 10%, 20%, 30%, 40%, 50%, and 60% phr; the process is shown in Figure 3. Once the leather fibers were mixed and evenly dispersed in the natural rubber, vulcanization activators were introduced. The composite was then mixed in a two-roll rubber mill at 60 °C for 20 min, with a friction ratio of 1.0:1.25. The leather fiber ratios were selected based on previous studies involving wet blue leather residues.

Figure 3. From left to right: skins on tanning drum, milled residues, mixing process residues with natural rubber in an open roller mixer, and samples obtained for the characterization process. Source: Authors' own work.

The activated mixtures were stored at room temperature for 24 h to allow for sufficient reaction time and to eliminate residual gasses, according to ASTM 3182 [43]. The formulations used in this work are shown in Table 1. Initially, the natural rubber was chewed for 10 min, so that a band would form around the roller and there would be a momentary loss of the elastic characteristic through the shearing of the polymer chains, making it plastic with low viscosity, facilitating the incorporation of filler and reagents and also their reaction with the rubber. After being chewed and plasticized, the reagents and filler were inserted until they were fully incorporated. The zinc oxide and stearic acid were incorporated simultaneously, followed by the leather residue until they were completely homogenized.

This facilitated the action of the accelerators and sulfur that would be added after this period. After the accelerators and sulfur were added, the compound was left to rest for another 12 h at room temperature so that the reaction between the zinc stearate and the accelerators could take place. In the next step, the composites were subjected to rheometric testing (ASTM D 5289) [44] to determine the optimum vulcanization time (T90) used in thermo-pressing, maximum torque (MH), and minimum torque (ML). The composites were molded by compression in a hydraulic press at a temperature of 150 °C for times equal to T90. After thermo-pressing, the composites obtained at different phr were subjected to mechanical, thermal, and morphological tests. In this final stage, the physical, mechanical, and thermal properties of the mixtures were evaluated in accordance with the international standard ASTM D 3184-07 [45].

Table 1. Vulcanization system formulation (phr).

Function	Materials	phr
Matrix	Natural rubber (NR)	100
Filler	Leather residue (LR)	0, 10, 20, 30, 40, 50, 60
Activator	Zinc oxide (ZnO)	5
Activator	Stearic acid	2
Curing agent	Sulfur (S)	2.5
Lubricant	* Oleic acid	6
Secondary accelerator	** TMTM	0.5
Primary accelerator	*** MBTS	1.25

* (9Z)-9-octadecenoic acid; ** Tetramethylthiuram monosulfide; *** Dibenzothiazole disulfide.

Rheological Properties

To gain insight into the vulcanization behavior, rheometry experiments were conducted using a rheometer (Team Equipment Ltd., Sao Pablo, Brazil) with an isothermal system set at 150 °C and a disk oscillation of 1°. The curing parameters were determined based on the curves obtained in accordance with ASTM D2084 [46]. Among the rheological properties, viscosity and elasticity are key factors related to structural characteristics such as weight and molecular weight distribution. In the case of blends, molecular weight distribution and phase morphology are particularly important, as the overall process outcome depends on the understanding of these variables. Studying the mixture's characteristics allows for the observation of polymer conditions, including degradation and crosslink formations, which manifest as a reduction or increase in torque.

The swelling method was employed to ascertain the crosslinking density of the composites. The samples, with a mass of approximately 0.25 ± 0.05 g each, were first weighed and then immersed in toluene for 5 days. After this period, the specimens were removed, surface-dried with absorbent paper, and weighed again. They were then placed in an oven at 60 °C for 24 h and weighed one final time. The crosslinking density was calculated using Equation (1), as suggested in [47,48].

$$\eta = \frac{-(\mathrm{Ln}(1 - V_b) + V_b + \mathrm{X}(V_b)^2)}{(p_b)(V_0)(V_b^{\frac{1}{3}} - \frac{V_b}{2})} \quad (1)$$

Here, V_b is the volume fraction of the polymer in the swollen gel at equilibrium; X, the polymer–solvent interaction parameter; P_b, the density of the polymer; and V_0, the molar volume of the solvent. The relative density was determined using a Sartorius BA210® gravimeter, and the densities of the various samples were measured following the specifications outlined in the UNE-ISO 2781:2015 Method A standard [49].

The mechanical properties of the composites were evaluated through tensile and deformation tests, including calculations for Young's modulus and elongation at break. Measurements were conducted using universal analysis equipment in accordance with ASTM D 412-06a [50]. The tests were performed using a C-type mold, with a displacement speed of 500 mm/min. In addition, abrasion tests were conducted in triplicate using a rubber abrasion tester featuring a revolving drum, in accordance with ASTM D 5963 [51]. The samples were subjected to a friction distance of 40 m, equivalent to 84 drum rotations, under a force of 5 $\pm$ 0.2 N (1.125 $-$ 0.02 lbf).

Hardness tests were performed using a Kiltler durometer (Sao Pablo, Brazil) with a Shore A scale, in accordance with ASTM D 2240 [52]. Mass loss measurements were obtained through thermogravimetric analyses, which also provided insights into the thermal stability of the materials—a critical property for this study. Analyses were carried out on a TA Instruments Q500 analyzer, covering a temperature range between 25 °C and 500 °C. It is important to clarify that the thermal degradation behavior of collagen fibers varies depending on the modifications applied to the fibers from different hides, due to their inherent structural complexity [53].

The derivative thermogravimetry (DTG) curve served to determine the temperature at which material loss occurred. Scanning electron microscopy (SEM) was employed to further evaluate the interaction between the constituents, specifically regarding morphology, homogeneity, and dispersion. Moreover, the energy-dispersive X-ray (EDX) technology integrated into the SEM was used to identify regions in the images that could be associated with chemical elements that did not react during the vulcanization process or flaws in the physical dispersion of the mixture components. When the electron fringe is incident on the composite, the outermost electrons are exited, causing changes in energy levels. As these electrons return to their original state, energy is released and captured by the detector. Since each chemical element has a unique energy spectrum, it can be accurately identified. The equipment used for this analysis was the Philips® XL-30 FEG microscope, equipped with energy-dispersive spectroscopy (EDS).

3. Results

In the preparation of the natural rubber composites, oleic acid—a long-chain unsaturated fatty acid derived from animal and vegetable fats through saponification and acidification—was added after incorporating the leather fibers during open roller milling [54]. This addition aimed to improve the interaction between the fibers and the natural rubber, thus promoting good dispersion and enhancing the physical and mechanical properties of the composites.

3.1. Rheometry and Crosslinking Density of Composites

To gain a more detailed understanding of the variables influencing the characteristics of the composites compared to vulcanized rubber without fillers, the rheometric parameters are presented in Table 2. Notably, the optimum cure time (t90) gradually increases for samples containing 10 to 20 phr of leather residues (see Figure 4). This can be attributed to the proportion of fibers added and the acidic nature of leather, which inhibits the activity of the vulcanization system accelerators. The most salient characteristic is the linear increase in torque, which correlates with the addition of leather fibers to the samples.

The gradual increase in optimum cure time for samples with 10 to 20 phr of leather residues can be attributed to the acidic nature of the leather residues. This acidity inhibits the action of the vulcanization system accelerators, as these interact with the free functional groups of the leather [55,56]. However, this inhibition results in improved physical and mechanical properties of the composite, which is analyzed in subsequent sections.

For samples containing 30 to 60 phr of residue filler, a gradual reduction in cure time was observed, with the 60% leather residue sample showing a shorter cure time compared to pure rubber. These results suggest that the addition of higher filler proportions promotes the vulcanization processes, likely due to the active groups present in the leather, as suggested in [57]. These findings are consistent with those of crosslinking density analyses, where a shorter cure time indicates a faster reaction rate and reduced crosslinking production time [58].

Table 2. Rheometric parameters of NR/LR composites.

Composite (100 phr)	MH (dNm)	ML (dNm)	ΔM (dNm)	t90 (dNm)
NR 0	0.68 ± 0.12	0.05 ± 0.12	0.63 ± 0.12	6.336 ± 0.12
NR/LR 10	0.88 ± 0.12	0.09 ± 0.14	0.79 ± 0.13	9.65 ± 0.12
NR/LR 20	1.27 ± 0.13	0.12 ± 0.15	1.15 ± 0.14	10.618 ± 0.14
NR/LR 30	1.03 ± 0.14	0.14 ± 0.13	0.89 ± 0.11	8.952 ± 0.13
NR/LR 40	1.49 ± 0.13	0.24 ± 0.11	1.25 ± 0.12	8.216 ± 0.13
NR/LR 50	1 ± 0.14	0.2 ± 0.12	0.8 ± 0.13	6.554 ± 0.12
NR/LR 60	1.77 ± 0.13	0.26 ± 0.11	1.51 ± 0.11	6.098 ± 0.14

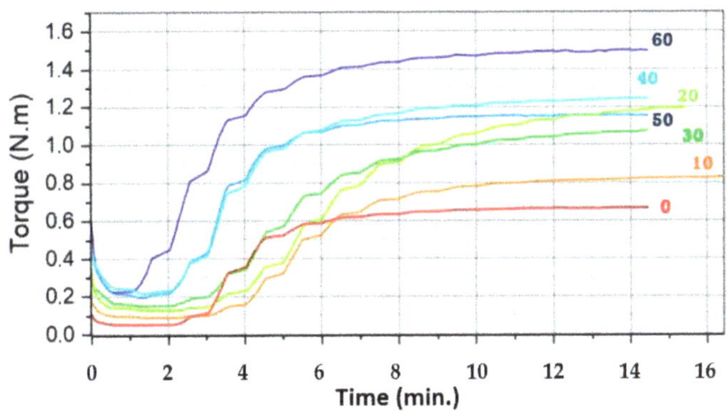

Figure 4. Rheological behavior curves for NR/LR composites.

The minimum torque (ML) values were observed during the pre-cure stage, when the accelerators and zinc stearate initiated the reaction process but no crosslinking had yet occurred. ML helps determine the degree of processability, i.e., the viscosity of the unvulcanized composites [59]. In contrast, the maximum shear strength (MH) values are associated with material stiffness. An increase in MH in the rheometer's oscillating disk may suggest an increase in matrix stiffness, necessitating greater rotational stress.

A gradual increase in minimum and maximum torque values was observed for all samples containing residues. This indicates that the viscosity and stiffness of the composites increase with the incorporation of leather residues, which are inherently stiffer due to their fibrous structure. The increase in MH can be attributed to the reduced macromolecular mobility of the matrix resulting from the incorporation of leather fibers [60]. The slight reduction in maximum torque for the composites with 30 and 50 phr can be attributed to the limited amount of chemicals available for vulcanization, as reported in [56]. The maximum torque values were considered together with the results of the hardness analyses, which will be discussed below. The obtained values are of great importance because they orientate the possibilities of saving in the cost of equipment in vulcanizing; considering some similar investigations, where the T90 value was obtained after 16 min of processing with NBR rubber [61], in this research, the T90 value was favorable as it was less than 12 min.

3.2. Determining Crosslinking Density

Figure 5 illustrates the crosslinking density of the vulcanized composites containing post-industrial leather residues. A linear increase in crosslinking density is observed as the leather residue content in the samples rises, compared to vulcanized rubber without filler.

This increase may be due to the formation of interfacial networks and enhanced bonding between the rubber matrix and the leather fibers.

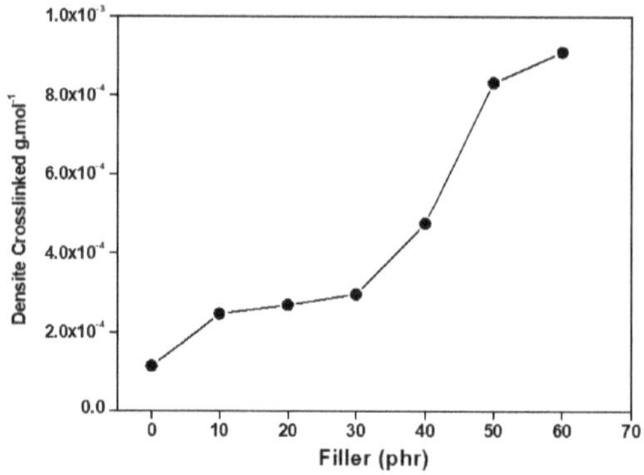

Figure 5. Crosslinking density of NR/LR composites.

The correlation of the number of crosslinks formed is evident in the results of the tensile strength and abrasion loss tests. Up to 20 phr of leather residue, the composites exhibited high resistance, but as the leather content increased beyond 30 phr, the stiffness compromised the flexibility of the samples, resulting in a fragile behavior. Nevertheless, this does not diminish the system's potential for industrial applications.

3.3. Analysis of Rubber/Fiber Interactions Using the Lorenz–Park Equation

The interaction between the leather residue fibers and the rubber matrix was determined employing the method developed by Lorenz and Park [62] and the parameters obtained from the solvent swelling tests used in Equation (2) [21]:

$$\frac{Q_l}{Q_r} = ae^{-z} + b \tag{2}$$

where

- Q = the weight of toluene absorbed per gram of rubber, where subscripts l and r denote the composites vulcanized with a loading and pure rubber, respectively;
- z = the ratio of filling mass per unit mass of rubber;
- a and b = constants.

The value of Q was calculated using Equation (3):

$$Q = \frac{w_s - w_d}{w_r \times 100/w_F} \tag{3}$$

where

- Ws = the weight of the swollen composite when the balance between the organic solvent and the polymer is achieved;
- wd = the weight of the dry composite;
- wr = the weight of the rubber in the dry composite;
- wF = the total weight of the formulation.

3.4. Analysis of Rubber/Loading Interactions Using the Lorentz–Park Equation

The swelling data of the composites in an organic solvent (Flory–Rehner) were used in the calculation of the Qf/Qg ratio applying Equation (2). The Qf/Qg ratio reflects the restriction to swelling by the rubber matrix in the vicinity of the loaded fibers. When rubber is immersed in the organic solvent, the latter tends to penetrate and degrade the regions that are not firmly bound to the loadings or crosslinked. Although rubber expands in response to the solvent, firmly adhered loaded particles can limit this expansion. The higher the Qf/Qg value, the weaker the load/matrix interaction because the latter tends to be load/load. And z is the ratio between the amount of loading and the amount of rubber (phr). Hence, an increase in the amount of loading in the composite generates an exponential reduction in z [63].

Figures 5 and 6 show a linear relationship between the curve of Qf/Qg and that of e^{-z}. As more loading is added, the Qf/Qg ratio increases as a function of e^{-z}. This behavior is due to the fact that the values are related to the swelling of the loaded rubber. Therefore, loading works as an obstruction to solvent penetration due to a good matrix/loading interaction. Parameters a and b are 1.3529 and 0.4967, respectively, with a correlation coefficient (R) of 0.8601 for leather residue fibers. According to Lorenz and Park, since the value of parameter a is greater than 0.7 and the slope of a is higher, the rubber/loading interaction is strong. The value of said parameter (much higher than 0.7) denotes a firm rubber/fiber interaction, related to the stress and strain response, respectively. Although all composites have the same formulation and crosslinking system, Figures 6 and 7 show a decrease in the Qf/Qg ratio with increasing load; in other words, less solvent penetrated the more loaded composites. This is associated with improved mechanical properties and is attributed to a strong interfacial interaction between the elastomer chains and the fibers, generated by a good dispersion of said fibers in the matrix.

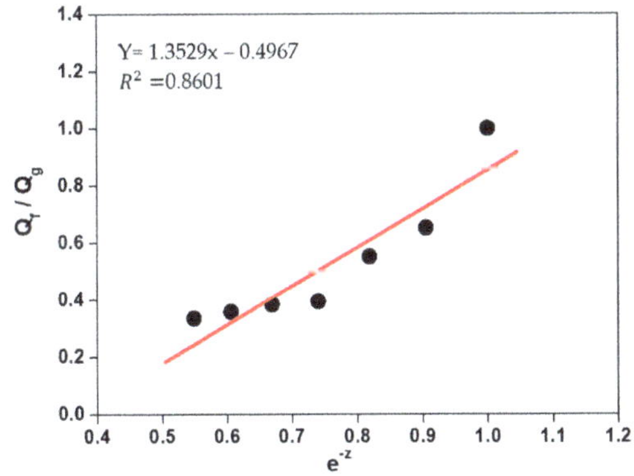

Figure 6. Variation in Qf/Qg versus e^{-z} in composites vulcanized with leather fiber residues.

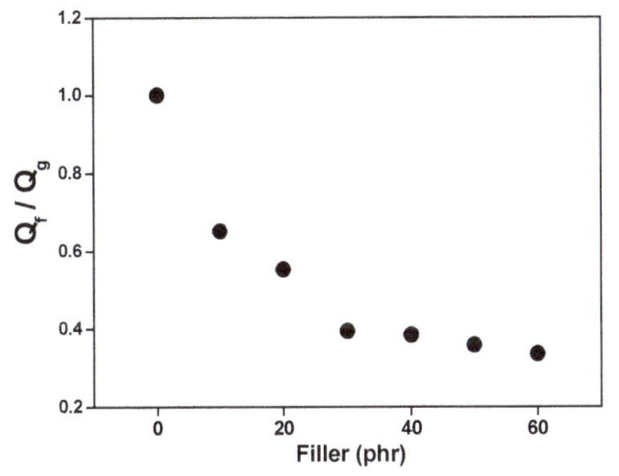

Figure 7. Effect of filler loading on the Qf/Qg of composites vulcanized with leather fiber residues.

Figure 7 shows the Qf/Qg ratio values as a function of filler content. It can be said that the ratio values decrease as a function of the increase in the filler in the matrix, and this behavior corroborates the strong interfacial interaction between filler and rubber.

3.5. Relative Density

Leather's density varies according to the stage of the production process [64]. As leather undergoes mechanical and chemical treatments, its density increases due to material compaction, which reduces, among other things, the fat content. Table 3 presents the density values of the composites as a function of the leather residue content. The addition of finished leather waste particles has been observed to proportionally increase the density of the composites. This change in density modifies physical properties such as elasticity and hardness [65], which aligns with the results of this study.

Table 3. Density for NR/LR composites.

Composite (phr)	Density (g/cm^3)
Pure gum (NR 0)	0.98 ± 2
NR/LR 10	1.0 ± 2
NR/LR 20	1.01 ± 2
NR/LR 30	1.02 ± 2
NR/LR 40	1.04 ± 3
NR/LR 50	1.06 ± 3
NR/LR 60	1.07 ± 3

This better performance points to a competitive relationship between rubber and leather waste. The samples' density was 10% greater than that of polybutadiene elastomers and 10% greater than that of natural leather [66], low-density EVA, medium-density soft PVC, and flexible PU, while the hardness was comparable to that of PVC, which is frequently utilized in the design of general-purpose soles [67].

3.6. Hardness (Shore A)

Figure 5 illustrates the crosslinking density values for the vulcanized composites. All composites display an increase in crosslinking density compared to natural rubber without filler. Shore A hardness values rise proportionally with the leather residue content in the samples. This increase can be attributed to the greater viscosity provided by the fibrous structure of leather, which is naturally stiffer than elastomeric materials [68,69].

The composite containing 60 phr of leather residue exhibited the highest crosslinking density among all samples, which coincides with the lowest swelling ratio observed. An increase in crosslinking density results in a decrease in the composite's molecular weight, which leads to the formation of additional bridges between the collagen fibers of leather and the natural rubber matrix. Additionally, it enhances the composite's resistance to solvents and increases the number of bonds during vulcanization. These results are consistent with those of previous studies [55,56].

The composites containing 40 and 50 phr of leather residue exhibited similar responses, shown in Figure 8. This similarity can likely be attributed to the heterogeneous dispersion of the fibers during the manual fabrication process of the samples. This suggests that a strong interphase relationship was achieved, promoting effective interaction and adhesion between the leather particles and the natural rubber matrix. When comparing the hardness parameters with the maximum torque values shown in Table 2, it is evident that both results support that the addition of leather residues increases the viscosity and hardness of the material. These findings are consistent with those reported in similar studies found in the specialized literature [55,70].

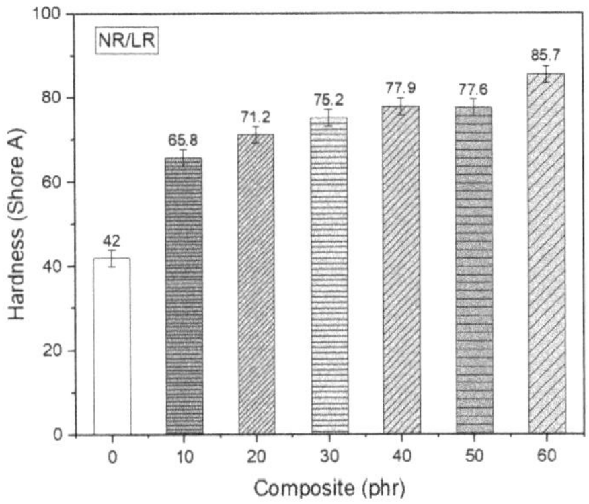

Figure 8. Hardness values of NR/LR composites.

3.7. Abrasion Resistance

The results of the abrasion tests performed on the samples are presented in Table 4. Abrasion resistance increased for all composites, except for the one with the highest percentage of filler (60 phr), which showed a 36.8% decrease. This reduction can be attributed to structural and physical variations, such as filler–filler interactions, which may create voids in the matrix due to the high leather residue content.

Table 4. Abrasion resistance.

Composite (phr)	Loss Volume (mm^3)	Abrasion Resistance (%)
NR	264.2 ± 2.4	59 ± 3
NR/LR 10	214.2 ± 1.4	72.6 ± 2
NR/LR 20	213.6 ± 2.8	72.9 ± 2
NR/LR 30	251.6 ± 9.6	61.9 ± 3.3
NR/LR 40	245.5 ± 8.4	63.5 ± 2.3
NR/LR 50	258.4 ± 8.9	60.4 ± 3.9
NR/LR 60	422.5 ± 7.6	36.8 ± 2.3

The composites containing 10 and 20 phr of leather residue exhibited the highest abrasion resistance, with minimal loss volume values of 213 mm^3 and 214 mm^3, respectively. These results suggest the effective dispersion of the leather residues within the rubber matrix, possibly due to the method of adding fibers through sieving during the lamination process. Moreover, the incorporation of oleic acid may have improved the distribution of the leather particles and contributed to composite stability [68]. Considering that the abrasion resistance values for materials used in daily wear soles are typically < 300 mm^3 [71], the composites containing up to 50 phr of leather residues are deemed suitable for such applications.

Other studies have demonstrated the potential to further improve abrasion resistance by incorporating higher amounts of leather residues into the rubber matrix, which may be achieved through the chemical treatment of the leather. Nevertheless, this process increases the complexity, cost, and energy consumption of the production process [71].

3.8. Tensile Strength/Deformation Test

The parameters obtained from the tensile and deformation tests of the composites containing finished leather residues are presented in Table 5 and Figure 9. The results show an increase in stress values for the composites with fillers, which indicates a high level of interaction between the rubber macromolecules and the leather fibers, which are more rigid. This interaction enables the composite to withstand greater tensile stress, effectively reinforcing the material [57,72]. The small size of the leather fibers (≤600 μm) also contributes to the improved tensile strength, as smaller fibers are better dispersed within the rubber matrix, enhancing the reinforcing effect [60,65]. Notably, the composite containing 20 phr of reinforcement exhibited the highest tensile strength values (5.18 MPa), representing a 128.19% increase compared to the natural rubber without reinforcement. However, composites with more than 10 phr of leather residue demonstrated a reduction in tensile strength values compared to the NR/LR 10 composite. This decline may be attributed to the physical interference between the rubber matrix and the leather fiber chains.

Table 5. Tensile strength/deformation test results.

Composite (phr)	Tensile Strength (MPa)	Deformation (%)
NR	2.27 ± 0.41	212 ± 0.39
NR/LR 10	4.38 ± 1.16	168 ± 1.12
NR/LR 20	5.18 ± 0.27	54 ± 0.08
NR/LR 30	3.72 ± 0.69	76 ± 0.46
NR/LR 40	3.76 ± 0.86	67 ± 0.15
NR/LR 50	4.2 ± 0.4	45 ± 0.08
NR/LR 60	3.85 ± 0.29	23 ± 0.05

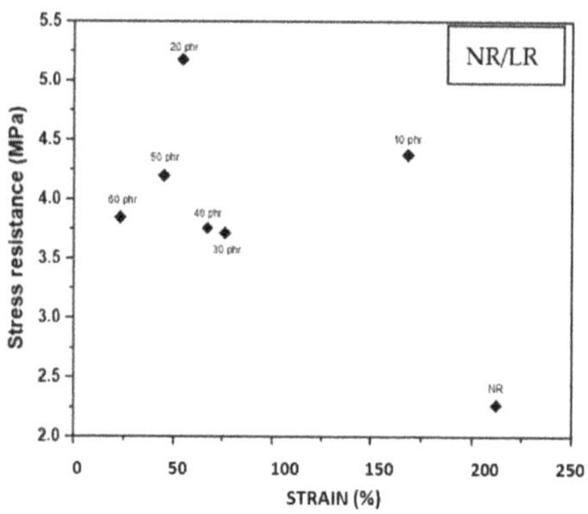

Figure 9. Tensile strength (stress resistance)/deformation (strain) of the NR/LR composites.

This reduction in flexibility is attributed to the increased crosslinking density and stiffness of the composite. When correlating the tensile test results with the morphological analysis from the SEM images in Figure 11, particularly for samples containing 10 phr and 20 phr of leather residues, a heterogeneous and partially amorphous morphology is observed in the NR/LR 30 phr sample. These observations confirm that the reduction in ductility is consistent with both the surface morphology and the mechanical properties of the composite.

The NR/LR 10 phr composite shows a notable increase in tensile strength, with a 92.9% improvement compared to natural rubber without fillers. However, this is accompanied by a 44% decrease in deformation. The addition of leather residues increases the stiffness of the material and influences the sulfur vulcanization system, promoting the formation of polysulfide bonds that enhance the elasticity of the rubber matrix. This means that the addition of leather residues in this proportion increases the mechanical performance of the composite, making it more resistant to tensile strength (4.38 MPa) than pure natural rubber. Additionally, the composites with less than 30 phr of fibers show higher ductility than those with greater fiber content; a similar effect was found in mixtures of natural rubber with leather, but previously neutralized with chemical agents; mixtures with urea 10% had lower tensile strength, but with a higher effect in hardness, orienting this more natural compound towards products with industrial applications, including the footwear sector [71].

3.9. Thermogravimetric Analysis

Thermogravimetric analyses were performed to assess the thermal stability of leather residues, natural rubber, and NR/LR composites through the decomposition of the material. The sample mass for this analysis ranged from approximately 14 to 23 mg. Figure 10 shows the thermogravimetry (TG) and DTG curves for (A) vulcanized natural rubber and (B) leather residue samples.

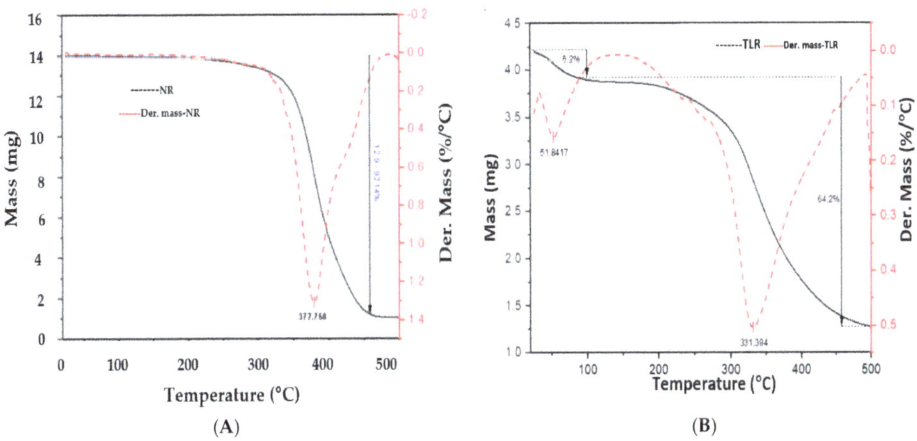

Figure 10. TG/DTG curves of (**A**) vulcanized natural rubber and (**B**) leather residues.

For natural rubber, a mass loss of 92.61% is observed between 280 °C and 480 °C, with an abrupt mass loss peaking at 377 °C, attributed to the degradation of the polymer chains in the rubber [73]. A lower shoulder is observed in the DTG curve (red) between 430 °C and 500 °C, characteristic of the degradation of higher-molecular-weight polymer chains, which degrade more slowly. This can also be linked to vulcanizing agents such as sulfur, which have low thermal stability and a slower degradation process [74], as observed in SEM images.

The pure natural rubber samples and the composites containing leather residues in proportions ranging from 10 to 60 phr display similar mass loss patterns, irrespective of the addition of leather residues (Figure 10A). The sample containing 60 phr of leather residues shows a mass loss of 21.6% at a maximum temperature of 300 °C (Figure 10B). This initial mass loss decreases as the leather residue content in the composite is reduced. It is noteworthy that the leather samples contained approximately 18% to 25% water, which correlates with the observed percentage of initial mass loss.

The residual mass percentage increases proportionally to the amount of leather residue in the mixture, which can be attributed to elements such as zinc oxide and sulfur from upwelling processes. However, unlike sulfur, zinc oxide requires temperatures exceeding 900 °C for degradation. Notably, the composite containing 60 phr of leather residue exhibits the lowest mass loss, with a maximum degradation temperature like that of pure natural rubber. The residual mass of this sample is higher due to the amount of leather residue, in which nitrogen is abundant, forming char and improving thermal stability.

Figure 11 demonstrates that the maximum degradation temperature for the NR/LR composites is comparable to that of the natural rubber sample without leather, suggesting a possible rubber coating that degrades at a higher temperature (378.8 °C) than leather

(331.4 °C), thereby reducing mass loss and causing single-stage decomposition similar to that of natural rubber.

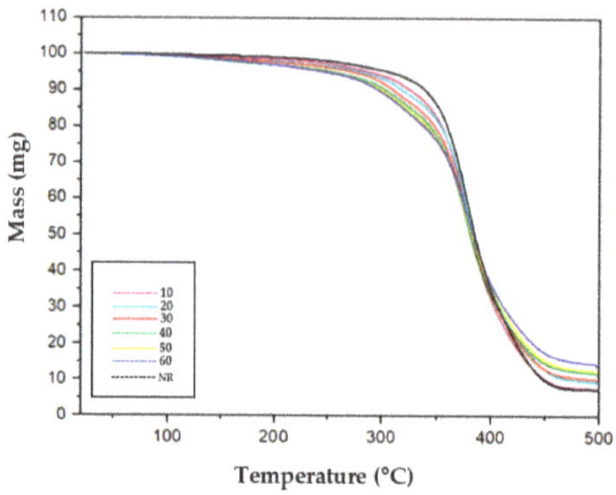

Figure 11. Thermogravimetric curves of the NR/LR composites.

The degradation process of composites with higher leather residue content occurs over a wider temperature range, indicating a slower degradation rate [75]. The reduction in the intensity of the maximum degradation temperature peak and the increase in residual mass for composites containing 20 to 60 phr of leather residues demonstrate improved thermal stability. This contrasts with the composite with 10 phr and the pure natural rubber sample, which show residual masses of 7.48% and 7.39%, respectively.

In conclusion, the incorporation of leather residues through the developed methods enables the fabrication of composites with mechanical properties suited for industrial applications such as footwear and thermal insulation. Moreover, leather residues can serve not only as filler material, but also as reinforcing material, offering versatile industrial applications.

Morphological Evaluation: Fiber–Rubber Interaction

Micrographs of the finished leather residues were randomly captured to analyze their structure and identify the relationship and distribution of the fibers prior to their incorporation into the natural rubber matrix. Figure 12 shows micrographs of leather fibers at various magnifications: (a) ×200, (b) ×500, (c) ×1000, and (d) ×5000. The interlacing of fibrils (fibrous structure) provides the material with flexibility and strength, confirming its composite nature [70,76,77].

The observed fibers have diameters ranging from 96.3 to 143 µm, as shown in Figure 12a, where individual fibrils are also visible. A 0.2 mm cross-section can contain up to 300 fibrils, each with a diameter of approximately 5 µm, and can aggregate up to 1000 filaments [78]. Figure 12b–d provide further detail of the fibrous network, revealing a matrix of intertwined fibers that form larger fibers, resulting in a high-strength material. This strength is attributed to the tanning process with chromium (III), which reacts with the amino acid chains to stabilize collagen, contributing good physical properties to the material [79].

The fibers are separated and well defined as a result of the tanning process applied to the material, which enhances its mechanical properties and makes it suitable for use in industry. This process, which improves the material softness, flexibility, and permeability, also results in the isolation of fibrils through the creation of voids that render the structure more porous and striated [80]. These properties improve the interaction of the rubber matrix with the leather fibers and are characteristic of low-density materials.

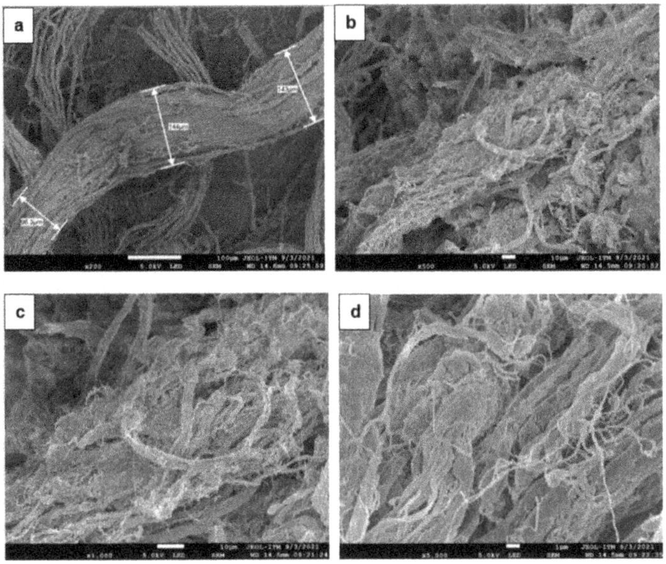

Figure 12. Micrographs of leather fibers at magnifications of (**a**) ×200, (**b**) ×500, (**c**) ×1000, and (**d**) ×5000.

Figure 13 presents micrographs of the composites with different filler proportions: (a and b) 10 phr, (c) 20 phr, and (d) 30 phr. The composites exhibit a predominant dark coloration, which is consistent across all samples regardless of the amount of leather added. This darkening can be explained by the natural color of the filler as well as the heating process, which causes the fibers to become more opaque due to dehydration. There is also evidence of increased surface roughness; however, no exposed leather fibers are observed, indicating the good coverage of the leather fibers by the rubber matrix.

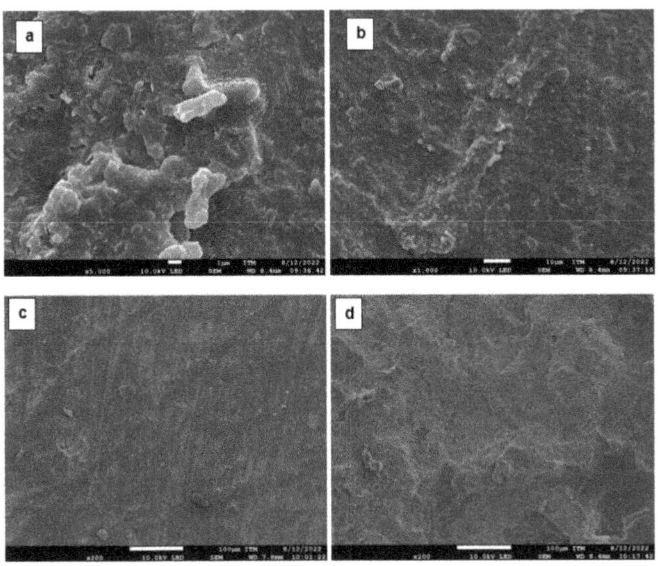

Figure 13. Composite micrographs: (**a**) NR/LR 10 at ×5000, (**b**) NR/LR 10 at ×1000, (**c**) NR/LR 20 at ×200, (**d**) NR/LR 30 at ×200.

The composite containing 20 phr fibers, shown in Figure 13c, displays a smoother surface compared to that in images a and b. This difference is due to the texture of the aluminum plate used during thermoforming, whereas the two first images represent cross-sectional views. In Figure 14, image (a) shows the process of mixing rubber with recycled leather and image (b) shows the compound obtained with 10 phr of material; as can be seen, it is a homogeneous product, which will have many applications in the footwear manufacturing sector, for example.

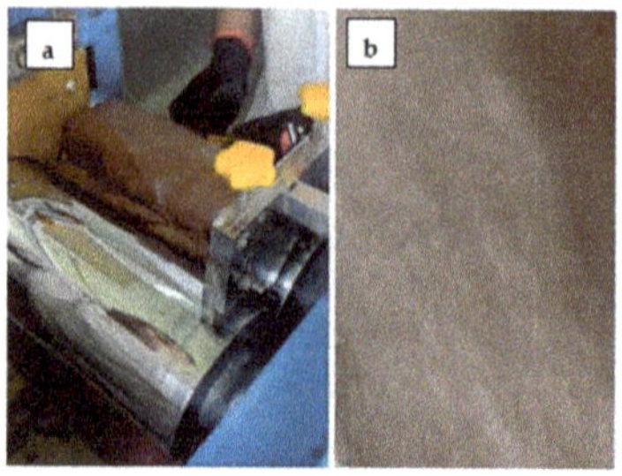

Figure 14. Composite obtained.

4. Final Observations

The reuse of leather waste for the development of new products through the traditional process of mixing rubber in roller cylinders, as outlined and proposed at the beginning of this article, has the potential to contribute significantly to the adoption of sustainable technologies. By collaborating with companies from other sectors and replicating programs from developed countries, this approach can become a driving force in transforming a highly productive sector, aligning it with sustainable practices and clean production methods.

By-products such as trimmings, sawdust, and shavings, which originate from the tanning and manufacturing processes, are currently considered non-usable waste by the company and are consequently disposed of with solid waste in a landfill. In contrast, wet blue leather, due to its chemical content, is separately disposed of by an external company. However, creating partnerships between the tannery and regional craft sector associations would offer the possibility for mutual benefits: the tannery could dispose of its leather waste and the craft sector would gain access to valuable raw materials.

5. Conclusions

A sustainability program must be a continuous effort rather than a one-time initiative. Once a plan is put in place, it is imperative to identify improvements that prioritize both environmental and financial sustainability. This is particularly important for Colombia's leather tanning companies, which are predominantly family-owned businesses with limited or inexistent environmental awareness in their practices.

It is crucial to inform small and medium-sized producers about available alternatives for utilizing waste through tools like clean production and design engineering. This study demonstrates that it is possible to carry out collaborative processes across different production areas, contributing to the achievement of Sustainable Development Goals 12 (Responsible Consumption and Production), 9 (Industry, Innovation, and Infrastructure), and 17 (Partnership for the Goals). Moreover, these efforts strengthen the communities where they take place.

To boost sustainability in Colombia's leather tanning industry, ongoing education for small producers is vital. Promoting waste utilization and exploring innovative materials with leather waste and natural rubber will enhance clean practices and community resilience.

Utilizing leather waste with alternative materials opens up opportunities to create new and innovative materials. Mixing leather waste with natural rubber provides a sustainable option for creating value-added products with positive environmental impact.

Author Contributions: Conceptualization, C.M.G.A., B.E.Á.Á., G.B.T. and J.A.J.C.; methodology, G.B.T., C.M.G.A. and B.E.Á.Á.; software, R.J.S. and G.B.T.; validation, G.B.T. and J.C.S.; formal analysis, R.J.S. and G.B.T.; investigation, E.R.L. and M.J.T.M.; resources, J.A.J.C., J.C.S. and G.B.T.; data curation, December 2024; writing—original draft preparation, G.B.T., C.M.G.A., R.J.S. and J.A.J.C.; writing—review and editing, G.B.T. and C.M.G.A.; visualization, C.M.G.A., B.E.Á.Á. and G.B.T.; supervision, G.B.T.; project administration, G.B.T.; funding acquisition, J.A.J.C., J.C.S. and G.B.T. All authors have read and agreed to the published version of the manuscript.

Funding: This research was funded by the Instituto Tecnológico Metropolitano (Grant number P20203) and the Centro para la Manufactura Avanzada, SENA, Medellin, Antioquia.

Institutional Review Board Statement: Not applicable.

Data Availability Statement: The original contributions presented in this study are included in the article. Further inquiries can be directed to the corresponding authors.

Acknowledgments: The authors thank the Instituto Tecnológico Metropolitano (Research code PCI 23205); the Agencia de Servicios Lingüísticos ITM; the Centro para la Manufactura Avanzada, SENA, Antioquia; and the Industrial Engineering Unit, Universidad Pontificia Bolivariana.

Conflicts of Interest: The authors declare no conflicts of interest.

References

1. Elías, C.; Jiménez, J.; Montón, J.; Muñoz, P.; Serrano, J.P.Y.F. Ciencias para el mundo contemporáneo 1 de 5. In *Impacto Ambiental. El Planeta Herido*; McGraw-Hill: Madrid, Germany, 2010.
2. Chojnacka, K.; Skrzypczak, D.; Mikula, K.; Witek-Krowiak, A.; Izydorczyk, G.; Kuligowski, K.; Bandrow, P.; Kułażyński, M. Progress in sustainable technologies of leather wastes valorization as solutions for the circular economy. *J. Clean. Prod.* **2021**, *313*, 127902. [CrossRef]
3. Yu, Y.; Wang, H.; Zeng, Y.; Zhou, J.; Zhang, Y.; Shi, B.; Wang, Y.N. Biomass-derived polycarboxylate–aluminum–zirconium complex tanning system: A sustainable and practical approach for chrome-free eco-leather manufacturing. *J. Clean. Prod.* **2024**, *452*, 142261. [CrossRef]
4. Shi, J.; Sheng, L.; Salmi, O.; Masi, M.; Puig, R. Life cycle assessment insights into nanosilicates-based chrome-free tanning processing towards eco-friendly leather manufacture. *J. Clean. Prod.* **2024**, *434*, 139892. [CrossRef]
5. Zhang, B.; Guan, X.; Han, Q.; Guo, H.; Zheng, S.; Sun, X.; El-Gowily, A.H.; Abosheasha, M.A.; Zhu, Y.; Ueda, M.; et al. Rational design of natural leather-based water evaporator for electricity generation and functional applications. *J. Energy Chem.* **2024**, *96*, 129–144. [CrossRef]
6. ONU. Transforming Our World: The 2030 Agenda for Sustainable Development. A/RES/70/1. Available online: https://sdgs.un.org/sites/default/files/publications/21252030%20Agenda%20for%20Sustainable%20Development%20web.pdf (accessed on 12 June 2024).
7. OECD Due Diligence Guidance for: Responsible Supply Chains in the Garment and Footwear Sector. 2017. Available online: https://mneguidelines.oecd.org/OECD-Due-Diligence-Guidance-Garment-Footwear.pdf (accessed on 12 June 2024).
8. European Commission. Final Report. Study on Due Diligence Requirements Through the Supply Chain. 2020. Available online: https://www.europarl.europa.eu/meetdocs/2014_2019/plmrep/COMMITTEES/DROI/DV/2020/06-22/Studyonduediligencerequirements_pdf (accessed on 10 January 2025).
9. Organización de las Naciones Unidas. *Cumbre de Rio*; united nations organization: New York, NY, USA, 1992.
10. Organización de las Naciones Unidas para el Desarrollo Industrial—ONUDI. *Transferencia o Desarrollo de Tecnología de la Industria 4.0 Para Establecer Modelos de Producción más Limpia*; UEESMUN: Guayaquil, Ecuador, 2015.
11. Buenos Aires produce más limpio. *Producción más Limpia, Concepto y Antecedentes*; Buenos Aires, Argentina, 2016.
12. Lee, S.Y.; Rhee, S.K. From end-of-pipe technology towards pollution preventive approach: The evolution of corporate environmentalism in Korea. *J. Clean. Prod.* **2003**, *13*, 387–395. [CrossRef]
13. Eisen, C.; Schenten, J.; Theis, A.; Rehn-Groenendijk, J.; Helferich, M.; Müller, H.; Hanss, D. Toward system innovation for more sustainable chemistry: Insights into consumers' perceptions, knowledge, and behavior related to traceability and product design strategies along leather supply chains. *Front. Sustain.* **2024**, *5*, 1351638. [CrossRef]

14. Shi, L.; Liu, J.; Wang, Y.; Chiu, A. Cleaner production progress in developing and transition countries. *J. Clean. Prod.* **2021**, *278*, 123763. [CrossRef]
15. ONU. Sustainable Development Goals (ODS). Available online: https://www.un.org/sustainabledevelopment/ (accessed on 12 June 2024).
16. Available online: https://mahileather.com/blogs/news/where-does-leather-come-from (accessed on 13 January 2025).
17. Shi, J.; Zhang, R.; Mi, Z.; Lyu, S.; Ma, J. Engineering a sustainable chrome-free leather processing based on novel lightfast wet-white tanning system towards eco-leather manufacture. *J. Clean. Prod.* **2021**, *282*, 124504. [CrossRef]
18. Kanagaraj, J.; Velappan, K.C.; Babu, N.K.; Sadulla, S. Solid wastes generation in the leather industry and its utilization for cleaner environment—A review. *J. Sci. Ind. Res.* **2006**, *65*, 541–548.
19. Koppiahraj, K.; Bathrinath, S.; Saravanasankar, S. Leather Waste Management Scenario in Developed and Developing Nations. *Int. J. Eng. Adv. Technol.* **2019**, *9*, 852–857. [CrossRef]
20. Rippel, M.M.; Bragança, F.D.C. Bragança Borracha natural e nano compósitos com argila. *Química Nova* **2009**, *32*, 818–826. [CrossRef]
21. Torres, G.B.; Hiranobe, C.T.; da Silva, E.A.; Cardim, G.P.; Cardim, H.P.; Cabrera, F.C.; Lozada, E.R.; Gutierrez-Aguilar, C.M.; Sánchez, J.C.; Carvalho, J.A.J.; et al. Eco-Friendly Natural Rubber–Jute Composites for the Footwear Industry. *Polymers* **2023**, *15*, 4183. [CrossRef] [PubMed]
22. Santos, R.J.; Hiranobe, C.T.; Dognani, G.; Silva, M.J.; Paim, L.L.; Cabrera, F.C.; Torres, G.B.; Job, A.E. Using the Lorenz–Park, Mooney–Rivlin, and dynamic mechanical analysis relationship on natural rubber/leather shavings composites. *J. Appl. Polym. Sci.* **2021**, *139*, 51880. [CrossRef]
23. Hiranobe, C.T.; Ribeiro, G.D.; Torres, G.B.; Reis, E.A.P.D.; Cabrera, F.C.; Job, A.E.; Paim, L.L.; Santos, R.J.D. Cross-Linked Density Determination of Natural Rubber Compounds by Different Analytical Techniques. *Mater. Res.* **2021**, *24* (Suppl. S1), e20210041. [CrossRef]
24. Cavalcante, D.G.; Gomes, A.S.; Santos, R.J.; Kerche-Silva, L.E.; Danna, C.S.; Yoshihara, E.; Job, A.E. Composites Produced from Natural Rubber and Chrome-Tanned Leather Wastes: Evaluation of their In Vitro Toxicological Effects for Application in Footwear and Textile Industries. *J. Polym. Environ.* **2018**, *26*, 980–988. [CrossRef]
25. Xu, Y.; Xu, Z.D.; Guo, Y.Q.; Huang, X.H.; Hu, Z.W.; Dong, Y.R.; Shah, A.A.; Dai, J.; Xu, C. Investigation of the Dynamic Properties of Viscoelastic Dampers with Three-Chain Micromolecular Configurations and Tube Constraint Effects. *J. Aerosp. Eng.* **2025**, *38*, 04024122. [CrossRef]
26. Masłowski, M.; Miedzianowska, J.; Strąkowska, A.; Strzelec, K.; Szynkowska, M.I. The Use of Rye, Oat and Triticale Straw as Fillers of Natural Rubber Composites. *Polym. Bull.* **2018**, *75*, 4607–4626. [CrossRef]
27. Peças, P.; Carvalho, H.; Salman, H.; Leite, M. Natural Fibre Composites and Their Applications: A Review. *J. Compos. Sci.* **2018**, *2*, 66. [CrossRef]
28. Chang, B.P.; Mohanty, A.K.; Misra, M. Studies on Durability of Sustainable Biobased Composites: A Review. *RSC Adv.* **2020**, *10*, 17955–17999. [CrossRef] [PubMed]
29. Barrera Torres, G.; Dognani, G.; da Silva Agostini, D.L.; dos Santos, R.J.; Camargo Cabrera, F.; Gutierrez Aguilar, C.M.; de Paiva, F.F.G.; Rainho Teixeira, S.; Job, A.E. Potential Eco-Friendly Application of Sugarcane Bagasse Ash in the Rubber Industry. *Waste Biomass Valorization* **2021**, *12*, 4599–4613. [CrossRef]
30. Masłowski, M.; Miedzianowska, J.; Strzelec, K. Influence of Wheat, Rye, and Triticale Straw on the Properties of Natural Rubber Composites. *Adv. Polym. Technol.* **2018**, *37*, 2866–2878. [CrossRef]
31. Roy, K.; Debnath, S.C.; Potiyaraj, P. A Review on Recent Trends and Future Prospects of Lignin Based Green Rubber Composites. *J. Polym. Environ.* **2020**, *28*, 367–387. [CrossRef]
32. Ministerio del Medio Ambiente—MMA. *Política Nacional de Producción Más Limpia*; Ministerio del Medio Ambiente—MMA: Bogotá, DC, USA, 1997.
33. Razzaq, M.A.; Lyzu, C.; Parveen, S.; Uddin, M.T.; Shaikh, M.A.A.; Chowdhury, M.J.; Jamal, A.S.I.M.; Al-Mansur, M.A. Fatliquor for fungus resistant leather-a sustainable ecofriendly approach. *Heliyon* **2024**, *10*, e31598. [CrossRef] [PubMed]
34. Rebelato, M.G.; Rodrigues, A.M.; Ferreira, B.S. Proposal of an environmental performance index to assess leather-manufacturing companies. *Green Technol. Sustain.* **2024**, *1*, 100106. [CrossRef]
35. Senthil, R. Regenerated products from leather industrial solid waste: Future perspective and current ad vances. *J. Hazard. Mater. Lett.* **2024**, *5*, 100112. [CrossRef]
36. Ma, S.; Ding, W.; Liu, Y.; Zhang, Y.; Ren, S.; Kong, X.; Leng, J. Industry 4.0 and cleaner production: A comprehensive review of sustainable and intelligent manu facturing for energy-intensive manufacturing industries. *J. Clean. Prod.* **2024**, *467*, 142879. [CrossRef]
37. La Grega, M.D.; Buckingham, P.L.; Evans, J.C. The Environmental Resources Management Group. In *Hazardous Waste Management*; McGraw-Hill: Singapore, 1994.

38. Al-Jabari, M.; Sawalha, H.; Pugazhendhi, A.; Rene, E.R. Cleaner production and resource recovery opportunities in leather tanneries: Technological applications and perspectives. *Bioresour. Technol. Rep.* **2021**, *16*, 100815. [CrossRef]
39. Zhang, Y.; Wang, Y.; Xiang, N.; Xu, F.; Li, X. Dynamic Optimized Cleaner Production Strategies to Improve Water Environment and Economic Development in Leather Industrial Parks: A Case Study in Xinji, China. *Sustainability* **2019**, *11*, 6828. [CrossRef]
40. Gonçales Filho, M.; Nunhes, T.V.; Barbosa, L.C.F.M.; de Campos, F.C.; de Oliveira, O.J. Opportunities and challenges for the use of cleaner production to reduce water consumption in Brazilian sugar-energy plants. *J. Clean. Prod.* **2018**, *186*, 353–363. [CrossRef]
41. Ribeiro, G.D.; Hiranobe, C.T.; da Silva Araújo, S.; da Silva Filgueira, M.; Rocha, J.A.; Mukuno, J.S.; Salmazo, L.O.; Gomes, A.S.; Tolosa, G.R.; Gennaro, E.M.; et al. Sustainable construction materials for low-cost housing: Thermal insulation potential of expanded SBR composites with leather waste. *J. Mater. Res. Technol.* **2024**, *30*, 5590–5604. [CrossRef]
42. Hiranobe, C.T.; Tolosa, G.R.; de Almeida Santos, G.T.; de Oliveira, J.P.J.; Budemberg, E.R.; da Silva, M.J.; Cabrera, F.C.; Job, A.E.; Paim, L.L.; Torres, G.B.; et al. Recycling waste polyurethane from the refrigeration industry as filler in SBR/NR composites for industrial applications. *J. Appl. Polym. Sci.* **2023**, *140*, e53709. [CrossRef]
43. *ASTM 3182*; Designation: D 3182-89 (Reapproved 1994) Standard Practice for Rubber—Materials, Equipment, and Procedures for Mixing Standard Compounds and Preparing Standard Vulcanized Sheets. ASTM: West Conshohocken, PA, USA, 2021.
44. *ASTM 5289*; Standard Test Method for Rubber Property—Vulcanization Using Rotorless Cure Meters. ASTM: West Conshohocken, PA, USA, 2019.
45. *ASTM D 3184-07*; Standard Test Methods for Rubber—Evaluation of NR (Natural Rubber). ASTM: West Conshohocken, PA, USA, 2007.
46. *ASTM D 2084-01*; Standard Test Method for Rubber Property-Vulcanization Using Oscillating Disk Cure Meter. American Society for Testing Materials (ASTM): West Conshohocken, PA, USA, 2020.
47. Vieyres, A.; Pérez-Aparicio, R.; Albouy, P.A.; Sanseau, O.; Saalwächter, K.; Long, D.R.; Sotta, P. Sulfur-Cured Natural Rubber Elastomer Networks: Correlating Cross-Link Density, Chain Orientation, and Mechanical Response by Combined Techniques. *Macromolecules* **2013**, *46*, 889–899. [CrossRef]
48. Flory, P.J.; Rehner, J. Statistical Mechanics of Cross-Linked Polymer Networks I. Rubberlike Elasticity. *J. Chem. Phys.* **1943**, *11*, 512–520. [CrossRef]
49. *UNE-ISO 2781:2015*; Method A Standard. Rubber, Vulcanized or Thermoplastic. Determination of Density. Asociacion Espanola de Normalizacion: Madrid, Spain, 2015.
50. *ASTM D 412-06a*; Standard Test Methods for Vulcanized Rubber and Thermoplastic Elastomers-Tension. American Society for Testing Materials (ASTM): West Conshohocken, PA, USA, 2019.
51. *ASTM D5963*; Test Method for Rubber Property—Abrasion Resistance (Rotary Drum Abrader). American Society for Testing Materials (ASTM): West Conshohocken, PA, USA, 2010.
52. *ASTM D2240*; Test Method for Rubber Property—Durometer Hardness. American Society for Testing Materials (ASTM): West Conshohocken, PA, USA, 2010.
53. Valeika, V. Low-pickle processing of leather: Assessment of leather tanning quality by methods of thermal analysis. *Medziagotyra* **2020**, *26*, 333–336. [CrossRef]
54. Bellucci, F.S.; Salmazo, L.O.; Budemberg, E.R.; Da Silva, M.R.; Rodríguez-Pérez, M.A.; Nobre, M.A.L.; Job, A.E. Preparation and structural characterization of vulcanized natural rubber nanocomposites containing nickel-zinc ferrite nanopowders. *J. Nanosci. Nanotechnol.* **2012**, *12*, 2691–2699. [CrossRef] [PubMed]
55. Santos, R.J.D.; Agostini, D.L.S.; Cabrera, F.C.; Budemberg, E.R.; Job, A.E. Recycling leather waste: Preparing and studying on the microstructure, mechanical, and rheological properties of leather waste/rubber composite. *Polym. Compos.* **2015**, *36*, 2275–2281. [CrossRef]
56. Ravichandran, K.; Natchimuthu, N. Vulcanization characteristics and mechanical properties of natural rubber-scrap rubber compositions filled with leather particles. *Polym. Int.* **2005**, *54*, 553–559. [CrossRef]
57. Zeng, S.; Ren, Y.; Liu, P.; Wang, Q.; He, L.; Wang, Q. Facile fabrication of chrome-tanned leather wastes/natural rubber composite: Mechanochemical de-crosslinking effect on collagen fibers and chrome complexation enabled in-situ compatibilization. *Compos. Sci. Technol.* **2021**, *214*, 108998. [CrossRef]
58. de Paiva, F.F.G.; de Maria, V.P.K.; Torres, G.B.; Dognani, G.; dos Santos, R.J.; Cabrera, F.C.; Job, A.E. Sugarcane bagasse fiber as semi-reinforcement filler in natural rubber composite sandals. *J. Mater. Cycles Waste Manag.* **2019**, *21*, 326–335. [CrossRef]
59. Malini, K.A.; Mohammed, E.M.; Sindhu, S.; Kurian, P.; Anantharaman, M.R. Dielectric and mechanical properties of rubber ferrite composites. *Plast. Rubber Compos.* **2002**, *31*, 449–457. [CrossRef]
60. El-Sabbagh, S.H.; Mohamed, O.A. Recycling of chrome-tanned leather waste in acrylonitrile butadiene rubber. *J. Appl. Polym. Sci.* **2011**, *121*, 979–988. [CrossRef]
61. Hang, L.T.; Do, Q.-V.; Hoang, L.; Nguyen, L.T.; Linh, N.P.D.; Doan, V.A. Mechanical Properties of Ternary Composite from Waste Leather Fibers and Waste Polyamide Fibers with Acrylonitrile-Butadiene Rubber. *Polymers* **2023**, *15*, 2453. [CrossRef] [PubMed]

62. Lorenz, O.; Parks, C.R. The Crosslinking Efficiency of Some Vulcanizing Agents in Natural Rubber. *J. Polym. Sci.* **1961**, *50*, 299–312. [CrossRef]
63. Haseena, A.P.; Dasan, K.P.; Namitha, R.; Unnikrishnan, G.; Thomas, S. Investigation on Interfacial Adhesion of Short Sisal/Coir Hybrid Fibre Reinforced Natural Rubber Composites by Restricted Equilibrium Swelling Technique. *Compos. Interfaces* **2004**, *11*, 489–513. [CrossRef]
64. Kanagy, J.R.; Wallace, E.L. Density of Leather and Its Significance. 1943. Available online: https://archive.org/details/jresv31n3p169/page/n1/mode/2up (accessed on 30 July 2024).
65. Guo, J.; Dai, R.; Chen, H.; Liang, Y.; Shan, Z. Research on the composite and functional characteristics of leather fiber mixed with nitrile rubber. *J. Leather Sci. Eng.* **2021**, *3*, 10. [CrossRef]
66. Papagiannis, P.; Azariadis, P.; Papanikos, P. Evaluation and optimization of footwear comfort parameters using finite element analysis and a discrete optimization algorithm. *IOP Conf. Ser. Mater. Sci. Eng.* **2017**, *254*, 162010. [CrossRef]
67. Golubeva, O.; Pogorelova, A. Analysis of the dependence of stresses and deformations on the surfaces of shoe soles of different thicknesses and materials. *E3S Web Conf.* **2021**, *273*, 05001. [CrossRef]
68. Ferreira, M.J.; Almeida, M.F.; Freitas, F. Formulation and characterization of leather and rubber wastes composites. *Polym. Eng. Sci.* **2011**, *51*, 1418–1427. [CrossRef]
69. Ferreira, M.J.; Freitas, F.; Almeida, M.F. The Effect of leather fibers on the properties of rubber-leather composites. *J. Compos. Mater.* **2010**, *44*, 2801–2817. [CrossRef]
70. Garcia, N.G.; dos Reis, E.A.P.; Budemberg, E.R.; Agostini, D.L.D.S.; Salmazo, L.O.; Cabrera, F.C.; Job, A.E. Natural rubber/leather waste composite foam: A new eco-friendly material and recycling ap proach. *J. Appl. Polym. Sci.* **2015**, *132*, 41636. [CrossRef]
71. Ravichandran, K.; Natchimuthu, N. Natural rubber—Leather composites. *Polímeros Ciência E Tecnol.* **2005**, *15*, 102–108. Available online: https://www.redalyc.org/articulo.oa?id=47015207 (accessed on 12 July 2024). [CrossRef]
72. Sui, G.; Zhong, W.H.; Yang, X.P.; Yu, Y.H.; Zhao, S.H. Preparation and properties of natural rubber composites reinforced with pretreated carbon nanotubes. *Polym. Adv. Technol.* **2008**, *19*, 1543–1549. [CrossRef]
73. Thomas, S.; Grohens, Y.; Jyotishkumar, P. *Characterization of Polymer Blends*; Wiley-VCH Verlag GmbH & Co. KgaA: Weinheim, Germany, 2014. [CrossRef]
74. Terada, N.; Kouge, K.; Komaguchi, K.; Hayakawa, S.; Tsutsumi, H. Thermal Stability Change of Insoluble Sulphur by a Heat Treatment and Its Mechanism Study. *Anal. Sci.* **2020**, *36*, 75–79. [CrossRef] [PubMed]
75. Bañón, E.; García, A.N.; Marcilla, A. Thermogravimetric analysis and Py-GC/MS for discrimination of leather from different animal species and tanning processes. *J. Anal. Appl. Pyrolysis* **2021**, *159*, 105244. [CrossRef]
76. Islam, M.H.; Islam, M.R.; Dulal, M.; Afroj, S.; Karim, N. The effect of surface treatments and graphene-based modifications on mechanical properties of natural jute fiber composites: A review. *iScience* **2022**, *25*, 103597. [CrossRef]
77. Teklay, A.; Gebeyehu, G.; Getachew, T.; Yaynshet, T.; Sastry, T.P. Preparation of value added composite boards using finished leather waste and plant fibers—A waste utilization effort in Ethiopia. *Clean Technol. Env. Policy* **2017**, *19*, 1285–1296. [CrossRef]
78. Falkiewicz-Dulík, M. leather and leather products. In *Handbook of Material Biodegradation, Biodeterioration and Biostablization*; Elsevier: Amsterdam, The Netherlands, 2015; pp. 133–256. [CrossRef]
79. Hoinacki, E. *Peles e Couros: Origens, Defeitos e Industrialização*, 2nd ed.; CIENTC/Novo Hamburgo: Hamburgo, Germany, 1989; Available online: https://books.google.com.co/books/about/Peles_e_couros.html?id=iinOZwEACAAJ&redir_esc=y (accessed on 22 September 2023).
80. Gutterres, M.; Da, T.; Osório, S. Métodos analíticos especiales aplicados al cuero. *Bol. De La Asoc. Química Esp. De La Ind. Del Cuero* **2005**, *56*, 71–80.

polymers

MDPI

Article

Tribological Performance of Short Fibers Reinforced Thermoplastic Polyurethane Composite Materials Under Water-Lubricated Condition

Yicong Yu [1,2,3], Pan Jiang [1,*], Wei Yu [4] and Zhiwei Guo [1,2,3,*]

1 School of Transportation and Logistics Engineering, Wuhan University of Technology, Wuhan 430063, China; yicongyu@whut.edu.cn
2 State Key Laboratory of Maritime Technology and Safety, Wuhan University of Technology, Wuhan 430063, China
3 Reliability Engineering Institute, National Engineering Research Center for Water Transportation Safety, Wuhan 430063, China
4 Wuhan Branch of China Classification Society, Wuhan 430022, China; 3220245434@bit.edu.cn
* Correspondence: jiangpan1@sina.com (P.J.); zwguo@whut.edu.cn (Z.G.)

Abstract: The water-lubricated bearing plays a crucial role in the ship propulsion system, significantly impacting vessel safety. However, under the harsh working conditions of low-speed and heavy-load, the lubrication state of water-lubricated bearings is usually poor, leading to serious friction and wear. To improve the tribological performance of composites and reduce friction, three short fibers (ultra-high-molecular-weight polyethylene fibers, basalt fibers, and bamboo fibers) with the same mass fraction (5%) were added into the melted thermoplastic polyurethane (TPU). The tribological behavior of these three composites under different loads and rotation speeds was investigated using the CBZ-1 friction and wear tester. Through the comprehensive analysis of the friction coefficient, the wear mass loss, and the surface morphology, it was confirmed that the filled fiber positively affected the tribological performance of thermoplastic polyurethane materials. The experimental results indicated that basalt fiber significantly improved the tribological performance of TPU, and the friction coefficient of the sample was only 0.088 under the working conditions of 0.5 MPa and 250 r/min, which was 70.57% lower than that of pure TPU material. And in all the tests, the minimum wear of the basalt fiber-reinforced composite is only 0.4 mg, which is also the smallest of all the materials under all conditions, and a decrease of 98.69% compared to TPU. Under high loads, ultra-high-molecular-weight polyethylene fiber and bamboo fiber-reinforced composites have smoother surfaces and exhibit better tribological properties. This study provides an experimental foundation for tribological performance enhancement for environmentally friendly, water-lubricated bearing composites.

Keywords: water-lubricated bearing; fiber-reinforced composites; thermoplastic polyurethane; tribological performance

1. Introduction

It is necessary to protect the environment and resources that oil-lubricated bearings need to be replaced because they might consume massive mineral oil and precious metal resources. It has been a trend to replace oil-lubricated bearings with water-lubricated bearings in the marine field because they meet the requirements of green shipping [1]. However, the stern tube journal bearing of a ship operates under frequent start–stop operation due to ship steering [2], which makes the bearing bushing in the unstable stress environment. At the same time, the viscosity of water is very low, and it is difficult to form a stable lubricating water film under harsh conditions, resulting in poor carrying capacity [3]. These harsh operating environments will cause obvious friction and wear of water-lubricated bearings, which directly affects the service life of bearings and the safety of ships. Therefore, it is of great significance to study the friction and wear properties of bearing materials under water lubrication for improving the lubrication conditions of friction pairs.

Polymers possess properties such as low friction, good wear resistance, and corrosion resistance, making it desirable for stern tube bearing applications [4]. Common polymer substrates are ultra-high-molecular-weight polyethylene (UHMWPE) [5], polytetrafluoroethylene (PTFE) [6], nitrile butadiene rubber (NBR) [7], and other materials, which provide the necessary strength for the bearing. Among these polymer matrix alternatives, thermoplastic polyurethane (TPU) has been studied widely due to its characteristic structure of soft/hard segments, providing excellent and tunable mechanical properties [8].

TPU is a multiblock copolymer whose molecular chain consists of a "hard segment" (HS) composed of diisocyanate and short-chain diol and a "soft segment" (SS) composed of macromolecular diols. Their molecular formula is shown in Figure 1b. The hard segment and soft segment are arranged alternately on the molecular chain [9], as shown in Figure 1a. SS exhibits strong mobility; the higher its content and the longer the chain segment, the better the elasticity and toughness of TPU. Conversely, when the content of HS is higher, the chain segment becomes less mobile, increasing the hardness and rigidity of TPU. The unique structure of TPU combines characteristics of both plastic and rubber. This dual nature enhances the tribological and mechanical performance of water-lubricated bearings made from TPU. Therefore, TPU is also widely used to improve the tribological properties of composites. The results show that TPU and polyamide (PA) composites exhibit good tribological properties under different sliding conditions. The toughness of PA6-TPU composites is significantly improved, and the friction coefficient stabilizes rapidly at low values, but the wear rate increases under high-load conditions [10]. In the biomedical field, biocompatible TPU composites containing titanium particles show excellent wear resistance. With the increase in titanium particle content, the hardness of the composite gradually increased, and the wear coefficient and wear rate reached the lowest at 2 wt%, but increased at 4 wt% due to the increase in volume loss [11]. The microfiber network of TPU and polytetrafluoroethylene (PTFE) composite was prepared by melting and mixing, and the friction coefficient was significantly reduced. It was found that the size of PTFE fiber is closely related to the shear stress, and the higher shear stress helps to form a thinner fiber, thereby improving the tribological properties of the composite [12].

In order to meet the demand for lightweight, high-strength, and durable materials in the high-tech field, high-performance fiber composites with various high-performance fibers as reinforcement materials have been rapidly developed. With the advancement of blending modification technology, new fiber-reinforced thermoplastic resin composites with new functions and high quality and low price are emerging one after another. UHMWPE fibers are known for their high strength-to-weight ratio, excellent abrasion resistance, and good impact resistance, which makes them ideal for enhancing the durability and mechanical strength of TPU composites [13]. Basalt fibers provide excellent thermal stability, high strength, and good chemical resistance. They contribute to the structural integrity and overall performance of TPU-based composites, especially in high-temperature applications [14]. Bamboo fibers are a sustainable and eco-friendly option that offers good mechanical properties, such as high tensile strength and flexibility. They also contribute to the biodegradability of the composite, making it a more environmentally conscious choice [15].

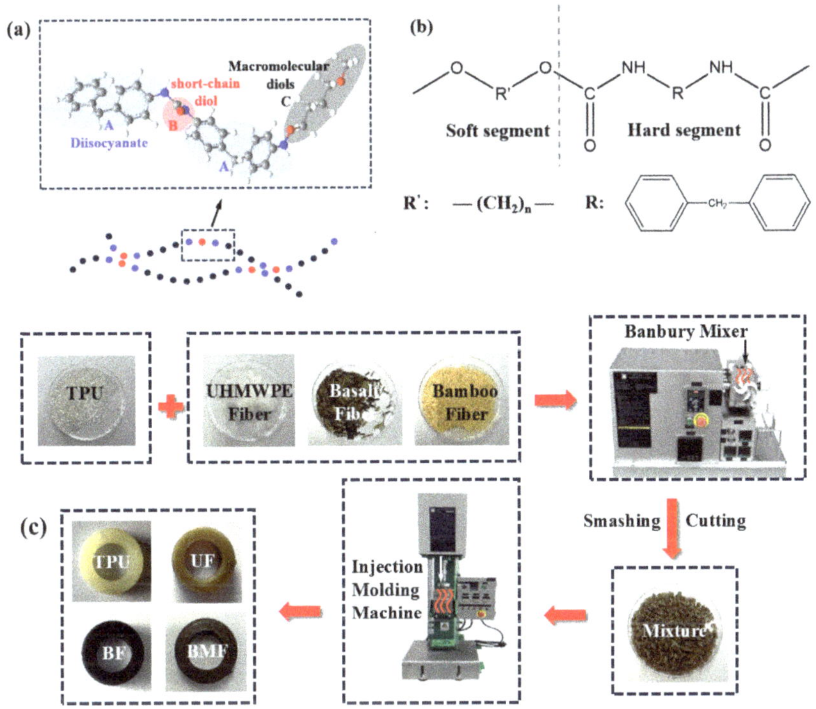

Figure 1. (**a**) Molecular chain of TPU, (**b**) structural formula of TPU, and (**c**) preparation process of composite material sample.

In this paper, three kinds of short fiber were selected as the reinforcing phase of the composites, and the TPU-based fiber-reinforced composites were prepared by injection molding. Under the condition of water lubrication, the CBZ-1 Marine shafting friction and wear testing machine was used to compare different fiber-reinforced composites (FRCs) under different working conditions, and the wear mechanism was analyzed by the surface morphology after the test. This study provides an experimental and theoretical basis for enhancing the tribological performance of environment-friendly, water-lubricated bearing composites.

2. Materials and Methodology

2.1. Materials

Neat thermoplastic polyurethane (Polyester TPU S172DL) was purchased from KOSLEN (Fujian, China) Co., Ltd. Ultra-high-molecular-weight polyethylene fibers were purchased from Sinopec Yizheng Chemical Fibre (Zhejiang, China) Co., Ltd. Basalt fibers were purchased from NCE Composites (Jiangsu, China) Co., Ltd. Bamboo fibers were purchased from Rui Mei bamboo technology (Jiangxi, China) Co., Ltd. Professional distilled water was purchased from Watsons (Beijing, China) Co., Ltd.

2.2. Specimen Preparation

In this study, all TPU-based samples were obtained by melt injection molding, which is convenient and cost-effective. First, three fibers and TPU of the same quality are added to the Banbury mixer, which is stirred at 180 °C for a period of time. The mixture is then crushed and cut into small particles, which are finally placed in an injection molding machine to obtain a test sample at a certain pressure and temperature. The preparation process is shown in Figure 1c.

Four types of samples were prepared as follows: pure thermoplastic polyurethane (TPU), TPU+5 wt% UHMPWPE fiber (named UF), TPU+5 wt% basalt fiber (named BF), and TPU+5 wt% bamboo fiber (named BMF). These three fibers were chosen as reinforcement phases because they mostly have good mechanical properties and a positive impact on the environment and are potential materials for water-lubricated bearings [13–15].

The test simulated the real working environment of the ship stern shaft and bearing under the condition of water lubrication through the contact between the tin bronze plate and sample. The sample is made of a ring shape with an outer diameter of 30 mm, an inner diameter of 18 mm, and a height of 10 mm. In addition, QSn7-0.2 is a commonly used material for shaft bushings in ship shafting, so tin bronze of the same material was selected as the grinding part of FRC in the test [16]; the element composition is shown in Table 1. In order to ensure the fully stable contact of the friction pair in the tribological test, the surface size of the tin bronze plate is designed to be slightly larger than the sample surface size. Among them, the outer diameter of the tin bronze plate is 32 mm, the inner diameter is 16 mm, and the thickness is 6mm. The morphology and size of the friction pair are shown in Figure 2. The mechanical properties of TPU and tin bronze plates used in the test are shown in Tables 2 and 3, respectively.

Table 1. Qsn7-0.2 tin bronze disc element composition.

Element	Cu	Zn	Sn	Ni	Al	Pb	Others
Proportion	90–92%	0.3%	6–8%	0.2%	0.01%	0.02%	0.15%

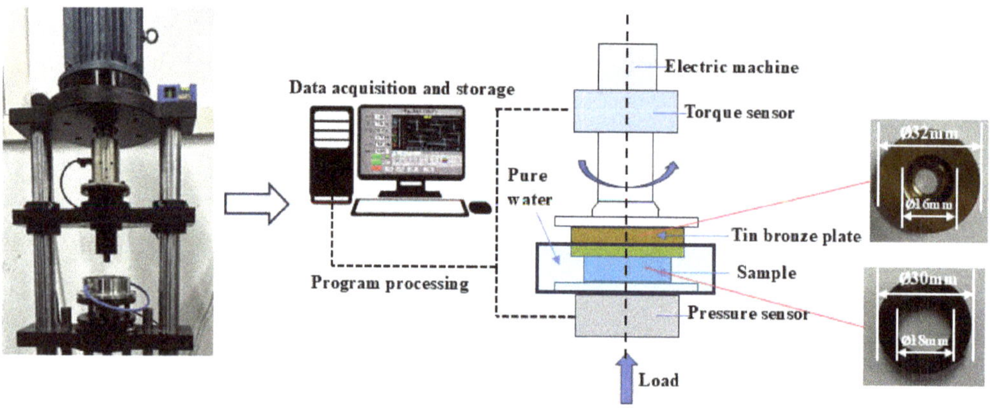

Figure 2. CBZ-1 friction and wear testing machine and schematic diagram.

Table 2. Properties of TPU.

	Density (g/cm ^{3})	Tensile Strength (MPa)	Elongation at Break (%)	Poisson Ratio	Hardness (Shore D)
TPU(S172DL)	1.23	34.1 (23 °C)	274 (23 °C)	0.38	72

Table 3. Qsn7-0.2 Mechanical properties of tin bronze plates.

	Material	Yield Strength (MPa)	Tensile Strength (MPa)	Hardness (HB)
Tin bronze plates	Qsn7–0.2	≥170	≥355	≥70

2.3. Materials Characterization Techniques

A universal material testing machine (Instron 5967, Co., Ltd., Boston, MA, USA) was used to measure the stress–strain curve of the material in a uniaxial tensile test according to the testing standard ISO 527-1 [17]. Shore hardness of samples was tested by a Shore hardness tester (HLX-AC, Adelberg Handpi Instrument Co., Ltd., Wenzhou, China). The surface morphology of the polymer disk was observed by scanning electron microscope (VEGA3, Tescan Co., Ltd., Brno, Czech Republic). Before scanning electron microscopy (SEM) measurements, all samples were gold-plated by sputtering. Laser scanning confocal microscope (VK-X2000, Keyence Ltd., Osaka, Japan) was used to observe the wear surface of the grinding pair.

2.4. Tribological Test

The CBZ-1 friction and wear testing machine (Haima Co., Ltd., Wuhan, China) was used to conduct a tribological performance test. The testing machine was composed of the host rotating system, data acquisition, and processing system. The physical drawing, working principle, and test details were shown in Figure 2.

The sample is fixed between the testing machine base and the tin bronze plate and is completely immersed in pure water. The tin-bronze grinding piece is driven by a motor to slide on the surface of the sample, and the experimental load and speed are adjusted by the control device of the friction and wear testing machine, so that the sample ring and the tin bronze plate are subjected to disc grinding tests. During the experiment, different sensors were used to collect the pressure, torque, and other parameters, and the friction coefficient was calculated in real time by the LabVIEW system in the computer. Calculate the real-time friction coefficient of the CBZ-1 friction and wear testing machine according to Equation (1) as follows:

$$\mu = \frac{T}{rF} \tag{1}$$

where μ represents the friction coefficient; T stands for torque, N · m; F stands for load, N; r represents the sample rotation radius, m.

The test conditions were selected according to the U.S. Navy military standard MIL-DTL-17901C [18]. The speed of the test spindle is set to 50, 150, and 250 r/min to simulate the working condition of the ship at low speed and normal operation. Since the nominal pressure of the ship's water-lubricated tail bearing is below 0.55 MPa [9], the load applied on the friction pair is set to 0.5 and 0.7 MPa to simulate the normal load and heavy load conditions, respectively, so as to explore the water lubrication performance of FRC under normal and harsh conditions. In order to ensure the reliability of the test results, the test is repeated three times under each friction and wear test condition to eliminate accidental errors.

Before the test, the surfaces of the tin bronze plate and composite material samples were polished with 240 mesh, 600 mesh, and 1000 mesh sandpaper on the polishing machine, respectively, so that the surface roughness of the friction pair was the same, about 0.3 μm. The polished material is rinsed with distilled water, placed in the oven for 12 h, and weighed. Each tin bronze plate and composite material sample is tested 3 times to calculate the average mass. In the test, when changing the material or changing the working condition, the tank was cleaned with distilled water, and the grinding was continued for 7200 s under each working condition. After the test, the tin bronze plate and composite material samples were washed again with distilled water, dried in the oven for 12 h, and weighed. Each tin bronze plate and composite material sample was recorded 3 times, and the average value was taken. By comparing the sample quality before and after the test, the wear data of the sample during the friction and wear test can be calculated. After the friction and wear test, the wear morphology of FRC material was observed by scanning electron microscope (SEM) and compared with that of pure TPU.

3. Results and Discussion

3.1. Water Absorption

Considering that there are organic fibers in the reinforced fibers added in this study, the addition of these fibers may cause large expansion in water, which may have an impact on the development of materials used in water-lubricated bearings [19]. Therefore, it is necessary to study the water absorption of the four different material samples in this test. Four samples with dimensions of 30 mm × 10 mm × 10 mm were soaked in distilled water in a 1 L beaker at 25 ° C. Vernier calipers were used to measure and record the average volume change of the sample under different soaking times. Table 4 shows the average volume changes in samples soaked in distilled water for different times.

Table 4. The average volume changes in samples for different soaking times.

Sample	TPU	UF	BF	BMF
Volume growth after 12 h	0.303%	0.301%	0.210%	0.405%
Volume growth after 24 h	0.548%	0.547%	0.318%	0.504%
Volume growth after 7 d	0.576%	0.551%	0.403%	0.950%

According to the US military standard MIL-DTL-17901C (SH) [19], for water-lubricated bearings made of plastic material as a substrate, the change in volume of the sample made of this material after soaking in water for 7 days should be less than 5%, and if the volume increase is less than 1% and 2%, it is grade IV and grade V, respectively.

The test results show that the volume growth rate of BMF with the worst water absorption performance in this test is less than 1%, which is in line with military standards. After adding UHMWPE fiber and basalt fiber, the water absorption performance of TPU has been improved. BF has the best water absorption performance and can meet the requirements of grade IV standards and is beneficial to the safe navigation of ships as a water-lubricated bearing material.

3.2. Mechanical Properties

Mechanical properties are one of the most important parameters to determine the applicability of engineering materials [20]. Moreover, in the field of tribology, the mechanical properties of materials are also one of the important reasons for the difference in tribological properties. Therefore, the hardness (Shore D) and tensile properties of TPU composites were measured in this paper. Figure 3 shows the test results of hardness and tensile strength of four different materials.

Figure 3a shows that the hardness of BF and BMF is higher than that of TPU, but the hardness of UF is lower than that of TPU. This shows that the addition of basalt fiber and bamboo fiber to the TPU matrix has a positive effect on the hardness of the composite; however, the addition of UHWMPE fiber will reduce the hardness of the composite, which may be due to the phase separation phenomenon between TPU and UHWMPE. In addition, UHMWPE has excellent self-lubricating properties, which, in some cases, reduce the friction coefficient and hardness of the material, making the material more susceptible to deformation under external forces [21]. At the same time, the hardness of BF is relatively highest, not only because of the high hardness of the basalt fiber itself but also because the presence of the fiber improves the structural integrity of the material and reduces the generation and expansion of cracks; the overall performance of the composite material is improved [22].

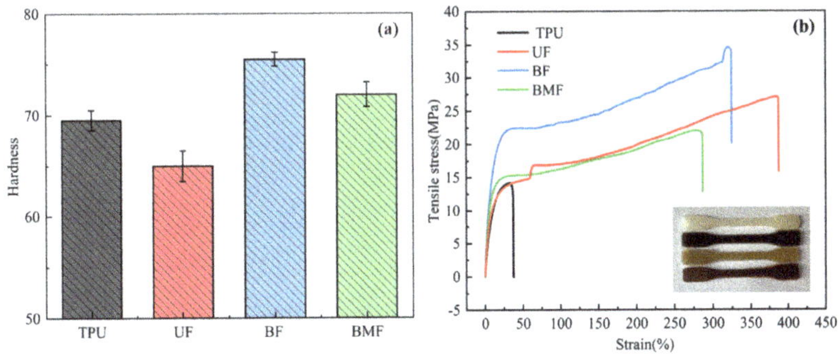

Figure 3. Mechanical properties of TPU− based samples: (**a**) hardness and (**b**) tensile stress−strain curve. The illustrations in (**b**) are dumbbell-shaped samples for tensile testing.

The stress–strain curves of the four materials can be seen in Figure 3b. The elongation at break of the materials in descending order is UF (390%) > BF (325%) > BMF (290%) > TPU (50%). However, the tensile strength of the material is BF (35 MPa) > UF (27 MPa) > BMF (22 MPa) > TPU (14 MPa) from large to small. The elongation at break and tensile strength of UF, BF, and BMF are higher than those of TPU, which indicates that the addition of three kinds of short fibers to the TPU matrix will lead to the improvement in the ductility and tensile strength of the composite. Among them, UF has the highest elongation at break, while BF has the highest tensile strength, which indicates that UF is very flexible and can have huge deformation before fracture, but the stress it can withstand is less than BF. This causes the UF to begin to deform at lower stresses, but because the structure is able to absorb energy, the material does not break until it deforms [23]. For BMF, after adding bamboo fiber, the hardness, tensile strength, and elongation at break of TPU material were slightly improved, but the effect was much weaker than that of UHMWPE fiber and basalt fiber, which may be related to the strength of plant fiber itself.

3.3. Friction Coefficient Analysis

Friction coefficient is one of the most important reference parameters that describes the tribological properties of material [24]. In order to better describe the tribological properties of materials, the numerical value, stability and variation trend of friction coefficients under different test conditions will be comprehensively considered in this section. In the test, the friction coefficients of different samples were recorded in real time to explore the tribological properties of short fiber-reinforced TPU materials under water lubrication, as shown in Table 5. In order to ensure the rigor of the test, it is necessary to consider the friction pair in the initial running-in stage of the test, and calculate the average friction coefficient of the sample under different loads starting from the friction coefficient after the 2000′s, as shown in Figure 4.

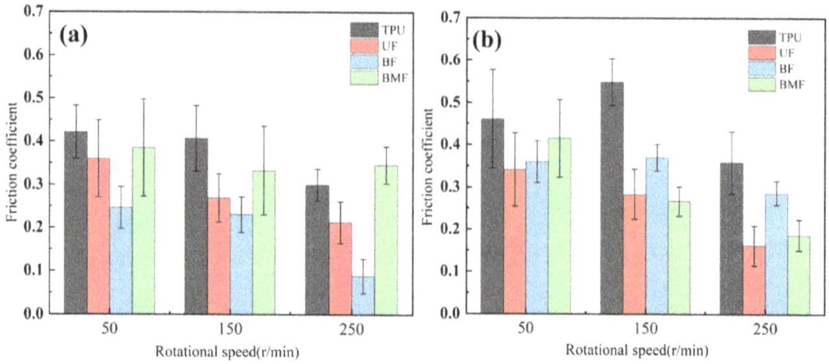

Figure 4. Friction coefficients of specimens under different loads: (**a**) 0.5 MPa and (**b**) 0.7 MPa.

It can be seen from Figure 4 that the friction coefficient of pure TPU samples is the largest under most working conditions of this test, which proves that the addition of three staple fibers can improve the tribological properties of TPU composites to varying degrees. At 0.5 MPa pressure, the friction coefficient value of BF is the smallest, which is 70.57% lower than that of pure TPU. Interestingly, however, UF and BMF may show smaller friction coefficient values at 0.7 MPa pressure. This is related to the properties of the fiber itself and the interface bonding between the fiber and the TPU matrix. Under the pressure of 0.5 MPa, basalt fiber provides good support for the sample with its high hardness and stiffness, reduces the contact area between the friction pairs, and has good lubrication status, which helps to improve the wear resistance of the composite material and reduce the friction coefficient. UHMWPE fiber can also reduce the friction coefficient of composite materials, but the effect is not as significant as basalt fiber under relatively low load conditions. However, when the pressure is increased to 0.7 MPa, huge pressure and constant relative rotation of friction pairs will accumulate a large amount of friction heat on the surface of the sample. When the surface temperature of the TPU composite is close to or exceeds the glass transition temperature (Tg) of the hard segment (HS), the hard domain will be oriented into type III crystals [25], which is conducive to a longer extension of the soft segment (SS). It is easy to cause the deformation of TPU, which affects the stick-slip phenomenon [26]. At this time, a large amount of material will be spalling off on the surface of the pure TPU sample, tearing and pits of the material will occur on the surface of the sample, and a large number of abrasive particles will exist in the lubricating medium, which will be discussed in the following. During the friction process, the contact temperature of TPU composites will rise due to frictional heating, and this temperature rise may affect its tribological properties because the thermal resistance of TPU is limited [27]. Moreover, the wear behavior of high-performance polymers at different temperatures shows that temperature changes will affect the brittle-ductility transition behavior of materials [28] and thus affect their wear properties. For BF samples, the basalt fibers were completely pulled out of the matrix by extreme shear force, and many structural defects appeared on the surface of the samples. When subjected to great pressure, UHMWPE fiber has a highly linear molecular chain structure and thus has good self-lubricating properties [29]. At the same time, bamboo fiber also contains natural vegetable lubrication components [30], so UF and BMF show better tribological properties. Under high load, the TPU on the surface of UHMWPE fiber is worn off, and the UHMWPE fiber with better wear resistance is exposed, but it is not pulled out, and the tribological properties of the overall structure are improved. Under the test condition of 0.7 MPa, the UF sample shows the best performance. In addition, for most samples under the same pressure, the average friction coefficient will decrease with the increase in rotational speed, which may be because when the rotational speed increases, the flow rate of distilled water between the two interfaces will accelerate, the lubrication pressure between the two interfaces will decrease, and the internal and external lubrication pressure difference will appear. Under the action of external liquid pressure, more distilled water enters the gap [31], the water film thickness increases, the lubrication state becomes better, and the friction coefficient decreases.

Table 5. Real-time friction coefficients under different test conditions.

Rotational Speed \ **Load**	**0.5 MPa**	**0.7 MPa**
50 r/min	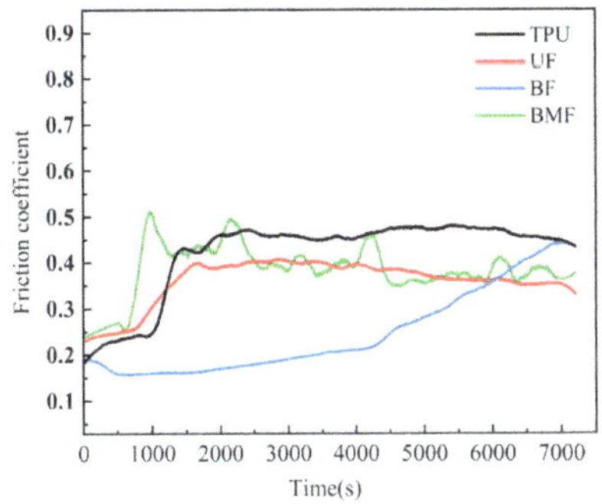	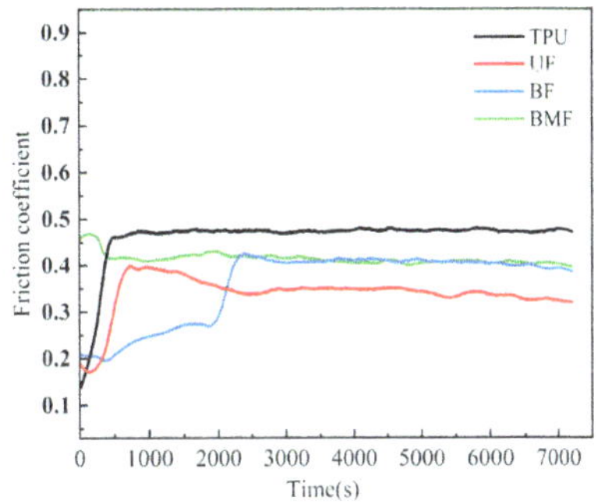
150 r/min	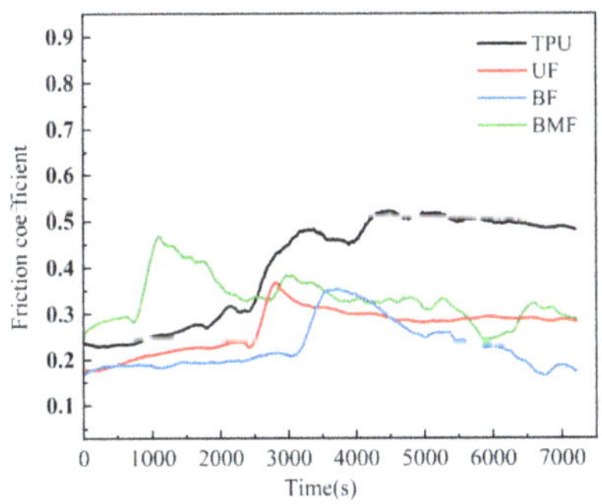	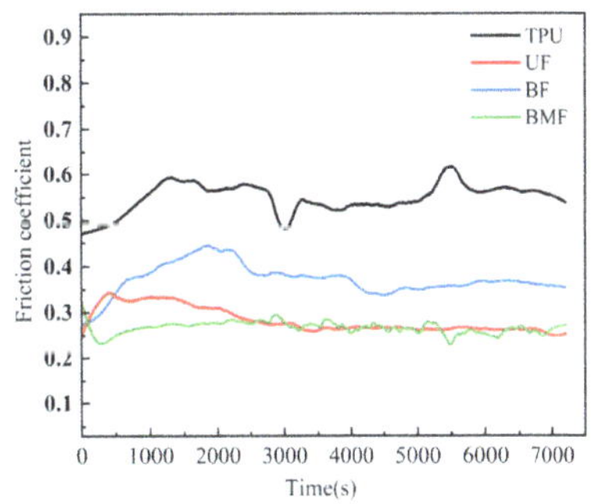
250 r/min	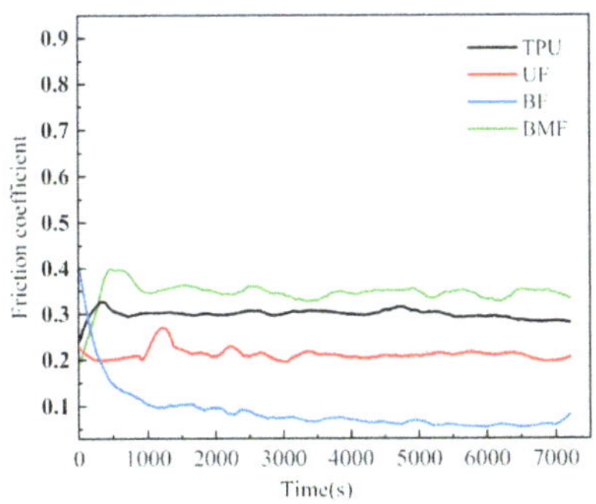	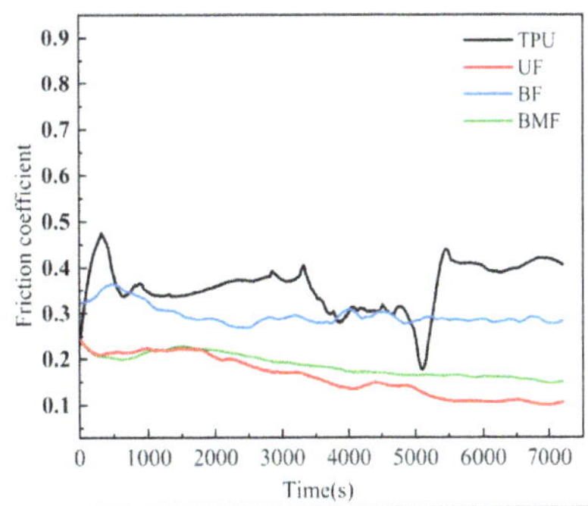

Table 5 shows the real-time changes in friction coefficients of four materials under six different working conditions, which can more directly observe the changes in friction coefficients of materials over time and under different working conditions. The overall value comparison is similar to the average friction coefficient in this section. It is worth noting that the real-time friction coefficient of polymer materials under high-load conditions is more stable than that under lower load conditions, especially when fiber is added as the reinforcement phase of the sample. This may be because the high load will make the friction pair fit more tightly, and it is easier to produce fiber fragments and friction films [32], and the lubrication conditions are improved. As can be seen from Table 5, when the load increases from 0.5 MPa to 0.7 MPa, the overall friction coefficient of TPU and BF samples increases, while that of UF and BMF samples decreases, which is consistent with the previous explanation. At the same time, under the condition of high load (0.7 MPa), the friction coefficients of the four materials are significantly compared. The real-time friction coefficient of the TPU sample remained at the highest level and fluctuated the most. The friction coefficient curve of TPU may fluctuate greatly because a large number of materials on the surface are stripped, the surface is pitted, a large number of abrasive particles exist in the lubricating water, and the friction process is extremely unstable. However, the friction coefficient of UF sample is the smallest among the four materials most of the time, showing a trend of continuous decline during the test. One of the most important reasons is that after TPU is worn off by the tin bronze plate, UHMWPE fiber, which has better wear resistance than TPU [33], plays a major role, and the friction coefficient continues to decrease. The friction coefficients of UF and BMF samples are maintained at a good level and are accompanied by a slow decline trend, which can also prove that UHMWPE fiber and bamboo fiber have better self-lubrication ability under high load, thus improving the lubrication environment. Occasionally, the friction coefficient of BMF samples can be found to fluctuate slightly, which indicates that the mechanical properties of bamboo fiber need to be improved. It should also be noted that by observing the friction coefficient curve of BF sample under 0.7 MPa test conditions, it can be found that after 7200 s of friction testing, the friction coefficient of BF does not decrease but will continue to rise most of the time. Combining the surface microstructure and wear mechanism of the sample, we can find that the basalt fiber in the BF sample has poor compatibility with the TPU matrix. As a result, many fibers are completely pulled out of the matrix, and the friction interface conditions are very bad.

3.4. Wear Mass Loss Analysis

The wear mass loss of materials is also one of the important indexes to measure the tribological properties of polymers. Figure 5 shows the wear mass loss of the samples under different test conditions. It can be found that the wear mass loss of the TPU composite material is significantly reduced after the addition of three kinds of staple fibers, which indicates that the addition of three kinds of short fibers has a positive impact on the wear resistance of TPU material.

It is worth noting that in all friction tests, the wear quality of TPU samples is the largest, which indicates that the addition of fibers has a positive effect on the wear resistance of TPU composites. By observing the surface of the sample and analyzing the wear mechanism (Figures 6 and 7), we can find that a large number of materials on the surface of pure TPU are spalling, which inevitably makes the wear quality extremely high. In addition, the accumulation of friction charges may increase friction and wear. The mechanical action on the contact surface can lead to the formation of free radicals and electron transfer, resulting in cationic and anionic polymer fragments, forming charge accumulation [34], and this charge distribution will increase the electrostatic interaction between the surfaces, thereby increasing the coefficient of friction and aggravating the wear of the material [35]. The minimum wear of the basalt fiber-reinforced composite was only 0.4mg, which is 98.7% lower than the pure TPU. This may be because the high hardness and strength of the basalt fiber enable it to effectively share the load during the friction process while improving the fatigue resistance of the material, reducing the microscopic crack growth during the friction process, and thus reducing the wear of the TPU matrix material [36]. At the same time, at 0.5 MPa, the fluctuation amplitude of BF wear mass is the smallest when the load and speed change, and the influence of test conditions on the wear quality is the least. This may be because basalt fibers enhance the mechanical strength and fatigue resistance of TPU materials. However, under high-load conditions, UF samples performed best, followed by BMF samples, and BF samples did not perform as well as when subjected to relatively low loads. This is because in the high-stress field, the basalt fibers are uprooted from the matrix by extreme shear forces, and the abrasive particles increase along with the deteriorating lubrication conditions. However, UHMWPE fiber has better compatibility with the matrix; only the TPU and a small amount of fiber body are worn off, and the wear mass loss is minimal in general. The BMF sample is between BF and UF.

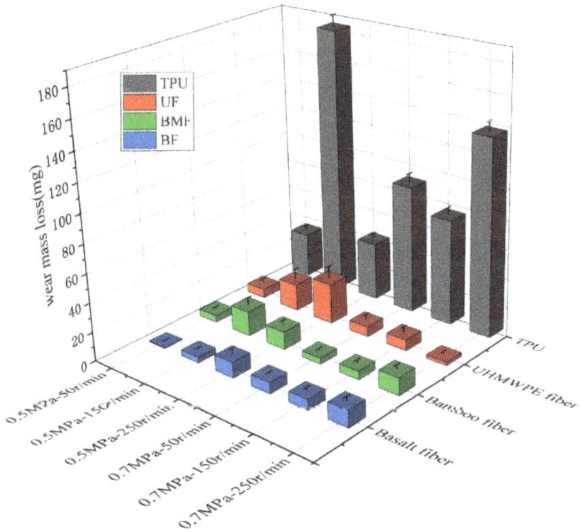

Figure 5. Wear mass loss of samples under different test conditions.

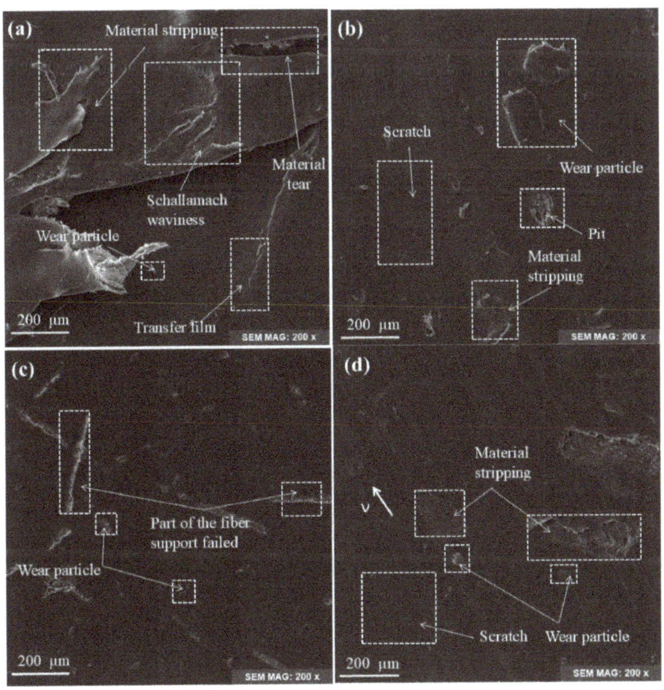

Figure 6. Surface microwear morphology of TPU-matrix composites: (**a**) TPU, (**b**) UF, (**c**) BF, and (**d**) BMF.

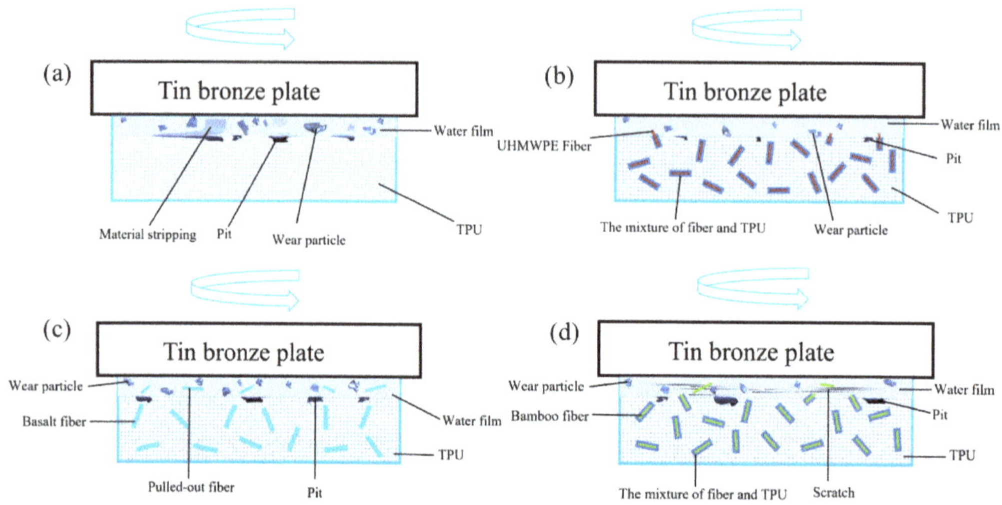

Figure 7. Wear mechanisms: (**a**) TPU, (**b**) UF, (**c**) BF, and (**d**) BMF.

3.5. Wear Surface Topography of the Specimens

In order to better evaluate the tribological properties of four kinds of short fiber-reinforced composites and more accurately describe the friction and wear of the materials, scanning electron microscopy (SEM) was used to observe the surface of the tested composites. According to previous studies, the actual maximum wear of water-lubricated bearings occurs at low-speed and heavy-load conditions, so SEM analysis mainly focuses on the morphology of samples tested at 0.7 MPa pressure [37].

Figure 6 shows the surface microstructure of TPU-based composites. On the whole, the surface wear condition of the three FRC samples is better than that of the pure TPU sample, which indicates that the three reinforced fibers can effectively improve the tribological properties of TPU materials. It can be observed from Figure 6a that the surface of TPU produces a thermal effect under the condition of low speed and heavy load, resulting in deformation of the TPU surface and thus affecting the sticking-slip phenomenon. For transfer film, large pieces of material spalling and tearing are generated on the surface of the sample, accompanied by a large amount of wear chips. At the same time, the Schallamach waves are caused by the periodic ripple deformation caused by the adhesive force and shear force on the contact surface, which indicates the existence of adhesive wear and abrasive wear, and the friction environment of the specimen surface is poor, and the friction mass loss is great. By observing Figure 6b–d, it can be found that there are slight scratches and pits on the UF surface, there are wear particles on the surface, and the surface roughness increases significantly, showing the characteristics of abrasive wear. At the same time, part of the fiber connected to the matrix can be observed from the pits, which indicates that the TPU near the fiber has been worn off, while a part of the wear-resistant UHMWPE fiber is still left in the matrix, as shown in Figure 7b. The BF surface has a small amount of abrasive particles and more fiber support failures, which are signs of high-stress areas during friction, indicating that the material has significant plastic deformation and breakage during friction. Combined with the previous analysis of BF friction coefficient and wear mass loss, we can see that these linear grooves indicate that the basalt fibers have been completely pulled out of the matrix, and the structural integrity of the sample is insufficient. The deep scratches present on the BMF surface extend along the direction of friction, showing the action of high stress and shear forces. In addition, there are tiny cracks and particle residue on the surface, and a small number of fiber bundles can be seen in the cracks, indicating that the base material and some fibers have been worn and produced fragments.

3.6. Wear Morphologies of Tin Bronze Plates

The friction pair of the water-lubricated bearing in this test is composed of a metal drive shaft and a polymer bearing [38]. Due to the complexity of the preparation process and the fine size requirements, the drive shaft is much more expensive than the bearing. Therefore, the wear morphology of the tin bronze plate used in the test is compared and analyzed in this paper. As with Section 3.5, considering that the greatest wear may occur under the highest load, tin bronze plates at 0.7 MPa were selected for analysis. Figures 8 and 9 are the wear surface topography of the tin bronze plate and the comparison of selected characteristic parameters, respectively.

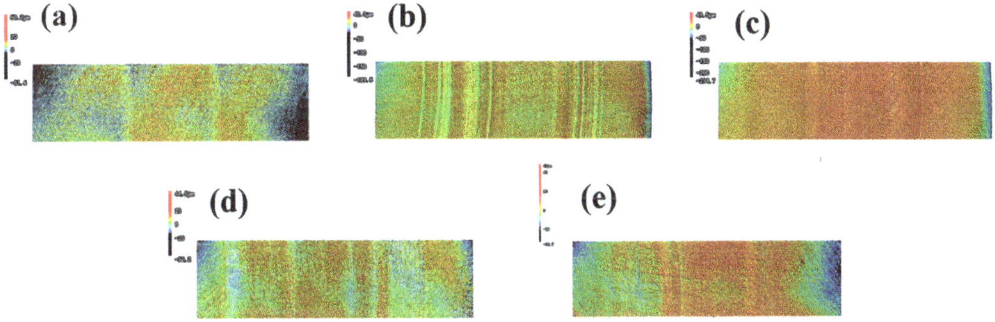

Figure 8. Wear morphologies of the tin bronze plates at 0.7 MPa contact specimens: (**a**) untested, (**b**) UF, (**c**) TPU, (**d**) BF, and (**e**) BMF.

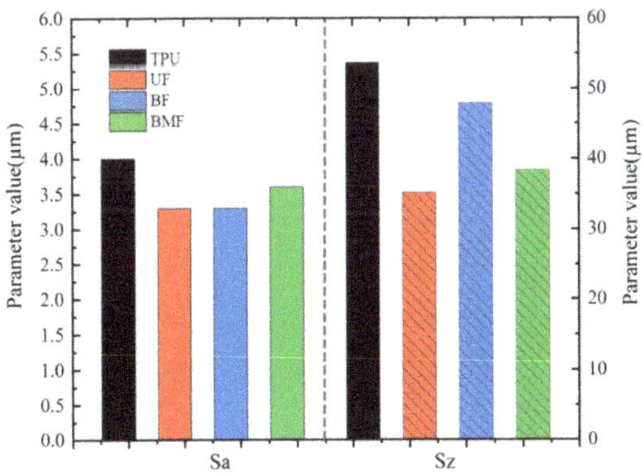

Figure 9. Surface characteristic parameters of tin bronze plate.

It can be seen from the surface morphology that the surface of the untested tin bronze plate is relatively flat, and there are only uniform marks caused by the killing in the polishing process. Other tin bronze plates tested all show abrasion marks of varying shades. Compared with the surface of the tin bronze plate that has not been tested, the roughness peak height difference of the tin bronze plate grinded with pure TPU is the most obvious, and there are a few scratches and grooves on the surface of the tin bronze plate grinded with BF. However, due to the high hardness of BF itself, it can be observed that the surface of the tin bronze plate presents a large number of microplastic convex structures [39], but it remains relatively flat on the macro level. The surface of the tin bronze plate milled with

UF and BMF showed more uniform scratches and adhesion, and it was also observed that part of the wear chips and adhesion filled the cracks of the scratches, improving the surface smoothness and friction properties. In Figure 9, Sa represents the average roughness, Sz represents the maximum peak and valley height, and both the average roughness and maximum peak and valley height of TPU remain at the highest level. Based on the above analysis, it may be because the deformed TPU is attached to the surface of the copper disk. The test results show that all three staple fibers can improve the friction environment of the materials. The surface roughness of the tin bronze plate ground with UF and BF is the smallest, and the surface peak of the tin bronze plate ground with UF and BMF is relatively gentle. The test results are consistent with the surface images of the samples taken by SEM. The wear debris and stick-slip TPU on the surface of the UF and BMF samples are embedded in the surface of the tin bronze plate, which has a positive influence on its roughness and surface peak. At the same time, the depression caused by the failure of many fibers to support the BF sample surface may also lead to sharp surface peaks.

4. Conclusions

In this study, the properties of thermoplastic polyurethane (TPU)-based composites were investigated under conditions similar to those of water-lubricated stern tube plain bearings. Three short fibers were added to TPU, and the tribological behavior of these three composites was studied under different loads and rotational speeds. Through the comprehensive analysis of mechanical properties, friction coefficient, wear mass loss, and surface topography, the conclusions drawn from this investigation are shown as follows:

1. In this paper, BF shows excellent mechanical properties, and the hardness and tensile strength are relatively the largest among the four materials. With the addition of UHMWPE fiber, the hardness of TPU will decrease, but the elongation at break will increase significantly, which indicates that UF is very flexible and can produce huge deformation. In addition, the water swelling of the three fiber-reinforced composites (FRCs) meets military standards, with BF providing the best performance.
2. The different speeds and loads applied in the test will affect the friction and wear properties of the composites. Under boundary lubrication, the frictional properties of TPU matrix composites are affected by contact mechanics and viscoelasticity. TPU increases friction and wear, but reinforced fillers improve wear resistance. The friction coefficient of BF was only 0.088 under the working conditions of 0.5 MPa and 250 r/min, which was 70.57% lower than that of pure TPU material. And in all the tests, the minimum wear of the BF is only 0.4 mg, which is also the smallest of all the materials under all conditions, and a decrease of 98.69% compared to TPU. However, under the test conditions of high load and low speed, UF samples showed lower friction coefficients and better wear resistance, which may be related to the good compatibility between UHMWPE fiber and TPU. The increase in sliding speed helps to form a water lubrication film, while the pure TPU surface appears to have a large number of material spalling and a large number of abrasive particles in the lubricating water, resulting in high wear and friction fluctuations, and the surface produces Schallamach waves.
3. Under the normal load (0.5 MPa), the superior mechanical properties of basalt fiber, such as high hardness, bear most of the load of the BF sample, reducing the contact area and improving the friction performance. However, under high load (0.7 MPa), the basalt fiber support partially fails, resulting in obvious stress concentration, plastic deformation, and fracture, and the whole fiber is pulled out from the matrix, destroying the structural integrity of the material. The self-lubricity of the relatively soft UHMWPE fiber and bamboo fiber plays a major role, and their compatibility with

the TPU matrix is relatively better. The fiber is fixed in the matrix and hardly pulled out. At the same time, part of the grinding chips are filled into the copper tray groove, which improves the surface finish, and the friction properties of UF and BMF are also improved.

4. This study proves that three different short fibers can improve the tribological properties of the TPU matrix in different degrees, which provides an experimental basis for improving the tribological properties of environmentally friendly water-lubricated bearing composites. It is necessary to further study the binding force between the reinforced fiber and the matrix in order to further improve the comprehensive properties of the material.

Author Contributions: Validation, W.Y.; writing—original draft, Y.Y.; writing—review and editing, Z.G.; supervision, P.J. All authors have read and agreed to the published version of the manuscript.

Funding: This research was funded by the National Natural Science Foundation of China (grant no.52171319) and, The Young Top-notch Talent Cultivation Program of Hubei Province and Hubei Provincial Chutian Talent Plan Science and Technology Innovation Team Project. APC was funded by the National Natural Science Foundation of China (grant no.52171319).

Institutional Review Board Statement: Not applicable.

Data Availability Statement: The data presented in this study are available on request from the corresponding author.

Acknowledgments: The authors would like to express their sincere gratitude to the National Natural Science Foundation of China (grant no. 52171319) and The Young Top-notch Talent Cultivation Program of Hubei Province and Hubei Provincial Chutian Talent Plan Science and Technology Innovation Team Project.

Conflicts of Interest: Mr. Wei Yu was employed by Wuhan Branch of China Classification Society. The remaining authors declare that the research was conducted in the absence of any commercial or financial relationships that could be construed as a potential conflict of interest.

References

1. Zhang, Z.; Ouyang, W.; Liang, X.X.; Yan, X.P.; Yuan, C.Q.; Zhou, X.C.; Guo, Z.W.; Dong, C.L.; Liu, Z.L.; Jin, Y.; et al. Review of the evolution and prevention of friction, wear, and noise for water-lubricated bearings used in ships. *Friction* **2024**, *12*, 1–38. [CrossRef]
2. Jiang, S.L.; Wong, J.S.S.; Puhan, D.; Yuan, T.; Bai, X.Q.; Yuan, C.Q. Tribological evaluation of thermoplastic polyurethane-based bearing materials under water lubrication: Effect of load, sliding speed, and temperature. *Friction* **2024**, *12*, 1801–1815. [CrossRef]
3. Han, H.S.; Lee, K.H. Experimental verification of the mechanism on stick-slip nonlinear friction induced vibration and its evaluation method in water-lubricated stern tube bearing. *Ocean Eng.* **2019**, *182*, 147–161. [CrossRef]
4. Litwin, W. Properties comparison of rubber and three layer PTFE-NBR-bronze water lubricated bearings with lubricating grooves along entire bush circumference based on experimental tests. *Tribol. Int.* **2015**, *90*, 404–411. [CrossRef]
5. Guo, Z.; Xie, X.; Yuan, C.Q.; Bai, X.Q. Study on influence of micro convex textures on tribological performances of UHMWPE material under the water-lubricated conditions. *Wear* **2019**, *426*, 1327–1335. [CrossRef]
6. Lin, B.; Wang, A.Y.; Sui, T.Y.; Wei, C.B.; Wei, J.H.; Yan, S. Friction and wear resistance of polytetrafluoroethylene-based composites reinforced with ceramic particles under aqueous environment. *Surf. Topogr.* **2020**, *8*, 015006. [CrossRef]
7. Xiao, B.; Zheng, X.L.; Zhou, Y.; Yao, D.; Wan, Y. Tribological behaviors of the water-lubricated rubber bearings under different lubricated conditions. *Ind. Lubr. Tribol.* **2021**, *73*, 260–265. [CrossRef]
8. Jiang, S.L.; Puhan, D.; Huang, J.; Yang, Z.R.; Zhang, L.Y.; Yuan, T.; Bai, X.Q.; Yuan, C.Q. Tribological properties of graphene-reinforced polyurethane bearing material. *Tribol. Int.* **2024**, *193*, 109481. [CrossRef]
9. Yuan, Y.; Wang, S.Q.; Tan, P.; Zhu, H.P. Mechanical performance and shear constitutive model study of a new high-capacity polyurethane elastomeric bearing. *Constr. Build. Mater.* **2020**, *232*, 117227. [CrossRef]
10. Zhou, S.F.; Huang, J.; Zhang, Q.X. Mechanical and tribological properties of polyamide-based composites modified by thermoplastic polyurethane. *J. Thermoplast. Compos. Mater.* **2014**, *27*, 18–34. [CrossRef]

11. Banoriya, D.; Purohit, R.; Dwivedi, R.K. Tribological behaviour of biocompatible TPU composites with reinforced titanium. *Adv. Mater. Process. Technol.* **2022**, *8* (Suppl. S2), 802–813. [CrossRef]
12. Rohm, K.; Amirkhosravi, M.; Manas-Zloczower, I. Microstructure and Tribological Properties of TPU/Fluoropolymer Composites. *Int. Polym. Process* **2020**, *35*, 415–421. [CrossRef]
13. Abdul Samad, M. Recent advances in UHMWPE/ UHMWPE nanocomposite/ UHMWPE hybrid nanocomposite polymer coatings for tribological applications: A comprehensive review. *Polymers* **2021**, *13*, 608. [CrossRef] [PubMed]
14. Yan, L.; Chu, F.L.; Tuo, W.Y.; Zhao, X.B.; Wang, Y.; Zhang, P.Q.; Gao, Y.B. Review of research on basalt fibers and basalt fiber-reinforced composites in China (I): Physicochemical and mechanical properties. *Polym. Polym. Compos.* **2021**, *29*, 1612–1624. [CrossRef]
15. Xu, D.D.; He, S.; Leng, W.Q.; Chen, Y.H.; Wu, Z.X. Replacing plastic with bamboo: A review of the properties and green applications of bamboo-fiber-reinforced polymer composites. *Polymers* **2023**, *15*, 4276. [CrossRef]
16. Huang, Q.R.; Guo, Z.W.; Wu, Z.M.; Yuan, C.Q. Insight into the tribological performance of polyurethane composites under high temperature water lubrication. *Tribol. Int.* **2021**, *155*, 106784. [CrossRef]
17. *BS EN ISO 527-5:1997*; Plastics-Determination of Tensile Properties. British Standard: London, UK, 1997.
18. Jiang, S.L.; Guo, Z.W.; Yuan, C.Q.; Liu, A.X.; Bai, X.Q. Study on the tribological properties of modified polyurethane material for water-lubricated stern bearing. *J. Appl. Polym. Sci.* **2018**, *135*, 46305. [CrossRef]
19. He, Y.; Guo, Z.W.; Wu, Z.M.; Yuan, C.Q. Effect of polyester fiber orientation on the tribological properties of ultrahigh molecular weight polyethylene composites for water lubrication. *Tribol. Int.* **2021**, *158*, 106935. [CrossRef]
20. Wang, C.B.; Bai, X.Q.; Guo, Z.W.; Dong, C.L.; Yuan, C.Q. Friction and wear behaviours of polyacrylamide hydrogel microsphere/UHMWPE composite under water lubrication. *Wear* **2021**, *477*, 203841. [CrossRef]
21. Wang, X.; Mu, B.; Wang, H.G. Preparation and Properties of Thermoplastic Polyurethane/Ultra High Molecular Weight Polyethylene Blends. *Polym. Compos.* **2015**, *36*, 897–906. [CrossRef]
22. Tomina, A.M.V.; Yeromenko, A.V.; Makarov, V.S. Influence of basalt fiber on tribological properties of secondary polyethylene terephthalate. *Prob. Tribol.* **2021**, *26*, 58–64. [CrossRef]
23. Li, J.C.M. *Microstructure and Properties of Materials*; World Scientific: London, UK, 1996.
24. Wu, Z.M.; Guo, Z.W.; Yuan, C.Q. Influence of polyethylene wax on wear resistance for polyurethane composite material under low speed water-lubricated conditions. *Wear* **2019**, *426*, 1008–1017. [CrossRef]
25. Li, X.; Wang, H.; Xiong, B.; Pöselt, E.; Eling, B.; Men, Y.F. Destruction and reorganization of physically cross-linked network of thermoplastic polyurethane depending on its glass transition temperature. *ACS Appl. Polym. Mater.* **2019**, *1*, 3074–3083. [CrossRef]
26. Dong, C.L.; Yuan, C.Q.; Bai, X.Q.; Qin, H.L.; Yan, X.P. Investigating relationship between deformation behaviours and stick-slip phenomena of polymer material. *Wear* **2017**, *376*, 1333–1338. [CrossRef]
27. Pinedo, B.; Hadfield, M.; Tzanakis, I.; Conte, M.; Anand, M. Thermal analysis and tribological investigation on TPU and NBR elastomers applied to sealing applications. *Tribol. Int.* **2018**, *127*, 24–36. [CrossRef]
28. Kurdi, A.; Kan, W.H.; Chang, L. Tribological behaviour of high performance polymers and polymer composites at elevated temperature. *Tribol. Int.* **2019**, *130*, 94–105. [CrossRef]
29. Kumar, R.M.; Sharma, S.K.; Kumar, B.V.M.; Lahiri, D. Effects of carbon nanotube aspect ratio on strengthening and tribological behavior of ultra high molecular weight polyethylene composite. *Compos. Part A Appl. Sci. Manuf.* **2015**, *76*, 62–72. [CrossRef]
30. Sumithra, H.; Sidda Reddy, B. A review on tribological behaviour of natural reinforced composites. *J. Reinf. Plast. Compos.* **2018**, *37*, 349–353. [CrossRef]
31. Yang, Z.R.; Guo, Z.W.; Yuan, C.Q. Tribological behavior of polymer composites functionalized with various microcapsule core materials. *Wear* **2019**, *426*, 853–861. [CrossRef]
32. Birleanu, C.; Pustan, M.; Pop, G.; Cioaza, M.; Popa, F.; Lăzărescu, L.; Conțiu, G. Experimental investigation of the tribological behaviors of carbon fiber reinforced polymer composites under boundary lubrication. *Polymers* **2022**, *14*, 3716. [CrossRef]
33. Hussain, M.; Naqvi, R.A.; Abbas, N.; Khan, S.M.; Nawaz, S.; Hussain, A.; Zahra, N.; Khalid, M. Ultra-high-molecular-weight-polyethylene (UHMWPE) as a promising polymer material for biomedical applications: A concise review. *Polymers* **2020**, *12*, 323. [CrossRef] [PubMed]
34. da Silveira Balestrin, L.B.; Del Duque, D.; da Silva, D.S.; Galembeck, F. Triboelectricity in insulating polymers: Evidence for a mechanochemical mechanism. *Faraday Discuss.* **2014**, *170*, 369–383. [CrossRef] [PubMed]
35. Luo, N.; Feng, Y.; Zhang, L.; Sun, W.X.; Wang, D.A.; Sun, X.J.; Zhou, F.; Liu, W.M. Controlling the tribological behavior at the friction interface by regulating the triboelectrification. *Nano Energy* **2021**, *87*, 106183. [CrossRef]
36. Wang, B.W.; Yu, S.D.; Mao, J.; Wang, Y.H.; Li, M.J.; Li, X.P. Effect of basalt fiber on tribological and mechanical properties of polyether-ether-ketone (PEEK) composites. *Compos. Struct.* **2021**, *266*, 113847. [CrossRef]
37. Guo, Z.W.; Yuan, C.Q.; Liu, A.X.; Jiang, S. Study on tribological properties of novel biomimetic material for water-lubricated stern tube bearing. *Wear* **2017**, *376*, 911–919. [CrossRef]

38. Xu, Q.R.; Guo, Z.W.; Huang, Q.R.; Yuan, C.Q. Effects of soft chain materials on the tribological performances of thermosetting polyurethane under water-lubrication. *Tribol. Int.* **2024**, *196*, 109692. [CrossRef]
39. Dehm, G.; Jaya, B.N.; Raghavan, R.; Kirchlechner, C. Overview on micro-and nanomechanical testing: New insights in interface plasticity and fracture at small length scales. *Acta Mater.* **2018**, *142*, 248–282. [CrossRef]

polymers

MDPI

Article

An OpenSees Surrogate Constitutive Model for High-Damping Rubber Based on Machine Learning

Feng Li [1] **and Tianbo Peng** [1,2,*]

[1] College of Civil Engineering, Tongji University, Shanghai 200092, China; lifeng19@tongji.edu.cn
[2] State Key Laboratory of Disaster Reduction in Civil Engineering, Tongji University, Shanghai 200092, China
* Correspondence: ptb@tongji.edu.cn

Abstract: The complex mechanical properties of high-damping rubber (HDR), a commonly used seismic isolation material in buildings and bridges, present a significant challenge in civil engineering. In a previous study, the authors proposed an HDR constitutive model that combines a Gated Recurrent Unit (GRU) and an attention mechanism, offering novel insights into the mechanical properties of HDR. The constitutive model was simplified first to facilitate the deployment of the proposed constitutive model within the finite element analysis environment. Then, the simplified constitutive model was converted into a uniaxial material format suitable for use within the open system for earthquake engineering simulation (OpenSees). In OpenSees, the uniaxial material was named HDRGA material, and the code for the HDRGA material header and source files was written. Finally, an HDR surrogate constitutive model was developed in OpenSees. To validate the precision of the HDRGA material in characterizing the mechanical attributes of HDR, a two-node model and a single-pier model were devised, and their accuracy was verified through a comparative analysis of test results and nonlinear time history calculation results, respectively. The results demonstrate that the developed HDRGA material is capable of performing well under earthquakes.

Keywords: high-damping rubber; machine learning; OpenSees; surrogate constitutive model; finite element analysis

1. Introduction

The rapid improvement of computer processing capabilities has facilitated substantial advancements in the domain of finite element simulation, enabling the analysis of intricate structures with greater precision and efficiency. Nevertheless, the advancement of high-fidelity and accurate simulations remains constrained by the limitations of computational resources and efficiency. The advantage of machine learning technology lies in its capacity to address the issue of scale in numerical solutions. Theoretically, a structure with millions of degrees of freedom can be regarded as a black box, capable of resolving issues that cannot be addressed in real-time due to the constraints of computer performance. The surrogate model based on machine learning algorithms is capable of not only providing the same computational results as finite element algorithms but also significantly improving computational efficiency and reducing computational costs.

In recent years, machine learning has emerged as a prominent research topic in a range of academic disciplines. Nguyen [1] et al. proposed an efficient method based on subset simulation and a surrogate model to study the reliability of building structures under earthquakes. Moustafa [2] et al. utilized Long Short-Term Memory (LSTM) techniques to simulate nonlinear numerical substructures. Guo [3] et al. and Yan [4] et al. put forth a seismic intensity index identifying and evaluation method based on a generalized linear regression model and a Gaussian process regression, respectively. Jia [5] et al. put forth a rapid assessment method for bridge seismic damage based on a random forest algorithm and an artificial neural network. Hwang [6] et al. and Kiani [7] et al. put forth a machine learning-based methodology for the prediction of seismic response. Alwanas [8] et al. employed an extreme learning machine model to predict the bearing capacity and failure mode of concrete in beam–column joints. Lei [9] et al. developed a Bayesian-optimized interpretable ensemble learning surrogate model to predict and interpret the essential seismic requirements of urban highway bridges. Wang [10] et al. employed machine learning algorithms to reliably estimate the bearing deformation and column offset ratio response of bridges.

To accurately and effectively simulate the hysteresis behavior of materials and components, numerous scholars have conducted extensive research based on machine learning.

Regarding the hysteresis behavior of seismic isolation devices, Zhang [11] et al. conducted static cyclic tests on laminated rubber bearings, taking into account the initial stiffness, friction coefficient, cross-sectional area, height, loading speed, vertical load, and aging time of the bearings. A novel constitutive model for bearings was devised through the application of artificial neural network technology. An artificial neural network-based bridge earthquake demand model was developed and implemented for the expeditious evaluation of bridge damage. The findings demonstrate that the artificial neural network earthquake demand model is capable of accurately fitting the complex functional relationship between a multitude of factors and bridge seismic response, thereby facilitating a rapid assessment of bridge seismic damage. Nasab [12,13] et al. developed a kriging model to address the issue of traditional modeling methods being unable to accurately simulate the mechanical properties of viscoelastic dampers that vary with different loading conditions, including input frequency, amplitude, and temperature. The model was utilized for the fragility analysis of structures reinforced with viscoelastic dampers. The findings demonstrate that the model exhibits high accuracy in forecasting experimental outcomes and is capable of effectively simulating the seismic performance of viscoelastic dampers with uncertain parameters. Mekaoui [14,15] et al. put forth a hybrid seismic analysis method for calculating the fully nonlinear response of building structures, which they validated using deep learning. The objective is to predict the nonlinear hysteresis response of seismic isolation devices with deformation and velocity-related characteristics. The results demonstrate that the integration of mechanics-based and data-driven methodologies through a hybrid analysis approach represents an effective methodology for simulating the response of buildings.

Regarding the hysteresis behavior of structural components and materials, Ni [16] et al. employed an enhanced bidirectional LSTM and bidirectional Gated Recurrent Unit (GRU) to assess the hysteresis behavior of HRB600 reinforced concrete columns subjected to repeated loading. Their analytical capabilities were then compared with those of finite element analysis. The findings demonstrate that the selected methodology is effective in simulating hysteresis curves. However, the GRU approach demonstrates the most precise predictive capability, as evidenced by the highest R^2 and lowest MAE, MSE, and RMSE values. Xu [17] et al. put forth a deep learning-based predictive simulation framework that offers precise and expedient hysteresis models for structural analysis. The analysis of the improved steel support model and the Bouc–Wen hysteresis model demonstrates that the model is both highly accurate and computationally efficient. Gu [18] et al. put forth a sensitivity-guided LSTM neural network methodology for the precise and expedient extraction of structural behavior features and the prediction of pivotal parameters associated with explicit hysteresis models. The findings demonstrate that this approach can accurately predict structural behavior, is more effective than manual construction, and exhibits superior generalization ability compared to classical data-driven hysteresis models. Xu [19] et al. constructed a deep neural network model comprising four hidden layers and one hundred units per layer to predict the complete residual stress–strain response of ultra-high performance concrete materials after exposure to high temperatures. The findings demonstrate that the stress–strain curve predicted by the deep neural network model exhibits a high degree of consistency with the UHPC test curve. The above studies show that the existing research has not yet integrated machine learning-based hysteresis behavior studies with finite element calculations.

High-damping rubber (HDR) bearings have been extensively utilized in the seismic design of building and bridge engineering. The mechanical properties of HDR are primarily influenced by temperature and loading cases. In the field of constitutive model research for HDR, several scholars have established constitutive models based on different theoretical frameworks.

Although hyperelastic constitutive models have been studied for nearly eighty years, choosing a model that accurately describes the mechanical response of rubber remains a challenge. He [20] et al. developed a fitting algorithm to quantitatively evaluate the ability of each strain energy function to reproduce experimental data of unfilled and highly filled rubber nanocomposites by reviewing eighty-five isotropic strain energy functions based on phenomenological theory and micromechanical network theory proposed from the 1940s to 2019. Xiang [21] et al. reviewed the constitutive models of physics-based soft materials to explain hyperelasticity, viscoelasticity, and damage phenomena. The quantitative comparison of hyperelastic models guides the selection of appropriate constitutive models.

Arruda [22] et al. proposed a constitutive model for the deformation of rubber materials based on the eight-chain representation of the underlying macromolecular network structure of rubber and the non-Gaussian behavior of individual chains in the proposed network. The model successfully represented the responses of these materials in uniaxial tension, biaxial tension, uniaxial compression, plane strain compression, and pure shear. Ogden [23] et al. believed that the common practice of writing strain energy as a function of two independent strain invariants often complicates the related mathematical analysis. Therefore, by fully utilizing the inherent simplicity of rubber's isotropic elasticity, a strain energy function was constructed. This function (i) provides a sufficient representation of the mechanical response of rubbery solids, and (ii) is simple enough for mathematical analysis. Han [24] et al. studied the complete constitutive relationship of hyperelastic materials within the theoretical framework of continuum mechanics, starting from the strain energy function. The research results indicate that based on experimental curves of the full deformation range under multiple deformation modes, the complete constitutive relationship of hyperelastic materials can be obtained, which has guiding significance for theoretical research and engineering applications of complex practical problems such as the fracture of hyperelastic materials. Wei [25] et al. conducted multi-step relaxation tests and cyclic shear tests under different compression forces and loading rates to investigate the effects of loading rates and compression forces on the mechanical behavior of HDR bearing. An HDR-bearing rate-dependent constitutive model considering the influence of compressive force was proposed, and its effect on structural seismic response was studied. Wei [26] et al. proposed a new elastic incompressible isotropic constitutive model for rubber-like materials. This model is an isotropic incompressible model defined based on the first and second principal stretches and is applicable to describe the elasticity of rubber-like materials in general deformation states. Wang [27] et al. proposed a multi-axial compressible strain energy function and directly and explicitly simulated the stress–strain hysteresis loop generated by the Mullins effect in rubber-like materials during loading and unloading cycles. Reese [28] et al. proposed a finite deformation viscoelastic model that employs nonlinear evolution laws and compared it with other models. Ghosh [29] et al. provided two potential frameworks for the constitutive modeling of dielectric elastomers, considering deformation-enhanced shear thinning due to viscosity dissipation.

It shows that the majority of existing studies have concentrated on monotonic loading test data under tension or compression. The findings of this study indicate that the strain history of HDR has a significant impact on its mechanical properties. It is a challenging task to establish a constitutive model that can accurately simulate the influence of various main factors on HDR. Furthermore, a comprehensive consideration of the Mullins effect represents a significant challenge. Accordingly, in the authors' previous research [30], an HDR constitutive model combining a GRU and an attention mechanism was proposed.

In this study, we initially present the HDR constitutive model, which combines a GRU and an attention mechanism. We then proceed to further simplify this constitutive model. Subsequently, the simplified constitutive model is transformed into a new open system for earthquake engineering simulation (OpenSees) uniaxial material HDRGA through format conversion. This enables the accurate description of the force–displacement relationship of HDR bearings when they are incorporated into finite element calculations.

2. Materials and Methods

2.1. Test Specimen and Cases

In engineering applications, HDR bearing is primarily responsible for bearing the vertical force transmitted by the upper structure and the horizontal force under earthquakes. Accordingly, this study employed multi-case compression shear testing on HDR specimens.

The HDR specimen is composed of two steel plates measuring 250 mm × 250 mm × 20 mm, with a single HDR layer measuring 250 mm × 250 mm × 5 mm sandwiched between them. The steel plates and HDR layer are bonded together through vulcanization, as illustrated in Figure 1. For temperature control, the high and low temperature alternating environment test chamber is used. The 100 t electro-hydraulic servo actuator is used as the horizontal loading equipment, while the 200 t vertical actuator is used as the vertical loading equipment. The cases for HDR specimens are shown in Table 1. Among them, the training cases include five temperatures and five strain rates. The testing cases include four temperatures and one strain rate. Each test case undergoes two types of strain amplitude loading processes: amplitude-increasing cyclic loading and amplitude-decreasing cyclic loading.

Table 1. The cases for the HDR specimen.

Case	Temperature/K	Strain Rate/s^{-1}	Strain Amplitude/%
	253.15	0.2	
	253.15	0.4	
	253.15	0.8	
	253.15	1.6	
	253.15	3.2	
	268.15	0.2	
	268.15	0.4	
	268.15	0.8	
	268.15	1.6	
	268.15	3.2	
	283.15	0.2	
	283.15	0.4	
Training case	283.15	0.8	40 → 80 → 120 → 160 → 200 and 200 → 160 → 120 → 80 → 40
	283.15	1.6	
	283.15	3.2	
	298.15	0.2	
	298.15	0.4	
	298.15	0.8	
	298.15	1.6	
	298.15	3.2	
	313.15	0.2	
	313.15	0.4	
	313.15	0.8	
	313.15	1.6	
	313.15	3.2	
	260.65	2.4	
	275.65	2.4	
Testing case	290.65	2.4	
	305.65	2.4	

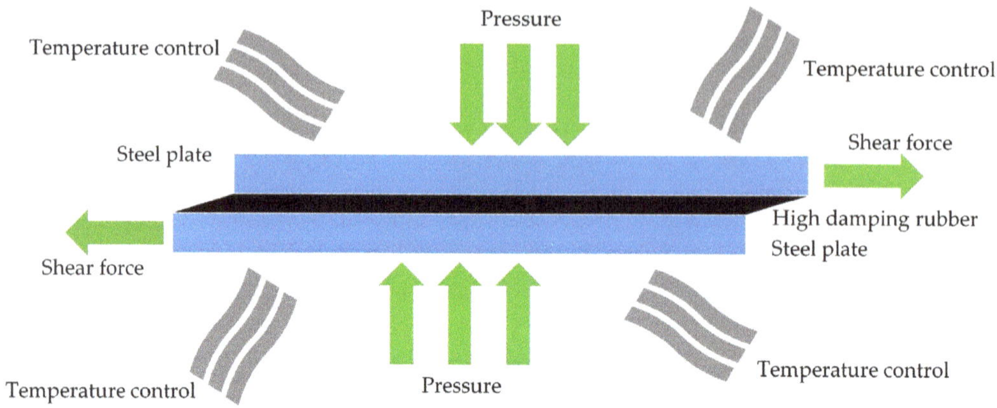

Figure 1. The HDR specimen.

2.2. LSTM and GRU

HDR has complex nonlinear constitutive relationships. However, explicit formulas are typically constrained by limited parameters, and constitutive models established through explicit formulas ultimately suffer from inaccurate descriptions. The constitutive model of HDR exhibits evident time-series characteristics, necessitating comprehensive investigation through techniques capable of integrating previous information, which has a profound impact on the current state.

A recurrent neural network (RNN) is a special type of large model neural network that is particularly suitable for processing and predicting temporal dependencies and temporal information in sequential data. Unlike traditional feedforward neural networks, an RNN has connected nodes between hidden layers, allowing them to store and transmit information from previous time steps to the current time step. An RNN generally takes sequence data as input and effectively captures the relationship features between sequences through the internal structure design of the network. It is usually output in the form of sequences. The recurrent mechanism of an RNN enables the results generated in the previous time step of the model's hidden layer to be used as part of the input for the current time step (the input for the current time step includes not only the normal input but also the output of the previous hidden layer), which affects the output of the current time step. However, an RNN has not been able to address the issue of long-term dependencies [31].

The LSTM [32] is a special type of RNN. LSTM aims to solve the problem of gradient vanishing or exploding encountered by traditional RNNs when processing long sequence data. Compared to standard RNNs, LSTM introduces more complex structures to maintain and update internal states. The GRU is a simplified adaptive approach derived from the LSTM. It has been demonstrated to achieve performance comparable to that of the LSTM [33].

The components of the LSTM [34] are shown in Figure 2. The yellow box represents the learning neural network layer comprising the σ network layer and the tanh network layer. The activation functions associated with these layers are the Sigmoid and tanh functions, represented by Equations (1) and (2), respectively. The pink circle signifies the point-by-point operation, denoting vector multiplication as "$\otimes$" and vector addition as "$\oplus$". Each line carries an entire vector from the output of one node to the inputs of others. The merging of lines indicates the concatenation of information, while the bifurcated line signifies the copy of the information, which is subsequently passed to the corresponding location. These components form the input gate, forget gate, and output gate, with the respective calculation equations outlined in Equations (3)–(8).

$$\text{Sigmoid}(x) = \frac{1}{1+e^{-x}} \tag{1}$$

$$\tanh(x) = \frac{e^{x}-e^{-x}}{e^{x}+e^{-x}} \tag{2}$$

$$f_t = \text{Sigmoid}\left(W_f \cdot [h_{t-1}, x_t] + b_f\right) \tag{3}$$

$$i_t = \text{Sigmoid}(W_i \cdot [h_{t-1}, x_t] + b_i) \tag{4}$$

$$\widetilde{C}_t = \tanh(W_C \cdot [h_{t-1}, x_t] + b_C) \tag{5}$$

$$C_t = f_t \times C_{t-1} + i_t \times \tilde{C}_t \tag{6}$$

$$o_t = \text{Sigmoid}(W_o \cdot [h_{t-1}, x_t] + b_o) \tag{7}$$

$$h_t = o_t \times \tanh(C_t) \tag{8}$$

where f_t, i_t, and o_t are the activation values of the forget gate, input gate, and output gate, respectively; $\widetilde{C}_t$ and C_t are the candidate memory unit state and updated memory unit state, respectively; h_t is the final hidden state output; W_f, W_i, W_C, and W_o are the weight matrix, respectively; b_f, b_i, b_C, and b_o are the bias term, respectively; h_{t-1} and x_t are the hidden state of the previous time step and the input of the current time step, respectively; C_{t-1} is updated memory unit state of the previous time step.

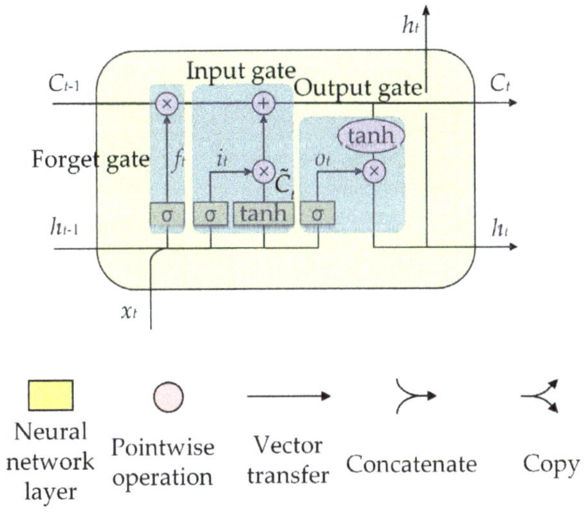

Figure 2. The LSTM structure.

Building upon the LSTM architecture, the GRU simplifies the gating mechanism by condensing the input gate, forget gate, and output gate of LSTM into a reset gate and an update gate. The structure of GRU [34] is illustrated in Figure 3. Each component is identical to its LSTM counterpart, and the corresponding calculation equations are presented as Equations (9)–(12). In comparison to LSTM, GRU exhibits a reduction in the number of parameters, making it more manageable for training and tuning while also enhancing computational efficiency.

$$z_t = \text{Sigmoid}(W_z \cdot [h_{t-1}, x_t]) \tag{9}$$

$$r_t = \text{Sigmoid}(W_r \cdot [h_{t-1}, x_t]) \tag{10}$$

$$\widetilde{h}_t = \tanh(W \cdot [r_t \times h_{t-1}, x_t]) \tag{11}$$

$$h_t = (1 - z_t) \times h_{t-1} + z_t \times \widetilde{h}_t \tag{12}$$

where z_t and r_t are the activation values of the update gate and reset gate, respectively; $\widetilde{h}_t$ is the candidate memory unit state; h_t is the final hidden state output; W_z, W_r, and W are the weight matrix, respectively; h_{t-1} and x_t are the hidden state of the previous time step and the input of the current time step, respectively.

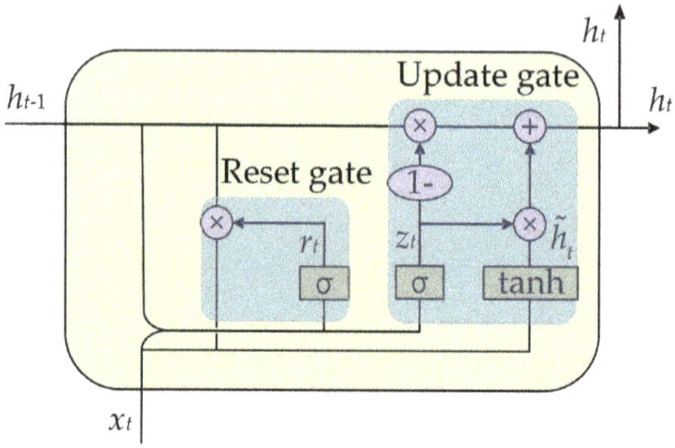

Figure 3. The GRU structure.

2.3. GRU + Attention Model Simplification

The attention mechanism is a commonly used technique in deep learning that plays a crucial role in effectively using input information by applying different weights based on specific requirements [35]. By incorporating the attention mechanism into the GRU model, it becomes possible to fully explore the distinctive characteristics of data and use the most significant components within the time series data for modeling purposes. This integration enhances the model performance by emphasizing relevant information. By incorporating the attention mechanism, the GRU model becomes capable of selectively attending to critical features, improving its ability to capture complex characteristics and enhance prediction accuracy.

Details regarding the specific model structure and prediction performance can be found in reference [30].

The primary objective of model simplification is to streamline the processing of data with a length that is less than the sliding window length. The multi-layer perceptron (MLP) is one of the most basic neural network models, with a relatively simple structure that is easy to understand and implement. It also has good scalability and universality and can be applied to various tasks such as classification and regression. In reference [30], an MLP is employed for training and prediction concerning this particular subset of the data. The MLP structure comprises three hidden layers, each comprising 16 units, as illustrated in Figure 4.

While the aforementioned method can also yield satisfactory prediction outcomes, it is necessary to employ an MLP to generate data of an equivalent length to the sliding window before utilizing the GRU + attention model for prediction. From an examination of the overall model structure and prediction process, it can be seen that the method is somewhat cumbersome.

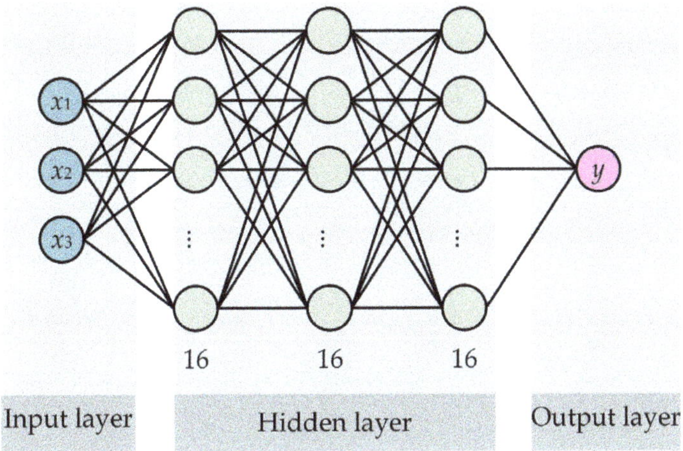

Figure 4. The MLP structure.

Accordingly, in the present study, the MLP component will be eliminated, and the prediction issue when the data length is insufficient for the sliding window length will be addressed by appending "0" to the test data. The specific approach is to insert an equal number of "0"s as the sliding window length at the beginning of each training case data, thereby enabling the GRU + attention model to be trained on a sequence comprising solely "0"s.

2.4. *OpenSees Uniaxial Material Development*

OpenSees is an open-source software framework [36] that is primarily utilized for the simulation and analysis of earthquake engineering and structural systems. The software includes a comprehensive material library, offering a variety of material models to address diverse engineering requirements. The models encompass a range of behaviors, from simple elastic behavior to complex nonlinear and damage behavior, and are suitable for a variety of structural simulations. Additionally, OpenSees enables users to develop their material models tailored to specific requirements, offering considerable flexibility.

The existing uniaxial material used to describe the constitutive relationship of HDR in OpenSees is KikuchiAikenHDR material. However, this material is unable to account for the influence of external factors on the constitutive relationship of HDR or to incorporate the Mullins effect of HDR.

2.4.1. KikuchiAikenHDR Material

The KikuchiAikenHDR material in OpenSees uniaxial material is used to simulate the nonlinear hysteresis behavior of HDR, and its material command is as follows:

uniaxialMaterial KikuchiAikenHDR $*matTag* $*tp* $*ar* $*hr*

where $*matTag* is the material number, $*tp* is the HDR type, $*ar* is the HDR area, and $*hr* is the total HDR thickness.

From the material command, it can be seen that the KikuchiAikenHDR material can only be used for simulating the hysteresis behavior of HDR simplification. It is not capable of considering the effects of temperature, loading rate, and other factors, nor can it describe the Mullins effect of HDR.

To address the shortcomings of the KikuchiAikenHDR material and to capitalize on the benefits of the simplified GRU + attention model, an HDR OpenSees surrogate constitutive model was developed. The surrogate constitutive model was named HDRGA material.

2.4.2. HDRGA Material

The HDRGA material has been developed based on the latest version of the OpenSees open-source code, version 3.6.0. As OpenSees is primarily developed in the C++ language, machine learning models trained in Python 3.9 cannot be directly applied within OpenSees. Accordingly, the initial step is to train the GRU + attention model, simplified as detailed in Section 2.3, in Python to generate a "model.h5" file. This is then converted to a "model. pb" file, and, finally, C++ code is written within the OpenSees. The specific development process is illustrated in Figure 5. The final result is OpenSees.exe, which can be directly executed for structural calculations or to replace the file with the same name in OpenSeespy with OpenSees.pyd. Subsequently, structural calculations can be performed through Python calls.

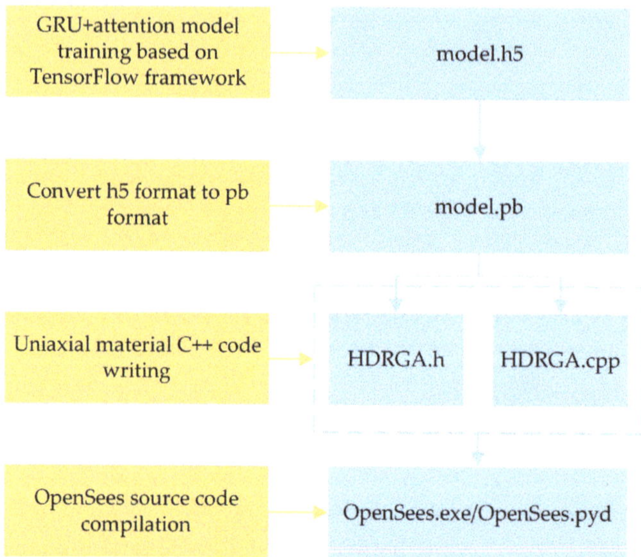

Figure 5. The HDRGA material development process.

In OpenSees, the command for accessing the HDRGA material is as follows:

uniaxialMaterial HDRGA $*matTag* $*nnid* $*temp*

where $*matTag* is the material number, $*nnid* is the neural network number, and $*temp* is the ambient temperature.

Next, the accuracy of the HDRGA material will be verified.

3. Results and Discussion

3.1. Test Results Verification

To verify the accuracy of the description of the HDR mechanical properties provided by the HDRGA material, a two-node model was constructed in OpenSees. Node 1 was designated as a fixed node, while node 2 was configured to release one degree of freedom in a single direction, as illustrated in Figure 6. The two nodes are connected by a two-node link element, which is assigned the HDRGA material type, as illustrated in Figure 7. Subsequently, the time history of shear displacement corresponding to the specified test cases is applied to node 2 through displacement loading to verify whether the shear force generated by the HDRGA material is in alignment with the test results.

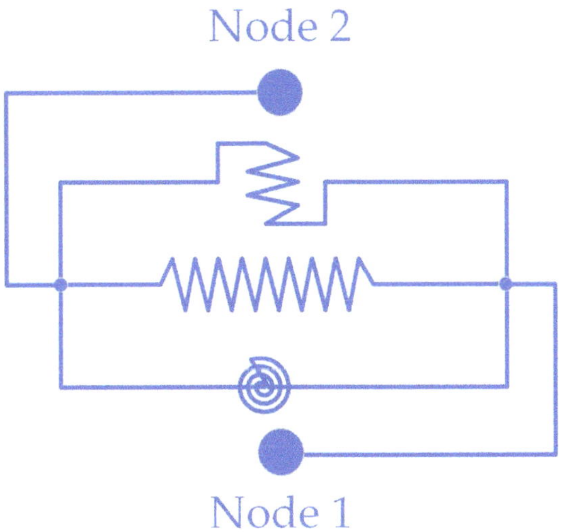

Figure 6. The two-node model.

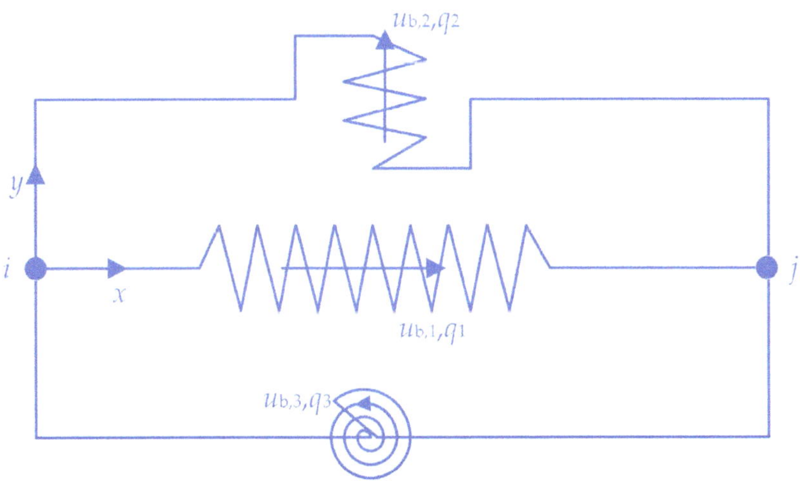

Figure 7. The two-node link element [36]. Where x, y are the local coordinate axes of the element, i, j are the nodes at both ends of the element, and $u_{b,i}$, q_i are the displacement and force in the corresponding direction of the element, respectively.

The accuracy of the HDRGA material was demonstrated by taking the temperature of 260.65 K amplitude-increasing loading and 305.65 K amplitude-decreasing loading as examples in the testing cases. The specific comparison results are shown in Figure 8. As illustrated in Figure 8, the shear force results calculated through OpenSees are found to be essentially consistent with the test results. Furthermore, the GRU+attention model is demonstrated to be capable of accurately describing the shear force–displacement relationship under both amplitude-increasing loading and amplitude-decreasing loading. This demonstrated the feasibility of introducing the GRU + attention model into OpenSees.

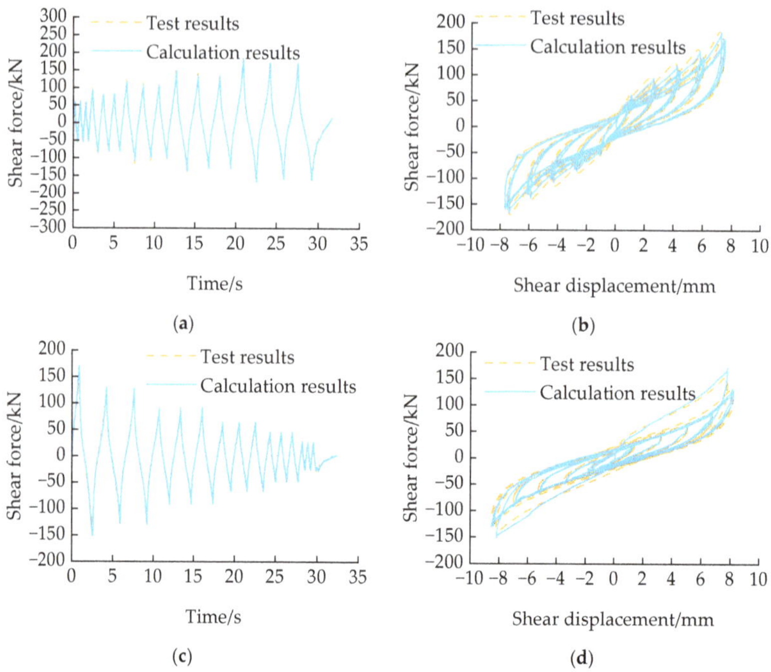

Figure 8. The verification results. (**a**) Shear force time history of amplitude-increasing loading at a temperature of 260.65 K; (**b**) shear force–displacement relationship of amplitude-increasing loading at a temperature of 260.65 K; (**c**) shear force time history of amplitude-decreasing loading at a temperature of 305.65 K; (**d**) shear force–displacement relationship of amplitude-decreasing loading at a temperature of 305.65 K.

3.2. Nonlinear Time History Verification

To ascertain whether the HDRGA material is capable of functioning correctly in the event of an earthquake, a single-pier model was constructed and simulated using an elastic beam column element. The lumped mass at the top of the pier is connected by a two-node link element, which is assigned the HDRGA material type, as illustrated in Figure 9. The lumped mass is 518.4 t, the pier body has an equal cross-section with a cross-sectional area of 17.05 m^2, a pier height of 5 m, an elastic modulus of 3.15×10^{10} Pa, a cross-sectional moment of inertia of 11.27 m^4, and the mass per unit length of the pier body is 40.75 t/m. In this structure, the HDRGA material corresponds to an HDR bearing area of 720 mm × 720 mm, a total HDR thickness of 140 mm, and a temperature of 313.15 K. An artificial seismic wave was generated using SeismoSignal software 2024, as illustrated in Figure 10. The peak acceleration of the seismic wave is 5.74 m/s^2, with a duration of 20 s.

The hysteresis curve of the HDRGA material under seismic wave action was calculated through nonlinear time history analysis (NTHA), as illustrated by the black dashed line in Figure 11. To ascertain the extent to which the constitutive behavior of the HDRGA material accurately reflects the results of the NTHA, the shear displacement of the HDRGA material obtained from the NTHA was applied to node 2 in Figure 6. The corresponding hysteresis curve was calculated and is shown by the red solid line in Figure 11. As illustrated in Figure 11, the hysteresis curve calculated through NTHA is in complete alignment with the hysteresis curve validated by the two-node model. This demonstrates that the HDRGA material is fully capable of meeting the structural calculation requirements under earthquakes.

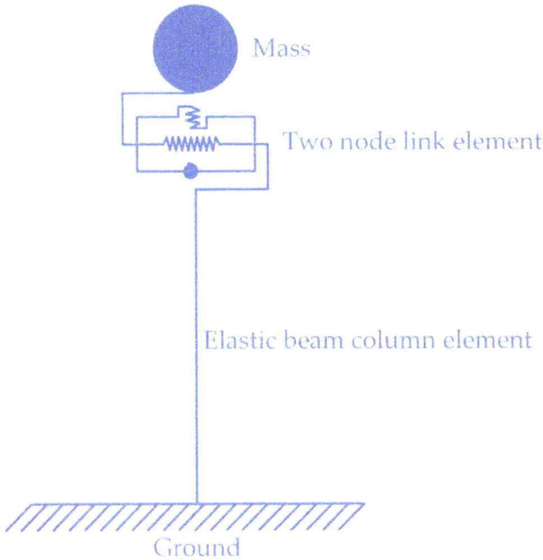

Figure 9. The single-pier model.

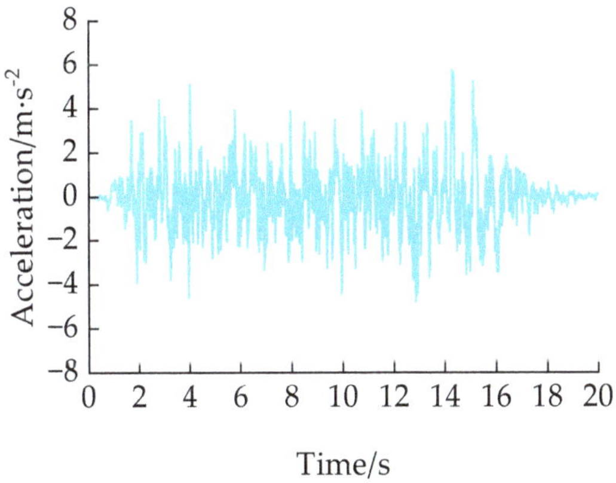

Figure 10. The artificial seismic wave.

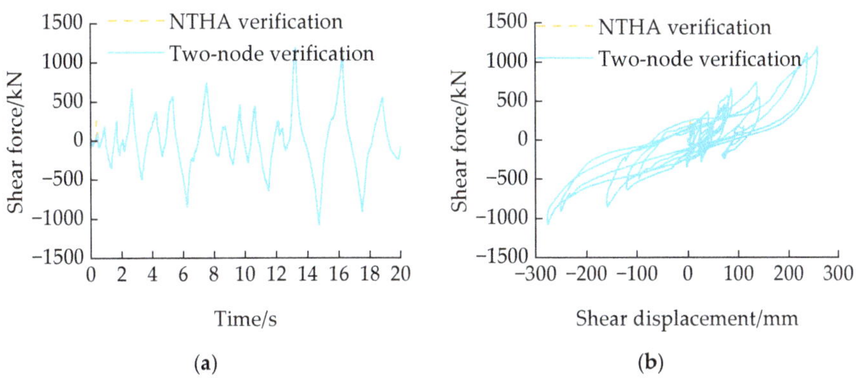

Figure 11. The nonlinear time history verification results. (**a**) Shear force time history; (**b**) shear force–displacement relationship.

4. Conclusions

An OpenSees surrogate constitutive model for HDR based on machine learning was developed by simplifying the GRU + attention constitutive model and deploying it to OpenSees. The following conclusions were drawn:

(1) The constitutive relationship of HDR materials has complex nonlinear characteristics, and existing HDR constitutive models make it difficult to fully describe the influence of multiple factors on their stress–strain relationship. The GRU + attention approach can effectively solve the problems in existing HDR constitutive models.

(2) The development of a material surrogate constitutive model has the potential to markedly reduce the consumption of computational resources, considerably enhance computational efficiency, and reduce computational costs while ensuring the expeditious and precise simulation of material constitutive relationships.

(3) The development of HDRGA material provides a novel approach for expediently comprehending materials with complex properties and implementing them in finite element calculations.

Author Contributions: Conceptualization, F.L.; data curation, F.L.; formal analysis, F.L.; funding acquisition, T.P.; investigation, F.L.; methodology, F.L.; project administration, T.P.; software, F.L.; supervision, T.P.; validation, F.L.; writing—original draft, F.L.; writing—review and editing, T.P. All authors have read and agreed to the published version of the manuscript.

Funding: This research was funded by the Independent Project of the State Key Laboratory of Civil Engineering Disaster Prevention, grant number SLDRCE19-B-21, and the National Natural Science Foundation of China, grant number 51278372.

Institutional Review Board Statement: Not applicable.

Data Availability Statement: The raw data supporting the conclusions of this article will be made available by the authors on request.

Conflicts of Interest: The authors declare no conflicts of interest.

References

1. Nguyen, T.T.; Ha, M.H.; Nguyen, T.P.; Dang, V.H. Seismic reliability analysis of building structures using subset simulation coupled with deep learning-based surrogate model. *Adv. Struct. Eng.* **2022**, *25*, 2301–2318. [CrossRef]
2. Bas, E.E.; Moustafa, M.A. Communication Development and Verification for Python-Based Machine Learning Models for Real-Time Hybrid Simulation. *Front. Built Environ.* **2020**, *6*, 574965. [CrossRef]
3. Guo, J.; Alam, M.S.; Wang, J.; Li, S.; Yuan, W. Optimal intensity measures for probabilistic seismic demand models of a cable-stayed bridge based on generalized linear regression models. *Soil Dyn. Earthq. Eng.* **2020**, *131*, 106024. [CrossRef]
4. Yan, Y.; Xia, Y.; Yang, J.; Sun, L. Optimal selection of scalar and vector-valued seismic intensity measures based on Gaussian Process Regression. *Soil Dyn. Earthq. Eng.* **2022**, *152*, 106961. [CrossRef]
5. Jia, H.; Lin, J.; Liu, J. Bridge Seismic Damage Assessment Model Applying Artificial Neural Networks and the Random Forest Algorithm. *Adv. Civ. Eng.* **2020**, *2020*, 6548682. [CrossRef]
6. Hwang, S.H.; Mangalathu, S.; Shin, J.; Jeon, J.S. Machine learning-based approaches for seismic demand and collapse of ductile reinforced concrete building frames. *J. Build. Eng.* **2021**, *34*, 101905. [CrossRef]
7. Kiani, J.; Camp, C.; Pezeshk, S. On the application of machine learning techniques to derive seismic fragility curves. *Comput. Struct.* **2019**, *218*, 108–122. [CrossRef]
8. Alwanas, A.A.H.; Al-Musawi, A.A.; Salih, S.Q.; Tao, H.; Ali, M.; Yaseen, Z.M. Load-carrying capacity and mode failure simulation of beam-column joint connection: Application of self-tuning machine learning model. *Eng. Struct.* **2019**, *194*, 220–229. [CrossRef]
9. Lei, X.; Feng, R.; Dong, Y.; Zhai, C. Bayesian-optimized interpretable surrogate model for seismic demand prediction of urban highway bridges. *Eng. Struct.* **2024**, *301*, 117307. [CrossRef]
10. Wang, X.; Li, Z.; Shafieezadeh, A. Seismic response prediction and variable importance analysis of extended pile-shaft-supported bridges against lateral spreading: Exploring optimized machine learning models. *Eng. Struct.* **2021**, *236*, 112142. [CrossRef]
11. Zhang, B.; Wang, K.; Lu, G.; Guo, W. Seismic Response Analysis and Evaluation of Laminated Rubber Bearing Supported Bridge Based on the Artificial Neural Network. *Shock Vib.* **2021**, *2021*, 5566874. [CrossRef]
12. Nasab, M.S.E.; Guo, Y.-Q.; Kim, J. Seismic retrofit of a soft first-story building using viscoelastic dampers considering inherent uncertainties. *J. Build. Eng.* **2022**, *47*, 103866. [CrossRef]
13. Eskandari Nasab, M.S.; Kim, J. Fuzzy analysis of a viscoelastic damper in seismic retrofit of structures. *Eng. Struct.* **2022**, *250*, 113473. [CrossRef]

14. Mekaoui, N.; Hada, S.; Saito, T. Deep learning model for prediction of non-linear cyclic hysteresis of seismic isolation devices: Full-scale experimental validation. *Soil Dyn. Earthq. Eng.* **2023**, *169*, 107873. [CrossRef]
15. Mekaoui, N.; Saito, T. A Deep Learning-Based Integration Method for Hybrid Seismic Analysis of Building Structures: Numerical Validation. *Appl. Sci.* **2022**, *12*, 3266. [CrossRef]
16. Ni, X.; Xiong, Q.; Kong, Q.; Yuan, C. Deep HystereticNet to predict hysteretic performance of RC columns against cyclic loading. *Eng. Struct.* **2022**, *273*, 115103. [CrossRef]
17. Xu, Y.; Fei, Y.; Huang, Y.; Tian, Y.; Lu, X. Advanced corrective training strategy for surrogating complex hysteretic behavior. *Structures* **2022**, *41*, 1792–1803. [CrossRef]
18. Gu, Y.; Lu, X.; Xu, Y. A deep ensemble learning-driven method for the intelligent construction of structural hysteresis models. *Comput. Struct.* **2023**, *286*, 107106. [CrossRef]
19. Xu, L.; Yang, Y.; Zhang, Y.; Xue, Y.; Yu, Y.; Hao, N. Estimation of stress-strain constitutive model for ultra-high performance concrete after high temperature with a deep neural network based method. *Constr. Build. Mater.* **2023**, *408*, 133690. [CrossRef]
20. He, H.; Zhang, Q.; Zhang, Y.; Chen, J.; Zhang, L.; Li, F. A comparative study of 85 hyperelastic constitutive models for both unfilled rubber and highly filled rubber nanocomposite material. *Nano Mater. Sci.* **2022**, *4*, 64–82. [CrossRef]
21. Xiang, Y.; Zhong, D.; Rudykh, S.; Zhou, H.; Qu, S.; Yang, W. A Review of Physically Based and Thermodynamically Based Constitutive Models for Soft Materials. *J. Appl. Mech.* **2020**, *87*, 110801. [CrossRef]
22. Arruda, E.M.; Boyce, M.C. A three-dimensional constitutive model for the large stretch behavior of rubber elastic materials. *J. Mech. Phys. Solids* **1993**, *41*, 389–412. [CrossRef]
23. Ogden, R.W.; Hill, R. Large deformation isotropic elasticity on the correlation of theory and experiment for incompressible rubberlike solids. *Proc. R. Soc. London. A. Math. Phys. Sci.* **1972**, *326*, 565–584. [CrossRef]
24. Han, L.; Wang, X.; Li, L. Complete constitutive relation of hyperelastic materials for Treloar's experimental data. *Chin. J. Theor. Appl. Mech.* **2022**, *54*, 3444–3455. (In Chinese)
25. Wei, W.; Yuan, Y.; Tan, P.; Zhu, H. A rate-dependent constitutive model of high damping rubber bearings considering the influence of compressive load and its effect on the seismic response. *China Civ. Eng. J.* **2020**, *53*, 23–32. (In Chinese)
26. Wei, Z.; Chen, H. A new elastic model for rubber-like materials. *Chin. J. Theor. Appl. Mech.* **2019**, *51*, 473–483. (In Chinese)
27. Wang, X.; Wu, R.; Xiao, H. Explicit modeling the hysteresis loops of the Mullins effect for rubber-like materials. *Chin. J. Theor. Appl. Mech.* **2019**, *51*, 484–493. (In Chinese)
28. Reese, S.; Govindjee, S. A theory of finite viscoelasticity and numerical aspects. *Int. J. Solids Struct.* **1998**, *35*, 3455–3482. [CrossRef]
29. Ghosh, K.; Lopez-Pamies, O. On the two-potential constitutive modeling of dielectric elastomers. *Meccanica* **2021**, *56*, 1505–1521. [CrossRef]
30. Li, F.; Peng, T. Developing a New Constitutive Model of High Damping Rubber by Combining GRU and Attention Mechanism. *Polymers* **2024**, *16*, 567. [CrossRef]
31. Bengio, Y.; Simard, P.; Frasconi, P. Learning long-term dependencies with gradient descent is difficult. *IEEE Trans. Neural Netw.* **1994**, *5*, 157–166. [CrossRef] [PubMed]
32. Hochreiter, S.; Schmidhuber, J. Long short-term memory. *Neural Comput.* **1997**, *9*, 1735–1780. [CrossRef] [PubMed]
33. Cho, K.; Van Merriënboer, B.; Gulcehre, C.; Bahdanau, D.; Bougares, F.; Schwenk, H.; Bengio, Y. Learning phrase representations using RNN encoder-decoder for statistical machine translation. *arXiv* **2014**, arXiv:1406.1078.
34. Olah, C. Understanding LSTM Networks. 2015. Available online: https://colah.github.io/posts/2015-08-Understanding-LSTMs (accessed on 1 June 2024).
35. Niu, Z.; Zhong, G.; Yu, H. A review on the attention mechanism of deep learning. *Neurocomputing* **2021**, *452*, 48–62. [CrossRef]
36. Mazzoni, S.; McKenna, F.; Scott, M.H.; Fenves, G.L. OpenSees command language manual. *Pac. Earthq. Eng. Res. (PEER) Cent.* **2006**, *264*, 137–158.

polymers

MDPI

Article

Application of Antimicrobial Rubber-Coated Cotton Gloves for Mangosteen-Peel-Extract-Mediated Biosynthesis of Ag–ZnO Nanocomposites

Montri Luengchavanon [1,2,*], Ekasit Anancharoenwong [3], Sutida Marthosa [4], Theerakamol Pengsakul [5] and Jidapa Szekely [6]

1 Sustainable Energy Management Program, Wind Energy and Energy Storage Systems Centre (WEESYC), Faculty of Environmental Management, Prince of Songkla University, Hatyai 90110, Thailand
2 Centre of Excellence in Metal and Materials Engineering (CEMME), Engineering Faculty, Prince of Songkla University, Hatyai 90110, Thailand
3 Materials and Renewable Energy Research Group, Faculty of Science and Industrial Technology, Prince of Songkla University, Surat Thani Campus, Surat Thani 84000, Thailand; ekasit.a@psu.ac.th
4 Centre of Excellence in Membrane Science and Technology, Faculty of Science and Industrial Technology, Prince of Songkla University, Surat Thani Campus, Surat Thani 84000, Thailand; sutida.m@psu.ac.th
5 Health and Environmental Research Centre, Faculty of Environmental Management, Prince of Songkla University, Hatyai 90110, Thailand; theerakamol.p@psu.ac.th
6 Faculty of Medical Technology, Prince of Songkla University, Hatyai 90110, Thailand; jidapa.sz@psu.ac.th
* Correspondence: montri.su@psu.ac.th

Abstract: Nanocomposites based on metal nanoparticles (MNP) prepared with mangosteen ($_{mgt}$) peel extract-mediated biosynthesis of Ag_{mgt}/Zn_{mgt} have attracted considerable interest due to their potential for various practical applications. In this study, their role in developing antibacterial protection for rubber cotton gloves is investigated. The process of mangosteen-peel-extract-mediated biosynthesis produced Ag_{mgt}/Zn_{mgt} nanocomposites with respective diameters of 23.84 ± 4.08 nm and 30.99 ± 5.73 nm, which were assessed in the context of antimicrobial rubber-coated gloves. The rubber glover surface exhibited a very dense deposition of the Ag+Zn_{mgt} nanocomposite, which subsequently demonstrated level 4 resistance to punctures under the ANSI-ISEA 105-2016 standard. This could be attributed to the Zn-cellulose double formation on the rubber surface. Notably, on testing the inhibition of bacterial growth, the extract with the Ag_{mgt} nanoparticles presented the least concentration capable of growth inhibition in comparison to the extracts with Zn_{mgt} and Ag+Zn_{mgt} nanoparticles. Each of the mangosteen extracts was shown to inhibit bacterial growth when tested against both Gram-positive cocci and Gram-negative bacilli, with MIC in the range 40–320 µg/mL. The growth of drug-resistant bacteria (MRSA) could also be inhibited with an MIC value of 160 µg/mL, and with 30 min of contact, gloves with respective coatings of Zn_{mgt} and Ag+Zn_{mgt} extract nanocomposites were shown to inhibit *K. pneumoniae* and MRSA. However, while effective bacterial inhibition occurred with the suspensions, the coatings on glove surfaces required a lengthy incubation period (contact time) of at least 30 min for efficacy.

Keywords: mangosteen peel; Ag nanoparticles; Zn nanoparticles; plant-extract-mediated biosynthesis; rubber gloves; antimicrobial activity

1. Introduction

Mangosteen fruits in Thailand have mainly been produced for exports, which benefit the farmers financially. Mangosteens are among the major fruits generating profits when exported from Thailand to many other countries, such as Japan, Taiwan, and China. Currently, there is a big problem with an imbalance of supply and demand [1]. The quantity of mangosteens that cannot be exported to other countries should instead be consumed domestically, when there is over supply. The mangosteen peel waste from restaurants, famers, and markets has been increasing, and there is a need for treating this waste to create added value.

In rubber glove production, estimating an appropriated curing degree involves the consideration of many mechanical properties fundamental to rubber gloves. Generally, rubber gloves have been tested for conventional measures of trust in experiments for protection, to determine time and temperature in the rubber-curing process to meet the required properties [2]. The typical physical resistance tests are conducted to assess abrasion, cut, and punch as well as tear resistances. The rubber gloves use natural rubber (NR) that can be combined with metal nanoparticles (MNPs) for added value in various applications. The linear polymer chains of NR must be cross linked with covalent bonds, forming a vulcanised molecular network that is mechanically strong [3]. The physical resistance tests follow standards such as ANSI-ISEA 105-2016 [4].

Metal nanoparticle technology has been widely developed for application in several areas, such as anticancer or antibacterial applications, catalysis, photoelectric phenomena, bioengineering [5], and medical imaging, based on the special physical and chemical properties of the nanoparticles. Normally, the general synthesis methods of MNPs involve toxic chemicals that are used as stabilising and reducing agents, but this process is complex to operate and excessively environmentally unfriendly. The plant-extract-mediated biosynthesis of NMPs can be provide a number of advantages, such as ease of operation, low cost, sustainability, and environmental friendliness. Additionally, the synthesised MNPs are nontoxic, and the process produces nanoparticles with better uniformity of size than is the case with the traditional method [6]. The silver (Ag) nanoparticles have served as bactericidal agents and plasmonic nanoparticles in environmental applications. For example, Ag nanoparticles can generate reactive oxygen species (ROS), offering specific reaction selectivities and surface plasmon resonance (SPR). Actually, Ag nanoparticles have antimicrobial capabilities [7]. Ag nanoparticles had significantly higher antimicrobial activities against *Bacillus* sp. and *Escherichia coli* when compared to $AgNO_3$. Furthermore, a toxicity evaluation of these Ag nanoparticles in suspension was carried out on seeds of chickpeas (*Cicer arietinum*) and mung beans (*Vigna radiata*) [8]. The zinc oxide (ZnO) nanoparticles have been reported to offer high antibacterial potential in the context of both Gram-negative and Gram-positive bacteria. The antibacterial effect of ZnO nanoparticles was demonstrated against *Staphylococcus aureus* and *E. coli* [9]. Recently, surface modifications have been applied in antibacterially coated gloves, which are a useful alternative for protecting health and public care workers from the spread of pathogenic microorganisms. The antimicrobial gloves can have advanced features in inhibiting growth or killing off microorganisms once they touch the glove surface [10].

This investigation studied the mangosteen-peel-extract-mediated biosynthesis of Ag–ZnO nanocomposites for application to antimicrobial rubber-coated gloves. The rubber gloves were tested for physical resistance under ANSI-ISEA 105-2016 standard as regards abrasion, cutting, tear, and puncture resistance, while the preparation of the liquid-natural-rubber-compound-dipped cotton gloves used a spray technique that provided a coating of Ag, Zn, or Ag+Zn nanoparticles. Subsequent antibacterial testing utilised challenges with *Pseudomonas aeruginosa*, *Klebsiella pneumonia*, *E. coli*, *Enterococcus faecalis*, Methicillin-resistant *S. aureus* (MRSA), and *S. aureus*.

2. Experimental

The chemical materials for the extract-mediated biosynthesis process were purchased as follows: 99.9% silver nitrate ($AgNO_3$) from Fisher Chemical (Loughborough, UK), 98% zinc nitrate hexahydrate ($N_2O_6Zn_6H_2O$) from Loba Chemie PVT. Ltd. (Mumbai, India), and 89% sodium hydroxide pellets (NaOH) from Loba Chemie PVT. Ltd. (Mumbai, India). The mangosteen peel was collected from Phatthalung province, Thailand. The selected mangosteen peel was only black in colour since this is suitable for extract-mediated biosynthesis.

Other chemical materials and composites included natural latex rubber from a local factory in Surat Thani province, Thailand; 85% potassium hydroxide pellets (KOH) from Loba Chemie PVT. Ltd. (Mumbai, India); 98% potassium laurate ($C_{12}H_{23}KO_2$) from Bio-Furan (Pittsburgh, PA, USA); 99.8% sulphur (S) from Loba Chemie PVT. Ltd. Mumbai, India); 97% zinc diethyldithiocarbamate (ZDC) from Sigma-Aldrich (Darmstadt, Germany); Wingstay ® L from Shuguang Chemical General Co., Ltd. (Anqing, China); 99% calcium carbonate ($CaCO_3$) from Sigma-Aldrich (Germany); 99% zinc oxide (ZnO) from Sigma-Aldrich (Germany); and cotton gloves from PPE Mate Company Ltd. (Don Ki Di, Thailand).

For physical resistance testing, the abrasion experiments used SODEMAT, AG04-Martindale, originally from France. The cut resistance experiments were run with SODEMAT, Coup Test, originally from France. The tear resistance and puncture resistance experiments were run with SHIMADZU, AG-X Plus, originally from Japan. All the samples were tested on the rubber-coated side of gloves (palm side).

For mangosteen-peel-extract-mediated Ag nanoparticle surface modification and its application for antibacterial purposes, the biosynthesis involved 30 g of mangosteen mixed with deionised water and boiled for 20 min at 80 °C. This mixture was then filtered with Whatman no. 1 (11 μm pore size) filter paper, and the filtrate was allowed to cool. Next, a 1-millimolar (mM) solution of silver nitrate was prepared in deionised water and mixed with 40 mL of mangosteen peel extract, while 0.1 M solution of sodium hydroxide (NaOH) was added to maintain pH 6 (indicated by the presence of dark reddish-brown coloured Ag nanoparticles [11] as the effect of pH; the organic material extracts were obtained using the silver nitrate solution preparation [12]), followed by vigorous stirring until the colour changed to dark brown, using a magnetic stirrer bar on a hot plate at 100 °C. The resulting product was a fine Ag nanocomposite prepared using mangosteen peel (Ag_{mgt}).

For mangosteen-peel-extract-mediated ZnO nanoparticle surface modification and its application for antibacterial purposes, the biosynthesis involved 30 g of mangosteen mixed with deionised water and boiled for 20 min at 80 °C. This mixture was then filtered with Whatman no. 1 (11 μm pore size) filter paper, and the filtrate was allowed to cool. Further, 2 g of zinc nitrate hexahydrate solution was prepared in 100 mL of deionised water and mixed with 100 mL of mangosteen peel extract, while 0.1 M solution of sodium hydroxide (NaOH) was added to maintain pH 8 (the pH was maintained at 7.4–8 during the analysis; cotton webs were used for the seeding of isolated organism agar plates that depend on the organic material extracts obtained using the silver nitrate solution [13,14]), followed by vigorous stirring until the colour changed to white using a magnetic stirrer bar on a hot plate at 60 °C. The resulting product was a fine Zn nanocomposite prepared using mangosteen peel (Zn_{mgt}).

The $Ag{+}Zn_{mgt}$ nanocomposite was prepared by mixing 50% Ag_{mgt} and 50% Zn_{mgt} (mass %) and then ball milling for 12 h. The resulting product was a fine Ag + Zn nanocomposite prepared with mangosteen peel extract ($Ag{+}Zn_{msg}$).

As regards the spraying technique for coating the gloves, Ag_{mgt}, Zn_{mgt}, and $Ag{+}Zn_{mgt}$ nanocomposites were first sieved using a 100 μm sieve. The mixer combined 80 mL ethanol and 1.813 g of Ag_{mgt}, Zn_{mgt}, or $Ag{+}Zn_{mgt}$ nanocomposite to spray on the rubber gloves.

The air pressure pump was set to 7 bars, and the nozzle was located at a 12 cm distance from the gloves, with a 10 cm radius.

The liquid rubber compound and chemicals were prepared as shown in Table 1. Natural latex rubber (2000 g) was placed in the container for 1–2 min, to allow for ammonium to evaporate. After that, the 60% natural latex was poured to form a mixture with 10% KOH, 20% $C_{12}H_{23}KO_2$, 50% S, 50% ZDC, 40% Wingstay@L, 50% ZnO, and 1% distilled water. This mixture was continuously stirred (60 rpm) as slurry for 48 h. The cotton gloves were inserted onto a solid mould for controlling the finger and palm shape. They were then dipped into the liquid rubber compound that covered the palm for 10 s, and then, the rubber gloves were sprayed with Ag_{mgt}, Zn_{mgt}, and $Ag{+}Zn_{mgt}$ nanocomposites as shown in Figure 1, before all gloves samples were baked in the oven at 70 °C for 3 h. The analysis of crystallisation, quantity (wt%), and structure of the samples used scanning electron microscopy (SEM, EDX, Quanta 400, Prague, Czech Republic), with XRD (PANalytical Empyrean, Almelo, The Netherlands).

Figure 1. The Ag_{mgt}, Zn_{mgt}, and $Ag{+}Zn_{mgt}$ nanocomposites were prepared and sprayed onto rubber gloves.

Table 1. The ratios of materials and chemicals which were mixed for production of the antimicrobial rubber-coated gloves.

Material and Chemical Concentrations	Typical Rubber Compound Proportions by Weight (phr)
60% Natural rubber latex (NR)	100.00
10% KOH	0.30
20% Potassium laurate	0.20
50% Sulphur	0.50
50% ZDC	0.75
40% Wingstay@L	0.75
50% $CaCO_3$	5.10
50% ZnO	0.40

Concerning the challenge organisms, six isolates of bacteria were used for bactericidal activity testing of the mangosteen extracts, which were common human pathogens and drug-resistant bacteria, namely *S. aureus*, *K. pneumoniae*, *E. coli*, *P. aeruginosa*, *E. faecalis*, and methicillin-resistant *S. aureus* (MRSA). The bacterial culture was grown on nutrient agar and incubated at 35 °C for 24 h, then 2–3 single colonies were selected and kept as stock cultures with nutrient broth containing 20% glycerol at −20 °C until required for testing.

In the antimicrobial susceptibility testing, the antifungal activity of the mangosteen extracts was tested using the broth microdilution standard method following the Clinical and Laboratory Standards Institute guidelines (CLSI) outlined in document CLSI M100 [15]. Briefly, the stock suspensions of mangosteen-extract-coated Zn and Ag nanoparticles were prepared by dissolving in dimethyl sulfoxide (DMSO) and performing serial dilutions to final concentrations in the range from 5120 μg/mL to 10 μg/mL, prepared in a sterile cation-adjusted Mueller Hinton broth (CAMHB). The inoculum preparation was performed following CLSI M100 to achieve the final inoculum with 5×10^6 CFU/mL.

Antimicrobial activity testing was performed using a microtiter plate. One hundred microliters of each dilution of mangosteen extract were added into wells 1 to 10. The drug-free growth control and sterility control were added to wells 11 and 12, respectively. One hundred microliters of prepared inoculum were added into wells 1 to 11. The plate was incubated at 37 °C for 24 h. The tests were performed in duplicate. Resazurin was used as a growth indicator [16]. The minimal inhibitory concentration (MIC) of each mangosteen extract was determined.

For the bactericidal activity testing of the glove surfaces, samples of nanoparticle coated gloves were prepared by cutting the coated area of the gloves into rectangular pieces of size 2×2.5 cm, for 5 cm^2 area. Uncoated gloves were used as control cases. Pieces of the test and control samples were sterilised by autoclaving before testing.

Bactericidal activity testing was carried out following the ASTM D7907-14 (2019) standard method [17]. ASTM D7907 is a standard for determining whether medical examination gloves kill bacteria effectively on their surfaces. Briefly, bacterial suspensions containing approximately 10^8 CFU/mL were prepared in sterile phosphate buffer without any interfering substances. A 10 μL bacterial suspension was distributed over the surface of the coated or uncoated glove samples and covered with a cover slip to allow contact with the glove's surface for 0, 10, 20, or 30 min. Thereafter, the remaining bacteria on the glove's surface were extracted with 10 mL of sterile phosphate buffer with 0.1% Tween 20. Serial dilutions were prepared and plated on tryptic soy agar (TSA) plates (BD Difco, MD, New York, USA) and incubated for 24 h at 35 ± 2 °C. The tests were conducted in triplicate, and the average bacterial colonies were calculated.

3. Results and Discussion

Figure 2 shows SEM image samples of Ag_{mgt} and Zn_{mgt} nanocomposites. Figure 2A shows the Ag_{mgt} nanocomposite that was interspersed on the fine carbon substrate and may be generated using mangosteen peel. The fine carbon substrate was of approximately 0.5–1.0 μm particle size. Figure 2B shows the Ag_{mgt} nanocomposite at 200,000× magnification for measuring the average 23.84 nm diameter, as presented in Table 2. Figure 2C shows the Zn_{mgt} nanocomposite revealing connections of many bubbles and high porosity, combined with the white fine carbon substrate, also generated using mangosteen peel. Figure 2D shows the Zn_{mgt} nanocomposite at 200,000× magnification for measuring the average 30.99 nm diameter, as presented in Table 2. When a nanocomposite film was composed of silver nanograins embedded in a carbonaceous matrix, created via physical vapour deposition (PVD) from separate sources of fullerene and silver acetate powders, these powders gave Ag nanograins with an average size of about 12–20 nm [18]. When an electrospray exposure system was operated at a high voltage to produce ZnO nanoparticles, they were about 20 nm in diameter [19].

Plant biosynthesis is generally related to bringing silver nitrate ($AgNO_3$) into contact with the biomass/extract of plants. The manifestation of a brownish-yellow colour shade after a short duration of contact indicates the formation of Ag nanocomposite in line with the following chemical reaction [20]:

$$Ag^{+}\,NO_3^{-} + \text{Plant molecule (OH, C=H, etc.)} \rightarrow Ag^{o}\ \text{Nanocomposites} \quad (1)$$

This equation describes Ag nanocomposite synthesis conducted using plants, as reported by Shankar et al. [21]. Meanwhile, zinc oxide (ZnO) nanocomposite can act as an attractive semiconductor particle for nano-electronic and photonic applications. Highly stable and spherical 25–40 nm ZnO nanoparticles [22] had particle size that could be regulated by varying the concentration of the organic material solution [23].

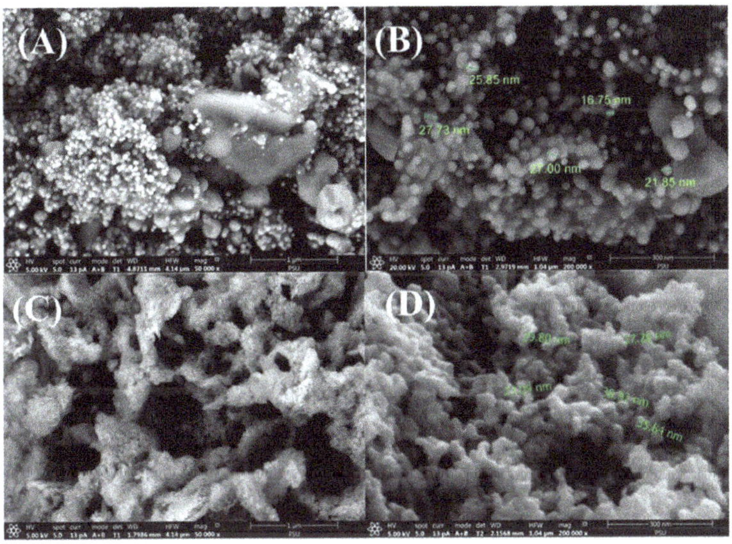

Figure 2. SEM image samples of Ag_{mgt} and Zn_{mgt} nanocomposites. (**A**,**B**) show the SEM images for Ag_{mgt} at 50,000× and 200,000× magnifications. (**C**,**D**) show the SEM images of Zn_{mgt} at 50,000× and 200,000× magnifications.

Table 2. The particle diameters in Ag $_{mgt}$ and Zn $_{mgt}$ nanocomposites.

Nanocomposites	Diameter (nm)
Ag_{mgt}	21.85
	16.75
	27.00
	27.73
	25.85
Average	23.84 ± 4.08
Zn_{mgt} *	29.80
	23.09
	27.28
	38.93
	35.84
Average	30.99 ± 5.73

* The Zn_{mgt} produced consisted of ZnO as indicated by XRD in Figure 3.

Table 2 allows for a comparison of diameters between Ag_{mgt} and Zn_{mgt} nanocomposites. Notably, the Zn_{mgt} nanocomposite had ZnO as indicated by XRD in Figure 3. Based on the SEM images, diameters of Ag_{mgt} and Zn_{mgt} nanocomposites were randomly sampled as 21.85, 16.75, 27.00, and 25.85 nm and as 29.80, 23.09, 27.28, 38.93, and 35.84 nm, respectively. Hence, the average diameters of the Ag_{mgt} and Zn_{mgt} nanocomposite particles were around 23.84 $\pm$ 4.08 nm and 30.99 $\pm$ 5.73 nm, respectively. When Ag nanoparticles were formed by the reaction of 1 mM silver nitrate and 5% leaf extract of banana, Neem (Indian plants), or Tulsi leaves (Indian plants), they could be fabricated in sizes of 50 nm, 20 nm, and 50 nm, respectively [8]. Therefore, the extract-mediated biosynthesis creates particles whose diameter depends on the biomaterials. Meanwhile, the biosynthesis of *Ocimum basilicum*-leaf-extract-mediated ZnO nanoparticles has produced sizes of 20.07 $\pm$ 0.34 nm [24].

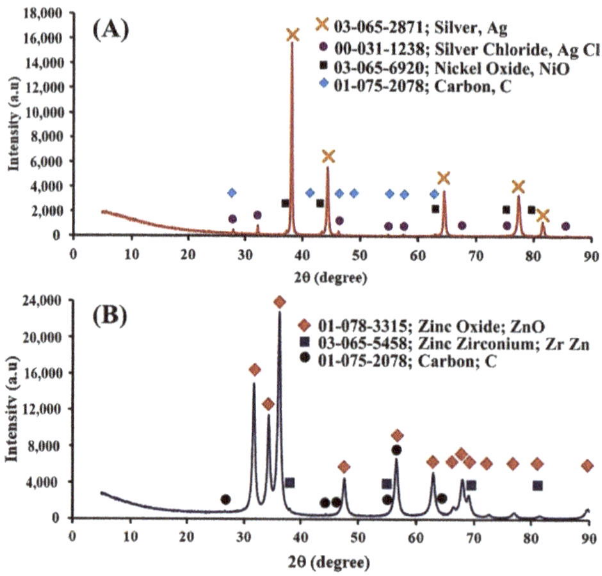

Figure 3. The XRD patterns of Ag_{mgt} and Zn_{mgt} nanocomposites: (**A**) the crystalline peaks of Ag_{mgt} nanocomposite and (**B**) the crystalline peaks of Zn_{mgt} nanocomposite.

Figure 3 shows the XRD patterns of Ag_{mgt} and Zn_{mgt} nanocomposites. Figure 3A for Ag_{mgt} nanocomposite revealed a highly crystalline peak of Ag with the serial number in the XRD standard being 03-065-2871, AgCl with the number 00-031-1238, NiO with the number 03-065-6920, and C with the number 01-075-2078. Figure 3B for Zn_{mgt} nanocomposite indicated a highly crystalline peak of ZnO with the serial number in XRD standard being 01-078-3315, Zr Zn with the number 03-065-5458, and C with the number 01-075-2078.

The peaks exhibited for metallic Ag revealed five specific diffraction peaks at 2θ = 38.2, 44.4, 64.5, 77.5, and 81.6°, which indicated the (111), (200), (220), (311), and (222) reflections of metallic Ag [25]. The peaks of XRD have indicated the C element from mangosteen-peel-derived porous carbon when activated by a temperature of 800 °C [26]. The NiO and Cl elements may have been affected by the mangosteen peel to be highly biosorptive in removal of Ni [27]. The elemental concentration of mangosteen peel extract analysed with X-ray fluorescence (in wt%) has revealed 13.45% Cl, so the black-coloured peel has a high Cl fraction [28]. Regarding the ZnO, the diffraction peaks at 31.84°, 34.52°, 36.33°, 47.63°, 56.71°, 62.96°, 68.13°, and 69.18° degrees were assigned to hexagonal wurtzite phases of ZnO [29]. Generally, the mangosteen peel has obviously contained Zn, Zr, and C elements, and the black-coloured peel has high amounts of Zn, Zr, and C [28]. Therefore, the Zn elements were sourced from the mangosteen peel and the extract-mediated biosynthesis process.

Figure 4 shows SEM image samples of the surfaces of rubber gloves, namely Ag_{mgt}-coated glove, Zn_{mgt}-coated glove, and Ag+Zn_{mgt}-coated glove. Figure 4A,B show the SEM images of the surface of the rubber glove at 200× and 1000× magnifications, for the control sample without coating. Figure 4C,D show the SEM images of the surface of the Ag_{mgt}-coated rubber glove at 200× and 1000× magnifications, whereby the Ag_{mgt} nanoparticles were spherical in agglomerates, deposited on the surface of the rubber and immersed in the rubber. They were deposited at a low density on the rubber surface. Figure 4E,F show the SEM images of the surface of the Zn_{mgt}-coated rubber glove at 200× and 1000× magnifications, whereby the Zn_{mgt} nanoparticles were elongated with a high aspect ratio in agglomerates, deposited on the surface of the rubber and immersed in the rubber. The deposition was of a higher density than of the Ag_{mgt} nanoparticles. Figure 4G,H show the SEM images of the surface of the Ag+Zn_{mgt}-coated rubber glove at 200× and 1000× magnifications, whereby the Ag+Zn_{mgt} nanoparticles aggregates were of various shapes and also deposited on the surface of the rubber and immersed in the rubber due to the fact that when using the spray technique, the rubber still had a soft texture. This deposition was of a very high density compared to Ag_{mgt} or Zn_{mgt} nanoparticles on the rubber surfaces. The electrospinning generated 8.9–9.89 × 10^4 g/mol ultrafine fibres with diameters of 70–175 nm and the particles were coated on the nitrile gloves [30]. Antimicrobial agents in the form of chemicals, whether naturally or chemically obtained from microorganisms, can be amalgamated into synthetic rubber (SR) or natural rubber (NR) films by coating or dispersion techniques [10].

Table 3 summarises the outcomes from physical testing of rubber gloves under the ANSI-ISEA 105-2016 standard to evaluate resistance to abrasion, cuts, tears, and punctures. Testing for abrasion resistance achieved level 1 at a load of 500 g for 100 cycles. Testing for cut resistance achieved level 3 at 1000 g, while testing for tear resistance achieved level 3 at 50 Newtons. Finally, testing for puncture resistance achieved level 4 at 100–149 Newtons. A spraying technique was used to apply coatings with Ag+Zn_{mgt}, Ag_{mgt}, or Zn_{mgt} on the rubber gloves, as shown in Figure 5. Figure 5A shows the Ag+Zn_{mgt} combined with carbon deposited on the rubber glove with Ag+Zn_{mgt} sharply formed, which impaled easily into the rubber glove texture. Figure 5B shows the Ag_{mgt} combined with carbon deposited on the rubber glove with Ag_{mgt} that stuck on the surface of the rubber glove. Figure 5C shows Zn_{mgt} combined with carbon deposited on the rubber glove, where Zn_{mgt} also stuck on the surface of the rubber glove. Regarding the Ag+Zn_{mgt} shape, it may be shaped in the ball milling process for 12 h.

The main considerations when measuring the strength of the rubber gloves were the resistance properties against abrasion, cuts, tears, and punctures, measured for each of the different rubber composite textures seen in Table 1. When the resistance to abrasion was compared among NR, NR/OC/MNR (OC = organoclay; MNR = maleic anhydride grafted natural rubber), the composite using only NR had a relatively poor resistance. When eight parts of MNR were introduced, this increased the filler–rubber interactions, preventing nano layer aggregation. Moreover, the use of OC improved the rubber layer dispersion, leading to a 46% reduction in NR chains peeling from the sample surface in comparison to pure NR [31]. The findings for resistance to cuts have shown that a mixed calcium carbonate ($CaCO_3$) composite exhibits greater hardness and Young's modulus [32], leading to increased resistance to cuts. The observed resistance to tears might be attributable to an elevated crosslink concentration resulting from additional sulphur crosslinks along with the stronger modulus due to the reinforcing filler. Meanwhile, the sulphur-vulcanised NR films contain polysulphidic linkages, whereas in the peroxide-vulcanised samples, they are predominantly C to C type links (the bond strengths have rank order C-C > C-S-C > C-S-S-C) [33]. Notably, in this current study, the best resistance to punctures was exhibited by the Ag+Zn_{mgt} and Zn_{mgt} composite, where the ZnO was embedded in the rubber, thus enhancing both suspension and dispersion. It was apparent that the number of activated rubber sites increased, leading to mixing sulphur consumption in the partial interparticle crosslinking of the rubber. The production of Zn–cellulose complexes along with the elastomeric matrix might serve to support zinc dispersion in the rubber, allowing for greater resistance to punctures in the composite [34,35]. Furthermore, ZDC composite coated with nano Zn_{mgt} was used in this current study, ensuring Zn–cellulose double formation on the rubber glove surfaces.

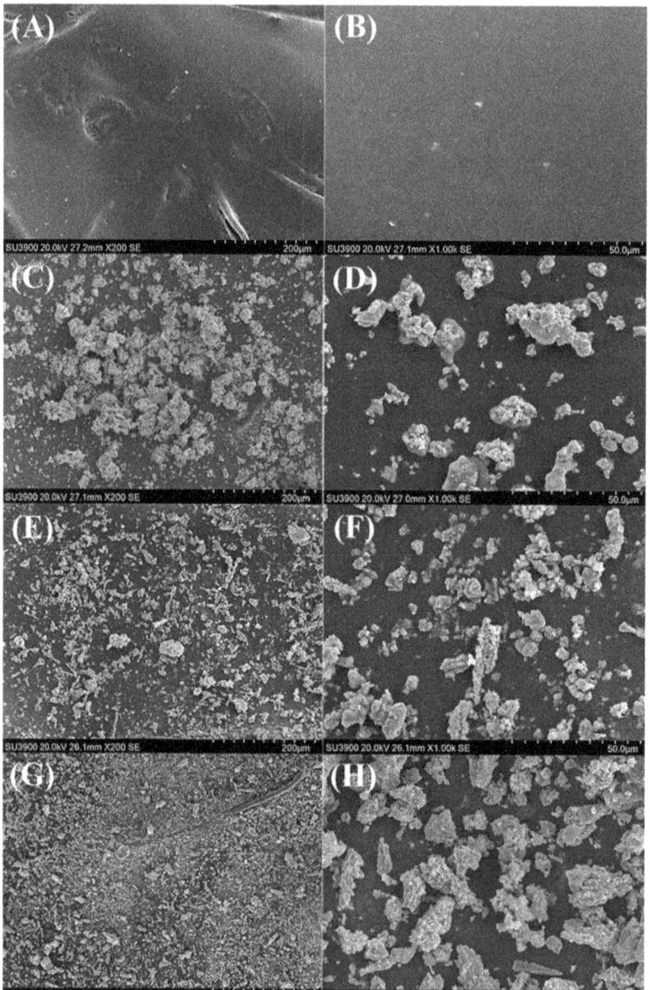

Figure 4. SEM image samples of the surfaces of rubber gloves; Ag_{mgt}-coated rubber glove, Zn_{mgt}-coated rubber glove, and Ag+Zn_{mgt}-coated rubber glove. (**A**,**B**) show the SEM images of the surface of the rubber glove at 200× and 1000× magnifications. (**C**,**D**) show the SEM images of the surface of the Ag_{mgt}-coated rubber glove at 200× and 1000× magnifications. (**E**,**F**) show the SEM image of the surface of the Zn_{mgt}-coated rubber glove at 200× and 1000× magnifications. (**G**,**H**) show the SEM image of the surface of the Ag+Zn_{mgt}-coated rubber glove at 200× and 1000× magnifications.

Table 3. Physical test results for the rubber gloves.

Physical Test	Result	Index	Reference
Abrasion	Level 1 (ANSI-ISEA 105-2016)	- Gram load (500) - ≥100 cycles	[36]
Cut	Level 3 (ANSI-ISEA 105-2016)	- ≥ 1000 g	[36]
Tear	Level 3 (ANSI-ISEA 105-2016)	- ≥50 Newton	[37]
Puncture	Level 4 (ANSI-ISEA 105-2016)	- 100–149 Newton	[36]

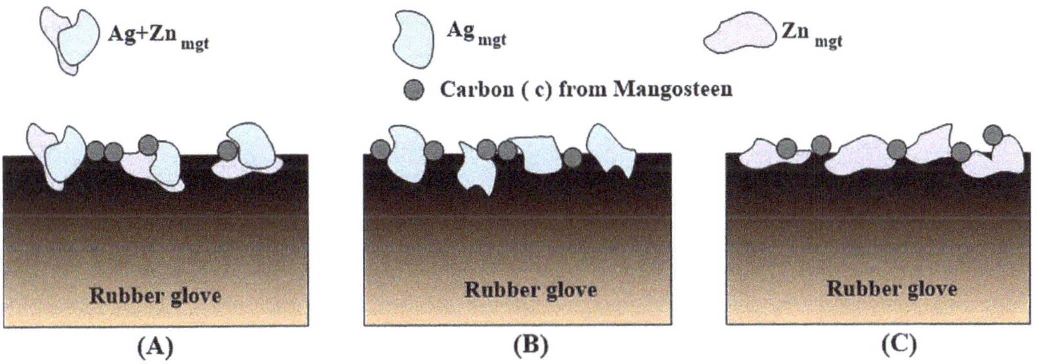

Figure 5. Schematic cross-sections of rubber gloves when sprayed with Ag+Zn$_{mgt}$, Ag$_{mgt}$, and Zn$_{mgt}$, to coat the surface of rubber gloves related to Figures 3 and 4. The XRD patterns indicated the Ag$_{mgt}$ and Zn$_{mgt}$ nanocomposites, while SEM exhibited the surfaces of the Ag$_{mgt}$ coated rubber glove, Zn$_{mgt}$ coated rubber glove, and Ag+Zn$_{mgt}$-coated rubber glove.

The minimal inhibitory concentration of mangosteen extracts and antibacterial activities of with Ag+Zn$_{mgt}$, Zn$_{mgt}$, and Ag$_{mgt}$ nanoparticles are presented in Table 4 and Figure 6. All bacterial tested isolates were susceptible to extract with Ag+Zn$_{mgt}$, Zn$_{mgt}$, and Ag$_{mgt}$ nanoparticles, except for *E. faecalis*, which was not susceptible to extract with ZnO. This finding is similar to a study by Zhu et al. (2021), who reported that mangosteen extract exhibited bactericidal activity, effectively inhibiting Gram-positive bacteria, such as *S. aureus* and *Micrococcus* [38]. The potential of α-mangostin (AMG) derived from mangosteen extract may play an important role in antibacterial activity [39].

The MICs of mangosteen extracts with Ag+Zn$_{mgt}$, Zn$_{mgt}$, and Ag$_{mgt}$ nanoparticles to susceptible bacteria ranged from 80 to 160, 40 to 320, and 40 to 160 μg/mL, respectively. All types of the mangosteen extract present bacterial inhibitory activity against both Gram-positive cocci and Gram-negative bacilli with MIC ranging in 40–320 μg/mL and against drug resistant bacteria (MRSA), with a similar MIC of 160 μg/mL. This is also corroborated by another study by Song et al. (2021), who assessed the combination of antibiotics and AMG on multidrug-resistant (MDR) bacteria. They suggested that AMG compound is promising candidate against MRSA, Vacomycin-resistant *Enterococcus* (VRE) and MDR pathogens by targeting bacterial inner membrane [30]. In our current study, we found that MICs of mangosteen extracts with metallic nanoparticles were higher than those reported earlier for a combination of antibiotics and AMG [40], and this finding could be explained by the effect of solvents on AMG yields in the extracts, for Soxhlet and ethanol solvent would provide better levels of α-mangostin compound than water [41].

Therefore, improving mangosteen extraction process may enhance antimicrobial efficacy of the coated gloves.

Table 4. Minimum inhibitory concentrations of mangosteen peel extracts against pathogenic bacteria.

Organism	Minimum Inhibitory Concentration (MIC) of Mangosteen Extract with Nanoparticles (μg/mL)		
	$Ag+Zn_{mgt}$	Zn_{mgt}	Ag_{mgt}
Pseudomonas aeruginosa	80	80	40
Klebsiella pneumoniae	160	320	80
Escherichia coli	80	160	40
Enterococcus faecalis	320	>5120	160
Methicillin-resistant *Staphylococcus aureus* (MRSA)	160	160	160
Staphylococcus aureus	80	40	160

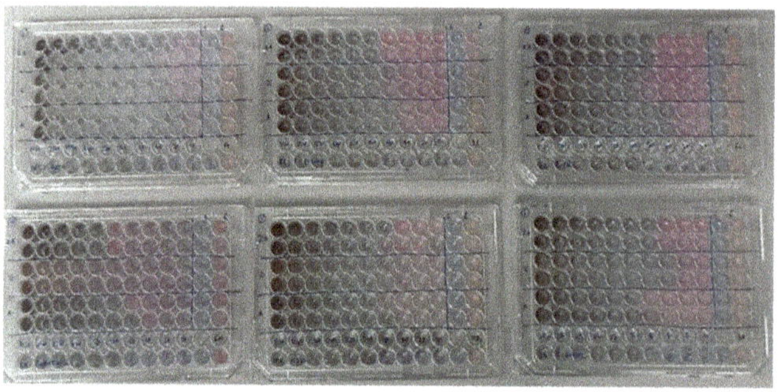

Figure 6. Antimicrobial activities at different concentrations of $Ag+Zn_{mgt}$, Zn_{mgt}, and Ag_{mgt} nanoparticles.

For bacterial inhibition activity of the coated glove surfaces, it was found that after 30 min contact time, the numbers of *K. pneumoniae* and MRSA were significantly reduced when tested on gloves coated with Zn_{mgt} and $Ag+Zn_{mgt}$ nanoparticles, respectively, compared to the uncoated control sample. However, the bactericidal effect was mostly not different between the control glove and the gloves sprayed with $Ag+Zn_{mgt}$, Zn_{mgt}, and Ag_{mgt} nanoparticles, as shown in Figure 7. Based on these results, the use of antibacterial gloves may be applicable to prevent or reducing cross-contamination and indirect transmission of bacterial pathogens in general use.

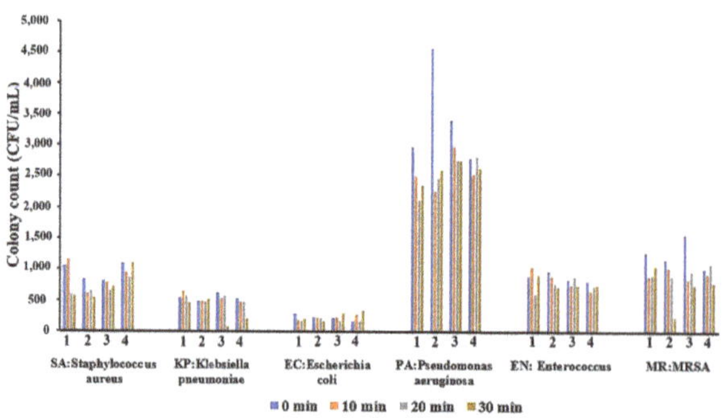

Figure 7. A comparison of (1) the bactericidal activity of the untreated glove, (2) $Ag+Zn_{mgt}$, (3) Zn_{mgt}, and (4) Ag_{mgt} spray coated on gloves, at contact times of 0, 10, 20, and 30 min.

4. Conclusions

Mangosteen-peel-extract-mediated biosynthesis of Ag_{mgt} and Zn_{mgt} nanocomposites for antimicrobial rubber-coated gloves generated particles with diameters of 23.84 ± 4.08 nm and 30.99 ± 5.73 nm, respectively. The elements Zn, Zr, and C were found in the nanocomposites. The $Ag+Zn_{mgt}$ nanocomposite was shown to be the most densely deposited one on the surface of rubber gloves.

The rubber-coated cotton glove showed powerful, level 4 puncture resistance under ANSI-ISEA 105-2016, which is caused by Zn–cellulose complex generation supporting the dispersion of zinc in the rubber, which contained ZDC composite and was coated with nano Zn_{mgt} using a double formation on the surface of the rubber glove, thus generating stronger puncture resistance.

From the antimicrobial susceptibility testing, we found that all types of extracts effectively inhibited all 5 species of human pathogenic bacteria and one strain of important drug-resistant bacteria (MRSA). In particular, the extract with Ag_{mgt} nanoparticles presented the lowest concentration against the growth of the tested bacteria, when compared to extracts with Zn_{mgt} or $Ag+Zn_{mgt}$ nanoparticles. All types of the mangosteen extracts presented bacterial inhibitory activity against both Gram-positive *cocci* and Gram-negative *bacilli*, with MIC ranges in 40–320 µg/mL, and against drug resistant bacteria (MRSA), with a similar MIC of 160 µg/mL.

However, when the nanoparticles were spray-coated on gloves, we found that the effectiveness in inhibiting bacteria contacting gloves surfaces had decreased. After a 30 min contact time, inhibition of *K. pneumoniae* and MRSA was found when using gloves coated with Zn_{mgt} and $Ag+Zn_{mgt}$, respectively. It seems that the extracts actively inhibited bacteria in suspension form, while as coatings on gloves their nanoparticles needed a long incubation period of at least 30 min.

Author Contributions: Conceptualization, M.L.; methodology, E.A. and J.S.; validation, S.M.; formal analysis, T.P.; investigation, M.L and J.S.; resources, E.A. and S.M.; data curation, E.A., S.M. and T.P.; writing—original draft preparation, M.L.; writing—review and editing, J.S.; visualization, M.L.; project administration, M.L. and T.P.; funding acquisition, M.L. All authors have read and agreed to the published version of the manuscript.

Funding: This research was financially supported by Prince of Songkla University under grant number SIT6601271c and by Prince of Songkla University, Surat Thani Campus, 2024.

Institutional Review Board Statement: Not applicable.

Data Availability Statement: No data were used for the research described in the article.

Acknowledgments: The authors extend their appreciation to the Fundamental Fund 2023 (FF66) at Prince of Songkla University, and thank you to Prapas Muangjunburee for supported laboratory room.

Conflicts of Interest: The authors declare no conflicts of interest.

Abbreviations

Ag_{mgt}	Ag nanocomposite prepared with mangosteen peel
Zn_{mgt}	Zn nanocomposite prepared with mangosteen peel
$Ag+Zn_{mgt}$	50% Ag_{mgt} + 50% Zn_{mgt} nanocomposite prepared with mangosteen peel
Ag	Silver
$AgNO_3$	Silver nitrate
AMG	α-mangostin
C	Carbon

$CaCO_3$	Calcium carbonate
CAMHB	Cation-adjusted Mueller Hinton broth
CLSI	Clinical and Laboratory Standards Institute guidelines
$C_{12}H_{23}KO_2$	Potassium laurate
DMSO	Dissolving in dimethyl sulfoxide
KOH	Potassium hydroxide
MDR	Multidrug-resistant
MNP	Metal nanoparticle
MRSA	Methicillin-resistant *S. aureus*
MIC	Minimal inhibitory concentration
$N_2O_6Zn_6H_2O$	Zinc nitrate hexahydrate
NaOH	Sodium hydroxide
NR	Natural rubber
S	Sulphur
SEM	Scanning electron microscope
SR	Synthetic
TSA	Tryptic soy agar
VRE	Vacomycin-resistant *Enterococcus*
XRD	X-ray diffraction
ZDC	Zinc diethyldithiocarbamate
ZnO	Zinc oxide
Zr	Zirconium

References

1. Ongkunaruk, P.; Piyakarn, C. Logistics Cost Structure for Mangosteen Farmers in Thailand. *Syst. Eng. Procedia* **2011**, *2*, 40–48. [CrossRef]
2. Loksupapaiboon, K.; Suvanjumrat, C. Curing analysis of rubber film on a Hand-Shaped former in the manufacturing of rubber gloves through novel OpenFOAM solver. *Int. J. Heat Fluid Flow* **2024**, *108*, 109476. [CrossRef]
3. Riyajan, S. Maleated epoxidized natural rubber grafted cassava starch/silver composite coating for cloth glove: Preparation and physical properties. *Ind. Crops Prod.* **2024**, *220*, 119245. [CrossRef]
4. Theiler, G.; Cano Murillo, N.; Halder, K.; Balasooriya, W.; Hausberger, A.; Kaiser, A. Effect of high-pressure hydrogen environment on the physical and mechanical properties of elastomers. *Int. J. Hydrog. Energy* **2024**, *58*, 389–399. [CrossRef]
5. Jha, P.K.; Chawengkijwanich, C.; Techato, K.; Limbut, W.; Luengchavanon, M. Callistemon viminalis Leaf Extract Mediated Biosynthesis of Ag, rGO-Ag-ZnO Nanomaterials for Catalytic PEM Fuel Cell Application. *Trends Sci.* **2022**, *19*, 493. [CrossRef]
6. Bao, Y.; He, J.; Song, K.; Guo, J.; Zhou, X.; Liu, S. Plant-extract-mediated synthesis of metal nanoparticles. *J. Chem.* **2021**, *2021*, 6562687. [CrossRef]
7. Mo, F.; Zhou, Q.; He, Y. Nano–Ag: Environmental applications and perspectives. *Sci. Total Environ.* **2022**, *829*, 154644. [CrossRef]
8. Banerjee, P.; Satapathy, M.; Mukhopahayay, A.; Das, P. Leaf extract mediated green synthesis of silver nanoparticles from widely available Indian plants: Synthesis, characterization, antimicrobial property and toxicity analysis. *Bioresour. Bioprocess.* **2014**, *1*, 3. [CrossRef]
9. Gudkov, S.V.; Burmistrov, D.E.; Serov, D.A.; Rebezov, M.B.; Semenova, A.A.; Lisitsyn, A.B. A mini review of antibacterial properties of ZnO nanoparticles. *Front. Phys.* **2021**, *9*, 641481. [CrossRef]
10. Hassan, A.A.; Singh, M. *Quick Insight on the Emergence of Antimicrobial Gloves*; Department of Statistics (DOS), Malaysia: Putrajaya, Malaysia, 2018; pp. 60–61.
11. Abaid, R.; Malik, M.; Iqbal, M.A.; Malik, M.; Shahwani, Z.; Ali, T.Z.; Morsy, K.; Capangpangan, R.Y.; Alguno, A.C.; Choi, J.R. Biosynthesizing Cassia fistula extract-mediated silver nanoparticles for MCF-7 cell lines anti-cancer assay. *ACS Omega* **2023**, *8*, 17317–17326. [CrossRef]
12. Ansari, M.; Ahmed, S.; Abbasi, A.; Khan, M.T.; Subhan, M.; Bukhari, N.A.; Hatamleh, A.A.; Abdelsalam, N.R. Plant mediated fabrication of silver nanoparticles, process optimization, and impact on tomato plant. *Sci. Rep.* **2023**, *13*, 18048. [CrossRef] [PubMed]
13. Javed, Z.; Tripathi, G.D.; Mishra, M.; Gattupalli, M.; Dashora, K. Cow dung extract mediated green synthesis of zinc oxide nanoparticles for agricultural applications. *Sci. Rep.* **2022**, *12*, 20371. [CrossRef] [PubMed]

14. Alnehia, A.; Al-Sharabi, A.; Al-Odayni, A.B.; Al-Hammadi, A.H.; Al-Ostoot, F.H.; Saeed, W.S.; Abduh, N.A.Y.; Alrahlah, A. Lepidium sativum Seed Extract-Mediated Synthesis of Zinc Oxide Nanoparticles: Structural, Morphological, Optical, Hemolysis, and Antibacterial Studies. *Bioinorg. Chem. Appl.* **2023**, *2023*, 4166128. [CrossRef] [PubMed]
15. Wu, A.Y.-J.; Chang, H.; Wang, N.-Y.; Sun, F.-J.; Liu, C.-P. Clinical and molecular characteristics and risk factors for patients acquiring carbapenemase-producing and non-carbapenemase-producing carbapenem-nonsusceptible-Enterobacterales bacteremia. *J. Microbiol. Immunol. Infect.* **2022**, *55*, 1229–1238. [CrossRef]
16. Borra, R.C.; Lotufo, M.A.; Gagioti, S.M.; Barros, F.d.M.; Andrade, P.M. A simple method to measure cell viability in proliferation and cytotoxicity assays. *Braz. Oral Res.* **2009**, *23*, 255–262. [CrossRef]
17. How, S.W.; Low, D.Y.S.; Leo, B.F.; Manickam, S.; Goh, B.H.; Tang, S.Y. A critical review on the current state of antimicrobial glove technologies: Advances, challenges and future prospects. *J. Hosp. Infect.* **2023**, *137*, 24–34. [CrossRef]
18. Terpilowska, S.; Gluszek, S.; Czerwosz, E.; Wronka, H.; Firek, P.; Szmidt, J.; Suchanska, M.; Keczkowska, J.; Kaczmarska, B.; Kozlowski, M. Nano-Ag Particles Embedded in C-Matrix: Preparation, Properties and Application in Cell Metabolism. *Materials* **2022**, *15*, 5826. [CrossRef]
19. Wu, B.; Wang, Y.; Lee, Y.-H.; Horst, A.; Wang, Z.; Chen, D.-R.; Sureshkumar, R.; Tang, Y.J. Comparative Eco-Toxicities of Nano-ZnO Particles under Aquatic and Aerosol Exposure Modes. *Environ. Sci. Technol.* **2010**, *44*, 1484–1489. [CrossRef]
20. Vijayaraghavan, K.; Ashokkumar, T. Plant-mediated biosynthesis of metallic nanoparticles: A review of literature, factors affecting synthesis, characterization techniques and applications. *J. Environ. Chem. Eng.* **2017**, *5*, 4866–4883. [CrossRef]
21. Shankar, S.S.; Ahmad, A.; Sastry, M. Geranium leaf assisted biosynthesis of silver nanoparticles. *Biotechnol. Prog.* **2003**, *19*, 1627–1631. [CrossRef]
22. Sangeetha, G.; Rajeshwari, S.; Venckatesh, R. Green synthesis of zinc oxide nanoparticles by aloe barbadensis miller leaf extract: Structure and optical properties. *Mater. Res. Bull.* **2011**, *46*, 2560–2566. [CrossRef]
23. Rajiv, P.; Rajeshwari, S.; Venckatesh, R. Bio-Fabrication of zinc oxide nanoparticles using leaf extract of Parthenium hysterophorus L. and its size-dependent antifungal activity against plant fungal pathogens. *Spectrochim. Acta Part A Mol. Biomol. Spectrosc.* **2013**, *112*, 384–387. [CrossRef] [PubMed]
24. Aggarwal, N.; Mehta, P.; Bhambra, J.; Sharma, A.; Ali, D.; Sharma, N.K.; Sharma, S.; Kumar, G.; Kumar, N. Biosynthesis of Ocimum basilicum leaf extract mediated zinc oxide nanoparticles and their cytotoxic activity. *Biomass Convers. Biorefin.* **2023**, *13*, 8373–8380. [CrossRef]
25. Corsino, D.C.; Balela, M.D.L. Room temperature sintering of printer silver nanoparticle conductive ink. *IOP Conf. Ser. Mater. Sci. Eng.* **2017**, *264*, 012020. [CrossRef]
26. Xue, M.; Chen, C.; Tan, Y.; Ren, Z.; Li, B.; Zhang, C. Mangosteen peel-derived porous carbon: Synthesis and its application in the sulfur cathode for lithium sulfur battery. *J. Mater. Sci.* **2018**, *53*, 11062–11077. [CrossRef]
27. Anitha, D.; Ramadevi, A.; Seetharaman, R. Biosorptive removal of Nickel(II) from aqueous solution by Mangosteen shell activated carbon. *Mater. Today Proc.* **2021**, *45*, 718–722. [CrossRef]
28. Gondokesumo, M.E.; Pardjianto, B.; Sumitro, S.B.; Widowati, W.; Dimyati, A. Microstructural characterization of the garcinia mangostana fruit at different maturity level. *J. Nat. Remedies* **2018**, *18*, 63–70.
29. Talam, S.; Karumuri, S.R.; Gunnam, N. Synthesis, characterization, and spectroscopic properties of ZnO nanoparticles. *Int. Sch. Res. Not.* **2012**, *2012*, 372505. [CrossRef]
30. Vongsetskul, T.; Wongsomboon, P.; Sunintaboon, P.; Tantimavanich, S.; Tangboriboonrat, P. Antimicrobial nitrile gloves coated by electrospun trimethylated chitosan-loaded polyvinyl alcohol ultrafine fibers. *Polym. Bull.* **2015**, *72*, 2285–2296. [CrossRef]
31. Nematollahi, M.; Jalali-Arani, A.; Golzar, K. Organoclay maleated natural rubber nanocomposite. Prediction of abrasion and mechanical properties by artificial neural network and adaptive neuro-fuzzy inference. *Appl. Clay Sci.* **2014**, *97–98*, 187–199. [CrossRef]
32. Kropidłowska, P.; Jurczyk-Kowalska, M.; Irzmańska, E.; Płociński, T.; Laskowski, R. Effects of Composite Coatings Functionalized with Material Additives Applied on Textile Materials for Cut Resistant Protective Gloves. *Materials* **2021**, *14*, 6876. [CrossRef] [PubMed]
33. Ikram, A.; Ma Azam, M.; Amir Hashim, M.; Fauzi, M.; Shamsul Bahri, A.; Kamaruzaman, S. Effect of Antioxidants and Latex Vulcanising Agents on the Environmental Degradation of Latex Films. *J. Rubber Res.* **2005**, *8*, 220–240.
34. Ogunsona, E.; Hojabr, S.; Berry, R.; Mekonnen, T.H. Nanocellulose-triggered structural and property changes of acrylonitrile-butadiene rubber films. *Int. J. Biol. Macromol.* **2020**, *164*, 2038–2050. [CrossRef] [PubMed]
35. Silva, F.L.; Correia, C.A.; Oliveira, L.M.; Ribeiro, H.; Valera, T.S. Nano cellulose-crystals: Isolation and their promising application as reinforcement in vulcanized natural rubber compound. *Ind. Crops Prod.* **2024**, *222*, 120023. [CrossRef]
36. Burns, D. Grip Characterization of Protective Gloves. Ph.D. Thesis, University of Nebraska Medical Center, Omaha, NE, USA, 2022.
37. Aly, N.M.; Hamouda, T. Stab, spike and knife resistant textiles. In *Advances in Healthcare and Protective Textiles*; Elsevier: Amsterdam, The Netherlands, 2023; pp. 355–385.

38. Zhu, J.; Liu, Q.; Wang, Y.; Zhu, K.; Guo, J.; Jin, Y.; Liu, Y. Mangosteen extract reduces the bacterial load of eggshell and improves egg quality. *Heliyon* **2024**, *10*, e35857. [CrossRef]
39. Felix, L.; Mishra, B.; Khader, R.; Ganesan, N.; Mylonakis, E. In Vitro and In Vivo Bactericidal and Antibiofilm Efficacy of Alpha Mangostin Against Staphylococcus aureus Persister Cells. *Front. Cell. Infect. Microbiol.* **2022**, *12*, 898794. [CrossRef]
40. Song, M.; Liu, Y.; Li, T.; Liu, X.; Hao, Z.; Ding, S.; Panichayupakaranant, P.; Zhu, K.; Shen, J. Plant Natural Flavonoids Against Multidrug Resistant Pathogens. *Adv. Sci.* **2021**, *8*, 2100749. [CrossRef]
41. Yuvanatemiya, V.; Srean, P.; Klangbud, W.K.; Venkatachalam, K.; Wongsa, J.; Parametthanuwat, T.; Charoenphun, N. A Review of the Influence of Various Extraction Techniques and the Biological Effects of the Xanthones from Mangosteen (*Garcinia mangostana* L.) Pericarps. *Molecules* **2022**, *27*, 8775. [CrossRef]

polymers

MDPI

Article

Influence of Kazakhstan's Shungites on the Physical–Mechanical Properties of Nitrile Butadiene Rubber Composites

Kanat Beknazarov [1], Rustam Tokpayev [1], Abdirakym Nakyp [1,2,*], Yulia Karaseva [2], Elena Cherezova [2], Miroslawa El Fray [3], Svetoslav Volfson [2] and Mikhail Nauryzbayev [1]

1 Center of Physical Chemical Methods of Research and Analysis, Faculty of Chemistry and Chemical Technology, Al-Farabi Kazakh National University, Al-Farabi Ave. 71, Almaty 050038, Kazakhstan; kanat.beknazarov@kaznu.edu.kz (K.B.); rustamtokpaev@mail.ru (R.T.); nauryzbaev@cfhma.kz (M.N.)

2 Institute of Polymers, Kazan National Research Technological University, 68 K. Marx Str., 420015 Kazan, Russia; karaseva_j@mail.ru (Y.K.); cherezova59@mail.ru (E.C.); svolfson@kstu.ru (S.V.)

3 Department of Polymer and Biomaterials Science, Faculty of Chemical Technology and Engineering, West Pomeranian University of Technology in Szczecin, Al. Piastów 45, 70-311 Szczecin, Poland; mirfray@zut.edu.pl

* Correspondence: abdirakym1994@mail.ru; Tel.: +7-778-666-48-00

Abstract: This study presents data on the use of shungite ore (the Bakyrchik deposit, Kazakhstan) and its concentrate as fillers in elastomer composites based on nitrile butadiene rubber. In addition to carbon, these shungite materials contain oxides of Si, Fe, K, Ca, Ti, Mn, and Al. The shungite concentrate was obtained through a flotation process involving five stages. The chemical composition analysis of these natural fillers revealed that during flotation, the carbon content increased 3.5 times (from 11.0 wt% to 39.0 wt%), while the silicon oxide content decreased threefold (from 49.4 wt% to 13.6 wt%). The contents of oxides of K, Ca, Ti, Mn, and Al decreased by less than 1%, and iron oxide content increased by 40% (from 6.7 wt% to 9.4 wt%). The study explored the impact of partial or full replacement of carbon black (CB) of P 324 grade with the shungite ore (ShO) and the shungite concentrate (ShC) on the vulcanization process and the physical–mechanical properties of the rubber. It was found that replacing CB with ShO and ShC reduces Mooney viscosity ML (1 + 4) 100 °C of the rubber compounds by up to 29% compared to the standard CB-filled sample. The use of the shungite fillers also increased scorch time (t_s) by up to 36% and cure time (t_{90}) by up to 35%. The carbon content in the shungite fillers had little influence on these parameters. Furthermore, it was demonstrated that replacing 5–10 wt% of CB with ShO or ShC improves the tensile strength of the rubber. The results of the flotation enrichment process enable the assessment of how these shungite fillers affect the properties of the composites for producing rubbers with specific characteristics. It was also found that substituting CB with ShO or ShC does not significantly affect the rubber's resistance to standard oil-based media. The findings indicate that Kazakhstan's shungite materials can be used as fillers in rubber to partially replace CB.

Keywords: shungite ore; flotation enrichment; shungite concentrate; carbon black; acrylonitrile butadiene rubber; vulcanization; physical–mechanical properties

1. Introduction

Shungite rocks can serve as a source for obtaining carbon-containing materials. Shungite is found in Russia, Ukraine, India, the Congo, the USA, Austria, and Kazakhstan [1].

Shungite rocks are a unique natural composite in composition and structure, composed of amorphous carbon and highly dispersed crystalline silicate material. Based on the composition of the mineral matrix, shungites are classified into siliceous, aluminosilicate, and carbonate varieties [2]. Shungite carbon is uniformly distributed within a silicate framework of fine-dispersed quartz crystals [3,4].

Shungite rocks from the deposits in Eastern Kazakhstan (the Bakyrchik deposit), like those from the Zazhoginsky deposit in Karelia (Russia), are classified as anthraxolites (Figures 1 and 2) [5,6].

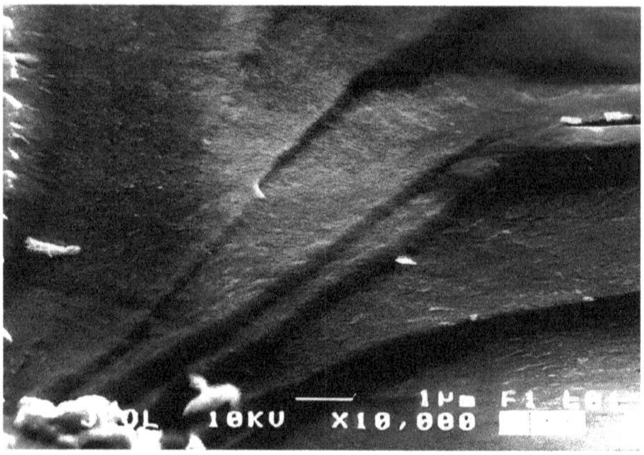

Figure 1. SEM images of the stepped micromorphology of fractures in high-grade anthraxolites from Kazakhstan.

Figure 2. High-resolution SEM images of the globular structure of Kazakhstan's anthraxolite.

Anthraxolites are a natural system of solid bitumens, whose chemical characteristics vary widely depending on the mineral phase and the degree of metamorphism. Anthraxolite contains up to 30 wt% of an allotropic form of metastable carbon in the pre-graphite stage of carbonification, with fullerene-like regular structures comprising up to 0.001 wt% [7]. Kazakhstan's shungite (the Bakyrchik deposit, 19.1 wt% of carbon) exhibits a turbostratic structure, similar to carbon black (CB) crystallites but with more disordered carbon. It is characterized by an interlayer spacing of approximately 0.35 nm, akin to shungite from the Zazhoginsky deposit in Russia [8–10].

Shungite rocks are of interest as sorbent materials for the purification of wastewater from petroleum products and phenols. Shungite possesses a broad spectrum of bactericidal properties; it exhibits adsorptive activity toward certain bacterial cells, phages, and pathogenic saprophytes [11].

Shungite from the Bakyrchik deposit is known to be used as a sorbent for oil extraction and determination of synthetic dyes in liquids [12–17], as well as electrode materials [18] and carriers of carbon–metal catalysts [19].

Crystals of crushed, finely ground shungite exhibit pronounced bipolar properties. As a result, shungite has a high level of adhesion and the ability to mix with virtually all organic and inorganic substances. This property allows shungite to be used as a filler for polymer and other composite materials [20–22].

In the work [23], elastomer compositions based on acrylonitrile butadiene rubber of BNR-40 grade and polybutadiene rubber BR with the addition of shungite (carbon content 28–32 wt%) were described. The authors [24–27] found that the addition of ground shungite filler (carbon content 30–33 wt%) gives elastomer composites increased strength properties and frost resistance.

The addition of shungite fillers in the manufacture of latex gloves makes it possible to obtain products with properties characteristic of gloves manufactured by standard technology [28]. In the study [29], shungite (carbon content 20–30 wt%) with a specific surface area of 20 m^2/g was used in rubber compounds. It has been shown that the introduction of shungite into rubber compounds reduces the rubber content in the bitumen-rubber binder system without a loss of quality [30].

A streamlined method to covalently bond hydroxylated carbon nanotubes (CNOH) within a polyphenol matrix, achieved through a direct, solvent-free process, was presented in paper [31]. This allows achieving a significant improvement in the mechanical properties of composites using low concentrations of CNOH along with topologically contrasting linkers.

In [32], the authors investigated the effects of interactions between nitrile butadiene rubber (NBR) and silica on the development of agglomerates by silica particles and of bound rubber in NBR/silica composites. The study showed that hydrogen bonding between nitrile and silanol groups suppresses the development of agglomerates by silica particles and affects the mechanical properties of NBR/silica composites by reducing agglomeration and improving strength properties.

The authors of works [33,34] used shungite to produce oil–petrol- and seawater-resistant rubbers. Studies [35–39] have shown that nanostructured shungite materials can increase the shielding efficiency of electromagnetic shielding composites due to their high absorption coefficient, as well as their ability to interact with the thermoplastic matrix, improving their thermophysical and electrophysical properties. The proposed composites have high chemical and thermal resistance.

Shungite rocks remain in the dumps of several deposits after the mining and enrichment of polymetallic ores. These shungite rocks are subjected to carbon enrichment and activation (thermal, mechanochemical) to increase the specific surface area and pore volume [40,41]. The concentrates can then be used as sorbents and fillers.

In particular, the concentrate of Kazakhstan's shungite (carbon content 45 wt%), obtained through flotation enrichment for carbon, is used as a sorbent for the capture of organic compounds and CO_2, demonstrating high speed and good selectivity [42,43].

Carbonate shungite of the "Taurite" grade ("Koksu" mining company" LLP, Kazakhstan) was used as a filler for bitumen after mechanochemical activation [44]. A "green" method for obtaining graphene from shungite is also known [45]. The synthesis of carbon materials from wastes of various industries fits into the paradigm of promising models of "clean energy" development.

After the mining of polymetallic ores at the Bakyrchik gold-sulfide deposit, three types of shungite-bearing rocks remain in the dumps. The main type is characterized as medium-carbon shungite (5–25% grade).

This work describes the method of obtaining shungite concentrate (ShC) by flotation enrichment to increase the carbon content from shungite ore (ShO) of the Bakyrchik deposit. The purpose of this work is to study the possibility of using ShO and ShC as a filler for rubber as an alternative to carbon black.

From the literature data presented, it follows that natural amorphous-crystalline carbon-silicate composites of various deposits (shungite ores), containing a maximum of 28 wt% of carbon, are successfully used as fillers for polymer materials. At the same time, CB is partially replaced by shungite. Shungite increases the plasticity and operational properties of rubbers and improves the technological characteristics of rubber compounds.

Replacing CB with shungite in rubber compositions is economically feasible because shungite has a lower cost compared to CB, allowing for a reduction in the total cost of obtaining polymer composites. In addition, the use of shungite improves the sanitary and hygienic conditions at the factories, as shungite is less "dusty" compared to CB.

The novelty of this work is that, for the first time, the shungite ore from the Bakyrchik deposit in the Abay Region, which is related to anthraxolites, and its concentrate enriched with carbon obtained by the flotation method, were investigated as fillers replacing carbon black in elastomer composites based on nitrile butadiene rubber. In this case, it was believed that increasing the carbon content of the shungite filler would improve the physical–mechanical and operational properties of rubber.

2. Materials and Methods

2.1. Materials

Nitrile butadiene rubber of BNKS-28 AMN grade (technical specifications TU 38.30313-2006) (2nd group, Krasnoyarsk Synthetic Rubber Plant JSC, Krasnoyarsk, Russia) was used in this work. Technical sulfur (1st class, 9995 grade, SERA CJSC, Orenberg, Russia) was used as a vulcanizing agent. The vulcanization accelerator 2-mercaptobenzothiazole (MBTS) (1st grade, Volzhsky Orgsintez JSC, Volgograd, Russia) was used. The vulcanization activator zinc oxide (ZnO) (A grade, Empils-Zinc LLC, Rostov-on-Don, Russia) and stearic acid (T-32 grade, Nefis Cosmetics JSC, Kazan, Russia) were used. The fillers included CB of P 324 grade (Sterlitamak Petrochemical Plant JSC, Sterlitamak, Russia), ShO (the Bakyrchik deposit, Almaty, Kazakhstan), and ShC obtained by flotation enrichment from ShO. Carbon Tetrachloride (CCl_4) was used as the solvent (State Standard GOST 20288-74, EKOS-1 JSC, Moscow, Russia). Kerosene TS-1 and pine oil Flotol 5219 (Clariant, Muttenz, Switzerland) were used as flotation reagents. The petroleum liquid of SZhR-1 grade was used as the standard hydrocarbon media in determining the properties of rubbers and rubber-technical products (technical specifications TU 38 10195-86, Gazpromneft-Omsk Refinery JSC, Omsk, Russia).

2.2. Preparation of Shungite Fillers

The preliminary preparation of shungite fillers was carried out on a grinding complex (GC) by grinding shungite ore to powder with particle sizes of 1–20 μm.

Flotation enrichment of the shungite ore was carried out in a laboratory flotation unit FML-3 (NPK Mechanobr-Technika, St. Petersburg, Russia) with a chamber volume of 3.0 L in a reagent mode at a pulp temperature of 15 ± 1 °C, with a solid-to-liquid ratio of 1:6. During flotation enrichment, reagents that selectively interact with the minerals constituting shungite are used. In the work, we used kerosene TS-1—1134 g/t, and pine oil Flotol 5219—1260 g/t. The duration of the main and control flotation was 10 min. Additional stages of recleaning flotation were conducted in three phases to increase the carbon content in the concentrate. The enrichment scheme is shown in Figure 3.

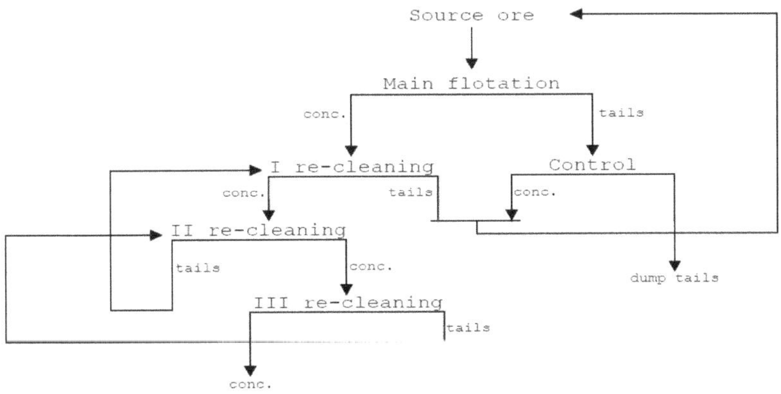

Figure 3. Technological scheme of flotation enrichment of the shungite ore from the Bakyrchik deposit.

2.3. *Preparation of Rubber Compounds*

The ingredients of the rubber compound were mixed in a closed laboratory Plasti-corder internal mixer Lab. Station Plasti-Corder, W50 (Brabender, Duisburg, Germany) at a temperature of 60 °C and a rotor rotation speed of 60 rpm according to the standard formulation for nitrile butadiene rubber according to State Standard GOST 34754-2021* (* State Standard) (Table 1). The type and concentration of ingredients in the formulation are specified in the standard. The formulation chosen for the analysis was standard, not industrial, because industrial formulations usually contain a larger number of ingredients, which can make it difficult to assess the impact of the tested fillers on the properties of rubbers.

Table 1. Formulation of the rubber compound.

No	Ingredients	Content, phr **
1	BNKS-28 AMN	100.0
2	Sulfur	1.5
3	ZnO	5.0
4	Stearic acid	1.5
5	MBTS	1.5
6	Filler	45.0

** Parts per hundred of rubber.

The ratios of the fillers are shown in Table 2.

Table 2. The ratio of the fillers in the formulation of the rubber compound.

No	Quantity, wt%		
	CB P 324	ShO	ShC
1 (control)	100	0	0
2	95	5	0
3	90	10	0
4	85	15	0

Table 2. *Cont.*

No	Quantity, wt%		
	CB P 324	**ShO**	**ShC**
5	80	20	0
6	75	25	0
7	50	50	0
8	25	75	0
9	0	100	0
10	95	0	5
11	90	0	10
12	85	0	15
13	80	0	20
14	75	0	25
15	50	0	50
16	25	0	75
17	0	0	100

2.4. Research Methods

2.4.1. Determination of the Chemical Composition of Shungite Materials

The chemical composition of the shungite fillers was determined using an electron probe microanalyzer Superprobe-733 (JEOL, Tokyo, Japan) and an energy-dispersive spectrometer INCA Energy (Oxford Instruments, Oxfordshire, UK) at an accelerating voltage of 25 kV and a probe current of 25 nA.

2.4.2. Determination of the Specific Surface Area of Fillers

The specific surface area of the fillers was determined by the method of thermodesorption of the gas adsorbate according to the BET theory using a "Sorbtemeter" (Katakon CJSC, Cheboksary, Russia) under the influence of liquid nitrogen on the test material.

2.4.3. Determination of the Surface Morphology of Fillers

The surface morphology of ShO and ShC was studied with a scanning electron microscope Quanta 3D 200i Dual system (FEI, Hillsboro, OR, USA) at the National Nanotechnology Laboratory of Al-Farabi Kazakh National University.

2.4.4. Determination of Mooney Viscosity

The Mooney viscosity of the rubber compounds was determined on a Mooney Viscozimeter-UGT7080S2 (Gotech, Taichung City, Taiwan) at a temperature of 100 °C (material preheating period 1 min, the rotor rotation duration 4 min).

2.4.5. Study of the Cure Characteristics

The rheometric characteristics of the rubber compounds were determined using an MDR (Moving Die Rheometer) MD-3000A (Gotech, Taichung City, Taiwan) in accordance with State Standard GOST 12535-84 at a chamber temperature of 150 °C for 30 min.

2.4.6. Vulcanization

The vulcanization process was carried out in a laboratory vulcanization press with induction heating of plates 100-400-2E (Krasin Plant CJSC, Kirov, Russia) at a temperature of 150 °C according to State Standard GOST 269-66.

2.4.7. Determination of the Physical–Mechanical Properties of Vulcanizates

The physical–mechanical properties of the rubber samples were evaluated according to State Standard GOST 270-75 by tensile strength, elongation at break, and residual elongation, which were determined on the universal electromechanical testing machine TRM-P 50 C1 Tochline (NPO Tochpribor LLC, Rostov-on-Don, Russia). Shore A hardness of the samples was measured on a portable hardness tester TH-200 (Time Group, Beijing, China) according to State Standard GOST 263-75. Rebound resilience was determined on a Shoba-type rebound resilience tester UMR-1 (Polymermash Group LLC, St. Petersburg, Russia) according to State Standard GOST 27110-86.

The oil and petroleum resistance of the rubber samples was determined by the change in mass of samples immersed in standard petroleum liquid of SZhR-1 grade for 72 h at 23 °C according to the formula:

$$\alpha = \frac{m_1 - m_0}{m_0} \times 100\%,$$

where, m_1—mass of the swollen sample, m_0—mass of the original sample.

2.4.8. Determination of Crosslink Density

For the testing, six specimens from each rubber sample of different shapes (square (10 mm × 10 mm × 2 mm), triangle (10 mm × 10 mm × 10 mm × 2 mm), circle (diameter 10 mm, thickness 2 mm)) were prepared. The samples were weighed and placed in a glass vessel (separately for each vulcanizate) and immersed in tetrachloromethane (CCl_4). The experiments were conducted at room temperature for 7 days. After 7 days, the samples were removed, dried with filter paper, and weighed. The number-average molecular mass of the polymer chain segments between adjacent nodes of the vulcanizate network was determined using Equation (1) based on the Flory–Rehner theory [46].

$$M_c = -\frac{\rho_1}{\rho_2} \cdot \frac{M_2\left(\varphi^{\frac{1}{3}} - \frac{2}{f \cdot \varphi_1}\right)}{\ln(1-\varphi) + \varphi + \chi \cdot \varphi^2} \quad (1)$$

where, ρ_1, ρ_2—density of nitrile butadiene rubber (0.96 g/cm^3) and density of the solvent CCl_4 (1.595 g/cm^3); χ—polymer-solvent interaction parameter (0.478); M_2—molecular weight of CCl_4 (153.8 g/mol); f—functionality of network nodes (4); φ—volume fraction of elastomer in the swollen sample.

The experimentally determined value of the volume fraction of elastomer, denoted as φ, is calculated from the mass of the original (m_0) and swollen (m_s) elastomer samples using Equation (2).

$$\varphi = \frac{1}{\left(1 + \left(\frac{m_s}{m_0} - 1\right) \cdot \frac{\rho_1}{\rho_2}\right)} \quad (2)$$

3. Results and Discussion

For fillers used in the rubber industry, important parameters include chemical composition, specific surface area, size of primary aggregates, and structure, which together determine the intensity of interaction between the filler and rubber macromolecules. On the basis of the above-mentioned, in the first stage of the work, the differences in the surface morphology of the studied shungite fillers and their chemical compositions were determined.

The surface morphology of ShO and ShC was studied by scanning electron microscopy (SEM) (Figures 4 and 5). A detailed study of the images shows an insignificant difference between the sizes of crystalline particles of the shungite fillers, which mainly have oblong shapes. This may influence the flow properties of the compound and the degree of interaction between the fillers and the rubber macromolecules.

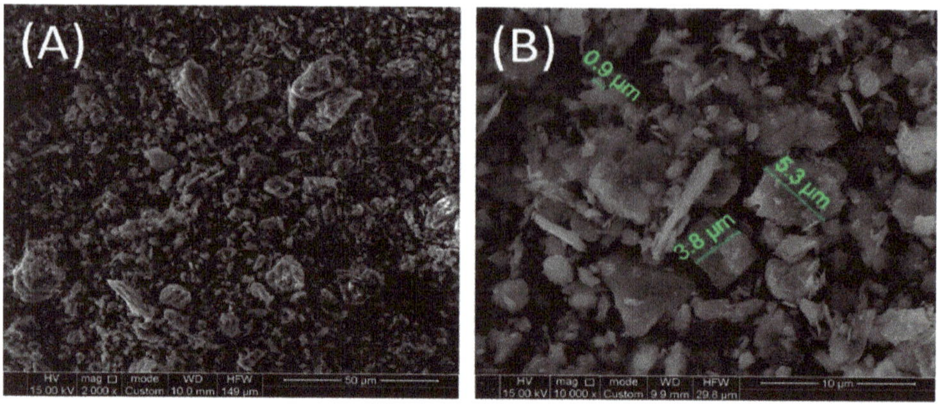

Figure 4. SEM images of ShO at different magnifications: (**A**) ×2000 and (**B**) ×10,000.

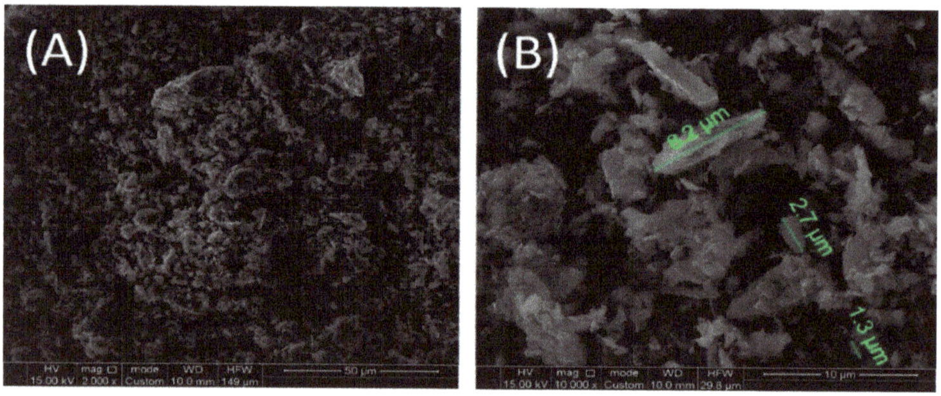

Figure 5. SEM images of ShC at different magnifications: (**A**) ×2000 and (**B**) ×10,000.

The most important factor determining the reinforcing effect of fillers is their dispersity, characterized by particle size or specific surface area. Fillers with particle sizes ranging from 5 to 1000 nm can be used to reinforce rubber, and fillers with particle sizes from 10 to 50 nm have the greatest reinforcing properties. This is confirmed by the results of the work [47,48].

The results of the X-ray spectral analysis of shungite fillers indicated (Table 3) that the main components of ShO and ShC are carbon, silicon dioxide, and aluminum oxide. During the flotation process for obtaining ShC, the carbon content increased more than 3.5 times (from 11 wt% to 39 wt%), while the proportion of silicon dioxide decreased more than 3 times (from 49.44 wt% to 13.64 wt%). The content of aluminum oxide practically did not change (21.13 wt% and 20.26 wt%, respectively). The proportion of iron oxide increased by 30% (from 6.75 wt% to 9.37 wt%) during the flotation of ShO. The amounts of K, Ca, Ti, and Mn oxides changed less than 1%.

The analysis of literature data on the influence of the chemical composition of mineral fillers on the physical–mechanical properties of rubbers suggests the following. Increasing the carbon content in ShC (39 wt%) enhances the mechanical properties of rubbers, such as compressive strength and elasticity. These parameters are of particular importance for rubbers that need to retain properties over a wide range of operating temperatures. The high content of SiO_2 in ShO (49.44 wt%) contributes to increased strength and wear resistance, which is important for rubber applications under friction and stress [49]. The reduction of SiO_2 in ShC (13.64 wt%) may limit some of the positive effects. Aluminum oxide (Al_2O_3) increases the heat resistance and tensile strength of rubbers [50,51]. The relatively stable content of Al_2O_3 in ShO and ShC (21.13 wt% and 20.26 wt%, respectively) allows us to expect good strength characteristics of rubbers during thermal aging. The oxides of several metals (Na_2O, CaO, MgO, K_2O) may act as vulcanization activators [52]. Each of these factors may ultimately affect the technological and operational properties of the resulting rubber products.

Table 3. Results of the X-ray spectral analysis of the shungite fillers.

Chemical Element, Component	Content, wt%	
	Shungite Ore	Shungite Concentrate
C	11.0 ± 0.5	39.0 ± 0.5
Na_2O	1.77 ± 0.1	1.39 ± 0.1
MgO	1.31 ± 0.1	1.57 ± 0.1
Al_2O_3	21.13 ± 0.1	20.26 ± 0.1
SiO_2	49.44 ± 0.1	13.64 ± 0.1
P_2O_5	0.22 ± 0.01	<0.10 ± 0.01
K_2O	3.42 ± 0.1	3.50 ± 0.1
CaO	2.47 ± 0.1	1.16 ± 0.1
TiO_2	3.06 ± 0.1	2.17 ± 0.1
MnO	0.20 ± 0.01	<0.10 ± 0.01
Fe_2O_3	6.75 ± 0.1	9.37
Specific surface area, m^2/g	9	17

During the enrichment of ShO, an increase in the specific surface area of the mineral material was observed. The specific surface area of ShO and ShC was 9 m^2/g and 17 m^2/g, respectively (Table 3). The specific surface area of shungite affects the vulcanization process in the following way: the developed surface of shungite leads to increased sorption of the rubber compound's ingredients, including the vulcanizing group. This prevents the formation of a sufficiently dense and strong vulcanization network and, consequently, reduces the strength characteristics of vulcanizates [49].

Viscosity is considered an important characteristic affecting the uniformity of ingredient distribution in a rubber compound. It was found that at partial or complete replacement of CB with ShO and ShC in the rubber compound, there is a decrease in Mooney viscosity ML (1 + 4) 100 °C in comparison with the standard sample filled only with CB (Figure 6).

When replacing less than 50 wt% of CB with the shungite fillers, no significant difference in viscosity between ShO and ShC was observed. However, when replacing more than 50 wt% of CB with the shungite fillers, the viscosity of the samples decreases to 61–70 Mooney units (Figure 6). At the same time, the viscosity of the rubber compound containing ShC is higher by 4–6 Mooney units. That is, the use of ShO and ShC leads to an increase in the plasticity of the rubber compounds, which in turn facilitates processing and allows for a reduction in the share of plasticizer. This is probably due to the presence of some fullerenes in the composition of the shungite fillers, which act as a kind of "lubricant" [53].

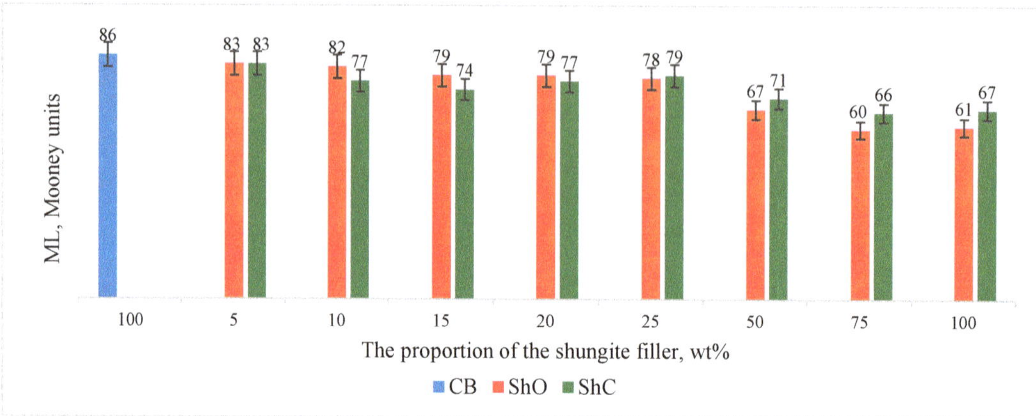

Figure 6. The influence of the proportion of the shungite filler (wt%) on Mooney viscosity of the rubber compounds.

The analysis of the curing characteristics of the rubber compounds showed that regardless of the differences in the chemical composition of the shungite fillers, the introduction of both ShO and ShC leads to an increase in the scorch (t_s) and cure (t_{90}) times (Table 4).

Table 4. Parameters of the rheometric curves of vulcanization.

Composition, No.	t_s, min	T_{min}, dN·m	T_{max}, dN·m	$t_{(90)}$, min
1	1.8	8	41	13.8
2	2.4	9	45	15.5
3	2.5	10	44	15.7
4	2.5	10	43	15.7
5	2.4	9	42	15.6
6	2.5	9	40	15.6
7	2.5	8	36	16.0
8	2.5	8	33	16.8
9	3.8	9	31	18.7
10	2.0	12	50	15.0
11	2.1	11	46	16.0
12	2.0	11	45	15.8
13	2.1	12	46	16.1
14	2.2	12	44	16.2
15	2.4	8	29	16.6
16	2.5	8	29	16.8
17	3.2	9	33	18.2

Apparently, the decrease in the vulcanization rate is due to the adsorption of the accelerator, activators, and other components of the vulcanizing group on the surface of the shungite fillers. This leads to the need to adjust the formulations to increase the content of the vulcanizing group.

In addition, the slowing of the vulcanization process may be due to the influence of metal ions contained in the shungite filler (Na^+, Mg^{+2}, Ca^{+2}, K^+). It is known that in a complex system consisting of rubber, sulfur, accelerator, and activator, coordination interactions occur, leading to the formation of intermediate components with various types of bonds. Coordination complexes are better dissolved in rubber than all initial components, which provides better distribution in the rubber matrix and more uniform vulcanization. This manifests the activation of the process. The structure of the complexes depends on the type of accelerator, but in general, it can be represented by the following Scheme 1 [54].

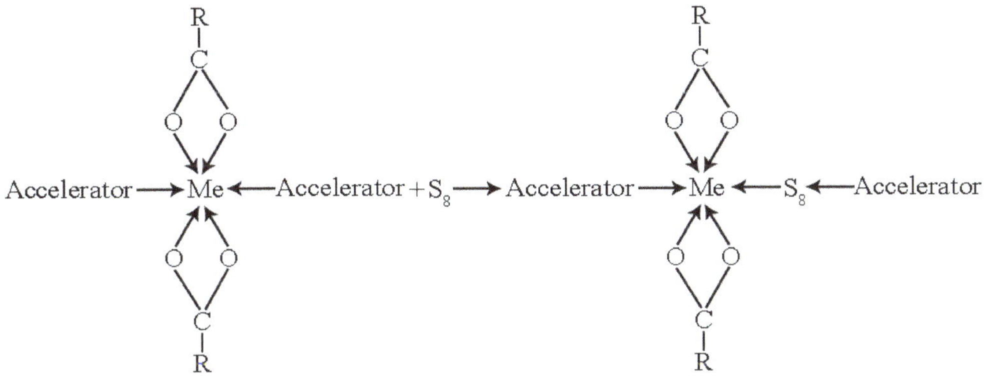

Scheme 1. Coordination complexes of the vulcanizing group (sulfur, accelerator, activator).

Metal ions present in the shungite filler can form less active complexes with the vulcanization accelerator that compete with zinc oxide.

The values of the maximum torque (T_{max}) of the samples containing ShC and ShO decrease with the increase in their proportion in the filler composition, which correlates with the obtained values of Mooney viscosity of the studied samples (Figure 6, Table 4). This result once again highlights the influence of the metal oxides present in the composition of the shungite fillers on the vulcanization process, as noted in several studies [55–58].

The results of determining the density of the network chains of the samples (Figure 7) containing the shungite fillers indicated, as expected from the results of rheometric tests, the formation of a more sparse three-dimensional network when replacing more than 5–10 wt% of CB with the shungite fillers. These data are consistent with the physical–mechanical properties of the rubbers presented in Table 5.

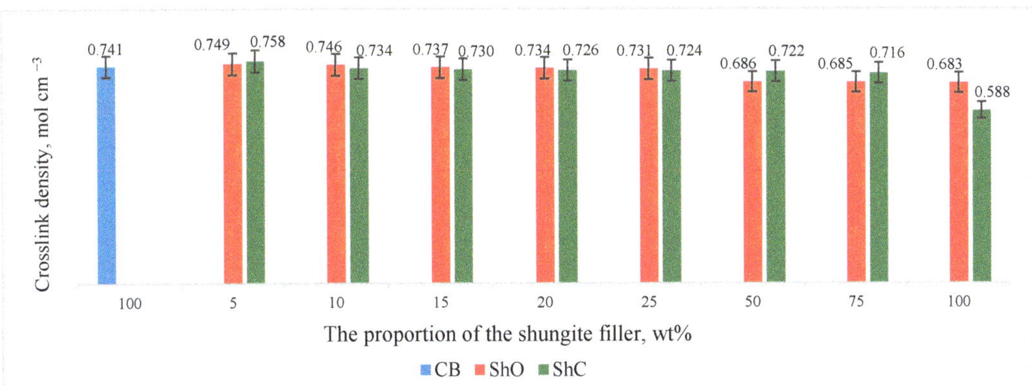

Figure 7. Influence of the proportion of the shungite filler on crosslink density of the vulcanizates.

Table 5. Results of the testing of vulcanizates.

No.	Composition Ratios (CB:ShO:ShC)	ε, %	ε′, %	Δm, %:
1	100:0:0	520	18	+0.1
2	95:5:0	580	18	+0.1
3	90:10:0	620	20	+0.1
4	85:15:0	650	20	+0.1
5	80:20:0	680	22	+0.1
6	75:25:0	680	23	+0.1
7	50:50:0	760	25	+0.3
8	25:75:0	870	37	+0.1
9	0:100:0	1010	49	+0.4
10	95:0:5	540	16	+0.1
11	90:0:10	580	19	+0.1
12	85:0:15	600	20	+0.1
13	80:0:20	620	22	+0.2
14	75:0:25	640	23	+0.2
15	50:0:50	660	27	+0.3
16	25:0:75	680	30	+0.4
17	0:0:100	700	32	+0.5

Note: ε, %—elongation at break; ε′, %—residual elongation; Δm, %—change in mass in the petroleum liquid of SZhR-1 grade.

The tensile strength of the samples in which 5–10 wt% of CB was replaced by ShC and ShO (samples 2–3, 10–11) remains at the level of the control sample (Figure 8). Further increase in the amount of the shungite fillers leads to a decrease in this parameter. The decrease in strength characteristics may be due to the fact that the shungite fillers, having their unique texture and structure, may not provide the same degree of compaction in the rubber composite as the traditional filler CB, which leads to a lower tensile strength of the rubber. Replacing 75 wt% of CB with ShO and ShC and completely replacing CB with ShO and ShC resulted in a dramatic decrease in tensile strength of the rubbers to 10.8 and 7.0 MPa for ShO, and to 11.6 and 7.9 MPa for ShC. The differences in tensile strength between the rubbers containing ShO and ShC indicate that the concentration of shungite ore to increase the carbon content positively affects the properties of the rubber. In general, the decrease in tensile strength of the samples with an increase in the proportion of the shungite filler by more than 10 wt% in the formulation is probably due to a decrease in the density of the vulcanization network chains of the samples, as well as a decrease in the total amount of the filler—carbon black.

It is logical that the elongation at break of the samples increases with increasing the content of the shungite fillers. Complete replacement of CB with ShO (sample 9, Table 5, Figure 9) leads to an increase in elongation at break by 94%, whereas complete replacement of CB with ShC (sample 17, Table 5, Figure 9) leads to an increase of 30% in comparison with the control sample. This correlates with the data on tensile strength and is associated with the change in the degree of crosslinking of the samples. Thus, the concentration of ShO leads to a lower dependence of elongation at break on the proportion of the shungite filler. The obtained data do not contradict the results of studies published by the authors [25,26].

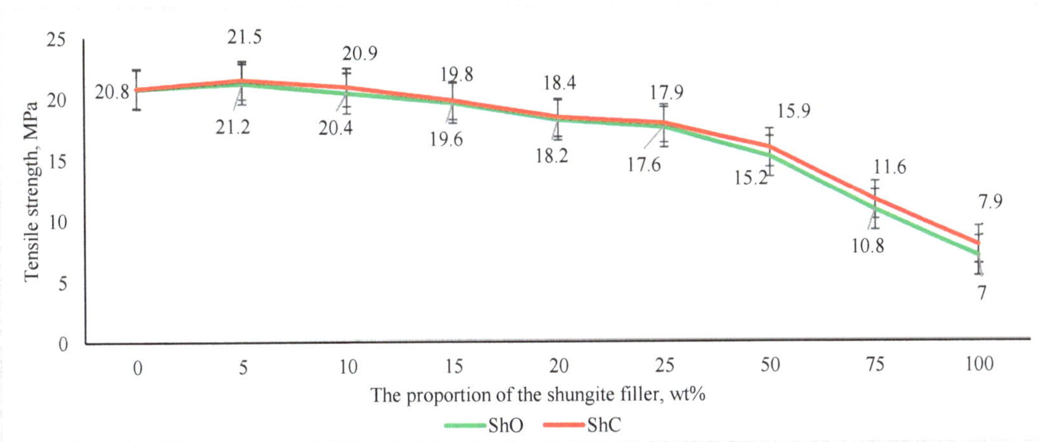

Figure 8. Influence of the proportion of the shungite filler on tensile strength.

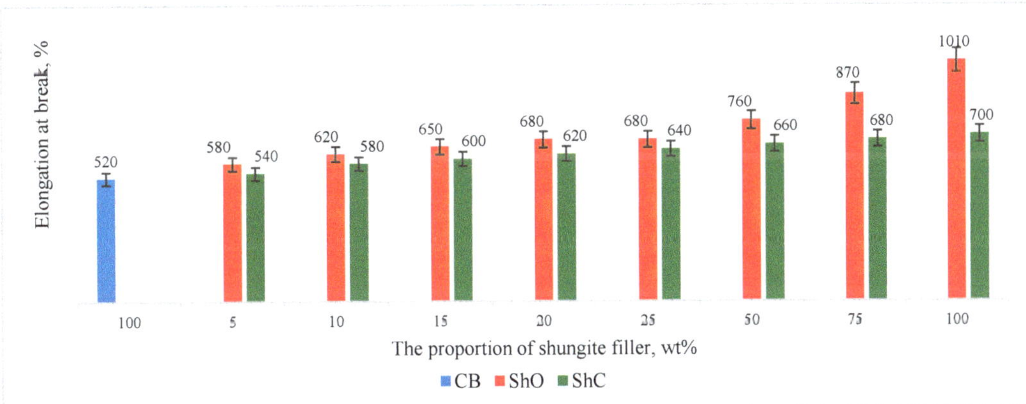

Figure 9. Influence of the proportion of the shungite filler on elongation at break.

The results of the determination of the residual elongation of the rubber samples are of particular interest. They show that with increasing the amount of the shungite fillers, the residual elongation of the samples also increases (Table 5). This is likely related to the influence of the chemical composition of ShO and ShC on the vulcanization network of the rubbers. It is important to note that the samples filled with ShO exhibit higher residual elongation compared to those filled with ShC (samples 2–9, Table 5). Consequently, rubbers filled with ShC possess better properties.

Upon partial or complete replacement of CB with ShO, a slight decrease in Shore A hardness (samples 2–9, Figure 10) is observed compared to the control sample filled with CB (sample 1, Table 5, Figure 10). As evident from the presented data, as the proportion of ShC in the filler increases from 5 wt% to 25 wt% (samples 5–14, Table 5, Figure 10), the Shore A hardness decreases linearly from 81 to 61 units. When the proportion of ShC is increased to 50 wt% or more, as well as in the case of complete replacement of CB with ShC (samples 16–17, Table 5, Figure 10), the hardness increases slightly. This can be explained by the fact that with an increase in the ShC content, the amount of silicon dioxide in the rubber matrix increases, which correlates with the data on the influence of SiO_2 on the hardness of rubbers [59].

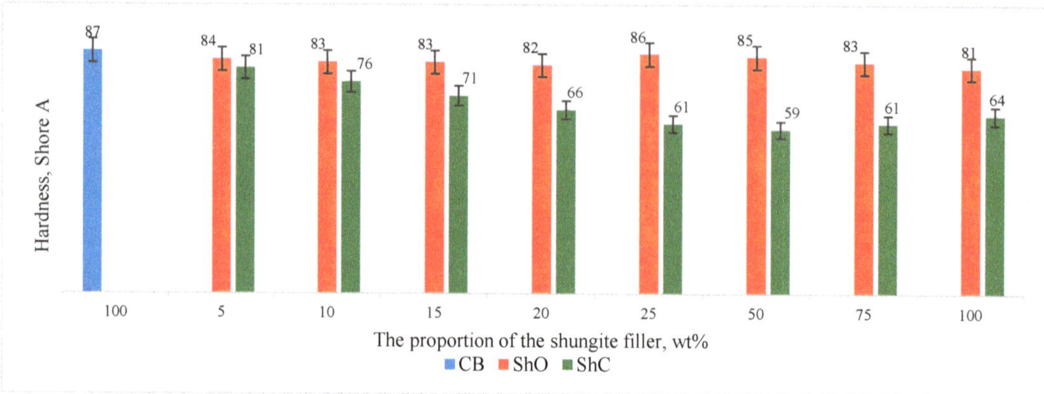

Figure 10. Influence of the proportion of the shungite filler on Shore A hardness.

According to the results of the study, the shungite fillers do not have a significant effect on rebound elasticity compared to the control sample filled with CB (Figure 11).

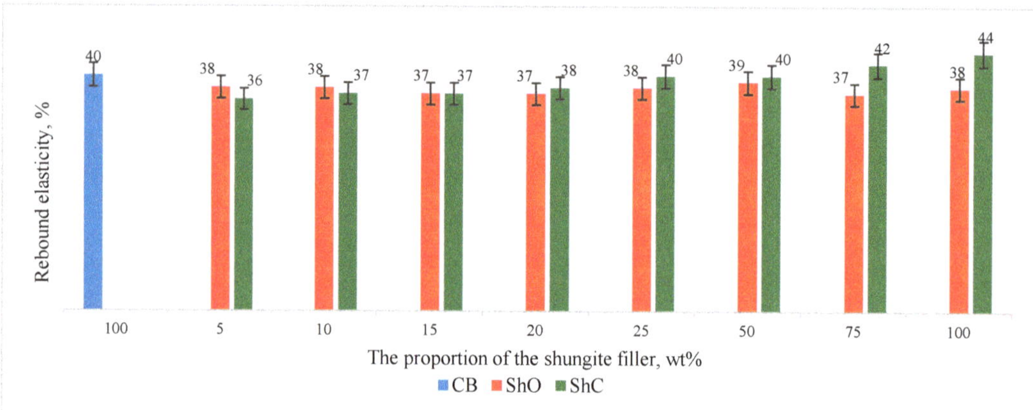

Figure 11. Influence of the proportion of the shungite filler on rebound elasticity.

When determining the resistance of rubbers to aggressive media, it was found that after exposure of the samples with different contents of ShO and ShC in the petroleum liquid of SZhR-1 grade for 72 h at 23 ± 1 °C, the change in mass does not exceed +0.5%. This allows us to assert that ShO and ShC do not impair the resistance of rubbers to SZhR-1. The results obtained from the testing of samples in which up to 10 wt% of CB was replaced with ShO and ShC correlate with the data from the study by Ushmarin et al. [35], in which Kazakhstan's shungite of the brand "Taurite TS-D" was used.

4. Conclusions

The influence of natural carbon-containing fillers—shungite ore (ShO) (the Bakyrchik deposit, Kazakhstan) and shungite concentrate (ShC), obtained through a five-stage flotation enrichment process—on the properties of elastomer composites based on nitrile butadiene rubber of BNKS-28 AMN grade has been investigated. It was found that replacing carbon black (CB) with ShO and ShC reduces Mooney viscosity ML (1 + 4) 100 °C by up to 29% compared to the standard sample filled only with CB. Additionally, the viscosity of the rubber compound containing ShC was 4–6 Mooney units higher. Rheometric analysis of the rubber compounds revealed that the use of the shungite fillers increased scorch time (t_s) by up to 36% and cure time (t_{90}) by up to 35%.

The main difference in the shungite fillers is that in the chemical composition of ShC, the carbon content is 3.5 times higher than in ShO (39 wt% and 11 wt%, respectively), but silicon oxide is 3 times lower (49.4 wt% and 13.6 wt%, respectively).

Complete replacement of CB with the shungite fillers reduces tensile strength of the rubbers by up to 66%. In the case of ShO, elongation at break increases by 94%, and in the case of ShC—by 30% compared to the control sample filled with CB. As the proportion of the shungite fillers increases, residual elongation of the samples rises, Shore A hardness decreases by up to 26%, and rebound elasticity remains at the control level. Complete replacement of CB with the shungite fillers results in a mass change of up to +0.5% after 72 h of exposure in the petroleum liquid of SZhR-1 grade at room temperature. The best results were obtained for the samples where 5–10 wt% of CB was replaced with the shungite fillers. At this ratio, it was possible to maintain tensile strength at the level of the control sample, and the change in mass of the samples after 72 h of exposure in SZhR-1 at 23 ± 2 °C was no more than +0.1%.

Author Contributions: Writing—original draft, visualization, formal analysis, methodology, resources, K.B.; conceptualization, formal analysis, writing—review, R.T.; formal analysis, methodology, conducting experiments, resources, writing—review and editing, A.N.; conceptualization, writing—review and editing, Y.K. and E.C.; conceptualization, writing—review, M.E.F.; writing—review and editing, S.V.; project administration, funding acquisition, M.N. All authors have read and agreed to the published version of the manuscript.

Funding: This research has been funded by the Science Committee of the Ministry of Science and Higher Education of the Republic of Kazakhstan (Grant No. AP14871403).

Institutional Review Board Statement: Not applicable.

Data Availability Statement: Data are contained within the article.

Conflicts of Interest: The authors declare no conflicts of interest.

References

1. Buryakov, N.P.; Sycheva, L.V.; Trukhachev, V.I.; Zaikina, A.S.; Buryakova, M.A.; Nikonov, I.N.; Petrov, A.S.; Kravchenko, A.V.; Fathala, M.M.; Medvedev, I.K.; et al. Role of Dietary Inclusion of Phytobiotics and Mineral Adsorbent Combination on Dairy Cows' Milk Production, Nutrient Digestibility, Nitrogen Utilization, and Biochemical Parameters. *Vet. Sci.* **2023**, *10*, 238. [CrossRef]
2. Medvedev, P.; Philippov, M.; Romashkin, A.; Vávra, N. Primary Organic Matter and Lithofacies of Siliceous Shungite Rocks from Karelia. 2001. Available online: https://repository.geologyscience.ru/handle/123456789/24983 (accessed on 30 October 2024).
3. Ignatov, I.; Mosin, O. Composition and structural properties of fullerene analogous mineral shungite. Mathematical model of interaction of shungite with water molecules. *Adv. Phys. Theor. Appl.* **2014**, *28*, 10–21.
4. Yushkin, N.P. Globular Supramolecular Structure of Shungite: Data from Scanning Tunneling Microscopy. *Reports USSR Acad. Sci.* **1994**, *337*, 800–803.
5. Golybev, E.A.; Glebashev, S.G.; Ignat'ev, S.V.; Filippov, V.N. Supramolecular Structure of Anthraxolite from the Bakyrchik Deposit, Eastern Kazakhstan. *Vestn. IG KNC UrO RAN [Bulletin IG KNC UrO RAS]* **2006**, *4*, 4–7.
6. Glebashev, S.V.; Ignatiev, S.V.K. Formation and Placement of Shungite Rocks of the Kyzylov Zone (Eastern Kazakhstan). *Sov. Geol. Soviet Geol.* **1989**, 33–42.
7. Volkova, I.B.; Bogdanova, M.V. Petrology and Genesis of Karelian Shungite—High Rank Coal. *Int. J. Coal Geol.* **1986**, *6*, 369–379. [CrossRef]
8. Kazankapova, M.; Bekzhanova, A.; Efremov, S.; Nauryzbayev, M. Study of Physical Chemical Characteristics of a Shungite. *Chem. Bull. Kazakh Natl. Univ.* **2012**, *66*, 149–157. [CrossRef]
9. Efremov, S.A.; Nechipurenko, S.V.; Kazankapova, M.K.; Washington, B.; Tassibekov, K.S.; Nauryzbaev, M.K. Physico-Chemical Characteristics of Shungite Rock of Kazakhstan. *Eurasian Chem. J.* **2013**, *15*, 241. [CrossRef]
10. Kazankapova, M.; Yefremov, S.; Nechipurenko, S.; Nauryzbayev, M. Study of Physical Chemical Characteristics of Shungite Rocks from Kazakhstan and Russia Using High-Resolution Physical Chemical Methods. *Chem. Bull. Kazakh Natl. Univ.* **2013**, *72*, 43–48. [CrossRef]
11. Kovalevski, V.V.; Reinikainen, S.P.; Reinikainen, V.; Rozhkova, V.S.; Sihvonen, T. Shungite rocks of varying genesis in innovative water treatment technologies. *Proc. Karelian Res. Cent. Russ. Acad. Sci.* **2020**, *6*, 97–105. [CrossRef]

12. Alham, A.; Ibraimov, A.; Alimzhanova, M.; Mamedova, M. Natural Material Shungite as Solid-Phase Extraction Sorbent for the Extraction of Red Synthetic Dye Ponceau 4R from Tap Water, Wine, and Juice. *Food Anal. Methods* **2022**, *15*, 707–716. [CrossRef]
13. Efremov, S.A. Shungite Rock Processing Technology. *Russ. J. Appl. Chem.* **2010**, *83*, 23–26. [CrossRef]
14. Yefremov, S.; Nechipurenko, S.; Atchabarova, A.; Kabulov, A.; Nauryzbayev, M. Obtaining Carbon-Mineral Sorbents Based on Shungite and Using Them to Clean Waste and Household Water. *Chem. Bull. Kazakh Natl. Univ.* **2011**, *64*, 85–88. [CrossRef]
15. Yefremov, S.A.; Kabulov, A.T.; Atchabarova, A.A.; Tokpayev, R.R.; Nechipurenko, S.V.; Nauryzbayev, M.K. Production of Shungite Concentrates—Multifunctional Fillers for Elastomers. In *XVIII International Coal Preparation Congress*; Springer International Publishing: Cham, Switzerland, 2016; pp. 1193–1196.
16. Cherezova, E.; Karaseva, Y.; Nakyp, A.; Nuriev, A.; Islambekuly, B.; Akylbekov, N. Influence of Partially Carboxylated Powdered Lignocellulose from Oat Straw on Technological and Strength Properties of Water-Swelling Rubber. *Polymers* **2024**, *16*, 282. [CrossRef] [PubMed]
17. Cherezova, E.N.; Galikhanov, M.F.; Karaseva, Y.S.; Nakyp, A.M. Impact of the Composition of Rubbers Filled with Carboxymethyl Cellulose on Their Properties. *Russ. J. Appl. Chem.* **2023**, *96*, 462–467. [CrossRef]
18. Atchabarova, A.A.; Tokpayev, R.R.; Kabulov, A.T.; Nechipurenko, S.V.; Nurmanova, R.A.; Yefremov, S.A.; Nauryzbayev, M.K. New Electrodes Prepared from Mineral and Plant Raw Materials of Kazakhstan. *Eurasian Chem. J.* **2016**, *18*, 141. [CrossRef]
19. Tokpayev, R.R.; Atchabarova, A.A.; Abdullayeva, S.A.; Nechipurenko, S.V.; Yefremov, S.A.; Nauryzbayev, M.K. New Supports for Carbon-Metal Catalytic Systems Based on Shungite and Carbonizates of Plant Raw Materials. *Eurasian Chem. J.* **2015**, *17*, 223. [CrossRef]
20. Rozhkova, N.N. Shungite—A Carbon-Mineral Filler for Polymeric Composite Materials. *Compos. Interfaces* **2001**, *8*, 307–312. [CrossRef]
21. Kovalevski, V.; Shchiptsov, V. Shungites and Their Industrial Potential. In *Proceedings of the 14th International Congress for Applied Mineralogy (ICAM2019), Belgorod, Russia, 23–27 September 2019*; Springer: Berlin/Heidelberg, Germany, 2019; pp. 201–204.
22. Melezhik, V.A.; Filippov, M.M.; Romashkin, A.E. A Giant Palaeoproterozoic Deposit of Shungite in NW Russia: Genesis and Practical Applications. *Ore Geol. Rev.* **2004**, *24*, 135–154. [CrossRef]
23. Solovieva, A.B.; Neschadina, L.E.; Rojkova, N.N.; Gorbatkina, J.A.; Kolbanev, I.V.; Wolfson, S.A. The Shungite Effect on the Physico-Mechanical Properties of Elastomers and Polypropylene. *Int. J. Polym. Mater.* **1997**, *36*, 11–17. [CrossRef]
24. Olewnik-Kruszkowska, E.; Adamczyk, A.; Gierszewska, M.; Grabska-Zielińska, S. Comparison of How Graphite and Shungite Affect Thermal, Mechanical, and Dielectric Properties of Dielectric Elastomer-Based Composites. *Energies* **2021**, *15*, 152. [CrossRef]
25. Kornev, Y.V.; Yanovskii, Y.G.; Boiko, O.V.; Semenov, N.A.; Chirkunova, S.V. Investigating the Influence of the Degree of Dispersion of Mineral Shungite on the Properties of Elastomeric Materials Based on Butadiene–Styrene Rubber. *Int. Polym. Sci. Technol.* **2013**, *40*, 27–32. [CrossRef]
26. Petrova, N.N.; Portnyagina, V.V.; Mukhin, V.V.; Timofeeva, E.N.; Matveeva, N.V.; Ivanova, S.F. Frost-Resistant Elastomeric Nanocomposites for Operation in the Far North Conditions. *Mater. Sci. Forum* **2019**, *945*, 412–416. [CrossRef]
27. Cherezova, E.N.; Karaseva, Y.S. Water-swelling rubbers filled with modified cotton powder. *ChemChemTech* **2022**, *65*, 71–78. [CrossRef]
28. Kravchenko, E.S.; Potapov, E.E.; Bobrov, A.P.; Smal, V.A.; Erokhin, A.I.; Plotnikova, M.F. A Study of the Possibility of Using Shungite in Latex Rubber Formulations for the Manufacture of Gloves with High Resistance to Aggressive Media. *Int. Polym. Sci. Technol.* **2016**, *43*, 21–24. [CrossRef]
29. Barashkova, I.I.; Komova, N.N.; Motyakin, M.V.; Potapov, E.E.; Wasserman, A.M. Shungite-Elastomer Interface Layers. *Dokl. Phys. Chem.* **2014**, *456*, 83–85. [CrossRef]
30. Yadykina, V.V.; Vyrodova, K.S.; Potapov, E.E. Study of the Impact of the Composite "Rubber:Shungite" on the Properties of Bituminous Rubber Binder. In *Innovations and Technologies in Construction*; Springer: Berlin/Heidelberg, Germany, 2021; pp. 302–308.
31. Kanbargi, N.; Damron, J.T.; Gao, Y.; Kearney, L.T.; Carrillo, J.M.; Keum, J.K.; Sumpter, B.G.; Naskar, A.K. Amplifying Nanoparticle Reinforcement through Low Volume Topologically Controlled Chemical Coupling. *ACS Macro Lett.* **2024**, *13*, 280–287. [CrossRef]
32. Suzuki, N.; Ito, M.; Ono, S. Effects of Rubber/Filler Interactions on the Structural Development and Mechanical Properties of NBR/Silica Composites. *J. Appl. Polym. Sci.* **2005**, *95*, 74–81. [CrossRef]
33. Egorov, E.N.; Kol'tsov, N.I.; Ushmarin, N.F.; Sandalov, S.I. Research of operational and dynamic properties of rubber for products working in sea water. *Izv. Vyss. Uchebnykh Zaved. Khimiya Khimicheskaya Tekhnologiya* **2020**, *63*, 96–102. [CrossRef]
34. Ushmarin, N.F.; Efimovskii, E.G.; Petrova, N.N.; Sandalov, S.I.; Kol'tsov, N.I. The influence of powder shungytes on properties of oil-and benzo resistant rubbers. *Izv. Vyss. Uchebnykh Zaved. Khimiya Khimicheskaya Tekhnologiya* **2018**, *62*, 54–60. [CrossRef]
35. Moshnikov, I.A.; Kovalevski, V.V. Composite Materials Based on Nanostructured Shungite Filler. *Mater. Today Proc.* **2018**, *5*, 25971–25975. [CrossRef]
36. Antonets, I.V.; Golubev, Y.A.; Shcheglov, V.I.; Sun, S. Electromagnetic Shielding Effectiveness of Lightweight and Flexible Ultrathin Shungite Plates. *Curr. Appl. Phys.* **2021**, *29*, 97–106. [CrossRef]
37. Danilova-Tret'yak, S.M.; Evseeva, L.E.; Kravtsevich, A.V.; Leshchenko, V.G.; Nikolaeva, K.V.; Rozhkova, N.N.; Shashura, L.I. Influence of the Concentration of the Filler and of the Type of Matrix on the Properties of Thermoplastics Filled with Shungite Nanoparticles. *J. Eng. Phys. Thermophys.* **2020**, *93*, 401–408. [CrossRef]

38. Podolsky, V.P.; Volkov, V.V.; Kukina, O.B.; Andreev, A.V. Substantiation of the possibility of using shungite as an effective radio-absorbing material. *Russ. J. Build. Constr. Archit.* **2022**, *3*, 74–80. [CrossRef]
39. Cherezova, E.N.; Karaseva, Y.S.; Nakyp, A.M. Evaluation of the Durability of Limited Swelling Rubber Filled with Modified Powdered Cellulose from Cotton Waste. *Polym. Sci. Ser. D* **2023**, *16*, 681–686. [CrossRef]
40. Kalugin, S.; Shildebaeva, Z.; Mametzhanova, M.; Yelibaeva, N.; Efremov, S.; Nechipurenko, S.; Omarova, D. Development of a Technology for Obtaining Flotation Reagent Oxane-3 for Carbon Mineral Raw Materials of Kazakhstan. *Chem. Bull. Kazakh Natl. Univ.* **2014**, *76*, 10–15. [CrossRef]
41. Kazankapova, M.K.; Nauryzbaev, M.K.; Efremov, S.A.; Ermagambet, B.T.; Nurgaliyev, N.U.; Nechipurenko, S.V. Preparation of Activated Shungite and Characterization of Its Chemical Composition and Adsorption Properties. *Solid Fuel Chem.* **2019**, *53*, 241–247. [CrossRef]
42. Kishibayev, K.K.; Voropaeva, N.L.; Kabulov, A.T.; Efremov, S.A.; Fernandez Sanjurjo, M.J.; Gur'yanov, V.V.; Nauryzbaev, M.K. Physicochemical Characteristics of Furfural-Based Activated Carbons and Application of These in Sorption of Cyclohexane. *Russ. J. Appl. Chem.* **2015**, *88*, 50–58. [CrossRef]
43. Kishibayev, K.K.; Serafin, J.; Tokpayev, R.R.; Khavaza, T.N.; Atchabarova, A.A.; Abduakhytova, D.A.; Ibraimov, Z.T.; Sreńscek-Nazzal, J. Physical and Chemical Properties of Activated Carbon Synthesized from Plant Wastes and Shungite for CO2 Capture. *J. Environ. Chem. Eng.* **2021**, *9*, 106798. [CrossRef]
44. Zhambolova, A.; Ongarbayev, Y.; Kenzhegaliyeva, A.; Abdikhan, D. Mechanochemically Activated Shungite as an Additive to Improve Bitumen Characteristics. *J. Ecol. Eng.* **2023**, *24*, 218–226. [CrossRef]
45. Novikova, A.; Karabchevsky, A. Green Extraction of Graphene from Natural Mineral Shungite. *Nanomaterials* **2022**, *12*, 4356. [CrossRef] [PubMed]
46. Flory, P.J.; Rehner, J. Statistical Mechanics of Cross-Linked Polymer Networks, I. Rubberlike Elasticity. *J. Chem. Phys.* **1943**, *11*, 512–520. [CrossRef]
47. Chuayjuljit, S.; Imvittaya, A.; Nuchanat, N.R.; Potiyaraj, P. Effects of Particle Size and Amount of Carbon Black and Calcium Carbonate on Curing Characteristics and Dynamic Mechanical Properties of Natural Rubber. *J. Met. Mater. Miner.* **2002**, *12*, 51–57.
48. Sirisinha, C.; Prayoonchatphan, N. Study of Carbon Black Distribution in BR/NBR Blends Based on Damping Properties: Influences of Carbon Black Particle Size, Filler, and Rubber Polarity. *J. Appl. Polym. Sci.* **2001**, *81*, 3198–3203. [CrossRef]
49. Salehi, M.M.; Khalkhali, T.; Davoodi, A.A. The Physical and Mechanical Properties and Cure Characteristics of NBR/Silica/MWCNT Hybrid Composites. *Polym. Sci. Ser. A* **2016**, *58*, 567–577. [CrossRef]
50. Xu, Y.; Gao, Q.; Ouyang, C.-F.; Guo, H.; Xiang, X. Study on the Properties of Nano-Al2O3/Phenyl Silicone Rubber Composites. In *Proceedings of the 2015 International Conference on Material Science and Applications (icmsa-15), Suzhou, China, 13–14 June 2015*; Atlantis Press: Amsterdam, The Netherlands, 2014.
51. Lin, J.; Su, S.; He, Y.; Kang, F. Improving Thermal and Mechanical Properties of the Alumina Filled Silicone Rubber Composite by Incorporating Carbon Nanotubes. *New Carbon Mater.* **2020**, *35*, 66–72. [CrossRef]
52. Heideman, G.; Noordermeer, J.W.M.; Datta, R.; van Baarle, B. Effect of Metal Oxides as Activator for Sulphur Vulcanisation in Various Rubbers. *Kautsch. Gummi Kunststoffe* **2005**, *58*, 30–42.
53. Rozhkova, N.N. Role of Fullerene-like Structures in the Reactivity of Shungite Carbon as Used in New Materials with Advanced Properties. In *Perspectives of Fullerene Nanotechnology*; Kluwer Academic Publishers: Dordrecht, The Netherlands, 2002; pp. 237–251.
54. Okhotina, N.A.; Kurbangaleeva, A.R.; Panfilova, O.A. *Syr'e i Materialy Dlya Rezinovoj Promyshlennosti (Raw and Other Materials for the Rubber Industry)*; Izd-vo Kazan. nats. isl. technol. un-ta; 2014.
55. Kabulov, A.T.; Beknazarov, K.I.; Tokpayev, R.R.; Atchabarova, A.A.; Bulybaev, M.E.; Khassanov, D.A.; Kubasheva, D.B.; Slyambayeva, A.K.; Kishibayev, K.K.; Nauryzbayev, M.K. Production of Dry Building Mixes Using Waste of the Mining and Metallurgical Complex of Kazakhstan. *Int. J. Biol. Chem.* **2022**, *15*, 83–90. [CrossRef]
56. Sombatsompop, N.; Thongsang, S.; Markpin, T.; Wimolmala, E. Fly Ash Particles and Precipitated Silica as Fillers in Rubbers. I. Untreated Fillers in Natural Rubber and Styrene–Butadiene Rubber Compounds. *J. Appl. Polym. Sci.* **2004**, *93*, 2119–2130. [CrossRef]
57. Sombatsompop, N.; Wimolmala, E.; Markpin, T. Fly-ash Particles and Precipitated Silica as Fillers in Rubbers. II. Effects of Silica Content and Si69-treatment in Natural Rubber/Styrene–Butadiene Rubber Vulcanizates. *J. Appl. Polym. Sci.* **2007**, *104*, 3396–3405. [CrossRef]
58. Sombatsompop, N.; Wimolmala, E.; Sirisinha, C. Fly Ash Particles and Precipitated Silica as Fillers in Rubbers. III. Cure Characteristics and Mechanical and Oil-resistance Properties of Acrylonitrile-butadiene Rubber. *J. Appl. Polym. Sci.* **2008**, *110*, 2877–2883. [CrossRef]
59. Ajinsamajan, A.; Rungvichaniwat, A.; Saetung, A. Utilization of Silicon Dioxide Powder from Industrial Wastes as Novel Filler in Rubber Isolator Application. *J. Met. Mater. Miner.* **2022**, *32*, 93–100. [CrossRef]

polymers

MDPI

Article

Hybrid Alumina–Silica Filler for Thermally Conductive Epoxidized Natural Rubber

Hassarutai Yangthong [1], Phattarawadee Nun-Anan [2], Apinya Krainoi [2], Boonphop Chaisrikhwun [2], Seppo Karrila [2] and Suphatchakorn Limhengha [2,*]

[1] Specialized Center of Rubber and Polymer Materials in Agriculture and Industry (RPM), Department of Materials Science, Faculty of Science, Kasetsart University, Chatuchak, Bangkok 10900, Thailand; fscihty@ku.ac.th

[2] Faculty of Science and Industrial Technology, Prince of Songkla University, Surat Thani Campus, Surat Thani 84000, Thailand; phattarawadee.anan@gmail.com (P.N.-A.); apinya_may2014@outlook.com (A.K.); boonphop.cs@gmail.com (B.C.); seppo.karrila@gmail.com (S.K.)

* Correspondence: suphatchakorn.l@psu.ac.th

Abstract: Thermally conductive composites were prepared based on epoxidized natural rubber (ENR) filled with alumina, silica, and hybrid alumina and silica. The thermal conductivity and mechanical properties were assessed. It was observed that the interactions of polar functional groups in the fillers and epoxy group in ENR supported a fine dispersion of filler in the ENR matrix. The mechanical properties were improved with alumina, silica, and hybrid alumina/silica loadings. The ENR/Silica composite at 50 phr of silica provided the highest 60 shore A hardness, a maximum 100% modulus up to 0.37 MPa, and the highest tensile strength of 27.3 MPa, while ENR/Alumina with 50 phr alumina gave the best thermal conductivity. The hybrid alumina/silica filler at 25/25 phr significantly improved the mechanical properties and thermal conductivity in an ENR composite. That is, the thermal conductivity of the ENR/Hybrid filler was 2.23 W/mK, much higher than that of gum ENR (1.16 W/mK). The experimental results were further analyzed using ANOVA and it was found that the ENR/Hybrid filler showed significant increases in mechanical and thermal properties compared to gum ENR. Moreover, silica in the hybrid composites contributed to higher strength when compared to both gum ENR and ENR/Alumina composites. The hybrid filler system also favors process ability with energy savings. As a result, ENR filled with hybrid alumina/silica is an alternative thermally conductive elastomeric material to expensive silicone rubber, and it could have commercial applications in the fabrication of electronic devices, solar energy conversion, rechargeable batteries, and sensors.

Keywords: thermal conductivity; natural rubber; alumina (Al_2O_3); silica; hybrid filler

1. Introduction

The demand for flexible conductive materials has motivated many studies on conductive polymer-based materials. The thermal conductivity of a material is given in W/mK (Watts per meter-Kelvin). Highly thermally conductive materials are needed to make heat sinks, while materials that transfer heat slowly act as insulators. For a variety of uses, including energy conversion, heat exchange, thermal insulation, and the thermal management of electronic and photonic devices, materials with ultrahigh and ultralow thermal conductivity are preferred. The majority of polymers have relatively low thermal conductivity, which could cause major issues in device applications including heat dissipation failure, particularly for organic electronic devices. One type of polymer that is of interest for combining thermal conductivity with flexibility is natural rubber (NR). Rubber is a raw material that possesses exclusive properties and has many industrial applications. Both natural and synthetic rubbers still need research to increase their production and to improve their quality, in order to match commercial demand. Moreover, when compared to other materials (in particular, metals), these alternatives combine light weight with good process ability. However, some properties of rubber materials need improvements in order to expand their applications. This especially applies to thermal conductivity [1,2] because of its importance in many functions, such as the protective covering of electronics and sensors [1], asphalt [2], dental gloves, and foodstuff conveyor belts [3]. Rubber composites prepared with fillers have been investigated in recent studies [4–6], demonstrating improved thermal conductivity upon using synthetic rubber (silicone) as the matrix, chosen because silicone rubber has an excellent high-temperature resistance due to its unique molecular structure with Si–O bonds [7]. Moreover, fillers have been added in order to increase the thermal conductance of silicone products. It was found that silicone rubber showed good thermal conductivity with hybrid alumina (Al_2O_3) powder as a filler [8,9]. Although silicone has the advantage of high thermal conductivity, it also has disadvantages. Namely, the effects of silicone on the environment are complex and include factors related to its manufacture, use, and disposal. The use of petroleum-derived hydrocarbons in the manufacture of silicone raises carbon emissions and has an effect on sustainability. In a case study on the degradation of room-temperature vulcanizable silicone elastomers, it was found that condensation catalysts play a key role at their end-of-life stage, in both chemical and biological degradation. These compounds have a significant effect on microbial communities [10]. In addition, often, silicone rubber is too costly, having a higher price than other rubbers [11]. Therefore, it is warranted to study other rubber composite candidates for thermally conductive elastomeric materials.

Epoxidized natural rubber (ENR) as a polymer matrix has attracted attention due to its good thermal resistance [4,12] and mechanical properties similar to natural rubber [13]. ENR is produced from natural rubber by reacting it with peroxycarboxylic acid to attach epoxide groups to the isoprene units of natural rubber [14]. The epoxidation percentage of ENR can be manipulated by controlling the proportions of the reagents [15]. Nowadays, ENR is commercially produced, especially with 25% and 50% epoxidation levels. Many ENR/filler composites have been found to demonstrate improved properties, such as ENR/CB [16–19], ENR/Silica [20–24], ENR/carbon nanotubes (CNTs) [25–27], ENR/rice husk ash [28–30], ENR/fiber [5], and ENR/geopolymer filler composites [31]. Moreover, ENR was also used as a matrix with ceramic fillers [4,32], that is, ENR/barium titanate ($BaTiO_3$) and lead titanate ($PbTiO_3$) were prepared. It was discovered that the mechanical properties of the gum ENR vulcanizates and ENR/$BaTiO_3$ composites were higher than those of the gum NR vulcanizate and $PbTiO_3$/NR composites [33]. However, the range of applications for ENR is significantly limited by its low heat conductivity. Incorporating inorganic thermally conductive ceramic filler into polymeric composites is an effective way to increase their thermal conductivity while preserving their insulating qualities. Various hybrid fillers have been evaluated in numerous studies to enhance the thermal characteristics of ENR composites, such as CNTs–titanium isopropoxide (TiO_2) [34] and magnesium carbonate ($MgCO_3$)–silica [3]. Thermally conductive fillers with alumina were chosen [11] to address the thermal management issues.

Aluminum oxide (Al_2O_3) or alumina has a variety of intriguing physical and chemical qualities, such as chemical inertness, suitable hardness, good thermal and mechanical capabilities, and a reasonable price [35,36]. Recently, Al_2O_3 was applied in styrene–butadiene rubber (SBR) composites [37] and in silicone rubber composites [4,38,39]. The outcomes show that the thermal conductivity of silicone rubber composite increased with the loading of Al_2O_3 powder [38]. Nevertheless, there is poor compatibility of Al_2O_3 with the organic matrix, making uniform filler dispersion challenging. This affected the mechanical properties of the composites and also caused other shortcomings [40]. Therefore, the use of inorganic fillers is limited by compatibility, and improved compatibility has been sought by modifying the filler surfaces [41], by choosing the appropriate inorganic matrix [42], or by using hybrid fillers. Improved thermal conductivity and mechanical properties of the composite have been demonstrated [38].

In this present study, we propose the novel combination of an ENR matrix with alumina and silica as thermally conductive fillers. In the first case, alumina alone or silica alone is used as a filler, and in the second case, alumina and silica are used as a blend, i.e., a hybrid filler. To the best of the authors' knowledge, this is the first instance in which a combination or hybrid filler of alumina and silica has been tested in an ENR matrix. The use of a silicone matrix or more expensive thermally conductive fillers was avoided in this study. The structure–property relationships were explored. The cure characteristics, structural and dispersion measurements (i.e., MDR, FTIR, and SEM), mechanical test results (i.e., tensile strength, 100% moduli, elongation at break, and hardness), and thermal conductivity were the evaluated properties of the ENR composites.

2. Materials and Methods

2.1. Materials

Epoxidized natural rubber with 50 mol% epoxide (ENR-50) used as the rubber matrix was manufactured by Muang Mai Guthrie Public (Surat Thani, Thailand). Aluminum oxide (Al_2O_3) or alumina (density = 3.94 g/cm^3) and silica (density = 1.36 g/cm^3), applied as fillers, were obtained from Sigma-Aldrich Chemical Co., Ltd. (Burlington, VT, USA) and from Evonik Industries (Essen, Germany). The particle sizes of ceramic materials or Al_2O_3 used in this work ranged from 0.4 μm to 2.4 μm. Bis(triethoxysilylpropyl)tetrasulfide, or TESPT (density of 1.09 g/cm^3), was used as a silane coupling agent, and it was produced by Sigma-Aldrich Co. Ltd. (Dorset, UK). Butylated Hydroxyl Toluene, or BHT (density = 1.05 g/cm^3), was used as an antioxidant and it was made by Sigma-Aldrich Chemical Co., Ltd. (St. Louis, MO, USA). Zinc oxide (ZnO) (density = 5.61 g/cm^3) and stearic acid (density = 0.85 g/cm^3) were utilized as activators, and were obtained from Global Chemical Co., Ltd. (Samut Prakarn, Thailand) and Imperial Industry Chemical Co., Ltd. (Pathum Thani, Thailand). Accelerators, including *N*-Cyclohexyl Benzothaizole-2-Sulfenamide or CBS (density = 1.34 g/cm^3) and tetramethylthiuram disulfide or TMTD (density = 1.39 g/cm^3), were manufactured by Xiamen AmoyChem Co., Ltd. (Xiamen, China) and TCI (Shanghai) Development Co., Ltd. (Shanghai, China). Sulfur (density = 2.07 g/cm^3) was used as a curing agent. It was manufactured by Ajax Chemical (Samutprakarn, Thailand).

2.2. Preparation of Epoxidized Natural Rubber Composites

Filled epoxidized natural rubbers (with the alternative fillers alumina, silica, and hybrid alumina/silica filler) were prepared with the formulations shown in Table 1. An internal mixer (Brabender GmbH & Co. KG, Duisburg, Germany) with a fill factor of 0.7 was used to mix ENR-50 with the filler and other chemicals at a rotor speed of 60 rpm, and the mixing steps are shown in Figure 1. First, ENR-50 was masticated at 130 °C, followed by adding filler and a coupling agent. It is noted that the filler loading was added in two parts, each providing half of the total content. After that, the masterbatch of ENR was dumped at room temperature. In the second step, masterbatch ENR was placed into the mixing chamber at 60 °C and ZnO and stearic acid were added. Then, CBS and BHT were added, and the blend was removed from the mixing chamber and allowed to cool to room temperature. In the third step, the filler dispersion was improved by passing the ENR compound ten times through the nip of a two-roll mill (Yong Fong machinery Co, Ltd., Samutsakorn, Thailand) while also mixing in TMTD and sulfur. For unfilled ENR, similar mixing was performed without filler. It is noted that the mixing time of unfilled and filled cases was controlled to 13 min. Finally, ENR composite sheets were prepared at 150 °C with compression molding (PR1D-W400L450PM, Charon Tut Co., LTD., Bangkok, Thailand). The cure time from a moving die rheometer (MDR) was used to compress each sample compound. It is noted that the samples were labeled "ENR/Alumina, ENR/Silica, and ENR/Hybrid filler". In addition, the sample without filler was labeled "Unfilled ENR".

Table 1. Compounding formulations of epoxidized natural rubber composites.

Ingredient	Quantity (phr)			
	Unfilled ENR	**ENR/Alumina**	**ENR/Silica**	**ENR/Hybrid Filler**
ENR 50 [1]	100	100	100	100
Al_2O_3 [2]	-	50	-	25
Silica	-	-	50	25
Coupling agent	-	5	5	5
Stearic acid	1	1	1	1
ZnO [3]	3	3	3	3
BHT [4]	1	1	1	1
CBS [5]	1.5	1.5	1.5	1.5
TMTD [6]	0.3	0.3	0.3	0.3
Sulfur	1.5	1.5	1.5	1.5

[1] Epoxidized natural rubber with 50 mol% epoxide, [2] aluminum oxide, [3] zinc oxide, [4] Butylated Hydroxyl Toluene, [5] *N*-Cyclohexyl Benzothaizole-2-Sulfenamide, [6] Tetramethyl Thiuram Disulfide.

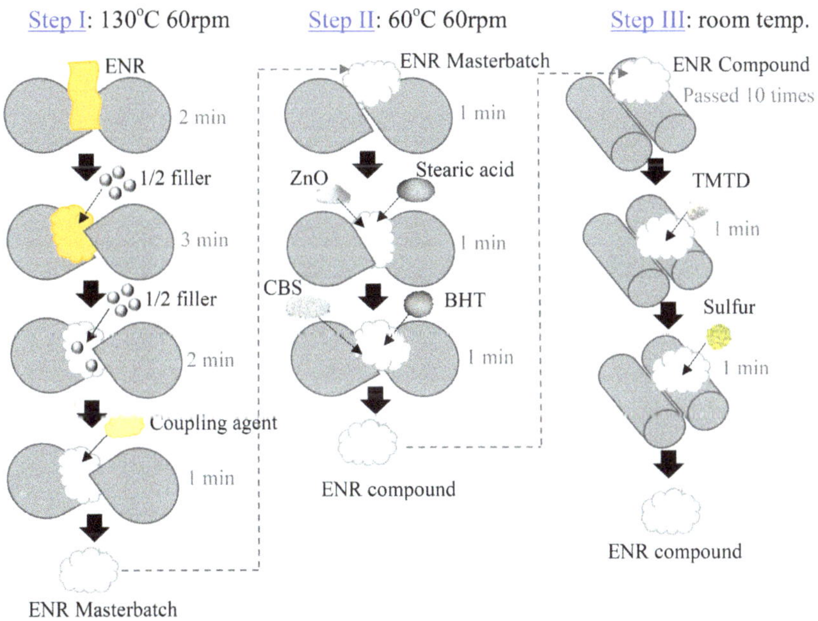

Figure 1. Mixing steps of ENR composites with alumina, silica, and hybrid fillers, following the formulations in Table 1.

2.3. *Testing and Characterization*

2.3.1. ATR-FTIR Spectroscopy

The functional groups and chemical characteristics of ENR composites with alumina, silica, and hybrid fillers were evaluated utilizing Attenuated Total Reflection Fourier Transform Infrared spectroscopy (ATR-FTIR) (FTIR, Vertex70, Bruker, Karlsruhe, Germany) in transmittance mode. The spectra were taken over the range of 4000–400 cm^{-1}, operated with a 4 cm^{-1} resolution and with 32 scans for each recording. The chemical alterations in the ENR composites were assessed by analyzing these spectra.

2.3.2. Mooney Viscosity

Mooney viscosity measurements of ENR compounds were conducted according to the procedures described in ASTM-D 1646-89 [43]. In this work, the samples were tested by using an MV-2020 Mooney viscometer (Montech, Tokyo, Japan). The testing was performed using a large rotor at 130 °C. The sample was preheated at the test temperature for one minute before the rotor was started, and then, the Mooney viscosity (ML1+4) was recorded as the torque after the rotor had rotated for 4 min at 2 rpm (average shear rate of about 1.6 s^{-1}).

2.3.3. Curing Properties

The cure characteristics of the ENR composites were tested with a moving die rheometer (MDR) (Monsanto Co., Ltd., Greenville, OH, USA), according to ASTM D5289, at 150 °C, with a measurement time of 20 min. The optimum cure time (t_c90), the scorch time (t_s2), the minimum torque (M_L), and the maximum torque (M_H) are reported. The ENR vulcanizates were prepared by compression molding with a PR1D-W400L450PM compression molding machine (Charon Tut Co., LTD., Bangkok, Thailand) at 150 °C, with a compression pressure of 150 kg/cm^2 and a time following the cure time. Finally, the ENR composite samples were conditioned in an ambient atmosphere for at least 24 h before further testing.

2.3.4. Morphological Characterization

The morphological characteristics were investigated as regards the sample surface topography and composition of the composites. Morphology evaluation is one of fastest methods available for the qualitative analysis of material composition and the level of homogeneity in the sample at a given scale. The dispersion states of alumina, silica, and hybrid fillers in ENR matrix were estimated by using a field emission scanning electron microscope (FEI-Czech Republic, Oxford, UK). The ENR composites were first cryogenically cracked in liquid nitrogen. The fresh fracture surfaces were then sputter-coated with a thin layer of gold for imaging to study the surface morphology and filler dispersion.

The elemental mapping was investigated to verify the distribution of hybrid filler in the ENR matrix by using an energy-dispersive X-ray fluorescence spectrometer (Oxford Instruments, Oxfordshire, UK), combined with FE-SEM apparatus. The samples were cryogenically cracked for the study of filler dispersion, but not coated with gold.

2.3.5. Mechanical Characterization

The mechanical properties (i.e., the tensile strength, modulus at 100%, and elongation at break) were investigated according to ISO 527. The test was performed using a 10ST tensile testing machine (Tinius Olsen, Salford, UK). The dumbbell shaped specimens were of the 5A-type. Testing was conducted with a crosshead speed of 200 mm/min. In addition, hardness was evaluated using a hardness testing machine (HT 3000, MonTech, Buchen, Germany). The test was run at room temperature in agreement with ISO 868. The scale of the indicating device was read after 15 s $\pm$ 1 s. Furthermore, the thermal aging properties of the ENR composites were estimated in terms of changes in the mechanical properties after aging at 100 °C for 48 h, according to ISO 188. The test results reported for each recipe are averages of five replicate samples.

2.3.6. Dynamic Mechanical Properties

The dynamic mechanical properties (i.e., storage modulus and tan delta) of the ENR composites were analyzed using a DMA 8000 (Perkin Elmer Co., Ltd., Waltham, MA, USA). The samples were rectangular with a 40 mm length, 10 mm width, and 2 mm thickness. The runs were performed in tension mode at a 1 Hz frequency. The dynamic strain amplitude was 0.1%, with temperature ranging from −100 to 70 °C at a heating rate of 10 °C/min.

2.3.7. Thermal Analysis

The thermal conductivity of the ENR composites was investigated. The investigation was performed using a thermal constant analyzer (TPS 2500S, Hot Disk Sweden, Goteborg, Sweden) with a Hot Disk 5465 sensor. Two samples were prepared per formulation, each of size 5 × 5 × 0.2 cm. The determinations used the transient plane source technique (TPS) [44].

2.3.8. Statistical Data Analysis

The mechanical (i.e., hardness, modulus at 100%, tensile strength, and elongation at break) and thermal property results obtained from Sections 2.3.5 and 2.3.7 were analyzed using ANOVA statistics in Excel, Microsoft 365 (Microsoft, Washington, DC, USA) with an alpha value of 0.05 (corresponding to 95% confidence level). The analysis of the data was performed from the repeated test data with 5 measurements per sample.

3. Results and Discussion

3.1. FTIR Spectroscopy for ENR Composites

Figure 2 and Table 2 show the FTIR spectra of the ENR composites with alumina, silica, and hybrid fillers. The spectrum of unfilled ENR exhibited peaks of isoprene structure at 2927, 2836, and 1463 cm^{-1} (C-H stretching vibrations), and at 1373 cm^{-1} (–CH_3 bending vibrations). The partly epoxidized rubber showed characteristic signals of epoxide units at 1260 cm^{-1} (C-O-C stretching) and 878 cm^{-1} (C-O-C ring vibration) [45]. In addition, the peaks at the wave numbers 1041, 641, and 578 for ENR/Alumina composite are characteristic of alumina, assigned to stretching modes of tetrahedral AlO_4, whereas those near 641 cm^{-1} and 578 cm^{-1} are attributed to stretching vibrations of octahedral AlO_6 [46,47]. For ENR/Silica composite, the relatively broad peak at 1068 cm^{-1} was for Si–O stretching vibrations of silica [48]. Moreover, the absorption peak at wavenumber 459 cm^{-1} represents amorphous silica [49], and was further investigated. The peaks in the ENR/Hybrid filler at 1068, 641, and 578 cm^{-1} were produced by the chemical properties of the ENR/Alumina/Silica mixture. In particular, the absorption peaks of the ENR/Hybrid filler composite at wavenumbers of about 1068 and 578 cm^{-1} were broader than the peaks in the spectrum of unfilled ENR. This might indicate chemical and physical interactions between the polar groups of silica and alumina particles with the oxirane groups or their opened ring products in the ENR matrix, as shown in Figure 3. In other words, during vulcanization, the silanol groups of silica interacted with the polar functional groups in epoxide rings as well as with the ring opening products of ENR [31], as shown in Figure 3a. Moreover, hydrogen bonding occurred between ENR and both fillers (Figure 3c,d), and in filler–filler interactions (Figure 3b) as well. The linkages between the ENR and functional groups of the hybrid filler were formed via rubber–filler interactions and H-bonding. The hybrid filler blend of alumina and silica had its dispersion in the ENR matrix supported by these interactions.

Table 2. Peak assignments in FTIR spectra for unfilled ENR, ENR/Alumina, ENR/Silica, and ENR/Hybrid fillers.

Wave Number (cm^{-1})	Assignment
2927	C-H stretching vibrations
2836	C-H stretching vibrations
1463	C-H stretching vibrations
1373	–CH_3 bending vibrations
1260	C-O-C stretching vibrations
1068	Si–O stretching vibrations
1041	AlO_4 tetrahedral stretching vibrations
878	C-O-C ring vibrations
641	AlO_6 octahedral stretching vibrations
578	AlO_6 octahedral stretching vibrations

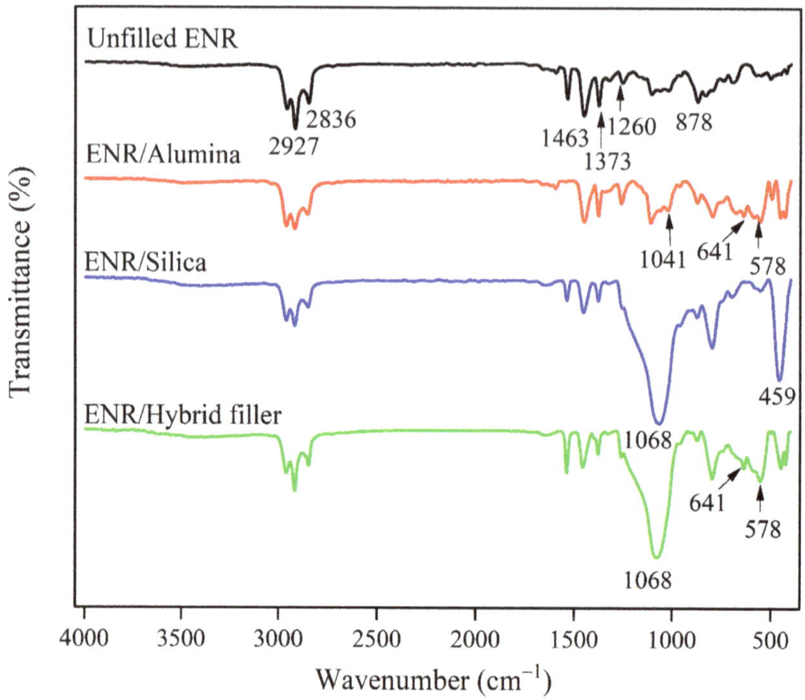

Figure 2. FTIR spectra of ENR composites with alumina, silica, and hybrid fillers.

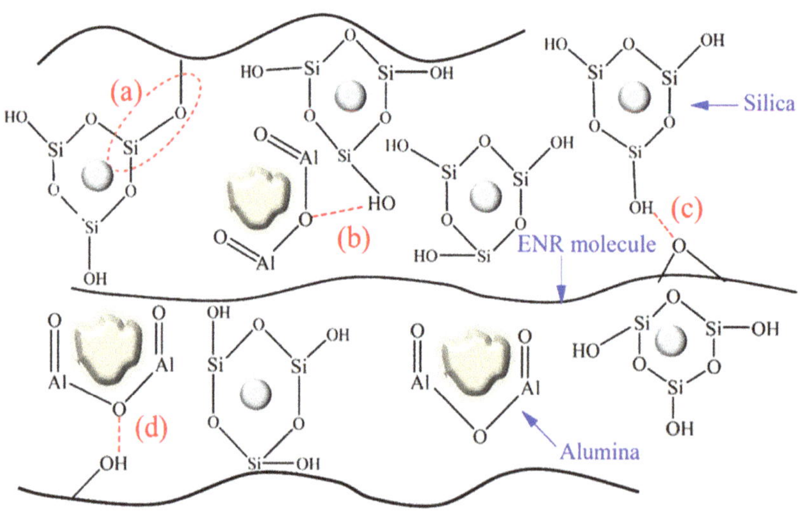

Figure 3. Possible chemical reactions between ENR and hybrid filler, (**a**) rubber-filler interaction, (**b**) hydrogen bonding between alumina and silica, (**c**) hydrogen bonding between ENR and silica, and (**d**) hydrogen bonding of ENR and alumina.

3.2. Morphological Properties

The SEM micrographs of unfilled ENR, ENR/Alumina, ENR/Silica, and ENR/Hybrid filler are shown in Figure 4a–d, respectively. These show a smooth surface and some small zinc oxide particles in Figure 4a. The filler particles (i.e., silica, alumina, and hybrid fillers) are more or less uniformly distributed in the ENR matrix. In Figure 4b, it is clear that the silica-filled ENR matrix showed good filler distribution with rubber-rich areas, indicating good rubber–filler interactions (Figure 3). This is due to the polar groups of 50 mol% epoxide (ENR-50) and silanol groups of silica, as well as polar groups of alumina. Therefore, the polarity of both the filler and the rubber promoted good compatibility, as shown in Figure 4b,c. Moreover, the coupling agent was introduced to favor the homogeneous dispersion of filler in the ENR composites. This is because the dual function coupling molecules' reactivity allows the ENR and filler to be coupled or bridged [10]. In addition, the ENR/Hybrid filler (Figure 4d) shows a fine dispersion of silica and alumina. This is attributed to the compatibility of materials and the functional groups of the coupling agent interacting with the polar groups on the hybrid filler and dominating the filler–filler interactions to prevent agglomeration. Such interactions were proven by the energy-dispersive X-ray fluorescence spectrometry.

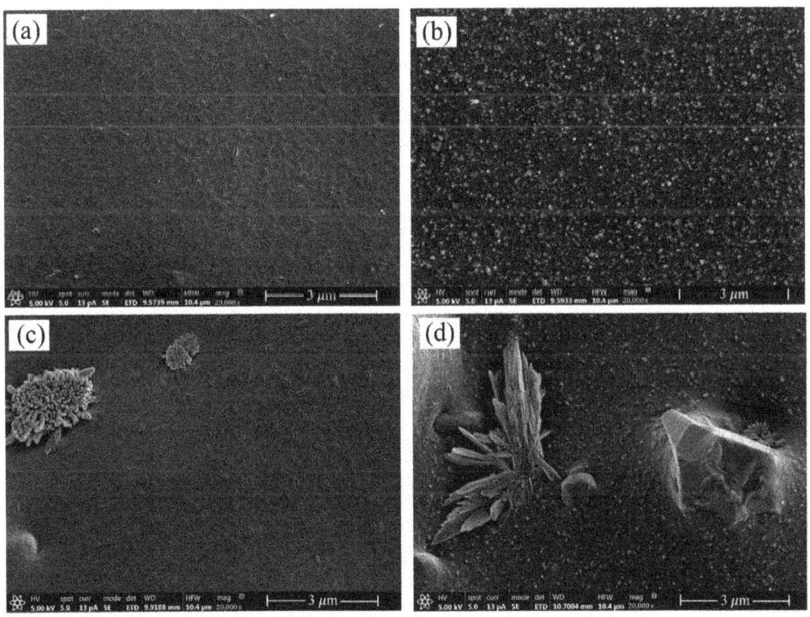

Figure 4. SEM micrographs of unfilled ENR (**a**) and ENR/Alumina (**b**), ENR/Silica (**c**), and ENR/Hybrid fillers (**d**) at 20,000× magnification.

Figure 5 shows BSE-SEM and SEM-EDX images of the ENR/Hybrid filler. It was proven that the hybrid filler was comparatively uniformly distributed in the ENR matrix. In Figure 5a, the matrix of the composite has three areas selected, as shown in the BSE-SEM images. The first area (point A) shows Al and Si as the main elements in the matrix, as does point B, while the spectrum of point C shows only Si as the main element. Therefore, points A and B indicate that silica and alumina were jointly dispersed and mutually compatible. Moreover, the SEM-EDX image (Figure 5b) was further analyzed through EDX attachment to understand the miscibility in the hybrid filler–rubber matrix. That is, the elemental distributions of silica and alumina were probed by using intensity maps of Si and Al, displayed in pink and blue colors. Si was well dispersed in the ENR matrix with small particles, as was the Al element, which showed larger sized particles. This clearly presents the dispersion states of fillers in the ENR matrix. Additionally, green was used to indicate the carbon of the matrix, and yellow for the sulfur of the vulcanizing agent in the rubber recipe.

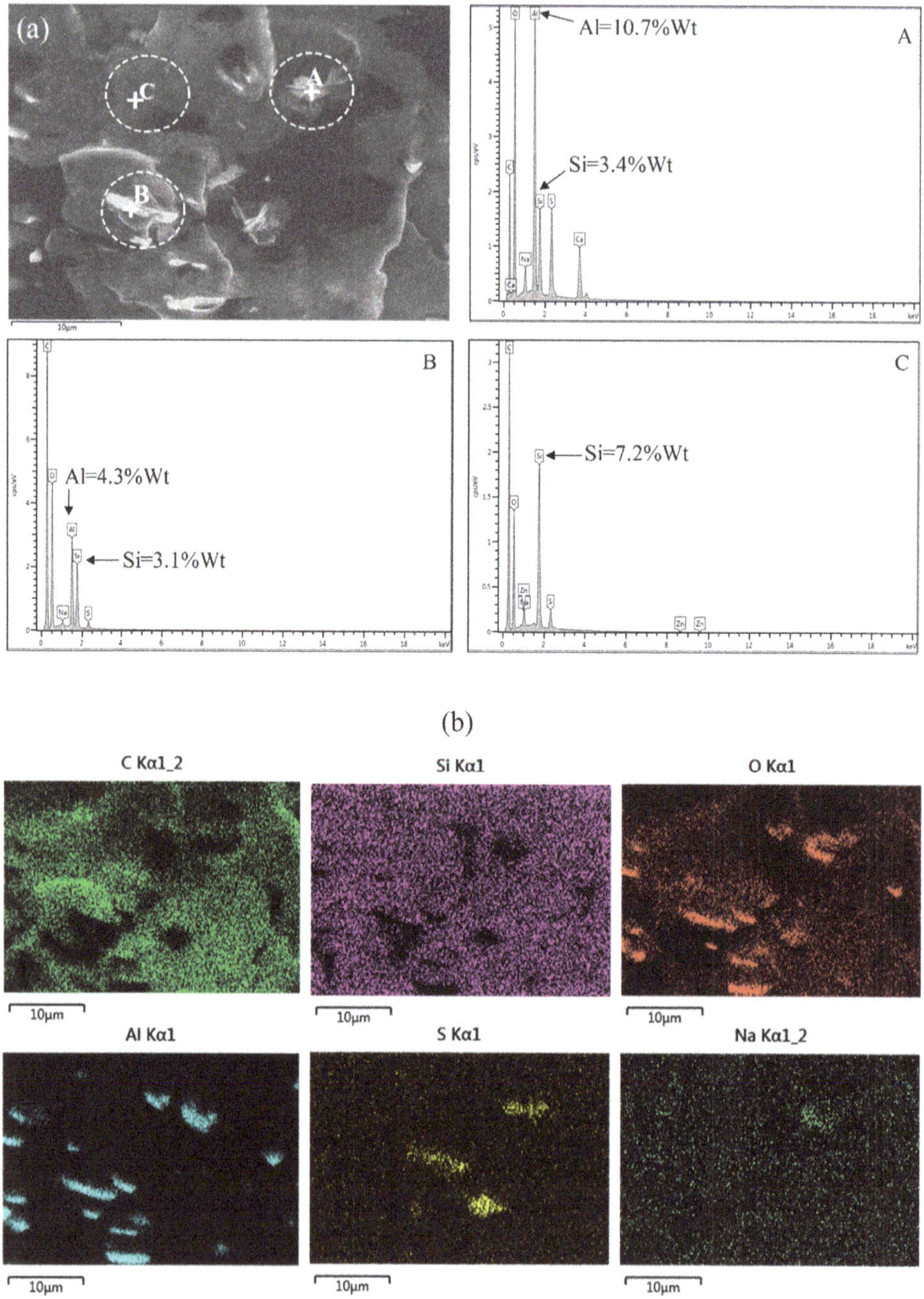

Figure 5. BSE-SEM (**a**) when (A)–(C) is elemental spectrum of each point on matrix, and SEM-EDX (**b**) images of ENR/Hybrid filler.

3.3. Cure Characteristics and Mooney Viscosity

The cure characteristics and Mooney viscosities are shown in Table 3. It was found that the compound without filler had the lowest Mooney viscosity and minimum, maximum, and delta torque, whereas the ENR composites filled with silica, alumina, and hybrid filler provided higher values. The reinforcing filler silica especially increased the torque difference. This is due to the filler (i.e., silica, alumina, or hybrid filler) increasing the stiffness and reducing the chain mobility of the ENR molecules. It should be mentioned that the crosslink density in rubber is reflected by the delta torque or torque differential [50]. In addition to crosslink density, filler–filler and filler–rubber interactions are also related to delta torque, as shown in Figure 3a–d. Therefore, increased torque suggests that there are good interactions between the filler and ENR molecules [51]. This can be mostly attributed to high compatibility between the alumina, silica, or hybrid filler and the ENR matrix, which contributed to reinforcement of the composites by the filler [52]. However, when comparing the delta of ENR/Silica and ENR/Alumina, it was found that ENR/Silica gave better delta than ENR/Alumina because of the particle size of silica. That is, silica had a larger specific surface area than alumina. Therefore, it provided more interactions with rubber molecules.

Table 3. Mooney viscosity and cure characteristics of epoxidized natural rubber composites.

Sample	Mooney Viscosity (ML1+4 100 °C)	Cure Characteristics					
		M_L [1] (dNm)	M_H [2] (dNm)	M_H-M_L [3] (dNm)	t_{s2} [4] (min)	t_{c90} [5] (min)	CRI (min^{-1})
Unfilled ENR	25.81 ± 1.31	1.05	14.24	13.19	0.30	1.23	107.5
ENR/Alumina	27.60 ± 0.92	1.92	16.63	14.71	0.29	1.11	121.95
ENR/Silica	50.20 ± 1.51	2.20	30.65	28.45	0.54	2.46	52.08
ENR/Hybrid filler	48.31 ± 1.32	2.82	27.98	25.16	0.54	2.06	65.78

[1] Minimum torque, [2] maximum torque, [3] delta torque, [4] scorch time, [5] cure time.

In Table 3, it is clear that the scorch times, cure times, and cure rate indexes of ENR/Alumina were close to those of unfilled ENR, and higher than those of ENR/Hybrid filler and ENR/Silica. When compared to the ENR/Silica composites, the inclusion of alumina significantly sped up the cure and scorch processes. Furthermore, it gave a very high cure rate index for the alumina-filled ENR and alumina/silica hybrid. This can be attributed to the metal oxides that, in the ENR, speed up crosslinking reactions and raise the cure rate index [53,54]. Therefore, the CRI of the ENR/Hybrid filler was between that of ENR/Silica and ENR/Alumina, with a convenient cure time. These results match the fine dispersion of both fillers, while the content of silica in the hybrid filler sample is lower than in ENR/Silica. As a result, the fillers are less likely to agglomerate.

3.4. Mechanical Properties

Figure 6 displays the hardness properties of ENR composites filled with alumina, silica, and hybrid fillers. It is clear that ENR/Silica composites showed the highest hardness, followed by the ENR/Hybrid filler, ENR/Alumina filler, and unfilled ENR, in that order. This shows the effects of filler particles and their interactions (silica, alumina, or hybrid filler) with the ENR matrix, as shown in Figure 3a–d. In addition, the mechanical properties, in terms of the 100% modulus, tensile strength, and elongation at break, are exhibited in Figure 7. It is seen that the 100% modulus increased with filler loading, as displayed in Figure 7a. The filler restricted the mobility of rubber chains and, hence, increased the material stiffness. Additionally, the hard filler particles in the soft matrix (the ENR matrix) were partly responsible for the rise in the modulus, whereas the interactions between the filler and rubber also contributed (Figure 3a), along with rubber–filler H-bonding (Figure 3c,d) and filler–filler interactions (Figure 3b). The silica in the ENR matrix had a smaller particle size and, accordingly, a larger effective surface area. This is consistent with the data in Table 3 showing a higher delta torque. Moreover, the tensile strength of the ENR composites showed the same trend with the 100% modulus. That is, ENR/Hybrid filler displayed lower tensile strength than the ENR/Silica composite, but higher than the ENR/Alumina composite, as seen in Figure 7b. This can be explained by the chemical interactions and hydrogen bonding between the ENR and filler, as shown in Figure 3. That is, ENR/Silica can have chemical interactions and hydrogen bonding. The presence of hydroxyl groups on the silica surfaces was detected, which can interact with the epoxide groups in ENR molecules, as shown in Figure 3a. The silanol group of silica can form an H-bond with the oxirane ring of ENR, as exhibited in Figure 3b. For ENR/Alumina, there can be H-bonding with the oxirane ring of ENR, as displayed in Figure 3d. Therefore, ENR/Hybrid filler is the most suitable formulation because a mix of silica and alumina gave no significant difference in tensile strength (i.e., 27 and 24 MPa of ENR/Silica and ENR/Hybrid filler). In addition, elongation at break, also known as the ratio of increased length to initial length at tensile failure, is depicted in Figure 7c. This shows how much bending and shaping a material can endure before breaking. It was found that the fillers decreased the maximal elongation. This may be due to the fillers decreasing the rubber content, which reduces elongation [55]. However, the filled loading in this work was controlled at 50 phr to balance the elasticity of the composites. It is seen that ENR/filler (i.e., silica, alumina, and hybrid fillers) showed lower elongation at break than unfilled ENR. This is because the filler particles reduce the chain mobility of rubber molecules and the chemical linkages between the filler items and the ENR matrix are intensified (Figure 3a), and H-bonding (Figure 3b–d) has a high reinforcing effect.

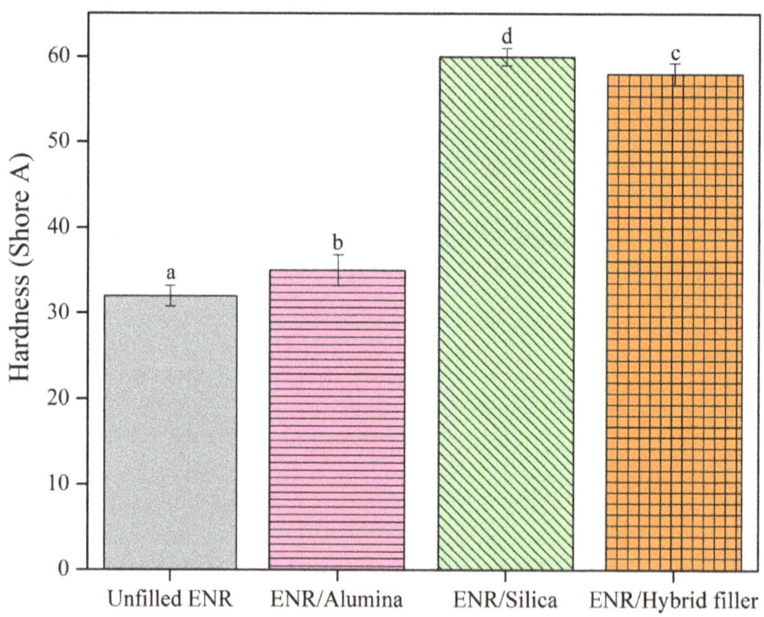

Figure 6. Hardness of ENR composites with alumina, silica, and hybrid fillers. a–d: Different letters within same picture indicate statistically significant differences at $p < 0.05$, based on one-way ANOVA.

In addition, the mechanical properties after aging are among the practically important properties of rubber vulcanizates [56]. It was found that ENR/Alumina and ENR/Silica exhibited an increased 100% modulus, as shown in Figure 7a. However, tensile strength and elongation at break after aging decreased in all cases (i.e., Unfilled ENR, ENR/Alumina, ENR/Silica, and ENR/Hybrid filler), as shown in Figure 7a,b. This is due to breaking crosslinks, especially those of the polysulphidic type [57]. These are then converted to di-

and mono-sulphidic bonds that are less flexible and also decrease elongation at break [58]. Moreover, the experimental results were further analyzed using ANOVA with p values less than 0.05 ($p < 0.05$). It was found that the hardness, 100% moduli, and tensile properties (before and after aging) of ENR/Hybrid filler exhibited significant differences. In addition, the elongation at break (before aging) of the ENR/Silica and ENR/Hybrid fillers showed no difference.

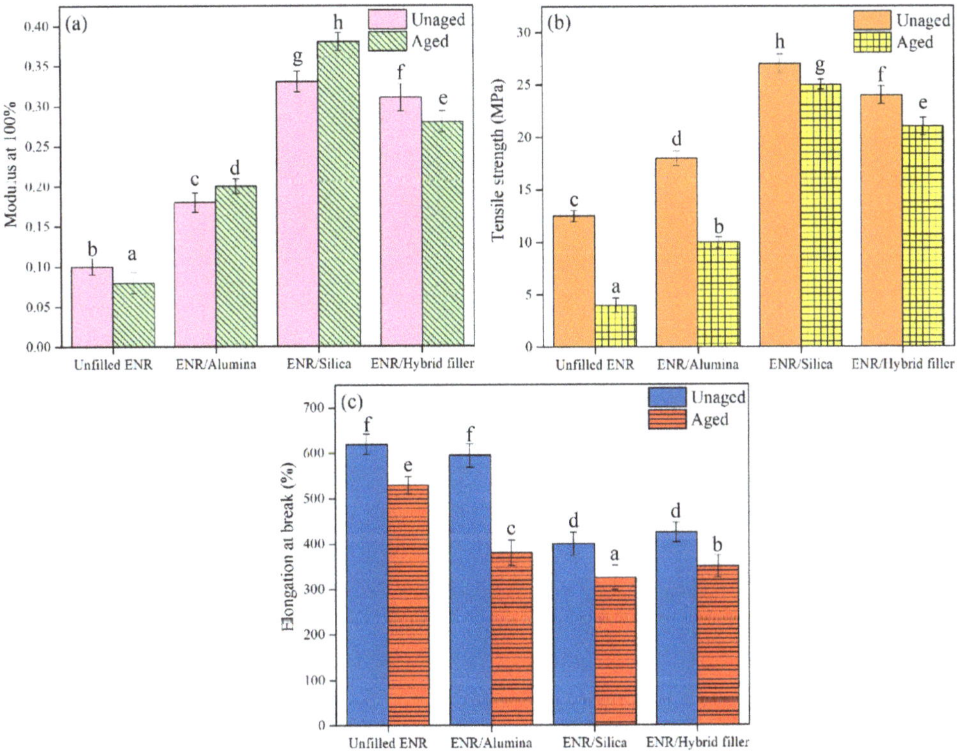

Figure 7. Mechanical properties: modulus at 100% (**a**), tensile strength (**b**), and elongation at break (**c**) of ENR composites, before and after aging at 100 °C for 48 h. a–h: Different letters within same picture represent statistically significant differences at $p < 0.05$, based on one-way ANOVA.

3.5. Thermal Properties

Figure 8a exhibits the thermal conductivity of ENR composites with alumina, silica, and hybrid fillers. It was observed that the filler enhanced the thermal conductivity of ENR composites from that of unfilled ENR. It is clear that the thermal conductivity increased significantly on adding filler, that is, from 1.16 W/mK of unfilled ENR to 2.22, 2.19, and 2.23 W/mK of ENR/Alumina, ENR/Silica, and ENR/Hybrid filler, respectively. The ANOVA showed that all recipes conducted heat significantly differently. The filled cases had far higher thermal conductivities than the rubber matrix. This is because NR is an insulator by nature, having poor thermal conductivity. Therefore, incorporating thermally conductive particles into the ENR matrix can improve the thermal conductivity of the composite. In addition, the composites ENR/Alumina, ENR/Silica, and ENR/Hybrid filler were compared. It can be seen that ENR/Alumina showed higher thermal conductivity compared to ENR/Silica composites. This is because the thermal conductivity of silica is in the range of 0.5 to 2.5 W/mK [59], while the thermal conductivity of alumina is about 25 W/mK [60]. Moreover, ENR/Hybrid filler composites achieved the highest thermal conductivity, indicating a synergistic effect of the fillers. In addition, this might be due to the dispersion of filler, which can be explained by the schematic diagram in Figure 8b. It can be seen that the thermally conductive paths (red dashed lines) formed by fillers can pass throughout the ENR matrix. However, filler agglomeration can lower the thermal conductivity performance of the composite. Therefore, the hybrid filler that had improved dispersion in the ENR matrix gave enhanced thermal conductivity. In this work, aluminum oxide was used as a thermally conductive filler due to the polarity of the filler. It can be finely dispersed in the ENR matrix to develop a polymer composite, and with this in mind, Al_2O_3 was chosen because of its comparatively low cost.

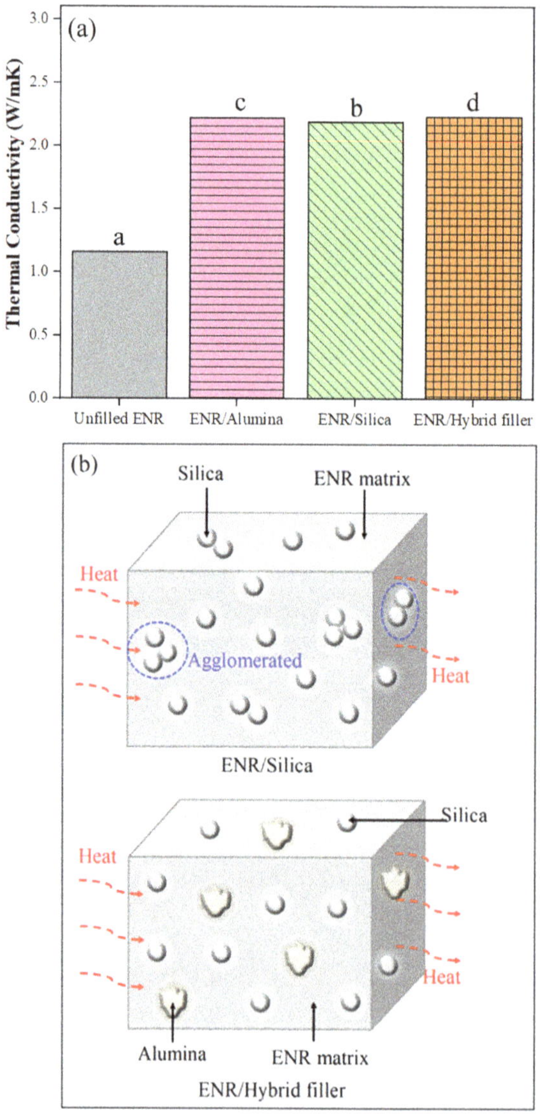

Figure 8. Thermal conductivity (**a**) and schematic diagram (**b**) of ENR composites. a–d: Different letters within same picture represent statistically significant differences at $p < 0.05$, based on one-way ANOVA.

3.6. Dynamic Properties

Figure 9 shows the temperature dependences of the storage modulus (a) and tan δ (b) for the ENR composites with alumina, silica, and hybrid fillers. In general, the storage modulus of the filled rubber composite is dependent on the physical interactions between rubber molecules and inorganic filler particles [31]. In Figure 9a, the ENR/Alumina, ENR/Silica, and ENR/Hybrid fillers display higher initial storage moduli than the unfilled ENR. The storage moduli give the rank order of unfilled ENR < ENR/Alumina < ENR/Hybrid filler < ENR/Silica. This is due to the physical and chemical interactions between alumina or silica and the ENR matrix, which can be explained by the possible chemical reactions in Figure 3.

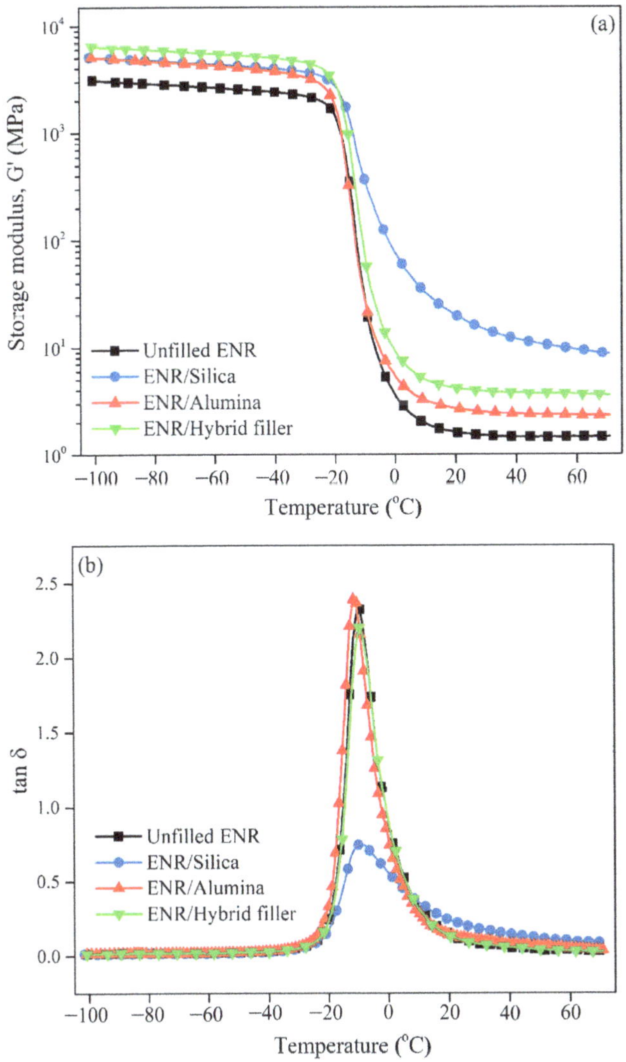

Figure 9. Temperature dependence of storage modulus (**a**) and of tan δ (**b**) for ENR composites with alumina, silica, and hybrid fillers.

Figure 9b displays the tan δ of the ENR composites with alumina, silica, and hybrid fillers. It was found that ENR/Silica showed the lowest tan δ. This may be attributed to the high surface area of silica increasing the filler–rubber interactions. The decreased value of tan δ suggests a comparatively high filler–rubber interaction [31]. The glass transition temperature (Tg) is traditionally defined as the temperature at which the tan peak appears. It can be observed that ENR/filler (i.e., silica, alumina, and hybrid fillers) composites displayed higher Tg than unfilled ENR (Table 4). That is, the filler shifted the Tg, especially in the case of silica filler. This is due to ENR/Silica chemical interactions, partly affected by particle size, being stronger than in the ENR/Alumina case. These chemical interactions restrict the molecular mobility of rubber [61]. However, it is clear that ENR/Hybrid filler showed Tg values between those of ENR/Silica and ENR/Alumina. This correlates well with the mechanical properties (Figure 7).

Table 4. Maximum tan δ (tan δ_{max}) and glass transition temperature (T_g) for epoxidized natural rubber composites.

Sample	Tan δ_{max} (dNm)	T_g (°C)
Unfilled ENR	2.2	−10.8
ENR/Alumina	2.3	−11.2
ENR/Silica	0.8	−8.7
ENR/Hybrid filler	2.0	−9.5

4. Conclusions

In this work, a hybrid filler blend of alumina and silica was investigated for use in an epoxidized natural rubber (ENR) matrix. The formulation was compared with alumina and silica as individual fillers, as well as with unfilled ENR. The FTIR results, SEM imaging, mechanical and dynamic properties, and thermal conductivity of the ENR composites were studied. The hybrid filler showed FTIR peaks of both silica and alumina. SEM micrographs showed the dispersions of alternative fillers in the ENR matrix. SEM-EDX especially proved that silica and alumina were compatible and well dispersed. In addition, the cure characteristics displayed that the scorch time, cure times, and cure rate index of ENR/Alumina closely matched those of unfilled ENR, while adding silica (ENR/Hybrid filler) showed a CRI between ENR/Alumina and ENR/Silica. The mechanical properties hardness, 100% modulus, and tensile strength, and the DMA results for ENR/Hybrid composites, were also between those obtained with singly used fillers. Moreover, ENR/Hybrid composites exhibited higher thermal conductivity than ENR/Alumina and ENR/Silica composites. Based on ANOVA, the thermal conductivity differences were statistically significant. Therefore, the combination of alumina and silica as a hybrid filler enhanced the mechanical properties of ENR and provided enhanced thermal conductivity superior to the ENR/Silica composite, with synergistic effects from alumina and silica. This study showed that hybrid alumina and silica filler in ENR composites can be used to prepare thermally conductive materials, which normally use silicone rubber/thermal conductive filler, and this can extend the practical applications of ENR. By varying the ratio of alumina to different ceramic materials, the ENR/Hybrid Alumina/Silica composite can be used in future studies to boost thermal conductivity. Furthermore, it is interesting to investigate how matrix polarity affects a composite's thermal conductivity by reducing the polarity of the matrix with a change from ENR-50 to ENR-25.

Author Contributions: Conceptualization, H.Y., P.N.-A., S.L., A.K., B.C. and S.K.; writing—original draft preparation, H.Y.; writing—review and editing, S.L. and S.K.; visualization, H.Y. and S.L.; supervision, S.L.; project administration, S.L. All authors have read and agreed to the published version of the manuscript.

Funding: No funding was received for conducting this study.

Institutional Review Board Statement: Not applicable.

Informed Consent Statement: Not applicable.

Data Availability Statement: Data are contained within the article, further request please contact the corresponding author.

Acknowledgments: This research applied testing tools supported by the Prince of Songkla University, Surat Thani Campus, 84000, Surat Thani, Thailand.

Conflicts of Interest: The authors declare no conflicts of interest.

References

1. Kampioti, K.; Matos, C.F.; Galembeck, F.; Jaillet, C.; Derré, A.; Zarbin, A.J.; Pénicaud, A. Highly conducting, sustainable, nanographitic rubber composites. *ACS Omega* **2018**, *3*, 1367–1373. [CrossRef] [PubMed]
2. Wen, Y.; Wang, Y.; Zhao, K.; Sumalee, A. The use of natural rubber latex as a renewable and sustainable modifier of asphalt binder. *Int. J. Pavement Eng.* **2017**, *18*, 547–559. [CrossRef]
3. Limhengha, S.; Sriseubsai, W. Optimizing the properties of vulcanized compounds for foodstuff conveyor belts in industrial microwave per-heating. *Int. J. Geomate* **2018**, *15*, 189–196. [CrossRef]
4. Xie, X.; Yang, D. Construction of thermal conduction networks and decrease of interfacial thermal resistance for improving thermal conductivity of epoxy natural rubber composites. *Ceram. Int.* **2023**, *49*, 17650–17659. [CrossRef]
5. Zhang, K.; Qiu, J.; Sakai, E.; Zhang, G.; Wu, H.; Guo, S.; Zhang, L.; Yamaguchi, H.; Chonan, Y. Preparation of continuous carbon fiber-filled silicone rubber with high thermal conductivity through wrapping. *Polym. Compos.* **2024**, *45*, 1437–1445. [CrossRef]
6. Shui, Y.; Huang, L.; Wei, C.; Sun, G.; Chen, J.; Lu, A.; Sun, L.; Liu, D. How the silica determines properties of filled silicone rubber by the formation of filler networking and bound rubber. *Compos. Sci. Technol.* **2021**, *215*, 109024. [CrossRef]
7. Xue, Y.; Wang, H.; Li, X.; Chen, Y. Exceptionally thermally conductive and electrical insulating multilaminar aligned silicone rubber flexible composites with highly oriented and dispersed filler network by mechanical shearing. *Compos.-A Appl. Sci. Manuf.* **2021**, *144*, 106336. [CrossRef]
8. Shit, S.C.; Shah, P. A review on silicone rubber. *Natl. Acad. Sci. Lett.* **2013**, *36*, 355–365. [CrossRef]
9. Ji, X.; Lu, Z.; Wang, J.; Ye, N.; Zhang, H.; Zhou, L.; Li, J.; Lu, Y. Construction of micro-nano hybrid structure based on carbon nanotube whisker and alumina for thermally conductive yet electrically insulating silicone rubber composites. *Compos. Sci. Technol.* **2024**, *249*, 110495. [CrossRef]
10. Laubie, B.; Ohannessian, A.; Desjardin, V.; Germain, P. Methodology to assess silicone (bio) degradation and its effects on microbial diversity. *J. Environ. Polym. Degrad.* **2012**, *20*, 1019–1026. [CrossRef]
11. Zhou, W.; Qi, S.; Tu, C.; Zhao, H. Novel heat-conductive composite silicone rubber. *J. Appl. Polym. Sci.* **2007**, *104*, 2478–2483. [CrossRef]
12. Shahamatifard, F.; Rodrigue, D.; Park, K.; Frikha, S.; Mighri, F. Natural rubber nanocomposites: Effect of carbon black/multi-walled carbon nanotubes hybrid fillers on the mechanical properties and thermal conductivity. *Polym.-Plast. Technol. Mater.* **2021**, *60*, 1686–1696. [CrossRef]
13. Saengdee, L.; Phinyocheep, P.; Daniel, P. Chemical modification of natural rubber in latex stage for improved thermal, oil, ozone and mechanical properties. *J. Polym. Res.* **2020**, *27*, 275. [CrossRef]
14. Yangthong, H.; Pichaiyut, S.; Jumrat, S.; Wisunthorn, S.; Nakason, C. Mechanical, thermal, morphological, and curing properties of geopolymer filled natural rubber composites. *J. Appl. Polym. Sci.* **2019**, *136*, 47346. [CrossRef]
15. Gelling, I. Modification of natural rubber latex with peracetic acid. *Rubber Chem. Technol.* **1985**, *58*, 86–96. [CrossRef]
16. Baker, C.S.L.; Gelling, I.R.; Newell, R. Epoxidized Natural Rubber. *Rubber Chem. Technol.* **1985**, *58*, 67–85. [CrossRef]
17. Song, W.; Wang, Z.; Xing, Y.; Zhang, G.; Zhang, X.; Lu, Y.; Tan, T.; Zhang, L. Green and catalyst-free cross-linking of bio-based elastomers toward robust, reprocessable abilities and improved thermal aging resistance enabled by β-hydroxyl esters. *ACS Sustain. Chem. Eng.* **2023**, *11*, 2784–2796. [CrossRef]
18. Amiruddin, M.A.A.; Sarkawi, S.S. A rheological characteristics and kinetic study for sulphur accelerated vulcanisation of epoxidized natural rubber. *J. Rubber Res.* **2023**, *26*, 363–372. [CrossRef]
19. Mensah, B.; Onwona-Agyeman, B.; Nyankson, E.; Bensah, D.Y. Effect of palm oil as plasticizer for compounding polar and non-polar rubber matrix reinforced carbon black composites. *J. Polym. Res.* **2023**, *30*, 67. [CrossRef]
20. Salaeh, S.; Nakason, C. Influence of modified natural rubber and structure of carbon black on properties of natural rubber compounds. *Polym. Compos.* **2012**, *33*, 489–500. [CrossRef]
21. Zhao, L.; Liu, L.; Gao, J.; Zhang, L. The structure and properties of bis (γ-triethoxysilylpropyl) tetrasulfide modified silica/epoxidized natural rubber nanocomposites as tire tread. *Polymer* **2024**, *311*, 127550. [CrossRef]
22. Abdulraman, D.; Tiensing, T.; Phinyocheep, P. A facile and green pre-dispersion method for the preparation of epoxide-functionalized natural rubber/silica nanocomposites with improved mechanical properties. *Iran. Polym. J.* **2024**, *33*, 1553–1567. [CrossRef]

23. Le, H.H.; Hoang, T.X.; Haider, S.B.; Subhradeep, M.; Reuter, U.; Dhakal, K.; Adhikari, R.; Reincke, K.; Saleah, S.; Wiessner, S. A new testing strategy based on the wetting concept for characterizing rubber-filler interaction in rubber compounds and its application to the study of the influence of epoxy groups and non-rubber components on rubber-filler interaction in natural rubber compounds. *Express Polym. Lett.* **2023**, *17*, 527–545. [CrossRef]
24. Xu, T.; Jia, Z.; Luo, Y.; Jia, D.; Peng, Z. Interfacial interaction between the epoxidized natural rubber and silica in natural rubber/silica composites. *Appl. Surf. Sci.* **2015**, *328*, 306–313. [CrossRef]
25. Damampai, K.; Pichaiyut, S.; Stöckelhuber, K.W.; Das, A.; Nakason, C. Ferric ions crosslinked epoxidized natural rubber filled with carbon nanotubes and conductive carbon black hybrid fillers. *Polymer* **2022**, *14*, 4392. [CrossRef]
26. Pichaiyut, S.; Kitisavetjit, W.; Nakason, C. Synergistic effects of graphite and carbon nanotube hybrid fillers on key properties of epoxidized natural rubber nanocomposites. *J. Polym. Res.* **2023**, *30*, 412. [CrossRef]
27. Adu, P.C.O.; Aakyiir, M.; Su, X.; Alam, J.; Tran, L.C.; Dai, J.; Meng, Q.; Kuan, H.C.; Ma, J. Challenges and advancements in Elastomer/CNT nanocomposites with mechanochemical treatment, reinforcement mechanisms and applications. *Smart Mater. Manuf.* **2024**, *2*, 100053. [CrossRef]
28. Wang, Y.; Liao, L.; Lin, H.; Zhang, F.; Zhong, J.; Xu, K.; Peng, Z. Damping properties of natural rubber/epoxidized natural rubber composites with different fillers. In Proceedings of the 2017 3rd International Forum on Energy, Environment Science and Materials (IFEESM 2017), Series of Advances in Engineering Research, Shenzhen, China, 25–26 November 2017; Atlantis Press: Amsterdam, The Netherlands, 2018; pp. 772–775. [CrossRef]
29. Dahham, O.S.; Noriman, N.; Sam, S.T.; Al-Samarrai, N.M.; Shayfull, Z.; Alakrach, A.; Elnaid, S.A. The influence of rice husk fiber on the properties of epoxidized natural rubber/rice husk compounds. In Proceedings of the MATEC Web of Conferences, 2nd International Conference on Green Design and Manufacture (IConGDM 2016), Phuket, Thailand, 1–2 May 2016; Volume 78, p. 01075. [CrossRef]
30. Attharangsan, S.; Ismail, H.; Bakar, M.A.; Ismail, J. The effect of rice husk powder on standard Malaysian natural rubber grade L (SMR L) and epoxidized natural rubber (ENR 50) composites. *Polym.-Plast. Technol.* **2012**, *51*, 231–237. [CrossRef]
31. Yangthong, H.; Wisunthorn, S.; Pichaiyut, S.; Nakason, C. Novel epoxidized natural rubber composites with geopolymers from fly ash waste. *Waste Manag.* **2019**, *87*, 148–160. [CrossRef] [PubMed]
32. Salaeh, S.; Boiteux, G.; Cassagnau, P.; Nakason, C. Flexible 0–3 Ceramic-Polymer Composites of Barium Titanate and Epoxidized Natural Rubber. *Int. J. Appl. Ceram. Technol.* **2015**, *12*, 106–115. [CrossRef]
33. Salaeh, S.; Muensit, N.; Bomlai, P.; Nakason, C. Ceramic/natural rubber composites: Influence types of rubber and ceramic materials on curing, mechanical, morphological, and dielectric properties. *J. Mater. Sci.* **2011**, *46*, 1723–1731. [CrossRef]
34. Jo, J.O.; Saha, P.; Kim, N.G.; Ho, C.C.; Kim, J.K. Development of nanocomposite with epoxidized natural rubber and functionalized multiwalled carbon nanotubes for enhanced thermal conductivity and gas barrier property. *Mater. Des.* **2015**, *83*, 777–785. [CrossRef]
35. Borisova, T.M.; Castro, R.A. Mechanism of charge transport in Si/Al_2O_3/Al structures. *J. Phys. Conf. Ser.* **2013**, *416*, 012017. [CrossRef]
36. Zhou, W.; Qi, S.; Tu, C.; Zhao, H.; Wang, C.; Kou, J. Effect of the particle size of Al_2O_3 on the properties of filled heat-conductive silicone rubber. *J. Appl. Polym. Sci.* **2007**, *104*, 1312–1318. [CrossRef]
37. Hoshikawa, Y.; Kawaguchi, R.; Nomura, K.; Akahane, H.; Ishii, T.; Ando, M.; Hoshino, N.; Akutagawa, T.; Yamada, H.; Kyotani, T. Quantitative analysis of the formation mechanism of tightly bound rubber by using carbon-coated alumina nanoparticles as a model filler. *Carbon* **2021**, *173*, 870–879. [CrossRef]
38. Lin, J.L.; Su, S.M.; He, Y.B.; Kang, F.Y. Improving thermal and mechanical properties of the alumina filled silicone rubber composite by incorporating carbon nanotubes. *New Carbon Mater.* **2020**, *35*, 66–72. [CrossRef]
39. Wang, Y.; Liu, T.; Zhang, H.; Luo, N.; Chen, F.; Fu, Q. Effect of spherical alumina crystalline phase content and particle size distribution polydispersity on the properties of silicone rubber composites. *Compos. Sci. Technol.* **2023**, *243*, 110273. [CrossRef]
40. Cheng, X.; Jiang, H. Surface modification of alumina and its application in thermal conductivity silicone potting. *Silicone Mater.* **2012**, *26*, 148–152. Available online: https://caod.oriprobe.com/articles/30438443/Surface_Modification_of_Alumina_and_its_Application_in_Thermal_Conduct.htm (accessed on 4 April 2024).
41. Presto, D.; Meyerhofer, J.; Kippenbrock, G.; Narayanan, S.; Ilavsky, J.; Moctezuma, S.; Sutton, M.; Foster, M.D. Influence of Silane Coupling Agents on Filler Network Structure and Stress-Induced Particle Rearrangement in Elastomer Nanocomposites. *ACS Appl. Mater. Interfaces* **2020**, *12*, 47891–47901. [CrossRef]
42. Ryu, G.; Kim, D.; Song, S.; Hwang, K.; Ahn, B.; Kim, W. Effect of Epoxide Content on the Vulcanizate Structure of Silica-Filled Epoxidized Natural Rubber (ENR) Compounds. *Polymers* **2021**, *13*, 1862. [CrossRef] [PubMed]
43. *D1646-89*; Viscosity and Vulcanization Characteristics (Mooney Voscometer). Annual Book of ASTM Standards: West Conshohocken, PA, USA, 1990; Volume 9, p. 307.
44. Solórzano, E.; Reglero, J.A.; Rodríguez-Pérez, M.A.; Lehmhus, D.; Wichmann, M.; De Saja, J.A. An experimental study on the thermal conductivity of aluminium foams by using the transient plane source method. *Int. J. Heat Mass Transf.* **2008**, *51*, 6259–6267. [CrossRef]
45. Samran, J.; Phinyocheep, P.; Daniel, P.; Kittipoom, S. Hydrogenation of unsaturated rubbers using diimide as a reducing agent. *J. Appl. Polym. Sci.* **2005**, *95*, 16–27. [CrossRef]

46. Padmaja, P.; Anilkumar, G.M.; Mukundan, P.; Aruldhas, G.; Warrier, K.G.K. Characterisation of stoichiometric sol–gel mullite by fourier transform infrared spectroscopy. *Int. J. Inorg. Mater.* **2001**, *3*, 693–698. [CrossRef]
47. Vasconcelos, D.C.L.; Nunes, E.H.M.; Vasconcelos, W.L. AES and FTIR characterization of sol–gel alumina films. *J. Non-Cryst. Solids* **2012**, *358*, 1374–1379. [CrossRef]
48. Innocenzi, P. Infrared spectroscopy of sol–gel derived silica-based films: A spectra-microstructure overview. *J. Non-Cryst. Solids* **2003**, *316*, 309–319. [CrossRef]
49. Sae-Oui, P.; Rakdee, C.; Thanmathorn, P. Use of rice husk ash as filler in natural rubber vulcanizates: In comparison with other commercial fillers. *J. Appl. Polym. Sci.* **2002**, *83*, 2485–2493. [CrossRef]
50. Wagner, M. Reinforcing silicas and silicates. *Rubber Chem. Technol.* **1976**, *49*, 703–774. [CrossRef]
51. Arroyo, M.; Lopez-Manchado, M.A.; Valentin, J.T.; Carretero, J. Morphology/behaviour relationship of nanocomposites based on natural rubber/epoxidized natural rubber blends. *Compos. Sci. Technol.* **2007**, *67*, 1330–1339. [CrossRef]
52. Varghese, S.; Karger-Kocsis, J.; Gatos, K.M. Melt compounded epoxidized natural rubber/layered silicate nanocomposites: Structure-properties relationships. *Polymer* **2003**, *44*, 3977–3983. [CrossRef]
53. Da Costa, H.M.; Visconte, Y.L.L.; Nunes, R.C.R.; Furtado, C.R.G. Rice husk ash filled natural rubber. III. Role of metal oxides in kinetics of sulfur vulcanization. *J. Appl. Polym. Sci.* **2003**, *90*, 1519–1531. [CrossRef]
54. Yangthong, H.; Pichaiyut, S.; Wisunthorn, S.; Kummerlöwe, C.; Vennemann, N.; Nakason, C. Role of geopolymer as a cure activator in sulfur vulcanization of epoxidized natural rubber. *J. Appl. Polym. Sci.* **2020**, *137*, 48624. [CrossRef]
55. Yangthong, H.; Nun-anan, P.; Faibunchan, P.; Karrila, S.; Limhengha, S. The enhancement of cure and mechanical properties of natural rubber vulcanizates with waste Aquilaria crassna wood. *Ind. Crops Prod.* **2021**, *171*, 113922. [CrossRef]
56. Alam, M.N.; Mandal, S.K.; Roy, K.; Debnath, S.C. Synergism of novel thiuram disulfide and dibenzothiazyl disulfide in the vulcanization of natural rubber: Curing, mechanical and aging resistance properties. *Int. J. Ind. Chem.* **2014**, *5*, 8. [CrossRef]
57. Tanrattanakul, V.; Wattanathai, B.; Tiangjunya, A.; Muhamud, P. In situ epoxidized natural rubber: Improved oil resistance of natural rubber. *J. Appl. Polym. Sci.* **2003**, *90*, 261–269. [CrossRef]
58. Mathew, N.; De, S.K. Thermo-oxidative ageing and its effect on the network structure and fracture mode of natural rubber vulcanizates. *Polymer* **1983**, *24*, 1042–1054. [CrossRef]
59. Ryningen, B.; Bellmann, M.P.; Kvande, R.; Lohne, O. The effect of crucible coating and the temperature field on minority carrier lifetime in directionally solidified multicrystalline silicon ingots. In Proceedings of the Conference 27th European Photovoltaic Solar Energy Conference and Exhibition, Frankfurt, Germany, 24–28 September 2012. [CrossRef]
60. Na Nakorn, P.; Jinawath, S.; Wada, S. Mechanical strength and thermal conductivity of high purity Al_2O_3 ceramics using AKP-30 Powder. *J. Sci. Res. Chula. Univ.* **2005**, *30*, 77–85. Available online: https://www.thaiscience.info/journals/Article/CJSR/10324248.pdf (accessed on 4 April 2024).
61. Jacob, M.; Francis, B.; Thomas, S.; Varughese, K.T. Dynamical mechanical analysis of sisal/oil palm hybrid fiber-reinforced natural rubber composites. *Polym. Compos.* **2006**, *27*, 671–680. [CrossRef]

polymers

MDPI

Article

Self-Oscillation of Liquid Crystal Elastomer Fiber-Slide System Driven by Self-Flickering Light Source

Dali Ge [1,2,3], Qingrui Hong [1], Xin Liu [1] and Haiyi Liang [1,2,3,*]

1 School of Civil Engineering, Anhui Jianzhu University, Hefei 230601, China
2 IAT-Chungu Joint Laboratory for Additive Manufacturing, Institute of Advanced Technology, University of Science and Technology of China, Hefei 241200, China
3 CAS Key Laboratory of Mechanical Behavior and Design of Materials, Department of Modern Mechanics, University of Science and Technology of China, Hefei 230026, China
* Correspondence: hyliang@ustc.edu.cn

Abstract: Self-oscillation, a control approach inspired by biological systems, demonstrates an autonomous, continuous, and regular response to constant external environmental stimuli. Until now, most self-oscillation systems have relied on a static external environment that continuously supplies energy, while materials typically absorb ambient energy only intermittently. In this article, we propose an innovative self-oscillation of liquid crystal elastomer (LCE) fiber-slide system driven by a self-flickering light source, which can efficiently regulate the energy input in sync with the self-oscillating behavior under constant voltage. This system primarily consists of a photo-responsive LCE fiber, a slider that includes a conductive segment and an insulating segment, a light source, and a conductive track. Using the dynamic LCE model, we derive the governing equation for the motion of the LCE fiber-slider system. Numerical simulations show that the LCE fiber-slide system under constant voltage exhibits two distinct motion phases, namely the stationary phase and the self-oscillation phase. The self-oscillation occurs due to the photo-induced contraction of the LCE fiber when the light source is activated. We also investigate the critical conditions required to initiate self-oscillation, and examine key system parameters influencing its frequency and amplitude. Unlike the continuous energy release from the static environmental field in most self-oscillation systems, our LCE fiber-slide self-oscillation system is driven by a self-flickering light source, which dynamically adjusts the energy input under a constant voltage to synchronize with the self-oscillating behavior. Our design features advantages such as spontaneous periodic lighting, a simple structure, energy efficiency, and ease of operation. It shows significant promise for dynamic circuit systems, monitoring devices, and optical applications.

Keywords: self-oscillation; photo-responsive; liquid crystal elastomer; fiber; dynamic circuit

1. Introduction

A diverse range of artificial materials has been meticulously engineered to respond effectively to various stimuli, including light [1,2], heat [3,4], chemical substances [5,6], magnetic fields [7], electric fields [8,9], humidity [10,11], and more. Many artificial materials, including liquid crystal elastomers (LCEs) [1,2], polyvinylidene fluoride (PVDF) [3], hydrogels [4,6], and carbon nitride polymers [11], possess a remarkable ability to undergo reversible shape changes when subjected to appropriate stimuli, enabling them to adapt dynamically to their environments [12,13]. In some cases, they can even exhibit controlled movement, which allows for complex interactions with surrounding conditions [14]. This unique feature not only enhances their functionality but also paves the way for their application in the development of intelligent and autonomous devices, including wearable health monitors, autonomous delivery robots, smart home assistants, and energy harvesters.

However, a significant challenge in the application of autonomous devices is that, in numerous instances, the associated external factors demand proper control strategies to achieve consistent and predictable dynamic behavior [15–17]. Conventional control strategies include electronic control, sensor-based control, external device control, and even manual control [18,19]. These strategies typically depend on external devices to send control signals, sensors to measure changes in physical quantities and provide real-time feedback, or manual switches to adjust the system. Nonetheless, conventional control strategies often have limitations, such as reliance on human intervention or the need for complex control systems. Additionally, they generally lack intelligent functionality. Given the diverse and complex real application scenarios, conventional control strategies for designing or utilizing autonomous devices may face numerous challenges.

The intelligent and autonomous features of biological organisms serve as a rich source of inspiration for designing autonomous devices. Self-oscillation in these organisms—exemplified by the involuntary beating of the heart, neural impulses, and cell cycling—demonstrates an autonomous, continuous, and regular response to constant external environmental stimuli, without the need for additional on/off switching signals [20,21]. The amplitude and frequency of their oscillations are determined by their intrinsic characteristics, exhibiting remarkable robustness [22,23]. Given these advantages, self-oscillation systems show great promise for use in autonomous robotics [24–26], energy-absorbing devices [27,28], mass transport systems [29,30], mechano-logistic devices [31], optical devices [32,33], and more.

To satisfy the multi-functional needs of different applications, a vast number of self-sustained motion modes have been constructed, such as bending [34–38], buckling [6,39–42], rocking [43–45], jumping [46,47], rolling [24,48–54], rotation [23,54–59], spinning [60,61], swimming [62,63], swinging [64,65], stretching and shrinking [34,66,67], snap-through [68,69], torsion [23,29,34,70], vibration [34,71–74], and even the harmonized motion of several interconnected self-oscillating systems [75,76]. Such self-sustained movements are often the result of nonlinear feedback processes, including phenomena like self-shadowing [40,44], the interaction between large deformations and chemical reactions [6], photothermal evaporation processes [37], and gradients in surface tension induced by photothermal effects [77].

Along with many other similar artificial materials, LCE is a type of responsive material that combines the elastic properties of long polymer chains with the ordered structure of liquid crystal (LC) mesogens. Due to the ability of these mesogens to reversibly switch between isotropic and anisotropic states, LCE can respond to various physical stimuli, including magnetic fields, heat, electricity, and light, resulting in significant macroscopic deformations [18,78]. Among these stimuli, optical excitation stands out due to its distinct benefits, including the ability to drive systems without physical contact, flexibility in programming, ease of control, high precision, and environmental sustainability [16,21,28]. When subjected to optical stimuli, optically responsive LCEs demonstrate rapid response times, substantial intrinsic deformation, and reversible shape changes. These distinctive characteristics facilitate self-oscillation in diverse forms, leading to considerable advancements in the development of LCE-based optically powered self-oscillation systems [24,29,32,33,40–42].

So far, most self-oscillation systems have been developed based on the continuous energy release from a static environmental field [40–42,52,56,62]. However, in these systems, materials tend to absorb ambient energy only intermittently, leading to considerable energy waste. To tackle this issue, we propose a novel self-oscillation of the LCE fiber-slide system driven by a self-flickering light source, which effectively regulates the energy source in harmony with the self-oscillating behavior. This system primarily consists of a photo-responsive LCE fiber, a slider that includes a conductive segment and an insulating segment, a light source, a conductive track, and direct current (DC) power supply. In contrast to the continuous energy release from the static environmental field in most self-oscillation systems, our system features a dynamic, self-flickering light source that interacts with the oscillation of the slider when exposed to a constant voltage, enabling periodic switching over time. This setup offers advantages such as spontaneous periodic lighting, a simple structure, energy efficiency, and ease of operation. It holds promise for constructing dynamic circuit systems, monitoring and warning devices, and optical devices, as well as other advanced technologies.

The paper is organized in the following way. In Section 2, a self-oscillating LCE fiber-slide system driven by a self-flickering light source is introduced, along with the formulation of its dynamic governing equations. In Section 3, numerical simulations are employed to explore two phases: the stationary phase and the self-oscillation phase, with a detailed revelation of the mechanisms underlying self-oscillation. In Section 4, the impact of system parameters on the self-oscillation behavior is extensively studied. Finally, conclusions are drawn in Section 5.

2. Theoretical Framework

In this section, we present an analysis of the self-oscillation in an LCE fiber-slide system powered by a self-flickering light source. Key topics include the theoretical framework for the self-oscillating LCE fiber-slider system, the dynamics of the cis molecule concentration evolution within the LCE fiber, the process of nondimensionalization, and the approach for solving the differential governing equations that feature variable coefficients.

2.1. Self-Oscillating Dynamic of the LCE Fiber-Slide System

Figure 1 illustrates a self-oscillation of the LCE fiber-slide system driven by a self-flickering light source. The system consists of a photo-responsive LCE fiber with an initial length of L_0 in a stress-free state, a long slider with a mass m that consists of a conductive segment and an insulating segment, a DC power supply, a light source, conductive wires, and a conductive track. In the reference state, the azobenzene molecules in the LCE fiber are aligned along the length of the fiber, maintaining a straight trans state, as illustrated in Figure 1a. It is widely recognized that exposure to light prompts the azobenzene molecules to shift from the straight trans state to a bent cis state, leading to reversible photo-induced contraction along the fiber's length. This contraction can subsequently return to its original state in the absence of light.

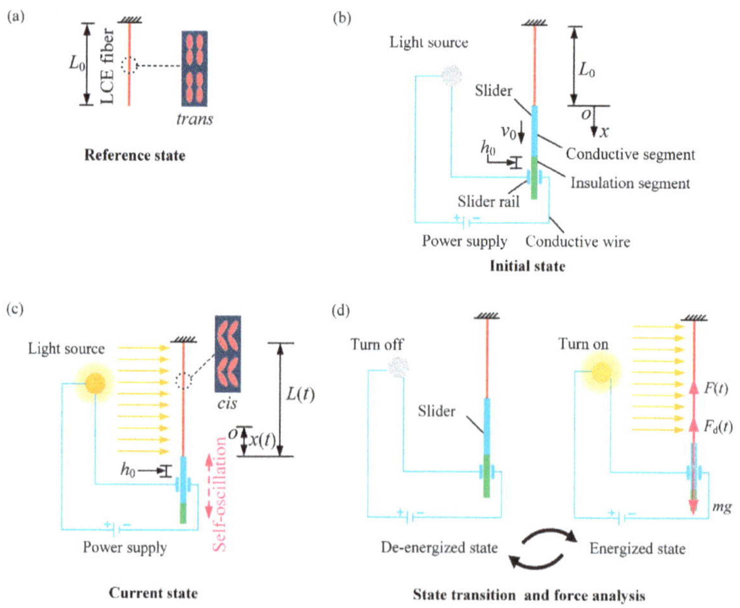

Figure 1. Schematic diagram of a self-oscillation of LCE fiber-slide system driven by self-flickering light source. (**a**) Reference state. (**b**) Initial state. (**c**) Current state. (**d**) State transition and force analysis.

One terminus of the LCE fiber is anchored at a suspension point, with its opposite extremity connected to a slider. The conductive segment of the slider is represented by the blue zone, and the insulating segment is highlighted in the green zone. The slider is situated within the conductive track. In a stress-neutral condition, the upper end of the insulating segment of the slider is positioned h_0 above the conductive track, as shown in Figure 1b. Conductive wires connect the power supply, light source, and both ends of the conductive track in series. Together, the power supply, light source, conducting wires, slider, and conductive track form a complete circuit system. The state of the circuit system, whether energized or de-energized, is determined by whether the conductive segment of the slider is in contact with the track or if the insulated segment of the slider is in contact with the track, as depicted in Figure 1d. In the starting configuration, the slider is set free under specific initial conditions. Influenced by gravity, the slider begins to move downward. To accurately describe the instantaneous displacement $x(t)$ of the slider at time t, we establish a coordinate system with the reference O positioned at the bottom terminus of the fiber in its stress-free state, as illustrated in Figure 1b. At this moment, the LCE fiber is stretched to a length $L(t)$ and experiences a tensile force $F(t)$. As the slider's displacement increases and surpasses h_0, its conductive segment comes into contact with the conductive track, energizing the circuit and activating the light source, which exposes the entire length of the LCE fiber to light, as shown in Figure 1c. Consequently, the LCE fiber contracts along its length, resulting in a reduction in its overall length. It is essential to understand that the tensile force $F(t)$, arising from the combined effects of the fiber's stretched state and the photo-induced contraction, causes the slider to decelerate and retract under the combined action of the damping force $F_d(t)$. As the slider moves upward and its insulating segment contacts the conductive track, the circuit enters a de-energized state, turning off the light source and allowing the LCE fiber to return to its original state in the dark. Concurrently, the decrease in tensile force $F(t)$ results in the slider accelerating downward, eventually causing it to move downward again. Thus, the slider may oscillate continuously over time, accompanied by the dynamic flickering of the light source.

During the motion of the slider, we assume that the frictional force generated between the slider and the track can be negligible. Under this assumption, the slider is subjected to the tensile force $F(t)$ of the LCE fiber, gravity mg, and the damping force $F_d(t)$, as illustrated in Figure 1d. Then, the equation governing the motion of the slider can be derived as follows:

$$m\ddot{x} = mg - F(t) - F_d(t), \tag{1}$$

with the initial conditions

$$x(t) = x_0 \text{ and } \dot{x} = v_0 \text{at } t = 0, \tag{2}$$

where g is the gravity acceleration, $\ddot{x}$ represents the acceleration of the slider, and $F_d(t)$ represents the damping force. It is essential to understand that, in practical situations, the damping force $F_d(t)$ is intricately related to the velocity of the system. To facilitate subsequent calculations, the damping force $F_d(t)$ is assumed to be directly proportional to the velocity $\dot{x}$ of the slider, with the force acting in the reverse direction to the motion, which can be given as following equation:

$$F_d(t) = \beta\dot{x}, \tag{3}$$

where β is the damping coefficient.

To determine the tensile force $F(t)$ in Equation (1), an assumption being made is that the tensile force is uniformly distributed and varies in direct proportional to the elastic strain $\varepsilon_e(t)$, i.e.,

$$F(t) = kL_0\varepsilon_e(t), \tag{4}$$

where k is the stiffness coefficient of the LCE fiber. Notably, the elastic strain $\varepsilon_e(t)$ in the LCE fiber is assumed to be homogeneous and can be represented as a linear combination of the overall strain $\varepsilon_{tot}(t)$ and the photo-induced contraction $\varepsilon(t)$. Considering the geometric relationship between the overall deformation of the LCE fiber and the displacement of the slider, the overall strain can be written as the following:

$$\varepsilon_{tot}(t) = x(t)/L_0. \tag{5}$$

Consequently, the tensile force can be given as the following:

$$F(t) = k[x(t) - \varepsilon(t)L_0]. \tag{6}$$

To determine the tensile force in Equation (6), it is essential to first estimate the photo-induced contraction $\varepsilon(t)$, the details of which are provided in the next section.

2.2. Dynamic LCE Model

The photo-induced contraction strain $\varepsilon(t)$ is closely related to the concentration $\varphi(t)$ of cis molecules within the LCE fiber. To simplify, we assume that the concentration is significantly smaller than 1. Consequently, by applying a Taylor expansion and considering only the first-order term, we can express the photo-induced contraction strain $\varepsilon(t)$ as being proportional to the cis molecule's concentration, i.e.,

$$\varepsilon(t) = -C_0\varphi(t), \tag{7}$$

where C_0 represents the contraction coefficient.

According to the previous work of Yu et al., trans-to-cis molecules in LCE can be induced by UV or laser light with wavelengths shorter than 400 nm [78]. Furthermore, the concentration $\varphi(t)$ of the cis molecules is influenced by thermal excitation from trans molecules to cis molecules, and the light-driven relaxation from cis to trans [20,79,80]. Typically, the thermal excitation from trans molecules to cis molecules is regarded as insignificant in comparison to the photo-induced processes [78,81]. To determine the concentration $\varphi(t)$ of cis molecules within the LCE fiber, we employ the well-established dynamic LCE model proposed by Finkelmann et al. [20,79,80]. The concentration $\varphi(t)$ is governed by the following equation:

$$\frac{d\varphi(t)}{dt} = \eta_0 I(t)[1 - \varphi(t)] - \tau^{-1}\varphi(t), \tag{8}$$

where τ is the thermal relaxation time from cis molecules to trans molecules, η_0 is the light absorption constant, and the light intensity $I(t)$ is determined by the current displacement of the slider. When the LCE fiber is illuminated due to the circuit being turned on (i.e., $x(t) > h_0$), the light intensity is set at $I(t) = I_0$. Clearly, when the circuit is turned off (i.e., $x(t) \leq h_0$), the light intensity can be set to $I(t) = 0$.

2.3. Nondimensionalization

To aid in exploring the dynamic characteristics of the self-oscillation of the LCE fiber-slider system, the following dimensionless quantities are defined: $\overline{I}_0 = \eta_0 I_0 \tau_0$, $\overline{I} = \eta_0 I \tau_0$, $\overline{t} = t/\tau$, $\overline{x} = x/L_0$, $\overline{x}_0 = x_0/L_0$, $\overline{v}_0 = \tau v_0/L_0$, $\overline{\dot{x}} = \dot{x}\tau/L_0$, $\overline{\ddot{x}} = \ddot{x}\tau^2/L_0$, $\overline{F} = F\tau^2/mL_0$, $\overline{F}_d = F_d\tau^2/mL_0$, $\overline{g} = g\tau^2/L_0$, $\overline{k} = k\tau^2/m$, $\overline{\beta} = \beta\tau/m$, and $\overline{h}_0 = h/L_0$. The governing Equations (1)–(6) for the slider can be rewritten in the following form:

$$\overline{\ddot{x}} = \overline{g} - \overline{F}(\overline{t}) - \overline{F}_d(\overline{t}), \tag{9}$$

with the initial conditions

$$\overline{x}(\overline{t}) = \overline{x}_0 \text{and } \overline{\dot{x}} = \overline{v}_0 \text{ at } \overline{t} = 0. \tag{10}$$

The dimensionless damping force can be calculated as

$$\overline{F}_d(\overline{t}) = \overline{\beta}\dot{\overline{x}}. \tag{11}$$

The tensile force of the LCE fiber can be determined as follows:

$$\overline{F}(\overline{t}) = \overline{k}[\overline{x}(\overline{t}) + C_0\varphi(\overline{t})]. \tag{12}$$

Similarly, the concentration can be determined by

$$\frac{d\varphi(\overline{t})}{d\overline{t}} = \overline{I}(\overline{t})[1 - \varphi(\overline{t})] - \varphi(\overline{t}), \tag{13}$$

where $\overline{I}(\overline{t}) = \overline{I}_0$ for $\overline{x}(\overline{t}) > \overline{h}_0$; otherwise, $\overline{I}(\overline{t}) = 0$.

Equations (7) and (11) govern the self-oscillating behavior of the LCE fiber-slider system. To tackle these complex differential equations with variable coefficients, we employ the Runge–Kutta method for numerical calculations in MATLAB 2023b software. At time $\overline{t}_i$, the concentration $\varphi(\overline{t}_i)$ of the *cis* molecules and the displacement $\overline{x}(\overline{t}_i)$ can be used to determine the current tensile force $\overline{F}(\overline{t}_i)$ in the LCE fiber according to Equation (12). Subsequently, the displacement $\overline{x}(\overline{t}_{i+1})$ of the slider at time $\overline{t}_{i+1}$ can be determined using Equation (9), while the light intensity $\overline{I}(\overline{t}_{i+1})$ is simultaneously estimated. Subsequently, the concentration $\varphi(\overline{t}_{i+1})$ of the *cis* molecules is derived from Equation (13). In summary, by iterating this process, the self-oscillation behavior of the LCE fiber-slider system can be numerically obtained for specified parameters: $\overline{I}_0$, C_0, $\overline{k}$, $\overline{x}_0$, $\overline{v}_0$, $\overline{h}_0$, $\overline{\beta}$, and $\overline{g}$. It is worth mentioning that $\overline{x}_0$ can be converted into the corresponding $\overline{v}_0$ by utilizing the system's energy relationship. To be concise, we fix $\overline{x}_0 = 0$ in the calculation.

3. Two Motion Phases and Mechanism of Self-Oscillation

In this section, we begin by discussing two distinct motion phases of the self-oscillation of the LCE fiber-slider system driven by a self-flickering light source, specifically the stationary phase and the self-oscillation phase. Following this, we offer a comprehensive explanation of the mechanism underlying the self-oscillation.

3.1. Two Motion Phases

To investigate the self-oscillation behavior of the self-oscillation of the LCE fiber-slider system, we should first establish the specific values of the dimensionless parameters in the model. Typical material properties and geometric parameters are provided in Table 1 [81–83]. The corresponding dimensionless parameters are also detailed in Table 2. These parameter values are utilized in the subsequent analysis of the self-oscillation behavior of the LCE fiber-slider system.

Table 1. Material properties and geometric parameters.

Parameter	Definition	Value	Units
C_0	Contraction coefficient	0~0.5	/
I_0	Light intensity	0~0.15	kW/m^2
τ	*trans*-to-*cis* thermal relaxation time	1~100	ms
η_0	Light absorption constant	0.0003	$m^2/(s{\cdot}W)$
L_0	Original length of LCE fiber	1~10	cm
k	Stiffness coefficient	0~20	N/m
g	Gravitational acceleration	10	m/s^2
β	Damping coefficient	0~0.6	mg·mm/s
m	Mass of the slider	0~10	g
h_0	Position of the conductive track	0~0.1	m
v_0	Initial velocity	0~0.1	m/s

Table 2. Dimensionless parameters.

Parameter	$\overline{I}_0$	$\overline{\beta}$	$\overline{k}$	$\overline{v}_0$	$\overline{g}$	$\overline{h}_0$
Value	0~0.5	0~0.5	0~10	0~0.2	0.1~2	0~0.3

Figure 2 illustrates two distinct motion phases of the self-oscillation in the LCE fiber-slider system: the stationary phase for $\overline{I}_0 = 0$, as shown in Figure 2a,b, and the self-oscillation phase for $\overline{I}_0 \neq 0$, as shown in Figure 2c,d. The other parameters are set as follows: $C_0 = 0.2$, $\overline{\beta} = 0.1$, $\overline{k} = 5.8$, $\overline{v}_0 = 0.1$, $\overline{g} = 1.2$, and $\overline{h}_0 = 0.2$. Through the numerical solution of Equations (7) and (11), we can derive both the temporal behavior of displacement and phase trajectory curves of the slider. In the scenario where $\overline{I}_0 = 0$, the slider experiences vertical oscillations driven by gravitational forces and the tension within the LCE fiber. However, due to damping effects, the amplitude and velocity of the oscillation gradually diminish, causing the slider to ultimately settle into a stationary equilibrium position, which we term the stationary phase, as illustrated Figure 2a,b. Conversely, when $\overline{I}_0 = 0.1$, the slider continues to oscillate due to the effects of gravity and the tensile force. During the alternating contact between the conductive and insulating segments of the slider and the conductive track, the dynamic circuit causes the light source to switch between the activated and deactivated states, resulting in self-flickering. The incorporation of a dynamic electrical circuit enables the LCE material to convert light energy from the self-flickering light source, effectively offsetting energy losses due to dissipation. As a result, the amplitude of the slider's vibrations increases over time until it stabilizes. With a constant voltage supplied by the DC power source, the slider eventually undergoes continuous periodic oscillations, entering what is known as the self-oscillation phase, as shown in Figure 2c,d.

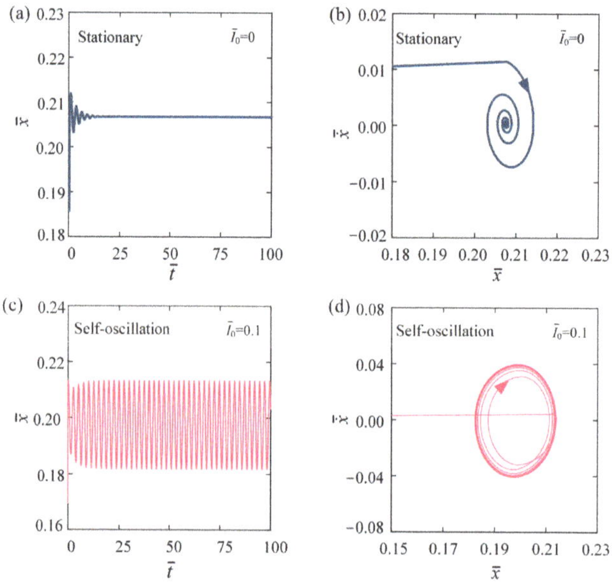

Figure 2. Temporal behavior of displacement and phase trajectory curves for two motion phases. (**a**) Displacement temporal behavior with $\overline{I}_0 = 0$; (**b**) phase trajectory with $\overline{I}_0 = 0$; (**c**) displacement temporal behavior with $\overline{I}_0 = 0.1$; and (**d**) phase trajectory with $\overline{I}_0 = 0.1$. Other parameters are $C_0 = 0.2$, $\overline{\beta} = 0.1$, $\overline{k} = 5.8$, $\overline{v}_0 = 0.1$, $\overline{g} = 1.2$, and $\overline{h}_0 = 0.2$. Under constant voltage, the LCE fiber-slide system driven by self-flickering light source exhibits two distinct motion phases: the stationary phase and the self-oscillation phase.

3.2. Mechanisms of Self-Oscillation

To elucidate the self-oscillation mechanism of the self-oscillation of the LCE fiber-slider system, Figure 3 presents several key physical quantities of the fiber-slider system for $C_0 = 0.2$, $\overline{I}_0 = 0.1$, $\overline{\beta} = 0.1$, $\overline{k} = 5.8$, $\overline{v}_0 = 0.1$, $\overline{g} = 1.2$, and $\overline{h}_0 = 0.2$. The green-highlighted regions in Figure 3a,b signify that the circuit is energized and the LCE fiber is subjected to light. As the slider moves downward from its initial state, the length of the LCE fiber increases, and then the slider's conductive segment makes contact with the conductive track, causing the circuit to be in an energized state. At this point, φ of the LCE fiber gradually increases over time, leading to a decrease in the length of the fiber and causing the slider to move upward. Once the insulating segment of the slider makes contact with the conductive track, the circuit enters a de-energized state. At this stage, φ gradually decreases over time, resulting in an elongation of the LCE fiber and causing the slider to move downward again. As a result, the concentration of the *cis* molecules exhibits a periodic change over time during the slider's motion, as shown in Figure 3a. Moreover, it is clear from Equation (4) that the tensile force $\overline{F}$ also exhibits a periodic variation over time, as depicted Figure 3b. This ultimately results in the displacement of the slider, which similarly demonstrates periodic changes.

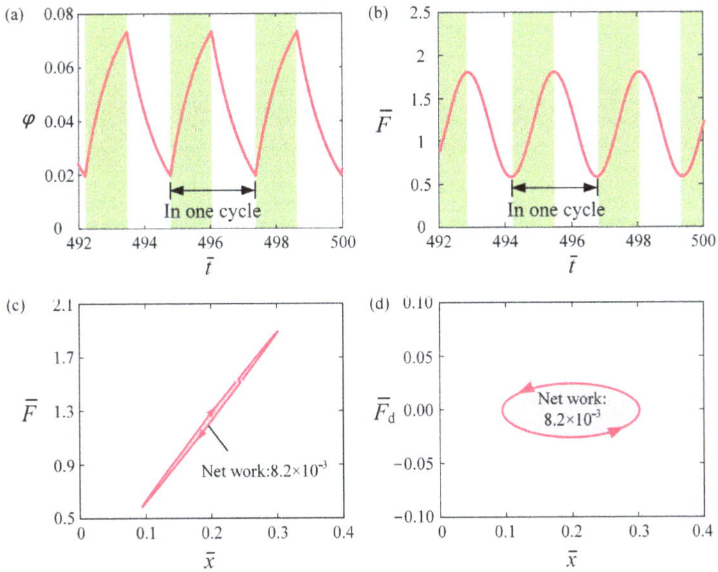

Figure 3. Mechanism of the self-oscillation of the LCE fiber-slider system driven by self-flickering light source for the typical case in Figure 2c,d. (**a**) Variation of concentration of *cis* molecules over $\overline{t}$; (**b**) Variation of LCE fiber $\overline{F}$ over $\overline{t}$. (**c**) The relationship between $\overline{F}$ and $\overline{x}$, and (**d**) relationship between $\overline{F}_d$ and $\overline{x}$. The damping dissipation is balanced by the energy input from the tensile force, thereby maintaining stable self-oscillation.

Figure 3c describes the variation of the tensile force with displacement. The curve representing the tensile force $\overline{F}$ against displacement $\overline{x}$ forms a closed loop in a clockwise direction. The area enclosed by this loop signifies the net work performed by tensile force on the system, representing the energy input, which is calculated to be 0.0082 over one cycle. Figure 3d depicts the variation of the damping force with displacement, forming a closed loop in a counterclockwise direction. The area within this loop corresponds to the damping dissipation, also calculated to be 0.0082—exactly matching the net work done by the tensile force. As a result, during the self-oscillation of the LCE fiber-slider system driven by the flickering light source, the damping dissipation is balanced by the energy input from the tensile force, thereby maintaining the stable self-oscillation of the system.

4. Parameter Analysis

In the preceding analysis of the self-oscillation in an LCE fiber-slide system powered by a self-flickering light source, seven dimensionless system parameters were identified: $\overline{I}_0$, C_0, $\overline{\beta}$, $\overline{k}$, $\overline{v}_0$, $\overline{g}$, and $\overline{h}_0$. This section aims to investigate how different system parameters influence the triggering conditions for self-oscillation, as well as their effects on the dimensionless amplitude A and frequency f of self-oscillation.

4.1. Impact of the Light Intensity

Figure 4 illustrates the impact of light intensity $\overline{I}_0$ on the self-oscillation behavior of the LCE fiber-slide system. We set $C_0 = 0.2$, $\overline{\beta} = 0.1$, $\overline{k} = 5.8$, $\overline{v}_0 = 0.1$, $\overline{g} = 1.2$, and $\overline{h}_0 = 0.2$. Figure 4a depicts the stable cycles for various $\overline{I}_0$. The threshold value for initiating self-oscillation in the system is approximately $\overline{I}_0 = 0.1$. Specifically, when $\overline{I}_0 < 0.1$, the system stays in a stationary phase, whereas at $\overline{I}_0 = 0.1$, $\overline{I}_0 = 0.12$, and $\overline{I}_0 = 0.14$, self-oscillation occurs. Figure 4b presents the relationship between the frequency f and amplitude A of self-oscillation and light intensity $\overline{I}_0$. As $\overline{I}_0$ increases, A tends to rise, while f remains relatively stable. As $\overline{I}_0$ increases, more energy is absorbed and converted into mechanical energy, leading to a rise in A. On the other hand, the stability in frequency can be attributed to the fact that the oscillation frequency of the LCE fiber-slide system is mainly governed by its inherent mechanical characteristics. Thus, increasing the $\overline{I}_0$ can enhance the practical use of self-oscillation in the LCE fiber-slide system.

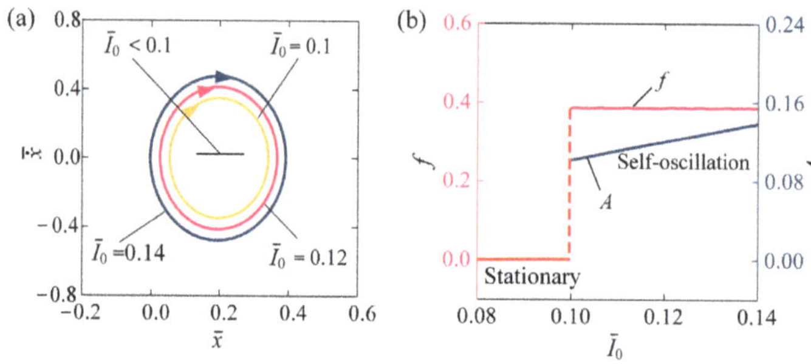

Figure 4. Influence of light intensity on self-oscillation of LCE fiber-slide system, for $C_0 = 0.2$, $\overline{\beta} = 0.1$, $\overline{k} = 5.8$, $\overline{v}_0 = 0.1$, $\overline{g} = 1.2$, and $\overline{h}_0 = 0.2$. (**a**) Stable cycles. (**b**) f and A. As $\overline{I}_0$ increases, the A tends to rise, while the f remains relatively stable.

4.2. Impact of the Contraction Coefficient

Figure 5 illustrates how the contraction coefficient C_0 impacts the self-oscillating behavior of the LCE fiber-slide system for $\overline{I}_0 = 0.1$, $\overline{\beta} = 0.1$, $\overline{k} = 5.8$, $\overline{v}_0 = 0.1$, $\overline{g} = 1.2$, and $\overline{h}_0 = 0.2$. Figure 5a shows the stable cycles for different C_0. The critical C_0 to trigger the self-oscillation of the LCE fiber-slide system is approximately $C_0 = 0.1$. That is, for $C_0 < 0.1$, the system is in a stationary phase, while for $C_0 = 0.1$, $C_0 = 0.2$, and $C_0 = 0.3$, the system exhibits a self-oscillation phase. Figure 5b demonstrates how the f and A vary with C_0. As C_0 rises, there is a noticeable increase in A, while the f remains relatively unchanged. Equation (5) supports the observation that as C_0 increases, the photo-induced contraction also increases, resulting in an increase in tensile force; consequently, the amplitude of self-oscillation increases. The findings indicate that boosting the contraction coefficient of the LCE materials can enhance the efficiency of converting light energy into mechanical energy.

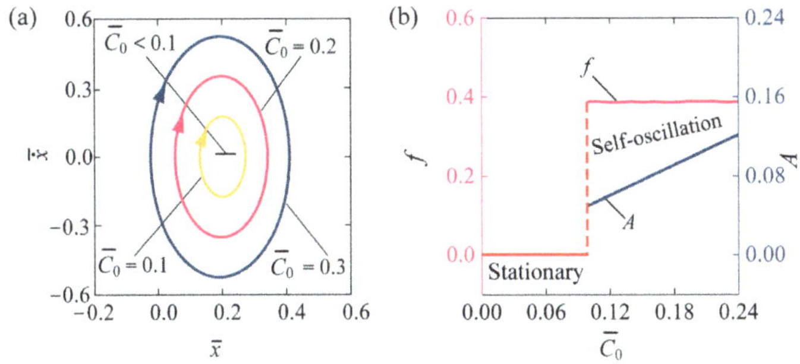

Figure 5. Influence of contraction coefficient on self-oscillation of LCE fiber-slide system, for $\overline{I}_0 = 0.1$, $\overline{\beta} = 0.1$, $\overline{k} = 5.8$, $\overline{v}_0 = 0.1$, $\overline{g} = 1.2$, and $\overline{h}_0 = 0.2$. (**a**) Stable cycles. (**b**) f and A. As the contraction coefficient increases, the A shows a distinct increase, while the f stays largely unaffected.

4.3. Impact of the Damping Coefficient

Figure 6 describes the influence of damping coefficient $\overline{\beta}$ on the self-oscillating behavior of the LCE fiber-slide system. In the analysis, we set $\overline{I}_0 = 0.1$, $C_0 = 0.2$, $\overline{k} = 5.8$, $\overline{v}_0 = 0.1$, $\overline{g} = 1.2$, and $\overline{h}_0 = 0.2$. Figure 6a illustrates the stable cycles associated with various damping coefficients. The threshold value for initiating self-oscillation is approximately $\overline{\beta} = 0.6$. In other terms, for $\overline{\beta} > 0.6$, the system is in a stationary phase, while for $\overline{\beta} = 0.1$, $\overline{\beta} = 0.2$, and $\overline{\beta} = 0.3$, the system is in a self-oscillation phase. Figure 6b shows how the f and A change with respect to $\overline{\beta}$. As $\overline{\beta}$ increases, A presents a decreasing trend, while f remains almost unchanged. When $\overline{\beta}$ surpasses 0.6, the energy input becomes inadequate to counterbalance the energy lost due to damping, resulting in the system reaching a stationary phase. Consequently, to enhance energy harvesting in engineering applications, a strategic reduction in the damping coefficient is an effective approach to improve light energy absorption during the self-oscillation of the LCE fiber-slide system.

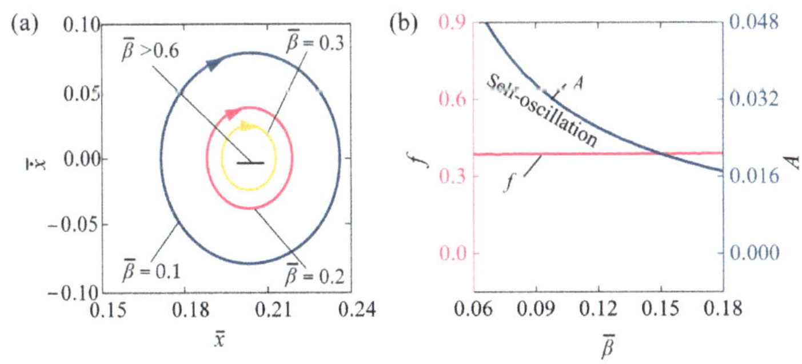

Figure 6. Influence of damping coefficient on self-oscillation of LCE fiber-slide system, for $\overline{I}_0 = 0.1$, $C_0 = 0.2$, $\overline{k} = 5.8$, $\overline{v}_0 = 0.1$, $\overline{g} = 1.2$, and $\overline{h}_0 = 0.2$. (**a**) Stable cycles. (**b**) f and A. With the increase in damping coefficient, the A presents a decreasing trend, while the f remains nearly constant.

4.4. Impact of the Stiffness Coefficient

Figure 7 demonstrates the influence of the stiffness coefficient $\overline{k}$ on the self-oscillating behavior of the LCE fiber-slide system, with parameters set to $\overline{I}_0 = 0.1$, $C_0 = 0.2$, $\overline{\beta} = 0.1$, $\overline{v}_0 = 0.1$, $\overline{g} = 1.2$, and $\overline{h}_0 = 0.2$. The stable cycles for various stiffness coefficients are depicted in Figure 7a. The critical threshold for initiating self-oscillation is approximately $\overline{k} = 4.6$. In this context, when $\overline{k} < 4.6$, the system remains static; however, at $\overline{k} = 5.0$, $\overline{k} = 5.4$, and $\overline{k} = 5.8$, the system exhibits self-oscillation. For lower values of $\overline{k}$, the tensile force is minimal, leading to insufficient net work done on the system. Consequently, the energy supplied is insufficient to counteract the damping losses required to maintain self-oscillation. Figure 7b illustrates the relationship between frequency, amplitude of self-oscillation, and the stiffness coefficient $\overline{k}$. As $\overline{k}$ increases, both the A and f exhibit an increasing pattern. Based on the derivation above, the tensile force of the LCE fiber rises with an increase in $\overline{k}$. As a result, the net work performed by this tensile force during each cycle also increases. A higher stiffness coefficient in the LCE fiber improves its ability to effectively convert light energy into mechanical energy within the self-oscillating system. Consequently, increasing the stiffness coefficient of the LCE material can enhance both the frequency and amplitude, making it better suited for a range of engineering applications.

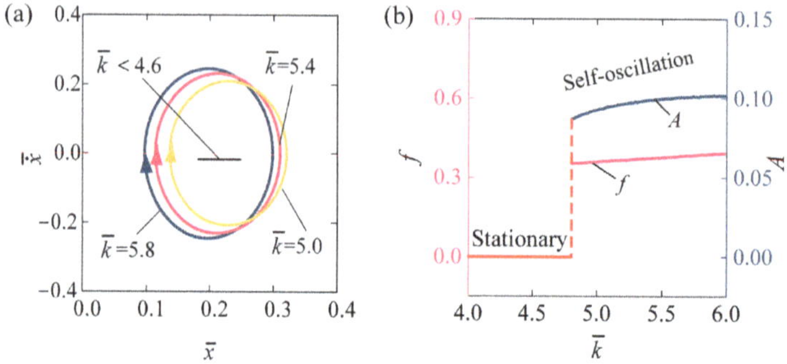

Figure 7. Influence of stiffness coefficient on self-oscillation of LCE fiber-slide system, for $\overline{I}_0 = 0.1$, $C_0 = 0.2$, $\overline{\beta} = 0.1$, $\overline{v}_0 = 0.1$, $\overline{g} = 1.2$, and $\overline{h}_0 = 0.2$. (**a**) Stable cycles. (**b**) f and A. As the stiffness coefficient increases, both the a f and A exhibit an upward trend.

4.5. Impact of the Initial Condition

Figure 8 illustrates the effect of initial velocity $\overline{v}_0$ on the self-oscillation of the LCE fiber-slide system for $\overline{I}_0 = 0.1$, $C_0 = 0.2$, $\overline{\beta} = 0.1$, $\overline{k} = 5.8$, $\overline{g} = 1.2$, and $\overline{h}_0 = 0.2$. The stable cycles for different initial velocities are presented in Figure 8a. A critical velocity $\overline{v}_0$ of approximately 0.01 marks the transition between the stationary phase and the self-oscillating phase. For $\overline{v}_0 < 0.01$, the system stays in the stationary phase because the energy input is insufficient to counterbalance the energy lost to damping. The self-oscillating phase is activated at $\overline{v}_0 = 0.01$, $\overline{v}_0 = 0.02$, and $\overline{v}_0 = 0.03$, each resulting in the same stable cycles, as shown in Figure 8a. Figure 8b displays the f and A of self-oscillation in relation to $\overline{v}_0$, respectively. Notably, both f and A remain unchanged with varying $\overline{v}_0$. Given that the effects of initial displacement $\overline{x}_0$ and initial velocity $\overline{v}_0$ can be equivalently transformed through energy conversion, we can conclude that the initial conditions only influence whether self-excited oscillation is triggered; however, they do not affect the amplitude and frequency of self-oscillation, which are intrinsic properties of the system [22].

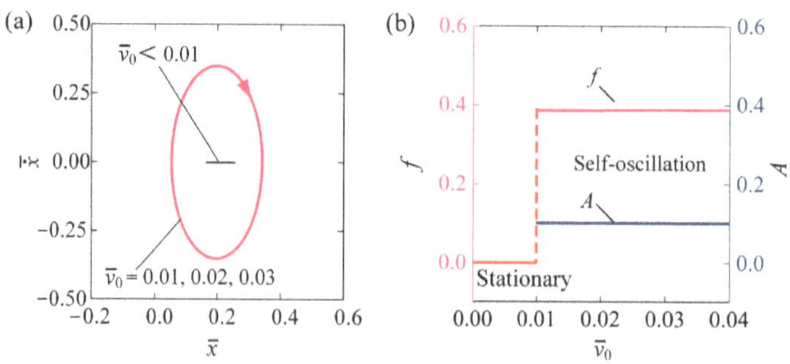

Figure 8. Influence of initial velocity on self-oscillation of LCE fiber-slide system, for $\overline{I}_0 = 0.1$, $C_0 = 0.2$, $\overline{\beta} = 0.1$, $\overline{k} = 5.8$, $\overline{g} = 1.2$, and $\overline{h}_0 = 0.2$. (**a**) Stable cycles. (**b**) f and A. Initial conditions do not influence the A and f of self-oscillation.

4.6. Impact of the Gravitational Acceleration

Figure 9 describes the influence of gravitational acceleration $\overline{g}$ on the self-oscillation of the LCE fiber-slide system. In the calculation, we set $\overline{I}_0 = 0.1$, $C_0 = 0.2$, $\overline{\beta} = 0.1$, $\overline{v}_0 = 0.1$, $\overline{k} = 5.8$, and $\overline{h}_0 = 0.2$. The stable cycles for various $\overline{g}$ are depicted Figure 9a. The range of gravitational acceleration required to initiate self-oscillation in the LCE fiber-slide system is approximately $0.9 \leq \overline{g} \leq 1.52$. In other terms, for $\overline{g} < 0.9$ or $\overline{g} > 1.52$, the system is in a stationary phase. For $\overline{g} = 1.0$, $\overline{g} = 1.2$, and $\overline{g} = 1.4$, the system exhibits a self-oscillation state. This outcome can be attributed to the balance between the energy input and the energy lost due to damping. Figure 9b illustrates how the f and A of self-oscillation vary with gravitational acceleration $\overline{g}$. As $\overline{g}$ increases, the A initially rises and then falls, while f has barely changed at all. It is noteworthy that $\overline{g}$ can be rewritten as $\overline{g} = (\tau/\tau_0)^2$, where $\tau_0 = \sqrt{L_0/g}$ represents the natural period of the LCE fiber-slider system. Thus, $\overline{g}$ reflects the relative speed of thermal relaxation compared to natural oscillation. For a small $\overline{g}$, thermal relaxation occurs too rapidly, causing the recovery of photo-induced contraction to happen almost exclusively at the moment the light source is turned off. Conversely, for a large $\overline{g}$, the thermal relaxation is too slow, resulting in a minimal recovery of photo-induced contraction when the light source is switched off. Consequently, when the light source is turned back on, the increase in photo-induced contraction is also insufficient. Therefore, whether $\overline{g}$ is too small or large, the net work done by the tensile force of the LCE fiber is inadequate to compensate for the energy dissipated, preventing sustained oscillation. These results suggest that a suitable increase in $\overline{g}$ within a specific range can enhance the oscillatory amplitude and improve the conversion efficiency of light energy into mechanical energy in engineering applications.

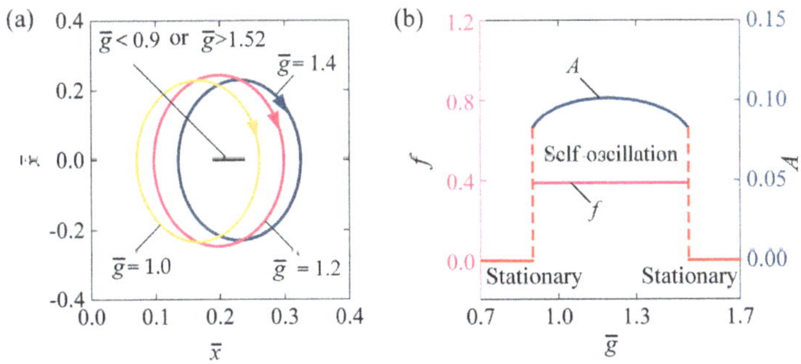

Figure 9. Influence of gravitational acceleration on self-oscillation of LCE fiber-slide system, for $\overline{I}_0 = 0.1$, $C_0 = 0.2$, $\overline{\beta} = 0.1$, $\overline{v}_0 = 0.1$, $\overline{k} = 5.8$, and $\overline{h}_0 = 0.2$. (**a**) Stable cycles. (**b**) f and A. As $\overline{g}$ increases, the A initially rises and then falls, while f has barely changed at all.

4.7. Impact of the Track Position

Figure 10 describes the influence of conductive track position $\overline{h}_0$ on the self-oscillation of the LCE fiber-slide system, with parameters set to $\overline{I}_0 = 0.1$, $C_0 = 0.2$, $\overline{\beta} = 0.1$, $\overline{v}_0 = 0.1$, $\overline{k} = 5.8$, and $\overline{g} = 1.2$. The stable cycles for various conductive track positions are shown in Figure 10a. The critical conductive track position $\overline{h}_0$ required to initiate the self-oscillation of the LCE fiber-slide system is approximately $\overline{h}_0 = 0.15$. In other terms, for $\overline{h}_0 < 0.15$, the system is in a stationary phase. For $\overline{h}_0 = 0.15$, $\overline{h}_0 = 0.16$, and $\overline{h}_0 = 0.21$, the system demonstrates a self-oscillation phase. Figure 10b illustrates how the f and A of self-oscillation and conductive track position $\overline{h}_0$ change. As $\overline{h}_0$ increases, the A initially rises and then falls, while f has barely changed at all. The principle of influence of $\overline{h}_0$ can be summarized as follows: for a small $\overline{h}_0$, the slider easily activates the light source in the conductive segment, resulting in significant contraction of the LCE fiber. However, it must travel a considerable distance before the insulating segment makes contact with the track to turn off the light and allow the fiber to recover, which is limited by the weight of the slider. For a large $\overline{h}_0$, the slider needs to move down a significant distance to activate the light source adequately. In this case, the tension in the LCE fiber also restricts how far the slider can move downward. Therefore, whether $\overline{h}_0$ is too small or large, the energy taken in by the LCE fiber-slider system is insufficient to offset energy loss, preventing sustained oscillation. Thus, adjusting $\overline{h}_0$ within a certain range appropriately can enhance the A, thereby improving the effectiveness of converting optical energy for practical engineering purposes.

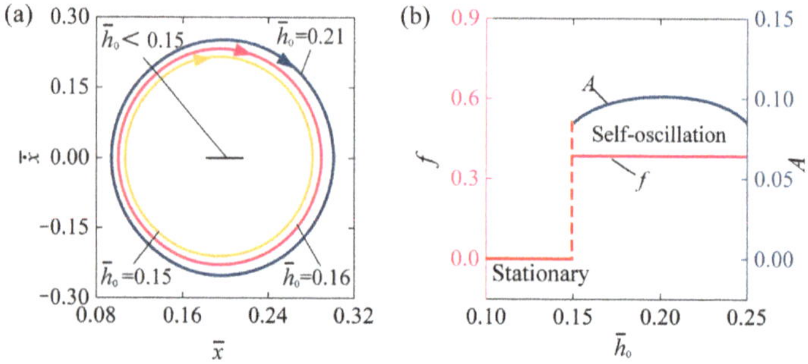

Figure 10. Influence of conductive track position on self-oscillation of LCE fiber-slide system, for $\overline{I}_0 = 0.1$, $C_0 = 0.2$, $\overline{\beta} = 0.1$, $\overline{v}_0 = 0.1$, $\overline{k} = 5.8$, and $\overline{g} = 1.2$. (**a**) Stable cycles. (**b**) f and A. As $\overline{h}_0$ increases, the A initially rises and then falls, while f has barely changed at all.

5. Conclusions

Self-oscillation efficiently transforms constant external inputs into mechanical energy, making it an excellent method for sustaining continuous motion. This characteristic forms the foundation for its broad application in engineering. Most self-oscillation systems have been developed by utilizing a static external environment that continuously releases energy. However, materials only intermittently absorb ambient energy, resulting in significant energy waste. To address this issue, we introduce a new self-oscillation of the LCE fiber-slide system driven by a self-flickering light source, which can regulate the energy source in accordance with the self-oscillating behavior. This system primarily consists of a photo-responsive LCE fiber, a slider that includes a conductive segment and an insulating segment, a light source, and a conductive track. By employing the dynamic LCE model, we derive the governing equation for the motion of the LCE fiber-slider system.

Through simulation, we identify two motion phases: the stationary phase and the self-oscillation phase. The self-oscillation of the LCE fiber-slider system arises from the photo-induced contraction of the LCE fiber when the light source is activated. We also investigate how the system parameters impact self-oscillation behavior, including critical conditions, frequency f, and amplitude A. The A primarily depends on parameters such as $\overline{I}_0$, C_0, $\overline{\beta}$, $\overline{k}$, $\overline{g}$, and $\overline{h}_0$. Increasing system parameters, including $\overline{I}_0$, C_0, and $\overline{k}$, can enhance both the f and A. As $\overline{g}$ increases, the A shows a pattern of gradually rising to a maximum point and then steadily decreasing, while the f has barely changed at all. Moreover, the change in A with increasing $\overline{h}_0$ follows a pattern identical to that of $\overline{g}$; conversely, an increase in $\overline{\beta}$ results in a decrease in A of self-oscillation. The f is primarily determined by $\overline{k}$. There exists a crucial $\overline{v}_0$ that triggers the self-oscillation, but $\overline{v}_0$ has no significant impact on the A and f.

Unlike conventional photo-induced self-oscillating systems with continuous illumination, our design features a dynamic self-flickering light source that interacts with the oscillation of the conductive slider under constant voltage, facilitating periodic switching over time. This configuration provides several benefits, including spontaneous periodic lighting, a streamlined structure, energy efficiency, and ease of use. It shows great potential for applications in dynamic circuit systems, monitoring and warning devices, optical devices, and more. It is important to acknowledge that the model proposed in this study has certain limitations, including the assumption of negligible friction between the slider and the track, a linear damping force, and a homogeneous tensile force proportional to the elastic strain, among others. In future work, we may investigate the effects of these factors or consider the use of more advanced models and illustrate the self-oscillation phenomenon through experimental validation to confirm the numerical calculations. We believe that this work has the potential to inspire new and diverse design concepts for dynamic circuit systems, monitoring and warning devices, optical devices, and more.

Author Contributions: Conceptualization, H.L.; Software, Q.H.; Data curation, X.L.; Writing—original draft, Q.H.; Writing—review & editing, D.G.; Visualization, X.L.; Supervision, H.L. All authors have read and agreed to the published version of the manuscript.

Funding: This research was funded by the Key Research and Development Program of Anhui Province (Grant No. 1804a09020001), University Natural Science Research Project of Anhui Province (Grant No. 2022AH020029), Anhui Provincial Natural Science Foundation (Grant No. 2208085Y01), National Natural Science Foundation of China (Grant No. 12172001) and Housing and Urban-Rural Development Science and Technology Project of Anhui Province (Grant No. 2023-YF140). Student Innovation and Entrepreneurship Training Program of Anhui Jianzhu University (No. S202410878074).

Institutional Review Board Statement: Not applicable.

Data Availability Statement: The original contributions presented in the study are included in the article, further inquiries can be directed to the corresponding author.

Conflicts of Interest: The authors declare that they have no known competing financial interests or personal relationships that could have appeared to influence the work reported in this article.

References

1. Wang, Y.; Dang, A.; Zhang, Z.; Yin, R.; Gao, Y.; Feng, L.; Yang, S. Repeatable and reprogrammable shape morphing from photo-responsive gold nanorod/liquid crystal elastomers. *Adv. Mater.* **2020**, *32*, 2004270. [CrossRef] [PubMed]
2. Wang, Y.; Yin, R.; Jin, L.; Liu, M.; Gao, Y.; Raney, J.; Yang, S. 3D-printed photo-responsive liquid crystal elastomer composites for free-form actuation. *Adv. Funct. Mater.* **2023**, *33*, 2210614. [CrossRef]
3. Wang, X.Q.; Tan, C.F.; Chan, K.H. In-built thermomechanical cooperative feedback mechanism for self-propelled multimodal locomotion and electricity generation. *Nat. Commun.* **2018**, *9*, 3438. [CrossRef] [PubMed]
4. Yang, Y.; Terentjev, E.M.; Zhang, Y. Reprocessable thermoset soft actuators. *Angew. Chem.* **2019**, *58*, 17474–17479. [CrossRef]
5. Techawanitchai, P.; Ebara, M.; Idota, N.; Asoh, T.A.; Kikuchi, A.; Aoyagi, T. Photo-switchable control of pH-responsive actuators via pH jump reaction. *Soft Matter* **2012**, *8*, 2844–2851. [CrossRef]
6. Hua, M.; Kim, C.; Du, Y.; Wu, D.; Bai, R.; He, X. Swaying gel: Chemo-mechanical self-oscillation based on dynamic buckling. *Matter* **2021**, *4*, 1029–1041. [CrossRef]
7. Deng, H.; Sattari, K.; Xie, Y.; Liao, P.; Yan, Z.; Lin, J. Laser reprogramming magnetic anisotropy in soft composites for reconfigurable 3D shaping. *Nat. Commun.* **2020**, *11*, 6325. [CrossRef]
8. Sun, J.; Wang, Y.; Liao, W.; Yang, Z. Ultrafast, high-contractile electrothermal-driven liquid crystal elastomer fibers towards artificial muscles. *Small* **2021**, *17*, 2103700. [CrossRef]
9. Liao, W.; Yang, Z. The integration of sensing and actuating based on a simple design fiber actuator towards intelligent soft robots. *Adv. Mater. Technol.* **2022**, *7*, 2101260. [CrossRef]
10. Shin, B.; Ha, J.; Lee, M. Hygrobot: A self-locomotive ratcheted actuator powered by environmental humidity. *Sci. Robot.* **2018**, *3*, eaar2629. [CrossRef]
11. Arazoe, H.; Miyajima, D.; Akaike, K. An autonomous actuator driven by fluctuations in ambient humidity. *Nat. Mater.* **2016**, *15*, 1084–1089. [CrossRef] [PubMed]
12. Yang, H.; Zhang, C.; Chen, B.; Wang, Z.; Xu, Y.; Xiao, R. Bioinspired design of stimuli-responsive artificial muscles with multiple actuation modes. *Smart Mater. Struct.* **2023**, *32*, 085023. [CrossRef]
13. Dai, L.; Wang, L.; Chen, B.; Xu, Z.; Wang, Z.; Xiao, R. Shape memory behaviors of 3D printed liquid crystal elastomers. *Soft Sci.* **2023**, *3*, 4. [CrossRef]
14. He, Q.; Yin, R.; Hua, Y.; Jiao, W.; Mo, C.; Shu, H.; Raney, J.R. A modular strategy for distributed, embodied control of electronics-free soft robots. *Sci. Adv.* **2023**, *9*, eade9247. [CrossRef] [PubMed]
15. Sun, Z.; Yamauchi, Y.; Araoka, F. An anisotropic hydrogel actuator enabling earthworm-like directed peristaltic crawling. *Angew. Chem.* **2018**, *57*, 15772–15776. [CrossRef]
16. Li, J.; Zhang, R.; Mou, L. Photothermal bimorph actuators with in-built cooler for light mills, frequency switches, and soft robots. *Adv. Funct. Mater.* **2019**, *29*, 1808995. [CrossRef]
17. Li, C.; Iscen, A.; Sai, H. Supramolecular-covalent hybrid polymers for light-activated mechanical actuation. *Nat. Mater.* **2020**, *19*, 900–909. [CrossRef]

18. Kularatne, R.; Kim, H.; Boothby, J.; Ware, T. Liquid crystal elastomer actuators: Synthesis, alignment, and applications. *J. Polym. Sci. Pol. Phys.* **2017**, *55*, 395. [CrossRef]
19. Xia, Z.; Deng, Z.; Fang, B.; Yang, Y.; Sun, F. A review on sensory perception for dexterous robotic manipulation. *Int. J. Robot. Res.* **2022**, *19*, 17298806221095974. [CrossRef]
20. Korner, K.; Kuenstler, A.S.; Hayward, R.C.; Audoly, B.; Bhattacharya, K. A nonlinear beam model of photomotile structures. *Proc. Natl. Acad. Sci. USA* **2020**, *117*, 9762–9770. [CrossRef]
21. Zibaei, R.; Zakerhamidi, M.S.; Korramab, S.; Ranjkesh, A. Effects of polarized light on the optical and self-oscillation behaviors of liquid crystal network polymers. *J. Mater. Chem. C* **2021**, *9*, 14908–14915. [CrossRef]
22. Jenkins, A. Self-oscillation. *Phys. Rep.* **2013**, *525*, 167–222. [CrossRef]
23. Nemati, Y.; Deng, Z.; Pi, H.; Guo, H.; Zhang, H. A scalable, incoherent-light-powered, omnidirectional self-oscillator. *Adv. Intell. Syst.* **2023**, *6*, 2300054. [CrossRef]
24. Kang, J.; Bai, C.; Liu, S.; Jia, Y. Light-induced nontethered rolling of liquid crystal elastomer and carbon nanotube composite ring. *ACS Appl. Polym. Mater.* **2024**, *6*, 2709–2718. [CrossRef]
25. Kotikian, A.; McMahan, C.; Davidson, E.C.; Muhammad, J.M.; Weeks, R.D.; Daraio, C.; Lewis, J.A. Untethered soft robotic matter with passive control of shape morphing and propulsion. *Sci. Robot.* **2019**, *4*, eaax7044. [CrossRef]
26. Zhu, Q.L.; Liu, W.; Khoruzhenko, O.; Breu, J.; Hong, W.; Zheng, Q.; Wu, Z.L. Animating hydrogel knotbots with topology-invoked self-regulation. *Nat. Commun.* **2024**, *14*, 300. [CrossRef]
27. Che, X.P.; Wang, T.; Zhang, B.L.; Zhai, Z.Z.; Chen, Y.J.; Pei, D.F.; Ge, A.L.; Li, M.J.; Li, C.X. Two-dimensionally nano-capsulating liquid metal for self-sintering and self-oscillating bimorph composites with persistent energy-harvest property. *Adv. Funct. Mater.* **2023**, *34*, 2307830. [CrossRef]
28. Liu, J.Q.; Xu, L.L.; Ji, Q.X.; Chang, L.F.; Hu, Y.; Peng, Q.Y.; He, X.D. A MXene-based light-driven actuator and motor with self-sustained oscillation for versatile applications. *Adv. Funct. Mater.* **2023**, *34*, 2310955. [CrossRef]
29. Nie, Z.-Z.; Wang, M.; Huang, S.; Liu, Z.; Yang, H. Multimodal self-sustainable autonomous locomotions of light-driven Seifert ribbon actuators based on liquid crystal elastomers. *Angew. Chem.* **2023**, *135*, e202304081. [CrossRef]
30. Zhai, Y.; Ng, T.N. Self-sustained robots based on functionally graded elastomeric actuators carrying up to 22 times their body weight. *Adv. Intell. Syst.* **2023**, *5*, 2100085. [CrossRef]
31. Helou, C.E.; Hyatta, L.P.; Buskohl, P.R.; Harne, R.L. Intelligent electroactive material systems with self-adaptive mechanical memory and sequential logic. *Proc. Natl. Acad. Sci. USA* **2024**, *121*, e2317340121. [CrossRef] [PubMed]
32. Jayoti, D.; Peeketi, A.R.; Kumbhar, P.Y.; Swaminathan, N.; Annabattula, R.K. Geometry controlled oscillations in liquid crystal polymer films triggered by thermal feedback. *ACS Appl. Mater. Interfaces* **2023**, *15*, 18362–18371. [CrossRef] [PubMed]
33. Zheng, R.; Ma, L.; Feng, W.; Pan, J.; Wang, Z.; Chen, Z.; Zhang, Y.; Li, C.; Chen, P.; Bisoyi, H.K.; et al. Autonomous self-sustained liquid crystal actuators enabling active photonic applications. *Adv. Funct. Mater.* **2023**, *33*, 2301142. [CrossRef]
34. Zeng, H.; Lahikainen, M.; Liu, L.; Ahmed, Z.; Wani, O.M.; Wang, M.; Yang, H.; Priimagi, A. Light-fuelled freestyle self-oscillators. *Nat. Commun.* **2019**, *10*, 5057. [CrossRef] [PubMed]
35. Hu, Y.; Ji, Q.; Huang, M.; Chang, L.; Zhang, C.; Wu, G.; Zi, B.; Bao, N.; Chen, W.; Wu, Y. Light-driven self-oscillating actuators with phototactic locomotion based on black phosphorus heterostructure. *Angew. Chem.* **2021**, *60*, 20511–20517. [CrossRef]
36. Yang, L.L.; Chang, L.F.; Hu, Y.; Huang, M.J.; Ji, Q.X.; Lu, P.; Liu, J.Q.; Chen, W.; Wu, Y.C. An autonomous soft actuator with light-driven self-sustained wavelike oscillation for phototactic self-locomotion and power generation. *Adv. Funct. Mater.* **2020**, *30*, 1908842. [CrossRef]
37. Li, J.; Mou, L.; Liu, Z.; Zhou, X.; Chen, Y. Oscillating light engine realized by photothermal solvent evaporation. *Nat. Commun.* **2022**, *13*, 5621. [CrossRef]
38. Ren, L.; He, Y.; Wang, B.; Xu, J.; Wu, Q.; Wang, Z.; Li, W.; Ren, L.; Zhou, X.; Liu, Q.; et al. 4D printed self-sustained soft crawling machines fueled by constant thermal field. *Adv. Funct. Mater.* **2024**, *34*, 2400161. [CrossRef]
39. Ge, D.; Li, K. Self-oscillating buckling and postbuckling of a liquid crystal elastomer disk under steady illumination. *Int. J. Mech. Sci.* **2022**, *221*, 107233. [CrossRef]
40. Gelebart, G.A.; Mulder, D.J.; Varga, M.; Konya, A.; Vantomme, G.; Meijer, E.W.; Selinger, R.L.B.; Broer, D.J. Making waves in a photoactive polymer film. *Nature* **2017**, *546*, 632–636. [CrossRef]
41. Shahsavana, H.; Aghakhania, A.; Zeng, H.; Guo, Y.; Davidson, Z.S.; Priimagi, A.; Sitti, M. Bioinspired underwater locomotion of light-driven liquid crystal gels. *Proc. Natl. Acad. Sci. USA* **2023**, *117*, 5125–5133. [CrossRef] [PubMed]
42. Zhao, T.; Fan, Y.; Lv, J. Photomorphogenesis of diverse autonomous traveling waves in a monolithic soft artificial muscle. *ACS Appl. Mater. Int.* **2022**, *14*, 23839–23849. [CrossRef] [PubMed]
43. da Cunha, M.P.; Peeketi, A.R.; Mehta, K.; Broer, D.J.; Annabattula, R.K.; Schenning, A.P.H.J.; Debije, M.G. A self-sustained soft actuator able to rock and roll. *Chem. Commun.* **2019**, *55*, 11029–11032. [CrossRef] [PubMed]
44. da Cunha, M.P.; Peeketi, A.R.; Ramgopal, A.; Annabattula, R.K.; Schenning, A.P.H.J. Light-driven continual oscillatory rocking of a polymer film. *ChemistryOpen* **2020**, *9*, 1149–1152. [CrossRef] [PubMed]
45. Ge, Y.; Cao, R.; Ye, S.; Chen, Z.; Zhu, Z.; Tu, Y.; Ge, D.; Yang, X. A bio-inspired homogeneous graphene oxide actuator driven by moisture gradients. *Chem. Commun.* **2018**, *54*, 3126. [CrossRef] [PubMed]
46. Xu, P.; Sun, X.; Dai, Y.; Li, K. Light-powered sustained chaotic jumping of a liquid crystal elastomer balloon. *Int. J. Mech. Sci.* **2024**, *266*, 108922. [CrossRef]

47. Hebner, T.S.; Korner, K.; Bowman, C.N.; Bhattacharya, K.; White, T.J. Leaping liquid crystal elastomers. *Sci. Adv.* **2023**, *9*, eade1320. [CrossRef]
48. Zhou, X.; Chen, G.; Jin, B.; Feng, H.; Chen, Z.; Fang, M.; Yang, B.; Xiao, R.; Xie, T.; Zheng, N. Multimodal autonomous locomotion of liquid crystal elastomer soft robot. *Adv. Sci.* **2024**, *11*, 2402358. [CrossRef]
49. Liang, Z.; Jin, B.; Zhao, H.; He, Z.; Jiang, Z.; Jiang, S. Rotini-like MXene@LCE actuator with diverse and programmable actuation based on dual-mode synergy. *Small* **2024**, *20*, 2305371. [CrossRef]
50. Wang, Z.; Bao, J.; Huang, R.; Song, C.; Shen, C.; Sun, J.; Lan, R.; Zhang, L.; Yang, H. Light- and humidity-driven fluorescence changeable soft robot enabled by water-gated photoinduced electron transfer pathway. *Sci. China Mater.* **2023**, *66*, 2445–2453. [CrossRef]
51. Zhao, Y.; Ch, Y.; Hong, Y.; Li, Y.; Yang, S.; Yin, J. Twisting for soft intelligent autonomous robot in unstructured environments. *Proc. Natl. Acad. Sci. USA* **2022**, *119*, e2200265119. [CrossRef] [PubMed]
52. Guo, K.; Yang, X.; Zhou, C.; Li, C. Self-regulated reversal deformation and locomotion of structurally homogenous hydrogels subjected to constant light illumination. *Nat. Commun.* **2024**, *15*, 1694. [CrossRef] [PubMed]
53. Wu, H.; Ge, D.; Chen, J.; Xu, P.; Li, K. A light-fueled self-rolling unicycle with a liquid crystal elastomer rod engine. *Chaos Solitons Fractals* **2024**, *186*, 115327. [CrossRef]
54. Ge, D.; Dai, Y.; Liang, H.; Li, K. Self-rolling and circling of a conical liquid crystal elastomer rod on a hot surface. *Int. J. Mech. Sci.* **2024**, *263*, 108780. [CrossRef]
55. Yu, Y.; Du, C.; Li, K.; Cai, S. Controllable and versatile self-motivated motion of a fiber on a hot surface. *Extrem. Mech. Lett.* **2022**, *57*, 101918. [CrossRef]
56. Li, K.; Qiu, Y.; Dai, Y.; Yu, Y. Modeling the dynamic response of a light-powered self-rotating liquid crystal elastomer-based system. *Int. J. Mech. Sci.* **2024**, *263*, 108794. [CrossRef]
57. Qiu, Y.; Ge, D.; Wu, H.; Li, K.; Xu, P. Self-rotation of a liquid crystal elastomer rod under constant illumination. *Int. J. Mech. Sci.* **2024**, *284*, 109665. [CrossRef]
58. Qiu, Y.; Li, K. Self-rotation-eversion of an anisotropic-friction-surface torus. *Int. J. Mech. Sci.* **2024**, *281*, 109584. [CrossRef]
59. Zhou, L.; Chen, H.; Li, K. Optically responsive liquid crystal elastomer thin film motors in linear/nonlinear optical fields. *Thin-Walled Struct.* **2024**, *202*, 112082. [CrossRef]
60. Li, K.; Qian, P.; Hu, H.; Dai, Y.; Ge, D. A light-powered liquid crystal elastomer semi-rotary motor. *Int. J. Solids Struct.* **2023**, *284*, 112509. [CrossRef]
61. Qiu, Y.; Dai, Y.; Li, K. Self-spinning of liquid crystal elastomer tubes under constant light intensity. *Commun. Nonlinear Sci. Numer. Simul.* **2024**, *133*, 108296. [CrossRef]
62. Zhu, Q.L.; Liu, W.; Khoruzhenko, O.; Breu, J.; Bai, H.; Hong, W.; Zheng, Q.; Wu, Z.L. Closed twisted hydrogel ribbons with self-sustained motions under static light irradiation. *Adv. Mater.* **2024**, *36*, 2314152. [CrossRef] [PubMed]
63. Deng, Z.; Li, K.; Priimagi, A.; Zeng, H. Light-steerable locomotion using zero-energy modes. *Nat. Mater.* **2024**. *Online ahead of print*. [CrossRef]
64. Liang, X.; Chen, Z.; Zhou, L.; Li, K. Light powered self-excited oscillation of a liquid crystal elastomer pendulum. *Mech. Syst. Signal. Pro.* **2021**, *163*, 108140. [CrossRef]
65. Bai, C.; Kang, J.; Wang, Y.Q. Light-induced motion of three-dimensional pendulum with liquid crystal elastomeric fiber. *Int. J. Mech. Sci.* **2024**, *266*, 108911. [CrossRef]
66. He, Q.; Wang, Z.; Wang, Y.; Wang, Z.; Li, C.; Annapooranan, R.; Zeng, J.; Chen, R.; Cai, S. Electrospun liquid crystal elastomer microfiber actuator. *Sci. Robot.* **2021**, *6*, eabi9704. [CrossRef]
67. Sun, X.; Zhou, K.; Xu, P. Chaotic self-beating of left ventricle modeled by liquid crystal elastomer. *Thin-Walled Struct.* **2024**, *205*, 112540. [CrossRef]
68. Ge, D.; Li, K. Pulsating self-snapping of a liquid crystal elastomer bilayer spherical shell under steady illumination. *Int. J. Mech. Sci.* **2022**, *233*, 107646. [CrossRef]
69. Zhao, Y.; Hong, Y.; Qi, F.; Chi, Y.; Hao, S.; Yin, J. Self-sustained snapping drives autonomous dancing and motion in free-standing wavy rings. *Adv. Mater.* **2023**, *35*, 2207372. [CrossRef]
70. Hu, Z.; Li, Y.; Lv, J. Phototunable self-oscillating system driven by a self-winding fiber actuator. *Nat. Commun.* **2021**, *12*, 3211. [CrossRef]
71. Xu, T.; Pei, D.; Yu, S.; Zhang, X.; Yi, M.; Li, C. Design of MXene composites with biomimetic rapid and self-oscillating actuation under ambient circumstances. *ACS Appl. Mater. Int.* **2021**, *13*, 31978–31985. [CrossRef]
72. Ge, D.; Liang, H.; Li, K. Self-oscillation of a liquid crystal elastomer string-mass system under constant gradient temperature. *Appl. Mech.* **2024**, *91*, 101001. [CrossRef]
73. Yu, Y.; Zhou, L.; Du, C.; Zhu, F.; Dai, Y.; Ge, D.; Li, K. Self-galloping of a liquid crystal elastomer catenary cable under a steady temperature field. *Thin-Walled Struct.* **2024**, *202*, 112071. [CrossRef]
74. Zhao, J.; Dai, C.; Dai, Y.; Wu, J.; Li, K. Self-oscillation of cantilevered silicone oil paper sheet system driven by steam. *Thin-Walled Struct.* **2024**, *203*, 112270. [CrossRef]
75. Vantomme, G.; Elands, L.C.M.; Gelebart, A.H.; Meijer, E.W.; Pogromsky, A.Y.; Nijmeijer, H.; Broer, D.J. Coupled liquid crystalline oscillators in Huygens' synchrony. *Nat. Mater.* **2021**, *20*, 1702–1706. [CrossRef]

76. Wu, H.; Zhang, B.; Li, K. Synchronous behaviors of three coupled liquid crystal elastomer-based spring oscillators under linear temperature fields. *Phys. Rev. E* **2024**, *109*, 024701. [CrossRef]
77. Kim, H.; Sundaram, S.; Kang, J.; Tanjeem, N.; Emrick, T.; Hayward, R.C. Coupled oscillation and spinning of photothermal particles in Marangoni optical traps. *Proc. Natl. Acad. Sci. USA* **2021**, *118*, e2024581118. [CrossRef]
78. Yu, Y.; Nakano, M.; Ikeda, T. Photomechanics: Directed bending of a polymer film by light-miniaturizing a simple photomechanical system could expand its range of applications. *Nature* **2003**, *425*, 145. [CrossRef]
79. Finkelmann, H.; Nishikawa, E.; Pereira, G.G.; Warner, M. A new opto-mechanical effect in solids. *Phys. Rev. Lett.* **2001**, *87*, 015501. [CrossRef]
80. Corbett, D.; Xuan, C.; Warner, M. Deep optical penetration dynamics in photobending. *Phys. Rev.* **2015**, *92*, 013206. [CrossRef]
81. Nagele, T.; Hoche, R.; Zinth, W.; Wachtveitl, J. Femtosecond photoisomerization of cis-azobenzene. *Chem. Phys. Lett.* **1997**, *272*, 489–495. [CrossRef]
82. Braun, L.B.; Hessberger, T.; Pütz, E.; Müller, C.; Giesselmann, F.; Serra, C.A.; Zentel, R. Actuating thermo- and photo-responsive tubes from liquid crystalline elastomers. *J. Mater. Chem. C* **2018**, *6*, 9093–9101. [CrossRef]
83. Camacho-Lopez, M.; Finkelmann, H.; Palffy-Muhoray, P.; Shelley, M. Fast liquid-crystal elastomer swims into the dark. *Nat. Mater.* **2004**, *3*, 307–310. [CrossRef] [PubMed]

polymers

MDPI

Article

Enhanced Dielectric Properties and Antibacterial Activity of Natural Rubber by Modification with Poly(Acrylic Acid-Co-Acrylamide) Incorporating Silver Nanoparticles and Titanium Dioxide

Supharat Inphonlek [1,2], Supawat Kotchapradit [3], Boonruang Marungsri [4], Yupaporn Ruksakulpiwat [1,2,*] and Chaiwat Ruksakulpiwat [1,2,*]

1 School of Polymer Engineering, Institute of Engineering, Suranaree University of Technology, Nakhon Ratchasima 30000, Thailand; supharat.inph@gmail.com
2 Research Center for Biocomposite Materials for Medical Industry and Agricultural and Food Industry, Suranaree University of Technology, Nakhon Ratchasima 30000, Thailand
3 School of Electronic Engineering, Institute of Engineering, Suranaree University of Technology, Nakhon Ratchasima 30000, Thailand; supawat@sut.ac.th
4 School of Electrical Engineering, Institute of Engineering, Suranaree University of Technology, Nakhon Ratchasima 30000, Thailand; bmshvee@sut.ac.th
* Correspondence: yupa@sut.ac.th (Y.R.); charuk@sut.ac.th (C.R.)

Abstract: This work aims to enhance natural rubber's dielectric properties and antibacterial activity by incorporating silver nanoparticles and titanium dioxide. Deproteinized natural rubber (DPNR) was modified through the graft copolymerization of acrylic acid and acrylamide using N′, N′-Methylenebisacrylamide as a crosslinking agent, resulting in poly(acrylic acid-co-acrylamide)-modified, deproteinized natural rubber (MDPNR). This modification facilitated coordination with silver ions and interaction with titanium dioxide. Silver nanoparticles were generated under heat and pressure. Modified natural rubber composites containing silver nanoparticles and titanium dioxide (MDPNR/Ag-TiO_2) were prepared. Scanning electron microscopy (SEM) revealed well-distributed silver in the modified natural rubber matrix, while agglomeration of titanium dioxide was observed at a high loading. Both MDPNR and MDPNR/Ag-TiO_2 showed high thermal stability compared to DPNR. The MDPNR/Ag-TiO_2 composites exhibited higher T_g and lower tan δ, indicating higher stiffness due to the restriction of chain movement compared to that in MDPNR. DPNR exhibited a low dielectric constant, enhanced by poly(acrylic acid-co-acrylamide) modification and silver nanoparticle/titanium dioxide incorporation. Incorporating 0.5 phr of $AgNO_3$ and 2.5 phr of TiO_2 in the composites increased the dielectric constant by 1.33 times compared to that of MDPNR. MDPNR showed no antibacterial activity, while the MDPNR/Ag-TiO_2 composites exhibited promising antibacterial activity against *Staphylococcus aureus* and *Escherichia coli*.

Keywords: natural rubber composites; silver nanoparticles; titanium dioxide; dielectric properties; antibacterial activity

1. Introduction

Composite materials with dielectric properties have garnered interest across various applications, including energy storage systems, flexible electronic equipment, sensors, and electrical robots [1–3]. A high dielectric constant allows for greater electrical energy storage, making materials more efficient for power storage systems. The development of energy storage devices with antibacterial properties represents a novel concept, integrating both energy storage capabilities and multifunctionality. Such systems can be highly beneficial for applications, such as healthcare storage devices and wearable electronics. Devices in healthcare are often exposed to bacteria, including *Staphylococcus aureus* (*S. aureus*) and *Escherichia coli* (*E. coli*). *S. aureus* is a Gram-positive bacterium, while *E. coli* is a Gram-negative bacterium. These bacteria are known to cause infections, particularly in vulnerable patients. Devices with antibacterial properties are safer because they reduce the risk of bacterial contamination that may come into contact with the skin or human body. Moreover, antibacterial materials prevent the growth of bacteria that can degrade the materials over time, thereby prolonging the shelf life of devices. Therefore, these devices require materials with biocompatibility, environmental friendliness, good thermal and mechanical stability, an enhanced dielectric constant, and antibacterial properties.

Among the polymers, natural rubber shows promise as a biopolymer matrix for preparing composite materials [4–6]. Natural rubber, derived from the rubber tree (*Hevea Brasiliensis*), exhibits high elasticity, flexibility, and tensile strength. It is easily processed and compounded with other materials, and is capable of forming films [7,8]. Moreover, it is biodegradable and renewable, making it more environmentally friendly than synthetic alternatives [9]. Considering these properties, natural rubber is an attractive choice for diverse applications. However, natural rubber has certain limitations, such as a low dielectric constant and susceptibility to bacterial contamination, which restrict its use in advanced applications like electronics, energy storage devices, healthcare, and environmental systems [10]. Enhancing the functional properties of natural rubber to meet these demands is an area of growing interest. One approach involves incorporating functional fillers, such as graphene and titanium dioxide [11–13]. To improve the compatibility of natural rubber with polar fillers, modified natural rubber composites have been developed. Previous work has demonstrated that graft the copolymerization of polar monomers, such as acrylic acid and acrylamide, onto deproteinized natural rubber can enhance its polarity and interactions with fillers [14]. This environmentally friendly modification method utilizes water as a medium, avoiding harmful organic solvents. The process involves using deproteinized natural rubber (DPNR), which suppresses side reactions during chemical modification, enhances biocompatibility, and reduces allergic reactions, making it particularly suitable for medical and sensitive applications [15]. Modified DPNR/filler composites exhibit improved mechanical and thermal properties. Furthermore, poly(acrylic acid-co-acrylamide) is capable of coordination with silver ions and interaction with titanium dioxide. Composites of poly(acrylic acid-co-acrylamide)-modified DPNR incorporating silver nanoparticles and titanium dioxide show enhanced compressive moduli, good structural stability, and effectiveness in dye removal [16]. Thus, the modification of deproteinized natural rubber by introducing polar functional components can enhance its compatibility with fillers, potentially improving the properties of modified natural rubber composites for a variety of applications.

Titanium dioxide (TiO_2), known for its favorable physical and chemical properties, including electronic properties, photocatalytic activity, chemical stability, and low toxicity, is particularly interesting [17,18]. Bunriw et al. prepared natural rubber/TiO_2 composites as environmentally friendly materials for triboelectric nanogenerator (TENG) applications, aiming to harvest mechanical energy into electricity [19]. The addition of TiO_2 improved the dielectric constant of the natural rubber composites. The enhancement of the TENG electrical output with the highest power density of 237 mW/m^2 was achieved. Therefore, these natural rubber/TiO_2 composites are promising for developing large-scale energy harvesting devices.

Silver-containing materials exhibit photocatalytic activity and antibacterial properties [20–23]. Silver nanoparticles may penetrate bacteria cells and release silver ions that disrupt metabolic processes and inhibit cell growth. Appamato et al. developed a triboelectric nanogenerator using a natural rubber–silver nanocomposite applied as a shoe insole to harvest human footstep energy [24]. These Ag nanoparticles created interfacial polarization between the conductive metal nanoparticle and the insulating polymer, thereby improving the dielectric constant and electrical output. Additionally, they exhibited antibacterial activity against *Staphylococcus aureus*, the bacterium responsible for foot odor. Zheyan Soo et al. prepared silver-doped TiO_2 nanofibers for antibacterial applications [25]. The silver-doped TiO_2 nanofibers demonstrated stronger antibacterial activity than bare TiO_2 nanofibers, indicating a synergistic enhancement in antibacterial performance through the combination of Ag nanoparticles and TiO_2. Among the nanofibers, those with 2% Ag content showed the highest antibacterial performance, reducing *Salmonella* Albany and *S. aureus* by 5.92 ± 0.00 and 1.38 ± 0.07 logCFU, respectively.

This study focuses on the preparation of poly(acrylic acid-co-acrylamide)-modified, deproteinized natural rubber composites incorporating silver nanoparticles and titanium dioxide (MDPNR/Ag-TiO_2), aiming to enhance the dielectric constant and antibacterial activity of natural rubber. Incorporating silver nanoparticles and titanium dioxide into the modified natural rubber matrix has emerged as a promising strategy for imparting additional functionalities. This work contributes to the development of modified natural rubber composites for advanced applications, such as energy storage devices in healthcare, where high dielectric constants are essential for effective energy storage and transfer. The biocompatibility, flexibility, and antibacterial properties of modified natural composites further enhance their potential for use in flexible and wearable medical devices. This research evaluates the physical, thermal, and dielectric properties of the composites, as well as their antibacterial performance against *Staphylococcus aureus* and *Escherichia coli.*

2. Materials and Methods

2.1. Materials

Natural rubber (NR), with a 60% dry rubber content and preserved using high ammonia, was received from Chemical and Materials Co., Ltd. (Bangkok, Thailand). Urea was purchased from RCI Labscan Limited (Bangkok, Thailand). Sodium dodecyl sulfate (SDS) and acrylamide (AM) monomer were acquired from Loba Chemie Pvt. Ltd. (Mumbai, India). Acrylic acid (AA) monomer and cumene hydroperoxide (CHP) initiator were supplied by Aldrich (St. Louis, MO, USA). The AA monomer was passed through a column packed with alumina adsorbent for purification. Tetraethylene pentamine (TEPA) was purchased from Acros Organics (Geel, Belgium), and Terric16A (10 wt%) was obtained from the Rubber Authority of Thailand (Bangkok, Thailand). N′, N′-Methylenebisacrylamide (MBA) was provided by Alfa Aesar (Ward Hill, MA, USA). Silver nitrate ($AgNO_3$) was obtained from Quality Reagent Chemical, controlled by QReC New Zealand (Rawang, Malaysia). Titanium dioxide (TiO_2), the anatase phase with an oil absorption value of 26 g/100 g and a residue from a 45 μm sieve of $\leq$0.1%, was obtained from Cernic International Co., Ltd. (Nakhon Pathom, Thailand). Throughout the study, deionized (DI) water was used.

2.2. Preparation of Poly(Acrylic Acid-Co-Acrylamide)-Modified, Deproteinized Natural Rubber Comprising Silver Nanoparticles and Titanium Dioxide (MDPNR/Ag-TiO_2) Composites

The poly(acrylic acid-co-acrylamide)-modified, deproteinized natural rubber composites containing silver nanoparticles and titanium dioxide (MDPNR/Ag-TiO_2) were prepared following the method outlined by Inphonlek et al. [16]. First, deproteinized natural rubber (DPNR) was obtained by removing proteins from NR latex, as described by Kawahara et al. [26]. The NR latex was mixed with 0.1%w urea and 1%w SDS under continuous magnetic stirring for 60 min. The mixture was then centrifuged at 15,000 rpm for 30 min to separate the rubber phase from the aqueous phase containing proteins and impurities. The washing process was repeated twice by redispersing the cream fraction in 1%w SDS, followed by centrifugation. The rubber phase was collected and dispersed in water in the presence of Terric16A as a stabilizer. Next, the poly(acrylic acid-co-acrylamide)-modified, deproteinized natural rubber (MDPNR) was synthesized via graft copolymerization of acrylic acid and acrylamide, using MBA as the crosslinking agent. Briefly, DPNR latex was transferred to a reactor and stirred mechanically at 100 rpm under a nitrogen atmosphere. The following chemicals were added in sequence: CHP, acrylic acid (50 mol% neutralized with 20%w NaOH solution), acrylamide, MBA, and TEPA. In this study, the monomer content was set at 20 phr, maintaining a 70:30 weight ratio of acrylic acid to acrylamide.

MBA was used at 1.00%w of the monomer. The CHP and TEPA contents were fixed at 1 phr. Polymerization was carried out for 6 h at 50 °C under a nitrogen atmosphere. Subsequently, the resulting MDPNR latex was mixed with a 1% *w*/*v* $AgNO_3$ solution and a 10% *w*/*v* TiO_2 dispersion. The composites were prepared with a fixed $AgNO_3$ concentration of 0.5 phr and varying TiO_2 contents of 1.0, 2.5, and 5.0 phr, denoted as MDPNR/Ag-1.0TiO_2, MDPNR/Ag-2.5TiO_2 and MDPNR/Ag-5.0TiO_2, respectively. The mixture was stirred at 600 rpm under dark conditions for 30 min and then subjected to an autoclave (HVA-110, Hirayama Manufacturing Corporation, Saitama, Japan) at 120 °C and 15 psi for 50 min. The resulting product was dried in a hot-air oven at 60 °C for 24 h. Additionally, samples containing either $AgNO_3$ or TiO_2 alone, referred to as MDPNR/Ag and MDPNR/5.0TiO_2 composites, respectively, were prepared for comparison. The prepared samples were stored for analysis, as shown in Figure 1.

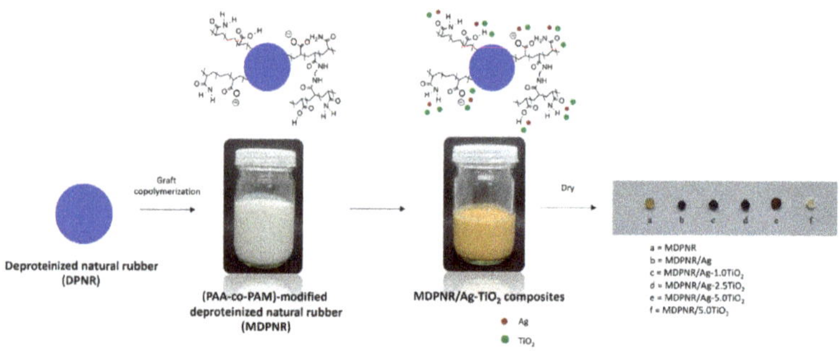

Figure 1. Schematic illustration of the preparation of the MDPNR/Ag-TiO_2 composites and the visual appearance of the obtained samples after drying.

2.3. *Fourier Transform Infrared Spectroscopy*

Attenuated total reflectance Fourier transform infrared spectroscopy (ATR-FTIR) was conducted to analyze the chemical structure of the prepared composites using an FTIR spectrophotometer (Tensor 27, Bruker, Billerica, MA, USA). Each spectrum was scanned from 4000 to 500 cm^{-1}, with a resolution of 4 cm^{-1}, and 64 scans were performed.

2.4. *Morphological Analysis*

The morphology of the prepared composites was observed using a scanning electron microscope (SEM). The dried samples were fixed on the stub and coated with gold under vacuum using a sputter coater (EM ACE600, Leica microsystems, Wetzlar, Germany). The surface morphology of samples was investigated using a JSM-7800F field emission scanning electron microscope (JEOL Ltd., Tokyo, Japan). Additionally, the element composition and distribution of the composites were determined by energy-dispersive spectroscopy coupled with SEM (SEM/EDS).

2.5. *Thermogravimetric Analysis*

The thermal stability of various types of composites was assessed using a thermogravimetric analyzer (TG 209 F3 Tarsus, Netzsch, Germany). For each sample, 10 mg of dried material was placed in a sample pan and heated from 50 to 600 °C at a heating rate of 10 °C/min. The measurement was carried out under a nitrogen atmosphere. The change in the remaining sample weight was continuously monitored throughout the measurement.

2.6. *Dynamic Mechanical Analysis*

The thermomechanical properties of the composites were evaluated using dynamic mechanical analysis (DMA) performed on a DMA850 instrument (TA Instruments, New Castle, DE, USA). A temperature sweep test was measured with a dynamic strain of 0.1% and a frequency of 1 Hz. The samples were tested over a temperature range from −80 to 100 °C, with a heating rate of 2 °C/min.

2.7. Dielectric Constant Testing

The dielectric constant was measured by using an impedance analyzer (Keysight E4294A, Agilent, Santa Clara, CA, USA) at room temperature. The samples were placed between the electrodes. The measurements were performed at varying frequencies ranging from 10^2 to 10^7 Hz.

2.8. Antibacterial Activity Evaluation

The antibacterial activity of the prepared composites was assessed using the disk diffusion method. *Staphylococcus aureus* TISTR 746 (*S. aureus*) and *Escherichia coli* TISTR 527 (*E. coli*) were chosen as representative Gram-positive and Gram-negative bacterial strains, respectively. The suspensions of microorganisms in CriterionTM Nutrient Broth (NB) were spread as thin layers on CriterionTM Mueller–Hinton (MH) agar in Petri dishes. Specimens measuring 6 mm in diameter were UV-sterilized for 1 h on each side prior to testing. Subsequently, the specimens were placed on top of the smeared agar, and then the plate was incubated at 37 °C for 24 h. Amoxicillin antibiotic disks were used as the positive control. The absence of bacterial growth, indicated by clear inhibition zones around the disk specimens, signified that those inhibitory concentrations had been achieved.

3. Results

3.1. FTIR Analysis

Modified natural rubber composites incorporating silver nanoparticles and titanium dioxide were prepared in this study. The deproteinized natural rubber was first modified by grafting with poly(acrylic acid-co-acrylamide) via emulsion graft copolymerization, acting as the polymeric matrix. The modified natural rubber was capable of forming coordination bonds with silver ions due to the presence of the carbonyl groups, carboxylate ions, and nitrogen atoms of amide groups in the polymeric matrix [27,28]. The silver ions were reduced to metallic silver atoms under a high temperature and applied pressure. These metallic silver atoms aggregate, forming clusters that grow into stable silver nanoparticles. Meanwhile, titanium dioxide can interact with modified natural rubber through hydrogen bonding [29]. As a result, the silver nanoparticles and titanium dioxide were distributed in the modified natural-rubber-based matrix.

The chemical functional groups in the samples were determined using FTIR analysis, as seen in Figure 2. In the FTIR spectrum of DPNR, peaks around 3000–2800 cm^{-1} were observed, attributed to the C-H stretching vibration of the polyisoprene backbone. Additionally, a peak at 1663 cm^{-1} was attributed to C=C stretching. Peaks at 1446 and 1375 cm^{-1} were observed, corresponding to $-CH_2$ and $-CH_3$ stretching vibration, respectively [30]. After modification through graft copolymerization, the FTIR spectrum of MDPNR exhibited a broad band around 3500–3000 cm^{-1}, indicating the presence of O-H and the N-H stretching of poly(acrylic acid-co-acrylamide). A peak at 1664 cm^{-1} was attributed to the C=O stretching. Another peak at 1561 cm^{-1} corresponded to carboxylate, resulting from the partial neutralization of acrylic acid. The presence of new peaks confirmed the success of the modification process [14,31,32]. In the case of the MDPNR/Ag, MDPNR/Ag-TiO_2, and MDPNR/TiO_2 composites, their spectra exhibited characteristics similar to those of modified natural rubber. The C=O stretching vibrations remained unchanged at 1664 cm^{-1}. However, the characteristic peaks of N-H stretching, O-H stretching, and carboxylate slightly shifted to a lower wavenumber. The peaks corresponding to N-H and O-H stretching vibrations shifted from 3396 to 3340 cm^{-1}, and carboxylate peaks shifted from 1561 to 1559 cm^{-1}. These corresponded to the results of the previous report by Anancharoenwong et al. [33]. These changes indicate the interactions of modified natural rubber with silver and titanium dioxide in the composites [34–36].

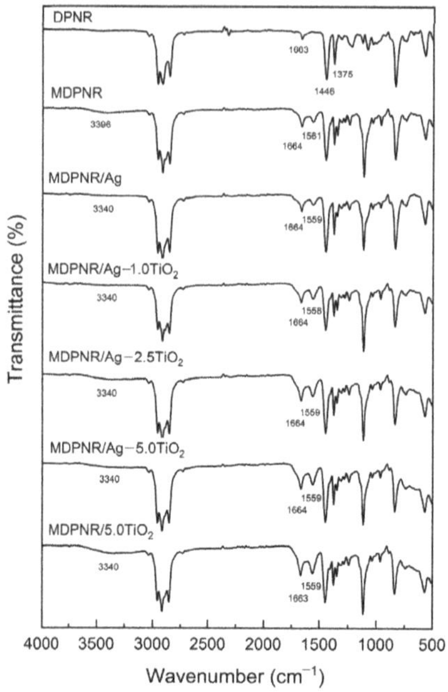

Figure 2. FTIR spectra of the DPNR, MDPNR, MDPNR/Ag, MDPNR/Ag-TiO_2 with various TiO_2 contents, and MDPNR/TiO_2 composites.

3.2. Morphology

The morphology of MDPNR and its composites, which include MDPNR/Ag, MDPNR/Ag-TiO_2 with varying TiO_2 contents, and MDPNR/TiO_2, is illustrated in Figure 3. SEM images revealed that the surface of MDPNR appeared relatively smooth, whereas the composites exhibited rougher surfaces due to the distribution of solid particles within the natural-rubber-based matrix. At higher magnification, the titanium-dioxide-containing composites exhibited solid particles approximately 240 $\pm$ 61 nm in size, as measured using ImageJ software (IJ 1.46r image analyzer software). The elemental composition and distribution of the samples were investigated by EDS analysis. Figure 4 depicts the composites' EDS spectra and mappings corresponding to Ag and Ti. The spectra revealed the presence of C, O, and Na in MDPNR. Additionally, a characteristic signal for Ag was observed in MDPNR/Ag [37], and signals corresponding to Ti appeared in MDPNR/5.0TiO_2 [38,39]. The MDPNR/Ag-TiO_2 composites with different TiO_2 contents exhibited signals indicative of modified natural rubber, Ag, and Ti, confirming the incorporation of these elements in the composites. However, the Ag peak appeared faint, likely due to the relatively low Ag content in the composites. EDS mappings demonstrated that Ag was well-distributed on the sample surface. The modified natural rubber can act as a stabilizing agent for the formation of stable silver nanoparticles with good dispersibility within the polymer matrix. Ti was also dispersed in the modified natural-rubber-based matrix. However, titanium tended to agglomerate at a higher loading, particularly evident when incorporating 5 phr of titanium dioxide, with agglomerate sizes ranging from 0.82 to 12.28 μm. The elemental compositions of the composites from EDS analysis are summarized in Table 1. The results indicated that carbon was the dominant element in the composites, ranging from a 66.44 to 84.67 weight percentage and a 75.06 to 88.19 atomic percentage, due to its presence as the primary component of the modified natural rubber matrix. Incorporating silver nanoparticles and titanium dioxide into the composites reduced the weight and atomic percentages of carbon. The composites contained a small amount of silver, with values ranging a from 0.61 to 1.08 weight percentage and a 0.08 to 0.13 atomic percentage.

An increase in the weight and atomic percentages of titanium was observed with higher titanium dioxide contents. The titanium weight and atomic percentages were increased from 1.53 to 8.21 and 0.42 to 2.36 for MDPNR/Ag-1.0TiO_2, MDPNR/Ag-2.5TiO_2, and MDPNR/Ag-5.0TiO_2, respectively.

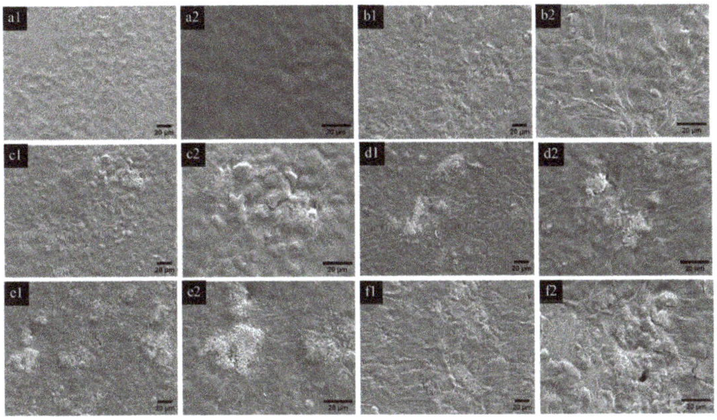

Figure 3. SEM images at ×500 and ×1000 magnifications of (**a1**,**a2**) MDPNR; (**b1**,**b2**) MDPNR/Ag; (**c1**,**c2**) MDPNR/Ag-1.0TiO_2; (**d1**,**d2**) MDPNR/Ag-2.5TiO_2; (**e1**,**e2**) MDPNR/Ag-5.0TiO_2; (**f1**,**f2**) MDPNR/5.0TiO_2.

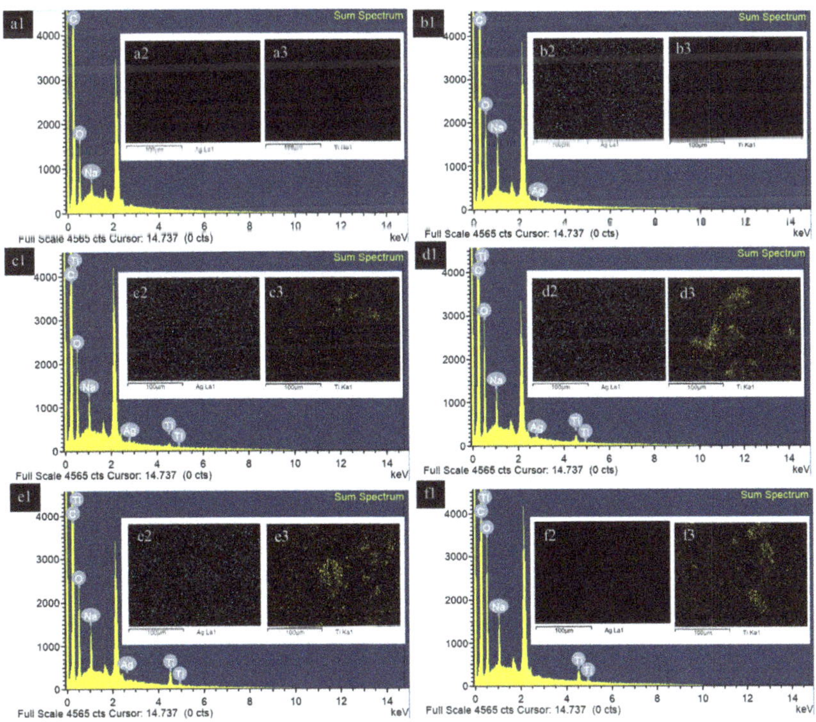

Figure 4. EDS spectra (first column) and EDS mappings, with green spots indicating silver (second column) and yellow spots indicating titanium (third column), of (**a1–a3**) MDPNR; (**b1–b3**) MDPNR/Ag; (**c1–c3**) MDPNR/Ag-1.0TiO_2; (**d1–d3**) MDPNR/Ag-2.5TiO_2; (**e1–e3**) MDPNR/Ag-5.0TiO_2; (**f1–f3**) MDPNR/5.0TiO_2.

Table 1. Elemental composition derived from EDS results of the composites.

Samples	Weight (%)					Atomic (%)				
	C	O	Na	Ag	Ti	C	O	Na	Ag	Ti
MDPNR	84.67	14.57	0.76	0.00	0.00	88.19	11.39	0.41	0.00	0.00
MDPNR/Ag	75.88	19.40	3.65	1.08	0.00	82.06	15.75	2.06	0.13	0.00
MDPNR/Ag-1.0TiO_2	75.63	19.65	2.23	0.96	1.53	82.17	16.03	1.26	0.12	0.42
MDPNR/Ag-2.5TiO_2	70.93	19.92	1.72	0.75	6.69	80.11	16.89	1.01	0.09	1.90
MDPNR/Ag-5.0TiO_2	69.26	19.69	2.22	0.61	8.21	79.31	16.93	1.33	0.08	2.36
MDPNR/5.0TiO_2	66.44	25.73	2.93	0.00	4.90	75.06	21.82	1.73	0.00	1.39

3.3. Thermal Properties

The thermal gravimetric analysis (TGA) and the first derivative of TGA (DTG) curves of the MDPNR/Ag-TiO_2 composites, compared to those of DPNR and MDPNR, are presented in Figure 5. The TGA thermograms indicate weight loss occurring within the temperature range of 50 to 600 °C, followed by residue formation. Detailed parameters, such as the temperature at 5% weight loss (T_5), temperature at maximum process rate (T_{max}), and residue content, are provided in Table 2. As a result, DPNR decomposition occurred between 336 and 476 °C, and almost no residue remained at 600 °C. After modification of DPNR via grafting with poly(acrylic acid-co-acrylamide), MDPNR exhibited initial weight loss between 70 and 170 °C due to the evaporation of absorbed and bound water. Since the natural rubber was modified with a hydrophilic polymer, the MDPNR could absorb water molecules in its structure. Subsequent weight loss between 180 and 291 °C was attributed to the decomposition of carboxylic acid and amide side groups. Principal decomposition, involving the natural rubber and the polymer backbone of poly(acrylic acid-co-acrylamide), occurred at 336–476 °C [40,41]. The shift in T_{max} to a higher temperature than DPNR suggests the improved thermal stability of MDPNR. From the result, the decomposition characteristics of the MDPNR/Ag-TiO_2 composites were similar to those of MDPNR. However, metallic silver and titanium dioxide in the composites reduced T_5 compared to MDPNR. This may be due to the heat conductivity of silver and titanium dioxide, accelerating the decomposition process of side groups [42]. Nonetheless, the decomposition of the main components of the composites, comprising natural rubber modified with poly(acrylic acid-co-acrylamide), exhibited no difference in T_{max} values, indicative of the high thermal stability of the composites. Furthermore, the residues of the composites were higher than those of MDPNR, and they increased with increasing silver and titanium dioxide contents in the composites.

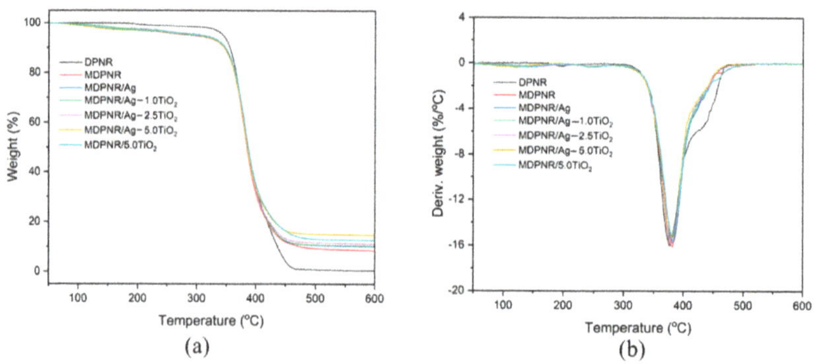

Figure 5. (**a**) TGA and (**b**) DTG thermograms of the DPNR, MDPNR, MDPNR/Ag, MDPNR/Ag-TiO_2 composites with various TiO_2 contents, and MDPNR/TiO_2.

Table 2. Thermal degradation temperature and residue from TGA curves of the DPNR, MDPNR, MDPNR/Ag, MDPNR/Ag-TiO_2 composites with various TiO_2 contents, and MDPNR/TiO_2.

Samples	T_5 (°C)	T_{max} (°C)	Residue at 600 °C (%)
DPNR	343.61	376.61	0.36
MDPNR	307.69	381.69	8.38
MDPNR/Ag	306.44	380.96	10.01
MDPNR/Ag-1.0TiO_2	302.27	383.27	10.28
MDPNR/Ag-2.5TiO_2	279.08	381.08	11.02
MDPNR/Ag-5.0TiO_2	291.98	380.98	14.57
MDPNR/5.0TiO_2	301.16	380.16	12.62

3.4. Dynamic Mechanical Properties

The viscoelastic behavior of the MDPNR/Ag-TiO_2 composites was investigated through temperature sweep testing. Figure 6a,b illustrate the storage modulus (E′) and loss tangent (tan δ) of the composites over a temperature range from −80 to 100 °C, respectively. All samples displayed a high storage modulus in the glassy state at low temperatures, as shown in Figure 6a. With increasing temperature, the storage modulus gradually decreased due to the increased mobility of the polymer chains [43]. The storage modulus decreased significantly due to the transition to a rubbery state. In dynamic mechanical analysis, the material's glass transition temperature (Tg) can be determined from the peak in the tan δ curve (Figure 6b). The viscoelastic properties are summarized in Table 3. As can be seen from the result, DPNR exhibited a single tan δ peak corresponding to the Tg of natural rubber, which was found to be −54.89 °C. The MDPNR showed two tan δ peaks. The first peak corresponded to the bulk of the modified natural rubber, while the second peak may be attributed to the polymeric chain network formed due to the modification. As observed in the bulk of the polymer, the Tg of MDPNR shifted to a lower temperature (−57.21 °C). This shift may be attributed to structural irregularities and intercalation between rubber chains, which decrease the number of chain entanglements and facilitate more effortless movement of the rubber chains. Interestingly, the Tg values of the MDPNR/Ag-TiO_2 composites appeared to increase compared to MDPNR. The Tg values of MDPNR/Ag-TiO_2 composites increased from −56.90 to −55.43 °C when TiO_2 was increased from 1.0 to 5.0 phr. Simultaneously, the peak height of tan δ decreased, suggesting increased stiffness and the restriction of polymer chain movement, probably due to the interaction of silver and titanium dioxide with a modified natural rubber matrix [44]. Silver can coordinate with active functional group-containing modified natural rubber [45], while titanium dioxide can interact with modified natural rubber through polar–polar interaction and hydrogen bonding between the hydroxylated titanium dioxide surface and modified natural rubber [46]. These findings suggest that the composites exhibit good stability against deformation.

Table 3. Viscoelastic properties of MDPNR/Ag-TiO_2 composites.

Samples	Tan δ Peak Position (°C)	Tan δ Peak Height
DPNR	−54.89	2.861
MDPNR	−57.21	1.045
MDPNR/Ag	−57.28	0.858
MDPNR/Ag-1.0TiO_2	−56.90	0.933
MDPNR/Ag-2.5TiO_2	−56.41	0.939
MDPNR/Ag-5.0TiO_2	−55.43	0.908
MDPNR/5.0TiO_2	−56.83	1.036

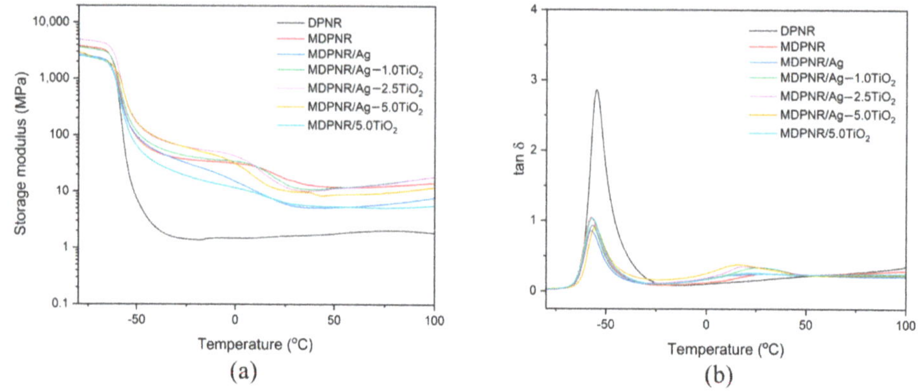

Figure 6. (**a**) Storage modulus and (**b**) loss tangent of composites as a function of temperature.

3.5. Dielectric Property

The materials' dielectric properties are important in studying their electrical response in an external electric field. The dielectric constant measures the ability of materials to store electrical energy from an electric field in their structure. Figure 7a illustrates the dielectric constants of DPNR, MDPNR, and MDPNR/Ag-TiO_2 composites across frequencies ranging from 10^2 to 10^7 Hz. DPNR exhibited a low dielectric constant due to the non-polar characteristic of natural rubber. The MDPNR and MDPNR/Ag-TiO_2 composites demonstrated a high dielectric constant at a low frequency. However, the dielectric constant decreased with increasing frequency, possibly due to the molecular movement's incapability and decreased orientation polarization at higher frequencies [47,48]. A comparison of dielectric constants of samples at 1 kHz is presented in Figure 7b. Notably, the dielectric constant of MDPNR was 74.54, which was higher than that of DPNR. This increase can be attributed to the modification of natural rubber with poly(acrylic acid-co-acrylamide), introducing a polar component to its structure [49–51].

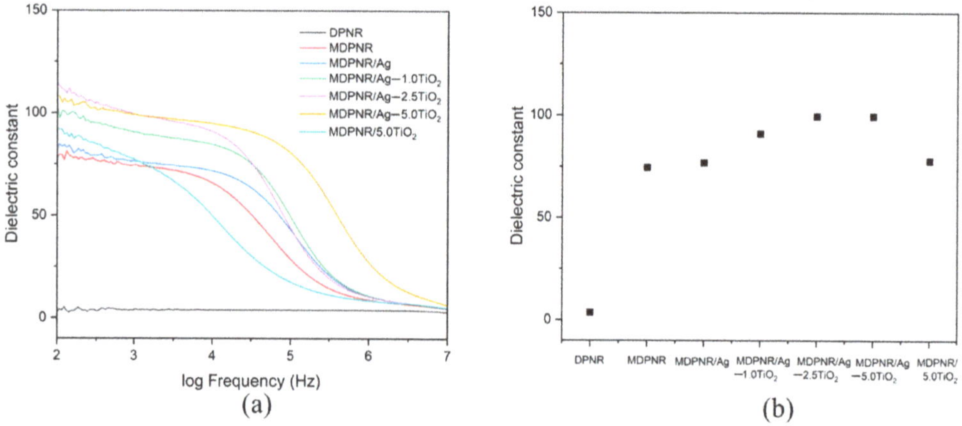

Figure 7. (**a**) Dielectric constant of the DPNR, MDPNR, and MDPNR/Ag-TiO_2 composites with various TiO_2 contents as a function of frequency and (**b**) dielectric constant of various types of composites at 1 kHz.

Furthermore, introducing silver nanoparticles and titanium dioxide enhanced the dielectric constant, suggesting that the composites possess more incredible electrical energy [52]. Silver and titanium dioxide act as fillers dispersed in the natural rubber-based matrix. Interfacial polarization occurred within the composites due to differences in the polarizations of the matrix and fillers, increasing the dielectric constant [53]. Indeed, the dielectric constant increases with higher titanium dioxide contents. Specifically, the dielectric constant values of MDPNR/Ag-TiO_2 composites increased to 91.03 and 99.46 when TiO_2 was added at 1.0 and 2.5 phr, respectively. At 5.0 phr, the dielectric constant does not differ significantly from that observed at 2.5 phr due to the agglomeration of titanium dioxide at a high loading. However, it exhibited a high dielectric constant in the broad frequency range. These correspond to the report by Sintharm et al., describing that the dielectric behavior of composites depends on the dielectric property of the polymer matrix and filler, chemical composition, chemical structure, and filler dispersion in the composites [54]. Thus, the enhancement of the dielectric constant of natural rubber through modification with poly(acrylic acid-co-acrylamide), incorporating silver nanoparticles and titanium dioxide, improved the dielectric constant, and made the composites more efficient for applications.

3.6. Antibacterial Activity

The antibacterial behavior of the MDPNR/Ag-TiO_2 composites against *S. aureus* and *E. coli*, compared to that of MDPNR, was examined, as shown in Figure 8. The results indicated that clear zones were not observed around the MDPNR samples on the agar plates against *S. aureus* and *E. coli*, suggesting no antibacterial activity. Interestingly, all composites demonstrated the ability to inhibit bacterial growth and form clear zones around the samples for both *S. aureus* and *E. coli*. The size of the clear zones for the samples was summarized in Table 4. It was observed that the composites composed of silver nanoparticles and/or titanium dioxide showed the inhibition zone, suggesting the effectiveness of silver nanoparticles and titanium dioxide for antibacterial activity. The antibacterial activity may be attributed to the penetration of silver particles and the release of silver ions, which interfere with bacterial cell membranes, disrupting their functions and leading to cell death [55]. Additionally, titanium dioxide may penetrate bacterial cell walls and disrupt essential cellular functions. As a result, the size of the clear zone for MDPNR/Ag composites against *S. aureus* and *E. coli* was measured as 12.0 ± 1.0 mm and 15.3 ± 0.6 mm, respectively, showing intense antibacterial activity. For the MDPNR/Ag-TiO_2 composites, the addition of titanium dioxide at 1.0 and 2.5 phr did not differ from the size of the clear zones compared to MDPNR/Ag. The inhibition zone was 11.7 ± 0.6, 12.0 ± 0.0 mm for *S. aureus* and 14.0 ± 0.0, 13.7 ± 0.6 mm for *E. coli* when titanium dioxide was added at 1.0 and 2.5 phr, respectively. However, upon increasing the titanium dioxide content to 5.0 phr, the inhibition zone against *S. aureus* and *E. coli* decreased to 10.7 ± 1.5 and 11.3 ± 2.5 mm, respectively. This reduction may have been due to the high titanium dioxide content, which could have resulted in poor distribution in the modified natural rubber matrix, as observed in SEM images, thereby reducing antibacterial efficacy. Nevertheless, the antibacterial efficiency of the MDPNR/Ag-5.0TiO_2 composites remained more remarkable than that of MDPNR/5.0TiO_2, which contains only titanium dioxide. It was indicated that the incorporation of silver nanoparticles and titanium dioxide promoted the antibacterial activity of natural rubber composites. Therefore, these MDPNR/Ag-TiO_2 composites exhibited antibacterial activity for both *S. aureus* and *E. coli* and could have potential for many applications.

Table 4. Antibacterial inhibition zone of MDPNR, MDPNR/Ag, MDPNR/Ag-TiO_2 with various TiO_2 contents, and MDPNR/TiO_2 composites.

Samples	Zone of Inhibition (mm)	
	S. aureus	*E. coli*
MDPNR	0.0 ± 0.0	0.0 ± 0.0
MDPNR/Ag	12.0 ± 1.0	15.3 ± 0.6
MDPNR/Ag-1.0TiO_2	11.7 ± 0.6	14.0 ± 0.0
MDPNR/Ag-2.5TiO_2	12.0 ± 0.0	13.7 ± 0.6
MDPNR/Ag-5.0TiO_2	10.7 ± 1.5	11.3 ± 2.5
MDPNR/5.0TiO_2	9.3 ± 1.2	8.7 ± 0.6

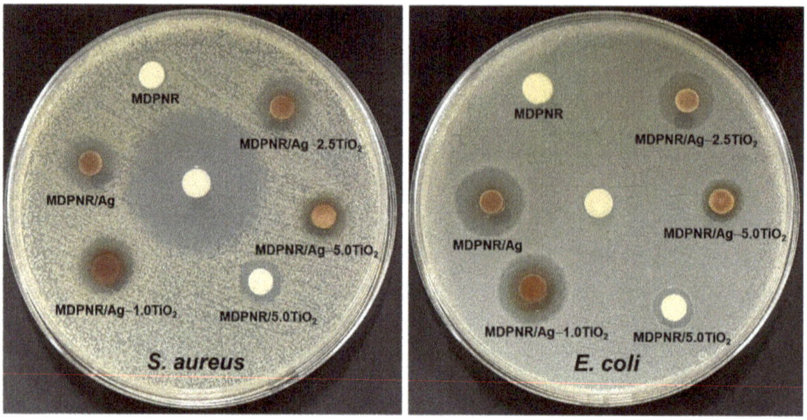

Figure 8. Disk diffusion test of MDPNR, MDPNR/Ag, MDPNR/Ag-TiO_2 with various TiO_2 contents, and MDPNR/TiO_2 composites.

4. Conclusions

The modification of deproteinized natural rubber (DPNR) through graft copolymerization facilitated the introduction of functional groups and enhanced interaction with silver nanoparticles and titanium dioxide. This modification was successfully achieved in this work, resulting in poly(acrylic acid-co-acrylamide)-modified, deproteinized natural rubber incorporating silver nanoparticles and titanium dioxide (MDPNR/Ag-TiO_2) composites. SEM-EDS mappings revealed that silver nanoparticles exhibited a uniform distribution in the modified natural rubber matrix. However, titanium dioxide tended to agglomerate at higher loadings, particularly noticeable at 5 phr of titanium dioxide. The MDPNR/Ag-TiO_2 composites demonstrated good thermal stability. The shift in Tg to a higher temperature and the decrease in the tan δ peak height indicate increased stiffness and resistance to deformation in the MDPNR/Ag-TiO_2 composites. The modification of DPNR with polar components, along with the presence of silver nanoparticles and titanium dioxide, improved the dielectric properties of the composites. Additionally, the existence of silver nanoparticles and titanium dioxide endowed the composites with antibacterial activity against *S. aureus* and *E. coli*. These composites, demonstrating good thermal stability, improved dielectric properties, and antibacterial activity, could have useful application in various fields.

Author Contributions: Conceptualization, S.I., S.K., B.M., Y.R. and C.R.; data curation, S.I. and C.R.; formal analysis, S.I., S.K., B.M., Y.R. and C.R.; funding acquisition, Y.R. and C.R.; investigation, S.I.; methodology, S.I., S.K. and C.R.; project administration, Y.R. and C.R.; resources, Y.R. and C.R.; supervision, C.R.; validation, Y.R. and C.R.; visualization, Y.R. and C.R.; writing—original draft, S.I.; writing—review and editing, S.I., S.K., B.M., Y.R. and C.R. All authors have read and agreed to the published version of the manuscript.

Funding: This work was financially supported by the Suranaree University of Technology Research and Development Fund (grant number IRD7-710-67-12-32).

Institutional Review Board Statement: Not applicable.

Data Availability Statement: Data are contained within the article.

Acknowledgments: The authors are grateful to Suranaree University of Technology Research and Development Fund for the financial support.

Conflicts of Interest: The authors declare no conflicts of interest.

References

1. Mahalingam, S.; Manap, A.; Lau, K.S.; Floresyona, D.; Medali Rachman, R.; Ayu Pradanawati, S.; Rabeya, R.; Chia, C.H.; Afandi, N.; Nugroho, A. Review of bioresource-based conductive composites for portable flexible electronic devices. *Renew. Sustain. Energy Rev.* **2024**, *189*, 113999. [CrossRef]
2. Zhang, L.; Shan, X.; Bass, P.; Tong, Y.; Rolin, T.D.; Hill, C.W.; Brewer, J.C.; Tucker, D.S.; Cheng, Z.Y. Process and Microstructure to Achieve Ultra-high Dielectric Constant in Ceramic-Polymer Composites. *Sci. Rep.* **2016**, *6*, 35763. [CrossRef] [PubMed]
3. Rubrice, K.; Castel, X.; Himdi, M.; Parneix, P. Dielectric Characteristics and Microwave Absorption of Graphene Composite Materials. *Materials* **2016**, *9*, 825. [CrossRef]
4. Kumar, A.; Ahmad, D.; Patra, K.; Hossain, M. Enhancement of electromechanical properties of natural rubber by adding barium titanate filler: An electro-mechanical study. *J. Appl. Polym. Sci.* **2021**, *138*, 50991. [CrossRef]
5. Kitisavetjit, W.; Nakaramontri, Y.; Pichaiyut, S.; Wisunthorn, S.; Kiatkamjornwong, S. Influences of carbon nanotubes and graphite hybrid filler on properties of natural rubber nanocomposites. *Polym. Test.* **2020**, *93*, 106981. [CrossRef]
6. Rajapaksha, N.; Perera, K.; Vidanapathirana, K. Characterization of a natural rubber based solid polymer electrolyte to be used for a magnesium rechargeable cell. *Polym. Bull.* **2022**, *79*, 1879. [CrossRef]
7. Yang, Y.; Zhao, G.; Cheng, X.; Deng, H.; Fu, Q. Stretchable and Healable Conductive Elastomer Based on PEDOT:PSS/Natural Rubber for Self-Powered Temperature and Strain Sensing. *ACS Appl. Mater. Interfaces* **2021**, *13*, 14599. [CrossRef]
8. Sarih, N.M.; Gwee, K.; Maher, S.; Rashid, A.A. Natural Rubber (NR) Latex Films with Antimicrobial Properties for Stethoscope Diaphragm Covers. *Materials* **2022**, *15*, 3433. [CrossRef]
9. Thomas, S.; Parameswaranpillai, J.; Krishnasamy, S.; Begam, P.M.; Nandi, D.; Siengchin, S.; George, J.; Hameed, N.; Salim, N.; Sienkiewicz, N. A comprehensive review on cellulose, chitin, and starch as fillers in natural rubber biocomposites. *Carbohydr. Polym. Technol. Appl.* **2021**, *2*, 100095. [CrossRef]
10. Pan, H.; Wang, Z.; Cui, Y.; Cao, L.; Zong, C. Influences of porous reduction graphene oxide/molybdenum disulfide as filler on dielectric properties, thermal stability, and mechanical properties of natural rubber. *J. Vinyl Addit. Technol.* **2021**, *27*, 868. [CrossRef]
11. Li, L.; Han, L.; Hu, H.; Zhang, R. A review on polymers and their composites for flexible electronics. *Mater. Adv.* **2023**, *4*, 726. [CrossRef]
12. Ni, Y.; Yang, D.; Wei, Q.; Yu, L.; Ai, J.; Zhang, L. Plasticizer-induced enhanced electromechanical performance of natural rubber dielectric elastomer composites. *Compos. Sci. Technol.* **2020**, *195*, 108202. [CrossRef]
13. Tian, M.; Zhang, J.; Zhang, L.; Liu, S.; Zan, X.; Nishi, T.; Ning, N. Graphene encapsulated rubber latex composites with high dielectric constant, low dielectric loss and low percolation threshold. *J. Colloid Interface Sci.* **2014**, *430*, 249. [CrossRef] [PubMed]
14. Inphonlek, S.; Bureewong, N.; Jarukumjorn, K.; Chumsamrong, P.; Ruksakulpiwat, C.; Ruksakulpiwat, Y. Preparation of Poly(acrylic acid-co-acrylamide)-Grafted Deproteinized Natural Rubber and Its Effect on the Properties of Natural Rubber/Silica Composites. *Polymers* **2022**, *14*, 4602. [CrossRef]
15. Pichayakorn, W.; Suksaeree, J.; Boonme, P.; Taweepreda, W.; Ritthidej, G.C. Preparation of Deproteinized Natural Rubber Latex and Properties of Films Formed by Itself and Several Adhesive Polymer Blends. *Ind. Eng. Chem. Res.* **2012**, *51*, 13393. [CrossRef]
16. Inphonlek, S.; Ruksakulpiwat, C.; Ruksakulpiwat, Y. The Effect of Silver Nanoparticles/Titanium Dioxide in Poly(acrylic acid-co-acrylamide)-Modified, Deproteinized, Natural Rubber Composites on Dye Removal. *Polymers* **2023**, *16*, 92. [CrossRef]
17. Niu, B.; Wang, X.; Wu, K.; He, X.; Zhang, R. Mesoporous Titanium Dioxide: Synthesis and Applications in Photocatalysis, Energy and Biology. *Materials* **2018**, *11*, 1910. [CrossRef]
18. Toro, R.G.; Diab, M.; de Caro, T.; Al-Shemy, M.; Adel, A.; Caschera, D. Study of the Effect of Titanium Dioxide Hydrosol on the Photocatalytic and Mechanical Properties of Paper Sheets. *Materials* **2020**, *13*, 1326. [CrossRef]
19. Bunriw, W.; Harnchana, V.; Chanthad, C.; Huynh, V.N. Natural Rubber-TiO_2 Nanocomposite Film for Triboelectric Nanogenerator Application. *Polymers* **2021**, *13*, 2213. [CrossRef]
20. Tan, X.; Liu, S.; Hu, X.; Zhang, R.; Su, X.; Qian, R.; Mai, Y.; Xu, Z.; Jing, W.; Tian, W.; et al. Near-Infrared-Enhanced Dual Enzyme-Mimicking Ag–TiO_2–x@Alginate Microspheres with Antibactericidal and Oxygeneration Abilities to Treat Periodontitis. *ACS Appl. Mater. Interfaces* **2023**, *15*, 391. [CrossRef]
21. Liao, C.; Li, Y.; Tjong, S.C. Visible-Light Active Titanium Dioxide Nanomaterials with Bactericidal Properties. *Nanomaterials* **2020**, *10*, 124. [CrossRef] [PubMed]
22. Zhang, Y.; Li, Q.; Gao, Q.; Wan, S.; Yao, P.; Zhu, X. Preparation of Ag/β-cyclodextrin co-doped TiO_2 floating photocatalytic membrane for dynamic adsorption and photoactivity under visible light. *Appl. Catal. B Environ.* **2020**, *267*, 118715. [CrossRef]
23. Ponja, S.D.; Sehmi, S.K.; Allan, E.; MacRobert, A.J.; Parkin, I.P.; Carmalt, C.J. Enhanced Bactericidal Activity of Silver Thin Films Deposited via Aerosol-Assisted Chemical Vapor Deposition. *ACS Appl. Mater. Interfaces* **2015**, *7*, 28616. [CrossRef] [PubMed]
24. Appamato, I.; Bunriw, W.; Harnchana, V.; Siriwong, C.; Mongkolthanaruk, W.; Thongbai, P.; Chanthad, C.; Chompoosor, A.; Ruangchai, S.; Prada, T.; et al. Engineering Triboelectric Charge in Natural Rubber–Ag Nanocomposite for Enhancing Electrical Output of a Triboelectric Nanogenerator. *ACS Appl. Mater. Interfaces* **2023**, *15*, 973. [CrossRef]
25. Soo, J.Z.; Chai, L.C.; Ang, B.C.; Ong, B.H. Enhancing the Antibacterial Performance of Titanium Dioxide Nanofibers by Coating with Silver Nanoparticles. *ACS Appl. Nano Mater.* **2020**, *3*, 5743. [CrossRef]
26. Kawahara, S.; Klinklai, W.; Kuroda, H.; Isono, Y. Removal of proteins from natural rubber with urea. *Polym. Adv. Technol.* **2004**, *15*, 181. [CrossRef]

27. Algethami, N.; Rajeh, A.; Ragab, H.; Tarabiah, A.; Gami, F. Characterization, optical, and electrical properties of chitosan/polyacrylamide blend doped silver nanoparticles. *J. Mater. Sci. Mater. Electron.* **2022**, *33*, 10645. [CrossRef]
28. Mkrtchyan, K.V.; Pigareva, V.A.; Zezina, E.A.; Kuznetsova, O.A.; Semenova, A.A.; Yushina, Y.K.; Tolordava, E.R.; Grudistova, M.A.; Sybachin, A.V.; Klimov, D.I.; et al. Preparation of Biocidal Nanocomposites in X-ray Irradiated Interpolyelectolyte Complexes of Polyacrylic Acid and Polyethylenimine with Ag-Ions. *Polymers* **2022**, *14*, 4417. [CrossRef] [PubMed]
29. Kanehira, K.; Banzai, T.; Ogino, C.; Shimizu, N.; Kubota, Y.; Sonezaki, S. Properties of TiO_2–polyacrylic acid dispersions with potential for molecular recognition. *Colloids Surf. B Biointerfaces* **2008**, *64*, 10. [CrossRef]
30. Joy, J.; Abraham, J.; Rajeevan, S.; George, S. Synthesis and characterization of natural rubber/graphene quantum dot nanocomposites. *J. Polym. Res.* **2021**, *28*, 358.
31. Mendoza, D.J.; Ayurini, M.; Raghuwanshi, V.S.; Simon, G.P.; Hooper, J.F.; Garnier, G. Synthesis of Superabsorbent Polyacrylic Acid-Grafted Cellulose Nanofibers via Silver-Promoted Decarboxylative Radical Polymerization. *Macromolecules* **2023**, *56*, 3497. [CrossRef]
32. Liu, Z.-K.; Deng, S.-S.; Zhou, Y.; Tong, Z.; Liu, J.-K.; Wang, Z.; Guo, M.-J.; Deng, L.; Zhen, Y.; Li, J.-T.; et al. On-Site Cross-Linking of Polyacrylamide to Efficiently Bind the Silicon Anode of Lithium-Ion Batteries. *ACS Appl. Mater. Interfaces* **2023**, *15*, 24416. [CrossRef] [PubMed]
33. Anancharoenwong, E.; Chueangchayaphan, W.; Rakkapao, N.; Marthosa, S.; Chaisrikhwun, B. Thermo-mechanical and antimicrobial properties of natural rubber-based polyurethane nanocomposites for biomedical applications. *Polym. Bull.* **2021**, *78*, 833. [CrossRef]
34. Yu, S.; Li, N.; Higgins, D.; Li, D.; Li, Q.; Xu, H.; Spendelow, J.S.; Wu, G. Self-Assembled Reduced Graphene Oxide/Polyacrylamide Conductive Composite Films. *ACS Appl. Mater. Interfaces* **2014**, *6*, 19783. [CrossRef] [PubMed]
35. Demchenko, V.; Rybalchenko, N.; Zahorodnia, S.; Naumenko, K.; Riabov, S.; Kobylinskyi, S.; Vashchuk, A.; Mamunya, Y.; Iurzhenko, M.; Demchenko, O.; et al. Preparation, Characterization, and Antimicrobial and Antiviral Properties of Silver-Containing Nanocomposites Based on Polylactic Acid–Chitosan. *ACS Appl. Bio Mater.* **2022**, *5*, 2576. [CrossRef]
36. Pai, K.R.N.; Anjusree, G.S.; Deepak, T.G.; Subash, D.; Nair, S.V.; Nair, A.S. High surface area TiO_2 nanoparticles by a freeze-drying approach for dye-sensitized solar cells. *RSC Adv.* **2014**, *4*, 36821. [CrossRef]
37. Bhardwaj, D.; Singh, R. Green biomimetic synthesis of Ag–TiO_2 nanocomposite using Origanum majorana leaf extract under sonication and their biological activities. *Bioresour. Bioprocess.* **2021**, *8*, 1. [CrossRef]
38. Vyatskikh, A.; Ng, R.C.; Edwards, B.; Briggs, R.M.; Greer, J.R. Additive Manufacturing of High-Refractive-Index, Nanoarchitected Titanium Dioxide for 3D Dielectric Photonic Crystals. *Nano Lett.* **2020**, *20*, 3513. [CrossRef]
39. Eddy, D.R.; Ishmah, S.N.; Permana, M.D.; Firdaus, M.L. Synthesis of Titanium Dioxide/Silicon Dioxide from Beach Sand as Photocatalyst for Cr and Pb Remediation. *Catalysts* **2020**, *10*, 1248. [CrossRef]
40. Beyler-Çiğil, A.; Şen, F.; Birtane, H.; Kahraman, M. Covalently bonded nanosilver-hydroxyethyl cellulose/polyacrylic acid/sorbitol hybrid matrix: Thermal, morphological and antibacterial properties. *Polym. Bull.* **2022**, *79*, 11353. [CrossRef]
41. Liu, Y.; Yan, Y.; Zhao, L.; Zhang, Y.; Zhang, L.; Zan, X. Thermally stable poly (acrylic acid-acrylamide-biomass-fly ash) composites with improved temperature resistance and salt resistance. *J. Appl. Polym. Sci.* **2022**, *139*, 51533. [CrossRef]
42. Noipitak, P.; Inphonlek, S.; Nillawong, M.; Sunintaboon, P.; Amornsakchai, T. Chitosan/alginate composite porous hydrogels reinforced with PHEMA/PEI core–shell particles and pineapple-leaf cellulose fibers: Their physico-mechanical properties and ability to incorporate AgNP. *J. Polym. Res.* **2021**, *28*, 182. [CrossRef]
43. Bashir, M.A. Use of Dynamic Mechanical Analysis (DMA) for Characterizing Interfacial Interactions in Filled Polymers. *Solids* **2021**, *2*, 108–120. [CrossRef]
44. Ren, X.; Barrera, C.S.; Tardiff, J.L.; Cornish, K. Sustainable Epoxidized Guayule Natural Rubber, Blends and Composites with Improved Oil Resistance and Greater Stiffness. *Materials* **2022**, *15*, 3946. [CrossRef] [PubMed]
45. Fahmy, A.; Agudo Jácome, L.; Schönhals, A. Effect of Silver Nanoparticles on the Dielectric Properties and the Homogeneity of Plasma Poly(acrylic acid) Thin Films. *J. Phys. Chem. C* **2020**, *124*, 22817. [CrossRef]
46. Liufu, S.; Xiao, H.; Li, Y. Adsorption of poly(acrylic acid) onto the surface of titanium dioxide and the colloidal stability of aqueous suspension. *J. Colloid Interface Sci.* **2005**, *281*, 155. [CrossRef]
47. Abdali, K.; Al-Bermany, E.; Haneen, K. Impact the silver nanoparticles on properties of new fabricated polyvinyl alcohol-polyacrylamide-polyacrylic acid nanocomposites films for optoelectronics and radiation pollution applications. *J. Polym. Res.* **2023**, *30*, 138. [CrossRef]
48. Mrudula, M.; Nair, M. Dielectric properties of natural rubber/polyethylene oxide block copolymer complexed with transition metal ions. *Polym. Bull.* **2020**, *77*, 6029. [CrossRef]
49. Liu, X.; Sun, H.; Liu, S.; Jiang, Y.; Yin, Z.; Yu, B.; Ning, N.; Tian, M.; Zhang, L. Dielectric elastomer sensor with high dielectric constant and capacitive strain sensing properties by designing polar-nonpolar fluorosilicone multiblock copolymers and introducing poly(dopamine) modified CNTs. *Compos. Part B Eng.* **2021**, *223*, 109103. [CrossRef]
50. Wang, Q.; Liu, X.; Qiang, Z.; Hu, Z.; Cui, X.; Wei, H.; Hu, J.; Xia, Y.; Huang, S.; Zhang, J.; et al. Cellulose nanocrystal enhanced, high dielectric 3D printing composite resin for energy applications. *Compos. Sci. Technol.* **2022**, *227*, 109601. [CrossRef]

51. Petchnui, K.; Uwanno, T.; Phonyiem Reilly, M.; Pinming, C.; Treetong, A.; Yordsri, V.; Moolsradoo, N.; Klamcheun, A.; Wongwiriyapan, W. Preparation of Chitin Nanofibers and Natural Rubber Composites and Their Triboelectric Nanogenerator Applications. *Materials* **2024**, *17*, 738. [CrossRef] [PubMed]
52. Suphasorn, P.; Appamato, I.; Harnchana, V.; Thongbai, P.; Chanthad, C.; Siriwong, C.; Amornkitbamrung, V. Ag Nanoparticle-Incorporated Natural Rubber for Mechanical Energy Harvesting Application. *Molecules* **2021**, *26*, 2720. [CrossRef] [PubMed]
53. Sreenivasan, D.P.; Sujith, A.; Asokan, A.N.; Rajesh, C. Dielectric Properties of Composites of Natural Rubber and Keratin Fibre from Chicken Feather. *Fibers Polym.* **2021**, *22*, 2588. [CrossRef]
54. Sintharm, P.; Phisalaphong, M. Green Natural Rubber Composites Reinforced with Black/White Rice Husk Ashes: Effects of Reinforcing Agent on Film's Mechanical and Dielectric Properties. *Polymers* **2021**, *13*, 882. [CrossRef]
55. Demchenko, V.; Mamunya, Y.; Kobylinskyi, S.; Riabov, S.; Naumenko, K.; Zahorodnia, S.; Povnitsa, O.; Rybalchenko, N.; Iurzhenko, M.; Adamus, G.; et al. Structure-Morphology-Antimicrobial and Antiviral Activity Relationship in Silver-Containing Nanocomposites Based on Polylactide. *Molecules* **2022**, *27*, 3769. [CrossRef]

polymers

MDPI

Article

Effect of Proteins on the Vulcanized Natural Rubber Crosslinking Network Structure and Mechanical Properties

Yueqiong Wang [1,2,*], Shiqi Su [1], Hongchao Liu [1,*], Rui Wang [1], Lusheng Liao [1], Zheng Peng [1], Jihua Li [1], Haijun Wu [3] and Dongning He [3]

[1] Guangdong Provincial Key Laboratory of Natural Rubber Processing, Agricultural Products Processing Research Institute, Chinese Academy of Tropical Agricultural Sciences, Zhanjiang 524001, China
[2] Key Laboratory of Advanced Materials of Tropical Island Resources, Ministry of Education, School of Materials Science and Engineering, Hainan University, Haikou 570228, China
[3] School of Chemistry and Chemical Engineering, Lingnan Normal University, Zhanjiang 524048, China
* Correspondence: yqw215@163.com (Y.W.); lhc2012168@163.com (H.L.)

Abstract: Proteins are important factors affecting the properties of natural rubber. Therefore, investigating the effect of free and bonded proteins on the structure and mechanical properties of the vulcanized crosslinking network of natural rubber would provide a theoretical basis for the production of high mechanical resistance natural rubber. Herein, natural rubbers with different protein contents and types were prepared by high-speed centrifugation. And, the effects on their network structure, vulcanization, tensile strength, tearing strength and dynamic mechanical properties were investigated. The results showed that the reduction in protein content led to the decrement in the entanglement networks, crosslinking density and tensile and tear strengths of the vulcanized natural rubber. Moreover, the bonded proteins had an obvious influence on the vulcanization process, while free proteins played an important role in the crosslink densities. These results reveal that both bonded and free proteins are involved in the vulcanization process and the construction of the vulcanized crosslinking network structure of natural rubber, which enhances the mechanical properties such as the modulus and tensile strength of vulcanized natural rubber.

Keywords: natural rubber; proteins; crosslinked network structure; mechanical properties

1. Introduction

Natural rubber (NR), an important engineering material, is a natural polymer synthesized from fresh latex collected from Brazilian three-leaf rubber trees. NR occupies an indispensable position in aerospace, rail transportation and medical and industrial fields for its excellent physical and mechanical properties [1,2]. The outstanding properties of NR are due to the non-rubber components that contain proteins and lipids. Both the proteins and lipids can affect the terminal molecular chains of NR. It is widely known that the non-rubber components have a significant effect on the network structure of NR. Tarachiwin et al. [3,4] demonstrated that proteins coupled to the ω-terminal group of the rubber chains, linked by hydrogen bonds, and phospholipids coupled to the α-terminal group either by hydrogen or ionic bonds. The aggregation of proteins and phospholipids then gradually formed a crosslinked network of natural rubber. Dixit et al. [5] showed that proteins in rubber were attached to each other in the form of hydrogen bonds to form crosslinking points to the ω-terminal end of the rubber molecular chain, while phospholipids formed branched structures at the α-terminal end of the rubber molecular chain. Former research proved that both proteins and lipids formed a "naturally occurring network" in unvulcanized NR [6,7].

In the non-rubber components, protein content is about 2.2% in NR latex [8], which is an important part. Research of proteins' effect on the network and mechanical properties of NR has been widely reported [6,9,10]. It has been found that the mechanical properties of NR dramatically decrease for the removal of proteins. Generally speaking, the proteins in NR are usually removed by centrifugation, proteases or urea. There are two main forms, called binding proteins and free proteins, in NR latex. The bonded proteins interact with the rubber particles and are distributed on the surface of the rubber particles, while the free proteins are mainly found in the serum of the natural latex. It is known that proteins affect the structure of the vulcanized crosslinking network of NR [11–13], which influences the properties of vulcanized NR [14–16]. Huang et al. [17] reported that protein content could be considered as the guidance for the selection of NR and pointed out the importance of proteins. There is also research focusing on the effect of exogenous proteins on the mechanical properties of NR. It has been found that adding exogenous proteins improves the mechanical properties and decreases the heat generation [18–20].

However, previous studies have illustrated the important influence of proteins on the structure and mechanical properties and network of NR. The contributions of the two types of endogenous proteins in NR are rarely discussed. As is known, the two types of proteins in NR are due to their types existing in NR. Proteins dispersed in NR latex that could be easily removed by a physical method are called free proteins. Those that are linked to the molecular chains of NR by functional groups are called bonded proteins. However, there is little research about the effects of bonded and free proteins on the structure and properties of NR. But, there are definitely different effects of the two types of proteins on the mechanical properties from our previous studies [21]. Free and bonded proteins and their contents have different influences on the structure and properties of the vulcanized crosslinking network of NR, but their influences have not been specially discussed. As is known, network structure is a crucial factor influencing NR's mechanical properties. Therefore, both the network structure and mechanical properties affected by the two types of proteins are necessary to study.

In this study, NR latex with different contents and existing types of proteins were prepared by high-speed centrifugation. Moreover, vulcanized NR was prepared by a mechanical mixing method, and the effects of the free and binding proteins on the crosslinking network and mechanical properties of the natural rubber were analyzed. The results will give advice on tuning the network structure and mechanical properties of NR.

2. Materials and Methods

2.1. Sample Preparation

2.1.1. Preparation of Natural Rubber Latex (NRL)

Fresh NRL (dry rubber content 35%) was provided by Guangdong Guangken Maoming Rubber Group Co., Ltd., Maoming, China. The NRL was added with 0.6% (*w*/*w*) ammonia at room temperature. Urea solution at 10% concentration and sodium dodecyl sulfate (SDS) solution at 10% concentration were prepared to be used with fresh latex. NRL with different protein contents were prepared using disc centrifugation or high-speed centrifugation, and the preparation process is shown in Figure 1. The samples were prepared as follows.

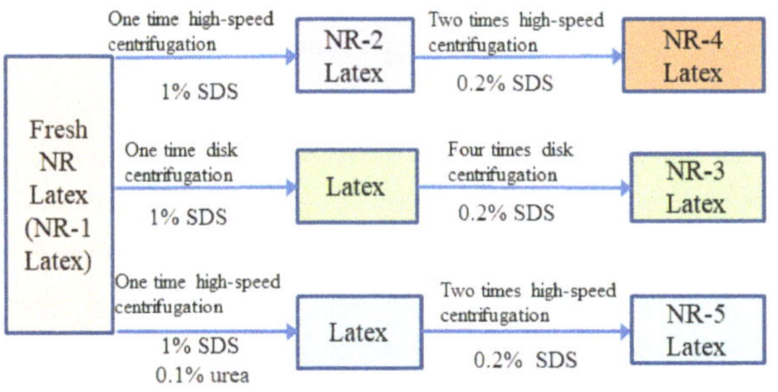

Figure 1. Preparation process of NRL with different protein contents.

NR-1. The NR-1 was the fresh NRL without centrifugation.

NR-2. For the NR-2, 1% (*w*/*w*) SDS was added to the fresh latex and diluted to a 30% dry rubber content with deionized water, and then the mixture was stirred for 2 h to mix it together. A high-speed freezing centrifuge (CR22N/21N, Hitachi, Tokyo, Japan) was used to centrifuge the paste at 8000 r/min for 30 min at 25 °C, and then the upper cream was collected and diluted to a 60% dry rubber content by adding deionized water and 0.6% (*w*/*w*) ammonia. NRL with high-speed centrifugation conducted once removed most of the free proteins in the natural latex and preserved the bonded proteins and some free proteins.

NR-3. Fresh NRL with 1% (*w*/*w*) SDS solution was centrifugated using a disc centrifuge (JRLDR200, Yixing Haide Separation Machinery Co., Ltd., Yixing, China) 5 times. Every cycle, the centrifugation diluted the NRL to a 30% dry rubber content, and 0.2% (*w*/*w*) SDS was added. After the fifth centrifugation, the protein content changed very little. Therefore, the free proteins were completely removed and only the bonded proteins were preserved in the NRL. The N content was 0.099%, which was similar to the previous report [22].

NR-4: The NR-4 was centrifuged one more time than the NR-2. After the first centrifugation of the NRL, the upper cream was diluted to a 30% dry rubber content and 0.2% (*w*/*w*) SDS was added. After the second centrifugation, the upper cream was diluted to a 60% dry rubber content by adding deionized water and 0.6% (*w*/*w*) ammonia.

NR-5: For the NR-5, 0.1% urea and 1% SDS were added to the fresh NRL; then, deionized water was added to dilute it to a 30% dry rubber content, and it was mechanically stirred for 2 h to mix it well. The urea-treated natural latex was centrifugated at 8000 r/min for 30 min at 25 °C using a high-speed refrigerated centrifuge, and the upper cream was diluted with 0.2% SDS and deionized water and centrifuged twice. Since urea-treated natural latex can remove bonded proteins, it can be compared with the samples treated three times by the high-speed centrifugation. The NR-4 can be used to compare the effects of different bonded protein contents on the properties of natural rubber.

2.1.2. Preparation of Vulcanized NR

The NR-1, NR-2, NR-3, NR-4 and NR-5 latex were poured into a glass mold and dried at 40 °C to a constant weight. The formula was as follows.

Zinc oxide (ZnO) at 6 parts per hundred of rubber (phr), sulfur (S) at 3.5 phr, stearic acid (SA) at 0.5 phr and accelerator NS at 0.5 phr were used. An open two roller mill was used to mix the natural rubber and the agents together, and then the uniform NR compounds were obtained.

A moving die rheometer (MDR-2000E, Huzhou Hongqiao Rubber Machinery Co., Ltd., Huzhou, China) was used to measure the t_{90} at 143 °C. After that, the NR compound was vulcanized for (t_{90} + 5) min on a plate vulcanizer to obtain the vulcanized NR.

2.2. *Sample Analysis*

2.2.1. Fourier Transform Infrared (FTIR) Spectroscopy

The NRL were dried on molds with small holes to obtain films with a thickness of 60 to 80 μm, and the molecular structure of the NR was tested using a FTIR spectrometer (Tensor 27, Bruker, Munich, Germany) in the wavelength range of 400 cm^{-1} to 4000 cm^{-1}.

2.2.2. Nitrogen Content

A fully automatic Kjeldahl nitrogen analyzer (Kjeltec 8400, Foss, Denmark) was used to determine the nitrogen content of the NR in accordance with the GB/T8088-2008 standard [23].

2.2.3. Mechanical Properties

A tensile machine (AI-7000-SUI, Gotech Co., Dongguan, China) was used to test the tensile properties at a rate of 500 mm/min in accordance with the GB/T528-2009 [24] standards. Five samples were repeated to obtain the strength parameters. The cyclic tensile test was operated at a rate of 200 mm/min.

2.2.4. Crosslinking Density

The equilibrium swelling method was used to determine the crosslinking density of the vulcanized NR. A sample of about 0.5 g was cut and weighed as m_0, the vulcanized rubber sheet was immersed in a 100 mL toluene solvent for 72 h and then the dissolved rubber was weighed and recorded as m_1. Subsequently, it was put in an oven at 80 °C and dried to a constant weight. Finally, the rubber sheet was removed and cooled down to room temperature and weighed again, recorded as m_2. The crosslinking density was calculated using the following formula.

$$v_r = \frac{\frac{m_2}{\rho}}{\frac{m_2}{\rho} + \frac{(m_1 - m_0)}{\rho_s}} = \frac{1}{Q} \quad (1)$$

$$V_c = -\frac{1}{V}\frac{ln(1 - v_r) + v_r + \chi v_r^2}{\sqrt[3]{v_r} - \frac{1}{2}v_r} \quad (2)$$

In Equations (1) and (2), Q is the swelling degree of the rubber; v_r is the volume fraction of the rubber after swelling; V_c is the average number of molecular chains per unit volume, i.e., in mol/cm^3; ρ and ρ_s are the densities of the rubber and toluene, respectively, which are 0.913 g/cm^3 and 0.865 g/cm^3; χ is the interaction parameter of the natural rubber and toluene, which is 0.393; and V is the molar volume fraction of the toluene, which is 106.3 cm^3/mol.

2.2.5. Dynamic Mechanical Analyzer (DMA)

The samples were cut down to long strip samples and tested using a dynamic thermo-mechanical analyzer (DMA Q800, TA, Newcastle, DE, USA). The temperature range was from −90 °C to 80 °C, and the rate was 3 °C/min.

3. Results and Discussion

3.1. FTIR Analysis

Figure 2 shows the infrared spectra of the NR gums with different protein contents. It can be seen in Figure 2a,b that the untreated fresh latex, NR-1, and the NR-2 treated one time with high-speed centrifugation have obvious peaks at 3319 cm^{-1} and 1553 cm^{-1}, which can be attributed to the N-H telescoping vibration and deformation vibration, respectively, which indicates that there are proteins in NR. As is seen, the 3319 cm^{-1} (Figure 2a) and 1553 cm^{-1} (Figure 2b) peaks in the NR-3, NR-4 and NR-5 almost disappear [21,25]. The peaks of the C=C stretching vibration and the C-H bending vibration are 1664 cm^{-1} and 837 cm^{-1}, respectively. As a result, the NR-1 contained all the proteins, which included both free and bonded proteins; the NR-2 lost most of the free proteins after the high-speed centrifugation process; the NR-3 retained all the bonded proteins, and the NR-4 and NR-5 contained only part of the bonded proteins, which also agreed with the reference [22].

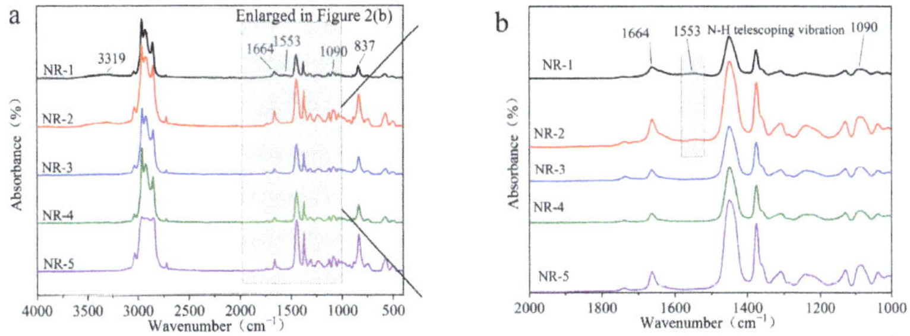

Figure 2. (**a**) Infrared spectrogram and (**b**) Enlarged infrared spectrogramof natural rubber with different protein content.

3.2. Nitrogen Content Analysis

The protein content in natural rubber can be calculated by multiplying the determined nitrogen content by a factor of 6.25 [26]. The nitrogen and protein contents of the NR are shown in Table 1. Free proteins are more easily removed during centrifugation, resulting in a decrease in protein content [6]. The protein content of the NR-2 was reduced from 3.356% to 0.984% after the high-speed centrifugation being conducted once. Disc centrifugation is not as effective as high-speed centrifugation because it is hard to separate the bonded proteins. Therefore, disc centrifugation was used to obtain the NR-3 sample with only bonded proteins, and the nitrogen content of the NR-3 was 0.099%, and the protein content was 0.619%. These data are consistent with the results in the literature [22], which indicates that the NR-3 had all the free proteins removed, and only the bonded proteins were preserved. High-speed centrifugation has a high separation force and can remove both free proteins and bonded proteins. For further removing proteins, additives should be used to depart the bonded proteins [27]. So, urea was used to depart the bonded proteins from the NR because urea can destroy the hydrogen bonding of proteins and increases the solubility of the hydrophobic side chains of proteins in water [28,29]. As a result, the protein content of the NR-5 was significantly decreased to 0.135%.

Table 1. Nitrogen content of natural rubber with different protein content.

Sample	Nitrogen Content %	Protein Content %
NR-1	0.537 ± 0.004	3.356 ± 0.025
NR-2	0.157 ± 0.001	0.984 ± 0.006
NR-3	0.099 ± 0.001	0.619 ± 0.006
NR-4	0.032 ± 0.001	0.198 ± 0.006
NR-5	0.022 ± 0.003	0.135 ± 0.017

3.3. *Effect of Proteins on the Vulcanization Properties of NR*

Table 2 shows the vulcanization parameters of the NR with different protein contents. It can be seen that the maximum torque of the NR decreased with the decrement in the protein content. Meanwhile, the scorch time t_{10} and optimum vulcanization time t_{90} increased with the decrement in the protein content as the amide group of the protein can enhance the vulcanization rate and promote the NR forming more crosslinking points that lead to the increment in the maximum torque [30,31]. Based on the existing forms of proteins, it can be seen that the free proteins had a significant effect on the maximum torque M_H of the natural rubber, which decreased from 7.14 dN·m to 6.12 dN·m after the removal of free proteins, as the protein content decreased from 3.356% to 0.619%. However, the M_H decreased from 6.12 dN·m to 6.04 dN·m after the bonded protein was reduced from 0.619% to 0.135%. In addition, the free proteins showed a significant effect on the t_{90}, which was prolonged from 12.2 min to 18.9 min, with a 2.737% (3.356% (the protein content of the NR-1) minus 0.619% (the protein content of the NR-3)) reduction in the free protein. The t_{90} was further prolonged from 18.9 min to 24.1 min after the reduction in the bonded proteins, while the content of the bonded proteins was reduced by 0.063% (0.198% (the protein content of the NR-4) minus 0.135% (the protein content of the NR-5)). Although the t_{90} reduction in the bonded proteins was lower than in the free proteins, the decreasing content of bonded proteins was much smaller than that of the free proteins. As a result, it is indicated that although the bonded protein content was lower in the NR, it had a more significant effect on the t_{90} than the free proteins.

Table 2. Vulcanization characteristic parameters of natural rubber with different protein content.

Samples	M_H/dN·m	M_L/dN·m	t_{10}/min	t_{90}/min
NR-1	7.14	0.72	1.7	12.2
NR-2	6.64	0.71	3.4	10.2
NR-3	6.12	0.55	8.8	18.9
NR-4	6.12	0.62	11.1	22.8
NR-5	6.04	0.63	11.8	24.1

3.4. Effect of Proteins on NR Mechanical Properties

Table 3 and Figure 3 show the tensile parameters and the stress–strain process and curves of the vulcanized NR. As can be seen in Table 3, the tensile strength of the NR tended to increase and then decrease as the protein content decreased, while the stress decreased. That is because the proteins had obvious influences on the NR crosslinking network. The tensile strength of the NR-2 increased after part of the free protein was removed. As free proteins can form crosslinking networks [32], there is an optimum crosslinking density in vulcanized NR [33]. If the crosslinking density is too large, the molecular weight between the crosslinking points decreases; short chains easily form defects, leading to a decline in tensile strength. After removing some free proteins, the density of the crosslinked network decreases and the number of defects decreases, and then the tensile strength increases. As a result, the tensile strength of the NR-2 was larger than that of the NR-1. With the further removal of the free proteins, the NR-3 contained only bonded proteins, and the crosslinking network formed by the proteins reduced, which led to a decrease in the tensile strength and stress. In addition, the strain-induced crystallization behavior of natural rubber also affects its mechanical properties, which can be reflected by the 500% stress, because vulcanized NR already crystallizes at the strain [34,35]. It is clear that the vulcanized NR exhibited a much higher 500% stress when it possessed more proteins, which was due to the higher crystallinity of the vulcanized NR. Moreover, our previous studies have shown that vulcanized natural rubber has the highest degree of crystallinity and the best tensile properties at a crosslinking density of 1.84×10^{-4} mol/cm^3. In brief, the effect of the proteins on the structure of the crosslinked network is in agreement with our former results [12].

Table 3. Mechanical property parameters of natural rubber with different protein content.

Projects	NR-1	NR-2	NR-3	NR-4	NR-5
Tensile strength (MPa)	27.80 ± 0.92	30.81 ± 1.57	27.45 ± 1.06	27.64 ± 2.08	26.70 ± 1.58
100% stress (MPa)	1.11 ± 0.02	0.86 ± 0.03	0.77 ± 0.01	0.70 ± 0.02	0.68 ± 0.03
300% stress (MPa)	2.82 ± 0.05	2.04 ± 0.06	1.77 ± 0.02	1.64 ± 0.06	1.58 ± 0.09
500% stress (MPa)	10.31 ± 0.4	5.03 ± 0.51	4.02 ± 0.07	3.7 ± 0.50	3.28 ± 0.43
Elongation at break (%)	656 ± 7	752 ± 20	738 ± 9	756 ± 26	753 ± 30

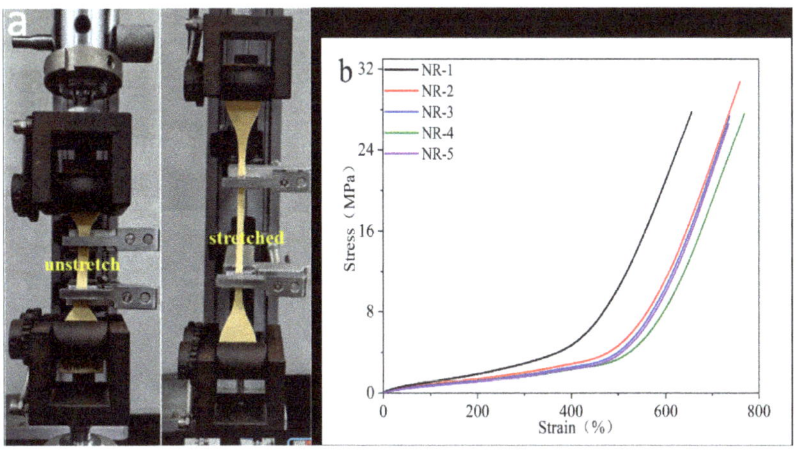

Figure 3. (**a**) Stress–strain process and (**b**) curves of rubber with different protein content.

3.5. Effect of Proteins on NR Crosslinking Network

Crosslinking density is an important type of data reflecting the structure of vulcanized networks [36]. Table 4 shows the crosslinking density V_c and average molecular weight between the crosslinking points M_c of the vulcanized NR. As can be seen in Table 4, the crosslinking density of the NR decreased with decrement in the protein content, while the average molecular weight between the crosslinking points increased. This is because the proteins in NR can participate in the construction of the crosslinking network by forming crosslinking points, which leads to an increase in the crosslinking density of the networks [32]. It is known that proteins exist in peptide bonds [37]; the carboxylic acid groups and amino groups in peptide bonds can react with zinc oxide to form coordination complexes, which improves the solubility of the zinc oxide in NR. Moreover, coordination complexes enhance the solubility of zinc salts in NR matrices and reduce the activation energy for forming crosslinking. Moreover, the unstable Zn-S bond of the coordination complexes effected by the amino or carboxyl groups can initiate a ring opening of the S_8 ring, which can act with the molecular chains of NR to form a crosslinking precursor [30]. Then, the crosslinking precursor and NR molecular chains react to form new crosslinking bonds, which promotes the NR vulcanization crosslinking reaction, so the crosslinking density of the NR increases. Therefore, as the protein content decreases, the carboxylic acid groups and amino groups contained in the NR decrease, and the coordination complexes formed by the proteins and zinc salts decrease, slowing down the vulcanization crosslinking reaction of the NR. And, the crosslinking structure of the NR reduces, leading to the crosslinking density decreasing. As can be seen from Table 4, the crosslinking density of the vulcanized NR decreased from 2.35×10^{-4} mol/cm^3 to 1.52×10^{-4} mol/cm^3 after the removal of the free proteins. The crosslinking density decreased from 1.52×10^{-4} mol/cm^3 to 1.46×10^{-4} mol/cm^3 after the bonded protein content was further reduced from 0.619% to 0.135%. The crosslinking density influenced by the bonded proteins was much lower than that influenced by the free proteins. This may be due to the fact that bonded proteins are involved in the construction of physical networks and are less likely to form ligand complexes with zinc salts during vulcanization crosslinking; thus, fewer crosslinking bonds are formed by the bonded proteins, which in turn has less effect on the crosslinking density.

Table 4. Crosslinking density of natural rubber with different protein content.

Samples	Q (%)	V_c (10^{-4} mol/cm^3)	M_C (kg/mol)
NR-1	4.06	2.35	3.88
NR-2	4.61	1.74	5.25
NR-3	4.89	1.52	6.02
NR-4	4.95	1.47	6.19
NR-5	4.97	1.46	6.24

The tube model theory was used to calculate the degree of contribution of the topological entanglement and chemical crosslinking network of the vulcanized NR. The effect of the proteins on the structure of the crosslinked network was analyzed using Equations (3) and (4) [38]. The diagrams of each sample based on these equations are shown in Figure 4.

$$\sigma_M = \frac{\sigma}{(\alpha - \alpha^{-2})} = G_c + G_e f(\alpha) \tag{3}$$

$$f(\alpha) = \frac{\beta}{2} \frac{\alpha^{\beta/2} - \alpha^{-\beta}}{\alpha^2 - \alpha^{-1}} \tag{4}$$

In Equations (3) and (4), σ_M is the reduced stress; G_c is the elastic modulus created by the chemical crosslinking network (i.e., the effective crosslinking); G_e is the elastic modulus created by the entanglement network; σ is the stress; α is the elongation ratio of the NR in tension; and β is the empirical parameter between the deformed pipeline in tension and the unstretched pipeline at the equilibrium, which is usually taken as 1 [39].

Figure 4 shows the f(α)-σ_m curves for each specimen, and linear fitting is operated in the range of 0.4–0.7 of the *X*-axis. The network structural parameters are shown in Table 5, which are the chemical crosslinking modulus G_C, the entanglement network modulus G_e, the effective crosslinking density V_c, the average chain length between the effective crosslinking points M_c, the molecular mean-square end distance between the crosslinking points R_c, the average number of chain segments between the crosslinking points N, the tube radius d_0 and the number of statistical chain segments ne between two neighboring effective entanglement points; the parameters were calculated using the same method as in our former research [36, 40]. As can be seen from the table, the chemical crosslinking modulus G_c and physical entanglement modulus G_e, as well as the effective crosslinking density V_c, of the NR show a decreasing trend as the content of protein decreases, while the related parameters of the rubber, such as M_c, R_c, N, d_0 and n_e, show an increasing trend. This is due to the fact that as the protein content decreases, the coordinated complexes formed by the protein and zinc salts decrease, and the crosslinking points of NR also reduce, resulting in a loosely arranged vulcanization network structure. Therefore, the decrease in the protein content of NR leads to a decrease in the chemical crosslinking modulus and physical entanglement modulus as well as the effective crosslinking density of vulcanized NR, whereas the related quantities such as the molecular weight between the effective crosslinking points of NR increase as the crosslinking points and entanglement points of NR decrease.

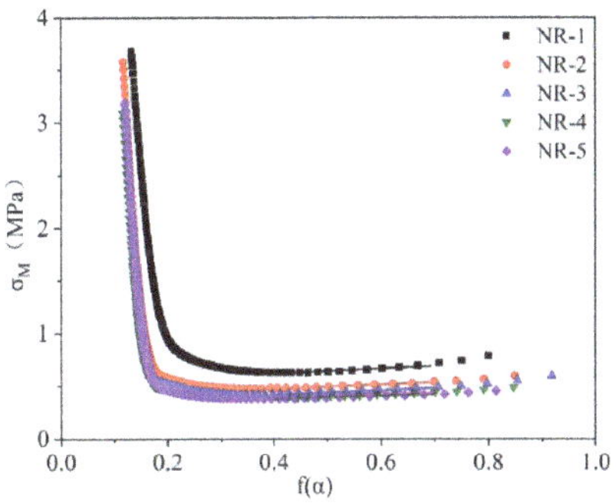

Figure 4. f(α)-σ_m diagram of each sample.

Table 5. Network structure parameters of each sample.

Samples	NR-1	NR-2	NR-3	NR-4	NR-5
G_c (MPa)	0.52	0.39	0.33	0.32	0.32
G_e (MPa)	0.24	0.20	0.22	0.18	0.14
$V_c \times 10^{-4}$ (mol/mL)	3.12	2.37	1.99	1.90	1.95
M_c (g/mol)	2978	3916	4664	4897	4776
R_c (nm)	4.05	4.64	5.07	5.19	5.13
N	28	37	44	46	45
d_0 (nm)	2.33	2.54	2.44	2.74	3.07
n_e	9	11	10	13	16

3.6. Hysteresis Loss

The hysteresis loss hysteresis loop is the energy loss due to the friction within the NR molecules during the stretch–recovery process [41]. NR hysteresis loops are operated by the stretch–recovery process, and the integral hysteresis loop area is the energy loss of the NR in the stretch–recovery process.

Figure 5 shows the tensile recovery curves of the vulcanized NR with different protein contents. From Figure 5a–e, it can be seen that as the strain increased, the hysteresis loop and its area became larger; as the protein content in the NR decreased, the area of the hysteresis loop of the vulcanized NR shrunk and the hysteresis loss decreased at the same strain. This is due to proteins forming network structures with molecular chains in NR, and because the breaking and rebuilding of ionic bonds during stretching causes more energy dissipation [22]. It also can be seen in Figure 5f that after the removal of all the free proteins, the hysteresis loss of the NR-3 vulcanized rubber decreased more significantly than the

NR-1 when the strain was more than 300%; after the further removal of the bonded proteins, the hysteresis loss of the NR-4 and NR-5 decreased more obviously than the NR-3 when the strain was more than 400%, and the difference between their hysteresis loss at 300% is not significant. It is indicated that the network structure formed by the free proteins breaks and slips at a 300% strain, whereas the crosslinked network structure formed by the bonded proteins does not break or slip until a 400% strain, resulting a different hysteresis loss.

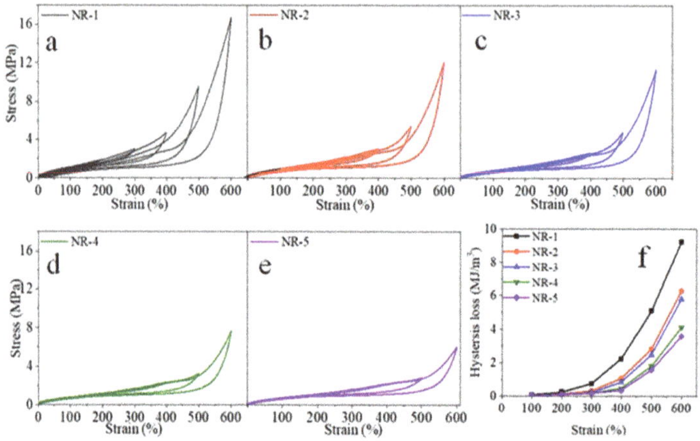

Figure 5. Hysteresis area of (**a**) NR-1, (**b**) NR-2, (**c**) NR-3, (**d**) NR-4, (**e**) NR-5 and (**f**) hysteresis loss of vulcanized NR with different protein content.

3.7. Effect of Protein on Dynamic and Mechanical Properties of Vulcanized NR

Figure 6 shows the energy storage modulus and loss factor versus temperature for the vulcanized NR with different protein contents. As can be seen in Figure 6, the storage modulus E′ of the NR-1 was much higher than that of other vulcanized NR. This suggests that proteins can participate in the crosslinked network construction, which increases the crosslinking density and thus improves the storage modulus and rigidity of vulcanized NR. From the loss factor curves, it can be seen that the T_g of the NR-3 was −43.4 °C, which was slightly lower than that of the NR-1, which was −42.9 °C, whereas after the bonded proteins were further removed, the T_g of the NR-5 was −43.9 °C, which was slightly lower than that of the NR-3. This indicates that both bonded and free proteins are involved in the construction of the crosslinked network, and the removal of these proteins decreases the crosslinking density and increases the mobility of the molecular chains, resulting in a slight decrement in the T_g. Both free and bonded proteins are involved in the construction of the crosslinking network of vulcanized NR.

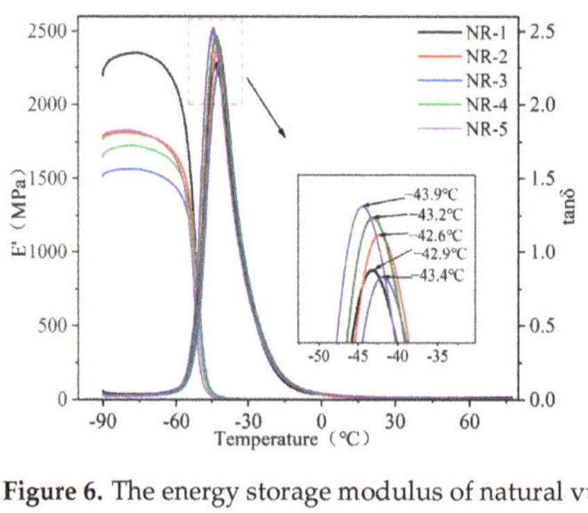

Figure 6. The energy storage modulus of natural vulcanized rubber with different protein content changes with temperature.

4. Conclusions

In this study, NRL with different forms and contents of proteins were prepared using centrifugation. The effects of the protein content and existing type on the vulcanization process, crosslinking network structure and mechanical properties of vulcanized NR were analyzed, and the following conclusions are drawn:

Both free and bonded proteins can promote NR vulcanization and reduce the vulcanization time; the bonded proteins have a more significant effect on the vulcanization time, although their content is lower than that of the free proteins. Both of them can increase the crosslinking density of vulcanized rubber, and the effect of the free proteins on the crosslinking density is more significant; the removal of the free proteins reduces the crosslinking density from 2.35×10^{-4} mol/cm^3 to 1.52×10^{-4} mol/cm^3, and the reduction in the bonded proteins has less of an effect on the crosslinking density, which is due to the fact that the bonded proteins are involved in the construction of the rubber gum network, leading to it being more difficult to form zinc salt coordinated complexes to promote the formation of crosslinking points. Both the free proteins and bonded proteins can participate in the construction of the crosslinking network of vulcanized NR, but the amount of the bonded proteins is not large enough to achieve the optimum crosslinking density, so the free proteins need to take part in the formation of the crosslinking network to achieve the optimum crosslink density; thus, the tensile properties of the vulcanized NR can be optimized. This conclusion can give guidance on preparing NR with different types of proteins and regulating NR properties by tuning proteins, which will promote the creation of vulcanized NR with excellent mechanical properties.

Author Contributions: Methodology, Y.W.; Validation, Y.W.; Formal analysis, H.L., R.W. and D.H.; Investigation, H.L. and H.W.; Writing—original draft, S.S.; Writing—review & editing, Y.W.; Visualization, J.L.; Supervision, L.L., Z.P., J.L. and D.H.; Funding acquisition, Y.W. and L.L. All authors have read and agreed to the published version of the manuscript.

Funding: This work was supported by the Central Public-interest Scientific Institution Basal Research Fund (No. 1630122024007), Earmarked Fund for China for Agriculture Research System (No. CARS-33-JG2), Hainan Provincial Natural Science Foundation of China (No. 521QN0929), Hainan Province Science and Technology Special Fund (No. ZDYF2024XDNY284), and Hainan Provincial Natural Science Foundation (No. 521CXTD438).

Institutional Review Board Statement: Not applicable.

Data Availability Statement: Data are contained within the article.

Conflicts of Interest: The authors declare no conflict of interest.

References

1. Xu, T.; Jia, Z.; Wu, L.; Chen, Y.; Luo, Y.; Jia, D.; Peng, Z. Influence of acetone extract from natural rubber on the structure and interface interaction in NR/silica composites. *Appl. Surf. Sci.* **2017**, *423*, 43–52. [CrossRef]
2. Kawahara, S. Discovery of island-nanomatrix structure in natural rubber. *Polym. J.* **2023**, *55*, 1007–1021. [CrossRef]
3. Tarachiwin, L.; Sakdapipanich, J.; Ute, K.; Kitayama, T.; Bamba, T.; Fukusaki, E.-I.; Kobayashi, A.; Tanaka, Y. Structural Characterization of α-Terminal Group of Natural Rubber. 1. Decomposition of Branch-Points by Lipase and Phosphatase Treatments. *Biomacromolecules* **2005**, *6*, 1851–1857. [CrossRef]
4. Tarachiwin, L.; Sakdapipanich, J.; Ute, K.; Kitayama, T.; Tanaka, Y. Structural Characterization of α-Terminal Group of Natural Rubber. 2. Decomposition of Branch-Points by Phospholipase and Chemical Treatments. *Biomacromolecules* **2005**, *6*, 1858–1863. [CrossRef]
5. Dixit, M.; Taniguchi, T. Effect of Impurities on the Formation of End-Group Clusters in Natural Rubber: Phenylalanine Dipeptide as an Impurity Protein. *Macromolecules* **2024**, *57*, 2588–2608. [CrossRef]
6. Huang, C.; Zhang, J.; Cai, X.; Huang, G.; Wu, J. The effects of proteins and phospholipids on the network structure of natural rubber: A rheological study in bulk and in solution. *J. Polym. Res.* **2020**, *27*, 158. [CrossRef]
7. Schlögl, S.; Trutschel, M.-L.; Chassé, W.; Riess, G.; Saalwächter, K. Entanglement Effects in Elastomers: Macroscopic vs Microscopic Properties. *Macromolecules* **2014**, *47*, 2759–2773. [CrossRef]
8. Kawahara, S.; Tanaka, Y. Structure of Natural Rubber. *Nippon Gomu Kyokaishi* **2009**, *82*, 417–423. [CrossRef]
9. Kawahara, S.; Nishioka, H.; Yamano, M.; Yamamoto, Y. Synthetic Rubber with the Tensile Strength of Natural Rubber. *ACS Appl. Polym. Mater.* **2022**, *4*, 2323–2328. [CrossRef]

10. Srisomboon, S.; Wadeesirisak, K.; Vaysse, L.; Sainte-Beuve, J.; Musigamart, N.; Liengprayoon, S.; Bonfils, F.; Rattanaporn, K.; Bottier, C. Optimization of a protein extraction method from natural rubber sheets made of Hevea brasiliensis latex. *J. Rubber Res.* **2021**, *24*, 27–39. [CrossRef]
11. Ariyawiriyanan, W.; Nuinu, J.; Sae-Heng, K.; Kawahara, S. The Mechanical Properties of Vulcanized Deproteinized Natural Rubber. *Energy Procedia* **2013**, *34*, 728–733. [CrossRef]
12. Lehman, N.; Keereerak, A.; Promsung, R.; Nakaramontri, Y.; Johns, J.; Songtipya, L.; Kalkornsurapranee, E. Influence of different protein contents from several clonal varieties of Hevea brasiliensis latex on the properties of cured natural rubber film using glutaraldehyde (GA) as a curing agent. *Ind. Crops Prod.* **2024**, *208*, 117868. [CrossRef]
13. Guo, Y.; Bao, K.; Wu, X.; Han, D.; Zhang, J. Morphology and aggregation process of natural rubber particles. *Ind. Crops Prod.* **2023**, *203*, 117153. [CrossRef]
14. Sadeghi, M.; Malekzadeh, M.; Taghvaei-Ganjali, S.; Motiee, F. Correlations between natural rubber protein content and rapid predictions of rheological properties, compression set and hardness of rubber compound. *J. Indian Chem. Soc.* **2021**, *98*, 100162. [CrossRef]
15. Zhou, Y.; Kosugi, K.; Yamamoto, Y.; Kawahara, S. Effect of non-rubber components on the mechanical properties of natural rubber. *Polym. Adv. Technol.* **2017**, *28*, 159–165. [CrossRef]
16. Jong, L. Influence of protein hydrolysis on the mechanical properties of natural rubber composites reinforced with soy protein particles. *Ind. Crops Prod.* **2015**, *65*, 102–109. [CrossRef]
17. Huang, S.-Q.; Zhang, J.-Q.; Zhu, Y.; Kong, L.-M.; Liao, L.-S.; Zhang, F.-Q.; Xie, Z.-T.; Wu, J.-R. Revealing the Structure-Property Difference of Natural Rubber Prepared by Different Methods: Protein and Gel Content are Key Factors. *Chin. J. Polym. Sci.* **2023**, *42*, 457–467. [CrossRef]
18. Zhan, Y.-H.; Wei, Y.-C.; Tian, J.-J.; Gao, Y.-Y.; Luo, M.-C.; Liao, S. Effect of protein on the thermogenesis performance of natural rubber matrix. *Sci. Rep.* **2020**, *10*, 16417. [CrossRef]
19. Jong, L. Reinforcement effect of soy protein nanoparticles in amine-modified natural rubber latex. *Ind. Crops Prod.* **2017**, *105*, 53–62. [CrossRef]
20. Yu, W.-W.; Xu, W.-Z.; Xia, J.-H.; Wei, Y.-C.; Liao, S.; Luo, M.-C. Toughening natural rubber by the innate sacrificial network. *Polymer* **2020**, *194*, 122419. [CrossRef]
21. Wang, Y.; Liu, H.; Yu, H.; Zhao, P.; Wang, Q.; Liao, L.; Luo, M.; Zheng, T.; Liao, S.; Peng, Z. New insight into naturally occurring network and entanglements induced strain behavior of vulcanized natural rubber. *Polymer* **2022**, *241*, 124545. [CrossRef]
22. Fu, X.; Huang, C.; Zhu, Y.; Huang, G.; Wu, J. Characterizing the naturally occurring sacrificial bond within natural rubber. *Polymer* **2019**, *161*, 41–48. [CrossRef]
23. *GB/T8088-2008*; Rubber, Raw Natural and Rubber Latex, Natural—Determination of Nitrogen Content. Chinese Standard Publisher: Beijing, China, 2008.
24. *GB/T528-2009*; Rubber, Vulcanized or Thermoplastic—Determination of Tensile Stress-Strain Properties. Chinese Standard Publisher: Beijing, China, 2009.
25. Rolere, S.; Liengprayoon, S.; Vaysse, L.; Sainte-Beuve, J.; Bonfils, F. Investigating natural rubber composition with Fourier Transform Infrared (FT-IR) spectroscopy: A rapid and non-destructive method to determine both protein and lipid contents simultaneously. *Polym. Test.* **2015**, *43*, 83–93. [CrossRef]
26. Yunyongwattanakorn, J.; Tanaka, Y.; Sakdapipanich, J.; Wongsasutthikul, V. Highly-Purified Natural Rubber by Saponification of Latex: Analysis of Residual Proteins in Saponified Natural Rubber. *Rubber Chem. Technol.* **2008**, *81*, 121–137. [CrossRef]
27. Sakdapipanich, J.; Kalah, R.; Nimpaiboon, A.; Ho, C.C. Influence of mixed layer of proteins and phospholipids on the unique film formation behavior of Hevea natural rubber latex. *Colloids Surf. A Physicochem. Eng. Asp.* **2015**, *466*, 100–106. [CrossRef]
28. Klinklai, W.; Saito, T.; Kawahara, S.; Tashiro, K.; Suzuki, Y.; Sakdapipanich, J.T.; Isono, Y. Hyperdeproteinized natural rubber prepared with urea. *J. Appl. Polym. Sci.* **2004**, *93*, 555–559. [CrossRef]
29. Yamamoto, Y.; Nghia, P.T.; Klinklai, W.; Saito, T.; Kawahara, S. Removal of proteins from natural rubber with urea and its application to continuous processes. *J. Appl. Polym. Sci.* **2008**, *107*, 2329–2332. [CrossRef]
30. Wei, Y.-C.; Liu, G.-X.; Zhang, L.; Zhao, F.; Liao, S.; Luo, M.-C. Exploring the unique characteristics of natural rubber induced by coordination interaction between proteins and Zn^{2+}. *Polymer* **2020**, *193*, 122357. [CrossRef]
31. Wei, Y.-C.; Zhu, D.; Xie, W.-Y.; Xia, J.-H.; He, M.-F.; Liao, S. In-situ observation of spatial organization of natural rubber latex particles and exploring the relationship between particle size and mechanical properties of natural rubber. *Ind. Crops Prod.* **2022**, *180*, 114737. [CrossRef]
32. Liu, X.-X.; He, M.-F.; Luo, M.-C.; Wei, Y.-C.; Liao, S. The role of natural rubber endogenous proteins in promoting the formation of vulcanization networks. *e-Polymers* **2022**, *22*, 445–453. [CrossRef]
33. Zhao, F.; Bi, W.; Zhao, S. Influence of Crosslink Density on Mechanical Properties of Natural Rubber Vulcanizates. *J. Macromol. Sci. Part B* **2011**, *50*, 1460–1469. [CrossRef]
34. Liu, H.; Huang, G.S.; Wei, L.Y.; Zeng, J.; Fu, X.; Huang, C.; Wu, J.R. Inhomogeneous Natural Network Promoting Strain-induced Crystallization:A Mesoscale Model of Natural Rubber. *Chin. J. Polym. Sci.* **2019**, *37*, 105–114. [CrossRef]
35. Huang, S.; Zhang, J.; Kong, L.; Li, W.; Xie, Z.-T.; Wu, J. Fully Biosourced, Vulcanization-Free, and Thermal-Responsive Natural Rubber Material. *Macromolecules* **2024**, *57*, 1642–1652. [CrossRef]

36. Wang, Y.; Liao, L.; Wang, R.; Yu, H.; Zheng, T.; Lian, Y.; Luo, M.; Liao, S.; Liu, H.; Peng, Z. Research of strain induced crystallization and tensile properties of vulcanized natural rubber based on crosslink densities. *Ind. Crops Prod.* **2023**, *202*, 117070. [CrossRef]
37. Wu, J.; Qu, W.; Huang, G.; Wang, S.; Huang, C.; Liu, H. Super-Resolution Fluorescence Imaging of Spatial Organization of Proteins and Lipids in Natural Rubber. *Biomacromolecules* **2017**, *18*, 1705–1712. [CrossRef]
38. Huang, C.; Huang, G.; Li, S.; Luo, M.; Liu, H.; Fu, X.; Qu, W.; Xie, Z.; Wu, J. Research on architecture and composition of natural network in natural rubber. *Polymer* **2018**, *154*, 90–100. [CrossRef]
39. Klüppel, M.; Menge, H.; Schmidt, H.; Schneider, H.; Schuster, R.H. Influence of Preparation Conditions on Network Parameters of Sulfur-Cured Natural Rubber. *Macromolecules* **2001**, *34*, 8107–8116. [CrossRef]
40. Wang, Y.; Liu, H.; Zheng, T.; Peng, Z.; Wang, R.; Yu, H.; Wang, Q.; Liao, S.; Liao, L. Strain-induced crystallization behavior and tensile properties of natural rubber with different vulcanization bond types. *Polym. Test.* **2023**, *129*, 108289. [CrossRef]
41. Shen, Q.; Wu, M.; Xu, C.; Wang, Y.; Wang, Q.; Liu, W. Sodium alginate crosslinked oxidized natural rubber supramolecular network with rapid self-healing at room temperature and improved mechanical properties. *Compos. Part A Appl. Sci. Manuf.* **2021**, *150*, 106601. [CrossRef]

polymers

MDPI

Article

The Use of Rubber-Polymer Composites in Bitumen Modification for the Disposal of Rubber and Polymer Waste

Anar Akkenzheyeva [1], Viktors Haritonovs [2], Akkenzhe Bussurmanova [3,*], Remo Merijs-Meri [4], Yerzhan Imanbayev [5], Akmaral Serikbayeva [1], Serik Sydykov [6], Yerbolat Ayapbergenov [7], Martynas Jankauskas [8], Anatolijs Trumpels [9], Murshida Aimova [3] and Maira Turkmenbayeva [1]

1 Engineering Faculty, Yessenov University, 32 Microdistrict, Aktau 130003, Kazakhstan; anar.akkenzheyeva@yu.edu.kz (A.A.); akmaral.serikbayeva@yu.edu.kz (A.S.); maira.turkmenbayeva@yu.edu (M.T.)
2 Faculty of Civil and Mechanical Engineering, Riga Technical University, 6A Kipsalas Street, LV-1048 Riga, Latvia; viktors.haritonovs@rtu.lv
3 Science and Technology Faculty, Yessenov University, 32 Microdistrict, Aktau 130003, Kazakhstan; murshida.aimova@yu.edu.kz
4 Faculty of Natural Sciences and Technologies, Riga Technical University, 3 Paula Valdena Street, LV-1048 Riga, Latvia; remo.merijs-meri@rtu.lv
5 Laboratory of Petrochemical Processes, RSE "Institute of Combustion Problems", Bogenbay Street, 172, Almaty 050012, Kazakhstan; erzhan.imanbayev@mail.ru
6 LLP "JV" CASPI BITUM Aktau Bitumen Plant, Industrial Zone 5, Building 65, Aktau 130000, Kazakhstan; s.sydykov82@mail.ru
7 Branch of LLP "KMG Engineering" "KazNIPImunaigas", 35 Microdistrict, Section 6/1, Aktau 130000, Kazakhstan; ayapbergenov_e@kaznipi.kz
8 JSC "Polylema", Jegaines Str. 8, Kauno District, LT-54469 Biruliskes, Lithuania; info@polylema.lt
9 LLC "Rubbintec", Drosmes 2-34, LV-2015 Jurmala, Latvia; atrumpels@gmail.com
* Correspondence: akkenzhe.bussurmanova@yu.edu.kz

Abstract: The use of rubber-polymer composites ELTC (End of Life Tire Compound) for bitumen modification was investigated. ELTC contains not only devulcanized rubber from used car tires, but also used plastics (polymers) such as polyethylene (PE) and polypropylene (PP). ELTC is obtained using the method of rubber devulcanization using a selective catalyst that allows selectively decomposing sulfide bonds at relatively low temperatures, while preserving most of the macromolecular chains. The characteristics of the asphalt binder improved after the modification of ELTC. After modification, the base asphalt binder became more homogeneous, and the thermal stability of the base asphalt binder increased. ELTC is evenly distributed, the compatibility between the components of the modified asphalt binder is good, which proves the uniformity of the modified asphalt binders. The results show that all ELTC formulations improve the softening temperature and increase their resistance to plastic deformation in the summer.

Keywords: petroleum bitumen; bitumen modification; crumb rubber; fluorescence microscopy; differential scanning calorimetry; thermogravimetric analyser

1. Introduction

Worldwide, sustainable development is an important need in the construction industry, and to this end, the use of waste in road construction is increasingly encouraged to reduce environmental impacts. A large number of original materials and technologies suitable for the construction and maintenance of road surfaces have been invented in the road infrastructure. Some of them are plastics and rubbers. Currently, the amount of various rubber waste and plastic waste is increasing. According to a study sanctioned by the World Business Council for Sustainable Development, in 2018, the United States and the EU produced about 3700 and 3400 kg of used tires, respectively [1]. It is predicted that by the end of 2030, the number of used tires worldwide will reach approximately 5 billion [2]. The current global annual accumulation of end-of-life plastic goods is approximately 150 million tons [3]. Their non-biodegradability and toxic components pose a serious threat to the environment [4–6]. The use of rubber and plastic waste for the production of modified asphalt plays an important role in resource conservation and environmental protection [7–9]. At the same time, the constant increase in the number of vehicles highlights the need for better quality roads and engineering design. Despite the many advantages offered by modern technologies for the production of modified binders, the materials obtained can increase the total cost of asphalt concrete mix by up to 40% [10,11].

Plastic and rubber waste can be used to partially replace conventional material to improve the desired mechanical characteristics for a particular road mix. Also, the active use of waste as bitumen modifiers is carried out in order to increase productivity, as well as reduce the cost of paving materials. Historically, crumb rubber has been used to produce modified asphalt for a very long time.

A large number of studies of rubber bitumen mixtures have shown that mixing rubber crumbs with petroleum asphalt can improve the characteristics of asphalt at high and low temperatures, viscoelastic characteristics and anti-aging properties, as well as rubber bitumen mixtures have a longer service life compared to conventional asphalt concrete mixtures, reduce traffic noise, slow down the appearance of reflective cracks, improve the adhesion of binders materials with fillers, improve the slip resistance of the road surface and wear resistance [12,13]. Many states around the world have already introduced the use of rubber crumbs for bitumen modification due to increased resistance to track and fatigue cracking of rubber-bitumen binders [14–18].

Nevertheless, the use of rubber-bitumen binders causes some problems, such as higher production and laying temperatures, additional energy consumption, increased greenhouse gas emissions, and the need for constant mixing to avoid phase separation [19,20]. To eliminate the above problems, various solutions have been investigated, such as the addition of additives and modifiers or the devulcanization of rubber [21]. Vulcanized rubber crumbs have attracted special attention because the structural transformation of rubber involved in the devulcanization process leads to the formation of a product that can simultaneously solve problems of both high viscosity and phase separation. However, the use of devulcanized rubber in asphalt binders has led to contradictory results in terms of mechanical properties and is therefore at the research stage [22–25].

On the other hand, with the development of the plastics industry, plastic-modified asphalt is already taking its place in the road materials market. Plastics can significantly improve the high-temperature characteristics of asphalt, but cannot improve the low-temperature characteristics of asphalt; instead plastics reduce the low-temperature characteristics of asphalt [26]. Polyethylene (PE) and polypropylene (PP) are the most attractive plastic wastes for the road industry and are among the most commonly used additives to reduce rut, and the effect is more obvious under heavy loads [27]. The authors [28] found that high-temperature characteristics, resistance to low-temperature cracking and resistance to deformation have significantly increased. Maharaj concluded that the particle size and polyethylene content are of great importance for resistance to fatigue cracking and rutting [29]. Modification with polyethylene reduced the residual deformation, and the modified asphalt mixture began to have excellent resistance to residual deformation [30,31]. The binder modified with PE and PP constantly demonstrates problems with phase stability due to reduced chemical compatibility between macromolecules and asphalt components [32–34]. Moreover, polyethylene particles will separate from the asphalt binder if stored improperly, which limits the use of polyethylene-modified asphalt binder [35]. In addition, there is no improvement in the low-temperature characteristics of the asphalt binder when mixed only with polyethylene particles [36].

In recent years, researchers have begun to study asphalt modified with rubber-plastic composites. The researchers found that some asphalt compositions modified with rubber-plastic composites had better characteristics at high temperatures than asphalt modified with SBS, but their addition reduced the low-temperature characteristics of asphalt [29,37,38]. Nizamuddin et al. studied how the use of innovative additives that improve compatibility can help stabilize recycled asphalt concrete mixtures of rubber and plastic during high-temperature storage [39]. Various scientists have studied the properties of bitumen binders modified with rubber-plastic composites [13,40–42] and some have found that asphalt modified with a rubber-plastic composite has no less high-temperature characteristics than asphalt modified with SBS, but also good low-temperature characteristics [43]. It was also found that when the ratio of rubber and plastic was 7:3, the powder of rubber waste and plastic waste had good compatibility; the modifier of the spent rubber-plastic composite has good compatibility with the main asphalt, significantly improving the road characteristics of the main asphalt [44]. Wang studied the combined modification of polyethylene and rubber. The result showed that after modification, the rheological properties of the asphalt binder improved in various temperature ranges. To obtain a stable modified asphalt binder, it is necessary to reduce the density difference and strengthen the interaction [45].

Some chemical agents used in the rubber devulcanization process, usually present in devulcanized rubber, may be useful to improve storage stability in the presence of plastics. Preliminary studies show that the use of devulcanized rubber in combination with plastic waste improves storage stability of bitumen modified by plastic waste, only due to the presence of plasticizers, compatibilizers and crosslinking agents [46].

Thus, it is necessary to further study the modification of bitumen with rubber-plastic composites or hybrid modifiers.

To assess the effect of the hybrid modifier on the modification of road bitumen, the results were compared with unmodified road bitumen, modified SBS road bitumen (the main competitor of rubber modifiers on the world market).

2. Materials and Methods

Modified bitumen B70/100 was used as the initial bitumen. The characteristics of petroleum bitumen B70/100 are shown in Table 1.

Table 1. Characteristics of petroleum road viscous bitumen grade B 70/100.

Bitumen Properties	Normative Indicators of the Road Bitumen B 70/100	Actual Value	Test Method
Penetration at 25 °C, not lower	87 ± 5	87.2	EN 1426:2015 [47]
Softening point °C, not below	45.8 ± 1.6	45.85	EN 1427:2015 [48]
Brittleness temperature on Fraas °C, not higher	–21 ± 3	–21	EN 12593:2015 [49]
Solubility %, not less	99.75 ± 0.1	99.75	EN 12592:2015 [50]
Flash point °C, not below	334 ± 4	335	EN 2592:2006 [51]
	Rolling Thin Film Oven Test (RTFOT)		
Change of Mass	−0.021 ± 0.01	−0.0212	EN 12607-1:2015 [52]
Increasing in Softening Point	7.0 ± 3.6	7.0	EN 1427:2015 [48]
Decreasing in Softening Point	52.8 ± 1.6	53.25	EN 1427:2015 [48]
Retained Penetration	52 ± 3	51.9	EN 1426:2015 [47]

In this work, the End of Life Tire Compound (ELTC) material from "Polylema" JSC manufacturer is used. "Rubbintec" LLC has proposed a method for the devulcanization of rubber using a selective catalyst that allows the selective decomposition of sulfide bonds at relatively low temperatures, while preserving most of the macromolecular chains. The resulting crumb from devulcanized rubber is intended for use both in rubber mixtures in the manufacture of new rubber products and as a bitumen modifier. A group of companies ("Rubbintec" and "Polylema") has developed and patented in the United States a unique tire recycling technology that allows you to completely extract rubber from tires and use it to produce almost any rubber products. A key part of the "Rubbintec" devulcanization process is the combination of a selective catalyst and a compatibilizer. The catalyst is designed to break sulfide bonds, and the compatibilizer helps it penetrate the rubber crumb. The second function of the catalyst is its assistance in "grafting" devulcanized rubber to other polymers and organic components.

"Rubbintec" technology allows starting the process of decomposition of rubber at a much lower temperature—thermomechanical cracking of rubber crumbs in a reactor with two screws. In addition, cracking at temperatures below 550 °C (preferably 350–500 °C) preserves a large number of the original double bonds present in diene rubbers.

ELTC technology is patented [53], and the copyright belongs to "Rubbintec" Ltd. The ELTC material is environmentally friendly. The strategy of the main enterprise of "Polylema" JSC is the development of a competitive material (increased flexibility at high and low operating temperatures will create a stable homogeneous bitumen and rubber system). Competitiveness must be achieved against alternatives available on the market, such as SBS (Styrene-butadiene-styrene) polymer, which is widely used to modify road bitumen. The cost of SBS polymer is about 3000 euros/ton, and ELTC is several times cheaper. The material produced by "Polylema" JSC is available in the form of granules with a diameter of up to 2.5 mm (see Figure 1).

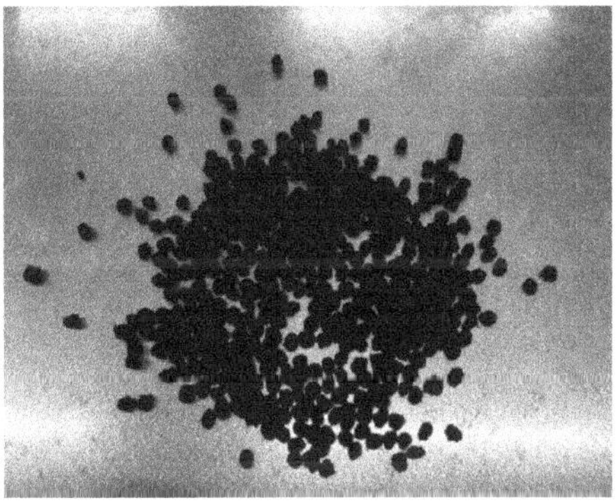

Figure 1. End of Life Tire Compound, developed by "Rubbintec" LLC and "Polylema" JSC.

The granules are black in color and consist of the following substances:

1. Crushed rubber from worn rubber tires 70–94%;
2. Thermoplastic polymer (waste), e.g., polypropylene or polyethylene 1–23%;
3. Additives 5–12%.

High Shear mixers are used in the laboratory to modify road bitumen. The production of modified road bitumen in the laboratory with the ELTC modifier of "Polylema" JSC consists of several stages:

1. Heating of road bitumen in an oven to 140 °C degrees;
2. The ELTC modifier is added (in parts) to 20% of the road bitumen mass and mixed at a speed of 6000 rpm and 170–180 °C degrees;
3. Mixing from 15 to 90 min;

A fluorescence microscopy (FM) Moticam PorS5Lite (OTICEUROPE, S.L.U., Barcelona, Spain) was used to scan the target using fluorescence excitation filter D480/30 nm to observe distribution properties of targets according to the fluorescence shape.

Differential scanning calorimetry DSC-1 (Mettler Toledo, Greifensee, Switzerland) was used to record heating thermograms from −100 °C to 200 °C at scanning speed 10 °C in air environment. Average sample mass was 10 mg.

Thermogravimetric analyser TGA-1 (Mettler Toledo) was used to record decomposition thermograms from 20 °C to 800 °C at scanning speed 10 °C in air environment. Average sample mass was 10 mg.

3. Results and Discussion

Using the innovative ELTC modifier, several modified bitumen compositions were created in the laboratory. As ELTC is made of not only rubber from used car tires, but also used thermoplastics such as PE or PP, together with customized devulcanization additives it belongs to the group of hybrid modifier. In this modifier, rubber provides flexibility over the broad temperature range, while thermoplastics provide rigidity and act as glue (1) helping hold rubber particles in one monolith during the production process, (2) giving the product the shape of granules and (3) making it more suitable for further introduction in bitumen.

One of the most important functions of the modifier is to increase the softening temperature of bitumen, making the binder more resistant at high exploitation temperatures, which becomes important due to global temperature increase. Figure 2 shows the results for the softening temperature as a function of the temperature and time of high-shear mixing of the bitumen with 20 wt.% of ELTC + 10% PP modifier.

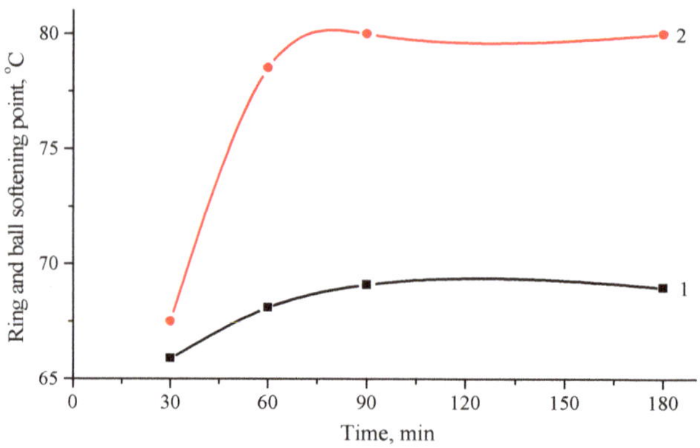

Figure 2. Change in softening temperature of ELTC + 10% PP modified bitumen from mixing time at different temperatures: 1—at 180 °C; 2—at 200 °C.

From the results shown in the Figure, it can be seen that all the investigated ELTC compositions improve the softening temperature. For the compositions obtained at 200 °C and mixing times over 60 min, softening temperatures are close to 80 °C, which means more than 20 °C increment compared to unmodified bitumen 70/100. Since the goal is to obtain modified bitumen with a softening temperature of at least 55 °C, the results show that the addition of the hybrid ELTC modifier not only allowed achieving this goal, but also exceeded it by a wide margin. It is important to note that in this study, the ELTC modifier was added to 20% of the bitumen mass. Also, in the previously reviewed literature [54,55], it was found that rubber improves the properties of bitumen by adding 15–25% by weight of bitumen. This denotes that large amount of rubber waste can be utilized in this way the cost of rubber-bitumen binders can be reduced. Also, the properties of modified bitumen were compared with the properties of bitumen modified with 4% SBS. The softening temperature with 4% modified SBS bitumen is 76.4 °C.

Bitumen modification was carried out at temperatures of 180 and 200 °C. These temperatures typical for bitumen modification at real production conditions. Figure 2 also shows that the modified bitumen was prepared in the laboratory using different mixing times within the range between 30 and 180 min. This was done in order to set a minimum modification time. It can be seen that the softening temperature rises to a stirring duration of 90 min, but then, when it reaches 180 min, there is no improvement. This means that it is enough to mix for 60–90 min. It is important to note that after 30 min, the softening temperature during mixing reaches 65.9 °C, which is a 17 °C degree higher than that of unmodified road bitumen B70/100.

The results show that a higher production temperature further improves the softening temperature, for example, when stirring at 180 °C for 90 min, the softening temperature is 69.1 °C and when stirring at 200 °C for 90 min–80 °C or almost 11 °C degrees higher. Figure 3 shows the effect of the PP content in the hybrid modifier on the softening temperature.

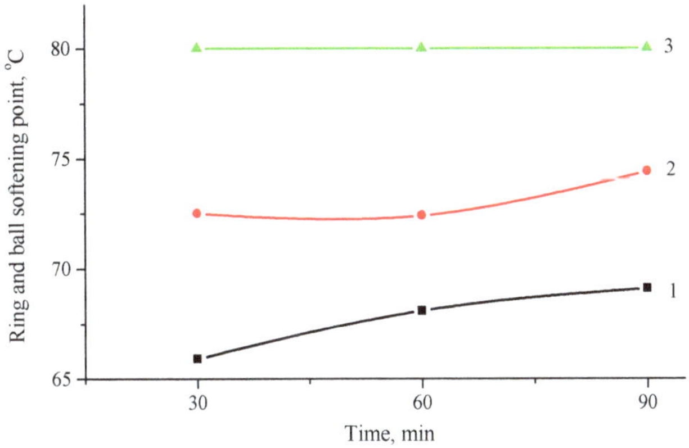

Figure 3. Change in softening temperature of ELTC modifier from mixing time at different concentration of PP: 1—at 10%; 2—at 20%; 3—at 30%.

It is clearly seen that with an increase in PP from 10% to 20%, the softening temperature increases by 4.5–6 °C. On the other hand, with an increase in the polypropylene content by another 10 percent, i.e., up to 30%, the softening temperature increases by another 5.5–7 degrees, reaching 80 °C. As indicated in [56], the softening point can reveal the high-temperature stability of the asphalt binder. As we can see, when modifying bitumen with the hybrid ELTC modifier, the high-temperature characteristics of bitumen improve, whereas the improvement is larger by increasing the concentration of more rigid component in the modifier.

The authors of [57] also studied the effect of the concentration of used rubber tires and recycled polyethylene on the softening temperature of the modified asphalt binder. The softening temperature increased with an increase in the amount of rubber tires and polyethylene added to the base asphalt binder and improved performance at high temperatures. It is assumed that the mechanism of bitumen modification with the ELTC modifier is a combination of physical and chemical processes. As mentioned above, a key part of the "Rubbintec" devulcanization process is the use of a selective catalyst and a compatibilizer. This helps in the "grafting" of devulcanized rubber crumbs to the polymer and bitumen fractions, which helps to improve distribution of the ELTC modifier particles within asphalt binder matrix. In addition, as shown in [57], the TGA weight loss curves of a bitumen binder modified with used tire rubber and recycled PE remained largely unchanged, which proves that thermal stability improved after modification.

PP and PE are the most common constituents or thermoplastic waste and both of these polymers usually are easily recyclable. However, due to their macromolecular structure both of these polymers do not have the necessary flexibility properties, especially at sub-zero temperatures, in contradiction to rubber or SBS polymers, which are widely used for bitumen modification. Therefore, in this study, PP and PE are used in combination with devulcanized rubber. The results show that there is not much difference between PE and PP as a constituent of ELTC modifier in terms of increment of softening temperature. However, somewhat higher increment of softening temperature is in the case of stiffer PP containing modifier. The softening temperature change is up to 1.8 °C (Figure 4).

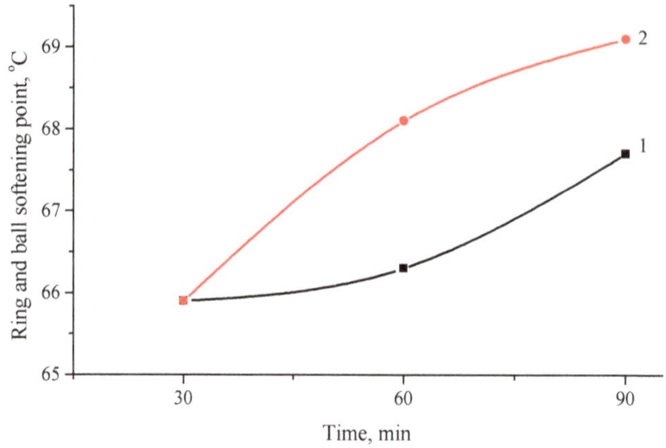

Figure 4. Dependence of softening temperature of ELTC modifier with 10 wt.% of PE (1) or PP (2) from mixing time.

Figure 5 shows the results of needle penetration at 25 °C of modified bitumen compositions, depending on the type of the polymer (PP and PE), present in the ELTC.

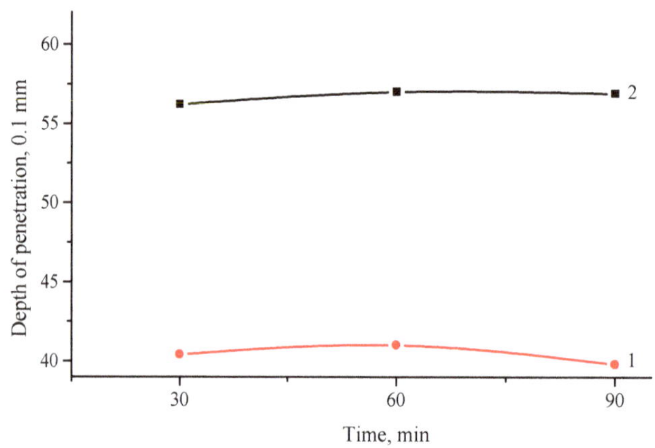

Figure 5. Change in penetration depth of ELTC modifier with 10 wt.% of PP (1) or PE (2) from mixing time.

Needle penetration value for ELTC compositions containing 10% PP is about 40, and for PE—more than 56. Thus, the obtained results show that under the same production conditions, road bitumen containing ELTC with 10% PE is softer than ELTC with 10% PP, which is in accordance with stiffer macromolecular structure of the latter.

Figure 6 shows the effect of the thermoplastics content in the composition of ELTC on the results of needle penetration.

From the results obtained, it can be seen that with an increase in the polymer content, the penetrating ability of the needle decreases for all the investigated ELTC modified binders, denoting to increased stiffness of the system. This trend is especially pronounced for the bitumen with PE containing ELTC binder: at 10% PE content needle penetration value is 56.2, whereas at 40% PE content it is 26.6. On the other hand, in the case of PP, a considerable decrease in needle penetration was not observed.

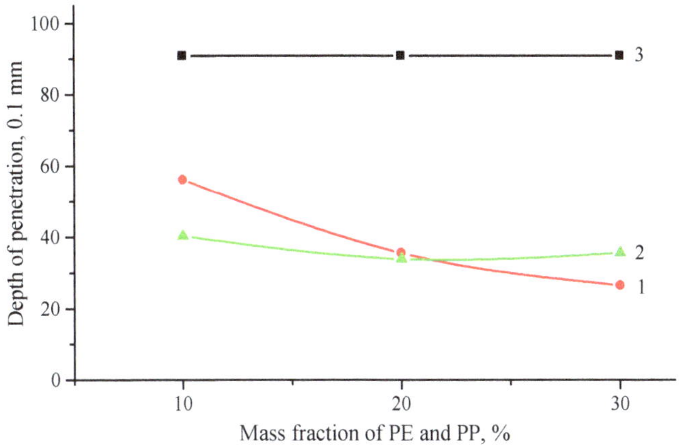

Figure 6. Change in penetration depth of ELTC modified bitumen binder (mixing time—30 min.; temperature—180 °C) with PE (1) and PP (2) in comparison to neat bitumen 70/100 (3) from the polymer concentration.

The results show that the addition of modifiers reduces the penetration value from 90.8 to 26.6. This means that, the modified bitumen becomes harder compared to unmodified bitumen. The traditional class of modified bitumen B45/80-55, where 45/80 is minimum and maximum needle penetration values, can be achieved using the ELTC + 10% PE modifier, for these compositions the needle penetration values are within the range of 56.2–57. Most compositions have a score below 45, which means that they correspond to the bitumen class 25/55.

The penetration value of the asphalt binder may reflect the consistency of the shows the effect of used rubber tires and recycled PE on the penetration of modified asphalt binder. Penetration decreased after the addition of used rubber tires and recycled PE. It has been shown that some molecular composition of the asphalt binder have changed after the addition of used rubber tire crumb and recycled PE, and some physical reactions have occurred, as a result of which the asphalt binder has hardened and its penetration has decreased. In addition, as a result of the introduction of polymer waste, the viscosity of the system becomes greater, which reduces the depth of penetration of the needle.

Figure 7 shows the results of elastic recovery. This is a very important property of modified bitumen, as this parameter indicates whether the added modifier improves flexibility.

Previous results have shown that the modifier affects the properties of bitumen, making it tougher. On the other hand, hard material can cause brittleness or premature cracking. The results obtained show that hybrid ELTC modifiers containing PP have larger elastic recovery at 25 °C. It is important to note that unmodified bitumen 70/100 does not show elastic recovery under these test conditions, while the 4% SBS modified bitumen shows a 60% elastic recovery. The polymer-modified road bitumen should have an elastic recovery of 50% at least. This figure is achieved using the ELTC + 10% PP modifier—54–55%. Other compositions show significantly lower elastic recovery value, that is, about 30–40%. The paper [57] shows the effect of the concentration of used rubber tires crumb and recycled PE on the plasticity of the modified asphalt binder, while the plasticity decreased as the PE content increased, confirming that thermoplastics cannot improve the low-temperature characteristics of asphalt, instead the low-temperature characteristics of asphalt are reduced [26,36]. Plasticity increased with an increase in the content of rubber crumbs, this may be due to the fact that with a higher content of polymer waste and rubber crumbs, the interaction between them and the light component of the asphalt binder was stronger.

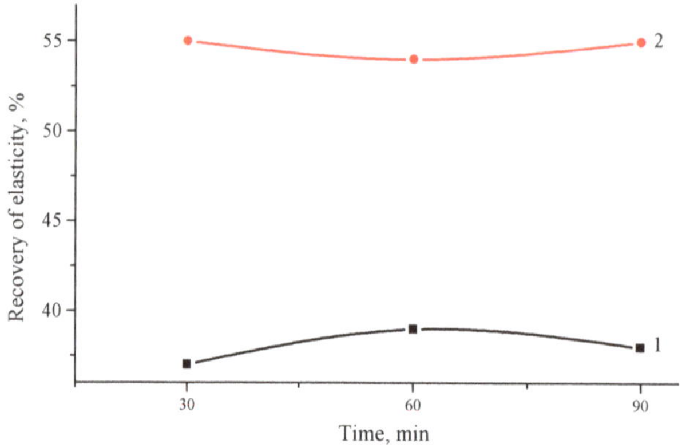

Figure 7. Change in of elastic recovery of ELTC modifier with 10 wt.% of PE (1) or PP (2) from mixing time.

Figures 8 and 9 show fluorescent micrographs of a modified asphalt binder, fluorescent micrographs can visually reflect the distribution of ELTC in asphalt binders.

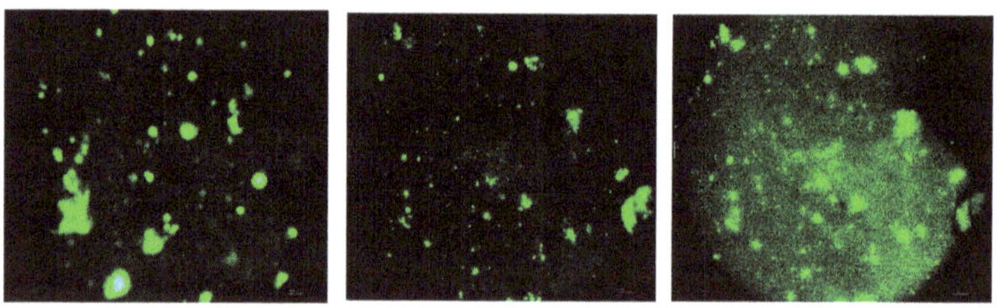

Figure 8. Morphology of modified bitumen binder with 20% of ELTC Mix 20/80PP mixed at 180 °C for 30 min (fracture surface of the sample, obtained by cooling the binder sample in liquid nitrogen).

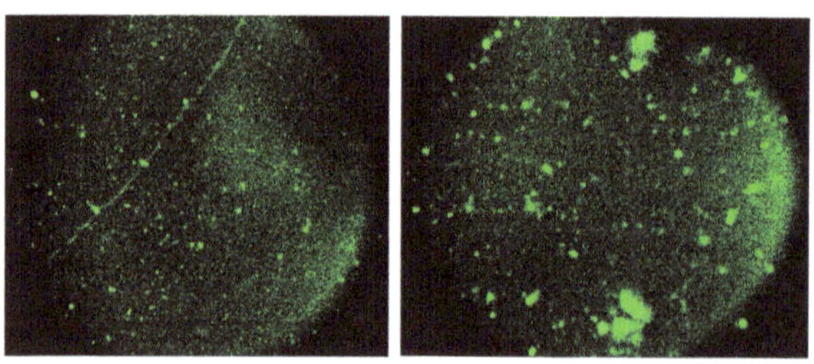

Figure 9. Morphology of modified bitumen binder with ELTC Mix 20/80PP 20% mixed at 180 °C for 90 min (fracture surface of the sample, obtained by cooling the binder sample in liquid nitrogen).

Polymer-bitumen binder is considered as a material in which the medium is bitumen, and the dispersed phase is a polymer. It surpasses the properties of conventional bitumen. Microscopy studies of the investigated bitumen binder show that with certain amounts of polymer additive in bitumen binder, it is able to "dissolve", but sometimes the polymer is distributed in bitumen in the form of separate, unregular particles. The effect of their action in the composition is similar to the effect of the filler. Mixing the composite for 30 min leads to distribution of ELTC binder in the bitumen in the form of separate particles. Mixing for 90 min leads to aggregation of particles and their fusion (Figures 8 and 9), this leads to increased strength and elasticity of the resulting product.

According to the mechanism of ELTC modified bitumen, the ELTC particles are distributed in the asphalt matrix during high-shear mixing. The fluorescence occurs if the ELTC particles are lighted by electromagnetic irradiation at certain wavelength. Bitumen/asphalt phase is black and opaque and have no significant fluorescence reaction. The fluorescence reaction difference between the ELTC particles and bitumen matrix can be used to present the distribution properties. In FM images, the light-green part shows the ELTC and the black part shows neat bitumen/asphalt.

It can be observed:

(1) ELTC is dispersed in bitumen/asphalt in the form of roundish particles forming non-continuous phase.
(2) Some poorly dispersed agglomerated ELTC particles also are observed within bitumen/asphalt matrix.
(3) Increased mixing/dispergation time reduce the formation of agglomerates.
(4) The largest particle images observed during investigation are around 200 μm.
(5) Finely, dispersed ELTC particles distributed in bitumen homogeneously can have a positive effect on storage stability.
(6) Coarser undispersed particles are unevenly distributed in the system and influence storage stability.

The authors of the paper [57] noted that when mixed with recycled PE, the mutual adhesion of polyethylene and rubber crumbs improved the compatibility between polyethylene and the asphalt binder phase, since the rubber crumbs absorbed light molecular components from asphalt binders when they were mixed [58]. It was found that with the correct content of the modifying agent, rubber crumbs and recycled PE merge with asphalt binders and are evenly distributed, compatibility between the components of the modified asphalt binder is good, which proves the uniformity of the modified asphalt binders.

By measuring the absorption and release of heat, DSC reveals the thermal characteristics of materials [34]. Table 2 demonstrates characteristic temperatures, whereas Figure 10 shows DSC thermograms of ELTC modifiers.

Table 2. Main characteristics of the polyolefine phase of the ELTC modifiers.

ELTC Modifier	Onset Melting Temperature, °C	Peak Melting Temperature, °C	Offset Melting Temperature, °C	Peak Width, °C
10%PE	67	126	136	69
20%PE	85	128	143	58
40%PE	94	131	146	52
PPMIX10/90	Peak overlapping	160	172	Peak overlapping
PPMIX20/80	Peak overlapping	161	174	Peak overlapping
PP30/70	Peak overlapping	161	173	Peak overlapping

As one can see melting peaks of PE based modifiers are considerably larger than in the case of PP based systems, which is explained by larger melting enthalpy of PE crystal (ca 290 J/g) in comparison of PP (ca 190 J/g). In the case of PE based modifier, widening of endothermic peak is observed by increasing the rubber content in the mixture denoting to restricted crystallization of the polyolefine phase. This complies well with decreased peak temperatures of PE crystalline phase by increasing the rubber content. In its turn, for PP based modifier, several endothermic effects are observed over the testing interval, especially for the composition with the highest rubber content. Consequently, the peaks centered around 0 °C and 93 °C are associated with rubber phase, denoting to presence of customized additives. It is evident that the endotherm with maxima around 93 °C, partly overlap with onset melting peak of PP crystalline phase. In spite of this overlap, maximum and offset melting peak temperatures of PP crystalline phase are not considerable affected by changing the rubber content. Most probably similar endothermic transitions of rubber phase are present also in the case of PE modifier; however, they are not observed due to peak overlapping and higher heat of fusion of PE crystal. Nevertheless, it is important to mention that melting of PE and PP crystalline fractions occurs below the used mixing temperature of ELTC with 70/100 grade bitumen.

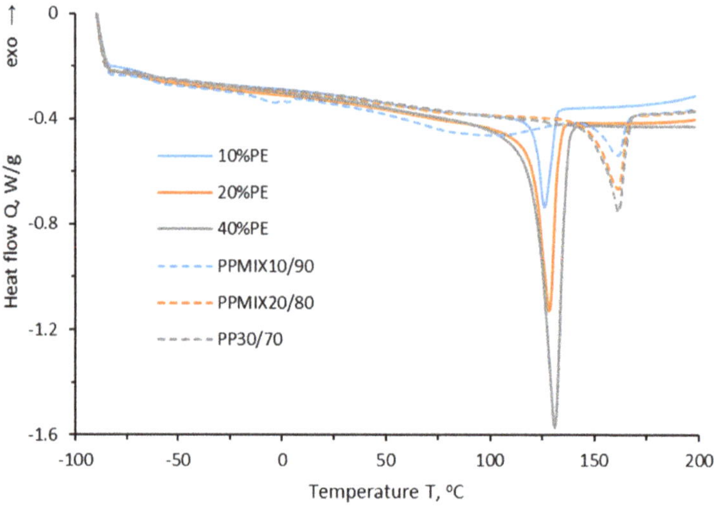

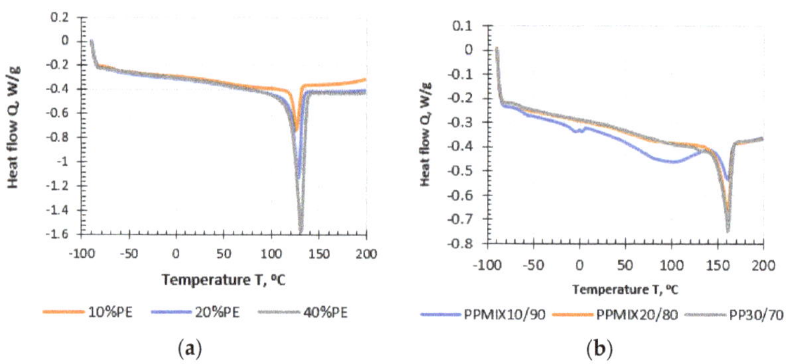

Figure 10. DSC thermograms of different ELTC modifiers based on PE (**a**) and PP (**b**).

TGA is a good indicator for assessing the thermal stability of an ELTC modifier [35]. Figure 11 depicts TGA thermograms, recorded by performing measurements under air atmosphere. Sufficient thermal stability of the investigated ELTC modifiers for compounding with bitumen or any other suitable constituent (e.g., thermoplastic) is observed by considering, if sufficient compatibility is ensured, i.e., main thermal destruction occurs above 250 °C.

According to the TGA graphs, ELTC modifiers decompose during several stages. Minor mass loss until 250 °C is associated with elimination of most volatile compounds. Considerable destruction of the binders starts only if heated above 300 °C. Comparatively thermal stability of PP based ELTC modifier is lower in comparison to than of PE based counterpart. This is evidently associated with the presence of tertiary carbon atoms in the PP macromolecule, which is more susceptible to thermal degradation. By increasing polyolefine concentration up to 40% onset of TGA curves is shifted towards direction of higher temperatures. Concomitant the first mass loss stage is shifted towards higher temperatures by increasing the polyolefine content. Increment in the first mass loss stage also may be observed by increasing the polyolefine concentration, which testify, that this mass loss stage is mainly associated with decomposition of PP or PE in the ELTC modifier. Consequently, decomposition rate of ELTC modifier becomes slower if rubber content is increased, resulting in larger char yields at the end of the experiment, especially in the case of PE-ELTC.

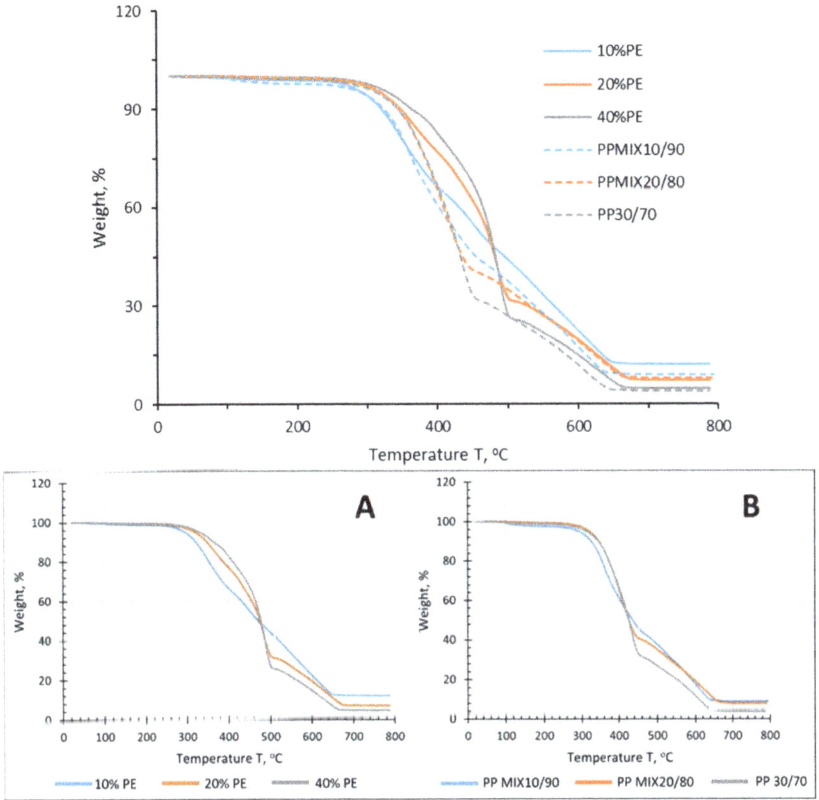

Figure 11. TGA thermograms of different ELTC modifiers based on PE (**A**) and PP (**B**).

4. Conclusions

Experimental results show that all ELTC compositions improve the softening temperature and increase their resistance to plastic deformation in the summer. Some compositions, such as ELTC + 20% PE and ELTC + 30% PE have a softening temperature of 80 °C, which means an increase in the softening temperature by more than 20° C compared to conventional unmodified bitumen B 70/100. The highest improvement is explained by more rigid macromolecular structure of PE and especially PP in comparison to partially devulcanized rubber phase.

The results of the needle penetration test at 25° C show that the addition of ELTC modifiers reduces the penetration value from 90.8 to 26.6 at the maximum. This means that, the modified bitumen becomes harder compared to unmodified bitumen, which is explained predominantly because of chain rigidity of the polyolefine phase.

The criteria of polymer modified road bitumen in respects of elastic recovery of 50% in accordance with EN 14023 can be achieved by adding 20% of the ELTC modifier with 10%PP to B 70/100 road bitumen, demonstrating elastic recovery value of 54–55%. Other compositions show significantly lower elasticity recovery values, i.e., about 30–40 percent.

Important, that by increasing mixing/dispergation time of bitumen with ELTC modifiers it is possible to reduce the formation of agglomerates, enabling to obtain more regular distribution of the binder within bituminous matrix.

It is also important to mention that mixing of bitumen with ELTC modifiers occurred above melting of crystalline phases of both thermoplastic polymers (PE and PP). Concomitant, no considerable thermal destruction of the ELTC modifiers occurs at the mixing temperatures (180 °C and 200 °C) used for preparation of the modified bitumen binder.

Author Contributions: Conceptualization, A.A., V.H. and A.B.; Data curation, A.B. and R.M.-M.; Formal analysis, R.M.-M., Y.I. and Y.A. Funding acquisition, A.A.; Investigation, A.B. and R.M.-M.; Methodology, A.S., S.S. and M.A.; Project administration, A.A.; Resources, A.S., S.S., M.J., A.T. and M.T.; Software, Y.I.; Visualization, R.M.-M.; Writing—original draft, A.A., V.H. and A.B.; Writing—review and editing, A.A., V.H., A.B., R.M.-M., M.J., A.T. and M.T. All authors have read and agreed to the published version of the manuscript.

Funding: This study has been funded by the Science Committee of the Ministry of Science and Higher Education of the Republic of Kazakhstan (grant No. AP19679081 "Modification of petroleum bitumen by industrial rubber waste").

Institutional Review Board Statement: Not applicable.

Data Availability Statement: Data are contained within the article.

Conflicts of Interest: Author Serik Sydykov is affiliated with LLP "JV" CASPI BITUM Aktau Bitumen Plant. Author Yerbolat Ayapbergenov is affiliated with Branch of LLP "KMG Engineering" "KazNIPImunaigas". Author Martynas Jankauskas is affiliated with JSC "Polylema". Author Anatolijs Trumpels. is affiliated with LLC "Rubbintec". The remaining of the authors declare that there is no conflicts of interest.

References

1. WBCSD & The Tire Industry Project, Global ELT Management-a Global State of Knowledge on Regulation, Management Systems, Impacts of Recovery and Technologies. 2019. Available online: https://docs.wbcsd.org/2019/12/Global_ELT_Management%E2%80%93A_global_state_of_knowledge_on_regulation_management_systems_impacts_of_recovery_and_technologies.pdf (accessed on 1 December 2019).
2. Grammelis, P.; Margaritis, N.; Dallas, P.; Rakopoulos, D.; Mavrias, G. A review on management of end of life tires (ELTs) and alternative uses of textile fibers. *Energies* **2021**, *14*, 571. [CrossRef]
3. Rahimi, A.R.; García, J.M. Chemical recycling of waste plastics for new materials production. *Nat. Rev. Chem.* **2017**, *1*, 46. [CrossRef]
4. Gondal, M.A.; Siddiqui, M.N. Identification of different kinds of plastics using laser-induced breakdown spectroscopy for waste management. *J. Environ. Sci. Health Part A Toxic/Hazard. Subst. Environ. Eng.* **2007**, *42*, 1989–1997. [CrossRef]
5. Hopewell, J.; Dvorak, R.; Kosior, E. Plastics recycling: Challenges and opportunities. *Philos. Trans. R. Soc. B Biol. Sci.* **2009**, *364*, 2115–2126. [CrossRef] [PubMed]
6. Wang, S.; Wang, Q.; Wu, X.; Zhang, Y. Asphalt modified by thermoplastic elastomer based on recycled rubber. *Constr. Build. Mater.* **2015**, *93*, 678–684. [CrossRef]
7. Wang, T.; Xiao, F.; Zhu, X.; Huang, B.; Wang, J.; Amirkhanian, S. Energy consumption and environmental impact of rubberized asphalt pavement. *J. Clean. Prod.* **2018**, *180*, 139–158. [CrossRef]
8. Imanbayev, Y.; Bussurmanova, A.; Ongarbayev, Y.; Serikbayeva, A.; Sydykov, S.; Tabylganov, M.; Akkenzheyeva, A.; Izteleu, N.; Mussabekova, Z.; Amangeldin, D.; et al. Modification of Bitumen with Recycled PET Plastics from Waste Materials. *Polymers* **2022**, *14*, 4719. [CrossRef]
9. Akkenzheyeva, A.; Haritonovs, V.; Bussurmanova, A.; Merijs-Meri, R.; Imanbayev, Y.; Riekstins, A.; Serikbayeva, A.; Sydykov, S.; Aimova, M.; Mustapayeva, G. Study of the Viscoelastic and Rheological Properties of Rubber-Bitumen Binders Obtained from Rubber Waste. *Polymers* **2024**, *16*, 114. [CrossRef]
10. Yildirim, Y. Polymer modified asphalt binders. *Constr. Build. Mater.* **2007**, *21*, 66–72. [CrossRef]
11. Brasileiro, L.; Moreno-Navarro, F.; Tauste-Martínez, R.; Matos, J.; Rubio-Gámez, M.d.C. Reclaimed polymers as asphalt binder modifiers for more sustainable roads: A review. *Sustainability* **2019**, *11*, 646. [CrossRef]
12. Li, J.; Xing, X.; Hou, X.; Wang, T.; Wang, J.; Xiao, F. Determination of SARA fractions in asphalts by mid-infrared spectroscopy and multivariate calibration. *Measurement* **2022**, *198*, 111361. [CrossRef]
13. Wang, T.; Xiao, F.; Amirkhanian, S.; Huang, W.; Zheng, M. A review on low temperature performances of rubberized asphalt materials. *Constr. Build. Mater.* **2017**, *145*, 483–505. [CrossRef]

14. Zanetti, M.C.; Fiore, S.; Ruffino, B.; Santagata, E.; Dalmazzo, D.; Lanotte, M. Characterization of crumb rubber from end-of-life tyres for paving applications. *Waste Manag.* **2015**, *45*, 161–170. [CrossRef] [PubMed]
15. Nejad, F.M.; Aghajani, P.; Modarres, A.; Firoozifar, H. Investigating the properties of crumb rubber modified bitumen using classic and SHRP testing methods. *Constr. Build. Mater.* **2012**, *26*, 481–489. [CrossRef]
16. Zhang, J.; Yao, Z.; Yu, T.; Liu, S.; Jiang, H. Experimental evaluation of crumb rubber and polyethylene integrated modified asphalt mixture upon related properties. *Road Mater. Pavement Des.* **2019**, *20*, 1413–1428. [CrossRef]
17. Sienkiewicz, M.; Borzędowska-Labuda, K.; Wojtkiewicz, A.; Janik, H. Development of methods improving storage stability of bitumen modified with ground tire rubber: A review. *Fuel Process. Technol.* **2017**, *159*, 272–279. [CrossRef]
18. Navarro, F.J.; Partal, P.; Martínez-Boza, F.; Gallegos, C. Thermo-rheological behaviour and storage stability of ground tire rubber-modified bitumens. *Fuel* **2004**, *83*, 2041–2049. [CrossRef]
19. Rodríguez-Alloza, A.M.; Gallego, J. Mechanical performance of asphalt rubber mixtures with warm mix asphalt additives. *Mater. Struct.* **2017**, *50*, 147. [CrossRef]
20. Liang, M.; Xin, X.; Fan, W.; Sun, H.; Yao, Y.; Xing, B. Viscous properties, storage stability and their relationships with microstructure of tire scrap rubber modified asphalt. *Constr. Build. Mater.* **2015**, *74*, 124–131. [CrossRef]
21. Hassan, M.M.; Aly, R.O.; Abdel Aal, S.E.; El-Masry, A.M.; Fathy, E.S. Mechanochemical devulcanization and gamma irradiation of devulcanized waste rubber/high density polyethylene thermoplastic elastomer. *J. Ind. Eng. Chem.* **2013**, *19*, 1722–1729. [CrossRef]
22. Qian, C.; Fan, W.; Liang, M.; He, Y.; Ren, S.; Lv, X.; Nan, G.; Luo, H. Rheological properties, storage stability and morphology of CR/SBS composite modified asphalt by high-cured method. *Constr. Build. Mater.* **2018**, *193*, 312–322. [CrossRef]
23. Dong, R.; Li, J.; Wang, S. Laboratory evaluation of pre-devulcanized crumb rubber–modified asphalt as a binder in hot-mix asphalt. *J. Mater. Civ. Eng.* **2011**, *23*, 1138–1144. [CrossRef]
24. Xu, O.; Rangaraju, P.R.; Wang, S.; Xiao, F. Comparison of rheological properties and hot storage characteristics of asphalt binders modified with devulcanized ground tire rubber and other modifiers. *Constr. Build. Mater.* **2017**, *154*, 841–848. [CrossRef]
25. Ghasemirad, A.; Asgharzadeh, S.M.; Tabatabaee, N. A comparative evaluation of crumb rubber and devulcanized rubber modified binders. *Pet. Sci. Technol.* **2017**, *35*, 1091–1096. [CrossRef]
26. Zhang, Z.; Sun, J.; Jia, M.; Ban, X.; Wang, L.; Chen, L.; Huang, T.; Liu, H. Effects of polyurethane thermoplastic elastomer on properties of asphalt binder and asphalt mixture. *J. Mater. Civ. Eng.* **2021**, *33*, 04020477. [CrossRef]
27. Punith, V.S.; Veeraragavan, A. Behavior of asphalt concrete mixtures with reclaimed polyethylene as additive. *J. Mater. Civ. Eng.* **2007**, *19*, 500–507. [CrossRef]
28. Fang, C.Q.; Yu, R.E.; Zhang, Y.; Hu, J.B.; Zhang, M.; Mi, X.H. Combined modification of asphalt with polyethylene packaging waste and organophilic montmorillonite. *Polym. Test.* **2012**, *31*, 276–281. [CrossRef]
29. Maharaj, C.; Maharaj, R.; Maynard, J. The effect of polyethylene terephthalate particle size and concentration on the properties of asphalt and bitumen as an additive. *Prog. Rubber Plast. Recycl.* **2015**, *31*, 1–23. [CrossRef]
30. Moghaddam, T.B.; Soltani, M.; Karim, M.R. Evaluation of permanent deformation characteristics of unmodified and polyethylene terephthalate modified asphalt mixtures using dynamic creep test. *Mater. Des.* **2014**, *53*, 317–324. [CrossRef]
31. Moghaddam, T.B.; Soltani, M.; Karim, M.R. Stiffness modulus of polyethylene terephthalate modified asphalt mixture: A statistical analysis of the laboratory testing results. *Mater. Des.* **2015**, *68*, 88–96. [CrossRef]
32. Sharma, S.; Sharma, S.; Upadhyay, N. Composition based physicochemical analysis of modified bitumen by HDPE/LDPE. *Orient. J. Chem.* **2019**, *35*, 1167–1173. [CrossRef]
33. Porto, M.; Caputo, P.; Loise, V.; Eskandarsefat, S.; Teltayev, B.; Rossi, C.O. Bitumen and bitumen modification: A review on latest advances. *Appl. Sci.* **2019**, *9*, 742. [CrossRef]
34. Fang, C.; Zhang, Y.; Yu, Q.; Zhou, X.; Guo, D.; Yu, R.; Zhang, M. Preparation, characterization and hot storage stability of asphalt modified by waste polyethylene packaging. *J. Mater. Sci. Technol.* **2013**, *29*, 434–438. [CrossRef]
35. Kim, Y.; Park, T.S. Reinforcement of recycled foamed asphalt using short polypropylene fibers. *Adv. Mater. Sci. Eng.* **2013**, *1*, 903236. [CrossRef]
36. Yan, K.Z.; Xu, H.B.; You, L.Y. Rheological properties of asphalts modified by waste tire rubber and reclaimed low density polyethylene. *Constr. Build. Mater.* **2015**, *83*, 143–149. [CrossRef]
37. Fathy, E.S.; Elnaggar, M.Y.; Raslan, H.A. Thermoplastic elastomer based on waste polyethylene/waste rubber containing activated carbon black: Impact of gamma irradiation. *J. Vinyl Addit. Technol.* **2018**, *25*, E166–E173. [CrossRef]
38. Wagner, M.A.; Hadian, A.; Sebastian, T.; Clemens, F.; Schweizer, T.; Rodriguez-Arbaizar, M.; Carreno-Morelli, E.; Spolenak, R. Fused filament fabrication of stainless steel structures—From binder development to sintered properties. *Addit. Manuf.* **2021**, *49*, 102472. [CrossRef]
39. Nizamuddin, S.; Giustozzi, F. The role of new compatibilizers in hybrid combinations of waste plastics and waste vehicle tyres crumb rubber-modified bitumen. In *Plastic Waste for Sustainable Asphalt Roads*; Woodhead Publishing: Cambridge, UK, 2022; pp. 165–178. [CrossRef]
40. Mulage, K.S.; Patkar, R.N.; Deuskar, V.D.; Pundlik, S.M.; Kakade, S.D.; Gupta, M. Studies on a Novel Thermoplastic Polyurethane as a Binder for Extruded Composite Propellants. *J. Energetic Mater.* **2007**, *25*, 233–245. [CrossRef]
41. Reddy, T.S.; Nair, J.K.; Satpute, R.S.; Gore, G.M.; Sikder, A.K. Rheological studies on energetic thermoplastic elastomers. *J. Appl. Polym. Sci.* **2010**, *118*, 2365–2368. [CrossRef]

42. Kobykhno, I.; Didenko, A.L.; Honcharenko, D.; Vasilyeva, E.; Kudryavtsev, V.; Tolochko, O. Development thermoplastic elastomer-based fiber-metal laminate for vibration damping application. *Mater. Today Proc.* **2020**, *30*, 393–397. [CrossRef]
43. Xu, P.; Zhu, Z.; Wang, Y.; Cong, P.; Li, D.; Hui, J.; Ye, M. Phase structure characterization and compatibilization mechanism of epoxy asphalt modified by thermoplastic elastomer (SBS). *Constr. Build. Mater.* **2022**, *320*, 126262. [CrossRef]
44. Wilkinson, P.J.; Weaver, M.C.; Kister, G.; Gill, P.P. Styrene-Ethylene/Butylene-Styrene (SEBS) Block Copolymer Binder for Solid Propellants. *Propellants Explos. Pyrotech.* **2021**, *47*, e202100142. [CrossRef]
45. Wang, S.F.; Yuan, C.H.; Deng, J.X. Crumb tire rubber and polyethylene mutually stabilized in asphalt by screw extrusion. *J. Appl. Polym. Sci.* **2014**, *131*. [CrossRef]
46. Haider, I.; Stefano, M.; Luca, D.; Michele, L. Recycled plastics and rubber for green roads: The case study of devulcanized tire rubber and waste plastics compounds to enhance bitumen performance. *Resour. Conserv. Recycl. Adv.* **2023**, *18*, 200157. [CrossRef]
47. *EN 12593:2015*; Bitumen and bituminous binders—Determination of the Fraass breaking point. SIST EN: Bratislava, Slovakia, 2015.
48. *EN 12592:2015*; Bitumen and bituminous binders—Determination of solubility. SIST EN: Nitra, Slovakia, 2015.
49. *EN ISO 2592:2017*; Petroleum and related products Determination of flash and fire points Cleveland open cup method. International Organization for Standardization: Geneva, Switzerland, 2017.
50. *EN 12607-1:2015*; Bitumen and bituminous binders—Determination of the resistance to hardening under influence of heat and air—Part 1: RTFOT method. SIST EN: Bratislava, Slovakia, 2014.
51. *EN 13398*; Bitumen and bituminous binders—Determination of the elastic recovery of modified bitumen. SIST EN: Bratislava, Slovakia, 2017.
52. *EN 14770*; Bitumen and bituminous binders—Determination of complex shear modulus and phase angle—Dynamic Shear Rheometer (DSR). SIST EN: Bratislava, Slovakia, 2012.
53. Vorobyev, L.R.; Bosnik, V.B. Method of Devulcanization of Sulfur-Cured Rubber. U.S. Patent 9,982,107 B1, 29 May 2018. Available online: https://patents.google.com/patent/US9982107B1/no (accessed on 31 May 2018).
54. Chavez FMarcobal, J.; Gallego, J. Laboratory evaluation of the mechanical properties of asphalt mixtures with rubber incorporated by the wet, dry, and semi-wet process. *Constr. Build. Mater.* **2019**, *205*, 164–174. [CrossRef]
55. Król, J.B.; Radziszewski, P.; Kowalski, K.J.; Sarnowski, M.; Czajkowski, P. Laboratory and field investigations of polymer and crumb rubber modified bitumen. *J. Civ. Eng. Archit.* **2014**, *8*, 1327–1334. [CrossRef]
56. Reyes-Ortiz, O.; Fuentes, L.; Alvarez, A. Mechanical response of asphalt mixtures modified with natural wax. In Proceedings of the Advanced Characterization of Asphalt and Concrete Materials, Geo-Hubei 2014, Yichang, China, 20–22 July 2014; pp. 58–66. [CrossRef]
57. Ge, D.; Yan, K.; You, Z.; Xu, H. Modification mechanism of asphalt binder with waste tire rubber and recycled polyethylene. *Constr. Build. Mater.* **2016**, *126*, 66–76. [CrossRef]
58. Ghavibazoo, A.; Abdelrahman, M. Composition analysis of crumb rubber during interaction with asphalt and effect on properties of binder. *Int. J. Pavement Eng.* **2013**, *14*, 517–530. [CrossRef]

polymers

MDPI

Article

On the Piezomagnetism of Magnetoactive Elastomeric Cylinders in Uniform Magnetic Fields: Height Modulation in the Vicinity of an Operating Point by Time-Harmonic Fields

Gašper Glavan *, Inna A. Belyaeva and Mikhail Shamonin

East Bavarian Centre for Intelligent Materials (EBACIM), Ostbayerische Technische Hochschule (OTH) Regensburg, Seybothstr. 2, 93053 Regensburg, Germany; inna.belyaeva@oth-regensburg.de (I.A.B.); mikhail.chamonine@oth-regensburg.de (M.S.)

* Correspondence: gasper.glavan@oth-regensburg.de

Abstract: Soft magnetoactive elastomers (MAEs) are currently considered to be promising materials for actuators in soft robotics. Magnetically controlled actuators often operate in the vicinity of a bias point. Their dynamic properties can be characterized by the piezomagnetic strain coefficient, which is a ratio of the time-harmonic strain amplitude to the corresponding magnetic field strength. Herein, the dynamic strain response of a family of MAE cylinders to the time-harmonic (frequency of 0.1–2.5 Hz) magnetic fields of varying amplitude (12.5 kA/m–62.5 kA/m), superimposed on different bias magnetic fields (25–127 kA/m), is systematically investigated for the first time. Strain measurements are based on optical imaging with sub-pixel resolution. It is found that the dynamic strain response of MAEs is considerably different from that in conventional magnetostrictive polymer composites (MPCs), and it cannot be described by the effective piezomagnetic constant from the quasi-static measurements. The obtained maximum values of the piezomagnetic strain coefficient ($\sim 10^2$ nm/A) are one to two orders of magnitude higher than in conventional MPCs, but there is a significant phase lag (35–60°) in the magnetostrictive response with respect to an alternating magnetic field. The experimental dependencies of the characteristics of the alternating strain on the amplitude of the alternating field, bias field, oscillation frequency, and aspect ratio of cylinders are given for several representative examples. It is hypothesized that the main cause of observed peculiarities is the non-linear viscoelasticity of these composite materials.

Keywords: magnetoactive; magnetorheological; elastomer; strain; magnetostriction; piezomagnetic; bias field; macroscopic deformation; time-harmonic field

1. Introduction

This paper concerns the concept of piezomagnetism (or linear magnetostriction) [1]. It is well known that the magnetostriction of conventional solid ferromagnetic materials is a second-order effect with respect to the applied magnetic field [2–4]. This means that the sign of the magnetostrictive (MS) strain does not change when the direction of the magnetic field strength is reversed. For example, a rod of a ferromagnetic material increases in length for both positive and negative values of the magnetic field strength. However, in conventional MS materials, magnetostriction "can be represented as effectively first-order when variations in the system parameters are small compared with the initial values of the parameters" [5]. Figure 1 illustrates the concept of the direct piezomagnetic effect in the context of employing an MS material as an actuator [5–7]. The effective piezomagnetic constant d can be calculated from the quasi-static field dependence of the magnetostrictive strain λ by calculating the first derivative of λ with respect to the applied magnetic field H_{DC}: $d = \partial\lambda/\partial H_{DC}$. Traditionally, it can be expected that the maximum piezomagnetic response is observed in the biased magnetic field, where the effective piezomagnetic constant d has its maximum.

For example, piezomagnetism is important for designing layered magnetoelectric (ME) sensors of AC magnetic fields comprising MS and piezoelectric layers [8–13]. In this context, the maximum ME voltage coefficient is expected in the bias magnetic field, where the magnitude of the effective piezomagnetic constant d of the MS layer has the highest value. Very recently, a new concept for designing layered multiferroic structures has been put forward in the literature [14–17]. The idea here is to depart from conventional rigid materials (ferromagnetic metals and alloys as well piezoelectric ceramics and single crystals) and to employ much softer polymer-based materials. The advantage of such an approach is the much lower electromechanical resonance frequency of the resulting heterostructures in comparison with conventional heterostructures [18,19]. The usage of polymer-based ferromagnetic and piezoelectric phases brings into play new physical effects related to their softness and viscoelasticity [20]. The ferromagnetic phase in layered polymer-based multiferroic heterostructures employed the magnetorheological or magnetoactive elastomers (MAEs), which consist of magnetic micro- or nanoparticles embedded into a polymer matrix [21–28].

However, our main motivation for investigating the piezomagnetic phenomena in MAEs is potential application of these materials in magnetically controlled soft actuators for soft robotics [29,30] owing to much higher deformations of MAEs (longitudinal strain up to 10^{-2}–10^{-1}) in technically feasible magnetic fields as compared to conventional (pure metals and alloys) MS materials (strain up to about 2×10^{-3}) [31–43].

Investigations of MS phenomena in composite ferromagnetic materials, which account for a strong viscoelastic response in the polymer matrix, are scarce and mostly of a theoretical nature [44–46]. It was reported that the inclusion of matrix viscosity results in an apparent hysteresis loop in magnetization and magnetostriction curves even though the model does not include magnetoelastic hysteresis in the magnetostrictive fibers [44]. Indeed, soft MAEs demonstrate a significant hysteresis of different physical properties, which is commonly attributed to the hysteresis of the consolidation of filler particles into elongated aggregates, i.e., the dependence of the internal microstructure of composite on the magnetization history [47–49]. In [50], it was pointed out that the dynamic behavior in certain scenarios of MAE-based actuators with soft-magnetic filling particles remains unexplored.

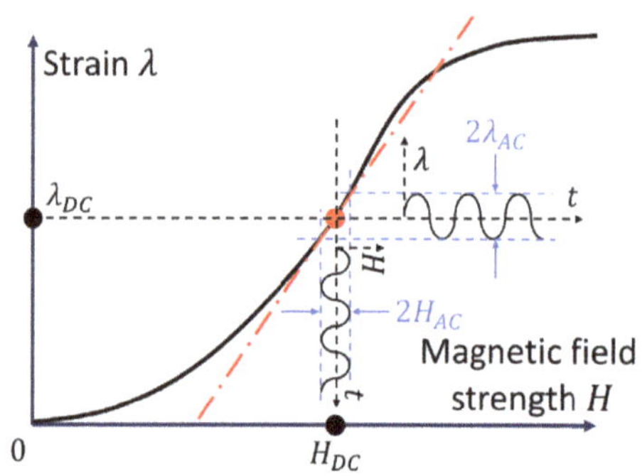

Figure 1. Schematic diagram of the dynamic behavior of a conventional MS actuator driven by a time-harmonic magnetic field in the vicinity of a bias point (H_{DC}, λ_{DC}). The red dash-dotted line denotes the tangent of the curve $\lambda(H)$ at point $H = H_{DC}$.

Hitherto, the effects of magnetic field variations around a biased magnetic field on magnetically induced strain in MAEs remain unstudied. Such a statement of the problem is similar to that of piezomagnetism in conventional ferromagnetic materials. The purpose of the present paper is to experimentally investigate the macroscopic deformation of MAE cylinders in the vicinity of an operation point upon superimposed time-harmonic magnetic-field excitation. The term time-harmonic indicates that only sinusoidal time-varying fields are considered [51]. The paper is organized as follows: The fabrication of MAE cylinders, the experimental setup, and the measurement protocol are described in Section 2. The results of measurements are presented and discussed in Section 3. In particular, the dependencies of the piezomagnetic strain coefficient d_{33} and the phase lag ϕ of the time-harmonic strain with respect to the time-harmonic magnetic field on the biased magnetic field, the cylinder's aspect ratio, the amplitude of the AC field, and its frequency are presented for the first time.

2. Materials and Methods

2.1. Synthesis of MAE Cylinders

This article is a follow-up paper to the research commenced in [41,52], where the fabrication of MAE materials and their characterization were described in detail. Below, we recall the main points. MAE composites were based on one and the same polydimethylsiloxane (PDMS) matrix with a stoichiometry ratio of the reaction (the ratio of the molar concentrations of hydride and vinyl reactive groups) $r \approx 1$ [41,52]. In the theoretical case, there should be only elastically active chains in the polymer network [53]. Carbonyl iron powder (CIP, type SQ, mean particle size of 3.9–5.0 μm, BASF SE, Ludwigshafen, Germany) was used as a soft-magnetic filling. The routine for synthesizing MAE materials was previously described in [54,55]. Liquid compound mixtures were poured into 3D printed moulds from a high-temperature-resistant material, acrylonitrile butadiene styrene (ABS) [41,52]. The curing took place in an oven with air circulation (Memmert model UF30, Memmert GmbH + Co. KG, Schwabach, Germany) for 1 h at 80 °C and next 24 h at 60 °C.

Cylinders differed in the anisotropy of particle arrangement (isotropic/anisotropic), the concentration of iron particles (70, 75 and 80 wt%, corresponding to three volume fractions of 22, 27 and 33 vol%), and the aspect ratio $\Gamma_0 = h_0/D_0$ (0.2, 0.4, 0.6, 0.8, 1.0 and 1.2), where the initial diameter D_0 was kept constant at 15 mm and the initial cylinder height h_0 was varied. Isotropic cylinders were cured in the absence of a magnetic field. To obtain anisotropic samples, two permanent samarium cobalt (SmCo) magnets (diameter of 25 mm and thickness of 5 mm, magnetized along the axis of the cylinder) were placed on the sides of the filled moulds [41,52].

Reference samples were produced for each material composition to analyze the rheological properties of MAE materials. Rheological measurement was conducted with a commercial rheometer (Anton Paar, model Physica MCR 301, with a magnetorheological device (MRD301) in the plate–plate (PP20/MRD/TI) geometry. The angular frequency $\omega = 10\ s^{-1}$ and the shear oscillation amplitude $\gamma = 0.01\%$ were constant through all the measurements. The results obtained and the chemical compositions are described in detail in [41]. The shear storage modulus of the PDMS matrix was about 7.7 kPa, and the shear loss modulus was roughly 0.3 kPa.

Magnetization curves of similar MAE materials have been published elsewhere [56]. Magnetization curves demonstrated a typical "pinched" hysteresis loop behavior. In comparison to magnetostrictive hysteresis loops, the magnetic hysteresis is barely pronounced. This means that the ascending and descending branches of magnetization versus external magnetic field were quite close to each other [56].

2.2. Experimental Setup

MAE cylinders were attached to a 3D printed holder made from polylactic acid (PLA) and positioned between the poles of an electromagnet (EM2 model, MAGMESS Magnetmesstechnik Jürgen Ballanyi e.K., Bochum, Germany), powered by a bi-polar power supply (FAST-PS 1k5, CAENels s.r.l., Basovizza, Italy). The magnetic field generated is known to possess high uniformity [56]. The deformation of the cylinders was recorded in side view using a CMOS camera (Alvium 1800 U-319 m, Allied Vision Technologies GmbH, Stadtroda, Germany). The experimental setup (Figure 2a) was enhanced in comparison with the previous works [41,52]. A lens with the decreased view area and thus increased accuracy of measurements (Kowa LM35JC10M 2/3″ 35 mm/F2.0 C-Mount, Kowa Optimed

Deutschland GmbH, Düsseldorf, Germany) was employed. The cylinders were backlit by a light-emitting diode (LED; Illuminant G4 Pen, Conrad Electronics, Hirschau, Germany) through a diffuser (Perspex diffuse, 2.5 mm, 3A Composites GmbH, Sins, Switzerland). The cylindrical samples were placed vertically between the electromagnet poles so that the cylinder axis was parallel to the magnetic field lines. The camera's vertical position was adjusted each time to maintain a constant angle between the camera and the top edge of a cylinder. The experiment was automated using LabVIEW software (version 2018, National Instruments, Austin, TX, USA) [41,52].

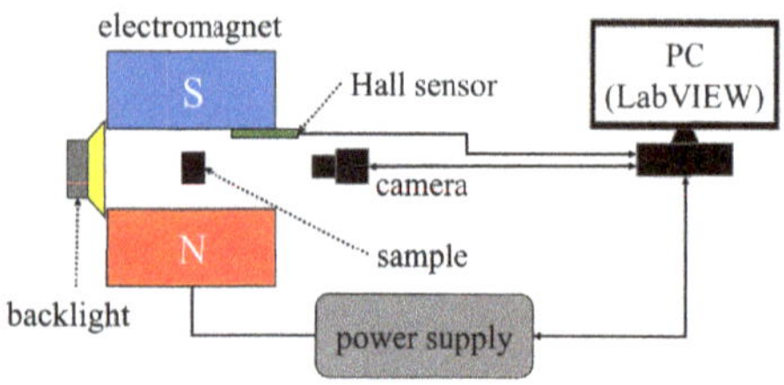

(**a**)

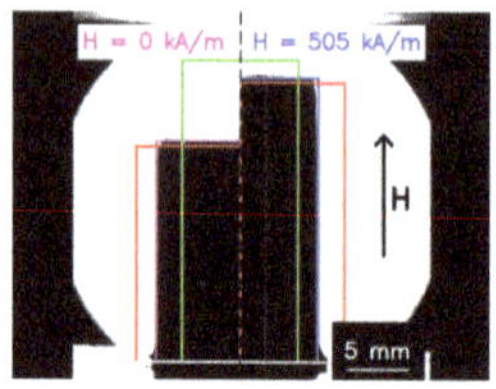

(**b**)

Figure 2. (**a**) Schematic diagram of the experimental setup. (**b**) Images of an isotropic MAE cylinder with an iron content 75 wt% and an aspect ratio Γ_0 of 1.2 in a zero magnetic field (**left** side) and in a maximum magnetic field (**right** side).

2.3. Image Processing

An example of obtained images is shown in Figure 2b, which compares images of an isotropic MAE cylinder in a zero magnetic field (left side of the figure) and a maximum magnetic field (right side of the figure). The left side of Figure 2b shows half of the cylinder in a zero magnetic field. Shown is a fragment of the entire image, which is cut along the cylinder´s axis. Similarly, the right side displays the image of the other half of the same cylinder in the maximum magnetic field. The changes in the cylinders' dimensions were obtained using the sub-pixel edge algorithm from [57], implemented in Python 3.12 The contact plane, indicated with the white horizontal line, was determined manually. The cylinder´s height h was determined inside a rectangular region of interest (ROI) with a width of 500 px as the difference of an average horizontal coordinate (x) of the top edge obtained with a sub-pixel edge detection algorithm and the manually determined contact plane. This ROI is shown as a green rectangle in Figure 2b. The pixel size was approximately 19×19 μm. According to [57], the uncertainty of edge detection was about ± 0.04 px. Two red ROIs (Figure 2b) were used to obtain the left and right edges of the cylinder. The differences between the averages of vertical coordinates (y) of these two edges return cylinders with a diameter of D. Dimensions of these two ROIs vary with the cylinders' deformation in such a way that the top edge is always 30 px below the calculated top edge obtained from green ROI. The magenta and blue lines on the edge of the MAE cylinder mark the calculated edges of the cylinder by the just-described algorithm. Apparent longitudinal strains λ were calculated from height changes: $\lambda = (h - h_0)/h_0$ [41,52].

2.4. Measurement Protocol

Figure 3a shows the measured dependencies of the longitudinal strain on the applied magnetic field strength H for an MAE cylinder. Typical magnetostrictive loops were observed. The magnetic field was gradually increased from zero to the maximum value of 500 kA/m with a step of 25 kA/m and then gradually decreased to zero. Each value of the magnetic field H was kept constant for 20 s, so that the steady state deformation of the cylinder could be reached [52]. Once the desired bias magnetic field (operating point) $H = H_{DC}$ was arrived at, after 20 s, the magnetic field was varied with time as

$$H = H_{DC} + H_{\sim}(t) = H_{DC} + H_{AC}\cos(2\pi f t), \quad (1)$$

where $H_\sim$ is the alternating magnetic field, H_{AC} is the oscillation amplitude, and f is the oscillation frequency. As an example, Figure 3b presents the resulting variation in the longitudinal strain with time. The duration of magnetic field oscillations was fixed at 30 s. The total duration of measurements at an operating point was, therefore, 50 s. The sampling rate of images in the course of magnetic field oscillations was 18 fps. The bias magnetic field H_{DC} was varied between 27 and 127 kA/m with a step of 25 kA/m, the oscillation amplitude of magnetic field H_{AC} was between 15 and 65 kA/m with a step of 12.5 kA/m, and the oscillation frequency f took six different values between 0.1 and 2.5 Hz (0.1, 0.5, 1.0, 1.5, 2.0 and 2.5 Hz). Note that the same value of a bias magnetic field (operating point) H_{DC} was set twice, for both ascending (increasing) and descending (decreasing) branches of the MS hysteresis loop. To ensure good repeatability and the same initial conditions, we allowed at least 24 h for the relaxation of samples between the measurements.

As an example, Figure 3b depicts the time response of the longitudinal strain to the applied oscillatory magnetic field. Some additional transient behavior is visible. To eliminate this transient effect, a sinusoidal function was fitted to the cylinder's magnetostrictive response during the last 15 s of magnetic field oscillations. The fitted sinusoidal function can be written as

$$\lambda = \lambda_{slow}(t) + \lambda_\sim(t) = \lambda_{slow}(t) + \lambda_{AC}\cos(2\pi f t - \phi), \tag{2}$$

where λ_{slow} is the transient longitudinal strain due to the step-like change in the bias field H_{DC}, $\lambda_\sim$ is the alternating longitudinal strain, λ_{AC} is the amplitude of the oscillatory response, and ϕ is the phase difference between the AC excitation field and the AC longitudinal strain. $\lambda_{slow}(t)$ was approximated as an exponential function $\lambda_{slow}(t) = \lambda_1 + \Delta\lambda(1 - e^{-at})$, with $\lambda_1 > 0$, $\Delta\lambda \lessgtr 0$ and $a > 0$ as fitting constants. The actual oscillation frequency f was determined by fitting a sinusoidal function onto the measured oscillation of a magnetic field. Examples of more detailed transient responses of the longitudinal strain to a step-like change in the DC magnetic field and a superimposed time-harmonic field are presented in Appendix A.

When a sinusoidal oscillation of the longitudinal strain (2) is plotted against the sinusoidal magnetic field (1), a Lissajous figure in the form of an ellipse is produced. Figure 3c presents a comparison between measured values and fitted functions. It is seen that the measured values fit the elliptic shape very well. The rotation direction of the point on a Lissajous ellipse is counterclockwise. This is to be expected because the longitudinal strain lags after the magnetic field [58]. Small experimental deviations of experimental points from the fitted ellipsoidal shape of the resulting loop in Figure 3c can be attributed to the transient response of magnetostriction to time-varying magnetic field [52] and the uncertainty of measurements. In experiments, it was not possible to keep the amplitude H_{AC} much smaller than H_{DC} because of the challenge of measuring small variations of the resulting strain due to the limited spatial resolution of the camera. In what follows, we will analyze fitted Lissajous ellipses without offset values due to the presence of H_{DC}. This means that $\lambda_\sim$ as functions of $H_\sim$ will be presented.

As an example, Figure 3d shows the magnetostrictive response of an MAE cylinder when the field oscillations have a positive or negative sign of H_{AC} in Equation (1). This means that a soft-magnetic MAE is insensitive to the direction of the AC magnetic field at the starting point and "forgets" minor details when attaining a steady cycle. It is visible that, although some differences did exist during the initial transient response of the longitudinal strain, the resulting Lissajous figures are practically the same. Investigations using the function (1) were sufficient to describe the relationships between the AC magnetic field and the AC longitudinal strain. For the sake of convenience for readers, Figure 3d is provided in more detail in Appendix B (Figure A2).

According to [59], the piezomagnetic strain coefficient,

$$d_{33} = \lambda_{AC}/H_{AC}, \tag{3}$$

can be used to characterize the dynamic properties of magnetostrictive polymer composites.

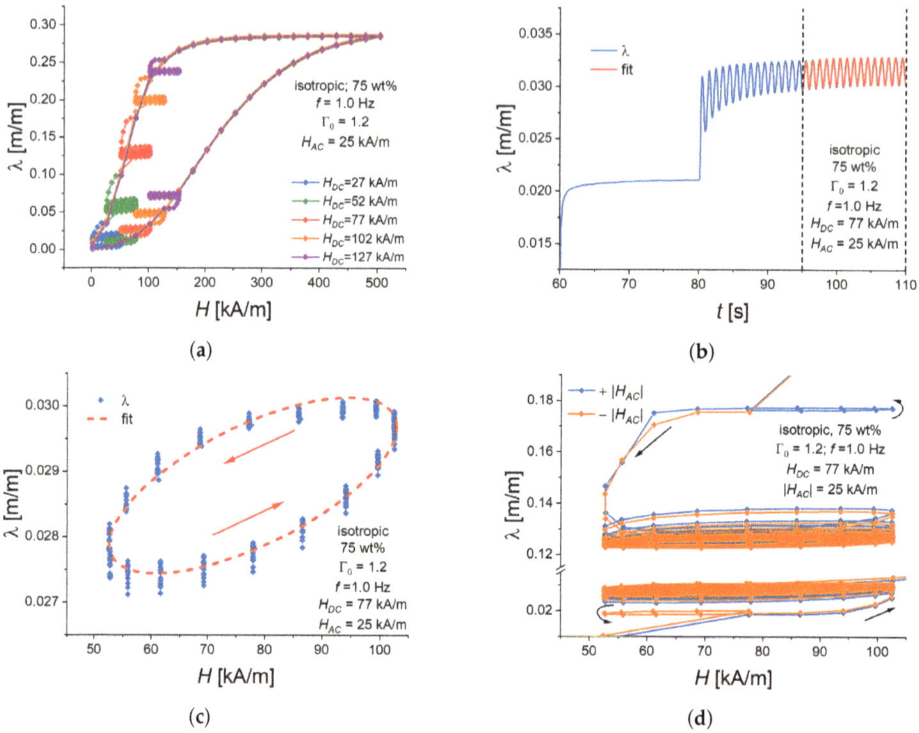

Figure 3. Macroscopic deformation of an isotropic MAE cylinder with 75 wt% of Fe and an aspect ratio $\Gamma_0 = 1.2$. Here and below, a line connecting experimental points serves as a guide to the eye. (**a**) Magnetostrictive hysteresis loops with additional minor loops due to superimposed oscillations of a magnetic field at different operation points. (**b**) Example of the transient response of magnetostrictive strain in a magnetic field (1). The measured values are shown in blue, and the fitted sinusoidal function is shown as a red curve. The bias field of 77 kA/m was set at the time point of $t = 60$ s. (**c**) Longitudinal strain λ versus the momentary value of magnetic field for the case (**b**). The red ellipse demonstrates the result from the fitted function (2). (**d**) Comparison of magnetostrictive response when magnetic field oscillations started with opposite phases (H_{AC} had the same magnitude, but it was either positive or negative). In (**b**–**d**), magnetic field oscillations had a frequency f of 1.0 Hz, an oscillation amplitude H_{AC} was 25 kA/m, and a bias magnetic field H_{DC} was 77 kA/m.

3. Results and Discussion

3.1. *Dependence of the Alternating Strain on the Magnitude of H_{AC}*

Figure 3b demonstrates that the magnetostrictive response of an MAE cylinder to a time-harmonic magnetic field $H_\sim$ in the vicinity of a bias field H_{DC} does resemble piezomagnetic behavior in conventional MS materials because there are positive and negative half waves of the longitudinal strain $\lambda_\sim$. However, the strain response is delayed with regard to the alternating magnetic field. This can be explained by the viscoelasticity of a polymer matrix and the entire composite material.

Furthermore, in the investigated range of the amplitude H_{AC} between 12.5 kA/m up to 65 kA/m in steps of 12.5 kA/m, the amplitude of alternating strain $\lambda_\sim$ depends on the amplitude of H_{AC} non-linearly. An example is shown in Figure 4a. For all isotropic samples with an aspect ratio of $\Gamma_0 = 1.2$, λ_{AC} monotonically increased with increasing amplitude of the oscillatory magnetic field. The resulting piezomagnetic coefficients d_{33}

also monotonically increased with increasing H_{AC} (Figure 4b). One of the reasons for the non-linear dependence of λ_{AC} on (H_{AC}) could be the Payne effect, which is characteristic of rubber composites containing fillers [60]. The Payne effect describes the decrease in the complex modulus of a filled, cross-linked elastomer system with increasing deformation amplitude. In MAEs, the Payne effect is also known to become well pronounced in magnetic fields, cf. [61–63] or Figure 1c,d in the very recent paper [50]. For MAEs considered in the present paper, an example of the Payne effect was given in Figure 4 of Ref. [41]. The highest sensitivity of λ_{AC} to H_{AC} was achieved for the sample with 75 wt of Fe, which could be expected from the previous investigation [41]. In general, for otherwise equal experimental parameters, d_{33} was somewhat lower for the ascending bias field than for the descending bias field. This also could be expected from the higher piezomagnetic constant d for the descending part of the MS hysteresis loop than for the ascending part of the loop at the same value of the bias field.

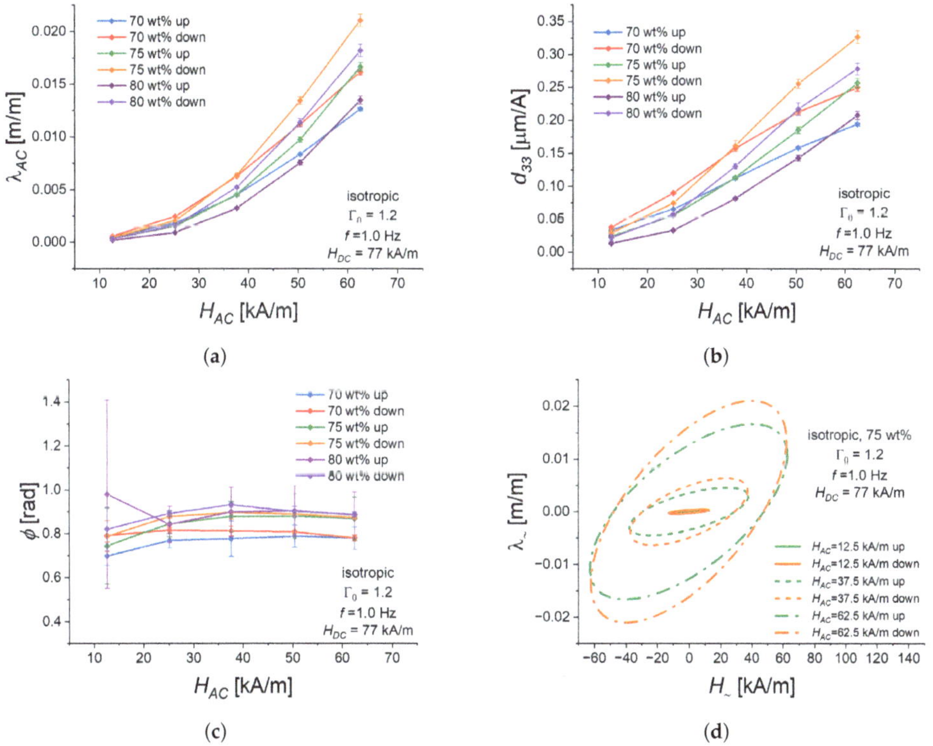

Figure 4. Dependencies of different characteristics of the alternating strain $\lambda_{\sim}$ with a frequency f = 1.0 Hz on the magnitude of H_{AC} for isotropic MAE cylinders with three different weight fractions of Fe (70, 75, 80 wt%) and $\Gamma_0 = 1.2$ in ascending ("up") or descending ("down") bias field $H_{DC} = 77$ kA/m of a quasi-static magnetostrictive hysteresis loop. In all sub-figures, a specific line color refers to the same value of iron content and the same part of hysteresis loop (ascending/descending). (**a**) The amplitude of harmonic strain oscillation λ_{AC} versus H_{AC}. (**b**) Piezomagnetic coefficient d_{33} versus H_{AC}. (**c**) Phase lag ϕ versus H_{AC}. (**d**) Lissajous figures for different values of H_{AC}. Continuous lines refer to $H_{AC} = 12.5$ kA/m, dashed lines refer to $H_{AC} = 37.5$ kA/m, and dash-dotted lines refer to $H_{AC} = 62.5$ kA/m.

The phase lag ϕ was weakly dependent on H_{AC} (Figure 4c) and had a tendency to moderately increase with increasing H_{AC}. The dependencies of ϕ on H_{AC} for samples with 80 and 75 wt% of Fe were within the uncertainty of measurements, while $\phi(H_{AC})$ for a cylinder with 70 wt% of Fe was about 0.1 rad lower. Figure 4d depicts Lissajous curves for the MAE cylinder with 75% of iron content and three representative values of H_{AC} (lowest, highest, and medium value). The features of the magneto-deformation response described above to the alternating magnetic field are clearly visible there.

The following results are presented for the fixed value $H_{AC} = 25$ kA/m. We have selected it as a compromise between a sufficiently high value of H_{AC} for the accurate measurements of AC strain and the increasing non-linearity of the piezomagnetic response with the increase in H_{AC}. As examples, we used isotropic MAE cylinders with 75 wt% of Fe because the samples with this filler concentration demonstrated the highest MS response in quasi-static experiments [41]. The aspect ratio Γ_0 of 1.2 was employed, since it resulted in the highest absolute elongation in quasi-static experiments [41].

3.2. Dependence of the Alternating Strain on the Bias Magnetic Field

As far as the usage of magnetostrictive polymer composites as actuators is concerned, the most interesting information is the value of the critical bias field at which the highest d_{33} is observed [59]. Figure 5a presents the measurement results for isotropic MAE cylinders with an aspect ratio $\Gamma_0 = 1.2$, where the bias magnetic field H_{DC} was varied, and the amplitude of field oscillations at a frequency f of 1.0 Hz was kept constant at $H_{AC} = 25$ kA/m.

As one can see from Figure 5a, the magnetostrictive response is qualitatively different for ascending and descending bias magnetic fields. The coefficient d_{33} decreased with increasing H_{DC} for all weight fractions of iron content in the ascending bias field. However, in the descending bias field, local maxima of d_{33} were observed at $H_{DC} = 52$ kA/m. The highest value of $d_{33} \approx 140$ nm/A was found for an isotropic cylinder with 75 wt% of Fe content followed by cylinders with 70 and 80 wt% of Fe, respectively. It is important that the observed dependencies of d_{33} on H_{DC} were qualitatively different from those expected from the dependencies of the quasi-static effective piezomagnetic constant d on bias field H_{DC} (Figure 5b). In the investigated range of H_{DC}, d increased with increasing H_{DC} for the ascending part of the hysteresis loop, while it had a local maximum around $\approx$100 kA/m (black dashed line in Figure 5a,b) for the descending part of the hysteresis loop. Furthermore, d_{33} was roughly one order of magnitude lower than d. It is obvious that the viscoelasticity of MAE materials and the non-linearity of the magnetostrictive response bring new effects, which were not observed in conventional magnetostrictive polymer composites. Additional theoretical research is required to explain our experimental results.

Due to the uncertainties of measurements, it can only be concluded that the phase lag ϕ (Figure 5c) was weakly dependent on the bias field and the dependencies of ϕ on H_{DC} for cylinders with 80 and 75 wt% of Fe were about 0.1 rad higher than those for a cylinder with 70 wt% of Fe.

Figure 5d depicts Lissajous curves for the MAE cylinder with 75% of iron content and three representative values of H_{DC}. The features described above of the magneto-deformation response to alternating magnetic field are clearly visible there.

Note that obtained maximum values of $d_{33} \sim 10^2$ nm/A are one to two orders of magnitude higher than those in conventional magnetostrictive polymer composites, while the critical magnetic fields where the maximum of d_{33} is observed are of the same order of magnitude (30–140 kA/m) [59].

Anisotropic cylinders with otherwise the same material parameters exhibited much lower magnetostrictive strains [41]; also, d_{33} was about 1.5-fold lower (Figure 6a). The phase lag ϕ was practically independent of H_{DC} (Figure 6b). Remarkably, d_{33} decreased with increasing H_{DC} for both ascending and descending bias fields (Figure 6a). Local maxima were absent.

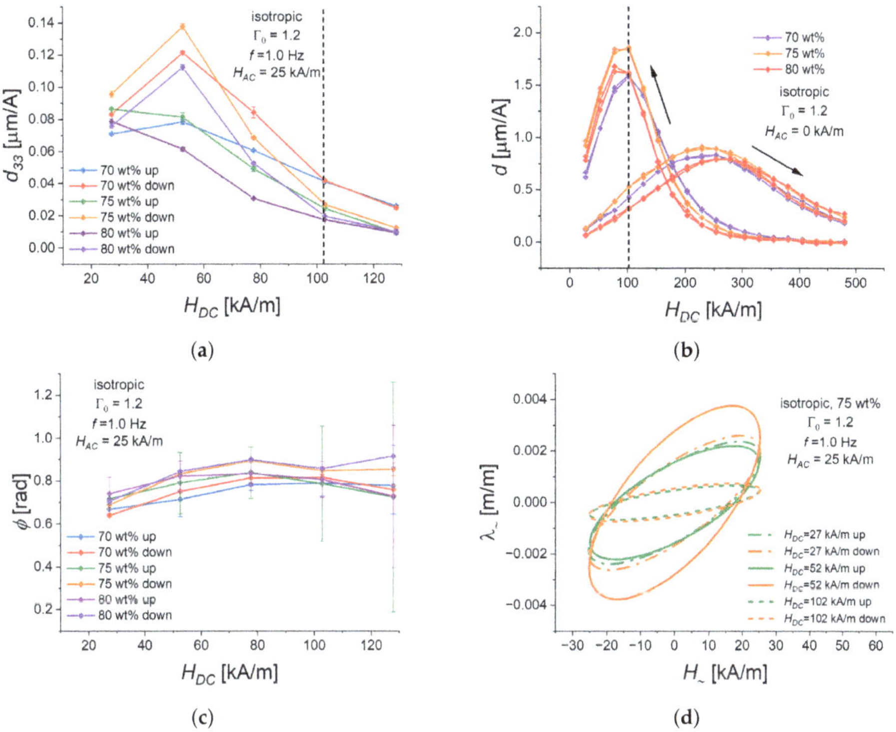

Figure 5. Dependencies of different characteristics of the alternating strain $\lambda_{\sim}$ with a frequency $f = 1.0$ Hz on the magnitude of H_{DC} for isotropic MAE cylinders with three different weight fractions of Fe (70, 75, 80 wt%), and $\Gamma_0 = 1.2$ in the ascending ("up") or descending ("down") bias field H_{DC} of a quasi-static magnetostrictive hysteresis loop. The value of H_{AC} is fixed at 25 kA/m. In all sub-figures, a specific line color refers to the same value of iron content and the same part of hysteresis loop (ascending/descending). (**a**) d_{33} versus H_{DC}. Dashed black line designates the position of the maximums of d for the descending part of the quasi-static hysteresis curves. (**b**) d versus H_{DC}, calculated as a finite central difference from experimental results of [41]. Black arrows designate the direction of field change. (**c**) Phase delay ϕ versus H_{DC}. (**d**) Lissajous figures for different values of H_{DC}. Continuous lines refer to $H_{DC} = 52$ kA/m, dashed lines refer to $H_{DC} = 102$ kA/m, and dash-dotted lines refer to $H_{DC} = 25$ kA/m.

Figure 6c,d depict Lissajous curves for the anisotropic MAE cylinder with 75 wt% of Fe and three representative values of H_{DC}, for both ascending (Figure 6c) and descending (Figure 6d) bias fields. The features described above of the magneto-deformation response to alternating magnetic field are clearly visible there. Since the piezomagnetic response of anisotropic cylinders was lower than that of their isotropic counterparts, they will be not be considered in the following.

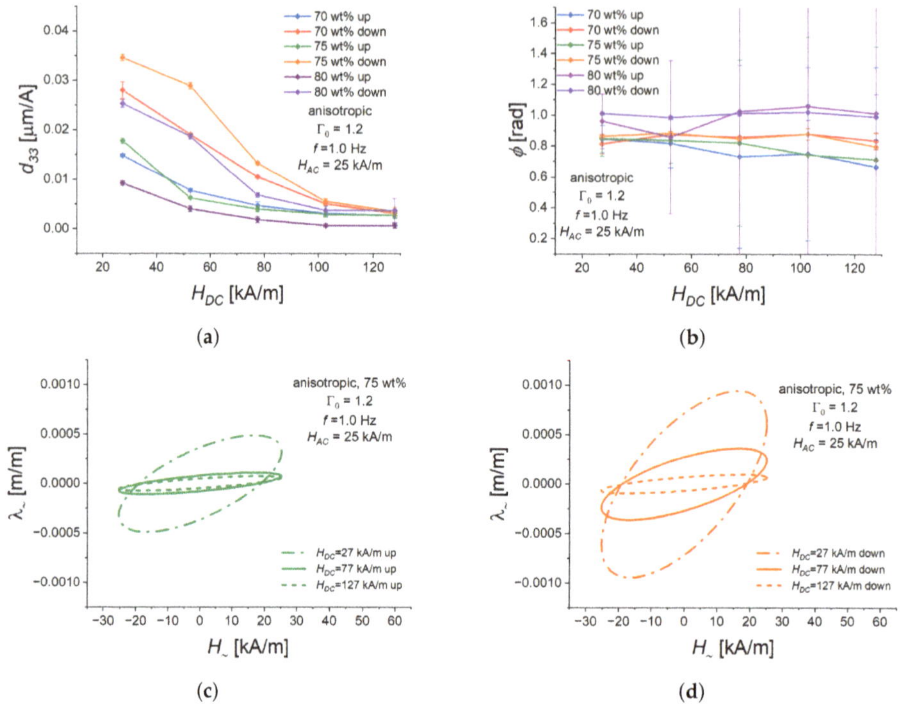

Figure 6. Dependencies of different characteristics of the alternating strain $\lambda_{\sim}$ with a frequency $f = 1.0$ Hz on the magnitude of H_{DC} for anisotropic MAE cylinders with three different weight fractions of Fe (70, 75, 80 wt%), and $\Gamma_0 = 1.2$ in ascending ("up") or descending ("down") bias field H_{DC} of a quasi-static magnetostrictive hysteresis loop. The value of H_{AC} is fixed at 25 kA/m. In all sub-figures, a specific line color refers to the same value of iron content and the same part of hysteresis loop (ascending/descending). (**a**) d_{33} versus H_{DC}. (**b**) Phase delay ϕ versus H_{DC}. (**c**) Lissajous figures for different values of ascending ("up") bias field H_{DC}. (**d**) Lissajous figures for different values of descending ("down") bias field H_{DC}.

3.3. Frequency Dependence of the Alternating Strain

In our next set of experiments, we investigated the dependence of the piezomagnetic characteristics on the oscillation frequency f in the range between 0.1 and 2.5 Hz for a isotropic MAE cylinders with an aspect ratio Γ_0 of 1.2. The frequency of 2.5 Hz was the highest achievable frequency, due to the experimentally limited sampling of the alternating magnetic field. The oscillation amplitude and the bias magnetic field were held constant at $H_{AC} = 25$ kA/m and $H_{DC} = 77$ kA/m, respectively.

The piezomagnetic coefficent d_{33} monotonically decreased with increasing frequency f (Figure 7a) for each particular concentration of filling particles and the direction of bias field change. For otherwise the same experimental parameters, d_{33} was somewhat higher for the descending bias field than for the ascending bias field. For each iron concentration and particular branch of hysteresis loop (up/down), the phase lag ϕ monotonically increased with increasing frequency f (Figure 7b). The decrease in d_{33} and the increase in ϕ with increasing frequency can be attributed to the viscoelasticity of the PDMS matrix and the composite material as a whole.

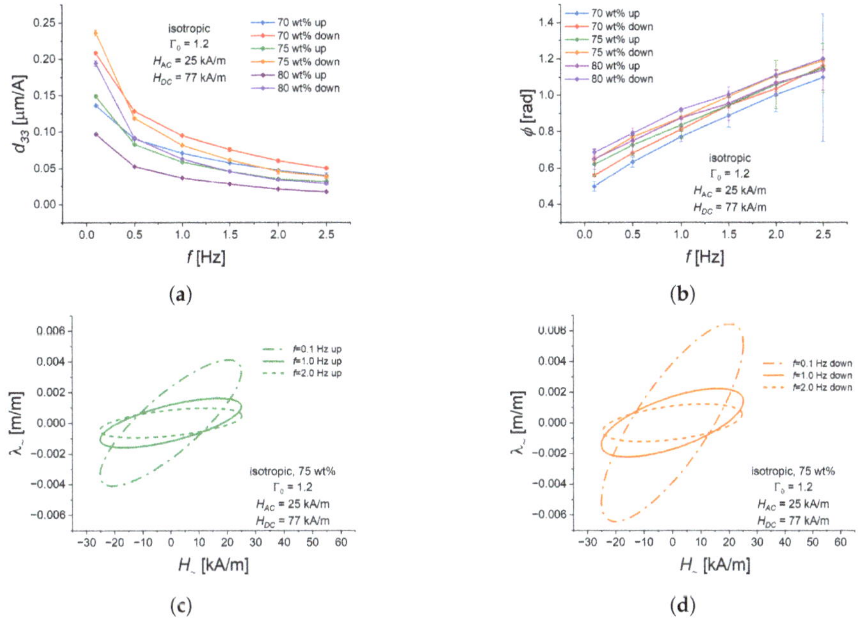

Figure 7. Dependencies of different characteristics of the alternating strain $\lambda_{\sim}$ on the oscillation frequency f for isotropic MAE cylinders with three different weight fractions of Fe (70, 75, 80 wt%) and $\Gamma_0 = 1.2$ in ascending ("up") or descending ("down") bias field of a quasi-static magnetostrictive hysteresis loop. The value of H_{AC} is fixed at 25 kA/m. The value of H_{DC} is fixed at 77 kA/m. In all sub-figures, a specific line color refers to the same value of iron content and the same part of the hysteresis loop (ascending/descending). (**a**) d_{33} versus f. (**b**) Phase delay ϕ versus f. (**c**) Lissajous figures for different values of frequency f in an ascending ("up") bias field H_{DC} of 77 kA/m. (**d**) Lissajous figures for different values of frequency f in a descending ("down") bias field H_{DC} of 77 kA/m.

Figure 7c,d depict Lissajous curves for the isotropic MAE cylinder with 75 wt% of Fe and three representative values of frequency f (0.1, 1.5, 2.5 Hz), both for ascending (Figure 7c) and descending (Figure 7d) bias fields, where the features described above of the magneto-deformation response to alternating magnetic field can be clearly observed.

3.4. Dependence of the Alternating Strain on the Aspect Ratio

Finally, we investigated how the piezomagnetic characteristics of isotropic MAE cylinders with 75 wt% of Fe depend on the aspect ratio Γ_0 (Figure 8). The amplitude of the alternating field H_{AC} was equal to 25 kA/m, and the bias magnetic field H_{DC} was 77 kA/m. The frequency f was 1.0 Hz. The primary effect of the aspect ratio should be the change in the demagnetizing factor [64]. Additionally, we have previously found that a concave dent is formed on the free circular base in magnetic fields [41]. The contribution of this dent to the apparent longitudinal strain increases with the decreasing aspect ratio [41].

Figure 8 presents the experimental results obtained. The piezomagnetic strain constant d_{33} was always somewhat lower for the ascending bias field than for the descending bias field (Figure 8a), in agreement with the observations described above. In general, d_{33} showed a peculiar dependence on the aspect ratio Γ_0 (Figure 8a), which currently have no explanation. We have verified that these results (Figure 8a) are repeatable. There seems to be a local minimum of d_{33} around $\Gamma_0 = 0.6$ and a local maximum between $\Gamma_0 = 0.8$ and 1.0. An increase in the aspect ratio should lead to an increase in the internal magnetic field strength with the cylinder, which should affect both DC and AC components of the internal magnetic field. Figures 4b and 5a demonstrate qualitatively different behavior of d_{33} on H_{AC} and H_{DC}, respectively. Could it be the interplay of non-linear dependencies of the coefficient d_{33} on the amplitudes of the overlapping DC and AC magnetic field, which lead to the appearance of the local maximum and minimum in dependence of d_{33} on the aspect ratio Γ_0? This question has to be investigated in detail in future works.

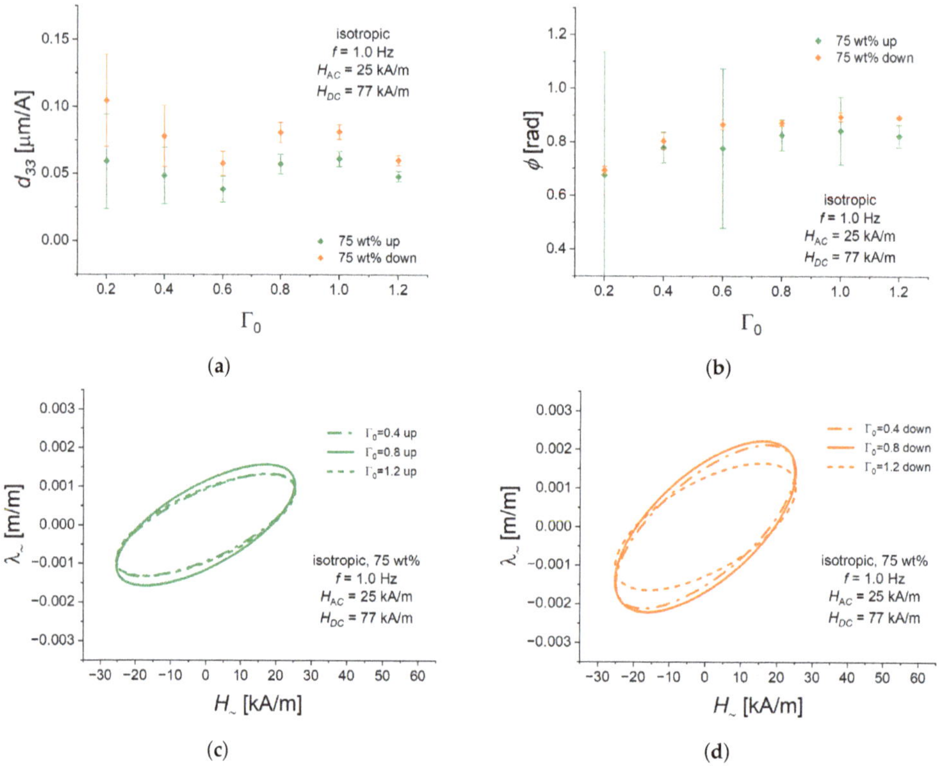

Figure 8. Dependencies of different characteristics of the alternating strain $\lambda_{\sim}$ on the aspect ratio Γ_0 for isotropic MAE cylinders with 75 wt% of Fe in ascending ("up") or descending ("down") bias field of a quasi-static magnetostrictive hysteresis loop. The value of H_{AC} is fixed at 25 kA/m. The value of H_{DC} is fixed at 77 kA/m. The frequency f is fixed at 1.0 Hz. In all sub-figures, a specific line color refers to the same part of the hysteresis loop (ascending/descending). (**a**) d_{33} versus Γ_0. (**b**) Phase delay ϕ versus aspect ratio Γ_0. (**c**) Lissajous figures for different aspect ratios Γ_0 in an ascending ("up") bias field H_{DC} of 77 kA/m. (**d**) Lissajous figures for different aspect ratios Γ_0 in a descending ("down") bias field H_{DC} of 77 kA/m. Dash-dotted lines correspond to $\Gamma_0 = 0.4$, continuous lines correspond to $\Gamma_0 = 0.8$, and dashed lines correspond to $\Gamma_0 = 1.2$.

Within the uncertainty of measurements, the phase lag ϕ was independent of the aspect ratio Γ_0 for both ascending and descending bias magnetic fields (Figure 8b).

Figure 8c,d provide the Lissajous figures of the alternating strain versus the alternating magnetic field for three selected values of the aspect ratio Γ_0. They present the results of Figure 8a,b in an alternative form.

4. Conclusions and Outlook

To summarize, we presented a comprehensive experimental study of the deformation response of MAE cylinders to time-harmonic magnetic fields in the vicinity of different operating points (bias magnetic fields). To the best of our knowledge, such an investigation was performed for the first time. It was found that the behavior of MAEs was different from conventional magnetostrictive polymer composites in many respects. The phenomenon resembles piezomagnetism in the sense that there are positive and negative half-cycles of alternating strain in response to a time-harmonic magnetic field, but there is a significant phase delay (≈0.6–1.0 rad) between the alternating magnetic field and alternating strain. Obviously, if time-harmonic quantities H_{AC}, λ_{AC} are written in the form of a phasor, d_{33} should be a complex number. The strain response to an oscillating magnetic field also involves a transient component (Figure 3d). The dependence of the piezomagnetic strain coefficient d_{33} on the amplitude of the alternating magnetic field H_{AC} was non-linear in the investigated range of H_{AC} (Figure 4b). The obtained values of d_{33} were one to two orders of magnitude higher than in conventional magnetostrictive polymer composites (Figures 4b and 5a). At a particular value of the bias field H_{DC}, d_{33} was higher when this value was reached at the descending part of the magnetostrictive hysteresis loop in comparison with the ascending part of the loop. In the investigated frequency range (0.1–2.5 Hz), the piezomagnetic strain coefficient d_{33} decreased with an increasing frequency f of alternating magnetic field.

It has been found that the strain response of MAE cylinders to a time-harmonic magnetic field superimposed on a biased magnetic is rather complex. It has several intriguing features not observed in conventional magnetostrictive polymer composites. It became also obvious that dynamic behavior of MAE cylinders cannot be simply derived from the quasi-static measurements. Unfortunately, there are currently no theories available to explain observed experimental dependencies. Given that MAE materials are considered to be promising as actuators in soft robotics [29,30], we hope that our experimental findings will encourage theoreticians to develop novel dynamic models incorporating non-linear viscoelasticity and inertia [50] for describing macroscopic deformations of MAE-based functional elements.

Author Contributions: Conceptualization, M.S.; methodology, M.S., G.G. and I.A.B.; software, G.G.; validation, G.G., I.A.B. and M.S.; investigation, G.G. and I.A.B.; resources, I.A.B. and M.S.; data curation, G.G.; writing—original draft preparation, G.G. and M.S.; writing—review and editing, M.S., G.G. and I.A.B.; visualization, G.G.; supervision, M.S.; project administration, M.S.; funding acquisition, M.S. All authors have read and agreed to the published version of the manuscript.

Funding: The research of G.G. and M.S. was funded by the Deutsche Forschungsgemeinschaft (DFG, German Research Foundation)—Project No. 466920132.

Institutional Review Board Statement: Not applicable.

Informed Consent Statement: Not applicable.

Data Availability Statement: The data supporting the findings of this paper ara available in Zenodo https://doi.org/10.5281/zenodo.13834054 (accessed on 24 September 2024).

Acknowledgments: We are grateful to Evonik Operations GmbH, Specialty Additives, Geesthacht, Germany for providing chemicals for synthesis of PDMS and to BASF SE, Ludwigshafen am Rhein, Germany, for providing CIP.

Conflicts of Interest: The authors declare no conflicts of interest.

Appendix A

Figure A1a shows in detail the transient behavior of the longitudinal strain when the DC magnetic field strength is increased from 52 kA/m to 77 kA/m and then overlapped by a sinusoidal magnetic field. Figure A1b presents in detail the transient behavior of the longitudinal strain when the DC magnetic field strength is decreased from 102 kA/m to 77 kA/m and then overlapped by a sinusoidal magnetic field. In both figures, the following step-like change in the external magnetic field is visible as well. For the sake of detailed presentation, the images for the calculation of the longitudinal strain were taken at 18 FPS. Notice that after the modulating AC magnetic field is switched off, the resulting longitudinal strain seems to be practically time independent, and it is not equal to the steady-state longitudinal strain in the DC field before the modulating AC magnetic field is switched on.

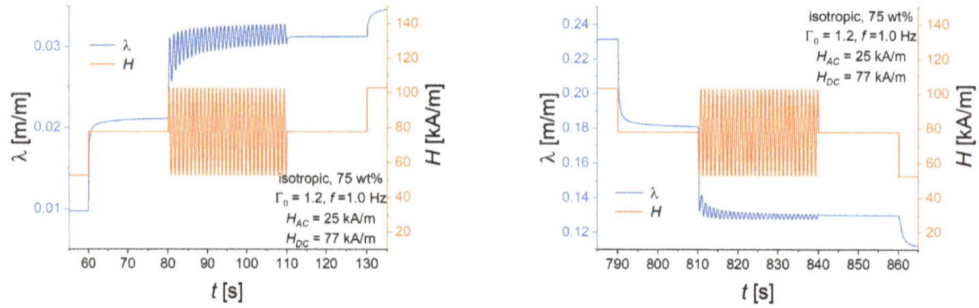

(a) (b)

Figure A1. Time dependencies of the longitudinal strain (blue curve) for harmonic magnetic field modulations (orange curve) when the DC magnetic field is increasing (**a**) and when it is decreasing (**b**). Parameters of the sample and measurement conditions are given within the figures.

Appendix B

Figure A2 presents enlarged views of the two curves in Figure 3d, separately.

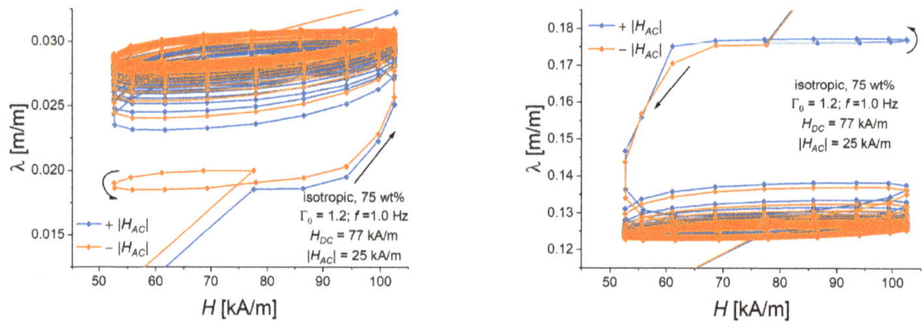

(a) (b)

Figure A2. Magnetostrictive response to oscillating magnetic field when the bias magnetic field is increasing (**a**) and when the bias field is decreasing (**b**). Blue dots and lines refer to the positive value of H_{AC} = 25 kA/m, and orange dots and lines refer to the negative value of H_{AC} = −25 kA/m. Parameters of the sample and measurement conditions are given within the figures.

References

1. Borovik-Romanov, A.S. Piezomagnetism, linear magnetostriction and magnetooptic effect. *Ferroelectrics* **1994**, *162*, 153–159. [CrossRef]
2. Piercy, A.R. The changing shape of magnetostriction. *Phys. Educ.* **1997**, *32*, 160. [CrossRef]
3. Ekreem, N.; Olabi, A.; Prescott, T.; Rafferty, A.; Hashmi, M. An overview of magnetostriction, its use and methods to measure these properties. *J. Mater. Process. Technol.* **2007**, *191*, 96–101. [CrossRef]
4. Giurgiutiu, V.; Lyshevski, S.E. *Micromechatronics: Modeling, Analysis, and Design with MATLAB*; CRC Press: Boca Raton, FL, USA, 2016. [CrossRef]
5. *IEEE Std 319-1990*; IEEE Standard on Magnetostrictive Materials: Piezomagnetic Nomenclature. IEEE Standard Association: Piscataway, NJ, USA, 1991. [CrossRef]
6. Magnetostriction and Piezomagnetism. Available online:. https://doc.comsol.com/6.1/doc/com.comsol.help.sme/sme_ug_theory.06.041.html (accessed on 11 July 2024).

7. Kim, Y.Y.; Kwon, Y.E. Review of magnetostrictive patch transducers and applications in ultrasonic nondestructive testing of waveguides. *Ultrasonics* **2015**, *62*, 3–19. [CrossRef]
8. Nan, C.W.; Bichurin, M.; Dong, S.; Viehland, D.; Srinivasan, G. Multiferroic magnetoelectric composites: Historical perspective, status, and future directions. *J. Appl. Phys.* **2008**, *103*, 1. [CrossRef]
9. Srinivasan, G. Magnetoelectric composites. *Annu. Rev. Mater. Res.* **2010**, *40*, 153–178. [CrossRef]
10. Zhou, Y.; Maurya, D.; Yan, Y.; Srinivasan, G.; Quandt, E.; Priya, S. Self-Biased Magnetoelectric Composites: An Overview and Future Perspectives. *Energy Harvest. Syst.* **2016**, *3*, 1–42. [CrossRef]
11. Leung, C.M.; Li, J.; Viehland, D.; Zhuang, X. A review on applications of magnetoelectric composites: From heterostructural uncooled magnetic sensors, energy harvesters to highly efficient power converters. *J. Phys. D Appl. Phys.* **2018**, *51*, 263002. [CrossRef]
12. Pradhan, D.K.; Kumari, S.; Rack, P.D. Magnetoelectric Composites: Applications, Coupling Mechanisms, and Future Directions. *Nanomaterials* **2020**, *10*, 2072. [CrossRef]
13. Liang, X.; Matyushov, A.; Hayes, P.; Schell, V.; Dong, C.; Chen, H.; He, Y.; Will-Cole, A.; Quandt, E.; Martins, P.; et al. Roadmap on Magnetoelectric Materials and Devices. *IEEE Trans. Magn.* **2021**, *57*, 1–57. [CrossRef]
14. Feng, J.; Xuan, S.; Ding, L.; Gong, X. Magnetoactive elastomer/PVDF composite film based magnetically controllable actuator with real-time deformation feedback property. *Compos. Part A Appl. Sci. Manuf.* **2017**, *103*, 25–34. [CrossRef]
15. Makarova, L.A.; Alekhina, Y.A.; Isaev, D.A.; Khairullin, M.F.; Perov, N.S. Tunable layered composites based on magnetoactive elastomers and piezopolymer for sensors and energy harvesting devices. *J. Phys. D Appl. Phys.* **2020**, *54*, 015003. [CrossRef]
16. Saveliev, D.V.; Glavan, G.; Belan, V.O.; Belyaeva, I.A.; Fetisov, L.Y.; Shamonin, M. Resonant Magnetoelectric Effect at Low Frequencies in Layered Polymeric Cantilevers Containing a Magnetoactive Elastomer. *Appl. Sci.* **2022**, *12*, 2102. [CrossRef]
17. Savelev, D.V.; Glavan, G.; Burdin, D.A.; Belyaeva, I.A.; Fetisov, L.Y.; Shamonin, M.; Fetisov, Y.K. Enhancement of magnetoelectric effect in polymer composites at low resonance frequencies by operation in the transverse-transverse mode. *J. Magn. Magn. Mater.* **2024**, *598*, 172020. [CrossRef]
18. Ju, S.; Chae, S.H.; Choi, Y.; Lee, S.; Lee, H.W.; Ji, C.H. A low frequency vibration energy harvester using magnetoelectric laminate composite. *Smart Mater. Struct.* **2013**, *22*, 115037. [CrossRef]
19. Marauska, S.; Jahns, R.; Greve, H.; Quandt, E.; Knöchel, R.; Wagner, B. MEMS magnetic field sensor based on magnetoelectric composites. *J. Micromech. Microeng.* **2012**, *22*, 065024. [CrossRef]
20. Moreno, M.; Gonzalez-Rico, J.; Lopez-Donaire, M.; Arias, A.; Garcia-Gonzalez, D. New experimental insights into magneto-mechanical rate dependences of magnetorheological elastomers. *Compos. Part B Eng.* **2021**, *224*, 109148. [CrossRef]
21. Ubaidillah; Sutrisno, J.; Purwanto, A.; Mazlan, S.A. Recent Progress on Magnetorheological Solids: Materials, Fabrication, Testing, and Applications. *Adv. Eng. Mater.* **2015**, *17*, 563–597. [CrossRef]
22. Menzel, A.M. Tuned, driven, and active soft matter. *Phys. Rep.* **2015**, *554*, 1–45. [CrossRef]
23. Lopez-Lopez, M.T.; Durán, J.D.G.; Iskakova, L.Y.; Zubarev, A.Y. Mechanics of Magnetopolymer Composites: A Review. *J. Nanofluids* **2016**, *5*, 479–495. [CrossRef]
24. Weeber, R.; Hermes, M.; Schmidt, A.M.; Holm, C. Polymer architecture of magnetic gels: A review. *J. Phys. Condens. Matter* **2018**, *30*, 063002. [CrossRef] [PubMed]
25. Bastola, A.K.; Paudel, M.; Li, L.; Li, W. Recent progress of magnetorheological elastomers: A review. *Smart Mater. Struct.* **2020**, *29*, 123002. [CrossRef]
26. Odenbach, S. (Ed.) *Magnetic Hybrid-Materials: Multi-Scale Modelling, Synthesis, and Applications*; Walter de Gruyter GmbH & Co KG: Berlin, Germany, 2021.
27. Nadzharyan, T.A.; Shamonin, M.; Kramarenko, E.Y. Theoretical Modeling of Magnetoactive Elastomers on Different Scales: A State-of-the-Art Review. *Polymers* **2022**, *14*, 4096. [CrossRef] [PubMed]
28. Kostrov, S.A.; Marshall, J.H.; Maw, M.; Sheiko, S.S.; Kramarenko, E.Y. Programming and Reprogramming the Viscoelasticity and Magnetic Response of Magnetoactive Thermoplastic Elastomers. *Polymers* **2023**, *15*, 4607. [CrossRef] [PubMed]
29. Boyraz, P.; Runge, G.; Raatz, A. An Overview of Novel Actuators for Soft Robotics. *Actuators* **2018**, *7*, 48. [CrossRef]
30. Bernat, J.; Gajewski, P.; Kołota, J.; Marcinkowska, A. Review of Soft Actuators Controlled with Electrical Stimuli: IPMC, DEAP, and MRE. *Appl. Sci.* **2023**, *13*, 1651. [CrossRef]
31. Domenjoud, M.; Berthelot, E.; Galopin, N.; Corcolle, R.; Bernard, Y.; Daniel, L. Characterization of giant magnetostrictive materials under static stress: Influence of loading boundary conditions. *Smart Mater. Struct.* **2019**, *28*, 095012. [CrossRef]
32. Romeis, D.; Toshchevikov, V.; Saphiannikova, M. Effects of local rearrangement of magnetic particles on deformation in magneto-sensitive elastomers. *Soft Matter* **2019**, *15*, 3552–3564. [CrossRef]
33. Stolbov, O.V.; Raikher, Y.L. Magnetostriction effect in soft magnetic elastomers. *Arch. Appl. Mech.* **2019**, *89*, 63–76. [CrossRef]
34. Sánchez, P.A.; Stolbov, O.V.; Kantorovich, S.S.; Raikher, Y.L. Modeling the magnetostriction effect in elastomers with magnetically soft and hard particles. *Soft Matter* **2019**, *15*, 7145–7158. [CrossRef]
35. Kalina, K.A.; Metsch, P.; Brummund, J.; Kästner, M. A macroscopic model for magnetorheological elastomers based on microscopic simulations. *Int. J. Solids Struct.* **2020**, *193–194*, 200–212. [CrossRef]

36. Saveliev, D.V.; Belyaeva, I.A.; Chashin, D.V.; Fetisov, L.Y.; Romeis, D.; Kettl, W.; Kramarenko, E.Y.; Saphiannikova, M.; Stepanov, G.V.; Shamonin, M. Giant Extensional Strain of Magnetoactive Elastomeric Cylinders in Uniform Magnetic Fields. *Materials* **2020**, *13*, 3297. [CrossRef] [PubMed]
37. Bastola, A.K.; Hossain, M. The shape—Morphing performance of magnetoactive soft materials. *Mater. Des.* **2021**, *211*, 110172. [CrossRef]
38. Silva, J.; Gouveia, C.; Dinis, G.; Pinto, A.; Pereira, A. Giant magnetostriction in low-concentration magnetorheological elastomers. *Compos. Part B Eng.* **2022**, *243*, 110125. [CrossRef]
39. Balogh, D.; Guba, S.; Horváth, B.; Szalai, I. Magnetic Field-Induced Deformation of Isotropic Magnetorheological Elastomers. *Magnetochemistry* **2022**, *8*, 146. [CrossRef]
40. Tasin, M.A.; Aziz, S.A.A.; Mazlan, S.A.; Johari, M.A.F.; Nordin, N.A.; Yusuf, S.Y.M.; Choi, S.B.; Bahiuddin, I. Magnetostriction Enhancement in Midrange Modulus Magnetorheological Elastomers for Sensor Applications. *Micromachines* **2023**, *14*, 767. [CrossRef]
41. Glavan, G.; Belyaeva, I.A.; Drevenšek-Olenik, I.; Shamonin, M. Experimental study of longitudinal, transverse and volume strains of magnetoactive elastomeric cylinders in uniform magnetic fields. *J. Magn. Magn. Mater.* **2023**, *579*, 170826. [CrossRef]
42. Roghani, M.; Romeis, D.; Saphiannikova, M. Effect of microstructure evolution on the mechanical behavior of magneto-active elastomers with different matrix stiffness. *Soft Matter* **2023**, *19*, 6387–6398. [CrossRef]
43. Goh, S.; Menzel, A.M.; Wittmann, R.; Löwen, H. Density functional approach to elastic properties of three-dimensional dipole-spring models for magnetic gels. *J. Chem. Phys.* **2023**, *158*, 054909. [CrossRef]
44. Hogea, C.S.; Armstrong, W.D. The time-dependent magneto-visco-elastic behavior of a magnetostrictive fiber actuated viscoelastic polymer matrix composite. *J. Acoust. Soc. Am.* **2002**, *112*, 1928–1936. [CrossRef]
45. Lin, C.h.; Lin, Y.Z. Analysis of non-linear piezomagnetism for magnetostrictive terfenol-D composites. *J. Magn. Magn. Mater.* **2021**, *540*, 168490. [CrossRef]
46. Lin, C.h.; Lin, C.J. Effective magnetoelastic responses of hybrid fiber composites with viscoelastic matrices. *Compos. Struct.* **2024**, *333*, 117874. [CrossRef]
47. Zubarev, A.; Chirikov, D.; Stepanov, G.; Borin, D.; Lopez-Lopez, M. On the theory of hysteretic magnetostriction of soft ferrogels. *Phys. A Stat. Mech. Its Appl.* **2018**, *498*, 86–95. [CrossRef]
48. Belyaeva, I.A.; Klepp, J.; Lemmel, H.; Shamonin, M. Feasibility of Probing the Filler Restructuring in Magnetoactive Elastomers by Ultra-Small-Angle Neutron Scattering. *Appl. Sci.* **2021**, *11*, 4470. [CrossRef]
49. Biller, A.M.; Stolbov, O.V.; Raikher, Y.L. Mesoscopic magnetomechanical hysteresis in a magnetorheological elastomer. *Phys. Rev. E* **2015**, *92*, 023202. [CrossRef]
50. Wang, B.; Kari, L.; Pang, H.; Gong, X. Modelling the dynamic magnetic actuation of isotropic soft magnetorheological elastomers. *Int. J. Mech. Sci.* **2024**, *266*, 108908. [CrossRef]
51. Fields, T.H.E. *Time Harmonic Electromagnetic Fields*; McGraw-Hill: New York City, NY, USA, 1961.
52. Glavan, G.; Belyaeva, I.A.; Shamonin, M. Transient Response of Macroscopic Deformation of Magnetoactive Elastomeric Cylinders in Uniform Magnetic Fields. *Polymers* **2024**, *16*, 586. [CrossRef]
53. Mazurek, P.; Vudayagiri, S.; Skov, A.L. How to tailor flexible silicone elastomers with mechanical integrity: A tutorial review. *Chem. Soc. Rev.* **2019**, *48*, 1448–1464. [CrossRef]
54. Sorokin, V.V.; Belyaeva, I.A.; Shamonin, M.; Kramarenko, E.Y. Magnetorheological response of highly filled magnetoactive elastomers from perspective of mechanical energy density: Fractal aggregates above the nanometer scale? *Phys. Rev. E* **2017**, *95*, 062501. [CrossRef]
55. Belyaeva, I.A.; Kramarenko, E.Y.; Shamonin, M. Magnetodielectric effect in magnetoactive elastomers: Transient response and hysteresis. *Polymer* **2017**, *127*, 119–128. [CrossRef]
56. Glavan, G.; Belyaeva, I.A.; Ruwisch, K.; Wollschläger, J.; Shamonin, M. Magnetoelectric Response of Laminated Cantilevers Comprising a Magnetoactive Elastomer and a Piezoelectric Polymer, in Pulsed Uniform Magnetic Fields. *Sensors* **2021**, *21*, 6390. [CrossRef] [PubMed]
57. Trujillo-Pino, A.; Krissian, K.; Alemán-Flores, M.; Santana-Cedrés, D. Accurate subpixel edge location based on partial area effect. *Image Vis. Comput.* **2013**, *31*, 72–90. [CrossRef]
58. Schmusch, W. *Elektronische Messtechnik: Prinzipien, Verfahren, Schaltungen*; Vogel: Würzburg, Germany, 2001; p. 139.
59. Elhajjar, R.; Law, C.T.; Pegoretti, A. Magnetostrictive polymer composites: Recent advances in materials, structures and properties. *Prog. Mater. Sci.* **2018**, *97*, 204–229. [CrossRef]
60. Payne, A.R. The dynamic properties of carbon black-loaded natural rubber vulcanizates. Part I. *J. Appl. Polym. Sci.* **1962**, *6*, 57–63. [CrossRef]
61. Ginder, J.; Clark, S.; Schlotter, W.; Nichols, M. Magnetostrictive phenomena in magnetorheological elastomers. *Int. J. Mod. Phys. B* **2002**, *16*, 2412–2418. [CrossRef]
62. Sorokin, V.V.; Ecker, E.; Stepanov, G.V.; Shamonin, M.; Monkman, G.J.; Kramarenko, E.Y.; Khokhlov, A.R. Experimental study of the magnetic field enhanced Payne effect in magnetorheological elastomers. *Soft Matter* **2014**, *10*, 8765–8776. [CrossRef]

63. Shamonin, M.; Kramarenko, E.Y. Chapter 7—Highly Responsive Magnetoactive Elastomers. In *Novel Magnetic Nanostructures*; Domracheva, N., Caporali, M., Rentschler, E., Eds.; Advanced Nanomaterials; Elsevier: Amsterdam, The Netherlands, 2018; pp. 221–245. [CrossRef]
64. Diguet, G.; Beaugnon, E.; Cavaillé, J. Shape effect in the magnetostriction of ferromagnetic composite. *J. Magn. Magn. Mater.* **2010**, *322*, 3337–3341. [CrossRef]

polymers

MDPI

Article

The Hydrogen Bonding in the Hard Domains of the Siloxane Polyurea Copolymer Elastomers

Ming Bao, Tianyu Liu, Ying Tao and Xiuyuan Ni *

State Key Laboratory of Molecular Engineering of Polymers, Department of Macromolecular Science, Fudan University, Shanghai 200438, China; 19110440004@fudan.edu.cn (M.B.); 21110440018@m.fudan.edu.cn (T.L.); 20110440021@fudan.edu.cn (Y.T.)
* Correspondence: xyni@fudan.edu.cn

Abstract: For probing the structure–property relationships of the polyurea elastomers, we synthesize the siloxane polyurea copolymer elastomer by using two aminopropyl-terminated polysiloxane monomers with low and high number-average molecular weight (M_n), i.e., L-30D and H-130D. To study the influence of the copolymer structures on the film properties, these films are analyzed to obtain the tensile performance, UV-vis spectra, cross-sectional topographies, and glass transition temperature (T_g). The two synthetic thermoplastic elastomer films are characterized by transparency, ductility, and the T_g of the hard domains, depending on the reacting compositions. Furthermore, the film elasticity behavior is studied by the strain recovery and cyclic tensile test, and then, the linear fitting of the tensile data is used to describe the film elasticity based on the Mooney–Rivlin model. Moreover, the temperature-dependent infrared (IR) spectra during heating and cooling are conducted to study the strength and recovery rate of the hydrogen bonding, respectively, and their influence on the film performance is further analyzed; the calculated M_n of the hard segment chains is correlated to the macroscopic recovery rate of the hydrogen bonding. These results can add deep insight to the structure–property relationships of the siloxane polyurea copolymer.

Keywords: siloxane polyurea copolymer; thermoplastic elastomer; hard domains; hydrogen bonding; temperature-dependent IR spectra

1. Introduction

The polymer elastomers can recover to the highest degree of strain after stretching for large strain, with a structure composed of soft chains as the reversible phase and cross-linked structures as the fixed phase. The formation of the fixed phase is contributed to by the chemical crosslinking, crystalline regions, and hard domains with the hydrogen bonding interactions [1,2]. The broadly used polymer elastomers include the styrene-butadiene rubber (SBR), ethylene propylene diene monomer (EPDM) rubber, polyolefin elastomer (POE), poly (ester-amine), and polyurethane elastomer that are determined by the phase structure [3,4]. The polyurethane or polyurea elastomers contain the soft domains composed of polyols or polyamines, and their hard domains consist of the chains reacting by isocyanates and extenders. Due to the low glass transition temperature (T_g) and low surface energy, the polysiloxane is introduced to the polyurethane elastomers, endowing the materials with excellent flexibility at low temperatures and smooth surfaces. Recently, the thermoplastic polyurea (TPU) elastomer containing polysiloxane has received rising attention for its melt processing capability and wider working temperature range.

The properties of the siloxane polyurea elastomer are influenced by many factors such as polysiloxane structures [5,6], diisocyanate structures and contents [7], and the types of chain extenders [8]. As reported in the work, the siloxane polyurea copolymers were synthesized by using aminopropyl terminated polydimethylsiloxane (PDMS) as soft segments with rather high M_n (10,800 g/mol and 31,500 g/mol). By comparison, the copolymer based on PDMS of higher M_n displays higher elongation at break and ultimate tensile strength [5]. Sheth et al. prepared segmented polyurea copolymers with three chain extenders including ethylene diamine (EDA), 1,6-diaminohexane, and 2-methyl-1,5-diaminopentane. The results show that the copolymers with shorter and symmetric chain extenders (EDA) display higher strength and wider temperature ranges [8]. To expand the end uses of the siloxane polyurea elastomers, it is vital to further study the copolymer structure–property relationships.

The fixed phases of the TPU are constructed by the hard domains through the hydrogen bonding interactions. In the conventional TPU using polyol monomers, the single ligands of hydrogen bonding are primarily attributed to the urethane N-H group as proton donors and the C=O or ester O groups as the proton acceptors [9]. The hydrogen bonding is influenced by the chemical structures and contents of hard segments [10], chain extenders [11], and M_w of the soft segments, thus impacting the microphase structures and physical properties of the material. For instance, the fluoridated TPU elastomers were synthesized utilizing 4,4′-[2,2,2-trifluoro-1-(trifluoromethyl)-ethylidene] bisphenol as the chain extender. It was found that the hydrogen bonding strength increased with the higher fluorine content, causing enhanced microphase separation and decreased transmittance for the elastomers [11]. In addition to polyol monomers, when introducing amine-terminated oligomers, the synthesized TPUs have different hard domains [12–14]. In detail, the double N-H groups in urea segments result in the presence of both single and ligands bidentate in the hydrogen bonding, implying stronger hard domain cohesion [15]. Incorporating amine-terminated polysiloxane into the copolymer may enhance the hydrogen bonding interaction through urea groups, leading to pronounced microphase separation, higher T_g, and better elasticity [16,17]. Further knowledge about the effects of the polysiloxane structure on the hydrogen bonding is crucial to preparing the copolymer elastomers with the controlled structures and properties.

In this work, the siloxane polyurea copolymer elastomers are synthesized by introducing two aminopropyl-terminated polysiloxane oligomers with low M_n (L-30D) and high M_n (H-130D). The effects of the copolymer structures on the film properties are further analyzed based on the performance including the tensile performance, transmittance of UV-vis spectra, cross-sectional topographies, and T_g. The two thermoplastic elastomer films are characterized by transparency, ductility, and the T_g of the hard domains. Furthermore, the elasticity behavior of the copolymer films is studied by the strain recovery and cyclic tensile test, and the linear fitting based on the Mooney–Rivlin model is conducted to analyze the film elasticity. Moreover, the temperature-dependent IR spectra during heating and cooling are utilized to monitor the shifts of carbonyl peaks, revealing the effects of the copolymer structures on the strength and recovery rate of the hydrogen bonding.

2. Experimental Section

2.1. Materials

Aminopropyl-terminated PDMS (L-30D: M_n is 2200 g/mol; H-130D: M_n is 9800 g/mol) was purchased from Waker Inc. (Munich, Germany), which was dried at 110 °C for 2 h before use. Isophorone diisocyanate (IPDI, 99%), and dibutyltin dilaurate (DBTDL, 95%) were bought from Aladdin Reagents Co. Ltd. (Shanghai, China). 1,3-Bis(3-aminopropyl) tetramethyldisiloxane (APTMS, 96%) was obtained from Rhwan Chemical Co. Ltd. (Shanghai, China) 1,4-Butanediol (BDO, purchased from Sinopharm Chemical Reagent Co. Ltd. (Shanghai, China)), Tetrahydrofuran (THF, 99.5%, anhydrous) was bought from Adamas Reagents Co. Ltd. (Shanghai, China).

2.2. Synthesis of the Siloxane Polyurea Copolymer Elastomers

A series of siloxane urea copolymers using aminopropyl-terminated PDMS were synthesized through two steps, as illustrated in Scheme 1. Initially, the prepolymer was prepared by charging a mixture of PDMS oligomer (L-30D or H-130D) and THF into the four-necked round bottom flask, which was equipped with a nitrogen venting device. Then, the diluent containing IPDI and THF was added dropwise to the flask, followed by stirring the mixture for 2 h at room temperature. In the second step, the THF solution containing APTMS as the chain extender was dropped into the reacting mixture for four batches, with the mass contents of approximately 50%, 20% twice, and 10%, respectively. The reaction was allowed to proceed for 5 h, and the copolymer solutions with about 25% solid contents were obtained. For the copolymer synthesized using L-30D and H-130D, the mass contents of IPDI were set as 11%, 13%, and 18%, respectively, with their reactant compositions listed in Table 1.

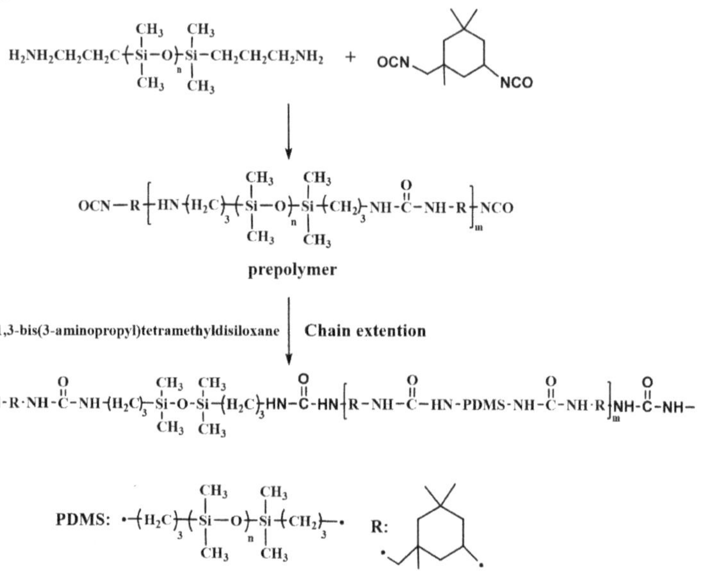

Scheme 1. The schematic representation of the synthesis of siloxane polyurea copolymers using aminopropyl-terminated PDMS.

Table 1. The reactant compositions of the copolymer using aminopropyl-terminated PDMS.

Samples	PDMS [mmol]	IPDI [mmol]	APTMS [mmol]	IPDI Content [wt%]	[a] Hard Segment content [wt%]
L-30D-11%	5.0	6.4	1.4	11	13
L-30D-13%	5.0	8.7	8.7	13	20
L-30D-18%	5.0	13.4	8.4	18	30
H-130D-11%	1.5	9.0	7.5	11	21
H-130D-13%	1.5	12.0	10.5	13	26
H-130D-18%	1.5	18.8	17.3	18	36

[a] Hard segment content is the mass percentage of IPDI and APTMS among the total components.

2.3. Preparation of the Copolymer Elastomer Films

The copolymer solid products were prepared by casting the copolymer solutions into polytetrafluoroethylene molds at room temperature for 24 h, followed by drying at 50 °C overnight. The dried solid products were hot-pressed at 110 °C for 5 min, and the films about 1 mm thick were obtained.

2.4. Characterization

The Fourier transform infrared (FTIR) spectra of the reacting products were recorded using a Nicolet 6700 from Thermo Fisher Scientific Inc. (Waltham, MA, USA) in ATR mode with the diamond as the crystal on the attachment, covering the range from 4000 cm^{-1} to 525 cm^{-1} with a resolution of 4 cm^{-1}. Meanwhile, the structure of the films was carried out in transmission mode. For the temperature scans, the thin films were placed in a heating cell that induced a hole to put the KBr tablet and maintain the transmittance. The heating rate was 5 °C per minute, followed by natural cooling at 20 °C. All measurements were conducted under a nitrogen atmosphere. The molecular weight of the copolymer was obtained using the gel permeation chromatography (GPC) on an Agilent 1206 instrument (Santa Clara, CA, USA). The samples were obtained by drying the copolymer solutions, and then solved in THF with the mass contents of 0.5%. The test was carried out using THF as the mobile phase with a flow rate of 1 mL/min.

The tensile tests were carried out using an Instron universal testing machine (Boston, MA, USA) with the specification of GBT 1040.3-2006. The dumbbell-shaped samples were utilized with the size of 50 mm × 4 mm × 1 mm, and the crosshead speed was 20 mm/min. Five samples were tested for each data point, and the average result was recorded. To evaluate the film resilience, the samples were stretched to strains of 300% or 200%. Then, one side of the samples was released from the clamp, followed by recording the strains versus time.

The cyclic tensile tests were performed at a maximum strain of 50% at a speed of 20 mm/min for 5 full cycles, with a 30 s rest between each cycle. The percentage of hysteresis was calculated by the curves of stress versus strain, as shown in the following formula [7].

$$\%hysteresis = \frac{A_{loading} - A_{recovery}}{A_{loading}} \times 100\% \quad (1)$$

where $A_{loading}$ is equal to the area increments under loading curves, and $A_{recovery}$ represents the area under recovery curves in the same cycle.

Dynamic mechanical analysis (DMA) was conducted using a QC800 analyzer from TA Company (Newcastle, WA, USA). The experiments were measured with a small tension clamp, heating the samples from −123 °C to 150 °C at a heating rate of 3 °C/min and a frequency of 1 Hz.

The film transmittance was obtained by Lambda750 ultraviolet-visible light (UV-vis) spectrometer from PerkinElmer (Waltham, MA, USA), with the wavelength ranging from 350 nm to 800 nm. The cross-sectional topography was imaged by a Tescan (Oberkochen, Germany) LaB_6 scanning electron microscope (SEM). The cross-sections of the samples were obtained by fracturing the films in liquid nitrogen. Before testing, the cross-sections were gold-plated.

3. Results and Discussion

3.1. The Structural Characterization of the Siloxane Polyurea Copolymer Films

Herein, the L-30D and H-130D oligomers were used to synthesize the siloxane polyurea copolymer, respectively. Scheme 1 gives the synthetic route of the copolymers. Figure 1a,b give the FTIR spectra of the copolymer films synthesized with L-30D and H-130D, respectively. The characteristic peaks of polysiloxane at 1263 cm^{-1} and 793 cm^{-1} are assigned to the in-plane bending vibration of the CH_3 connecting with Si and the stretching vibration of Si-O, respectively [18], and the peaks at 2000 cm^{-1} can be attributed to the group of Si-CH=CH_2, which might derive from the siloxane raw materials. For the copolymer films with the same monomer, the absorption peak intensities of the urea groups (3345 cm^{-1}, 1630 cm^{-1}, and 1570 cm^{-1}) rise with the increasing IPDI contents. The results show that the urea group ratio in the copolymer increases with higher IPDI concentration. In addition, the results of NMR spectra for the oligomers and copolymers are shown in Supplementary Figure S1, which agrees with the FTIR results. By comparison, at the same IPDI content, the copolymer using H-130D displays higher relative peak intensities of the urea groups, reflecting more hard segment contents in the copolymer (Table 1). For the copolymer films using L-30D, the M_n is 2.1 × 10^4 g/mol (11% IPDI), 1.6 × 10^4 g/mol (13% IPDI), and 1.6 × 10^4 g/mol (18% IPDI), respectively, with a distribution index ranging from 1.4 to 1.7. At the IPDI contents of 11%, 13%, and 18%, the M_n of the synthetic films using H-130D is 8.2 × 10^4 g/mol, 7.8 × 10^4 g/mol, and 5.2 × 10^4 g/mol, respectively, with a distribution index between 1.7 and 1.8. In comparison with the copolymers using L-30D, the M_n of the copolymers is significantly enhanced, which might be attributed to the introduction of more APTMS extenders to ensure the same IPDI contents. We can conclude that the siloxane polyurea copolymers have been synthesized.

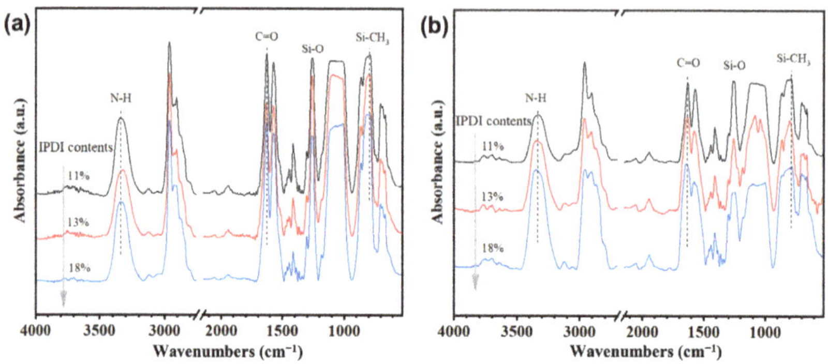

Figure 1. The FTIR spectra of the siloxane polyurea copolymer using L-30D (**a**) and H-130D (**b**).

3.1.1. The Mechanical Properties of the Copolymer Films

The films were prepared from the synthetic copolymer products to study their mechanical performance, transmittance, and morphology. Figure 2a,b show the stress–strain curves of the films using L-30D and H-130D, respectively. When the L-30D is used as the soft segment, the films display excellent ductility with an elongation at break of more than 700%, which is higher than that of urea elastomer in the previous works [19–21]. Notably, the films with an IPDI content of 11% exhibit an elongation at break of more than 4000%. The tensile strength of the films ranges from 0.08 MPa to 0.33 MPa. As for the films using H-130D, with the IPDI content increasing, the elongation at break of the films decreases from roughly 670% to 250%, with the tensile strength enhanced from 0.6 MPa to 1.3 MPa. Compared with the copolymers containing L-30D, at the equal IPDI contents, the films using H-130D have much higher tensile strength, which indicates that the films using H-130D might have more physical cross-linked structures.

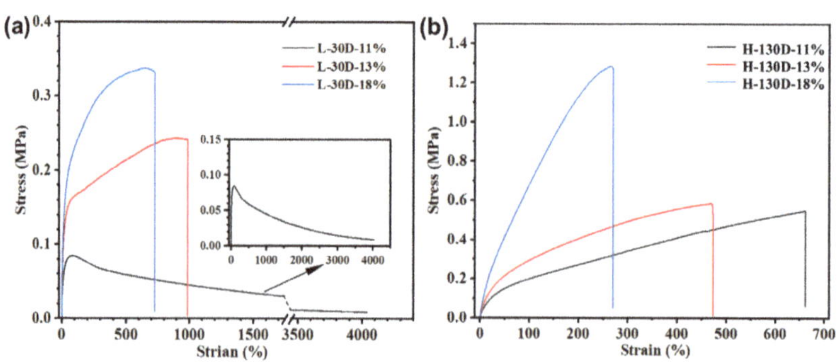

Figure 2. The tensile stress–strain curves of the copolymer films using L-30D (**a**) and H-130D (**b**).

The areas under the stress–strain curves before fracture can represent the work of the tension (W), reflecting the energy absorption and toughness of the material [22,23]. For the films synthesized with L-30D, when the IPDI contents of the films are 11%, 13%, and 18%, the work of tension (W) is 0.15 N·m, 0.20 N·m, and 0.21 N·m, respectively. The work of tension (W) for the films using H-130D exhibits the values of 0.29 N·m, 0.16 N·m, and 0.26 N·m, respectively, when the IPDI contents of 11%, 13%, and 18%. Clearly, the H-130D-11% films exhibit the highest toughness.

The T_g of the films can be analyzed through the DMA test. The results in Supplementary Figure S2 indicate that all samples have a rubbery plateau, followed by a rapid decline in the storage modulus with rising temperature, which agrees with the characterization of the thermoplastic elastomer. At the same IPDI content, the films using H-130D show a wider rubbery plateau, ranging from roughly −40 °C to around 70 °C (Supplementary Figure S2), indicating their wider working temperature ranges. The films have two T_g in total, as shown in Figure 3. The T_g at low temperature is approximately −115 °C, which is related to the soft segments of polysiloxane chains [7,8]. Another T_g at high temperature is ascribed to the hard domains. For the films using L-30D, the results in Figure 3a indicate that, when the IPDI content rises from 11% to 18%, the T_g of the hard domains increases from around 45 °C to 86 °C, which implies that the various IPDI contents can influence the hard domain structures. As shown in Figure 3b, the films using H-130D also display an increasing T_g of the hard domains ranging from 89 °C to 135 °C, as the IPDI content increases; moreover, the minor peaks at roughly −51 °C are attributed to the crystallinity of the polysiloxane that has high M_n, which is close to the reported melting temperature for the polysiloxane crystalline region [8]. In compassion, there were no obvious peaks at around −50 °C for the copolymer using L-30D due to the poor crystallizing ability of the L-30D with low M_n.

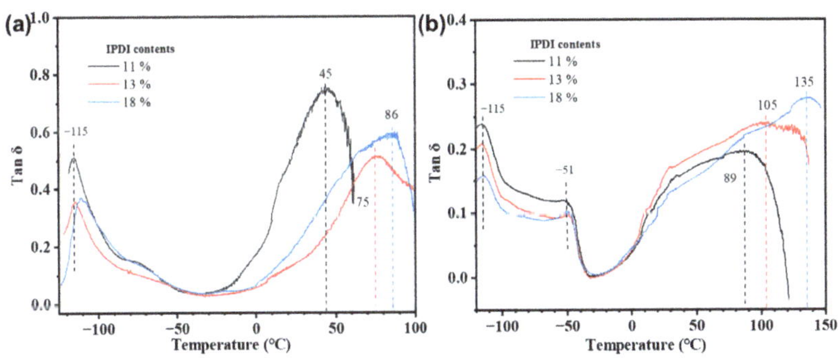

Figure 3. The loss factor (tan δ) versus temperature of the films using L-30D (**a**) and H-130D (**b**).

3.1.2. The Transparency of the Copolymer Films

Figure 4a gives the UV-vis absorption spectra of the films synthesized with L-30D. When the IPDI contents are 11% and 13%, the films show quite high transparency at visible light, with light transmittance of roughly 99% and 96% at the wavelengths from 500 nm to 800 nm, respectively, which is similar to the transmittance of non-siloxane block polyurethane coatings described in the literature [24,25]. As the IPDI content increases to 18%, the films display a light transmittance ranging from 80% to 68%. As presented in Figure 4b, the films using H-130D display the transmittance below 80%. For the L-30D films using 11% and 13% contents of IPDI, the characters on the bottom of the films can be seen clearly, which agrees with the results of the higher transmittance. It is known that the transmittance is affected by the phase separation structure of the material containing multiple compounds [26,27]. In the siloxane polyurea copolymer, there is a notable difference in the polarity among the soft domains of polysiloxane and the hard domains formed by hydrogen bonding. The solubility parameters for the soft and hard domains are 15.6 $(J{\cdot}cm^{-3})^{1/2}$ and 45.6 $(J{\cdot}cm^{-3})^{1/2}$, respectively, showing a phase-separated trend [28]. Conclusively, the copolymer films exhibit decreased transmittance when the H-130D is introduced, implying that the films might have higher degree of the phase separation that leads to more light scattering or reflection.

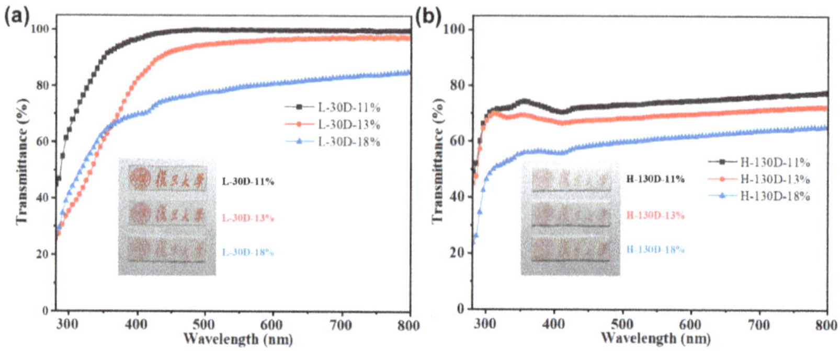

Figure 4. The UV-vis absorption spectra and photos of the films using L-30D (**a**) and H-130D (**b**). The characters covered by the films mean Fuand University in Chinese.

Furthermore, the SEM has been used to study the cross-sectional topographies of the synthetic films after brittle fracture in liquid nitrogen. Figure 5 images the cross-sectional topographies of the films using L-30D. When the IPDI content is 11%, the cross-section is flat and smooth; as the IPDI content increases to 13%, little particles appear on the cross-section. In contrast, when the IPDI content is raised to 18%, the topography changes distinctly, resulting in more obvious heterogeneous phase structures; specifically, little particles are distributed with the size from 1 μm to 2 μm. These results are consistent with the film transmittance test in Figure 4a. The SEM images in Figure 6 display the cross-sectional topographies of the films using H-130D. The results indicate that the films display various heterogeneous structures, characterized by multiple dispersed phases ranging scatter across the cross-sections, and the number of the dispersed phases increases with the higher IPDI content. By comparison, the thermoplastic copolymer films based on H-130D exhibit much lower transmittance and a higher degree of phase separation at the same IPDI content.

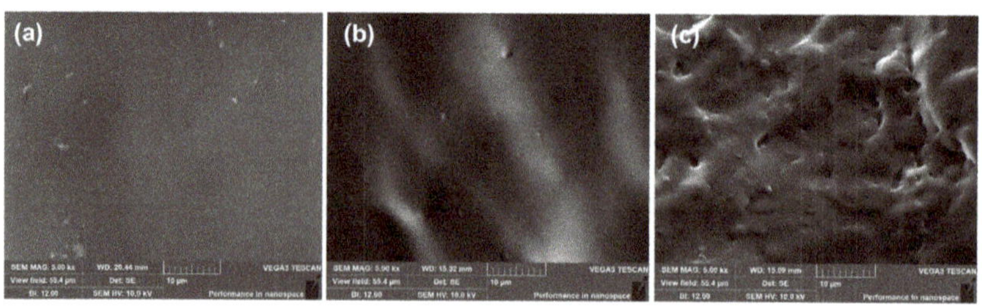

Figure 5. The cross-sectional topography of SEM for the films using L-30D, the IPDI contents: (**a**) 11%, (**b**) 13%, (**c**) 18%.

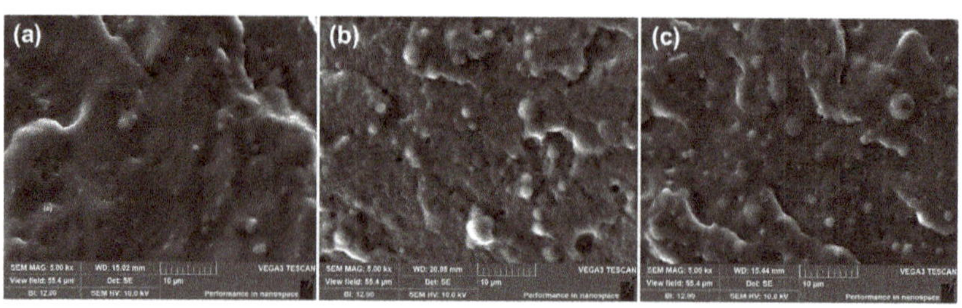

Figure 6. The cross-sectional topography of SEM for the films using H-130D, the IPDI contents: (**a**) 11%, (**b**) 13%, (**c**) 18%.

3.2. The Modeling Analysis of the High Elasticity for the Copolymer Films

The tensile recovery and cyclic tensile tests are utilized to characterize the elasticity of the copolymer films. The results in Supplementary Figure S3 indicate that the strain can recover to roughly 60% at the greatest level for the films using L-30D, while the films containing H-130D exhibit remarkable resilience with the ability to recover to less than 10% strain. The cyclic tensile curves are presented in Figure 7, with the hysteresis listed in Table 2. The hysteresis is calculated by the Formula (1) and represents the energy dissipation of the films. The results show that the films have the highest hysteresis in the first cycle, which is caused by the energy dissipation during the tensile process. The energy dissipation is primarily formed by the dissociation of the hydrogen bonding, the deformation of soft segments, and the interaction of the hard and soft segments [29,30]. It is seen that the film hysteresis tends to decrease with the increase of IPDI contents, indicating that the increasing IPDI contents may affect the film phase structure. Compared with the films using L-30D, the films containing H-130D display significantly lower hysteresis at the same IPDI contents, and their differences enlarge with the increasing IPDI contents. The result might be ascribed to fewer breakages of the hydrogen bonding owing to the hard domains connected by soft segments with higher M_n [25].

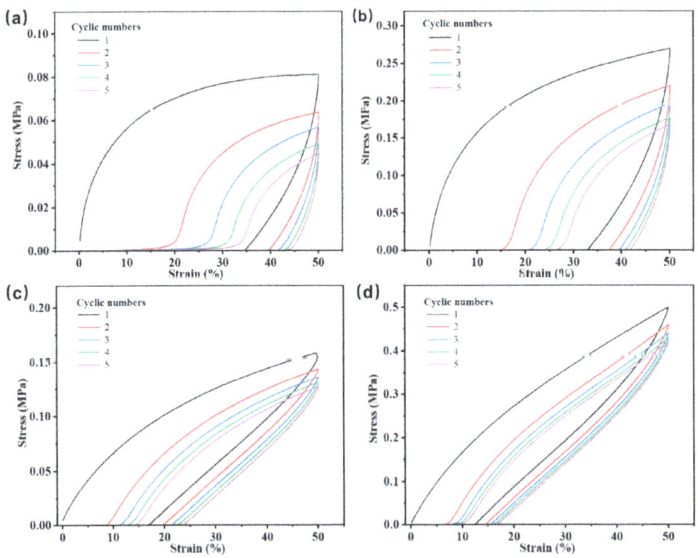

Figure 7. The cyclic tensile curves of the samples: (**a**) L-30D-11%, (**b**) L-30D-18%, (**c**) H-130D-11%, and (**d**) H-130D-18%.

Table 2. The hysteresis values of the cyclic tensile curves for the copolymer films.

Samples/Cycles	L-30D-11%	L-30D-18%	H-130D-11%	H-130D-18%
1	87.6	84.7	53.3	45.3
3	84.2	81.7	45.7	35.5
5	84.0	81.5	45.3	35.0

As for the copolymer films using the two polysiloxane monomers, the results mentioned before show notable distinctions in many aspects, such as the limited highest temperature (T_g), toughness (W), ductility, resilience, and the hysteresis of the cyclic tensile process. These results reflect the influence of the oligomer structures and isocyanate contents on the mechanical performance, thermal properties, and transparency of the copolymer films. Having realized that the copolymer properties are mostly affected by the hydrogen bonding interaction, we investigate the hydrogen bonding by the modeling fitting and the temperature-dependent IR spectroscopy to study the effect of the copolymer structures on the film properties.

For the ideal highly elastic materials that are isotropic in the undeformed state, the physical properties can be specified by the stored energy function W. Based on the Mooney–Rivlin model [29,30], the strain during the deforming process in a single direction can be described by the Formula (2).

$$\delta = (\lambda - \frac{1}{\lambda^2})(2C + 2k\frac{1}{\lambda}) \tag{2}$$

where λ represents the strain in the tensile direction ($\lambda \neq 0$), and δ is calculated through the W as a function of the strain invariants (λ). C and k are both the modeling parameters; the parameter C reflects the cross-linked degree of the material, and its value increases with the cross-linked degree.

The Formula (2) can be transformed according to the following formula:

$$[f^*] = \delta / \left(\lambda - \frac{1}{\lambda^2}\right) = 2C + 2k\frac{1}{\lambda} \tag{3}$$

where $[f^*]$ is named the contrastive strain.

As for the copolymer films using H-130D, based on the tensile strain–stress data within the strain range from 10% to 100%, the contrastive strain $[f^*]$ is calculated firstly, followed by obtaining the function curves of $[f^*]$ and λ^{-1} with the $[f^*]$ and λ^{-1} as Y-axis and X-axis, respectively, as shown in Figure 8. It is seen that the $[f^*]$ and λ^{-1} approximately conform to the linear relation, which indicates that the Mooney–Rivlin model can describe the elasticity of the siloxane polyurea copolymer. Furthermore, the linear fitting is conducted according to the Formula (3), with the calculated model parameters of C, k, and R^2 listed in Table 3. The R^2 of all the samples is higher than 0.92, certifying the reliability of the fitting results.

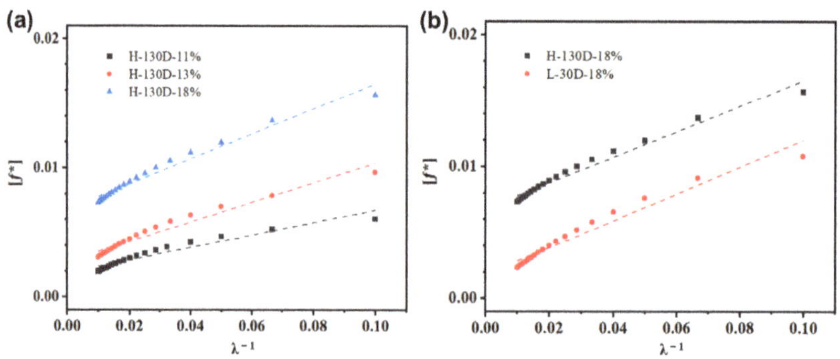

Figure 8. The scatters and fitted lines (dashes) of $[f^*]$ versus λ^{-1} for the copolymer films using H-130D (**a**) and the films by the oligomers of different M_n (**b**).

Table 3. The fitted parameters used for the Mooney–Rivlin model.

Samples	C (MP_a)	k (MP_a)	^{a} R^2
L-30D-18%	1.0×10^{-3}	0.018	0.95
H-130D-18%	3.5×10^{-3}	0.049	0.97
H-130D-13%	1.4×10^{-3}	0.038	0.96
H-130D-11%	1.0×10^{-3}	0.025	0.92

[a] R^2 represents the linear regression index of the fitting curves.

The parameter 2C is the Y-intercept for the fitted lines of the $[f^*]$ and λ^{-1}. The results in Figure 8a indicate that the model parameter C is enhanced with the increase of IPDI contents. Based on the model, the value of C is positively related to the elastomer cross-linked degree, which indicates that the synthetic polyurea copolymer films display a higher cross-linked degree as the IPDI content increases. In the TPU materials, the physical crosslinking is mostly composed of the interaction of hydrogen bonding. Therefore, the changing trend of the parameter C indicates that the hydrogen bonding interaction is improved with the IPDI content increasing.

Moreover, at the same experimental conditions with the IPDI contents of 18%, the tensile data for the films using two oligomers separately are also fitted and calculated, with the results shown in Figure 8 and Table 3. At the same IPDI content, when using the oligomer of low M_n (L-30D), the synthetic film displays a smaller parameter C, suggesting that the film displays fewer crosslinking due to the weaker hydrogen bonding interaction.

3.3. Analyzing the Hydrogen Bonding of the Hard Domains by Using Temperature-Dependent IR Spectroscopy

In the polyurea materials, the soft segments of the copolymer are typically composed of diamine oligomers, and the hard segments are formed by the chains synthesized by the isocyanate and extender. Due to the hydrogen bonding, the hard segments tend to aggregate as hard domains, resulting in the microphase separation structure. The FTIR spectra are an efficient tool for analyzing the hydrogen bonding interactions. For example, the formation of hydrogen bonding may cause a shift in the carbonyl peaks towards lower wavenumbers. According to the literature results [27,31], the polyurea copolymer contains three carbonyl states: free (~1670 cm^{-1}), disordered (~1645 cm^{-1}), and ordered (~ 1620 cm^{-1}). The free carbonyl is without the hydrogen bonding, while the disordered and ordered carbonyls are formed by the hydrogen bonding of single and bidentate ligands, respectively.

The peaks in the IR spectra exhibit different wavenumbers for the carbonyl with or without the hydrogen bonding. Therefore, as the samples are heated, the weakening of hydrogen bonding can result in shifts in the carbonyl peak [32]. Based on this theory, when the films were heated from room temperature to 150 °C, the IR spectra were tested at various temperatures, with the change in carbonyl peaks presented in Figure 9 and Supplementary Figure S4. The carbonyl peaks in all copolymer samples exhibit a shift towards higher wavenumbers. The results suggest that the hydrogen bonding is dissipated upon heating, leading to an increasing amount of the free carbonyls.

Considering that the carbonyl peaks associated with three states are detected at various wavenumbers, the carbonyl peaks can be peak fitted, as shown in Figure 9c. These results reveal that the free carbonyl peaks in all samples rise in percentage to various extents during heating. Notably, at the same temperature, the free carbonyl contents decline as the IPDI contents increase, certifying that the carbonyl with hydrogen bonding is enhanced in percentage. Furthermore, with the same IPDI amount (18%), the temperature-dependent IR spectra were detected for the samples using different oligomers. For the copolymer containing H-130D, the free carbonyl percentage at the same temperature is much lower than that of the copolymer using L-30D. It implies that the copolymers containing H-130D have stronger hydrogen bonding interaction.

The experimental evidence has exhibited the various performances of the films using L-30D and H-130D, which are correlated to the hydrogen bonding strength of the hard domains. The stronger hydrogen bonding may produce more separated hard domains, implying that the films using H-130D display more structures of heterogeneous phase (Figure 5), which can explain the relatively low visible-light transmittance of the films (seen in Figure 4b). As reported, the microphase separation of hard segments generally forms the physical crosslinking as the fixed phase in the TPU materials [1,2]. Obviously, the films containing L-30D display less cross-linking owing to the weaker interaction of the hydrogen bonding, which results in lower tensile strength and T_g, as well as better ductility with the elongation at break of more than 700%. Meanwhile, compared with the films containing H-130D, the films using L-30D show the visible stress yield within 100% strain (seen in Figure 2a), which may be attributed to the unstable hard domains due to the relatively few hydrogen bonding. Moreover, the more separated hard domains can contribute to the better toughness of the films with H-130D. Additionally, the toughness of the TPU is also correlated to the entanglement of polymer chains.

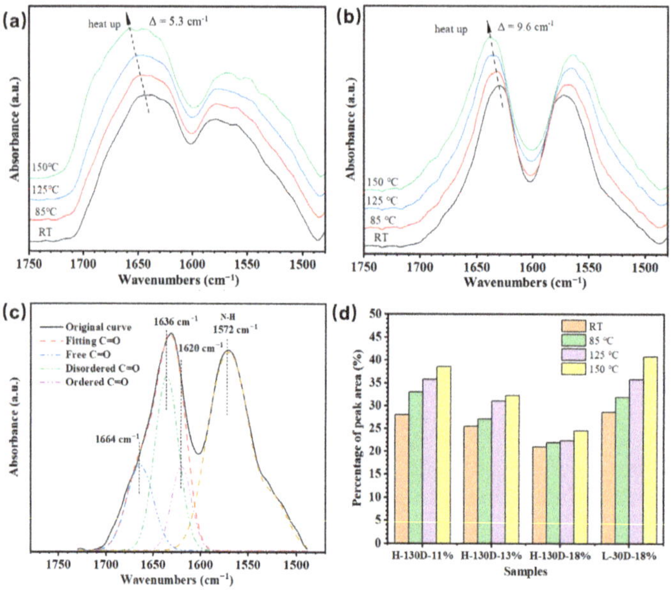

Figure 9. The FTIR spectra of the carbonyl during heating for the films using H-130D (**a**) and L-30D (**b**) at the IPDI content of 18%. (**c**) The fitting example of carbonyl peaks in IR spectra for the films. (**d**) The area percentage of free carbonyl peaks at rising temperatures for these samples.

The deformation of the hard domains may affect the elastic recovery rate and degree of the films. Meanwhile, heating the films can disrupt the hard domains by dissipating the hydrogen bonding. To study the recovery ability of the hydrogen bonding, the temperature-dependent IR spectra are utilized to analyze the changing rule of carbonyl peaks during cooling. The variation of carbonyl peaks is closely associated with the relaxation ability of the hard segments.

The films were heated to 150 °C under the same condition and then cooled naturally. The IR spectra were recorded at various cooling times to observe the change in carbonyl peaks, as shown in Figures 10 and 11. The results indicate that the carbonyl peak of each film shifts toward lower wavenumbers during cooling, implying the hydrogen bonding recovery. The recovery percentage (φ) of carbonyl peaks is defined using the following formula.

$$\varphi = \frac{\lambda_0 - \lambda_t}{\lambda_0 - \lambda_{RT}} \times 100\% \quad (4)$$

where λ_0, λ_t represent the wavenumbers of carbonyl peaks at the start of cooling and after t minutes of cooling, respectively, and λ_{RT} is the wavenumber of carbonyl peaks at room temperature. Herein, φ can qualitatively describe the recovery degree of hydrogen bonding.

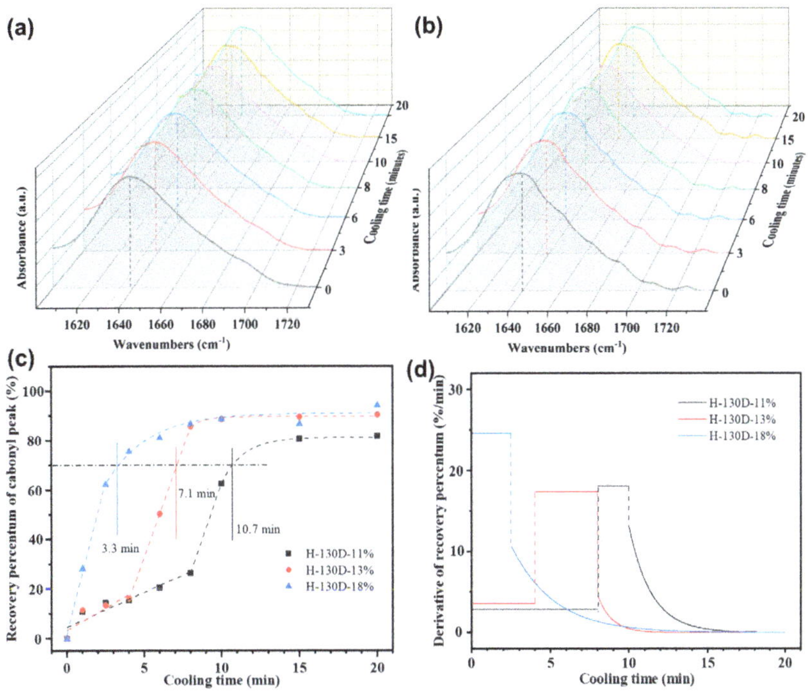

Figure 10. The FTIR spectra of the carbonyls during cooling for the films using H-130D, the IPDI contents: (**a**) 11%; (**b**) 13%. The recovery percentage curves (**c**) and the recovery rate (**d**) of the carbonyl peaks for the films using various IPDI contents.

Figure 10c presents the recovery percentage curves for the films using different IPDI contents, with the oligomer being H-130D. It is evident that φ increases with the higher IPDI contents, exhibiting more obvious differences at the beginning of cooling. Based on the time to need for reaching 70% (φ = 70%) of the original carbonyl peak ($t_{0.7}$), the macroscopic recovery rate of the hydrogen bonding can be compared quantitatively. As shown in Table 4, the required time ($t_{0.7}$) for the samples declines from 10.7 min to 3.3 min, as the IPDI content increases from 11% to 18%.

Table 4. The average M_n of hard segment chains and the required time for reaching φ = 70%.

Samples	$\overline{M}_H$ [g/mol]	$t_{0.7}$ [min]
H-130D-11%	2102	10.7
H-130D-13%	2651	7.1
H-130D-18%	3609	3.3

According to Formula (5) [6,33], we have calculated the $\overline{M}_n$ of the hard segment chains, i.e., $\overline{M}_H$.

$$\overline{M}_H = M_S \times \omega + 2M_0 \quad (5)$$

where M_s represents the M_w of the polysiloxane oligomers, and ω is the ratio of the mass of the hard segments consisting of isocyanates and extenders to the total mass of the reactants. M_0 takes the value of 16 g/mol, which indicates M_n of the amino (-NH_2) group in the urea group at the two ends of the hard segment chain.

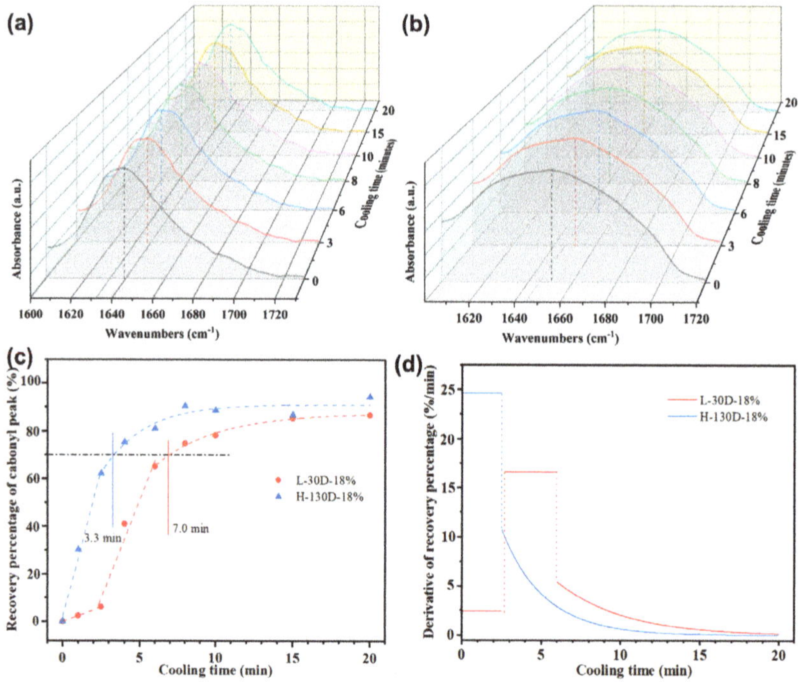

Figure 11. The FTIR spectra of the carbonyls during cooling for the films using L-30D (**a**) and H-130D (**b**). The recovery percentage curves (**c**) and the recovery rate (**d**) for the carbonyl peaks of these films.

Table 4 presents the $\overline{M}_H$ of the polyurea copolymers synthesized with H-130D and various IPDI contents. It is observed that the $\overline{M}_H$ is negatively related to the $t_{0.7}$, meaning that the required time of recovery declines with the higher $\overline{M}_H$. In theory, it is more possible to aggregate the polymer chains with higher M_n. Therefore, $\overline{M}_H$ is a significant factor that influences the hydrogen bonding recovery.

To further study the hydrogen bonding recovery, the derivative of the φ against t curves is calculated to obtain the function relation of φ' ($d\varphi/dt$) and cooling time, which represents the instantaneous recovery rate. As for the H-130D-18% film, the hydrogen bonding rapidly reverts during cooling, followed by a decrease in the recovery rate. Separately, the other copolymers containing H-130D display a very low φ' within 8 min, indicating minimal hydrogen bonding recovery at the start of cooling; afterward, a relatively high constant rate is observed, followed by a gradual slowdown. Based on these results, the hydrogen bonding recovery can be categorized into three stages: the initial stage, with a very low φ'; the second stage, having a constant rate; and the third state, with a declining rate. The duration of the initial stage varies for each film, and the retention time for the H-130D-18% films can be regarded as zero. Figure 10d illustrates that when the IPDI contents increase from 11% to 18%, the sample retention time in the initial stage decreases from 7.2 min to 0 min. Furthermore, the recovery rates at the second stage are nearly identical. Based on the above results, it can be concluded that the notable difference in the initial stage of φ' leads to variations in the recovery time ($t_{0.7}$). The polymers with higher molecular weight have a greater tendency to aggregate, thereby the retention time at the initial stage should be influenced by the $\overline{M}_H$ for the copolymer. Moreover, the T_g also impacts the relaxation time of the polymer chains, indicating that the recovery of the hydrogen bonding may be influenced by both the hard segments and T_g.

Additionally, at the same IPDI contents of 18%, the synthetic films using L-30D and H-130D were characterized by the temperature-dependent IR spectra during cooling. The relation curves of φ' and t are analyzed as before, with the results shown in Figure 11. The copolymer films with L-30D require a longer recovery time, with a difference in $t_{0.7}$ between the two samples of 4 min. Moreover, the calculated $\overline{M}_H$ of the L-30D copolymer is 701 g/mol, which is far less than that of the films using H-130D. The result still agrees with the negative correlation of the $\overline{M}_H$ and $t_{0.7}$.

During the tensile process at large strain, the breakage of hydrogen bonding occurs in the deformation of hard domains, leading to energy dissipation [34]. Similar to the heating and cooling process, the dissipation and recovery of the hydrogen bonding occur during the cyclic tensile process, indicating the deformation and reorganization of the hard domains. For the films containing H-130D, the stronger hydrogen bonding and the faster recovery rate reflect less irreversible deformation of the hard domains, reflecting less energy dissipation (Table 2). The rule is also suitable for the films using the same oligomer with higher IPDI contents. Moreover, the film resilience is determined by the relaxation of polymer chains, which are mainly influenced by the recovery rate of the deformed hard domains. Evidently, the films using H-130D show faster recovery rates of the hydrogen bonding, consistent with their better resilience (Supplementary Figure S3).

4. Conclusions

We have synthesized the siloxane polyurea copolymer elastomers using two oligomers with low M_n (L-30D) and high M_n (H-130D) as soft segments, respectively. Further, the influence of the copolymer structures on the film properties is studied by the Mooney–Rivlin model and the temperature-dependent IR spectroscopy. The two kinds of thermoplastic elastomer films are characterized by ductility, transparency, and T_g of the hard domains, respectively. Moreover, the film transmittance is correlated to the phase structures.

The elasticity behavior of the copolymer films is studied by the strain recovery and cyclic tensile test, and the Mooney–Rivlin modeling fitting based on tensile data can describe the film elasticity, correlated to the degree of physical cross-linking. The results of the temperature-dependent IR spectra during heating show that the copolymers display stronger hydrogen bonding strength when using the higher diisocyanate contents or the oligomer H-130D, which can explain the influence on the film properties. Further, the IR spectra during cooling are utilized to analyze the hydrogen bonding recovery. For the copolymers using H-130D, the $\overline{M}_H$ is negatively related to the recovery rate of the hydrogen bonding, which is also suitable for the films using two various monomers at the same IPDI contents. The hydrogen bonding interactions can explain the distinctions of the copolymer film properties; specially, the lower hysteresis of the H-130D films is attributed to the strong hydrogen bonding and the great stability of the hard domains. These results can be helpful in designing other TPU materials.

Supplementary Materials: The following supporting information can be downloaded at: https://www.mdpi.com/article/10.3390/polym16172438/s1, Figure S1: The 1H-NMR of the samples: (a) the oligomer L-30D, (b) the oligomer H-130D, (c, d) the copolymer L-30D-11% and L-30D-18%; (e, f) the copolymer H-130D-11% and H-130D-18%. Figure S2: The storage modulus versus temperature of the films using L-30D (a) and H-130D (b). Figure S3: The strain recovery curves of the films using L-30D (a) and H-130D (b). Figure S4: The FTIR spectra of the carbonyl during heating for the films using H-130D with the IPDI contents of 11% (a) and 13% (b).

Author Contributions: Conceptualization, X.N. and M.B.; methodology, X.N. and M.B.; software, M.B. and T.L.; validation, X.N., M.B. and T.L.; formal analysis, M.B.; investigation, X.N. and M.B.; resources, X.N. and M.B.; data curation, M.B. and Y.T.; writing—original draft preparation, M.B., Y.T. and T.L.; writing—review and editing, M.B. and X.N.; visualization, M.B.; supervision, X.N.; project administration, X.N.; funding acquisition, X.N. All authors have read and agreed to the published version of the manuscript.

Funding: This research received no external funding.

Institutional Review Board Statement: Not applicable.

Data Availability Statement: The raw data supporting the conclusions of this article will be made available by the authors on request.

Conflicts of Interest: The authors declare no conflict of interest.

References

1. Drobny, J.G. *Handbook of Thermoplastic Elastomers*; Elsevier: London, UK, 2007; pp. 1–435.
2. Tato, W.; Urchegui, M.A.; Martínez, A.B.; Aginagalde, A.; Bartolomé, L. Experimental Characterization and Modelling of Large-Strain Viscoelastic Behavior of a Thermoplastic Polyurethane Elastomer. *Rubber. Chem. Technol.* **2013**, *86*, 146–164.
3. Basak, S. Thermoplastic elastomers in biomedical industry—Evolution and current trends. *J. Macromol. Sci. A* **2021**, *58*, 579–593. [CrossRef]
4. Costa, N.L.; Hiranobe, C.T.; Cardim, H.P.; Dognani, G.; Sanchez, J.C.; Carvalho, J.J.; Torres, G.B.; Paim, L.L.; Pinto, L.F.; Cardim, G.P.; et al. A Review of EPDM (Ethylene Propylene Diene Monomer) Rubber-Based Nanocomposites: Properties and Progress. *Polymers* **2024**, *16*, 1720. [CrossRef] [PubMed]
5. Yilgor, I.; Eynur, T.; Yilgor, E.; Wilkes, G.L. Contribution of soft segment entanglement on the tensile properties of silicone–urea copolymers with low hard segment contents. *Polymer* **2009**, *50*, 4432–4437. [CrossRef]
6. Riehle, N.; Thude, S.; Götz, T.; Kandelbauer, A.; Thanos, S.; Tovar GE, M.; Lorenz, G. Influence of PDMS molecular weight on transparency and mechanical properties of soft polysiloxane-urea-elastomers for intraocular lens application. *Eur. Polym. J.* **2018**, *101*, 190–201. [CrossRef]
7. Sirrine, J.M.; Schexnayder, S.A.; Dennis, J.M.; Long, T.E. Urea as a monomer for isocyanate-free synthesis of segmented poly (dimethyl siloxane) polyureas. *Polymer* **2018**, *154*, 225–232. [CrossRef]
8. Sheth, J.P.; Aneja, A.; Wilkes, G.L.; Yilgor, E.; Atilla, G.E.; Yilgor, I.; Beyer, F.L. Influence of system variables on the morphological and dynamic mechanical behavior of polydimethylsiloxane based segmented polyurethane and polyurea copolymers: A comparative perspective. *Polymer* **2004**, *45*, 6919–6932. [CrossRef]
9. He, Y.; Xie, D.; Zhang, X. The structure, microphase-separated morphology, and property of polyurethanes and polyureas. *J. Mater. Sci.* **2014**, *49*, 7339–7352. [CrossRef]
10. Teo, L.S.; Chen, C.Y.; Kuo, J.F. Fourier transform infrared spectroscopy study on effects of temperature on hydrogen bonding in amine-containing polyurethanes and poly(urethane-urea)s. *Macromolecules* **1997**, *30*, 1793–1799. [CrossRef]
11. Yang, L.J.; Wang, Y.L.; Peng, X.H. Synthesis and characterization of novel fluorinated thermoplastic polyurethane with high transmittance and superior physical properties. *J. Macromol. Sci. A* **2017**, *54*, 516–523. [CrossRef]
12. Yilgor, E.; Yilgor, I. Hydrogen bonding: A critical parameter in designing silicone copolymers. *Polymer* **2001**, *42*, 7953–7959. [CrossRef]
13. Xu, B.W.; Yin, Q.C.; Su, C.; Cheng, J.; Zhang, J.Y.; Zhao, J.B. High-Performance Nonisocyanate Thermoplastic Polythiourethane with High Hydrogen Bond Content. *ACS Macro. Lett.* **2022**, *11*, 517–524. [CrossRef]
14. Toader, G.; Diacon, A.; Axinte, S.M.; Mocanu, A.; Rusen, E. State-of-the-Art Polyurea Coatings: Synthesis Aspects, Structure-Properties Relationship, and Nanocomposites for Ballistic Protection Applications. *Polymers* **2024**, *16*, 454. [CrossRef]
15. Paiksung, C.S.; Smith, T.W.; Sung, N.H. Properties of segmented polyether poly(urethaneureas) based on2,4-Toluene diisocyanate, 2. infrared and mechanical studies. *Macromolecules* **1980**, *13*, 117–121. [CrossRef]
16. Zhang, Y.; Wang, T.; Si, H.; Liu, Y. Superhydrophobic polyurea coating with excellent adhesion and abrasion resistance prepared via convenient microphase separation. *Prog. Org. Coat.* **2022**, *170*, 106994. [CrossRef]
17. Li, S.; Kong, X.; Feng, S. Preparation of uniform poly(urea–siloxane) microspheres through precipitation polymerization. *RSC Adv.* **2015**, *5*, 90313–90320. [CrossRef]
18. Pergal, M.V.; Džunuzović, J.V.; Poręba, R.; Steinhart, M.; Pergal, M.M.; Vodnik, V.V.; Špírková, M. Structure–property correlation study of novel poly(urethane–ester–siloxane) networks. *Ind. Eng. Chem. Res.* **2013**, *52*, 6164–6176. [CrossRef]
19. Cui, Y.; Wang, H.; Pan, H.; Yan, T.; Zong, C. The effect of mixed soft segment on the microstructure of thermoplastic polyurethane. *J. Appl. Polym. Sci.* **2021**, *138*, 51346. [CrossRef]
20. Que, Y.-H.; Shi, Y.; Liu, L.-Z.; Wang, Y.-X.; Wang, C.-C.; Zhang, H.-C.; Han, X.-Y. The crystallisation, microphase separation and mechanical properties of the mixture of ether-based TPU with different ester-based TPUs. *Polymers* **2021**, *13*, 3475. [CrossRef]
21. Xiang, N.; Zhang, X.; Zheng, M.; Xu, R.; Yan, Y. Investigation of tensile behavior and molecular structure of the thermoplastic polyurethane sheets injection molded at different mold temperatures. *J. Appl. Polym. Sci.* **2021**, *138*, 50959. [CrossRef]
22. Alexopoulos, N.D.; Tiryakioglu, M. Relationship between Fracture Toughness and Tensile Properties of A357 Cast Aluminum Alloy. *Metall. Mater. Trans. A* **2009**, *40A*, 702–716. [CrossRef]
23. Manikkavel, A.; Kumar, V.; Lee, D.-J. Simple fracture model for an electrode and interfacial crack in a dielectric elastomer under tensile loading. *Theor. Appl. Fract. Mech.* **2020**, *108*, 102626. [CrossRef]
24. Mooney, M. A theory of large elastic deformation. *J. Appl. Phys.* **1940**, *11*, 582–592. [CrossRef]
25. Bonart, R.; Müller-Riederer, G. Modellvorstellungen zur Molekülorientierung in gedehnten segmentierten Polyurethan Elastomeren. *Colloid Polym. Sci.* **1981**, *259*, 926–936. [CrossRef]
26. Rivlin, R.S.; Saunders, D.W. Large elastic deformations of isotropic materials VII. Experiments on the deformation of rubber. *Philos. Trans. R. Soc. Lond. Ser. A* **1951**, *243*, 251–288.

27. Gao, Y.; Lv, J.; Liu, L.; Yu, Y. Effect of diacylhydrazine as chain extender on microphase separation and performance of energetic polyurethane elastomer. *e-Polymers* **2020**, *20*, 469–481. [CrossRef]
28. Yilgör, I.; Yilgör, E.; Wilkes, G.L. Critical parameters in designing segmented polyurethanes and their effect on morphology and properties: A comprehensive review. *Polymer* **2015**, *58*, A1–A36. [CrossRef]
29. Qi, H.J.; Boyce, M.C. Stress–strain behavior of thermoplastic polyurethanes. *Mech. Mater.* **2005**, *37*, 817–839. [CrossRef]
30. Estes, G.M.; Seymour, R.W.; Cooper, S.L. Infrared studies of segmented polyurethane elastomers. II. Infrared dichroism. *Macromolecules* **1971**, *4*, 452–457. [CrossRef]
31. Mattia, J.; Painter, P. A comparison of hydrogen bonding and order in a polyurethane and poly(urethane-urea) and their blends with poly(ethylene glycol). *Macromolecules* **2007**, *40*, 1546–1554. [CrossRef]
32. Tareq, A.Z.; Hyder, M.; Merino, D.H.; Chippindale, A.M.; Kaur, A.; Cooper, J.A.; Hayes, W. Thermally and mechanically robust self-healing supramolecular polyurethanes featuring aliphatic amide end caps. *Polymer* **2024**, *302*, 127052. [CrossRef]
33. Falco, G.; Simonin, L.; Pensec, S.; Dalmas, F.; Chenal, J.-M.; Bouteiller, L.; Chazeau, L. Linear and nonlinear viscoelastic properties of segmented silicone-urea copolymers: Influence of the hard segment structure. *Polymer* **2020**, *186*, 122041. [CrossRef]
34. Lee, K.; Lee, B.; Kim, C.; Kim, H.; Kim, K.; Nah, C. Stress-strain behavior of the electrospun thermoplastic polyurethane elastomer fiber mats. *Macromol. Res.* **2005**, *13*, 441–445. [CrossRef]

polymers

MDPI

Article

The Effect of Rubber–Metal Interactions on the Mechanical, Magneto–Mechanical, and Electrical Properties of Iron, Aluminum, and Hybrid Filler-Based Styrene–Butadiene Rubber Composites

Md Najib Alam [1], Vineet Kumar [1], Seok-U Jeong [2] and Sang-Shin Park [1,*]

[1] School of Mechanical Engineering, Yeungnam University, 280, Daehak-ro, Gyeongsan 38541, Republic of Korea; mdnajib.alam3@gmail.com (M.N.A.); vineetfri@gmail.com (V.K.)
[2] Graduate School of Mechanical Engineering, Yeungnam University, 280, Daehak-ro, Gyeongsan 38541, Republic of Korea; fffu123@naver.com
* Correspondence: pss@ynu.ac.kr

Abstract: Multifunctional stretchable rubber composites are gaining attention due to their unique electrical, mechanical, and magnetic properties. However, their high production costs pose economic challenges. This study explores the use of cost-effective metal powders—iron, aluminum, and their 1:1 (vol/vol) hybrid filler—in styrene–butadiene rubber composites, varying from 10 to 20 vol%. The effects of these metal particles on the mechanical, electrical, morphological, and swelling properties were investigated. Metal particles generally act as non-reinforcing fillers but can significantly enhance the mechanical modulus, electrical, and magnetic properties based on the filler structure and the filler–rubber interactions. Iron-based composites exhibit significant electrical conductivity and excellent magnetic properties. Aluminum enhances the modulus, while the combination yields average mechanical properties with added magnetic characteristics. Iron demonstrates higher reactivity with sulfur-based crosslinking ingredients, adversely affecting the rubber matrix's crosslinks, as shown by swelling tests. This reactivity is attributed to iron's transition metal characteristics. At 20 vol%, iron-filled composites display the highest magnetic anisotropic effect on toughness (~25%) under a magnetic field by permanent magnets and excellent electrical conductivity (1.5×10^{-2} S/m). While iron significantly boosts the electrical and magnetic properties, higher filler amounts degrade the mechanical properties. These composites are currently suitable for electrical and smart mechanical applications, but incorporating reinforcing fillers could enhance their robustness for broader applications.

Keywords: styrene–butadiene rubber; metallic fillers; conducting rubber composites; magneto–mechanical properties; swelling properties

1. Introduction

Rubber composites are increasingly being utilized in various industrial and engineering applications [1,2]. Traditionally, fillers like silica and clay have been used for mechanical reinforcement, but incorporating fillers with superior electrical, thermal, and magnetic properties can enhance the performance of rubber composites in these areas [3–5]. Rubber inherently has poor thermal and electrical conductivity, but these can be significantly improved by adding functional fillers, expanding the applications to electrical, mechanical, and thermal management systems [6,7]. For instance, rubber composites with iron powder fillers exhibit excellent magnetic properties, making them suitable for advanced technologies such as sensors, actuators, and magnetorheological dampers [8–10]. Additionally, rubber composites with metal fillers offer microwave absorption properties, making them highly effective for radiation shielding [11]. Metal-based conductive rubber composites are valuable for discharging static electricity in triboelectric systems, which is crucial for safety in environments with flammable gases [12]. Furthermore, due to their stretchability, rubber–metal composites can serve as stretchable conductors in various applications [13–15].

Different metals, such as aluminum, nickel, cobalt, iron, and zinc, are commonly used to make rubber–metal composites. Vinod et al. [16] studied the effects of heat, ozone, and high-energy radiation on aluminum powder-filled natural rubber composites, finding that these composites exhibit better resistance in mechanical properties under such conditions. In another investigation, Vinod et al. [17] found that aluminum improves the thermal conductivity of nitrile rubber composites. Anuar et al. [18] examined the curing, mechanical, and electrical properties of aluminum and zinc-filled natural rubber composites. They observed that zinc-based composites have higher tensile strength and elongation properties but a lower modulus compared to aluminum-based composites. They also found that electrical resistance is influenced by the size of the filler materials, with aluminum-filled rubber composites showing lower electrical resistance due to the larger particle size compared to zinc-based composites. Nassar et al. [19] investigated the effect of aluminum content up to 60 wt% in styrene–butadiene rubber composites, finding enhanced dielectric constant and dielectric loss values. Due to their magnetic properties and electrical conductivities, iron, cobalt, and nickel are useful for fabricating conducting rubber composites with tunable mechanical properties in various magnetic fields, making them ideal for smart magnetorheological elastomeric applications [20]. Zainudin et al. [8] found that cobalt, as a magnetic filler in silicone rubber, improved the magnetorheological properties and enhanced the electrical conductivity, making it useful as a sensor composite. Metal powders with a low aspect ratio in the filler structure demonstrate excellent electrical conductivity in rubber composites. Due to their low aspect ratio, even small deformations lead to significant changes in electrical resistivity in the composite, allowing for extremely high piezoresistive strain sensitivity [15,21]. Various studies [22–25] indicate that rubber composites with magnetic metals can improve microwave absorption and electromagnetic interference shielding efficiencies, along with enhancing electrical and magnetic losses. Additionally, iron filler-based rubber composites can be used as stretchable electrical induction heating materials for various technological applications [26].

Previous studies [16–26] have primarily focused on the thermal, electrical, and mechanical properties of metal-based polymer composites. However, the different chemical characteristics of various metals can lead to distinct effects on the physical and chemical properties of rubber composites. For instance, metallic iron, such as carbonyl iron particles, exhibits strong physical interactions with benzene π-electrons and d-electron systems in SBR rubber composites, thereby improving the mechanical properties [27]. Besides these physical interactions, there may also be chemical interactions with rubber additives. Most rubbers require curing with additives such as sulfur and accelerators, which can chemically interact with metals and influence the final chemical crosslinks. Both physical interactions and chemical crosslinks between rubber chains significantly impact the ultimate mechanical properties of rubber composites. Therefore, it is essential to evaluate all possible interactions of metals in rubber composites to fully understand their effects on mechanical and electrical properties. However, no studies have yet explored physical and chemical interactions separately. By considering these interactions independently, we can gain a clearer understanding of the electrical, mechanical, and other properties of metal-based rubber composites.

Initially, three metal powders—iron, copper, and aluminum—were selected as fillers for styrene–butadiene rubber-based composites. However, after curing the rubber composite with copper powder, an unconventional sticky compound was formed. This could be due to the high reactivity of copper with the sulfur-based curing ingredients, which significantly affects the crosslinking formation. Consequently, further experiments with copper powder were not conducted.

In this study, two classes of metal elements based on their periodic properties—iron (d-block) and aluminum (p-block)—were investigated as fillers in a styrene–butadiene rubber system. Iron is known for its strong magnetic properties, while aluminum exhibits excellent thermal conductivity, making it suitable for magnetic applications requiring enhanced thermal management. Styrene–butadiene rubber was selected as the matrix due to its stretchability, high modulus, and effective physical interactions with various fillers, which are advantageous for applications in high-load-bearing mechanical systems. Comparative analyses were conducted to examine the influence of these different metals on the physical and chemical properties of the resulting rubber composites. Swelling studies were performed to assess the chemical effects of the metals on crosslink formations within the rubber matrix. Additionally, considering its electrical conductivity and magnetic properties, iron was evaluated for its magneto-mechanical characteristics, aiming at potential applications in conductive magnetorheological elastomers.

2. Materials and Methods

2.1. Materials

Styrene–butadiene rubber (SBR 1502, ~23.5% styrene content) was procured from Kumho Petrochemicals, Republic of Korea. A masterbatch rubber was prepared using the following curing ingredients in parts per hundred grams of rubber (phr): zinc oxide (5 phr), stearic acid (2 phr), accelerator N-tert-butyl-benzothiazole sulfonamide (1.75 phr), accelerator tetramethyl thiuram disulfide (1 phr), and sulfur (1.5 phr). The mixing process was carried out on an open two-roll mill. Electrolytic iron powder with an average particle size of 400 mesh was sourced from AO Metal Corporation Ltd., Republic of Korea. Aluminum powder, with an average particle size of 325 mesh, and the solvent toluene were obtained from Daejung Chemicals & Metals Co. Ltd., Siheung, Republic of Korea. The scanning electron micrographs of the two metals are shown in Figure 1a,b. It is evident from these figures that iron particles (Figure 1a) are relatively smaller in size with more uneven structures compared to the aluminum particles (Figure 1b).

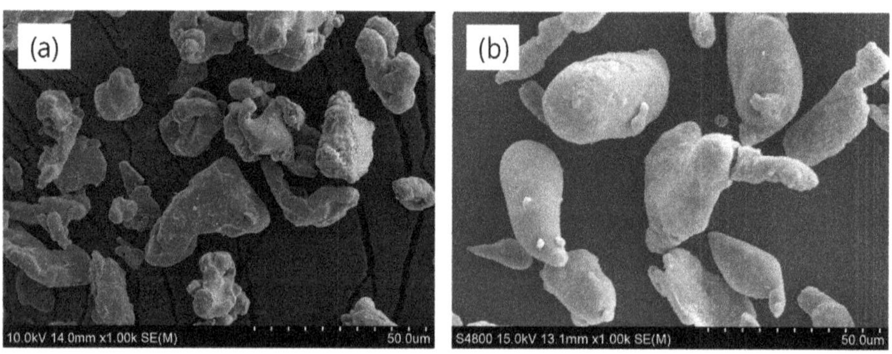

Figure 1. SEM images; (**a**) iron particles and (**b**) aluminum particles.

2.2. Fabrication of Rubber Composites

A specific amount of masterbatch rubber, as indicated in Table 1, was soaked in 100 mL of toluene for 24 h in a glass jar. The soaked rubber was then homogenized through mechanical stirring, resulting in a highly viscous slurry. In a separate container, varying amounts (0–20 vol%) of iron or aluminum powder, or their 1:1 (v/v) hybrid, were mixed with 20 mL of toluene and subjected to 30 min of sonication. This filler solution was subsequently added to the homogenized rubber slurry and mixed vigorously for about 10 min using mechanical stirring. The filler-mixed rubber slurry was then placed in an oven at 80 °C for 24 h to allow for solvent evaporation. To determine the volume percentage of filler in the rubber composites, the weight/density formula was applied, using theoretical density values of 7.87 g/cm^3 for iron and 2.7 g/cm^3 for aluminum. The volume fraction of rubber was calculated based on the measured density of unfilled rubber, which was 0.94 g/cm^3. The different mixing compositions and filler amounts are detailed in Table 1.

Table 1. Mix compositions of materials by actual weight taken in grams.

Formulation	Amount of Masterbatch Rubber (vol%)	Amount of Iron (vol%)	Amount of Aluminum (vol%)	Hybrid (1:1 *v*/*v*)	
				Amount of Iron (vol%)	Amount of Aluminum (vol%)
SBR-unfilled	23.5 (100) *	-	-	-	
SBR/10vol% Fe	21.15 (90)	19.68 (10)	-	-	
SBR/15vol% Fe	19.98 (85)	29.51 (15)	-	-	
SBR/20vol% Fe	18.8 (80)	39.35 (20)	-	-	
SBR/10vol% Al	21.15 (90)	-	6.75 (10)	-	
SBR/15vol% Al	19.98 (85)	-	10.13 (15)	-	
SBR/20vol% Al	18.8 (80)	-	13.5 (20)	-	
SBR/10vol% Hyb	21.15 (90)	-	-	9.84 (5)	3.38 (5)
SBR/15vol% Hyb	19.98 (85)	-	-	14.76 (7.5)	5.07 (7.5)
SBR/20vol% Hyb	18.8 (80)			19.68 (10)	6.75 (10)

* The values in parentheses indicate the volume fraction of each component in the compounded rubber.

After drying, the compounded rubber was vulcanized at 150 °C in a hot press molding machine into cylindrical or sheet-like shapes, as described in previous studies [27,28]. The colors and physical shapes of the unfilled and highest filler-loaded vulcanizates are illustrated in Figure 2a–d for both the unfilled and highest filler-loaded compounds. The vulcanized rubber samples were then stored in a refrigerator to control environmental effects. They were removed from the refrigerator one day before measuring their physical and mechanical properties.

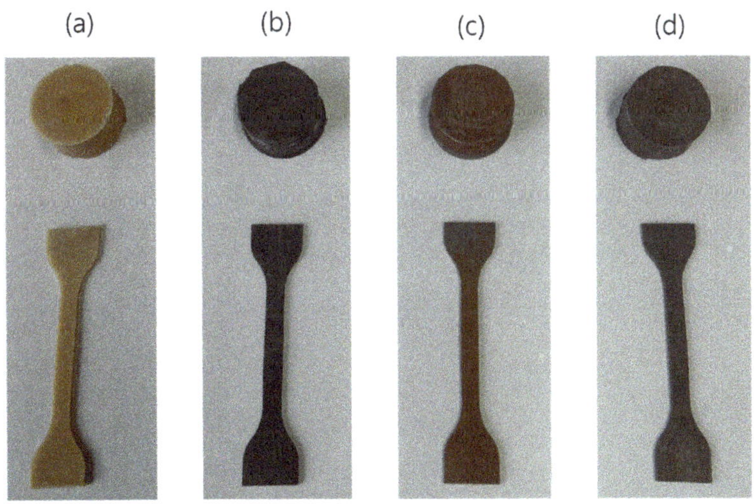

Figure 2. Camera images; (**a**) SBR-unfilled, (**b**) SBR/20vol% Fe, (**c**) SBR/20vol% Al, and (**d**) SBR/20vol% Hyb.

2.3. *Measurements of the Mechanical Properties*

Tensile, compressive, and stress relaxation tests were conducted to investigate the mechanical properties of the rubber composites using a universal testing machine (Lloyd UTM, Westminster, UK) equipped with a 1 kN load cell. For the tensile test, dumbbell-shaped test pieces were cut from the vulcanized sheet according to the ISO-37 [29], type-2 standard. The test was performed with a constant motor speed of 300 mm/min and a gauge length of 25 mm. For the compressive test, cylindrical samples with a diameter of 20 mm and a height of 10 mm were used. The motor speed was set at 2 mm/min, and the ultimate compressive strain was 30%. To investigate the stress relaxation behavior, the cylindrical sample was compressed to 30%, and the load relaxation was observed for up to 60 s.

2.4. Magneto-Mechanical Properties

The magneto-mechanical properties in the presence of neodymium permanent magnets (Grade: N35 NdFeB, Remanence ~1.2 T, dimensions: d = 30 mm and h = 10 mm) were studied similarly to the compressive mechanical properties. The compressive mechanical characteristics exhibited changes due to the magnetic field-induced anisotropic effect in the ferromagnetic rubber composites. The anisotropic effects on various mechanical properties were determined by comparing the mechanical properties in the presence and absence of the magnetic field.

2.5. Electrical Properties

The electrical properties, including the resistivity and conductivity, of the rubber composites were measured using cylindrical samples. The two-probe resistance measurement method was employed, with two copper electrodes placed on opposite sides of the cylinder. A digital multimeter (Keysight, 34461A, Santa Rosa, CA, USA) was used for these measurements. The multimeter's maximum resistance measurement limit was 100 megaohms.

2.6. Filler Dispersion

Field-emission scanning electron microscopy (FE-SEM, S-4800, Hitachi, Tokyo, Japan) was used to study the dispersion of the fillers in the rubber matrix. Prior to capturing the SEM images, the tensile fractured surfaces were coated with gold using a sputtering machine.

2.7. Swelling Studies and Crosslink Density Measurements

Rubber can swell in a solvent until it reaches thermodynamic equilibrium. It is known that as the chemical crosslinking density increases, the rubber becomes less swellable. Therefore, information about the chemical crosslinks achieved in the rubber composites in the presence of metallic fillers can be obtained. It is assumed that due to the strong metallic bonds in the metallic filler systems, there will be no solvent swelling of the fillers. Consequently, any observed solvent swelling must be attributed to the rubber phase, and the crosslink density in the rubber phase can be determined experimentally.

To study the swelling characteristics of the rubber compounds, cylindrical test pieces were immersed in toluene (solvent) for 7 days to reach a swelling equilibrium. From the swelling equilibrium data, two swelling index values were determined: one for the total composite and one for the fraction of rubber in the composite. The swelling of the solvent relative to the fraction of rubber was used to determine the crosslink density in the rubber matrix.

The swelling index values were calculated using the formula shown in Equation (1).

$$\text{Swelling Index} = \frac{\text{Swelled volume of solvent}}{\text{Volume of whole composite or volume of rubber fraction}} \tag{1}$$

The crosslink density in the rubber matrix was determined using the Flory–Rehner equation [30], as depicted in Equations (2) and (3).

$$V_c = \frac{-\left\{\ln(1 - V_r) + V_r + \chi V_r^2\right\}}{V_s d_r \left(V_r^{\frac{1}{3}} - \frac{V_r}{2}\right)} \tag{2}$$

In Equation (2), V_r represents the volume fraction of rubber in the swollen phase, and V_c denotes the crosslink density. The interaction parameter $\chi = 0.49$ applies to the

SBR/toluene (solvent) system, while V_s = 106.2 signifies the molar volume of toluene (solvent), and d_r = 0.94 g/cm^3 stands for the density of rubber.

$$V_r = \left(\frac{\frac{w_r}{d_r}}{\frac{w_r}{d_r} + \frac{w_s}{d_s}} \right) \tag{3}$$

In Equation (3), w_r and w_s denote the weights, and d_r and d_s represent the densities of the rubber and solvent, respectively. The density of toluene was assumed to be 0.87 g/cm^3.

3. Results and Discussion

3.1. Mechanical Properties

Figure 3a–c illustrates the tensile stress–strain curves for different rubber composites. At lower strains, the stress values increase with higher filler amounts, indicating improved mechanical strength. However, at higher strains, the stress decreases relative to the unfilled rubber, possibly due to diminished filler–polymer interactions, given that the fillers are micrometer-sized particles [27]. Notably, in Figure 3a, iron-based composites exhibit comparable or slightly higher toughness than unfilled rubber, suggesting distinct physical interactions with iron particles [27]. In contrast, such interactions appear absent in aluminum-based composites as shown in Figure 3b.

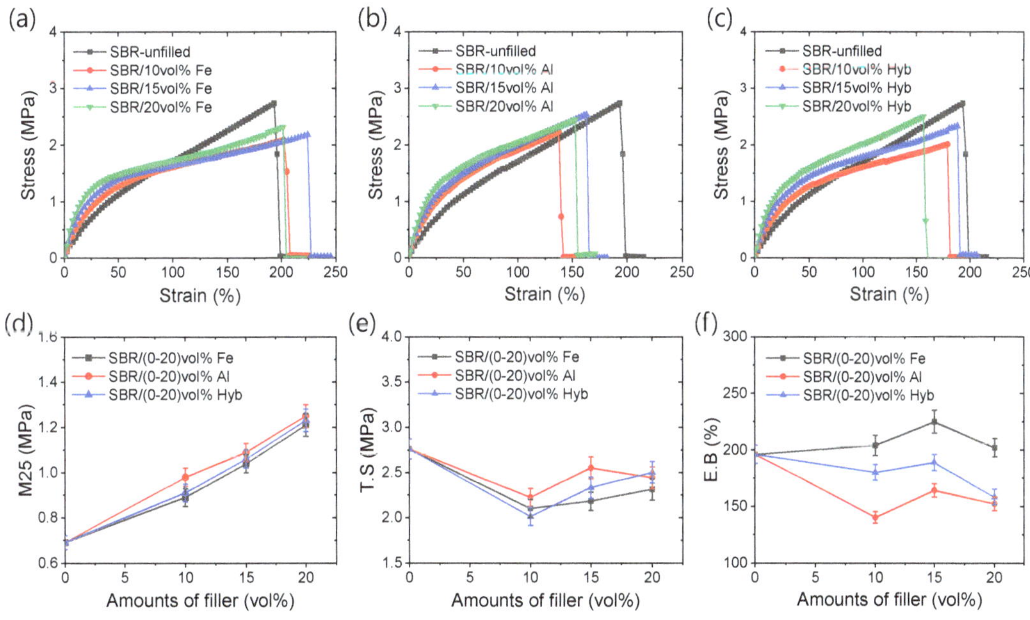

Figure 3. Tensile mechanical properties; (**a**) stress–strain of iron-based composites, (**b**) stress–strain of aluminum-based composites, (**c**) stress–strain of hybrid filler-based composites, (**d**) M25, (**e**) T.S, and (**f**) E.B.

Figure 3d displays the modulus at 25% strain (M25) values for various composites as a function of filler amounts. It is evident that the modulus increases with higher filler concentrations, with aluminum-based composites exhibiting higher modulus values compared to iron and hybrid filler-based composites. However, these differences diminish as the filler content increases. Fröhlich et al. [31] propose that the modulus of filled rubber depends on the filler–filler interactions, in-rubber structures, hydrodynamic effects, and the rubber network. Given that both iron and aluminum particles exceed micrometer sizes, their filler–filler interactions and in-rubber structures are expected to contribute less to the modulus values at lower filler contents. Instead, hydrodynamic effects likely play a major role due to the rigidity and size of the filler particles. Moreover, the rubber network significantly influences the modulus, as indicated by the stress–strain properties and modulus of the unfilled rubber. Aluminum particles, being larger, are likely to enhance the modulus values more through hydrodynamic effects. Additionally, aluminum's lower chemical influence on crosslink formation compared to iron, noted in swelling investigations, suggests another reason for their modulus enhancement. As filler content increases, the filler–filler interactions and in-rubber structures develop. Iron particles, more irregular and capable of strong interactions with SBR's benzene ring, may form more robust in-rubber structures at higher filler levels [27,32,33]. This is reflected in Figure 3d by similar modulus values at higher filler contents and the pronounced effects observed in compressive properties, where fillers aggregate rather than separate as seen in the tensile testing.

Figure 3e displays the tensile strength (T.S) values of various rubber composites. The figure indicates that the T.S initially decreases with 10 vol% filler content in the composites, followed by an increase with higher filler amounts. These findings suggest the formation of a potentially stronger filler network at an optimal filler concentration, likely due to enhanced filler–polymer interactions. However, the T.S value of the rubber composites does not match that of the unfilled rubber, possibly attributable to the micron size of the particles.

The maximum stretchability of rubber composites is influenced by factors such as the homogeneity of the filler distribution, the filler–rubber interactions, and the rigidity of the rubber networks. Figure 3f plots the elongation at break (E.B) values of the rubber composites. From this figure, it is evident that aluminum filler-based composites exhibit lower stretchability, whereas the iron-based composites demonstrate higher stretchability. This trend aligns with the observations of the tensile strength values, suggesting that more stretchable filler networks may form due to uniform filler dispersion and improved filler–rubber interactions up to an optimal filler amount. Further insights from swelling properties, discussed later, indicate that iron-based rubber composites, characterized by lower crosslink densities, enable higher stretchability compared to other filler systems.

Figure 4a–d presents various compressive mechanical properties. From the compressive stress–strain curves shown in Figure 4a–c, it is evident that the slope of the compressive stress–strain increases more rapidly with compression for the iron-based composites compared to the aluminum and hybrid filler-based composites. It is important to note that during compression, the filler particles come closer together unlike in the tensile deformation where they tend to separate. The irregular structure of iron particles allows them to potentially form in-rubber structures after reaching a certain filler content, effectively occluding the rubber within the filler matrix [31]. This shielding prevents rubber deformation, leading to an increase in the effective filler content and contributing to the compressive modulus. In Figure 4d, the Young's modulus values show that aluminum-based composites have a lower modulus compared to iron-based composites. At lower filler contents, hybrid filler-based composites exhibit better modulus values, possibly due to stronger polymer networks and the presence of iron particles that promote in-rubber structures. At the highest filler content, a significant number of in-rubber structures may form due to strong filler–polymer interactions with iron particles [27,31], resulting in the highest Young's modulus.

Figure 5a–c illustrates the stress relaxation behaviors of the rubber composites under a 30% applied strain. From the stress relaxation curves in Figure 5a–c, it is evident that relaxation rates increase with higher filler amounts. This phenomenon is likely due to the increased Young's modulus and enhanced rigidity upon the addition of fillers. It is noticeable from Figure 5a–c that significant stress relaxation occurs within 10 s of the relaxation time. Iron-based composites exhibit prolonged relaxation times, indicating higher viscoelastic behavior compared to aluminum-based composites. The increased viscoelasticity in iron-based composites may be attributed to physical interactions between the π-electrons of benzene in SBR and the d-electrons in iron [27,32,33]. In contrast, aluminum-based composites, lacking such interactions, relax more quickly than their iron-containing counterparts.

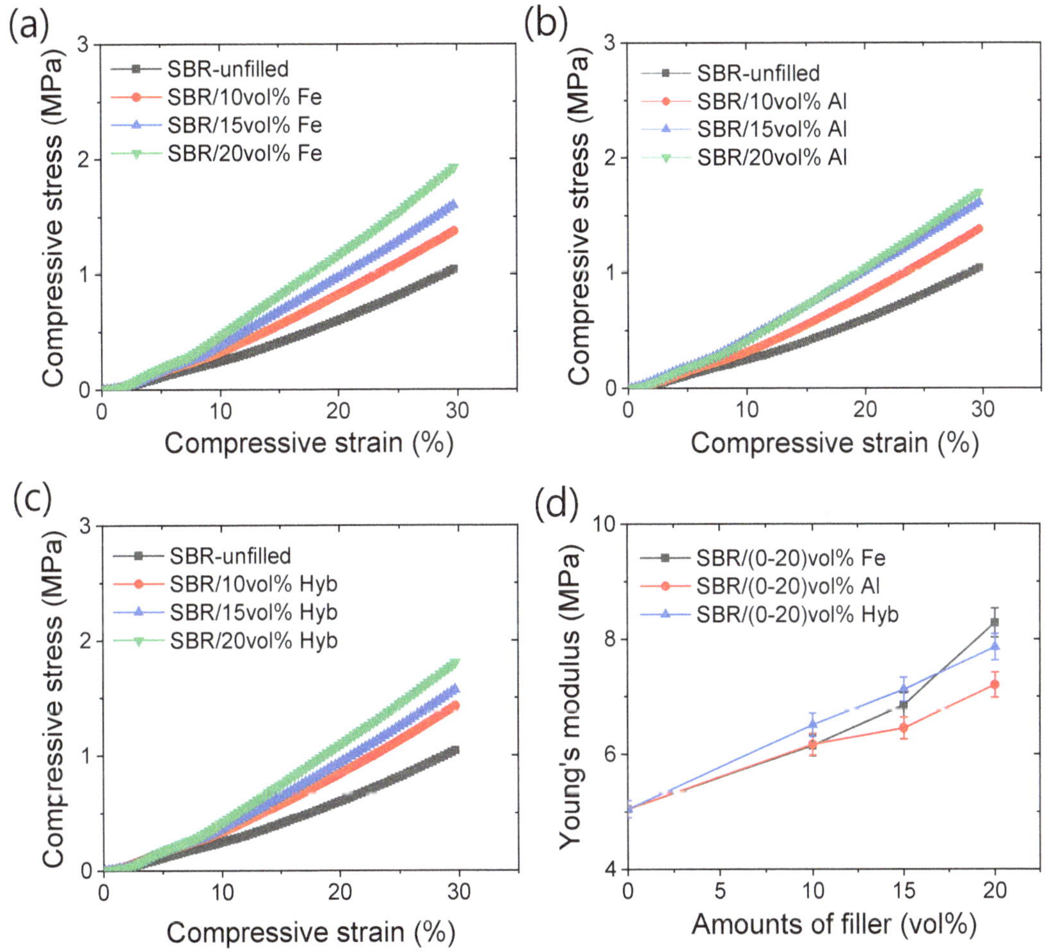

Figure 4. Compressive mechanical properties; (**a**) stress–strain of iron-based composites, (**b**) stress–strain of aluminum-based composites, (**c**) stress–strain of hybrid filler-based composites, and (**d**) Young's modulus.

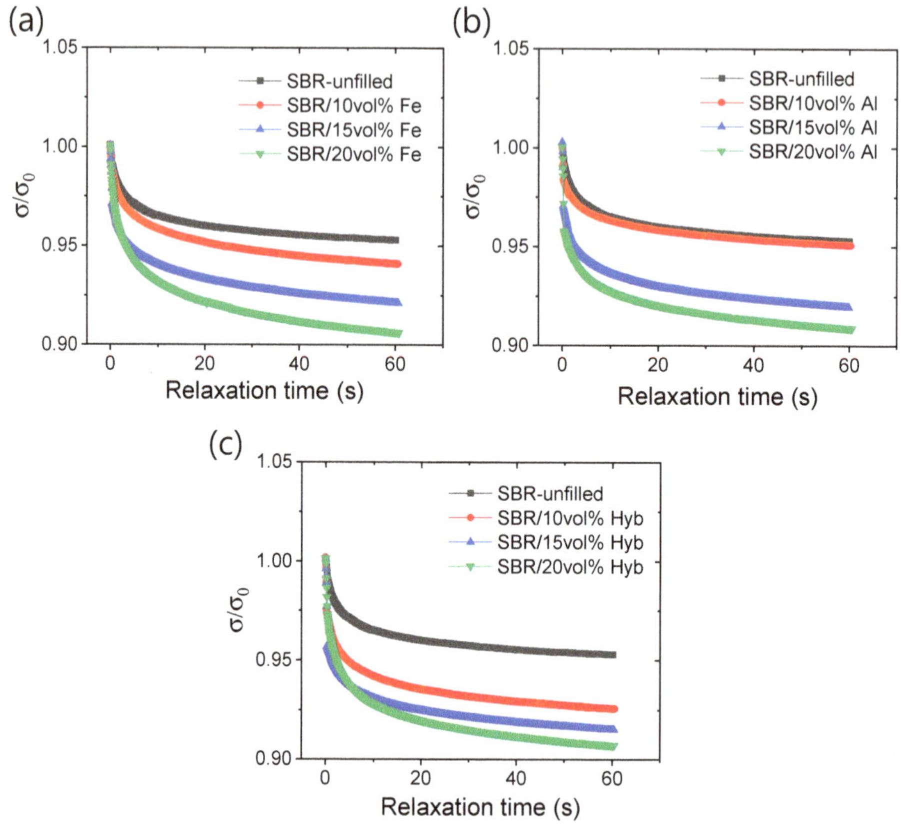

Figure 5. Stress relaxation behaviors; (**a**) iron-based composites, (**b**) aluminum-based composites, and (**c**) hybrid filler-based composites.

3.2. Magneto-Mechanical and Electrical Properties

Iron is a ferromagnetic material, imparting an anisotropic effect to iron-containing rubber composites when subjected to an external magnetic field [34–37]. This effect enhances the stiffness of magnetorheological elastomers, requiring more energy to deform or reducing the deformation amplitude. Consequently, they can effectively dampen significant forces and find utility as dampers in various mechanical systems [36,37]. Figure 6a–e present investigations into the anisotropic effects on different compressive mechanical properties. Figure 6a illustrates the typical compressive stress–strain curves in the presence and absence of magnetic fields, demonstrating the enhanced stiffness of magnetorheological elastomers and their potential in mechanical damping systems. Figure 6b,c depict changes in the stiffness and the relative changes induced by a permanent magnet in the composites with varying iron content. It is evident that the highest changes in stiffness and relative changes occur in composites with the highest iron content. For instance, the change in stiffness can exceed 25 kN/m for the 20 vol% iron-containing rubber composite with 1.2 T of magnetic field. Figure 6d,e illustrate changes in the toughness and relative toughness under magnetic field conditions. It shows that ΔToughness increases with the increasing iron content in the rubber composites. However, the relative toughness may decrease depending on the initial toughness of the composite. From Figure 6e, it can be observed that a composite with 20 vol% iron exhibits 25% higher relative toughness at 30% deformation. This suggests that applying a 1.2 T magnetic field can dampen 25% more energy due to the filler anisotropy. However, a higher damping efficiency is achieved at lower deformation levels, as the rubber composites become stiffer with increased deformation, leading to a reduced relative change in the stiffness values [28,38]. Significant changes in the stiffness and toughness values in the presence of a magnetic field, particularly at 20 vol% iron content, indicate enhanced filler percolation at this filler concentration.

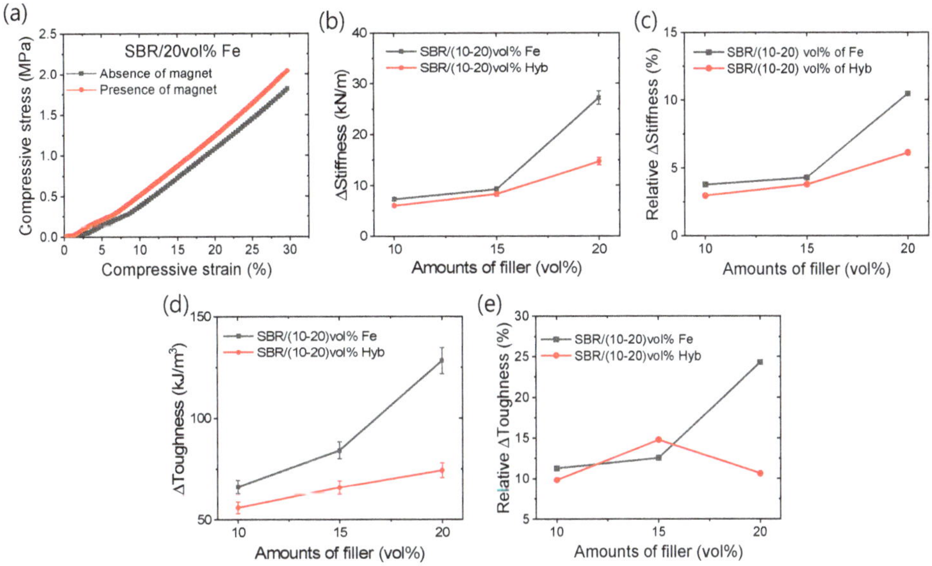

Figure 6. Magneto-mechanical properties; (**a**) typical stress–strain curves showing magnetic effect, (**b**) change in stiffness, (**c**) relative change in stiffness, (**d**) change in toughness, and (**e**) relative change in toughness.

Although all the composites contain metal fillers, they do not exhibit significant electrical conductivity until the filler particles percolate to form conductive paths. It is noteworthy that iron, as a single filler system, demonstrates notable electrical conductivity, as shown and measured in Figure 7a,b. From Figure 7a,b, it is evident that significantly lower resistivity and higher electrical conductivity in the rubber composite can be achieved with a 20 vol% iron content. This level of iron content indicates that filler particles are sufficiently close to form effective conductive paths, thereby enhancing both the magnetic and electrical properties. The improved electrical conductivity and magnetic properties of these magnetic rubber composites render them ideal for a diverse range of applications in advanced mechanical and electronic devices.

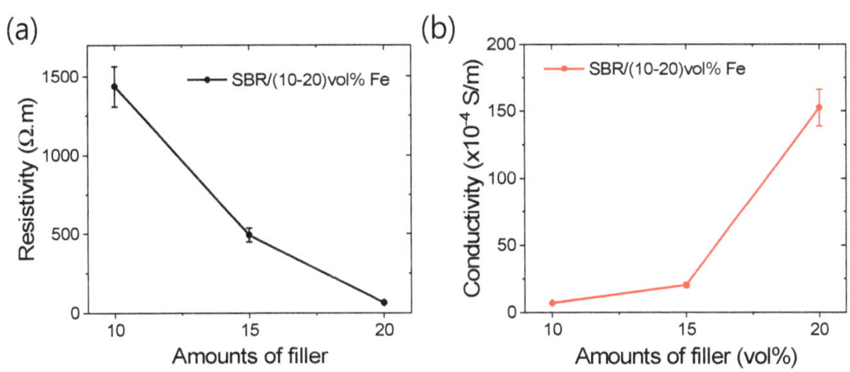

Figure 7. Electrical properties; (**a**) resistivity and (**b**) conductivity.

3.3. Morphology of Rubber Composites

Figure 8a–f presents SEM images of both unfilled and filled rubber composites, revealing critical insights into their microstructures. The unfilled rubber composites (Figure 8a) appear smooth without significant features. In contrast, the filled rubber composites exhibit numerous vacuoles, indicating weak filler–rubber interactions. These composites are prone to tearing under tensile loading and show reduced tensile strength compared to the unfilled rubber. In iron-containing composites (Figure 8b–d), aggregated filler structures are evident, promoting electrical conductivity. This characteristic enhances their magneto-mechanical properties and results in a higher anisotropic effect due to improved filler particle contacts. Conversely, aluminum-based composites (Figure 8e) and hybrid filler systems (Figure 8f) demonstrate filler particles that are either separated or fail to form proper conductive channels, even at high filler concentrations. Consequently, these composites exhibit weaker filler–rubber interactions and inferior mechanical properties compared to iron-only composites. Moreover, iron-only composites show in-rubber structures that confirm their superior compressive modulus at higher filler contents. To further verify the filler distributions in the hybrid composite, energy-dispersive X-ray spectroscopy was conducted, as shown in Figure 9a,b. Figure 9a confirms the presence of elements such as C, N, O, S, Zn, Al, and Fe in the composites, originating from either curatives or added fillers. Elemental mapping of the major elements in Figure 9b indicates that while there is no effective filler networking for electrical conductivity, the composite exhibits magnetic properties due to the iron particles. Overall, these SEM observations underscore the significant impact of filler distribution and interactions on the mechanical and electrical properties of rubber composites, highlighting the advantages of iron-based composites in magneto-mechanical applications.

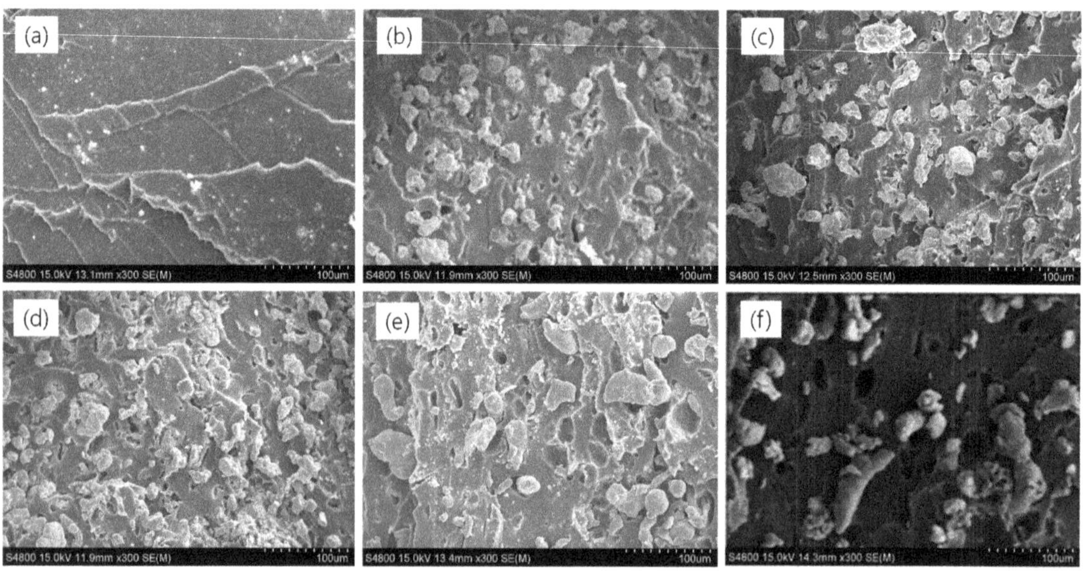

Figure 8. SEM images; (**a**) SBR-unfilled, (**b**) SBR/10vol% Fe, (**c**) SBR/15vol% Fe, (**d**) SBR/20vol% Fe, (**e**) SBR/20vol% Al, and (**f**) SBR/20vol% Hyb.

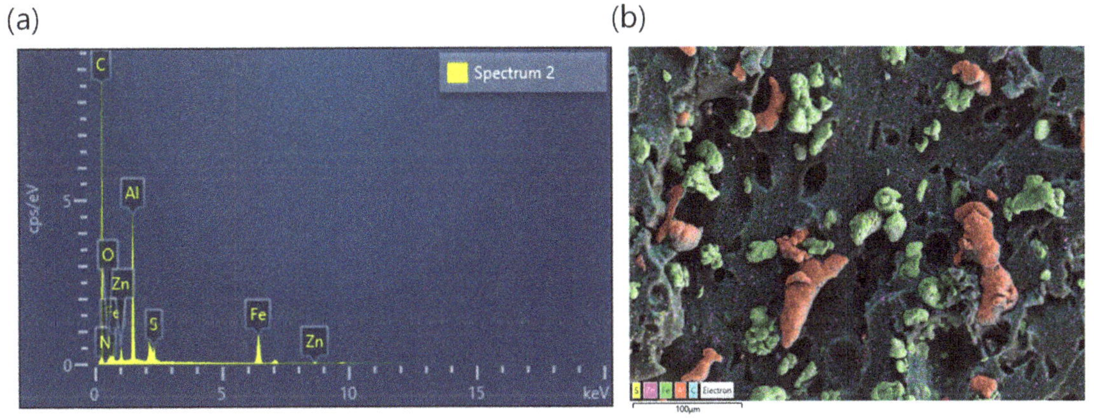

Figure 9. (**a**) Energy-dispersive X-ray spectroscopy of SBR/20vol% Hyb composite and (**b**) mapping of major elements in SBR/20vol% Hyb composite including iron (green) and aluminum (red) particles.

3.4. Swelling and Chemical Crosslink Densities

Figure 10a–c illustrates the swelling behavior and crosslink density values of different rubber composites. In Figure 10a,b, the swelling index values are depicted for the entire composites and for the rubber matrix alone, respectively. From Figure 10a, there appear to be no significant differences in the swelling index values with increasing filler content in specific filler systems. However, Figure 10b shows that swelling index values increase with higher filler amounts. These results suggest that the fillers themselves do not significantly swell in toluene. Moreover, increasing the filler content can reduce the number of crosslinks in the rubber matrix, as shown in Figure 10c. Figure 10c demonstrates that metal composites generally exhibit reduced crosslink density compared to unfilled rubber, with aluminum-based composites showing better crosslink densities than other composites. Mechanical properties, such as modulus values of the composites, heavily depend on factors like network density in the rubber matrix and physical or chemical interactions between rubber and fillers. Reinforcement in these properties, particularly modulus values, primarily occurs through physical interactions such as filler–filler and filler–rubber interactions. From Figure 10c, it is also evident that iron particles have a more detrimental effect on crosslinks compared to aluminum particles. Anuar et al. [18] similarly found that zinc-based rubber composites exhibit lower mechanical properties and thermal stability compared to aluminum-based natural rubber composites, suggesting a higher destructive effect on crosslinks in the former. It is plausible that d-block elements like iron and copper exhibit higher reactivity with vulcanizing ingredients compared to p-block metals like aluminum, resulting in reduced crosslink density. Additionally, the antioxidant behavior or higher thermal conductivity of metals may inhibit vulcanization or decompose certain curatives before the formation of active sulfurating complexes [16,39,40]. Since vulcanization follows first-order kinetics with respect to the active sulfurating complex, a reduced amount of this complex may lead to fewer sulfur crosslinks in the rubber matrix [39,40]. Consequently, rubber–metal composites typically exhibit lower crosslink density values compared to unfilled rubber compounds. Despite iron-based composites having lower crosslink density than aluminum-based composites, their tensile toughness values are significantly better, suggesting stronger rubber–filler physical interactions, possibly through π-electron systems of rubber with d-electrons of iron [27] as illustrated in Figure 11.

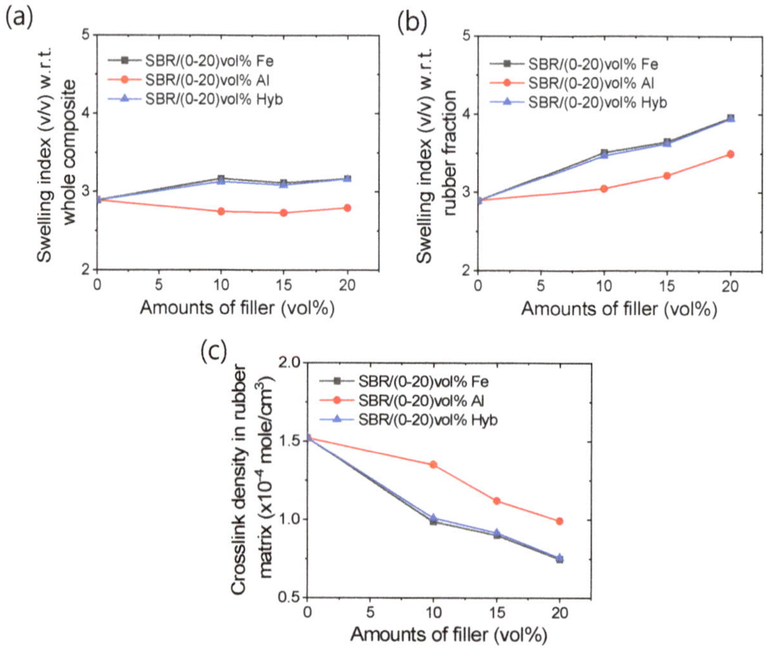

Figure 10. Swelling characteristics; (**a**) swelling index with respect to whole composites, (**b**) swelling index with respect to the rubber faction, (**c**) crosslink density in the rubber fraction.

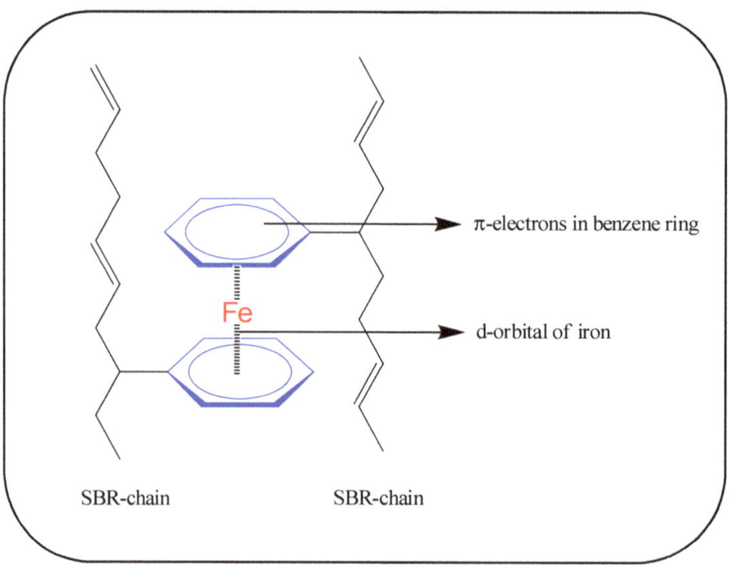

Figure 11. Possible interactions between π-electrons of the benzene ring in the SBR-chain and the d-orbital of iron.

4. Conclusions

Various metal powders, including iron, aluminum, and a 1:1 (v/v) hybrid, were studied as fillers in styrene–butadiene rubber (SBR). The mechanical properties of these composites indicate that all metal fillers can enhance the modulus values due to their physical interactions with the rubber. However, in many cases, metal fillers act as non-reinforcing fillers in SBR-based composites. This limitation is attributed to the micron size of the filler particles and their detrimental effects on chemical crosslink formation in the rubber matrix. Nevertheless, iron-based rubber composites exhibit exceptional magneto-mechanical properties and excellent electrical conductivity compared to other composites. At 20 vol% iron content, the composite exhibits a 25% relative change in toughness at 30% compression when exposed to a 1.2 T magnetic field, compared to the absence of the magnetic field. Additionally, this composite with 20 vol% iron content demonstrates electrical percolation, with an electrical conductivity of 1.5×10^{-2} S/m. It has been observed that iron, as a d-block element, has a more pronounced detrimental effect on crosslink formation than aluminum, which is a p-block element. The superior electrical and magnetic properties of iron-based rubber composites make them promising for applications in magnetorheology, sensors, electronics, and other advanced engineering systems. However, to address their poor mechanical properties, it is necessary to incorporate some amount of reinforcing filler such as carbon black or nano-silica to enhance overall performance.

Author Contributions: Conceptualization, M.N.A.; methodology, M.N.A.; validation, M.N.A., V.K., S.-U.J. and S.-S.P.; formal analysis, M.N.A., V.K. and S.-U.J.; investigation, M.N.A.; data curation, M.N.A.; writing—original draft preparation, M.N.A.; writing—review and editing, M.N.A., V.K., S.-U.J. and S.-S.P.; visualization, M.N.A. and S.-S.P.; supervision, S.-S.P.; project administration, S.-S.P.; funding acquisition, S.-S.P. All authors have read and agreed to the published version of the manuscript.

Funding: This research received no external funding.

Institutional Review Board Statement: Not applicable.

Data Availability Statement: Data are contained within the article.

Conflicts of Interest: The authors declare no conflicts of interest.

References

1. Pothen, L.A.; Chan, C.H.; Thomas, S. *Natural Rubber Materials, Volume 2-Composites and Nanocomposites*; Royal Society of Chemistry: London, UK, 2013.
2. Umoren, S.A.; Solomon, M.M. Protective Polymeric Films for Industrial Substrates: A Critical Review on Past and Recent Applications with Conducting Polymers and Polymer Composites/Nanocomposites. *Prog. Mater. Sci.* **2019**, *104*, 380–450.
3. Al-Sehemi, A.G.; Al-Ghamdi, A.A.; Dishovsky, N.T.; Malinova, P.A.; Atanasov, N.T.; Atanasova, G.L. Analysis of the Electrical and Magnetic Properties of Elastomeric Composites and Their Applicability in Small Flexible Wearable Antennas. *Mater. Res. Express* **2017**, *4*, 076304. [CrossRef]
4. Marin, C.N.; Malaescu, I.; Sfirloaga, P.; Teusdea, A. Electric and Magnetic Properties of a Composite Consisting of Silicone Rubber and Ferrofluid. *J. Ind. Eng. Chem.* **2021**, *101*, 405–413. [CrossRef]
5. Kruželák, J.; Kvasničáková, A.; Džuganová, M.; Dosoudil, R.; Hudec, I.; Krump, H. The Electrical Conductivity, EMI Absorption Shielding Performance, Curing Process, and Mechanical Properties of Rubber Composites. *Polymers* **2024**, *16*, 566. [CrossRef]
6. Boudenne, A.; Mamunya, Y.; Levchenko, V.; Garnier, B.; Lebedev, E. Improvement of Thermal and Electrical Properties of Silicone–Ni Composites Using Magnetic Field. *Eur. Polym. J.* **2015**, *63*, 11–19. [CrossRef]
7. Wu, S.; Hu, W.; Ze, Q.; Sitti, M.; Zhao, R. Multifunctional Magnetic Soft Composites: A Review. *Multifunct. Mater.* **2020**, *3*, 042003. [CrossRef] [PubMed]
8. Zainudin, A.A.; Yunus, N.A.; Mazlan, S.A.; Shabdin, M.K.; Abdul Aziz, S.A.; Nordin, N.A.; Nazmi, N.; Abdul Rahman, M.A. Rheological and Resistance Properties of Magnetorheological Elastomer with Cobalt for Sensor Application. *Appl. Sci.* **2020**, *10*, 1638. [CrossRef]
9. Böse, H.; Gerlach, T.; Ehrlich, J. Magnetorheological Elastomers—An Underestimated Class of Soft Actuator Materials. *J. Intell. Mater. Syst. Struct.* **2021**, *32*, 1550–1564. [CrossRef]
10. Bastola, A.K.; Hossain, M. A Review on Magneto-Mechanical Characterizations of Magnetorheological Elastomers. *Compos. Part B Eng.* **2020**, *200*, 108348. [CrossRef]
11. Sankaran, S.; Deshmukh, K.; Ahamed, M.B.; Pasha, S.K. Recent Advances in Electromagnetic Interference Shielding Properties of Metal and Carbon Filler Reinforced Flexible Polymer Composites: A Review. *Compos. Part A Appl. Sci. Manuf.* **2018**, *114*, 49–71. [CrossRef]

12. Wu, W.; Yang, T.; Zhang, Y.; Wang, F.; Nie, Q.; Ma, Y.; Cao, X.; Wang, Z.L.; Wang, N.; Zhang, L. Application of Displacement-Current-Governed Triboelectric Nanogenerator in an Electrostatic Discharge Protection System for the Next-Generation Green Tire. *ACS Nano* **2019**, *13*, 8202–8212. [CrossRef] [PubMed]
13. Lee, S.; Shin, S.; Lee, S.; Seo, J.; Lee, J.; Son, S.; Cho, H.J.; Algadi, H.; Al-Sayari, S.; Kim, D.E.; et al. Ag Nanowire Reinforced Highly Stretchable Conductive Fibers for Wearable Electronics. *Adv. Funct. Mater.* **2015**, *25*, 3114–3121. [CrossRef]
14. Tian, Z.; Zhao, Y.; Wang, S.; Zhou, G.; Zhao, N.; Wong, C.P. A Highly Stretchable and Conductive Composite Based on an Emulsion-Templated Silver Nanowire Aerogel. *J. Mater. Chem. A* **2020**, *8*, 1724–1730. [CrossRef]
15. Yun, G.; Tang, S.Y.; Lu, H.; Zhang, S.; Dickey, M.D.; Li, W. Hybrid-Filler Stretchable Conductive Composites: From Fabrication to Application. *Small Sci.* **2021**, *1*, 2000080. [CrossRef]
16. Vinod, V.S.; Varghese, S.; Kuriakose, B. Degradation Behaviour of Natural Rubber–Aluminium Powder Composites: Effect of Heat, Ozone and High Energy Radiation. *Polym. Degrad. Stab.* **2002**, *75*, 405–412. [CrossRef]
17. Vinod, V.S.; Varghese, S.; Kuriakose, B. Aluminium Powder Filled Nitrile Rubber Composites: Effect of Bonding Agents. *Kautsch. Gummi Kunstst.* **2004**, *57*, 641–644.
18. Anuar, J.; Mariatti, M.; Ismail, H. Properties of Aluminium and Zinc-Filled Natural Rubber Composites. *Polym. Plast. Technol. Eng.* **2007**, *46*, 667–674. [CrossRef]
19. Nassar, A.; Yehia, A.A.; El-Sabbagh, S.H. Evaluation of the Physico-Mechanical and Electrical Properties of Styrene-Butadiene Rubber/Aluminum Powder and Styrene-Butadiene Rubber/Cerium Sulfate Composites. *Polimery* **2015**, *60*, 100–108. [CrossRef]
20. Kang, S.S.; Choi, K.; Nam, J.D.; Choi, H.J. Magnetorheological Elastomers: Fabrication, Characteristics, and Applications. *Materials* **2020**, *13*, 4597. [CrossRef] [PubMed]
21. Mamunya, Y.P.; Davydenko, V.V.; Pissis, P.; Lebedev, E.V. Electrical and Thermal Conductivity of Polymers Filled with Metal Powders. *Eur. Polym. J.* **2002**, *38*, 1887–1897. [CrossRef]
22. Shen, X.Z.; Xie, S.M.; Guo, J.; Liu, Z.C. Microwave Absorbing Properties of Ternary Linear Low-Density Polyethylene/Carbonyl Iron Powder/Carbon Black Composites. *J. Appl. Polym. Sci.* **2009**, *114*, 3434–3439. [CrossRef]
23. Song, Z.; Xie, J.; Zhou, P.; Wang, X.; Liu, T.; Deng, L. Toughened Polymer Composites with Flake Carbonyl Iron Powders and Their Electromagnetic/Absorption Properties. *J. Alloys Compd.* **2013**, *551*, 677–681. [CrossRef]
24. Zhang, J.; Feng, Y.; Qiu, T.; Tang, C. Preparation and Characterization of Carbonyl Iron Powder/Millable Polyurethane Elastomer Microwave Absorbing Patch. *Polym. Compos.* **2014**, *35*, 1318–1324. [CrossRef]
25. Amutha Jeevakumari, S.A.; Indhumathi, K.; Arun Prakash, V.R. Role of cobalt nanowire and graphene nanoplatelet on microwave shielding behavior of natural rubber composite in high frequency bands. *Polym. Compos.* **2020**, *41*, 4362–4371. [CrossRef]
26. Darwish, M.S.; Al-Harbi, L.M. The Influence of Self-Heating Iron on the Thermal, Mechanical, and Swelling Properties of PDMS Composites for Organic Solvents Removal. *Polymers* **2021**, *13*, 4231. [CrossRef] [PubMed]
27. Alam, M.N.; Kumar, V.; Jo, C.R.; Ryu, S.R.; Lee, D.J.; Park, S.S. Mechanical and magneto-mechanical properties of styrene-butadiene-rubber-based magnetorheological elastomers conferred by novel filler-polymer interactions. *Compos. Sci. Technol.* **2022**, *229*, 109669. [CrossRef]
28. Alam, M.N.; Kumar, V.; Ryu, S.R.; Ko, T.J.; Lee, D.J.; Choi, J. Synergistic magnetorheological NR–NBR elastomer blend with electrolytic iron particles. *Rubber Chem. Technol.* **2021**, *94*, 642–656. [CrossRef]
29. *ISO-37*; Rubber, Vulcanized or Thermoplastic-Determination of Tensile Stress-Strain Properties. ISO: Geneva, Switzerland, 2017.
30. Flory, P.J.; Rehner, J., Jr. Statistical Mechanics of Cross-Linked Polymer Networks II. Swelling. *J. Chem. Phys.* **1943**, *11*, 521–526. [CrossRef]
31. Fröhlich, J.; Niedermeier, W.; Luginsland, H.D. The effect of filler–filler and filler–elastomer interaction on rubber reinforcement. *Compos. Part A Appl. Sci. Manuf.* **2005**, *36*, 449–460. [CrossRef]
32. Youn, I.S.; Kim, D.Y.; Singh, N.J.; Park, S.W.; Youn, J.; Kim, K.S. Intercalation of transition metals into stacked benzene rings: A model study of the intercalation of transition metals into bilayered graphene. *J. Chem. Theory Comput.* **2012**, *8*, 99–105. [CrossRef]
33. Demircan, C.A.; Bozkaya, U. Transition metal cation−π interactions: Complexes formed by Fe^{2+}, Co^{2+}, Ni^{2+}, Cu^{2+}, and Zn^{2+} binding with benzene molecules. *J. Phys. Chem. A* **2017**, *121*, 6500–6509. [CrossRef]
34. Liao, G.J.; Gong, X.L.; Xuan, S.H.; Kang, C.J.; Zong, L.H. Development of a real-time tunable stiffness and damping vibration isolator based on magnetorheological elastomer. *J. Intell. Mater. Syst. Struct.* **2012**, *23*, 25–33. [CrossRef]
35. Alam, M.N.; Kumar, V.; Lee, D.J.; Choi, J. Magnetically active response of acrylonitrile-butadiene-rubber-based magnetorheological elastomers with different types of iron fillers and their hybrid. *Compos. Commun.* **2021**, *24*, 100657. [CrossRef]
36. Chen, L.; Gong, X.L.; Li, W.H. Damping of magnetorheological elastomers. *Chin. J. Chem. Phys.* **2008**, *21*, 581–585. [CrossRef]
37. Li, Y.; Li, J.; Li, W.; Du, H. A state-of-the-art review on magnetorheological elastomer devices. *Smart Mater. Struct.* **2014**, *23*, 123001. [CrossRef]
38. Yang, J.; Gong, X.; Deng, H.; Qin, L.; Xuan, S. Investigation on the Mechanism of Damping Behavior of Magnetorheological Elastomers. *Smart Mater. Struct.* **2012**, *21*, 125015. [CrossRef]

39. Alam, M.N.; Debnath, S.C.; Choi, J. Nitrosamine-safe thiuram disulfide and benzothiazole sulfenamide as a synergistic pair of accelerators for the vulcanization of rubber. *J. Polym. Sci.* **2021**, *28*, 141. [CrossRef]
40. Alam, M.N.; Kumar, V.; Jeong, M.; Park, S.S. MgO as a cure modifier for reducing the conventional rubber vulcanization temperature at water's boiling point with improved vulcanization kinetics. *J. Polym. Res.* **2024**, *31*, 317. [CrossRef]

polymers

MDPI

Review

Research Progress of Natural Rubber Wet Mixing Technology

Qinghan Zhao, Fangyan Niu, Junyu Liu and Haishan Yin *

College of Electromechanical and Engineering, Qingdao University of Science and Technology, Qingdao 266100, China; zqhqust@163.com (Q.Z.); nfy5068@163.com (F.N.); ljy38324@126.com (J.L.)
* Correspondence: yinhais@163.com

Abstract: The performance of natural rubber (NR), a naturally occurring and sustainable material, can be greatly enhanced by adding different fillers to the NR matrix. The homogeneous dispersion of fillers in the NR matrix is a key factor in their ability to reinforce. As a novel method, wet mixing technology may effectively provide good filler dispersion in the NR matrix while overcoming the drawbacks of conventional dry mixing. This study examines the literature on wet mixing fillers, such as graphene, carbon nanotubes, silica, carbon black, and others, to prepare natural rubber composites. It also focuses on the wet preparation techniques and key characteristics of these fillers. Furthermore, the mechanism of filler reinforcement is also examined. To give guidance for the future development of wet mixing technology, this study also highlights the shortcomings of the current system and the urgent need to address them.

Keywords: wet compounding; fillers; natural rubber; research progress; rubber reinforcement; composites

1. Introduction

Extracted from the Brazilian rubber tree, natural rubber (NR) is a sustainable and renewable resource. Because of its exceptional physical qualities, including its exceptional elasticity, resilience, high tensile strength, and raw strength, it is one of the elastomers that is utilized in tire production the most. Because of its low heat buildup, resistance to fracture formation, and low hysteresis, NR is favored for tire production [1]. In most cases, the qualities of pure rubber are insufficient for practical uses; therefore, reinforcing fillers must be added to improve the properties of rubber [2]. The failure behavior of rubber composites is greatly influenced by the reinforcing fillers' particle size, form factor, surface activity, number of filler components, and dispersion condition in the rubber matrix [3].

Traditional mixing, sometimes referred to as dry mixing, is mostly accomplished by alternating or continuously mixing fillers with granular or solid rubber ingredients. Reinforcing fillers, such as carbon black, tend to clump when traditional dry mixing techniques are applied. This leads to inadequate filler dispersion in the rubber matrix, and the resulting rubber composite does not show the desired qualities.

The wet mixing technique is a novel approach to chemical solidification that involves combining filler solution and polymer latex (solution) following conventional dry mixing. Wet mixing can address issues with dust pollution, energy consumption, and other issues associated with traditional dry mixing since it uses a liquid phase to complete the mixing and dispersion of the filler and rubber. Wet mixing offers notable benefits, particularly for heavily packed rubber composites.

According to contemporary basic research on rubber composites, in order to manufacture rubber composites with outstanding overall performance, a thorough investigation of the issue of particle binding at the microscopic level is required. As illustrated in Figure 1, the three layers that make up natural latex's rubber particles are a viscous sol-gel layer made of rubber hydrocarbons with a small degree of polymerization in the innermost layer, a gel layer made of rubber hydrocarbons with a large degree of polymerization in the middle layer, and a protective layer made of proteins and lipids in the outermost layer.

Of these, the protective layer can help the rubber particles stay evenly distributed in the natural latex; however, because of its presence, it prevents the filler particles from coming into direct contact with the rubber particles in the natural latex, which in turn impacts the filler reinforcement's effectiveness.

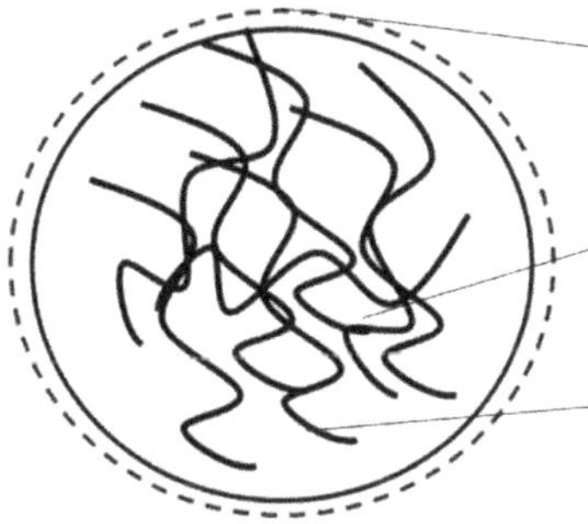

Figure 1. Schematic structure of natural latex rubber particles.

The filler particles must first have a high dispersibility to effectuate the reinforcement of natural latex by reinforcing filler in an emulsion condition and to obtain an outstanding reinforcing effect. To produce a uniform distribution of fillers in the natural latex film and improve the reinforcing effect, co-precipitation of the natural latex particles and reinforcing filler particles is required when coalescence occurs. Rubber hydrocarbon and reinforcing filler need to come into direct contact with one another; the greater their mutual contact area, the better and more excellent the reinforcing effect.

This research examines the wet fabrication and application of graphene (GO)/NR, carbon nanotubes (CNTs)/NR, silica (SiO_2)/NR, and carbon black (CB)/NR composites. While there are a lot of reviews on natural rubber composites, there are not many on the wet preparation of NR composites. This review aims to close that gap by offering a thorough analysis of the body of research as well as a forecast for the promising future possibilities in this area.

2. Wet Preparation of NR Composites

The two primary challenges of agglomeration and filler–matrix interaction must be overcome to produce NR composites with homogenous filler dispersion. Stronger filler–matrix contact and high filler dispersion are necessary for the composites to have a reinforcing effect. The three main wet preparation techniques for NR composites are sol-gel, latex, and solution mixing.

2.1. Latex Mixing Method

This approach uses natural latex as the matrix and stabilizes the reinforcing particles through a series of techniques that culminate in a homogenous mixture and agglomeration. Zhao et al. [4] prepared GO/NR nanocomposites by using the latex blending method. It was discovered that the latex mixing approach may be used to accomplish the homogenous dispersion of GO in the NR matrix. Without compromising the final strength, the uniformly distributed GO greatly raised the tensile strength and energy storage modulus of NR at a reduced filling percentage. This process is less harmful to the environment than the solution mixing method since it does not require costly and highly toxic organic solvents [5].

2.2. Solution Mixing Method

By combining the rubber solution and nanofiller dispersion and then draining the solvent, rubber composites were produced using this technique [6]. Yang et al. [7] prepared SiO_2/NR composites by accomplishing the mixing between silica and rubber molecular chains in an organic phase solution. The silica/natural rubber composites made via the solution approach presented superior filler dispersion when compared to those made by the dry mixing method. The main disadvantages of this technique, however, are its high cost, the need for a lot of solvents during the solvent removal process, and the environmental issues related to disposing of solvents [8]. The final properties of the rubber composites are greatly influenced by the solvent used during the preparation process [5].

2.3. Sol-Gel Method

Precursors like siloxanes or metal salts are added to the rubber matrix in the sol-gel process to facilitate hydrolysis and condensation processes, which produce uniformly distributed nanoparticles in situ. Tetraethoxysilane (TEOS) is a widely utilized precursor for the preparation of SiO_2/NR composites. Poompradub et al. [9] used ethyl orthosilicate (TEOS) as a silica precursor to create SiO_2/NR composites utilizing the sol-gel process. When compared to commercial (ectopic formation) SiO_2/NR vulcanizates, the mechanical and thermal characteristics of SiO_2/NR composites, including silica produced in situ using the sol-gel method, were greatly enhanced. While the non-in situ generated commercial silica agglomerated in the NR matrix, the in situ-created silica particles were evenly dispersed throughout the matrix.

3. Progress in the Application of Wet Mixing Process in Natural Rubber Composites

3.1. Carbon Black (CB)/Natural Rubber Composites

3.1.1. Mechanism of Carbon Black Reinforced Rubber

Carbon black is thought to be the most efficient filler material additive since it makes rubber-based polymers stronger and harder [10]. The binding effect between the carbon black particles and the rubber matrix is the primary mechanism by which carbon black is reinforced with rubber. Due to the small particle size of carbon black, its large contact area with the rubber matrix, and its ease of infiltration by rubber molecules, physical bonding mostly manifests as van der Waals force. Simultaneously, the presence of active sites on the surface of carbon black particles, which have a higher surface energy and are easier to firmly connect with the rubber matrix, is the primary cause of chemical adsorption. Numerous academics and researchers have investigated the workings of carbon black reinforcing rubber in great detail, and they have put forth the following models of reinforcement.

According to the principle of volumetric effect, scattered carbon black in the rubber matrix creates a solvent effect that raises the rubber's viscosity and has a strong reinforcing impact. The majority of the mechanical characteristics of carbon black rubber composites are associated with rubber's tendency to increase viscosity. Rubber's mechanical strength is increased because the molecular chains adsorbed on the surface of carbon black particles exhibit an orientated state. When there is a tensile tension applied to the rubber, if the carbon black adsorption on the rubber molecular chain is weak, the "gap phenomenon" will occur [11]. Rubber can more easily penetrate areas with higher carbon black particle surface adsorption energies. The degree of surface structure of carbon black can be improved by actively treating the particles' surfaces, which increases the material's ability to permeate the rubber matrix and distribute itself evenly.

However, the surface structure theory [12] contends that the active filler's surface is not smooth and that the filler's size range and surface roughness would affect the rubber's characteristics. Donnet et al. [13] first saw the surface of carbon black particles using a scanning tunneling electron microscope. They discovered that the surface of the carbon black is uneven and rough, with sharp edges separating the particles. Based on these observations, they proposed a model theory for carbon black. Later, Donnet et al. discovered numerous localized crystal structures on carbon black's surface and put forth the notion of the polymer adsorption carbon black surface model.

A molecular chain sliding model was proposed by Dannenberg et al. [14] in which diverse adsorption activities are exhibited by carbon black particles and rubber macromolecules gliding over the surface. Under tension, the rubber chains adsorbed on the carbon black surface lengthen and slide. Hysteresis energy loss into thermal energy occurs during the molecular chain sliding process, preventing the rubber substance from being destroyed. Most molecular chains do not require sliding when stretched back to their initial length, demonstrating the phenomena of stress relaxation.

The interaction between the filler and the rubber is crucial for the reinforcing of filled rubbers, as several investigations have demonstrated. Generally speaking, the quantity of binding gel that forms between the filler and the rubber is how much of an interaction there is. The inclusive rubber model, the rubber-shell model, and the glassy rubber-shell model are three of the most developed theoretical models of binding gel that have been put forth.

The inclusive rubber model, as described by Medalia [15] and Kraus [16], states that a certain amount of rubber is contained in the spaces of the filler network, whereas aggregates or agglomerates of active fillers can create a network of fillers due to their adsorption (Figure 2a). Due to its decreased mobility, the encapsulated rubber no longer contributes to the rubber's elastic behavior and joins the packing, increasing the packing's effective volume.

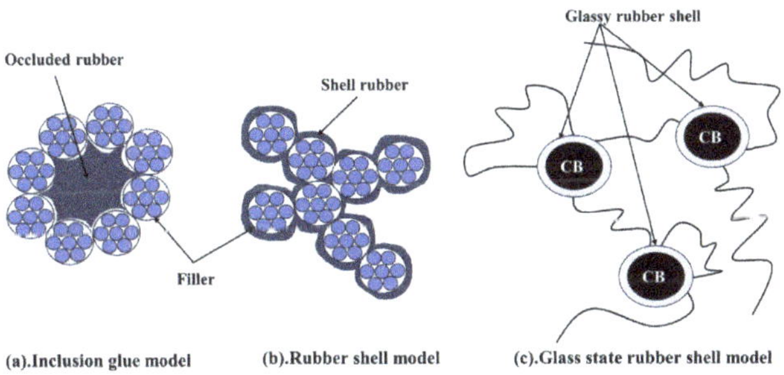

Figure 2. Binding gel theory model diagram.

Rubber macromolecular chains may be chemisorbed by the filler's surface active sites as a result of chemisorption around the filler particles, as demonstrated in Figure 2b, according to the rubber shell model put out by Smit [17] and Pliskin [18]. Using Figure 2c as support, O'Brien [19] further suggested that the rubber molecular chains adsorbed around the filler particles are practically glassy and have no motility. In this regard, LeBlanc [20] conducted numerous experimental investigations utilizing H1-NMR, and he suggested that rubber adsorbed in the immediate area of the filler particles as a layer that was strongly linked. The external force field is in a glassy state and does not affect the macromolecular chains, which interact significantly with the filler particles. When the adsorption force weakens, rubber molecule chains become more mobile, and the binding layer becomes comparatively free. As seen in Figure 3, the rubber's solvent can dissolve the opposing outer layer, or free rubber, which is not constrained by the filler particles.

An interfacial model based on the interaction of fillers and rubber macromolecules is called the binding rubber model. Based on this binding rubber model, Fukahori [21] suggested a new interfacial structure model based on the stress analysis results, accounting for the fact that the rubber's volume expands in a large deformation state. As seen in Figure 4a,b, the bonded rubber bilayer structure model is the core of this model. According to the model, the structure of the binding rubber adsorbed at the carbon black's periphery is made up of two parts: rubber layers with distinct moduli and an uncrosslinked structure. The inner structure is the glassy hard layer (GH) of the polymer in its glassy state, which has a thickness of around 2 nm, and the outer structure is the sticky hard layer (SH) of the polymer, which has a thickness of between 3 and 8 nm. The outermost layer, which has a thickness of 3–8 nm, is a sticky hard layer with limited polymer macromolecule mobility. The combined thickness of the two layers is around 5 nm for carbon black particles with a smooth surface and 10 nm for carbon black particles with a high degree of roughness on their outer layer.

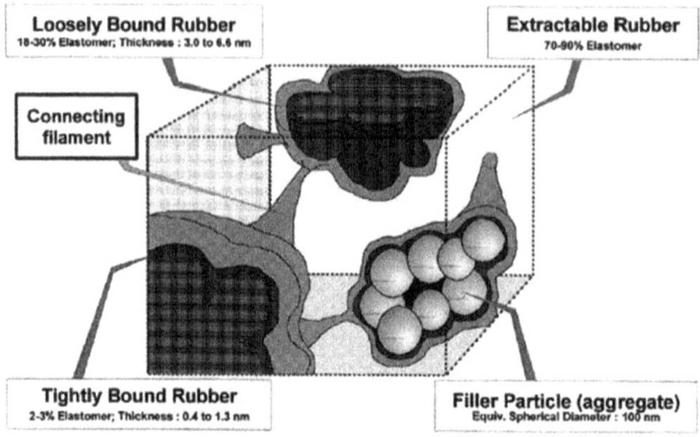

Figure 3. Theoretical diagram of rubber layer distribution around carbon black particles [20].

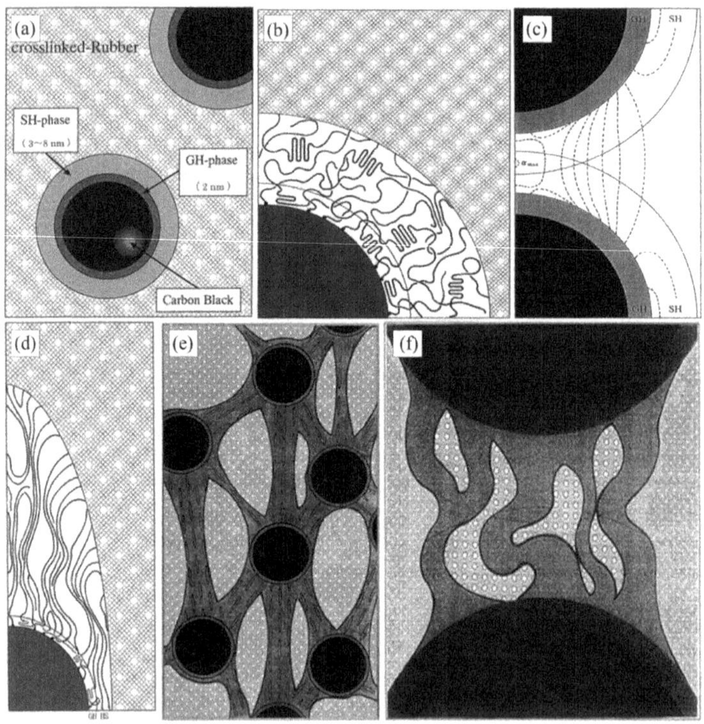

Figure 4. Diagrammatic representation of the binding gel bilayer model (**a**) A new double-layer interface model consisting of a GH layer and a SH layer. (**b**) Detailed molecular structures in the GH and SH layers as vulcanized. (**c**) Overlapped SH layers in triaxial expansion along extension direction at $\varepsilon_0 = 10\%$ and $\phi = 0.2$. (**d**) Molecular movements within the SH layer under large extension. (**e**) Supernetwork structure of carbon particles interconnected by strands of oriented molecules. (**f**) Buckling of extended and oriented molecular bundles [21].

The GH layer does not affect the sharp increase in stress under large magnitudes because the effect of the polymer glassy hard layer is limited to increasing the effective diameter of the carbon black particles, and its contribution to stress remains constant regardless of the amount of strain applied to it. While the GH layer cannot undergo orientation due to the poor mobility of the molecular chains, it does not contribute to the increase in the modulus of the rubber. At the same time, the viscous hard layer behaves similarly to the rubber matrix at small strains and contributes less to the modulus (Figure 4d). When the strain gradually increases, the SH layer undergoes orientation, a phenomenon that plays an important role in the increase in the rubber's modulus. Furthermore, at higher carbon black contents, the SH layers separating various binding rubber particles may overlap, creating a super-network structure (Figure 4c). This super-network structure changes orientation and hardens under high strains, which adds to the modulus. Little holes will grow between the bundles of rubber macromolecular chains as they are assembled, and these holes can absorb some energy, postponing the rubber molecular chains' eventual disintegration (Figure 4e). Simultaneously, the polymer's volume expands under tension due to the existence of these microscopic pores.

Haghgoo et al. [22] investigated how multiscale fillers affected the electrical conductivity and resistivity of multiscale nanocomposites made of reinforced polymer and CB/CF. The outcomes demonstrated that the conductivity qualities and interoperability in the direction of electron flow were enhanced by the multiscale fillers. These enhancements were in good agreement with the microstructure's multiscale filler modifications. Multiscale fillers can be added to polymer matrix composites' microstructure to improve their electrical characteristics and reduce percolation thresholds.

Drawing on the micro-level analysis of the reinforcing model discussed above, we can employ several techniques to enhance the surface quality and surface activity of carbon black, resulting in NR composites with consistent filler dispersion. To create high-performance natural rubber composites, we can simultaneously lessen the force between carbon black particles, increase the force between carbon black and the NR matrix, and encourage the combination of carbon black and latex particles.

3.1.2. CB/NR Composites Wet Mixing Preparation Process

Due to their hydrophobicity and/or high elemental carbon content (90~99%), carbon black particles agglomerate and produce unstable dispersions when dissolved in water at a pH close to the isoelectric point (IEP). Consequently, it is essential to regulate the surface charge of carbon black and limit its particle size to produce extremely stable carbon black aqueous dispersions [23]. An old issue in materials science is the creation of stable carbon black dispersions in water, either for use as pigments or as reinforcing filler particles in polymers. The application properties of carbon black are weakened due to their tendency to form agglomerates due to their tiny primary particle size and high inter-particle forces. Table 1 [24] illustrates the three primary categories into which the techniques employed in the literature to characterize the dispersion of carbon black can be separated.

Table 1. Carbon black water dispersion main method.

Form	Methodologies
1	Using water-dispersible amphiphilic polymer structures or traditional surfactants to disperse carbon black in water
2	Carbon black suspension in water, precipitation encapsulation, and emulsion polymerization
3	Methods like using the right oligomers to ball mill carbon black powder in the solid state

As was previously noted, carbon black was dispersed in water using conventional surfactants or water-dispersible amphiphilic polymer architectures. Fresh natural latex was combined and agitated with a carbon black slurry that had been made by Alex et al. [25] while surfactants (alkali metal salts of fatty acids) were present. To create carbon black wet master gum, drying and acid flocculation were applied in the end. The results demonstrated that protein displacement and high adsorption onto the rubber particles were caused by adding fatty acid soap-based surfactants to the carbon black slurry. This process converts the protein-stabilized latex into a system that is stabilized by surfactants. The adsorbed anions combine with the acid in the surfactant enclosing the latex to generate undissociated surfactants, which rob the latex particles of their stabilizer. In addition, the latex's Menni viscosity is low, and the surfactant might sensitize it to promote quick solidification.

Martínez-Pedrero et al. [26] studied the rheological properties of binary colloidal mixtures. During the sonication step, carbon black's dispersion state was improved with the addition of sodium dodecyl sulfate (SDS). It was then agitated and combined with natural latex to produce a gel. The findings of the experiment demonstrated that the carbon black particles experienced a structural transformation in the binary colloidal mixes, bridging the natural rubber particles together. This phenomenon of bridging allows a generative network of fractal clusters to emerge, which in turn forms an elastic solid. The interparticle attraction energy between CB particles and NR droplets, which is impacted by the surfactant concentration, is primarily responsible for controlling this transition. The homogeneity of the dispersion is controlled before it loses stability by the adsorption of surfactant molecules on the carbon black surface, which modifies the interaction between the two particles. Applying shear can overcome the energy barrier preventing the bridging effect in the more stable scenario.

Using a similar strategy, Dong et al. [27] were able to effectively diminish the interaction between carbon blacks and improve the interaction between carbon blacks and rubber molecule chains, hence improving the dispersion of carbon blacks. Furthermore, it was shown that while the hardness and elongation at break showed the opposite pattern, the mechanical characteristics increased first and then dropped as the size of the carbon black particles increased. This is because more active sites are present in carbon black particles with smaller particle sizes since they have bigger specific surface areas. As a result, there is a greater degree of chemical bonding and physical adsorption between the rubber molecular chain and the surface of carbon black, resulting in improved bonding. The performance of vulcanized rubber made by the latex co-sinking process is still superior to that of conventional dry mixing, even though too-small carbon black particles are easy to agglomerate in the rubber matrix and reduce its mechanical qualities.

As previously shown, surfactants or polymers grafted with carbon black have improved dispersion stability. Carbon black can also employ its hydroxyl group and other active sites as low molecules for cationic, anionic, and free radical polymerization processes. The inter-particle resistive effect of the carbon black is enhanced by the grafting of the polymer onto its surface, which can successfully stop the carbon black from clumping together in the aqueous phase. The production of new or modified polymers has made extensive use of ultrasonic, high-energy electron beam (EB), and C-ray radiation in recent years as effective and ecologically acceptable free radical polymerization techniques.

By grafting acrylic acid (PAA) onto the CB surface with high-energy electron beam irradiation (EB), Jiang et al. [28] created water-dispersible CB. When compared to C-ray irradiation, EB is more efficient [29]. In addition to increasing the CB's hydrophilicity and lowering its surface energy, grafting also improves the CB's average aggregated particle size and dispersity when compared to unmodified CB. Fu et al. [30] grafted polyethylene glycol 400 onto the surface of carbon black using an in situ liquid phase grafting approach. Better dispersion of the modified carbon black in the rubber system increases the mechanical characteristics and the interaction between the modified filler and rubber. The dispersibility of carbon black can be effectively increased by the polymer graft modification method; however, because linear polymers are susceptible to molecular chain entanglement, carbon black particles may reaggregate and form new aggregates.

Because of their distinct structure, high number of reactive functional groups, good solubility, low viscosity, non-entangled molecular chains, and simplicity of synthesis, hyperbranched polymers are of interest [31,32]. On the surface of carbon black, Han et al. [33] grafted end-carboxylated hyperbranched poly(2-hydroxypropane-1,2,3-tricarboxylic acid) (Figure 5). The modified CB greatly enhanced the dispersibility and wettability of the composite; its average particle size was significantly smaller, and its particle size distribution was narrower, according to the results. This could be explained by the hyperbranched polymer development on the CB surface, which raises the spatial site resistance and electrostatic repulsion between CB particles. This, in turn, increases the repulsive force between CB particles, making it more challenging for CB to assemble [34,35]. A significant amount of carboxyl groups, which are hydrophilic oxygen-containing functional groups, are present in the topmost layer of the hyperbranched polymer layer and may help to improve the dispersion of carbon black in water.

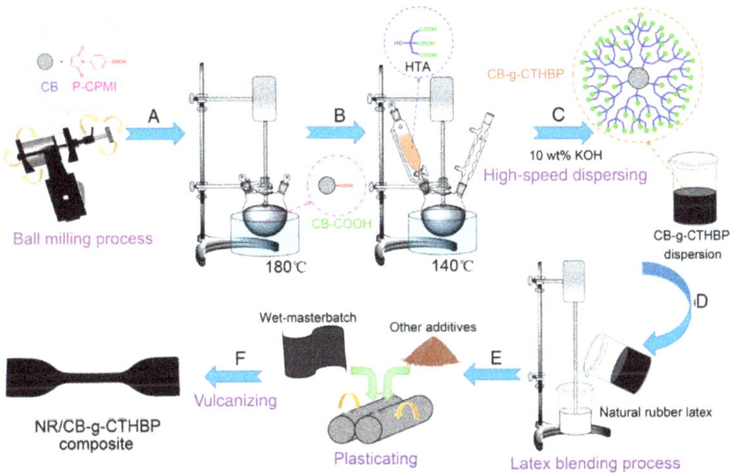

Figure 5. Schematic diagram illustrating the preparation route (**A**) Reaction to prepare CB-COOH. (**B**) Reaction preparation of CB-g-CTHBP. (**C**) Preparation of CB-g-CTHBP dispersions (**D**) Wet mixing of dispersion with latex. (**E**,**F**) Compounding to make rubber compounds [33].

As was previously indicated, the emulsion coalescence process can be used to create composites by adding coating resin to a mixed system of carbon black and latex to form a powder system [36] With the use of the emulsifier AEO-9, Lin et al. [37] generated the CB emulsion, which they subsequently mixed and combined with NR latex to create a powder system. After being heated in a water bath, carbon black and natural rubber thermally aggregate. Afterward, flocculation and drying were accomplished using a $CaCl_2$ solution. According to the findings, natural rubber composites filled with carbon black and powdered had superior mechanical and dynamic properties than those made by traditional dry blending. The improved composite properties were also attributed to the good dispersion of CB in the rubber matrix and the improvement in rubber–filler interactions.

Process approaches can also be used to obtain a homogenous mixing of carbon black with natural latex in addition to carbon black modification. The earliest method to homogeneously disperse carbon black in latex was developed by Cabot [38,39]. It involved using high-pressure jetting technology to quickly mix and flocculate natural latex with carbon black slurry, but the equipment and process requirements are complex. Afterward, Han et al. [40] used high-speed impact jet processing to directly combine natural rubber latex (NRL) with carbon black (CB) to create rubber composites. The filler dispersion and latex were combined using a side-mounted jet mixer to create a combination that was then baked in an oven to create master rubber. The findings demonstrated that, in comparison to the traditional dry procedure, the jet compounding technique allowed for a more uniform dispersion of CB into natural rubber. Furthermore, a comparison of the jet Reynolds number mixing forms revealed that the carbon black was more likely to agglomerate and have bigger particle sizes in the matrix and that the particles in laminar flow mixing were subjected to less shear stress. Turbulent jet mixing, on the other hand, successfully prevents particle agglomeration. It improves the interaction between rubber and fillers, diminishes the network of fillers, and improves filler dispersion in the rubber matrix, all of which increase load transfer efficiency.

Using ultrasonic technology to pretreat natural latex can boost the reinforcing effect by increasing the contact area between rubber hydrocarbon particles and filler particles and rupturing the protective barrier. Simultaneously, filler particle size can be decreased via ultrasonic treatment, improving filler dispersion uniformity. Galinovskiy et al. [41] explored the effect of ultrasound and ultrasonic jet treatment methods on the dispersion of nanosuspensions. The experimental results indicate that ultrasonic jet dispersion is a potential method for particle deagglomeration and a solution to the issue of achieving the required degree of dispersion in suspensions. Using a novel technique called field emission scanning electron microscopy (FESEM), Cattinari et al. [42] examined how the nanoscale structure of NR–CB concretions changed as the solvent evaporated. Osmium vapor was used to specifically chemically immobilize the sample, preventing it from submerging in the immobilizer-containing liquid solution while maintaining the colloidal structure of the NR particles. We looked into and examined the impact of external physical stressors (shear and ultrasonic) on the coagulation of NR latex and CB slurry. It was discovered that combining the two parts in a colloidal suspension while subjected to sonication kept the CB filler in the form of tiny aggregates (20–200 nm), with excellent filler homogeneity surrounding the NR pellets (Figure 6). It was determined through experimentation that the NR's sonication did not affect the structure (Figure 6g,h). The NR and CB were evenly distributed, and heterogeneity was mostly responsible for the clots' aggregation (Figure 6a,b). Some of the aggregates between the NR are highlighted in images captured at high magnification (Figure 6c,d). Some of the aggregates have sunk into the sphere, as shown by the red arrows in Figure 6f, and these aggregates formed a distinct contact on the sphere's surface (Figure 6e).

Using the grinding balls' inherent gravity and the impact of the balls colliding, ball milling technology breaks up aggregates and helps refine particle size. Yamamoto et al. [43] used a ball mill under various ball milling conditions to study the dispersion characteristics of carbon black in solvents. The results of the trials demonstrated that the dispersion performance rose as the stirrer speed and bead loading ratio increased and that there was a strong correlation between the two parameters and the impact energy of the beads. To prepare silica and carbon black dispersions for the creation of natural rubber composites, Hamran et al. [44] used ball milling sonication. The dispersion treated with ultrasonication followed by ball milling demonstrated the best dispersing ability, and the results indicated that this combination of treatment methods was the optimum way to minimize the particle size of silica and carbon black.

Sui et al. [45] suggested a completely formulated wet mixing approach for carbon black. Using all of the formulation's fillers, they ball-milled the pretreatment to create an aqueous dispersion slurry, which was then combined physically, via stirring, with the latex. Ultimately, a twin-screw was used to accomplish the flocculation and dewatering steps, and an oven was used to dry and further treat the rubber masterbatch. The generated rubber composites containing carbon black exhibited a homogeneous dispersion and distribution, tiny particle size, and no visible aggregation phenomenon due to the completely formulated wet continuous mixing procedure that combines wet and continuous mixing methods. The carbon black dispersion was enhanced in comparison to the conventional dry mixing, and the composites' tensile strength, elongation at break, and cut resistance all increased by 9.4%, 9.6%, and 35%, respectively. On the other hand, constant wet mixing permits the molecular chains in natural rubber to be evenly distributed throughout the rubber by minimizing chain breaking. Consequently, the fully formulated wet blend's Menni viscosity is higher than the dry blend's.

In summary, dispersant or surfactant can be used to modify carbon black, even though it is dense, hydrophobic, and settles a lot in aqueous dispersion. Furthermore, the modified carbon black frequently cannot match the latex's settling speed due to its differing polarity and density, which may have an impact on the carbon black filler's stability and dispersion in the rubber matrix. Since the rubber phase in the mixed product is not cross-linked and tends to aggregate due to the thermodynamics of carbon black, we should try to take steps to dehydrate and dry the product quickly to reduce its viscosity and lengthen its survival period. To minimize wastewater, waste gas, and solid waste, shorten the production cycle, reduce the three wastes, and improve product quality and performance, wet mixing technology's benefits for the economy and environment should also be carefully taken into account during the production process. This is because there is an antagonistic relationship between the search for wet mixing technology's advantages and the development of low-carbon, energy-saving, and environmentally friendly directions.

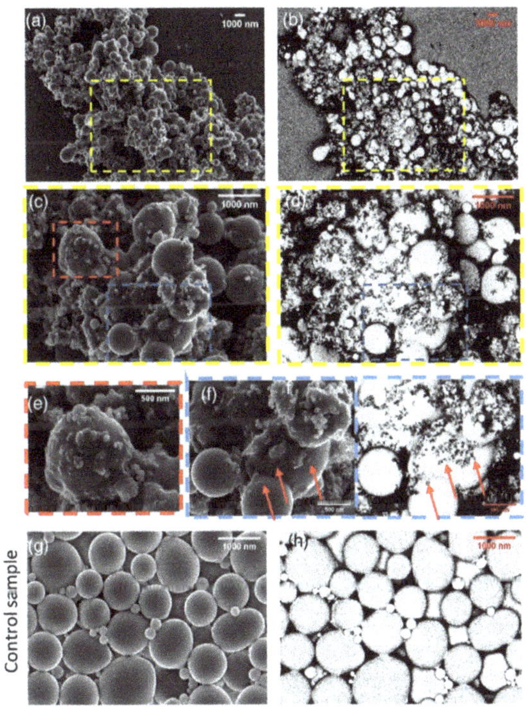

Figure 6. FESEM micrographs of the NR–CB heterocoagulum obtained with sonication during mixing: Mass ratio in the liquid suspension = 40% CB/rubber and the structure was fixed at t = 1 h. (**a**) Large view SE and (**b**) BSE images of the same region showing the homogeneous distribution of NRs and CB filler (magnification: ×5000). (**c**) Magnified SE and (**d**) BSE views corresponding to the yellow squared region in (**a**,**b**) (magnification: ×10,000), highlighting the partial coalescence between the NR globules that are surrounded by several small CB aggregates. (**e**) Magnified SE of the red boxed region in (**c**) (magnification: ×80,000). (**f**) Magnified SE and BSE views of the blue boxed region shown in (**c**,**b**) (magnification: ×40,000), showing that the CB filler partially sinks toward the inner part of the NR globules. (**g**) Magnified SE and (**h**) BSE views corresponding to the control sample composed only of sonicated NR globules [42].

3.2. NR/Silica (SiO_2) Composites

3.2.1. Mechanism of Silica Reinforced Rubber

Silica's reinforcing process is quite similar to that of carbon black. When compared to carbon black, silica's biggest distinguishing attribute is its surface activity. Silica is an amorphous structure of silicon dioxide, having silicon atoms in the center and oxygen atoms at the apex, forming an irregular tetrahedral shape. The silicon atoms on the surface are unevenly organized, and the hydroxyl groups that connect them change depending on the chemical reaction. The numerous hydroxyl groups on the surface of silica can be classified into three types: double hydroxyl, isolated hydroxyl, and neighboring hydroxyl.

The adsorption–slip theory, which states that fillers strengthen rubber primarily through molecular chain slip when the composite is stretched by external pressures, is now generally accepted. The addition of silica filler to rubber results in the binding of several silica particles to the rubber molecular chain, which in turn binds each silica particle to numerous rubber molecular chains. When an external force is applied, the rubber molecules will align themselves in the force's direction. This is because the silica particles are separated by several rubber molecular chains, each of which has a different length. The longer chains are forced to slip on the silica surface when the shorter chain segments restrict the orientation of the larger chains after they have finished their tensile orientation. The molecular chains connecting the silica particles are completely stretched as the orientation proceeds. To achieve a greater modulus, each rubber molecular chain segment is currently subjected to a more uniform external force thanks to the silica-reinforcing effect of the rubber. This reinforcing structure will be destroyed, the slip orientation will cease, and the composite material will fracture when the external force exceeds a particular critical amount.

The silica exhibits a high specific surface area and strong surface adsorption. It is typically presented in an aggregated state, which allows for a larger contact area with rubber and promotes the formation of a bond between the two materials. The crystallization effect of the particles causes the adsorption layer to rise, resulting in a smaller particle diameter than the spacing between them, which plays the role of physical reinforcing [46]. Chemical reinforcement can be created when the double bond of the rubber molecular chain reacts with the hydroxyl group present on the surface of silica [47]. Bonded rubber is created when the alkaline reactive groups of rubber react with the acidic hydroxyl groups on the surface of silica [48]. This reaction between the two groups creates silica-reinforced rubber.

3.2.2. SiO_2/NR Composites Wet Mixing Preparation Process

Silica (SiO_2) is the preferred filler for tire tread applications because it has lower rolling resistance and higher wet traction than carbon black. Utilizing silica as much as possible instead of traditional carbon black while preparing NR-based green tire rubber compounds is crucial for the automotive sector, given the current state of the world and environmental concerns [49]. Generally speaking, rubber's mechanical qualities are enhanced by silica addition. However, silica is more polar due to the abundance of hydroxyl groups on its surface, which reduces the interaction between silica and rubber. Furthermore, silica dispersion is typically poor in carbon black-filled compounds made using the traditional mixing method. The viscosity rises dramatically with substantial additions of silica, which complicates processing and puts undue strain on processing machinery. The strong contact between silica particles is responsible for the viscosity increase. Two approaches can be taken to address this issue: either the hydrophilic silica surface is modified, or the rubber's non-polar structure is functionalized to make it more polar [50]. Tables 2 and 3 include a list of papers on the wet filling of natural latex with silica.

Table 2. Summary of modified silica/natural rubber composite methods improved in this paper.

NR Composites	Modified Treatment Method/Reagent	Preparation Method	Year/Reference
NR/SiO_2	Methyltriethoxysilane, vinyltriethoxysilane, γ-aminopropyltrimethoxysilane	Sol-gel method	2014 [51]
NR/SiO_2/GE	Electron-beam irradiation	Latex mixing	2019 [52]
NR/SiO_2	Deep eutectic solvents (DES) 3phr	Mechanical mixing	2022 [53]

Table 2. *Cont.*

NR Composites	Modified Treatment Method/Reagent	Preparation Method	Year/Reference
NR/SiO_2	Cystamine dihydrochloride (CDHC)	Latex mixing	2020 [54]
NR/SiO_2	AEO-9, KH-590	Solution compounding	2022 [6]

Table 3. Summary of wet preparation methods for silica/rubber composites.

NR Composites	Synthesis	Flocculants/Methods	Year/Reference
NR/SiO_2	Latex compounding self-assembling techniques	$MgSO_4$	2019 [55]
NR/SiO_2	Latex co-coagulation method	3 wt% acetic acid	2020 [56]
NR/SiO_2	Latex co-coagulation method	Calcium chloride/acetic acid	2020 [57]
NR/ENR/SiO_2	Wet masterbatch technique	Ethanol	2016 [58]
NR/SiO_2	Solution compounding	Solvent removal	2022 [6]
NR/SiO_2	Latex co-coagulation method	Calcium chloride	2011 [59,60]
NR/SiO_2	Latex co-coagulation method	Acetic acid	2022 [61]

Silane coupling agents (SCAs) are used to alter the silica surface. Amphoteric surface modifiers and interfacial compatibilizers are silane coupling agents (SCA). Chemically speaking, SCA has at least one alkoxy group that, by dehydration condensation, can form stable Si-O-Si structures with SiO_2 particles. Reactive groups, particularly reactive sulfur atoms, are also present in SCA and can react chemically with diene rubber macromolecules.

The most practical and technologically advantageous silane coupling agents for silicone-based NR compounds are sulfide–alkoxy silane coupling agents, according to research by Kaewsakul et al. [62] on silane coupling agents with various particular functionalities. By comparatively offering higher filler–rubber contacts than single alkoxy or sulfide-based silanes, these chemicals can effectively lower the viscosity and filler–filler interactions of the compounds, leading to a notable increase in mechanical characteristics.

By altering SiO_2 with 3-mercaptopropylethoxybis(tridecylpentamethoxy)silane (Si-747), a long-arm silane coupling agent, Zheng et al. [63] created NR/SiO_2 composites. Si747 is a type of SCA with two long arms that can form abundant and concentrated hydrogen bonds with SiO_2, and the creation of an NR/NR masterbatch will be more satisfying when the total amount of Si747 is sufficient to reduce the electronegativity of SiO_2. When the total concentration of Si747 is sufficient to counteract the electronegativity of SiO_2, a more suitable silica/NR masterbatch will be prepared. Some papers on wet filling of natural latex with silica are listed in Tables 2 and 3.

The hydroxyl groups on the surface of silica react with conventional silane coupling agents to produce a lot of volatile organic compounds (VOCs), which degrade rubber composite performance and contaminate the environment [64]. By reacting fatty alcohol polyoxyethylene ether (AEO-X) with varying polyether lengths with bis[γ-(triethoxysilyl)propyl]-tetrasulphide TESPT (Si69), Li et al. [65] created a range of novel coupling agents with low VOC emission, which they then integrated into silica/natural rubber nanocomposites. The findings demonstrated that the new coupling agents could properly balance the tires' "magic triangle" performance and greatly enhance silica dispersion in the rubber matrix. SiO_2/NR composites showed improvements in wet slip resistance of 14.2%, abrasion resistance of 20.3%, and rolling resistance of 15.8%.

A set of low volatile organic compound (VOC) Mx-Si69 couplers (Figure 7) was applied to SiO_2/NR nanocomposites by Zhai et al. [66] These couplers correspond to the number of ethoxyl groups in bis-(γtriethoxysilylpropyl)-tetrasulphide (Si69) substituted by aliphatic polyether chains (x = 1, 2, 3, 4, 5, 6). Excellent performing tire treads were ready. First off, M1-Si69 is the optimal option for both Si69 and Mx-Si69. In addition to giving the tires exceptionally good anti-slip performance and very low energy loss, applying the recently developed M1-Si69 coupling agent to SiO_2/NR nanocomposites can also drastically cut down on VOC gas emissions.

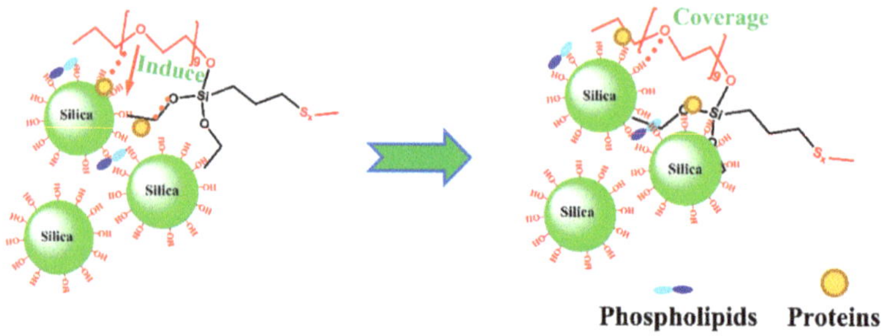

Figure 7. Schematic diagram of silica modified with Mx-Si69 induced by NRCs [66].

Different ratios of TWEEN-20 and TESPT were chosen by Xiao et al. [67] to alter silica and create NR nanocomposites. The lengthy fatty chain portion of TWEEN—20 is more suited for rubber, while the terminal hydroxyl group and polyether portion can interact chemically and physically with silica. Consequently, TWEEN-20 can significantly improve silica dispersion in NR composites. In addition to lowering volatile organic compounds (VOCs), the synergistic interaction of TWEEN—20 and TESPT can enhance the ideal dispersion of silica in the rubber matrix. When the ratio of TESPT-to-TWEEN-20 was 2:1, the rubber nanocomposites' dynamic and static characteristics were at their best.

Premature scorching during the blending process is a problem caused by TESPT because the silanization reaction demands a temperature range of 140 °C to 150 °C. Alternative silanes such as octanoylthiopropyltriethoxysilane and bis-triethoxysilylpropyl disulfide (TESPD) were introduced to get around this problem. Therefore, the effectiveness of silica-filled rubber products mostly depends on the proper arrangement of essential system components [68]. When Kaewsakul et al. [68] examined two silane coupling agents, TESPT produced better mechanical qualities, filler–rubber interaction, and dispersion stability. TESPD extended the rubber's coking period, with ideal loadings of roughly 9.0 weight percent (about silica) for both. Additionally, it was shown that DPG, when used as a catalyst, may enhance the silanization reaction between the silane coupling agent and silica, with the best loading being 2.0 weight percent.

To improve the interaction of silica-filled compounds, certain polar rubbers with functional groups that can interact with the silica surface have been studied as silane coupling agent alternatives. Using epoxidized natural rubber (ENR) as a bulking agent, Sengloyluan et al. [69] studied tire tread rubber made of silica-based natural rubber (NR). Epoxidized natural rubber was added, which enhanced the dispersion of silica in the rubber and decreased the rubber's Menni viscosity, Payne effect, flocculation rate constant, and packing network factor. Epoxidized natural rubber (ENR) was employed by Xu et al. [70] as an interfacial modifier to enhance the mechanical properties of SiO_2/NR composites. This resulted in the formation of covalent bonds between the epoxy groups on the ENR chain and the Si-OH groups on the surface of silica, which enhanced the dispersion of SiO_2 in the rubber matrix and reinforced the interfacial bond between NR and SiO_2. Not only does the ring opening response happen during vulcanization, but it also happens during compounding, and the latter is more significant because of the interaction of temperature and mechanical forces. He et al. [71] prepared a silica dispersion by a high shear process using liquid natural rubber (NRL) as a bulking agent and then blended it at low speed with latex. Finally, flocculation and acetic acid drying were used to create natural rubber–silica composites. The tensile and tear strengths increased by 37.9% and 18.9%, respectively, and the silica dispersion in the NR matrix was enhanced with the addition of NRL, which also encouraged vulcanization. Natural rubber masterbatch was created by Wang et al. [72] from NRL-treated silica. It had improved wet slip resistance, good compatibility with rubber, good dispersion in rubber matrix, and a diminished Payne effect of the composites. The performance of the rubber composites was significantly enhanced by the dispersion and interfacial adhesion of silica in the rubber matrix.

The most practical method for silica surface functionalization is to use silica silanol groups that have reacted through the sol-gel reaction with appropriate organoalkoxysilane reagents. However, because of spatial site resistance effects, the conversion of organosilanes to SiO_2 particles via the sol-gel process somewhat reduces as the size of the substituent groups on the organosilanes rises. To increase the compatibility of organosilanes with NR, Theppradit et al. [51] modified the SiO_2 surface silica hydroxyl groups with methyl, vinyl, and aminopropyl groups using MTES, VTES, and APTMS, respectively. This resulted in the modified SiO_2 particles having distinct surface properties depending on their hydrophobicity. The vulcanization features, as well as the physico-mechanical, thermal, and dynamic mechanical properties of the filled NR vulcanized rubber, were enhanced when compared to the unmodified SiO_2 particles due to a greater interfacial connection between the modified SiO_2 particles and the NR matrix.

To create NR composites, Jansomboon et al. [52] blended SiO_2/GE blends made using the hybrid approach with NR by latex blending. The products were then cured with EB radiation. The outcomes demonstrated that NR composites might be crosslinked using EB irradiation to enhance their tensile characteristics. The NR composites filled with 2.5 parts SiO_2/0.056% GE blend had the best tensile strength at an EB irradiation dose of 150 kGy. Additionally, the modulus and tear strength were significantly enhanced by irradiation.

These days, a lot of fresh additives are also applied to enhance the silica-based reinforcing of natural rubber. The impact of low eutectic solvent (DES) modified silica on the characteristics of natural rubber composites was examined by Yang et al. [53] Choline chloride and urea were combined in a 1:2 molar ratio to create DES. By forming hydrogen bonds with silica, the low eutectic solvent can enhance the compatibility of rubber and silica. The tensile strength, grip strength, crosslink density, and constant elongation stress of the composites were enhanced when the amount of DES was three parts. Additionally, the rolling resistance of vulcanized rubber was reduced while maintaining good wet slip resistance. Wang et al. [54] developed graphene oxide (rGO)/SiO_2/natural rubber (NR) composites by combining wet mixing with ultrasound-assisted latex mixing (as shown in Figure 8). Bladder cysteamine dihydrochloride (CDHC), a new interfacial modifier with coagulation capabilities, was added to the procedure; the latex particles, SiO_2, and CDHC's electrostatic interactions served as a coagulant. As a coagulant during preparation and an interfacial compatibilizer in the final composites, CDHC can establish potent electrostatic contacts with rGO, silica, and NR latex particles.

Xia et al. [73] used L-lysine as a catalyst in the hydrolytic condensation reaction of ethyl orthosilicate to create nanoscale autonomous monodisperse silica (AS) particles. Next, using the latex blending process, NR latex was combined with AS dispersion to create a silica/natural rubber (NR) masterbatch. The AS/NR composites demonstrated superior tensile strength, abrasion resistance, and resilience in comparison to PS/NR due to the presence of more bound rubber and a decrease in filler–filler interactions, as demonstrated by the use of commercially available precipitated silica (PS) as a reference.

Hydrogen bonding, covalent bonding, hydrophobic contacts, ligand bonding, electrostatic attraction, and charge-transfer interactions are the common forces used in the self-assembly process, which may also be used to synthesize NR/SiO_2. A novel technique was developed by Curk et al. [74] to produce 3D self-assembled SiO_2-NR nanocomposites in the presence of Mg^{2+} and in aqueous phase conditions. Negatively charged NR and

SiO_2 colloidal particles can be positioned on an advantageous NR-SiO_2 heterogeneous condensation by creating molecular bridges. Sattar et al. [55] used self-assembly and latex compounding procedures to manufacture SiO_2/NR nanocomposites (Figure 9). Ultrasonic dispersion was used to create the suspensions, which were subsequently combined with latex, acid flocculated, and dried. Moreover, emulsion blending and Mg^{2+} particle-induced heterogeneous coalescence self-assembly were utilized to achieve the homogeneous dispersion of SiO_2 particles in the NR matrix. In contrast to NR-NR or SiO_2 NP-SiO_2 NP self-aggregation, divalent metal ions were crucial in creating molecular bridges between negatively charged stable NR and SiO_2 colloidal particles under aqueous-phase conditions. It was shown that SiO_2 NP-SiO_2 NP self-aggregation or NR-NR hetero-aggregation were preferable.

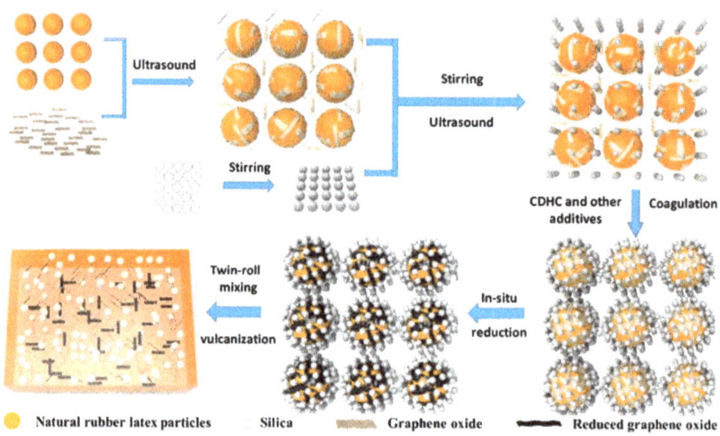

Figure 8. A wet compounding process combined with ultrasonically assisted latex mixing (WCL) method for the preparation of reduced graphene oxide (rGO)/silica/natural rubber (NR) composites [54].

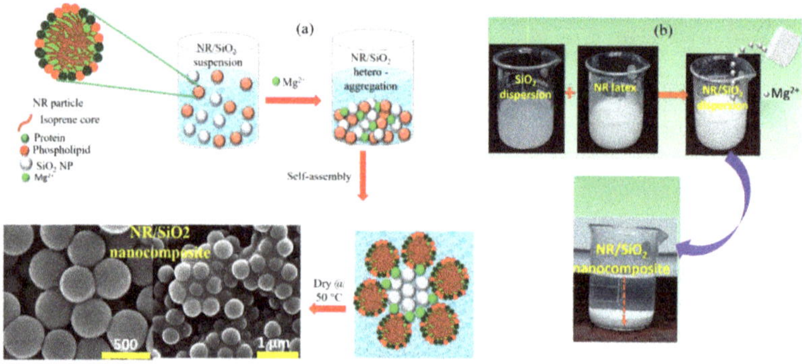

Figure 9. (**a**) Schematic of the synthetic pathway for the formation of 3D self-assembled NR-SiO_2 nanocomposite and its SEM images. (**b**) Depiction of nanoscopic aggregation in NR-SiO_2 nanocomposite in the presence of 100 mM Mg^{2+} [55].

Using liquid mixing of model NR latex–silica NPs, Cattinari et al. [56] employed nanofluorescence imaging to thoroughly analyze the structure and evolution of heterogeneous aggregates generated by colloidal instability. SiO_2NPs were found surrounding the NR spherical particles as a result of Mg^{2+} ion-mediated electrolyte-responsive contacts between NR and filler particles; however, this interaction did not cause a film-forming mechanism but rather stopped the aggregation between the spheres. Additionally, throughout the interaction, the distribution of proteins and lipids within the biofilm does not alter, which restricts the biofilm's ability to shift to a membrane state. Either the solvent evaporating or the liquid suspension being sheared can accomplish this change. Shearing has the advantage of quickening the complex's formation by accelerating the kinetics of heterogeneous aggregation and facilitating the polymerization between NR spheres. The composites had a network of elliptical ring-like structures grouped in a regular pattern that mimicked the original NR spherical appearance. As a result, the liquid state interactions established by NR are directly related to the dispersion of SiO_2NPs in the composites.

By mechanically combining silica solutions, blending them with NR latex, and then drying and agitating the mixture, Ryu et al. [57] created SiO_2/NR composites. Without compromising other rubber qualities, the wet masterbatch rubber's high silica dispersion, low Payne effect, and low loss modulus prevent silica aggregation and result in low viscosity, good starting safety, and low rolling resistance. Kim et al. [75] created silica slurry by stirring, flocculation by mixing with the latex plus aqueous acetic acid solution, and drying by mixing. They also prepared ESBR/SiO_2, BR/SiO_2, and NR/SiO_2 composites by wet and dry techniques, respectively. There was an 82% improvement in the dispersion of silica in the rubber matrix. The wet masterbatch showed good elongation, tensile strength, and abrasion resistance, along with low Menni viscosity and soft filler flocculation because of the silica's good dispersion.

NR/silica and NR/ENR/silica composites were made using the wet masterbatch method by Wang [58]. The silica suspensions were made by ultrasonic dispersion, blended with ENR and NR latex, and then solidified by adding ethanol. Wet masterbatch-based composites demonstrated comparable wet slip resistance with reduced rolling resistance, an important characteristic of materials used in high-performance green tire treads.

After processing the silica with a ball mill, Prasertsri et al. [59] created well-dispersed silica suspensions, as shown in Figure 10. They also created a silica/NR masterbatch by flocculating it with calcium chloride after blending it with natural latex. Because the moisture during the preparation process protected the curing adsorption on the silica surface, the wet preparation of silica/NR masterbatch did not experience the phenomenon of delayed vulcanization that is associated with conventional silica-filled rubber. This is in contrast to the conventional preparation method. Furthermore, the wet-preparation blend has a stronger reinforcing effect than the conventional dry blend when Si69 is absent. Subsequently, utilizing a latex system and fumed silica (FSi) and precipitated silica (PSi) suspensions as raw materials, Prasertsri et al. [60] developed a silica/natural rubber (NR) masterbatch. The findings demonstrated that while a stirred ball mill and an ultrasonic water bath could both produce well-dispersed FSi suspensions, only the latter could produce well-dispersed PSi suspensions. The loss of silica in the composite FSi/NR masterbatch was significantly lower than in the dry preparation of silica-filled NR, and the silica content of the FSi and PSi-filled samples gradually increased the composite viscosity, cross-linking initiation point, and stiffness. Phumnok et al. [61] used a response surface approach in their ball milling process to prepare silica dispersions. When the silica content was around 22% and the ball milling time was 62 h, the best dispersion of silica suspension was achieved. The wet method offers a longer coking and curing time than the dry method, which is beneficial for process control. The silica particle size was reduced to 4.9 mm. This is especially true at high silica loading. The rolling resistance is reduced and the mechanical qualities are superior to the dry process, making them appropriate for the manufacture of green tires. Neena et al. [76] examined how the processing technique affected the creation of SiO_2/NR composites and demonstrated that the technique had a significant impact on how fillers are distributed throughout the polymer matrix. A nano-silica network encircled the rubber microspheres, created by the NR/nano-silica composites made using the latex stage mixing-cast film forming technique. The enhanced crosslink density and solvent resistance of the nanocomposites suggest that the rubber–filler interactions are functioning well.

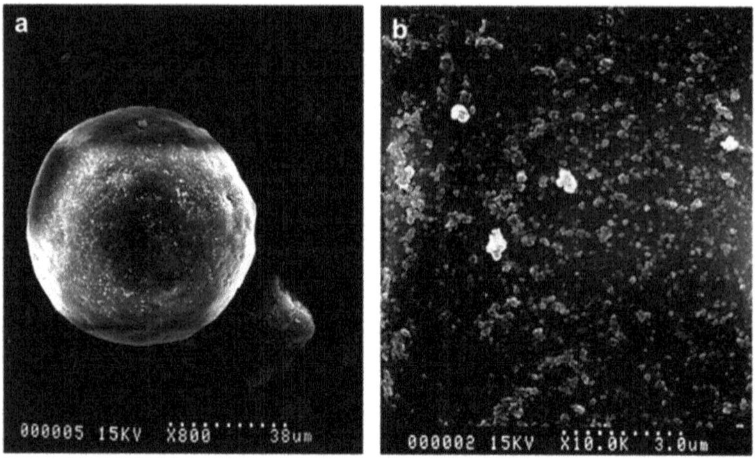

Figure 10. SEM micrographs of silica suspension; (**a**) before grinding and (**b**) after grinding [61].

Nano-SiO_2/NR composites were made by Lin et al. [77] via in situ production and coalescence. By adding Si69, the vulcanized rubber's performance was further enhanced. It offered reduced rolling resistance and increased wet resistance when compared to the vulcanized rubber without Si69. The mechanical and dynamic qualities outperformed those of SiO_2/natural rubber composites made using the traditional dry mixing technique.

A unique experimental technique based on sol-gel chemistry was proposed by Utrera-Barrios et al. [78] for the in situ synthesis of Si and Zi nanoparticles in previously reinforced nano-reinforced materials. The outcomes demonstrated that the Si particles produced in situ were reinforcing at low to medium strains (500%) and raised the rubber's stiffness because of their improved hardness and resilience to abrasion.

A novel technique called atomization drying allows for quick flocculation drying without the need for a flocculant. The latex mixture can be atomized to create tiny droplets with a high specific surface area. The high temperature can then be used to evaporate the moisture from the droplets quickly, allowing for drying. To create natural latex/silica composites, Bian et al. [79] combined silica and latex, sprayed it via atomization with a spray cannon onto a self-designed high-temperature circular roller for quick drying, and then scraped it off with a scraper. The composites made by atomization combined with the high-temperature sputtering and drying (A-HTS) method had the best physical and mechanical properties when compared to the strong acid flocculation method, the natural drying method, and the general mixing method of cigarette flake rubber. In the dynamic mechanical property analysis, these composites also demonstrated excellent rolling resistance and anti-slippery properties. A-HTS is able to ensure good dispersion and uniform distribution of fillers in the latex by preventing filler agglomeration after dispersion and promoting uniform dispersion. Graphene oxide/silica/natural latex composites were made by Wang et al. [80] using traditional drying methods, mechanical stirring, dry ice expansion pre-dispersion, and spray sputtering drying, respectively. The spray sputtering method produced greater silica dispersion in the rubber matrix and higher physico-mechanical characteristics when compared to the other three procedures. When compared to the dry approach, the composites made using the spray sputtering drying process had better tensile strength, tear strength, and abrasion resistance by 27.2%, 43.6%, and 24.6%, respectively. SiO_2/NR composites were made by Zhu et al. [81] using a continuous wet mixing method that included twin-screw mixing, high-temperature flash drying, and ball milling to treat the filler dispersion. The continuous wet mixing approach facilitated silica dispersion and distribution throughout the preparation process, in contrast to traditional dry mixing. The rubber materials showed improvements in tensile strength, abrasion resistance, and fracture energy index of 20.4%, 45.1%, and 14.3%, respectively.

As research progressed, scientists started examining how silica and natural latex self-flocculate to prevent flocculants from altering the characteristics of rubber materials. The promoter 1,3 diphenylguanidine (DPG), which has pro-sulfurization and pro-silanization properties, was loaded onto the green filler silica by Xiao et al. [82] using high-energy ball milling technology, as shown in Figure 11. Simultaneously, DPG destroyed the latex's ionization equilibrium and accomplished self-flocculation.By promoting the dispersion and silanization reaction of silica, the DPG-induced self-flocculation approach helped to produce a more comprehensive filler–rubber network and three-dimensional crosslinked network. The ball milling–autoflocculation technology raised the silanization index, tensile strength, and aging factor by 8%, 31%, and 19%, respectively, in comparison to the conventional dry mixing approach. Rolling resistance is reduced, and the pace of cure is accelerated.

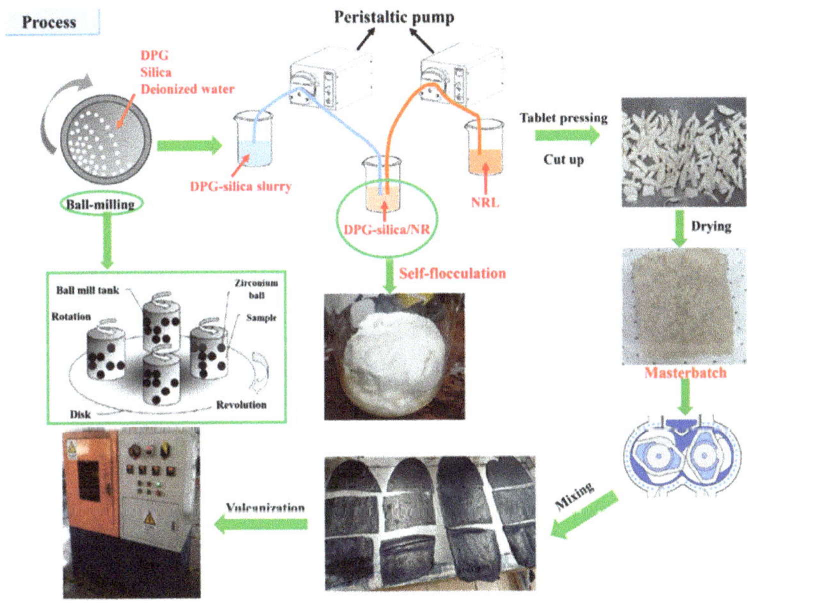

Figure 11. Silicon dioxide-loaded 1,3 diphenylguanidine (DPG) combined with self-flocculation technology process [82].

By using a wet mixing methodology instead of the conventional dry mixing method, SiO_2/NR composites can be prepared with better silica dispersibility, filling capacity, and reinforcing capacity. Processing is made extremely difficult by the wet masterbatch's high Menni viscosity and lack of mixing. To summarize, silica can be modified through surfactant treatment, which can help to partially restore the masterbatch's Mooney. Therefore, it is anticipated that wet blending of silica will have a big development space if the issue of wet masterbatch processing difficulties is successfully resolved.

3.3. NR/Carbon Nanotubes (CNTs) Composites

3.3.1. Mechanism of Carbon Nanotubes Reinforced Rubber

Carbon nanotubes are formed of one or more layers of graphite flakes twisted at a specific helix angle. They are seamless tubular materials with diameters that are nanometers. Consequently, based on the various layers of graphite flakes, carbon nanotubes can be divided into two groups: single-walled carbon nanotubes (SWCNTs) and multi-walled carbon nanotubes (MWCNTs).

When Nah et al. [83] compared the CNT-filled rubber composites' reinforcement mechanism to that of CB, they found that, for the same filler quantity, the mechanical properties of CNTs/NR composites were much better than those of CB/NR composites. Conversely, more stress softening at both low and high strains and volume expansion in tension were linked to the reduced amount of bonded rubber. These findings imply that there is not much interfacial contact between rubber and CNTs and that interfacial connections are easily broken when the materials are stretched. Therefore, rather than strong interfacial adhesion, the most significant element for the enhancement of CNTs appears to be their high aspect ratio, which causes physical entanglement and efficient load transfer.

The uniaxial tensile big deformation behavior of MWNT/NR nanocomposites with varying MWNT volume fractions was studied by Deng [84]. In the tensile strain range of 150% to 400%, the materials exhibited a highly nonlinear stress–strain relationship. To comprehend the possible enhancement process under massive deformation, a theoretical model of MWNT/NR composites was created. The model indicates that the macroscopic mechanical response is significantly influenced by the orientation evolution and persistence length of MWNTs.

A model for the "cellular structure" that forms in CNT/elastomer composites was proposed by Wang [85] (Figure 12). The mechanical interfacial zones of the CNT and the CNT itself are consistently and separately distributed in the NR matrix at low CNT loadings (Figure 12a). This composite has relatively low mechanical characteristics since there is no contact between the CNTs. The CNTs and the mechanical interface regions are tightly entwined into a network to generate a three-dimensional structure as the CNT loading increases. As a result, the mechanical interface area and the CNT itself make up the tiny unit that is known as the NR matrix. "Cytoarchitecture" refers to the three-dimensional linkages that exist between the CNT and the mechano-interfacial area. The "cell" is restricted to the NR matrix, and the mechanical interfacial zone functions as a cell wall. CNT/NR composites are essentially collections of cells. There is the formation of a cellular structure at high CNT loading. A portion of the cell-like arrangement developed when the filling amount of CNTs reached 10 phr, as seen in Figure 12b. A significant number of vesicles developed and dispersed throughout the matrix when the addition of CNTs reached 60 phr (Figure 12c). The significant improvement in mechanical characteristics can be attributed to the creation of the cell-like organization.

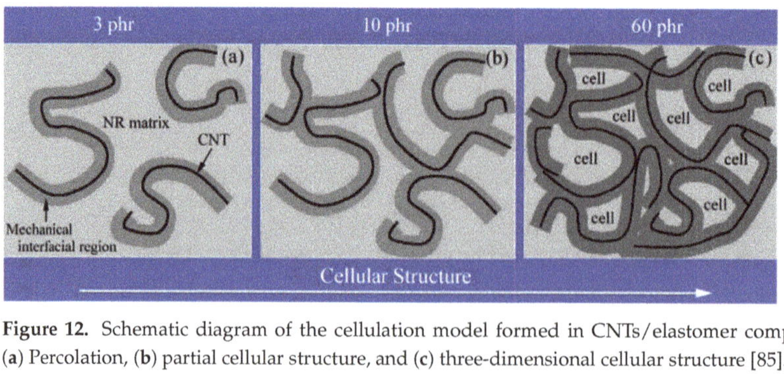

Figure 12. Schematic diagram of the cellulation model formed in CNTs/elastomer composites. (**a**) Percolation, (**b**) partial cellular structure, and (**c**) three-dimensional cellular structure [85].

To predict the creep behavior of CNT–polymer nanocomposites, Hassanzadeh-Aghdam et al. [86] proposed a new modeling approach based on micromechanics. They discovered that agglomeration of CNTs affects and reduces the creep resistance of CNT–polymer nanocomposites and that important factors affecting the composites include random dispersion, transverse anisotropy, and viscoelastic interfaces of CNTs.

Because of their exceptional qualities, which include high mechanical, high thermal, and high electrical conductivity, carbon nanotubes (CNTs) have garnered a lot of interest [87–89]. In the rubber business, adding carbon nanotubes (CNTs) to rubber products can enhance their mechanical, electrical, and thermal conductivity even further. Additionally, achieving the optimal reinforcing effect and maximizing the exceptional performance of rubber requires a good dispersion of CNTs. As a result, the rubber and CNT mixing procedure is essential. CNTs are quickly entangled with each other due to their wide aspect ratio and the strong van der Waals forces between two tubes [90,91], which makes it challenging for them to be evenly distributed throughout the matrix. Liquid-phase mixing has been demonstrated to further increase the dispersion of carbon nanotubes in the rubber matrix and boost their reinforcing capacity [92,93].

3.3.2. CNTs/NR Composites Wet Mixing Preparation Process

Many surface modification techniques (as shown in Table 4), based on mechanical milling, covalent bonding of functional groups, noncovalent encapsulation, or adsorption of medium molecules on the CNT surface, have been proposed to improve the dispersion of CNTs. These techniques are applied in solvent media like water or in polymer media [94]. It has been suggested that CNTs can be hydrophilically modified to improve their dispersion and compatibility with polymer matrices. To counteract van der Waals forces via physical or chemical methods, hydrophilic modification is mostly accomplished by adding hydrophilic specific groups to carbon nanotubes (CNTs) as a spatial site resistance. Physical modification techniques such as adsorption or encapsulation of organic molecules around carbon nanotubes by hydrophobic or van der Waals interactions, such as low molecular weight surfactants, supramolecules, or polymers, can accomplish this [95,96].

Table 4. Summary of methods mentioned in this paper on wet preparation of NR Carbon Nanotube composites.

NR Composites	Synthesis Method	Surfactants/Modifiers	Year/Reference
NR/MWCNT	Latex co-coagulation method	TCl_4	2014 [97]
NR/CNT	Latex co-coagulation method	SDS	2016 [98]
NR/MWCNTs	Latex co-coagulation method	Vulcastab VL	2017 [99]
NR/MWCNTs	Latex co-coagulation method	1-ethyl3-methylimidazolium bromide and 1-hexyl-3-methylimidazolium bromide	2018 [100]
NR/MWNTs	Latex co-coagulation method	K_2FeO_4	2024 [101]
NR/MWCNTs	Latex co-coagulation method	SDBS/AOTPh/TCPh	2015 [102]
NR/SWNTs	Latex co-coagulation method	NaDDBS	2009 [102]
NR/MWCNTs	Latex co-coagulation method	SDS	2010 [103]
NR/CNTs	Slurry blending method	SDS	2019 [104]
NR/CB/CNTs	Latex co-coagulation method	Emulsifier OP	2011 [105]

CNTs/NR composites were prepared by Gao et al. [104] using the slurry mixing method, which decreased carbon nanotube agglomeration and sedimentation when compared to the latex mixing method. This improved carbon nanotube dispersion in NR (Figure 13) led to a 15.2% increase in tensile strength, a more stable thermal conductivity, a weaker Payne effect, and a higher dielectric constant. NR's interfacial tension was first lowered by SDS, which also enhanced NR's attraction to CNTs. This enhanced electrostatic driving force between CNTs improved CNT dispersion in the latex. Second, the NR latex became a paste with a decreased interfacial tension as a result of the mixing ingredients increasing its viscosity. The CNTs were evenly distributed throughout the rubber matrix (Figure 13a,c), and the carbon black particles contained a large number of single CNTs (shown by the red arrows in Figure 13). In Figure 13b, some agglomerated CNTs were observed in the red dotted circles. This was most likely caused by adding more filler, which increased the combined latex's viscosity and decreased its fluidity. Therefore, to prevent subsequent agglomeration, the CNTs were cemented in place after being mixed with NR.

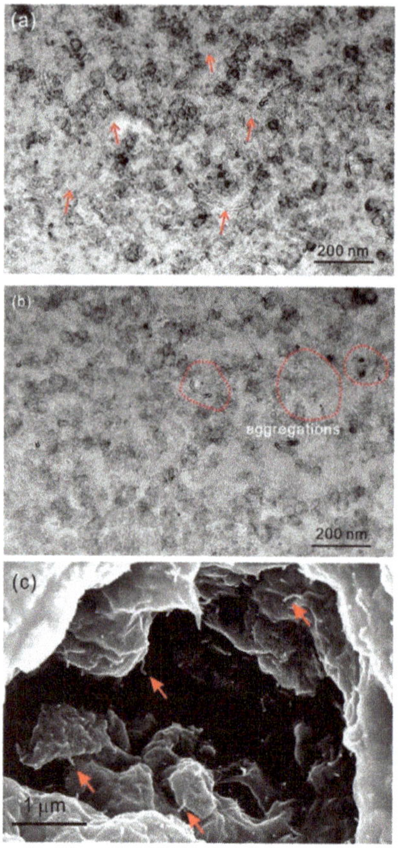

Figure 13. TEM images of NR/CNTs composites prepared by slurry blending method and latex blending method, (**a**) slurry blending method, (**b**) latex blending method. (**c**) The SEM image of NR/CNTs composites prepared by slurry blending method [104].

Anand K et al. [106] prepared SWNTs/NR nanocomposites using the latex stage mixing method. After using ultrasonication to disperse the SWNTs in water, surfactants were added to stabilize the mixture. Five distinct surfactants were examined for their dispersing power: sodium benzoate, polyvinyl alcohol (PVA), sodium dodecyl sulfate (SDS), sodium dodecylbenzene sulfonate (NaDDBS), and iso-octylphenoxypolyethoxyethanol (disinfectant) (Creighton X-100). The best dispersing active agent was determined to be NaDDBS, with a mass ratio of 1:5. After being combined with NR latex, the stabilized aqueous dispersion of SWNTs was dried and cured. The outcomes demonstrated that the strength and modulus of natural rubber were greatly enhanced by SWNTs. Electroosmotic flow may be routed to the insulating NR matrix at low-weight fractions.

Multi-walled carbon nanotube suspension in aqueous solutions of two anionic surfactants (SDS and SDBS) was studied by Bystrzejewski et al. [107]. Both dispersants were shown to be capable of forming stable suspensions of carbon nanotubes below their CMC limits. Compared to SDS, SDBS surfactant had a dispersion capacity that was 26% to 45% higher. While the diameter distribution of CNTs suspended in SDS solution was the same as that of the original CNTs, SDBS demonstrated a larger preference for narrower

CNTs. Mohammad et al. [102] used the emulsion mixing approach to treat surfactants and nanocomposites non-covalently, resulting in the successful preparation of homogeneously dispersed MWCNTs in NR latex (Figure 14). First, the surfactant solution was agitated for one hour using five weight percent MWCNTs, followed by three hours of sonication and dispersion. Subsequently, NR latex was introduced to enhance mixing and sonication. To obtain the composites, the materials were finally dried at 80 °C in an oven. The findings demonstrated that, in contrast to the commercially available SDBS surfactant with only one phenyl group, the application of a three-chain surfactant with three phenyl groups significantly stabilized the colloidal mixes of MWCNTs distributed in the NR latex matrix. Furthermore, advantageous π–π interactions between the surfactant tail chain and the MWCNT surface are produced by increasing the number of phenyl rings in the surfactant tail chain from one to three.

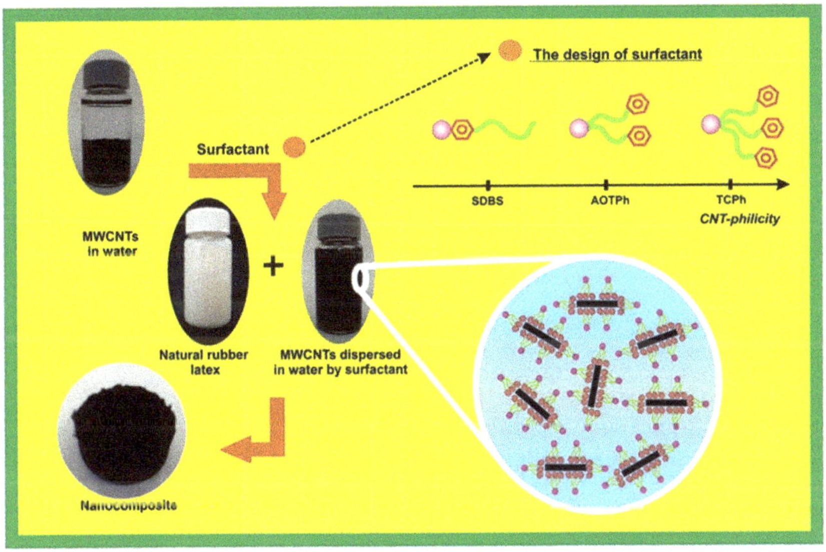

Figure 14. The general route for nanocomposite preparation [102].

The effectiveness of single-, double-, and triple-chain anionic sulfosuccinate surfactants for the dispersion of multi-walled carbon nanotubes (MWCNTs) in natural latex (NR-latex) was examined by Mohamed [97]. Hyperbranched triple-chain sulfosuccinate anionic surfactants, specifically sodium 1,4-bis(neopentyloxy)3-(neopentyloxycarbonyl)-1,4-dioxobutane-2-sulfonate ($TC1_4$), were utilized to effectively disperse MWCNTs in NR-latex. This method outperformed sodium dodecyl sulfate (SDS), a commercially available surfactant.

Carbon nanotube-filled natural rubber composites were created by Nakaramontri et al. [98] using melt mixing and latex mixing techniques, respectively. Furthermore, in situ, functionalization of carbon nanotubes using bis(triethoxysilylpropyl) tetrasulphide (TESPT), a silane coupling agent, was conducted to enhance the filler–rubber interactions between rubber molecules and carbon nanotube surfaces. The characteristics of NR–CNT composites are improved, and CNT dispersion is improved with in situ functionalization of CNTs through latex blending and TESPT.

Using Vulcastab VL (poly(ethylene oxide) condensate) as a nonionic surfactant, George et al. [99] were able to produce stable aqueous dispersions of multiwalled carbon nanotubes (MWCNTs). Emulsion mixing assisted by ultrasonography was used to make MWNR/CNTs composites. The best ratio of CNT-to-VL was found to be 1:1, and the lowest amount of ultrasonic energy needed to produce a stable MWCNT aqueous dispersion was 85,000 J.

Multi-walled carbon nanotubes (MWCNTs) were modified, and Ge et al. [100] examined the impact of the hydrophilic ionic liquids 1-ethyl-3-methylimidazolium bromide and 1-hexyl-3-methylimidazolium bromide on this process. After a month of ultrasonic treatment, the carbon nanotubes (CNTs) that were suspended in water remained evenly dispersed. Using a liquid latex mixing technique, high-performance composites comprising natural rubber latex (NRL) and carbon nanotubes modified by IL were created. The NRL/CNT—IL composites' fatigue resistance and mechanical qualities were enhanced by the homogeneous distribution of CNTs within the matrix.

Multi-walled carbon nanotubes (MWNTs) were oxidized by Xue et al. [101] using potassium persulfate (K_2FeO_4) in both neutral and acidic environments. A wet flash evaporation method was used to create the composites of natural rubber, MWNTs, and carbon black. The qualities of rubber products were finally improved by the rise in the oxidation degree of MWNTs, which also raised the number of oxygenated hydroxyl groups on the surface and encouraged their dispersion in the natural rubber matrix. The natural rubber created a suitable route for electron transport, while the effective dispersion of MWNTs decreased the electron transport barrier. According to the experimental findings, K_2FeO_4 performed best in an acidic environment and had the maximum degree of oxidation. In comparison to the untreated MWNTs rubber composites, the oxidized MWNTs rubber composites had a 34.9% reduced Payne effect, 15.7% lower wear, 12.7% lower rolling resistance, and 2 orders of magnitude higher electrical conductivity.

Liu et al. [108] functionalized multi-walled carbon nanotubes (MWNTs) with sodium lignosulfonate (SLS). A straightforward physical milling method (Figure 15) was able to evenly scatter the SLS-functionalized MWNTs in water, and the MWNTs' equivalent solubility reached 1.5 mg/mL. Furthermore, the high stability was sustained for over three months. Methylcellulose (MC) was employed by Du et al. [87] to stabilize MWCNT dispersions at a constant mass concentration of 0.08 weight percent. The outcomes demonstrated that MC could considerably raise MWCNT viscosity through the viscosity effect. The findings demonstrated that MC could dramatically improve the stability of MWCNT dispersions in an alkaline environment and significantly increase MWCNT dispersion through the viscosity effect. The addition of 0.18 weight percent MC to MWCNT suspension had the greatest impact on improving MWCNT dispersion.

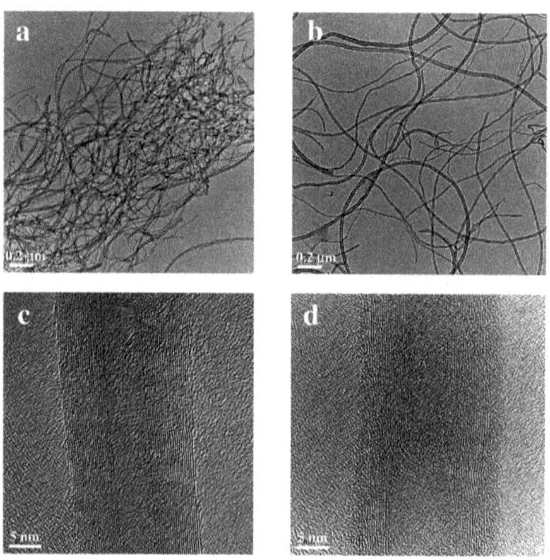

Figure 15. TEM and HRTEM images of aqueous dispersions of (**a**,**c**) pristine MWNTs and (**b**,**d**) SLS-functionalized MWNTs [108].

MWCNT/NR composites were made by Peng et al. [103] utilizing self-assembly and latex mixing methods. After being dissolved in water, the acid-treated MWCNTs and SDS were subsequently added dropwise to the negatively charged latex, dried, and combined. According to the results, MWCNTs were uniformly dispersed in NR latex as single carbon nanotubes, and their surface functionalization significantly inhibited the strong self-aggregation of MWCNTs. Additionally, the uniformly dispersed MWCNTs significantly improved the tensile strength of NR with better endogenous thermal and thermal stability, meaning that only 1 weight percent of the material was used.

To create CB/CNT/NR composites, Zhan et al. [105] also used an ultrasound-assisted latex mixing technique. The outcomes demonstrated that, in comparison to the traditional mixing approach, the ultrasound-assisted latex mixing procedure produced a more uniform dispersion of CNTs in the matrix, and the well-dispersed CNTs and CBs had a synergistic impact. Maximum mechanical characteristics were achieved at a mass ratio of 20:5 (per 100 g of rubber) for CB/CNTs.

To sum up, surfactants can be added or modified to distribute carbon nanotubes evenly throughout the rubber matrix. It is still possible to get a greater reinforcing effect with fewer filler elements. Through functional group modification and ultrasonication, carbon nanotubes' huge aspect ratio and strong van der Waals force can enhance aggregation.

3.4. Graphene (GO)/NR Composites

3.4.1. Mechanism of GO Reinforced Rubber

Li et al. [109] used the latex mixing method to manufacture GE/NR and GE/NR composites. The in situ chemical reduction method in the latex allowed for the conversion of GO to GE. Additionally, the enhancement mechanism was investigated using the entangled tube model (Figure 16), and the findings indicated that in contrast to the carbon black-reinforced rubber model, GE or GO nanosheets with high specific surface area and nanoscale surface roughness have a greater chance of interacting with the rubber. Due to the high aspect ratio of GE or GO nanosheets, even a tiny quantity of nanolayer dispersion is sufficient to interact with the rubber matrix, allowing only a small number of rubber chains to stay free. Compared to GO, GE is more likely to interact with rubber.

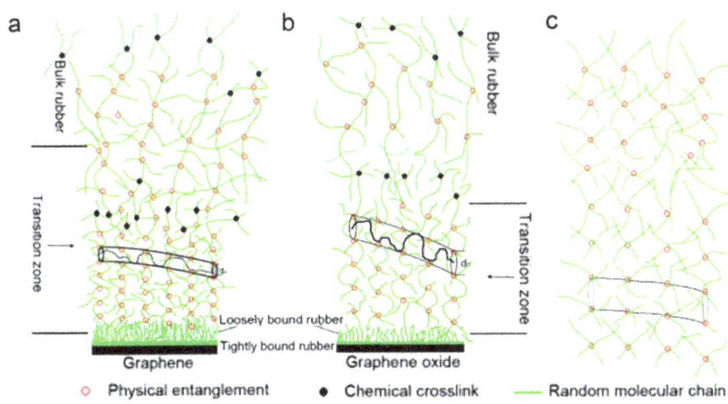

Figure 16. Schematic representation of the EBT model, (**a**) GE/NR (**b**) GO/NR (**c**) NR [109].

The microstructure and enhancement mechanism of nanocomposites containing graphene nanoparticles (GNPs) in natural rubber (NR) were examined by Li et al [110]. They discovered that the GNPs were folded and annular in the NR matrix and that multiple stacking of the nanoparticles was occasionally visible. Perhaps as a result of heat pressing during processing, the GNPs stayed intact and aligned parallel to the surface of the nanocomposites. The stress-induced Raman Pu band shift was used to quantify the stress transfer from natural rubber to graphene nanosheets.

Although graphene has a stronger reinforcing impact, its potential is still restricted by its aspect ratio, surface roughness, and other parameters, as indicated by the examination of the aforementioned process. Currently, research into the process by which graphene reinforces natural rubber is still needed to fully utilize graphene's reinforcing properties in rubber.

3.4.2. Graphene/NR Composites Wet Mixing Preparation Process

Graphene has garnered significant interest as a rubbery composite material due to its exceptional electrical, thermal, and mechanical properties [111]. As a result, there has been a gradual increase in wet preparation investigations (Table 5). Nevertheless, it is difficult and expensive to produce clean graphene on a big scale. Modifying graphene can prevent its uniform dispersion in the polymer matrix, which is still a significant issue. Three modes of analysis were used by Hassanzadeh-Aghdam et al. [112] to examine the impact of graphene dispersion on the creep properties of nanocomposites: agglomeration, homogeneous dispersion, and directional alignment. The results showed that the highest creep modulus was achieved by aligning GNPs, while the lowest was achieved by agglomeration. An increased creep resistance is a result of the interfacial phase's effect when its mechanical properties surpass those of the matrix. Compared to the randomly distributed nanocomposites, the orientated aligned GNPs nanocomposites showed superior creep properties.

Table 5. Summary of methods mentioned in this paper on wet preparation of NR/graphene composites.

NR Composites	Dispersal Methods	Improvement of Performance	Year/Reference
NR/GE	Ultrasonic dispersion	Mechanical properties	2011 [113]
NR/GE	Ultrasonic irradiation	Electrical conductivity	2012 [114]
NR/graphene	Ultrasonic irradiation	Mechanical properties	2013 [115]
NR/GO	Vibrate	Mechanical properties	2014 [116]
NR/GO	Mechanical stirring and bath	Electrical conductivity	2015 [117]
NR/rGO	Mechanical stirring	Electrical conductivity Thermal conductivity	2017 [118]
rGO/NR/CB	Ultrasonic dispersion	Mechanical properties	2018 [119]
NR/GO	Ultrasonic dispersion	Mechanical properties	2020 [120]
CB/GO/NR	Ultrasonic dispersion	Mechanical properties	2023 [121]

Hydrophilic groups in the structure of graphene oxide (GO), an oxidation product of graphene, facilitate its dispersion in aqueous solutions. Graphene oxide may be obtained in vast quantities at a lower cost than pristine graphene. Furthermore, GO has excellent mechanical characteristics and a large concentration of oxygen functional groups. When all these characteristics are combined, GO becomes a more potent reinforcing agent [122]. Reduced graphene oxide (rGO) is a structure that is created when reduced graphene is stripped of its oxygen-containing groups [123]. This is one of the derivatives of graphene, which are also known as reduced graphene, functionalized graphene, chemically modified graphene, and chemically converted graphene [124].

Zhan et al. [113] used in situ reduction techniques and ultrasonic assistance to manufacture NR/graphene (GE) composites. To create the NR/GE masterbatch, graphene oxide was distributed in NRL using ultrasonic, then in situ reduction and latex solidification. In comparison to traditional direct mixing, the technique improved GE's dispersion and exfoliation from the matrix and enhanced tensile strength, according to the data. With an increase in GE content, the NR/GE composites' maximum torque, crosslink density, modulus of elasticity, and thermal conductivity all increased.

To create cross-linked rubber composites with electrical conductivity, a water vapor barrier, and high tensile qualities, Zhan [114] et al. used self-assembly and static vulcanization in latex. The composites' electrical conductivity and electrical impermeability threshold were both significantly increased by the graphene network's creation. Five orders of magnitude more electrical conductivity was observed in the composites made using the self-assembly procedure in rubber latex with a GE content of 1.78 Vol% as compared to the composites made using conventional methods.

To manufacture innovative graphene/natural latex (NRL) composites, Li et al. [115] developed a stabilizer-free modification technique in which ammonium, the stabilizer, was dispersed during the final heat vulcanization process. The findings demonstrate the viability of the ammonium-assisted approach for the manufacture of graphene/natural latex composites and the superior qualities of the resulting materials. Strong p–p interactions and hydrogen bonding exist between graphene and NRL matrix even at modest loading because of the material's unique two-dimensional structure and good dispersion.

GO/NR nanocomposites were created by Stanier et al. [116] by combining NR latex with an aqueous dispersion of GO flakes. Rectangular specimens underwent tensile tests at various strain rates up to 600% distortion. Even with modest filler content, the tensile tests demonstrated a considerable rise in the composites' Young's modulus, confirming the strong interfacial contact between GO and natural rubber. As the GO content rises, Young's modulus rises with strain rate, and dissipation rises as a result of higher friction brought on by the high aspect ratio GO. These results indicate that, in comparison to the widely utilized carbon black or carbon nanotubes, considerable increases in mechanical characteristics can be obtained by employing smaller quantities of fillers.

Suriani et al. [117] created GO/NRL nanocomposites in a single step by combining GO and NRL in a 1:1 volume ratio. They also generated surfactant SDS at a concentration of 0.01 M by electrochemical stripping. The GO/NRL nanocomposites prepared in one step (103.7 Fg^{-1}) and the nanocomposites prepared in two steps (32.6 Fg^{-1}) had a lower specific capacitance. This method involved preparing the GO/NRL nanocomposites concurrently with the GO and NRL mixing product. Consequently, the synthesized GO/NRL nanocomposites may find use in energy storage devices as supercapacitors.

Sodium poly 4-styrene sulfonate was utilized by Li et al. [118] as a stabilizer for the chemical reduction of graphite oxide. The resulting graphene was then combined with NR using a latex process. It was demonstrated that adding surfactants to the graphene surface causes NR molecules to experience both spatial site resistance and electrostatic repulsion. Using the latex composite approach, high-performance NR/graphene nanocomposites were created in conformity with the morphological findings, and the generated nanocomposites' energy storage modulus was greatly increased.

Following the solidification and drying of rGO/NR latex mixtures made by ultrasound-assisted latex mixing and in situ reduction, Wang et al. [119] developed rGO/CB/NR composites. Both rGO and CB were evenly distributed throughout the rubber composites, as demonstrated by morphological observations. The inclusion of rGO significantly increased the composites' hardness, heat conductivity, and resistance to aging. When compared to normal latex mixing, the rGO/CB/NR composites made with ultrasonic assistance had superior mechanical characteristics.

Mao et al. [120] created GO/NR composites by using the latex co-coagulation process. By ultrasonically removing graphite oxide from water, combining the aqueous GO dispersion with the latex, and co-coagulating the mixes with the addition of a flocculant, an aqueous GO dispersion stabilized by electrostatic repulsion was produced. The outcomes demonstrated the strong interfacial contacts between GO and NR and the fine dispersion of the highly exfoliated GO flakes in the NR rubber matrix. The mechanical characteristics of the GO/NR composites were further assessed by the researchers. With a GO concentration of less than 2 phr, the results demonstrated a considerable improvement in the tensile strength, rip strength, and modulus. In particular, because of the stress-induced crystallization effect of NR, GO shows a unique strengthening mechanism in NR.

Using the latex co-precipitation approach, Li et al. [121] created CB/GO/NR composites with various crosslinked networks. In the meantime, research was done on how various crosslinking techniques affected fatigue life and crack extension resistance in various vulcanization systems. The findings demonstrated that, in the conventional vulcanization (CV) system, the polysulfone-based CB/GO/NR composites had the lowest crack extension rate (64.1 nm/cycle) and the highest tear strength (71.6 KN/m). The improvement in crack extension resistance was primarily attributed to the CV system's well-developed crosslinking network and polysulfone-based crosslinking structure.

A wide range of graphene can be modified to improve its dispersion in a rubber matrix or water. With the emergence of several of the aforementioned surfactants and processes, graphene dispersion acts as a potent aid, and when compared to conventional dry mixing, wet mixing can better capture the graphene's reinforcing properties.

4. Expectations and Conclusions

Due to its distinct benefits over dry mixing, wet mixing technology has drawn a lot of attention and has recently been a hot topic for research. The review above indicates that the primary issue with CB/NR composites is the high density and hydrophobicity of carbon black, which settles a lot in the aqueous dispersion and makes it difficult to co-settlement well with the latex. This requires modification with a dispersant or surfactant. The effective interface between particular active sites on the surface of carbon black and the rubber macromolecular chains is always reduced or hindered by the surface's preferential contact with and adsorption of surfactants.

A significant amount of research has also been conducted on wet mixing technology for SiO_2/NR composites. However, because of the conflict between silanization and the rubber material's characteristics and the high viscosity of the prepared masterbatch resulting from the presence of SiO_2, overloading the mixing equipment can easily occur, impeding appropriate mixing and processing. Wet mixing methods for graphene/NR and NR/CNT composites have been extensively studied, and related research is still ongoing.

In summary, weakening filler–filler interactions and strengthening filler–matrix interactions are necessary to greatly improve the characteristics of NR nanocomposites. However, the broad use of NR composite nanomaterials is restricted by the weaker interactions between nonpolar NR and other nanofillers. As this review points out, better dispersion mixing techniques and filler surface functionalization can help with this. In summary, the wet mixing method makes up for the drawbacks of conventional dry mixing and is anticipated to become a widely used emergent technology.

Author Contributions: Conceptualization, H.Y.; writing—original draft preparation, Q.Z.; writing—review and editing, F.N. and J.L.; supervision, H.Y.; All authors have read and agreed to the published version of the manuscript.

Funding: This research received no external funding.

Conflicts of Interest: The authors declare no conflicts of interest.

References

1. Hirata, Y.; Kondo, H.; Ozawa, Y. Natural rubber (NR) for the tyre industry. In *Chemistry, Manufacture and Applications of Natural Rubber*; Elsevier: Amsterdam, The Netherlands, 2014; pp. 325–352.
2. Liu, Y.; Zhou, Q.; Lu, Q.; Han, C.; Zhou, Z.; Liang, Z.; Liu, R.; Nie, Y. Reinforcement and toughening of rubber by bridging graphene and nanosilica. *J. Inorg. Organomet. Polym. Mater.* **2020**, *30*, 337–348. [CrossRef]
3. Dong, B.; Liu, C.; Zhang, L.; Wu, Y. Preparation, fracture, and fatigue of exfoliated graphene oxide/natural rubber composites. *Rsc Adv.* **2015**, *5*, 17140–17148. [CrossRef]
4. Zhao, L.; Sun, X.; Liu, Q.; Zhao, J.; Xing, W. Natural rubber/graphene oxide nanocomposites prepared by latex mixing. *J. Macromol. Sci. Part B* **2015**, *54*, 581–592. [CrossRef]

5. Malas, A. Rubber nanocomposites with graphene as the nanofiller. In *Progress in Rubber Nanocomposites*; Elsevier: Amsterdam, The Netherlands, 2017; pp. 179–229.
6. Sethulekshmi, A.; Saritha, A.; Joseph, K. A comprehensive review on the recent advancements in natural rubber nanocomposites. *Int. J. Biol. Macromol.* **2022**, *194*, 819–842. [CrossRef]
7. Yang, J.; Xian, B.; Li, H.; Zhang, L.; Han, D. Preparation of silica/natural rubber masterbatch using solution compounding. *Polymer* **2022**, *244*, 124661. [CrossRef]
8. Fawaz, J.; Mittal, V. Synthesis of polymer nanocomposites: Review of various techniques. In *Synthesis Techniques for Polymer Nanocomposites*; Wiley-VCH: Hoboken, NJ, USA, 2014; pp. 1–30.
9. Poompradub, S.; Thirakulrati, M.; Prasassarakich, P. In situ generated silica in natural rubber latex via the sol–gel technique and properties of the silica rubber composites. *Mater. Chem. Phys.* **2014**, *144*, 122–131. [CrossRef]
10. Fan, Y.; Fowler, G.D.; Zhao, M. The past, present and future of carbon black as a rubber reinforcing filler–A review. *J. Clean. Prod.* **2020**, *247*, 119115. [CrossRef]
11. Goudarzi, T.; Spring, D.W.; Paulino, G.H.; Lopez-pamies, O. Filled elastomers: A theory of filler reinforcement based on hydrodynamic and interphasial effects. *J. Mech. Phys. Solids* **2015**, *80*, 37–67. [CrossRef]
12. Huber, G.; Vilgis, T.A. Universal properties of filled rubbers: Mechanisms for reinforcement on different length scales. *Kautsch. Gummi Kunststoffe* **1999**, *52*, 102–107.
13. Donnet, J. Nano and microcomposites of polymers elastomers and their reinforcement. *Compos. Sci. Technol.* **2003**, *63*, 1085–1088. [CrossRef]
14. Dannenberg, E. Molecular slippage mechanism of reinforcement. *Trans. Inst. Rubber Ind.* **1966**, *42*, 26–32.
15. Medalia, A. Effective volume of aggregates of carbon black from electron microscopy; application to vehicle adsorption and to die swell of filled rubber. *J. Colloid Interface Sci.* **1970**, *32*, 115–131. [CrossRef]
16. Kraus, G. A carbon black structure-concentration equivalence principle. *Appl. Stress-Strain Relatsh. Fill. Rubbers. Rubber Chem. Technol.* **1971**, *44*, 199–213.
17. Smit, P. Glass transition in carbon black reinforced rubber. *Rubber Chem. Technol.* **1968**, *41*, 1194–1202. [CrossRef]
18. Pliskin, I.; Tokita, N. Bound rubber in elastomers: Analysis of elastomer-filler interaction and its effect on viscosity and modulus of composite systems. *J. Appl. Polym. Sci.* **1972**, *16*, 473–492. [CrossRef]
19. O'brien, J.; Cashell, E.; Wardell, G.; Mcbrierty, V. An NMR investigation of the interaction between carbon black and cis-polybutadiene. *Macromolecules* **1976**, *9*, 653–660. [CrossRef]
20. Leblanc, J.L. Rubber–filler interactions and rheological properties in filled compounds. *Prog. Polym. Sci.* **2002**, *27*, 627–687. [CrossRef]
21. Fukahori, Y. New progress in the theory and model of carbon black reinforcement of elastomers. *J. Appl. Polym. Sci.* **2005**, *95*, 60–67. [CrossRef]
22. Haghgoo, M.; Ansari, R.; Hassanzadeh-Aghdam, M.K. The effect of nanoparticle conglomeration on the overall conductivity of nanocomposites. *Int. J. Eng. Sci.* **2020**, *157*, 103392. [CrossRef]
23. Ridaoui, H.; Jada, A.; Vidal, L.; Donnet, J.-B. Effect of cationic surfactant and block copolymer on carbon black particle surface charge and size. *Colloids Surf. A Physicochem. Eng. Asp.* **2006**, *278*, 149–159. [CrossRef]
24. Tiarks, F.; Landfester, K.; Antonietti, M. Encapsulation of carbon black by miniemulsion polymerization. *Macromol. Chem. Phys.* **2001**, *202*, 51–60. [CrossRef]
25. Alex, R.; Sasidharan, K.; Kurian, T.; Chandra, A. Carbon black master batch from fresh natural rubber latex. *Plast. Rubber Compos. Sci. Technol.* **2011**, *40*, 420–424. [CrossRef]
26. Martínez-Pedrero, F.; Alousque, F.; De Gaudemaris, B.; Berriot, J.; Gaboriaud, F.; Bremond, N.; Bibette, J. Making an elastomeric composite material via the heteroaggregation of a binary colloidal dispersion. *Soft Matter* **2012**, *8*, 8752–8757. [CrossRef]
27. Dong, A.; Zhiyi, Z.; Haixiang, J.; Jinquan, S.; Huan, Z.; Yaqing, L. Effect of carbon black nature on dynamic mechanical properties and reinforcement of nr vulcanization by latex techniques. *Rubber Chem. Technol.* **2017**, *90*, 611–620. [CrossRef]
28. Jiang, Q.; Wang, S.; Xu, S. Preparation and characterization of water-dispersible carbon black grafted with polyacrylic acid by high-energy electron beam irradiation. *J. Mater. Sci.* **2018**, *53*, 6106–6115. [CrossRef]
29. Bo, Y.; Cui, J.; Cai, Y.; Xu, S. Preparation and characterization of poly (methyl methacrylate) and poly (maleic anhydride-co-diallyl phthalate) grafted carbon black through γ-ray irradiation. *Radiat. Phys. Chem.* **2016**, *119*, 236–246. [CrossRef]
30. Fu, W.; Liu, A.; Yang, J. Effects of in-situ grafting-modified carbon black on properties of natural rubber. *J. South China Univ. Technol. Nat. Sci. Ed.* **2014**, *42*, 7.
31. Zheng, Y.; Li, S.; Weng, Z.; Gao, C. Hyperbranched polymers: Advances from synthesis to applications. *Chem. Soc. Rev.* **2015**, *44*, 4091–4130. [CrossRef] [PubMed]
32. Zhang, T.; Howell, B.A.; Smith, P.B. Rational Synthesis of hyperbranched poly (ester) s. *Ind. Eng. Chem. Res.* **2017**, *56*, 1661–1670. [CrossRef]
33. Han, F.; Zhang, Z.; Ma, T.; Shou, C. Preparation and mechanical properties of water-dispersible hyperbranched polymer grafted carbon black/natural rubber composites by latex blending method. *Polym. Adv. Technol.* **2022**, *33*, 368–379. [CrossRef]
34. Ortega-Vinuesa, J.; Martin-rodriguez, A.; Hidalgo-Alvarez, R. Colloidal stability of polymer colloids with different interfacial properties: Mechanisms. *J. Colloid Interface Sci.* **1996**, *184*, 259–267. [CrossRef] [PubMed]

35. Adamczyk, Z.; Weroński, P. Application of the DLVO theory for particle deposition problems. *Adv. Colloid Interface Sci.* **1999**, *83*, 137–226. [CrossRef]
36. Zhang, A.; Wang, L.; Lin, Y.; Mi, X. Carbon black filled powdered natural rubber: Preparation, particle size distribution, mechanical properties, and structures. *J. Appl. Polym. Sci.* **2006**, *101*, 1763–1774. [CrossRef]
37. Lin, Y.; Zhang, A.; Wang, L.; Pei, C.; Gu, Q. Carbon black filled powdered natural rubber. *J. Macromol. Sci. Part B* **2012**, *51*, 1267–1281. [CrossRef]
38. Wang, M.-J.; Wolff, S.; Donnet, J.-B. Filler-elastomer interactions. Part I: Silica surface energies and interactions with model compounds. *Rubber Chem. Technol.* **1991**, *64*, 559–576. [CrossRef]
39. Wang, M.-J.; Wolff, S.; Donnet, J.-B. Filler—Elastomer interactions. Part III. Carbon-black-surface energies and interactions with elastomer analogs. *Rubber Chem. Technol.* **1991**, *64*, 714–736.
40. Han, J.; Zhang, Y.; Wu, C.; Ma, Y. Effect of reynolds number on the dispersion of carbon black in natural rubber latex and filler–rubber interfacial interaction in high-speed jet flow field. *Polym. Compos.* **2013**, *34*, 1071–1075. [CrossRef]
41. Galinovskiy, A.; Htet, K.M.; Provatorov, A. Ultra-Jet as a Tool for Dispersing Nanosuspensions. *Polym. Sci. Ser. D* **2020**, *13*, 209–213. [CrossRef]
42. Cattinari, G.; Steenkeste, K.; Le Bris, C.; Canette, A.; Gallopin, M.; Couty, M.; Fontaine-Aupart, M.P. Natural rubber-carbon black coagulation: Following the nanostructure evolution from a colloidal suspension to a composite. *J. Appl. Polym. Sci.* **2021**, *138*, 50221. [CrossRef]
43. Yamamoto, Y.; Soda, R.; Kano, J.; Saito, F. Dispersion of carbon black by a media mill and correlation of its rate constant with beads impact energy. *Powder Technol.* **2012**, *219*, 105–110. [CrossRef]
44. Hamran, N.; Rashid, A.A. Effect of combination ultrasonic and ball milling techniques of commercial fillers dispersion on mechanical properties of natural rubber (NR) latex films. *AIP Conf. Proc.* **2017**, *1865*, 40015.
45. Sui, Y.; Lai, Q.; Nie, W.; Tang, S.; Wang, C.; Bian, H. Wet continuous mixing technique based on full formula of carbon black. *J. Appl. Polym. Sci.* **2022**, *139*, e52971. [CrossRef]
46. Shu, H.; Liu, K.; Liu, F.; Zhang, Z.; Li, X. Improving mechanical properties of poly (vinyl chloride) by doping with organically functionalized reactive nanosilica. *J. Appl. Polym. Sci.* **2013**, *129*, 2931–2939. [CrossRef]
47. Zhong, B.; Jia, Z.; Luo, Y.; Jia, D. Surface modification of silica with N-cyclohexyl-2-benzothiazole sulfenamide for styrene–butadiene rubber composites with dramatically improved mechanical property. *Mater. Lett.* **2015**, *145*, 41–43. [CrossRef]
48. Law, Y.Y.; Feke, D.L.; Manas-Zloczower, I. Thermogravimetric analysis of the kinetics of bound-rubber formation on surface-modified silica. *Rubber Chem. Technol.* **2014**, *87*, 311–319. [CrossRef]
49. Bera, A.; Sarkar, K.; Ganguly, D.; Ghorai, S.; Hore, R.; Kumar, N.; Amarnath, S.; Chattopadhyay, S. Recent advancements in silica filled natural rubber composite: A green approach to achieve smart properties in tyre. *J. Polym. Res.* **2024**, *31*, 122. [CrossRef]
50. Phinyocheep, P. Chemical modification of natural rubber (NR) for improved performance. In *Chemistry, Manufacture and Applications of Natural Rubber*; Elsevier: Amsterdam, The Netherlands, 2014; pp. 68–118.
51. Theppradit, T.; Prasassarakich, P.; Poompradub, S. Surface modification of silica particles and its effects on cure and mechanical properties of the natural rubber composites. *Mater. Chem. Phys.* **2014**, *148*, 940–948. [CrossRef]
52. Jansomboon, W.; Loykulnant, S.; Kongkachuichay, P.; Dittanet, P.; Prapainainar, P. Electron beam radiation curing of natural rubber filled with silica-graphene mixture prepared by latex mixing. *Ind. Crops Prod.* **2019**, *141*, 111789. [CrossRef]
53. Yang, H.; Yang, L.; Guo, H.; Hu, W.; Du, A. The effect of silica modified by deep eutectic solvents on the properties of nature rubber/silica composites. *J. Elastomers Plast.* **2022**, *54*, 111–122. [CrossRef]
54. Wang, J.; Zhang, K.Y.; Fei, G.X.; De Luna, M.S.; Lavorgna, M.; Xia, H.S. High Silica Content Graphene/Natural Rubber Composites Prepared by a Wet Compounding and Latex Mixing Process. *Polymers* **2020**, *12*, 14. [CrossRef]
55. Sattar, M.A.; Nair, A.S.; Xavier, P.; Patnaik, A. Natural rubber–SiO 2 nanohybrids: Interface structures and dynamics. *Soft Matter* **2019**, *15*, 2826–2837. [CrossRef]
56. Cattinari, G.; Steenkeste, K.; Lévêque-Fort, S.; Cabriel, C.; Deniset-Besseau, A.; Canette, A.; Oberdisse, J.; Gallopin, M.; Couty, M.; Fontaine-Aupart, M.-P. Nanostructural Evolution of Natural Rubber/Silica Nanoparticle Coagulation from Binary Colloidal Suspensions to Composites: Implications for Tire Materials. *ACS Appl. Nano Mater.* **2021**, *4*, 6722–6733. [CrossRef]
57. Ryu, C.; Yang, J.K.; Park, W.; Kim, S.J.; Kim, D.I.; Seo, G. Silica-filled NR compounds prepared by dry and wet masterbatches with different mixing times. *J. Appl. Polym. Sci.* **2020**, *137*, 49548. [CrossRef]
58. Wang, Y.; Liao, L.; Zhong, J.; He, D.; Xu, K.; Yang, C.; Luo, Y.; Peng, Z. The behavior of natural rubber–epoxidized natural rubber–silica composites based on wet masterbatch technique. *J. Appl. Polym. Sci.* **2016**, *133*. [CrossRef]
59. Prasertsri, S.; Rattanasom, N. Mechanical and damping properties of silica/natural rubber composites prepared from latex system. *Polym. Test.* **2011**, *30*, 515–526. [CrossRef]
60. Prasertsri, S.; Rattanasom, N. Fumed and precipitated silica reinforced natural rubber composites prepared from latex system: Mechanical and dynamic properties. *Polym. Test.* **2012**, *31*, 593–605. [CrossRef]
61. Phumnok, E.; Khongprom, P.; Ratanawilai, S. Preparation of natural rubber composites with high silica contents using a wet mixing process. *ACS Omega* **2022**, *7*, 8364–8376. [CrossRef]
62. Kaewsakul, W.; Sahakaro, K.; Dierkes, W.K.; Noordermeer, J.W. Mechanistic aspects of silane coupling agents with different functionalities on reinforcement of silica-filled natural rubber compounds. *Polym. Eng. Sci.* **2015**, *55*, 836–842. [CrossRef]

63. Zheng, J.; Han, D.; Ye, X.; Wu, X.; Wu, Y.; Wang, Y.; Zhang, L. Chemical and physical interaction between silane coupling agent with long arms and silica and its effect on silica/natural rubber composites. *Polymer* **2018**, *135*, 200–210. [CrossRef]
64. Yao, B.B.; Xia, L.J.; Wang, H.; Kan, Z. The effects of natural astaxanthin-modified silica on properties of natural rubber. *J. Appl. Polym. Sci.* **2019**, *136*, 47287. [CrossRef]
65. Li, S.; Zhai, X.; Chen, Z.; Han, D.; Ye, X.; Zhang, L. Balancing the" magic triangle" performance of tire rubber nanocomposites with novel low VOC silane coupling agents. *Compos. Commun.* **2024**, *46*, 101836. [CrossRef]
66. Zhai, X.; Chen, X.; Zheng, F.; Han, D.; Zheng, J.; Ye, X.; Li, X.; Zhang, L. Designing a Coupling Agent with Aliphatic Polyether Chain and Exploring Its Effect on Silica/Natural Rubber Nanocomposites under the Action of Non-Rubber Contents. *Polymers* **2023**, *15*, 674. [CrossRef] [PubMed]
67. Xiao, Y.; Zou, H.; Zhang, L.; Ye, X.; Han, D. Surface modification of silica nanoparticles by a polyoxyethylene sorbitan and silane coupling agent to prepare high-performance rubber composites. *Polym. Test.* **2020**, *81*, 106195. [CrossRef]
68. Kaewsakul, W.; Sahakaro, K.; Dierkes, W.K.; Noordermeer, J.W. Optimization of rubber formulation for silica-reinforced natural rubber compounds. *Rubber Chem. Technol.* **2013**, *86*, 313–329. [CrossRef]
69. Sengloyluan, K.; Sahakaro, K.; Dierkes, W.K.; Noordermeer, J.W. Silica-reinforced tire tread compounds compatibilized by using epoxidized natural rubber. *Eur. Polym. J.* **2014**, *51*, 69–79. [CrossRef]
70. Xu, T.; Jia, Z.; Luo, Y.; Jia, D.; Peng, Z. Interfacial interaction between the epoxidized natural rubber and silica in natural rubber/silica composites. *Appl. Surf. Sci.* **2015**, *328*, 306–313. [CrossRef]
71. He, F.; Yuan, T.; Li, C.; Sun, L.; Liao, S. Interfacial interactions and properties of natural rubber–silica composites with liquid natural rubber as a compatibilizer and prepared by a wet-compounding method. *J. Appl. Polym. Sci.* **2018**, *135*, 46457. [CrossRef]
72. Wang, X.; Luo, Z.; Xu, Y.; Zhong, J.; Zhang, H.; Liang, J. Preparation and performance of natural rubber latex treatment for silica filled natural rubber composites. *J. Appl. Polym. Sci.* **2023**, *140*, e53966. [CrossRef]
73. Xia, L.; Song, J.; Wang, H.; Kan, Z. Silica nanoparticles reinforced natural rubber latex composites: The effects of silica dimension and polydispersity on performance. *J. Appl. Polym. Sci.* **2019**, *136*, 47449. [CrossRef]
74. Curk, T.; Dobnikar, J.; Frenkel, D. Rational design of molecularly imprinted polymers. *Soft Matter* **2016**, *12*, 35–44. [CrossRef]
75. Kim, W.; Yu, E.; Ryu, G.; Kim, D.; Ryu, C.; Seo, Y.; Kim, W. Silica dispersion and properties of silica filled ESBR/BR/NR ternary blend composites by applying wet masterbatch technology. *Polym. Test.* **2020**, *84*, 106350. [CrossRef]
76. Neena, G.; Amrutha, S.; Rani, J.; Jose, P.; Mathiazhagan, A. Nano-silica as reinforcing filler in NR latex: Role of processing method on filler morphology inside the rubber and properties of the nanocomposite. *Express Polym. Lett.* **2021**, *15*, 1101–1112. [CrossRef]
77. Lin, Y.; Zhang, A.; Sun, J.; Wang, L. Properties of natural rubber vulcanizates/nanosilica composites prepared based on the method of in-situ generation and coagulation. *J. Macromol. Sci. Part B* **2013**, *52*, 1494–1507. [CrossRef]
78. Utrera-Barrios, S.; Perera, R.; León, N.; Santana, M.H.; Martínez, N. Reinforcement of natural rubber using a novel combination of conventional and in situ generated fillers. *Compos. Part C Open Access* **2021**, *5*, 100133. [CrossRef]
79. Bian, H.; Chang, T.; Wang, C.; Zhu, D.; Ren, D.; Yao, X.; Tao, Z. Atomization combined with high-temperature sputtering method applied in latex compounding technology for preparation of natural rubber latex/silica composites. *Polym. Bull.* **2022**, *79*, 1–14. [CrossRef]
80. Wang, C.; Chang, T.; Bian, H.; Zhang, L. Study on the preparation of graphene oxide/silica/natural rubber latex composites by different processes. *Polym. Polym. Compos.* **2019**, *27*, 135–142. [CrossRef]
81. Zhu, D.; Wang, C.; Xiao, Y.; Wang, X.; Han, S.; Wang, Z.; Bian, H. Studying the continuous wet mixing process based on the silica formula. *J. Appl. Polym. Sci.* **2021**, *138*, 50726. [CrossRef]
82. Xiao, Y.; Xu, Z.; Gong, Z.; Li, B.; Huang, Y.; Bian, H.; Wang, C. Self-flocculation and its mechanism of natural rubber latex by accelerant-modified silica. *J. Appl. Polym. Sci.* **2023**, *140*, e54049. [CrossRef]
83. Nah, C.; Lim, J.Y.; Cho, B.H.; Hong, C.K.; Gent, A.N. Reinforcing rubber with carbon nanotubes. *J. Appl. Polym. Sci.* **2010**, *118*, 1574–1581. [CrossRef]
84. Deng, F.; Ito, M.; Noguchi, T.; Wang, L.; Ueki, H.; Niihara, K.-I.; Kim, Y.A.; Endo, M.; Zheng, Q.-S. Elucidation of the reinforcing mechanism in carbon nanotube/rubber nanocomposites. *ACS Nano* **2011**, *5*, 3858–3866. [CrossRef]
85. Wang, D.; Fujinami, S.; Nakajima, K.; Niihara, K.-I.; Inukai, S.; Ueki, H.; Magario, A.; Noguchi, T.; Endo, M.; Nishi, T. Production of a cellular structure in carbon nanotube/natural rubber composites revealed by nanomechanical mapping. *Carbon* **2010**, *48*, 3708–3714. [CrossRef]
86. Hassanzadeh-Aghdam, M.K.; Mahmoodi, M.J.; Ansari, R. Creep performance of CNT polymer nanocomposites-An emphasis on viscoelastic interphase and CNT agglomeration. *Compos. Part B Eng.* **2019**, *168*, 274–281. [CrossRef]
87. Du, M.R.; Jing, H.W.; Duan, W.H.; Han, G.S.; Chen, S.J. Methylcellulose stabilized multi-walled carbon nanotubes dispersion for sustainable cement composites. *Constr. Build. Mater.* **2017**, *146*, 76–85. [CrossRef]
88. Mei, H.; Xia, J.; Zhang, D.; Han, D.; Xiao, S.; Cheng, L. Highly conductive and high-strength carbon nanotube film reinforced silicon carbide composites. *Ceram. Int.* **2017**, *43*, 8873–8878. [CrossRef]
89. Seong, M.; Jeong, C.; Yi, H.; Park, H.-H.; Bae, W.-G.; Park, Y.-B.; Jeong, H.E. Adhesion of bioinspired nanocomposite microstructure at high temperatures. *Appl. Surf. Sci.* **2017**, *413*, 275–283. [CrossRef]
90. Chen, S.; Chen, J.; Xu, K.; Pan, R.; Peng, Z.; Ma, L.; Wu, S. Effect of coal gangue/carbon black/multiwall carbon nanotubes hybrid fillers on the properties of natural rubber composites. *Polym. Compos.* **2016**, *37*, 3083–3092. [CrossRef]

91. Basirjafari, S.; Khadem, S.E.; Malekfar, R. Radial breathing mode frequencies of carbon nanotubes for determination of their diameters. *Curr. Appl. Phys.* **2013**, *13*, 599–609. [CrossRef]
92. Zhou, X.; Zhu, Y.; Liang, J.; Yu, S. New fabrication and mechanical properties of styrene-butadiene rubber/carbon nanotubes nanocomposite. *J. Mater. Sci. Technol.* **2010**, *26*, 1127–1132. [CrossRef]
93. Wang, L.; Wang, X.; Li, Y. Relation between repeated uniaxial compressive pressure and electrical resistance of carbon nanotube filled silicone rubber composite. *Compos. Part A Appl. Sci. Manuf.* **2012**, *43*, 268–274. [CrossRef]
94. Kim, S.W.; Kim, T.; Kim, Y.S.; Choi, H.S.; Lim, H.J.; Yang, S.J.; Park, C.R. Surface modifications for the effective dispersion of carbon nanotubes in solvents and polymers. *Carbon* **2012**, *50*, 3–33. [CrossRef]
95. O'connell, M.J.; Bachilo, S.M.; Huffman, C.B. Band gap fluorescence from individual single-walled carbon nanotubes. *Science* **2002**, *297*, 593–596. [CrossRef] [PubMed]
96. Peng, F.; Pan, F.; Sun, H.; Lu, L.; Jiang, Z. Novel nanocomposite pervaporation membranes composed of poly (vinyl alcohol) and chitosan-wrapped carbon nanotube. *J. Membr. Sci.* **2007**, *300*, 13–19. [CrossRef]
97. Mohamed, A.; Anas, A.K.; Abu Bakar, S.; Aziz, A.A.; Sagisaka, M.; Brown, P.; Eastoe, J.; Kamari, A.; Hashim, N.; Isa, I.M. Preparation of multiwall carbon nanotubes (MWCNTs) stabilised by highly branched hydrocarbon surfactants and dispersed in natural rubber latex nanocomposites. *Colloid Polym. Sci.* **2014**, *292*, 3013–3023. [CrossRef]
98. Nakaramontri, Y.; Nakason, C.; Kummerlöwe, C.; Vennemann, N. Enhancement of electrical conductivity and filler dispersion of carbon nanotube filled natural rubber composites by latex mixing and in situ silanization. *Rubber Chem. Technol.* **2016**, *89*, 272–291. [CrossRef]
99. George, N.; Bipinbal, P.; Bhadran, B.; Mathiazhagan, A.; Joseph, R. Segregated network formation of multiwalled carbon nanotubes in natural rubber through surfactant assisted latex compounding: A novel technique for multifunctional properties. *Polymer* **2017**, *112*, 264–277. [CrossRef]
100. Ge, Y.; Zhang, Q.; Zhang, Y.; Liu, F.; Han, J.; Wu, C. High-performance natural rubber latex composites developed by a green approach using ionic liquid-modified multiwalled carbon nanotubes. *J. Appl. Polym. Sci.* **2018**, *135*, 46588. [CrossRef]
101. Xue, J.; Hao, Y.; Qu, S.; Wang, C.; Li, L. Exploring the reinforcement and conductivity mechanism of K_2FeO_4-modified multiwalled carbon nanotubes in carbon black/natural rubber composites. *Polym. Compos.* **2024**, *45*, 6252–6263. [CrossRef]
102. Mohamed, A.; Anas, A.K.; Bakar, S.A.; Ardyani, T.; Zin, W.M.W.; Ibrahim, S.; Sagisaka, M.; Brown, P.; Eastoe, J. Enhanced dispersion of multiwall carbon nanotubes in natural rubber latex nanocomposites by surfactants bearing phenyl groups. *J. Colloid Interface Sci.* **2015**, *455*, 179–187. [CrossRef]
103. Peng, Z.; Feng, C.; Luo, Y.; Li, Y.; Kong, L. Self-assembled natural rubber/multi-walled carbon nanotube composites using latex compounding techniques. *Carbon* **2010**, *48*, 4497–4503. [CrossRef]
104. Gao, J.-S.; Liu, Z.; Yan, Z.; He, Y. A novel slurry blending method for a uniform dispersion of carbon nanotubes in natural rubber composites. *Results Phys.* **2019**, *15*, 102720. [CrossRef]
105. Zhan, Y.; Liu, G.; Xia, H.; Yan, N. Natural rubber/carbon black/carbon nanotubes composites prepared through ultrasonic assisted latex mixing process. *Plast. Rubber Compos.* **2011**, *40*, 32–39. [CrossRef]
106. Anand, K.A.; Jose, T.S.; Alex, R.; Joseph, R. Natural rubber-carbon nanotube composites through latex compounding. *Int. J. Polym. Mater.* **2009**, *59*, 33–44. [CrossRef]
107. Bystrzejewski, M.; Huczko, A.; Lange, H.; Gemming, T.; Büchner, B.; Rümmeli, M. Dispersion and diameter separation of multi-wall carbon nanotubes in aqueous solutions. *J. Colloid Interface Sci.* **2010**, *345*, 138–142. [CrossRef] [PubMed]
108. Liu, Y.; Gao, L.; Sun, J. Noncovalent functionalization of carbon nanotubes with sodium lignosulfonate and subsequent quantum dot decoration. *J. Phys. Chem. C* **2007**, *111*, 1223–1229. [CrossRef]
109. Li, F.; Yan, N.; Zhan, Y.; Fei, G.; Xia, H. Probing the reinforcing mechanism of graphene and graphene oxide in natural rubber. *J. Appl. Polym. Sci.* **2013**, *129*, 2342–2351. [CrossRef]
110. Li, S.; Li, Z.; Burnett, T.L.; Slater, T.J.; Hashimoto, T.; Young, R.J. Nanocomposites of graphene nanoplatelets in natural rubber: Microstructure and mechanisms of reinforcement. *J. Mater. Sci.* **2017**, *52*, 9558–9572. [CrossRef]
111. Boland, C.S.; Khan, U.; Backes, C.; O'neill, A.; Mccauley, J.; Duane, S.; Shanker, R.; Liu, Y.; Jurewicz, I.; Dalton, A.B. Sensitive, high-strain, high-rate bodily motion sensors based on graphene–rubber composites. *ACS Nano* **2014**, *8*, 8819–8830. [CrossRef] [PubMed]
112. Hassanzadeh-Aghdam, M. Evaluating the effective creep properties of graphene-reinforced polymer nanocomposites by a homogenization approach. *Compos. Sci. Technol.* **2021**, *209*, 108791. [CrossRef]
113. Zhan, Y.; Wu, J.; Xia, H.; Yan, N.; Fei, G.; Yuan, G. Dispersion and exfoliation of graphene in rubber by an ultrasonically-assisted latex mixing and in situ reduction process. *Macromol. Mater. Eng.* **2011**, *296*, 590–602. [CrossRef]
114. Zhan, Y.; Lavorgna, M.; Buonocore, G.; Xia, H. Enhancing electrical conductivity of rubber composites by constructing interconnected network of self-assembled graphene with latex mixing. *J. Mater. Chem.* **2012**, *22*, 10464–10468. [CrossRef]
115. Li, C.; Feng, C.; Peng, Z.; Gong, W.; Kong, L. Ammonium-assisted green fabrication of graphene/natural rubber latex composite. *Polym. Compos.* **2013**, *34*, 88–95. [CrossRef]
116. Stanier, D.; Patil, A.; Sriwong, C.; Rahatekar, S.; Ciambella, J. The reinforcement effect of exfoliated graphene oxide nanoplatelets on the mechanical and viscoelastic properties of natural rubber. *Compos. Sci. Technol.* **2014**, *95*, 59–66. [CrossRef]
117. Suriani, A.; Nurhafizah, M.; Mohamed, A.; Zainol, I.; Masrom, A. A facile one-step method for graphene oxide/natural rubber latex nanocomposite production for supercapacitor applications. *Mater. Lett.* **2015**, *161*, 665–668. [CrossRef]

118. Li, C.; Wang, J.; Chen, X.; Song, Y.; Jiang, K.; Fan, H.; Tang, M.; Zhan, W.; Liao, S. Structure and properties of reduced graphene oxide/natural rubber latex nanocomposites. *J. Nanosci. Nanotechnol.* **2017**, *17*, 1133–1139. [CrossRef]
119. Wang, J.; Zhang, K.; Cheng, Z.; Lavorgna, M.; Xia, H. Graphene/carbon black/natural rubber composites prepared by a wet compounding and latex mixing process. *Plast. Rubber Compos.* **2018**, *47*, 398–412. [CrossRef]
120. Mao, Y.; Wang, C.; Liu, L. Preparation of graphene oxide/natural rubber composites by latex co-coagulation: Relationship between microstructure and reinforcement. *Chin. J. Chem. Eng.* **2020**, *28*, 1187–1193. [CrossRef]
121. Li, Z.; An, D.; He, R.; Sun, Z.; Li, J.; Zhang, Z.; Liu, Y.; Wong, C. The design of crosslinks in different vulcanized systems to improve crack growth resistance for carbon black/graphene oxide/natural rubber composites. *Adv. Compos. Hybrid Mater.* **2023**, *6*, 82. [CrossRef]
122. Zhu, Y.; Murali, S.; Cai, W.; Li, X.; Suk, J.W.; Potts, J.R.; Ruoff, R.S. Graphene and graphene oxide: Synthesis, properties, and applications. *Adv. Mater.* **2010**, *22*, 3906–3924. [CrossRef]
123. Park, S.; Ruoff, R.S. Chemical methods for the production of graphenes. *Nat. Nanotechnol.* **2009**, *4*, 217–224. [CrossRef]
124. Eda, G.; Chhowalla, M. Chemically derived graphene oxide: Towards large-area thin-film electronics and optoelectronics. *Adv. Mater.* **2010**, *22*, 2392–2415. [CrossRef]

polymers

MDPI

Article

Novel Reactive Polyhedral Oligomeric Silsesquioxane-Reinforced and Toughened Epoxy Resins for Advanced Composites

Weibo Liu [1], Caiyun Wang [1], Yu Feng [1], Yongfeng Chen [1], Liqiang Wan [1], Farong Huang [1,*], Zuozhen Liu [2], Jianhua Qian [2] and Weiping Liu [3]

[1] Key Laboratory for Specially Functional Polymeric Materials and Related Technology of (Ministry of Education), School of Materials Science and Engineering, East China University of Science and Technology, Shanghai 200237, China; y82210301@mail.ecust.edu.cn (W.L.); y10210074@mail.ecust.edu.cn (C.W.); fengyu10161366@163.com (Y.F.); 13918350379@163.com (Y.C.); wanliqiang@163.com (L.W.)

[2] HuaChang Polymers Co., Ltd., East China University of Science and Technology, Shanghai 200241, China; lzz@ecust.edu.cn (Z.L.); qjh@hchp.com.cn (J.Q.)

[3] Manufacturing Center of Composite Materials for Commercial Aircraft, Shanghai Aircraft Manufacturing Co., Ltd., COMAC, Shanghai 201324, China; liuweiping@comac.cc

* Correspondence: fhuanglab@ecust.edu.cn

Abstract: Most toughening methods for epoxy resins are usually used at the expense of other properties. Some polyhedral oligomeric silsesquioxanes (POSSs) with both a rigid Si-O-Si structure and flexible organic chain segments could be expected to be effective toughening agents. In this study, three reactive polyhedral oligomeric silsesquioxanes with a thiol group (OMPPS), a carboxyl group (OCOPS), and an epoxy group (OGCPS) were synthesized and characterized. They were utilized as modifiers to toughen 3-(oxiran-2-ylmethoxy)-*N*,*N*-bis(oxiran-2-ylmethyl)aniline (AFG-90MH)/4,4′-methylenebis(2-ethylaniline) (MOEA) (epoxy resin) with different molar ratios to obtain hybrid resins named OMPPS-EP-i, OCOPS-EP-j, and OGCPS-EP-k. The effects of the amount of modifier added and the length of the organic chain on the cage structure on various properties of the hybrid resins were investigated. The results show that all three modifiers show good compatibility with the epoxy resin. The hybrid resins have a low viscosity at 45~85 °C and can be cured at a low temperature (110 °C). The cured hybrid resins display improved toughness. Typically, the critical stress intensity factor (K_{IC}) and impact strength of OGCPS-EP-0.6-C are 2.54 $MPa \cdot m^{-1/2}$ and 19.33 $kJ \cdot m^{-2}$, respectively, which increased by 58.75% and 22.48% compared with the pristine epoxy resin, respectively. In addition, the glass transition temperature and flexural strength of the hybrid resins are basically unchanged.

Keywords: high-performance epoxy resin; polyhedral oligomeric silsesquioxane; toughness; liquid molding resin for advanced composites

1. Introduction

Aerospace technology, being a highly sophisticated and state-of-the-art field, serves as a significant emblem of advancements in modern science and technology. The realization of this technology is inseparable from the support of advanced materials. In the aerospace field, due to its extremely harsh operating environment, the material requirements of performance are more stringent. High-performance epoxy resins have found extensive applications as matrix resins in aerospace component composites owing to their good thermal stability [1], high glass transition temperature, excellent mechanical properties, and low curing shrinkage [2–6]. However, the high crosslinking density of the epoxy resins also results in low toughness, which significantly limits their usage in some situations. Therefore, enhancing the toughness of the epoxy resins has always been one of the key subjects for high-performance epoxy resins.

There are organic and inorganic toughening agents for toughening epoxy resins. Organic toughening agents include liquid rubbers [7,8], thermoplastic polymers [9–11], hyperbranched polymers [12,13], and liquid crystal polymers [14]. They have a good toughening effect, but usually at the expense of the mechanical and thermal properties of the epoxy resins. In addition, because the load amount of the agents is generally high, the viscosity of the modified epoxy resins will increase, which is not conducive to the processing of the resins. Inorganic toughening agents include graphene oxide [15–18], carbon nanotubes [19,20], nano-silica [21,22], and nano-clay [23]. They can maintain good mechanical properties while toughening epoxy resins. However, they are easy to aggregate in the resins because of the small size and large specific surface area of the nanoparticles, which form stress concentration points and reduce the performance of the resins.

Polyhedral oligomeric silsesquioxane (POSS), a kind of nano organic–inorganic hybrid molecule, is composed of a silicon and oxygen skeleton and has attracted extensive attention from researchers in recent years [24,25]. The general molecular formula of POSS can be abbreviated as $(RSiO_{1.5})_n$ (where R stands for an organic group and n could be 8, 10, or 12). The center of POSS is a rigid inorganic Si-O-Si core and the exterior is connected to soft organic groups. POSS can be diversified through different chemical reactions, which can make POSS meet various application requirements. According to different external organic groups, POSSs can be divided into reactive POSSs and non-reactive POSSs. As a toughening agent for an epoxy resin, reactive POSSs can participate in the curing reaction of the epoxy resin through organic groups [26–31], and non-reactive POSSs can be uniformly dispersed in the resin by physical blending [32]. The molecular dispersion of reactive POSSs can be achieved through participation in the curing reaction of the epoxy resin, thereby enhancing the compatibility between POSSs and the epoxy resin.

Liu et al. [25] toughened an epoxy resin with POSS-grafted polyurethane (POSS-PU). The impact strength of the toughened epoxy resin increased by 115%, but the viscosity of the resin increased and the glass transition temperature decreased. Yang et al. [33] mixed octavinyl polyhedral oligomeric silsesquioxane (EPOSS)-modified vinyl resin with epoxy resin to form an interpenetrating network structure to toughen the epoxy resin. The impact strength of the resin increased by 139.6%, but the glass transition temperature decreased. Mishra et al. [28] incorporated trisilanol phenyl POSS and glycidyl POSS separately into an epoxy resin (EPON 862). Obviously, the trisilanol phenyl POSS is a non-reactive POSS and the glycidyl POSS is a reactive POSS. The toughness of the incorporated epoxy resins is increased by 30% and 130%, respectively. The toughening efficiency of the reactive POSS is much higher than that of the non-reactive POSS. Han et al. [26] modified an epoxy resin with octa-glycidyl POSS (OG-POSS) as a toughening agent. OG-POSS participated in the curing reaction and was evenly dispersed in the epoxy resin. The modified resin not only maintained the original glass transition temperature, flexural strength, and thermal stability but also increased the impact strength by 56%.

In this study, in view of the nano-sized hollow structure of polyhedral oligomeric silsesquioxanes and their good compatibility with a resin, three reactive polyhedral oligomeric silsesquioxanes with a thiol group (OMPPS), a carboxyl group (OCOPS), and an epoxy group (OGCPS) were synthesized and incorporated into the blend of 3-(oxiran-2-ylmethoxy)-*N*,*N*-bis(oxiran-2-ylmethyl)aniline/4,4′-methylenebis(2-ethylaniline) (epoxy resin) to obtain OMPPS-EP-i, OCOPS-EP-j, and OGCPS-EP-k hybrid resins with good comprehensive properties. The effect of the structure and content of the three modifiers on the curing behavior of the hybrid resins, as well as the thermal and mechanical properties of the cured resins, was investigated and compared.

2. Experimental

2.1. Materials

(3-Mercaptopropyl) trimethoxysilane (analytical purity), 10-undecylenoic acid (analytical purity), and 2,2′-azobis(2-methylpropionitrile) (AIBN) (analytical purity) were purchased from Shanghai Titan Technology Co., Ltd. 4-Hydroxybenzaldehyde (analytical purity) was bought from Jiangsu Anhuai Chemical Technology Co., Ltd. (Nantong, China). Methyl triphenyl phosphonium bromide (analytical purity) was supplied by Shanghai Zhonghe Chemical Technology Co., Ltd. (Shanghai, China). Potassium tert-butanol (analytical purity) was purchased from Shanghai XINBO Chemical Technology Co., Ltd. (Shanghai, China). Sodium hydride (analytical purity), epichlorohydrin (analytical purity), tetrabutylammonium bromide (analytical purity), and sodium hydroxide (analytical purity) were bought from Shanghai Boer Chemical Technology Co., Ltd. (Shanghai, China). Hydrochloric acid was bought from Sinopharm Chemical Reagent Co., Ltd. (Shanghai, China). Epoxy resin AFG-90MH (industrial grade) was supplied by Shanghai Huayi Resin Co., Ltd. (Shanghai, China). 4,4″-Methylenebis(2-ethylbenzenamine) (MOEA) (analytical purity) was purchased from Shanghai Bide Pharmaceutical technology Co., Ltd. (Shanghai, China). The release agent (909A 45) was supplied by Dongguan Jiadan lubricating oil Co., Ltd. (Dongguan, China). All solvents (analytical purity) were purchased from Shanghai Titan Technology Co., Ltd. (Shanghai, China).

2.2. Synthesis of Octa (Mercaptopropyl) Polyhedral Oligomeric Silsesquioxane (OMPPS)

The synthetic route of OMPPS is shown in Scheme 1. First, 20 g of (3-Mercaptopropyl) trimethoxysilane (0.1 mol) and 25 mL of concentrated hydrochloric acid (36%) were added to a 1 L three-necked flask which contained 500 mL of methanol and stirred at 65 °C for 24 h under a N_2 atmosphere. At the end of the reaction, the precipitate at the bottom of the flask was collected, washed three times with methanol, and dissolved in dichloromethane. Then, the obtained solution was washed three times with deionized water and dried to remove the solvent, followed by vacuum drying in a vacuum oven at 40 °C for 8 h to obtain OMPPS with a yield of 75%.

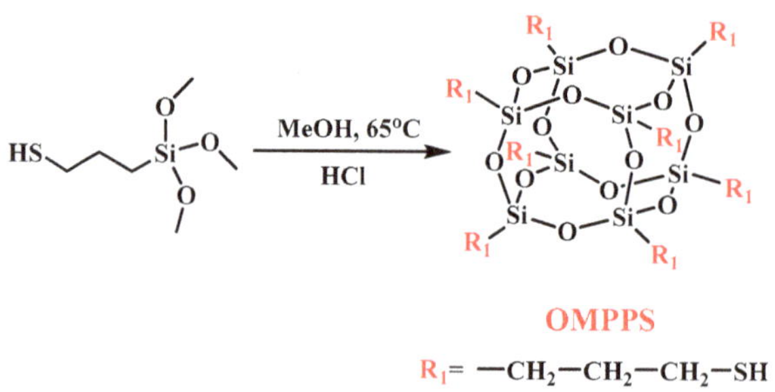

Scheme 1. Synthetic route of OMPPS.

2.3. Synthesis of Octa (Carboxyl Decyl Thiopropyl) Polyhedral Oligomeric Silsesquioxane (OCOPS)

The synthetic route of OCOPS is shown in Scheme 2. First, 18.4 g of undecylenic acid (0.1 mol) and 1.0 g of 2,2′-azobis(2-methylpropionitrile) (AIBN) (6 mmol) were added to a 250 mL three-necked flask which contained 70 mL of dried tetrahydrofuran. The mixture in the flask was stirred until the solids were completely dissolved and heated up to 60 °C under N_2. Then, a tetrahydrofuran solution (50 mL) containing 10.2 g (0.01 mol) OMPPS was added dropwise to the flask while stirring, and the addition time was controlled at 10~15 min. The reaction was stopped after maintaining at 60 °C for 6 h. After the reaction, the solution was concentrated and poured into 300 mL of methanol for sedimentation, and a white precipitate product was obtained. The product was placed in a vacuum oven and dried at 60 °C for 8 h to obtain OCOPS with a yield of 57%.

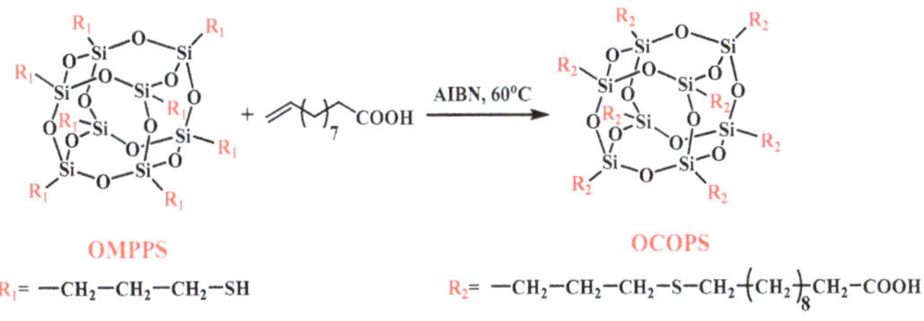

Scheme 2. Synthetic route of OCOPS.

2.4. Synthesis of Octa (Glycidyl Phenyl Ethyl Thiopropyl) Polyhedral Oligomeric Silsesquioxane (OGCPS)

The synthetic route of OGCPS is shown in Scheme 3. 4-Vinyl phenyl glycidyl ether (DGHDS) was synthesized from 4-hydroxy styrene (HDS) and epichlorohydrin, which is shown in the supporting information.

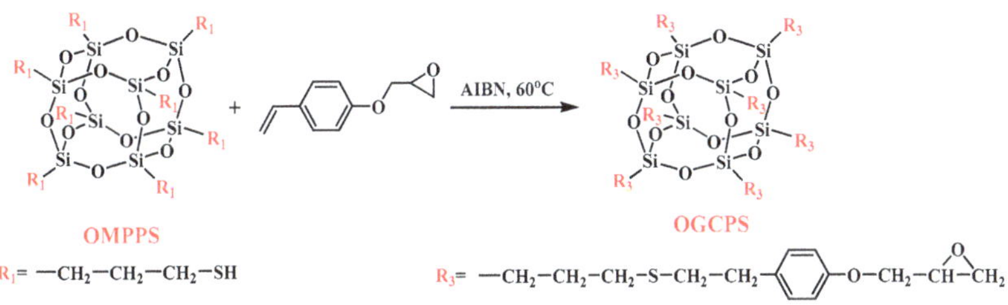

Scheme 3. Synthetic route of OGCPS.

First, 17.6 g of DGHDS (0.1 mol) and 1.0 g of 2,2′-azobis(2-methylpropionitrile) (AIBN) (6 mmol) were added to a 250 mL three-necked flask which contained 100 mL of dried tetrahydrofuran. The mixture was stirred until the solids were completely dissolved and heated up to 60 °C under N_2. Then, a tetrahydrofuran solution (50 mL) containing 10.2 g (0.01 mol) OMPPS was added dropwise to the flask while stirring, and the addition time was controlled at 10~15 min. The reaction was stopped after maintaining at 60 °C for 6 h. After the reaction, the solution was concentrated and poured into 300 mL of methanol for sedimentation, and a light-yellow oily precipitate was obtained. After washing with methanol three times, the light-yellow oily precipitate was placed in a vacuum oven and dried at 60 °C for 8 h to obtain OMPPS with a yield of 52%.

2.5. Preparation of Hybrid Resins

The epoxy resin used in this study is 3-(oxiran-2-ylmethoxy)-*N*,*N*-bis(oxiran-2-ylmethyl) aniline (AFG-90MH), and the curing agent is 4,4′-methylenebis(2-ethylaniline) (MOEA). The structural formulas of AFG-90MH and MOEA are shown in Figure 1. The pristine epoxy resin (EP) used as the control group consisted of AFG-90MH and MOEA. The obtained reactive polyhedral oligomeric silsesquioxanes (OMPPS, OCOPS, and OGCPS) were separately mixed into the EP to obtain the hybrid resins. The molar ratio of active hydrogen to the epoxy group in the hybrid resins was controlled at 1:1.

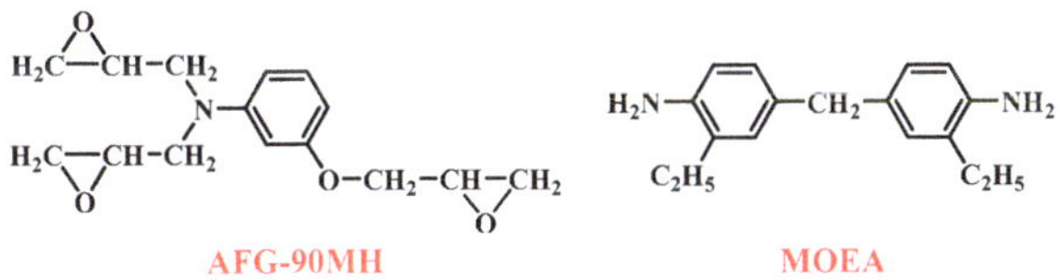

Figure 1. Structural formulas of AFG-90MH and MOEA.

2.5.1. Preparation of OMPPS- or OGCPS-Modified Epoxy Hybrid Resins

AFG-90MH, MOEA, and OMPPS were dissolved in THF in a flask under stirring. After removing THF using a vacuum at 60 °C, the OMPPS-modified epoxy hybrid resins were obtained and named as OMPPS-EP-i (i = 0.3, 0.8, 1.4, 1.9, 2.8). The OGCPS-modified epoxy resins were made in the same way and named as OGCPS-EP-j (j = 0.1, 0.3, 0.6, 0.8, 1.2). The i and j are the molar percentage of OMPPS and OGCPS in the hybrid resins, respectively.

2.5.2. Preparation of OCOPS-Modified Epoxy Hybrid Resins

First, OCOPS was dissolved in DMF in a flask, and the produced solution was mixed with AFG-90MH and prepolymerized at 110 °C for 0.5 h. Then, DMF was removed using a vacuum, and a homogeneous resin was obtained. Finally, MOEA was added to the resin to obtain the OCOPS-modified epoxy hybrid resins, named OCOPS-EP-k (k = 0.1, 0.3, 0.6, 0.8, 1.2, 1.6, 1.9), in which k is the molar percentage of OCOPS in the OCOPS-modified epoxy resins.

2.6. Preparation of Cured Hybrid Resins

The obtained hybrid resins were poured into the treated molds, respectively, and then the molds containing the resins were placed in a vacuum oven and evacuated at 90 °C for 0.5 h to remove the residual solvent. Afterward, the resins with molds were moved to an oven and heated to 110 °C, then kept at 110 °C for 2 h and 150 °C for another 2 h to obtain the cured resins. The obtained cured resins were abbreviated as EP-C, OMPPS-EP-i-C, OCOPS-EP-j-C, and OGCPS-EP-k-C, respectively.

2.7. Instruments and Measurements

The ^{1}H NMR spectrum was obtained on the AVANCE 400 III spectrometer (Bruker, Switzerland) at an operating frequency of 400 MHz with chloroform-D ($CDCl_3$) or DMSO-*d6* as the deuterated solvent and tetramethylsilane as the internal standard. The FT-IR spectrum was obtained using a Nicolet iS10 infrared spectrometer (Thermo Scientific, Madison, WI, USA). Elemental analysis tests were conducted on the VARIO EL CUBE elemental analyzer (ELEMENTAR, Hanau, Germany). The relative molecular masses of POSSs were obtained by 4800 plus time-of-flight mass spectrometry (ABS, MAS Building, Singapore). The ^{29}Si NMR spectrum was obtained on the Ascend 600 spectrometer at an operating frequency of 600 MHz with $CDCl_3$ or DMSO-*d6* as the deuterated solvent and chromium acetylacetonate as the relaxation reagent. Differential scanning calorimetry (DSC) was run on a DSC214 (NETZSCH, Selb, Germany) instrument at the heating rate of 10 °C/min under nitrogen flow from 40 °C to 300 °C. Dynamic thermomechanical analysis (DMA) was performed on a DMA 1 mechanical analyzer (Mettler Toledo, Greifensee, Switzerland) by using the three-point bending mode at a heating rate of 5 °C/min with a frequency of 1 Hz under nitrogen flow. The flexural performance and fracture toughness of the cured resin were measured using a SANS CMT 4204 universal testing machine (Sansi Material Testing Co., Ltd., Shenzhen, China) at a speed of 2.00 mm/min and 1.5 mm/min, respectively, at room temperature with the three-point bending method. The impact performances of the cured resins were obtained by a CEAST 9050 cantilever beam impact testing machine (CEAST, Torino, Italy). The fracture surface of the cured resins was characterized by S-4800/S-3400 scanning electron microscopy (HITACHI, Tokyo, Japan).

3. Results and Discussion

3.1. The Structural Characterization of OMPPS, OCOPS, and OGCPS

The structures of OMPPS, OCOPS, and OGCPS were characterized by FT-IR spectroscopy, and the spectra are shown in Figure 2a. The absorption peaks around 1100 cm^{-1} are the Si-O-Si stretching vibration peaks in OMPPS, OCOPS, and OGCPS. The peak at 2554 cm^{-1} corresponds to the -SH of OMPPS. The absorption peak of -SH disappears and the peak of -COOH at 1703 cm^{-1} for OCOPS appears, indicating that OCOPS has been successfully synthesized. In the FT-IR spectrum of OGCPS, the absorption peak of -SH also disappeared, and the absorption peak of the epoxy group appeared at 915 cm^{-1}, indicating that the -SH at the eight corners of OMPPS has completely reacted with the vinyl group of DGHDS to form OGCPS. Figure 2b shows the ^{1}H NMR spectra of OMPPS, OCOPS, and OGCPS. The solvent of the ^{1}H NMR analysis for OMPPS and OGCPS is $CDCl_3$ and the solvent for OCOPS is DMSO-*d6*. In the ^{1}H NMR spectrum of OMPPS, the resonance peaks for the hydrogens of three methylene in the mercaptopropyl group are found at chemical shifts of 0.77 ppm(a_1), 1.72 ppm(b_1), and 2.57 ppm(c_1), respectively, and the resonance peak of sulfhydryl hydrogen is found at chemical shift of 1.38 ppm(d_1). The area ratio of the peaks for the hydrogens at a_1, b_1, c_1, and d_1 is 2.07:2.14:1.98:1.0. In the ^{1}H NMR spectrum of OCOPS, the chemical shifts of the methylene hydrogens at a_2, b_2, and c_2 move to the high field under the influence of the long-chain alkane group, compared to those of the hydrogens of OMPPS at a_1, b_1, and c_1, which are 0.73 ppm, 1.60 ppm, and 2.50 ppm, respectively. The resonance peak of sulfhydryl hydrogen disappears, indicating that the sulfhydryl group has completely participated in the reaction, and the eight corners of POSS have become carboxyl-terminated groups. Under the influence of the thioether group, the chemical shifts of hydrogens for methylene-connected sulfur atoms are close, at about 2.50 ppm. The chemical shift of methylene hydrogens at d_2 is influenced by the thioether group and moves to the lower field, at 1.47 ppm, while the chemical shift of methylene hydrogens at e_2 is less affected and is 1.24 ppm. Due to the influence of the carboxyl group, the resonance peak of methylene hydrogens connected with the carboxyl group moves to the lower field and is 2.17 ppm. The peak area ratio is 2.00:2.05:4.07:3.98:12.14:2.0 for hydrogens at a_2, b_2, c_2, d_2, e_2, and f_2, respectively. In the ^{1}H NMR spectrum of OGCPS, the resonance peaks at a_3, b_3, and c_3 are slightly shifted to the lower field as compared to the hydrogens of OMPPS at a_1, b_1, and c_1, with chemical shifts of 0.74 ppm, 1.68 ppm, and 2.53 ppm, respectively. The resonance peak of sulfhydryl hydrogen disappears, indicating that the sulfhydryl group has completely reacted, and the eight corners of POSS have become hydrocarbon chains with a terminal epoxy group. In addition, the chemical shifts of other hydrogen atoms in OGCPS can also correspond with each other in the spectrum. The peak area ratio of the hydrogens at a_3, b_3, (c_3 + d_3 + j_3 + e_3), i_3, h_3, g_3, and f_3 is 2.14:2.18:7.92:1.00:2.14:2.08:2.0.

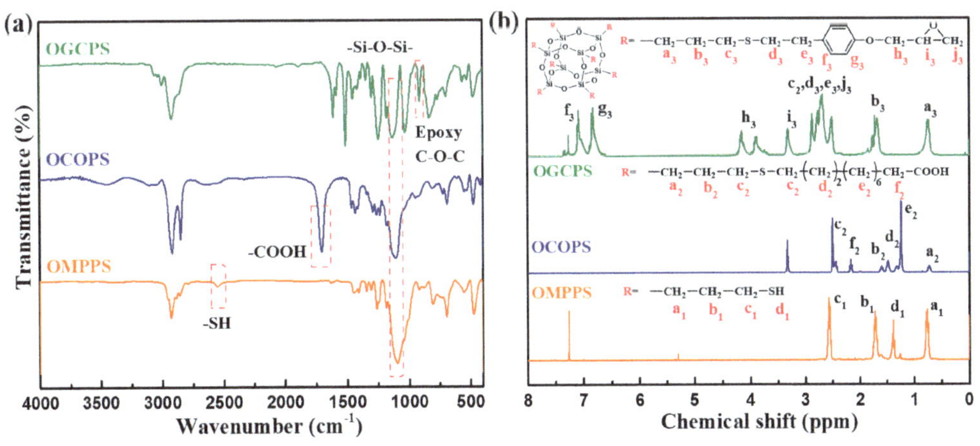

Figure 2. (**a**) FT-IR spectra of OMPPS, OCOPS, and OGCPS and (**b**) ^{1}H NMR spectra of OMPPS, OCOPS, and OGCPS.

The ^{29}Si NMR spectra of OMPPS, OCOPS, and OGCPS are shown in Figure 3. The solvent of the ^{29}Si NMR analysis of OMPPS and OGCPS is $CDCl_3$ and the solvent for OCOPS is DMSO-*d6*. The ^{29}Si NMR spectra of OMPPS, OCOPS, and OGCPS show a large single peak, indicating that the purity is high. In addition, the relative molecular masses of OMPPS, OCOPS, and OGCPS were determined by using MALDI-TOF-MS, and the results are shown in Figure S2. The analysis results are as follows: OMPPS (product + Na^+): 1039.20 Da (calculated: 1039.14), OCOPS (product + Na^+): 2512.33 Da (calculated: 2512.33), and OGCPS (product + Na^+): 2447.30 Da (calculated: 2447.27). These results show that the molecular weights of OMPPS, OCOPS, and OGCPS are 1016, 2489, and 2424, respectively. The above test results demonstrate that OMPPS, OCOPS, and OGCPS were successfully synthesized.

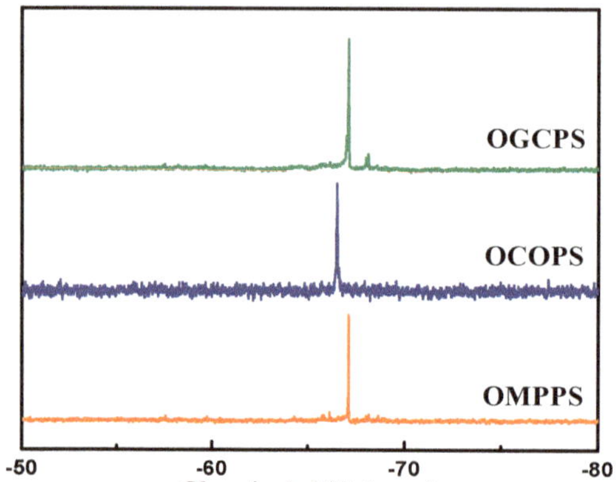

Figure 3. ^{29}Si NMR spectra of OMPPS, OCOPS, and OGCPS.

The solubility of OMPPS, OCOPS, and OGCPS has an influence on the subsequent application. The dissolution behaviors of OMPPS, OCOPS, and OGCPS in common organic solvents are listed in Table 1. OMPPS and OGCPS have good solubility and can be dissolved in dichloromethane, tetrahydrofuran, ethyl acetate, *N,N*-dimethylformamide, and dimethyl sulfoxide. However, OCOPS has poor solubility and can be dissolved in tetrahydrofuran, *N,N*-dimethylformamide, and dimethyl sulfoxide.

Table 1. Solubility of OMPPS, OCOPS, and OGCPS.

Solvent	Solubility		
	OMPPS	OCOPS	OGCPS
Methanol	−	−	−
Ethanol	−	−	−
Tetrahydrofuran	+	+	+
Ethyl acetate	+	−	+
Dimethylformamide	+	+	+
Dimethyl sulfoxide	+	+	+
Dichloromethane	+	−	+
Petroleum ether	−	−	−
Chloroform	+	−	+

+ soluble, − insoluble.

3.2. The Processability of Hybrid Resins

Viscosity has an important effect on the processability of the epoxy resin. Figure 4 shows the viscosity of different resins as a function of the temperature. Obviously, the addition of OMPPS, OCOPS, and OGCPS has no significant effect on the viscosity of the epoxy resin. The viscosity of hybrid resins is lower than 500 mPa·s at 45~85 °C, which is suitable for the liquid molding (RTM) process of composite materials.

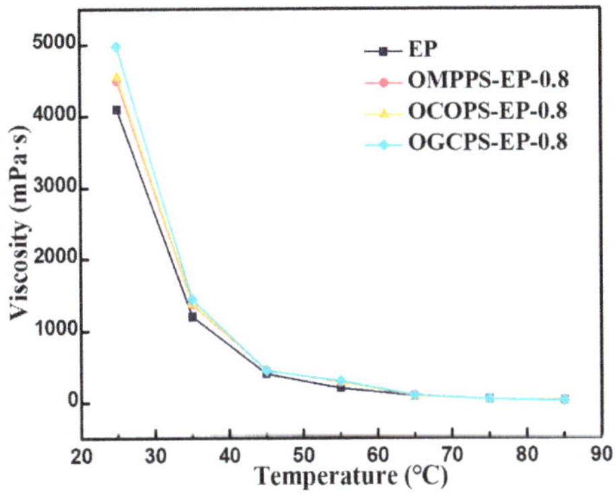

Figure 4. Viscosity of pristine epoxy resin, OMPPS-EP-0.8, OCOPS-EP-0.8, and OGCPS-EP-0.8.

3.3. Thermal Curing Behaviors of Hybrid Resins

In order to investigate the curing behavior of the hybrid resins, a differential scanning calorimeter (DSC) was used. Figure 5a–c shows the DSC curves of the OMPPS-EP-i, OCOPS-EP-j, and OGCPS-EP-k hybrid resins, and the detailed analysis data are listed in Table S1. It can be seen that the addition of OMPPS and OGCPS has no effect on the curing behaviors of the epoxy resin, but the addition of OCOPS can slightly reduce the curing temperature. This may be due to the fact that the carboxyl groups in OCOPS have a higher reactivity with epoxy resin than active groups in OMPPS and OGCPS. In order to compare the effects of different modifiers on the thermal curing behavior of the epoxy resin, we selected three hybrid resins with a modifier addition amount of 0.8 mol% for comparison. Figure 5d shows the DSC curves of the epoxy resin and hybrid resins with a 0.8 mol% load of the modifier. It can be seen that under conditions where the same molar amount is added, OCOPS is beneficial in promoting the thermal cross-linking reaction of epoxy resin.

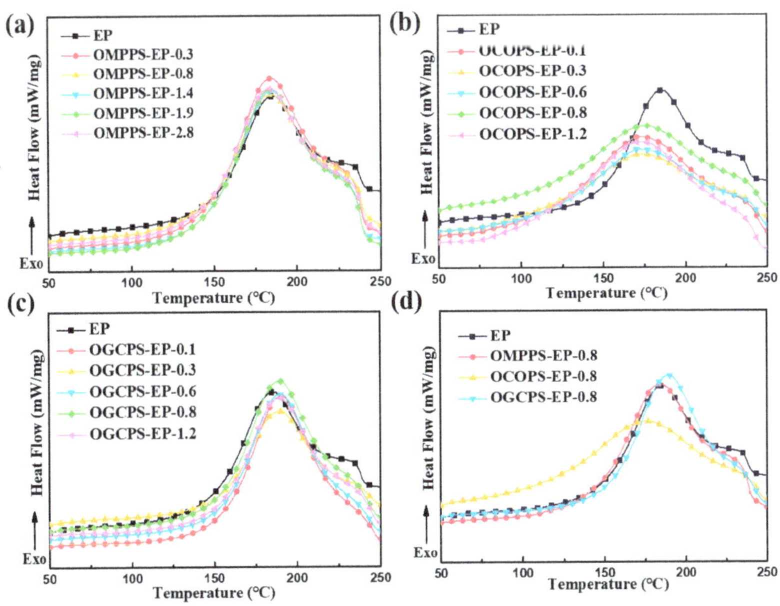

Figure 5. DSC curves of the OMPPS-EP-i (**a**), OCOPS-EP-j (**b**), and OGCPS-EP-k (**c**) hybrid resins. (**d**) The pristine epoxy resin and hybrid resins with the same load amount of modifier.

The T_i (initial curing temperature) and T_p (top curing peak temperature) of the modified resins usually vary with the heating rate. The higher the heating rate, the higher the T_i and T_p. We conducted DSC tests for the pristine epoxy resin at various heating rates. The T_{i0} and T_{p0} at the heating rate of zero are obtained through the linear extrapolation of the T_i and T_p with the heating rate (Figure S4), and are 110 °C and 155 °C, respectively. Therefore, 110 °C/2 h + 150 °C/2 h was selected as the curing process of the epoxy resin. Figure 6 shows the DSC curves of OMPPS-EP-0.8, OCOPS-EP-0.8, and OGCPS-EP-0.8 before and after curing. There is no obvious exothermic peak in the cured epoxy resins, indicating that the epoxy resins have been completely cured.

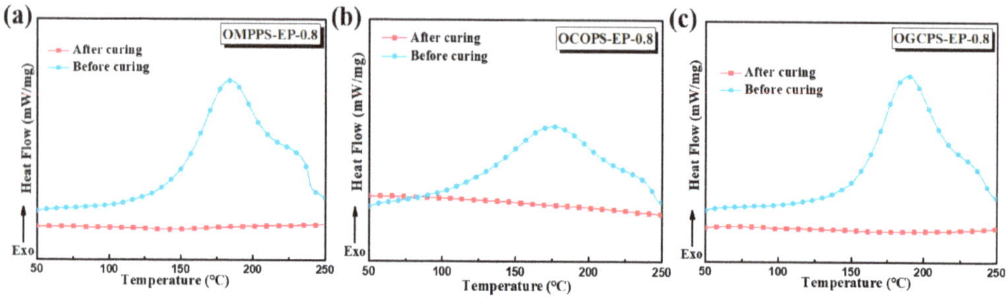

Figure 6. DSC curves of OMPPS-EP-0.8 (**a**), OCOPS-EP-0.8 (**b**), and OGCPS-EP-0.8 (**c**) before and after curing.

FT-IR spectroscopy was adopted to investigate the structural changes in the hybrid resins during the curing process. The FT-IR spectra of the hybrid resins with 0.8 mol% modifier content before and after curing are shown in Figure 7. As shown in Figure 7a, the absorption peaks at 3400~3500 cm^{-1} are assigned to the stretching vibration peak of the amino group in the curing agent (MOEA). The absorption bands at 2554 cm^{-1} and 1703 cm^{-1} are attributed to the stretching vibration of -SH and -COOH, respectively, indicating that the modifiers have been introduced into the epoxy resin. After curing, the absorption peaks of -NH_2 for the curing agent and the absorption peaks of the epoxy group (915 cm^{-1}) disappeared, and a new absorption peak of -OH appeared, indicating that the amino group and epoxy group have completely reacted. In addition, the absorption peak of -SH in OMPPS-EP-0.8 and the absorption peak of -COOH in OCOPS-EP-0.8 disappeared after curing, indicating that OMPPS and OCOPS have participated in the curing reaction of the epoxy resins.

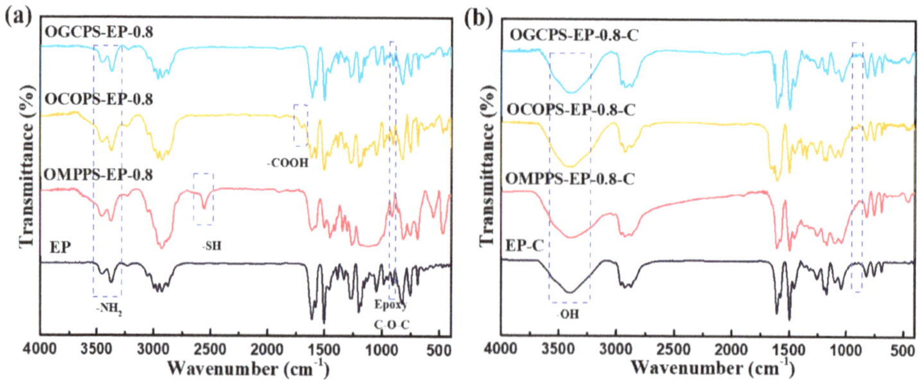

Figure 7. FT-IR spectra of the hybrid resins with 0.8 mol% modifier content before (**a**) and after (**b**) curing.

3.4. Mechanical Properties of Cured Hybrid Resins

3.4.1. Flexural and Impact Properties of Cured Hybrid Resins

The toughness and strength of resins have a significant influence on the performance of terminal fiber reinforced composites. The flexural properties of the cured hybrid resins were tested and the results are shown in Figure 8. When the amount of modifier added is small, the flexural properties of the cured hybrid resins increase slightly and then reach a peak value. However, as the amount of modifier added increases, the flexural properties begin to decline slightly. The reason for this phenomenon may be that at low loading, the rigid inorganic nano-scale Si-O-Si core is uniformly dispersed, which can disperse the stresses concentrated on the crack tip and thus increase the flexural properties of the cured hybrid resins. However, with the continuous addition of the POSS modifiers, POSS molecules self-assemble and aggregate to form the so-called "hard-core soft-shell particles", which play a plasticizing role, resulting in a decline in the flexural properties of cured hybrid resins [28].

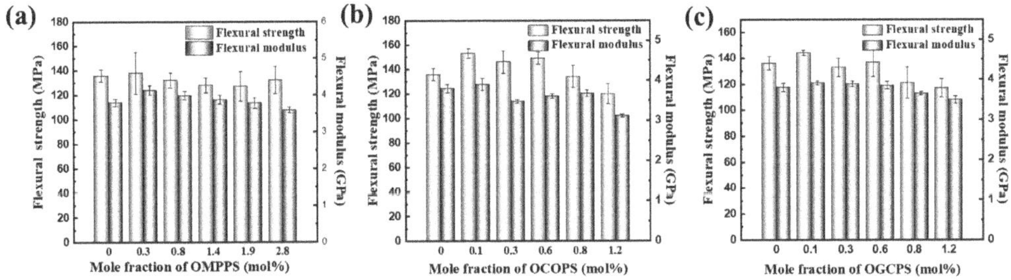

Figure 8. Flexural properties of OMPPS-EP-i-C (**a**), OCOPS-EP-j-C (**b**), and OGCPS-EP-k-C (**c**).

As shown in Figure 8, the flexural strength of OMPPS-EP-0.3-C reaches 138 MPa, compared with the flexural strength of 136 MPa for EP-C. Compared with OMPPS, OCOPS has longer and more flexible alkyl chain segments, the distance between the molecular chains is increased, and the free volume fraction in the cured resin is increased. Thus, the flexural strength of OCOPS-EP-0.6-C increases to 149 MPa, which is 10% higher than that of EP-C. As for OGCPS-EP-0.6-C with the same modifier addition ratio, the flexural strength of OGCPS-EP-0.6-C reaches 137 MPa, which is close to that of the pristine resin. This is mainly due to OGCPS's molecular structure, which contains rigid benzene rings, increasing the rigidity of the cured resins (flexural modulus: OGCPS-EP-0.6-C > OCOPS-EP-0.6-C) and restricting the movement of the molecular chain. In summary, these results show that OMPPS, OCOPS, and OGCPS can effectively maintain or improve the flexural properties of epoxy resins at a certain dosage.

Impact performance tests were conducted on the cured resins, and the results are shown in Figure 9. It can be seen that the impact strength of OMPPS-EP-i-C and OGCPS-EP-j-C increases first and then decreases with an increase in the amount of modifier added, while the impact strength of OCOPS-EP-k-C increases with an increase in the amount of OCOPS added.

OMPPS, OCOPS, and OGCPS contain functional groups that can participate in the curing reactions of the epoxy resin, which can cause them to become better dispersed in the resin. POSSs, with rigid -Si-O-Si- cage-like structures, are uniformly distributed in the epoxy networks. When the cured epoxy structures are subjected to impact, POSSs play a dual role. On the one hand, they are able to absorb some energy as stress concentration points and thereby facilitate the dispersion of the stress. On the other hand, they also prevent the formation and propagation of micro-cracks, while forming shear bands within the resins to absorb another portion of the impact energy [26,28,34]. Organic soft segments in POSSs can also increase the flexibility of the polymer chain [35–37] and thus increase the impact strength of the epoxy resins. However, the impact strength decreases with the further loading of the amount of POSSs. The decline is possibly related to the assembling of POSSs.

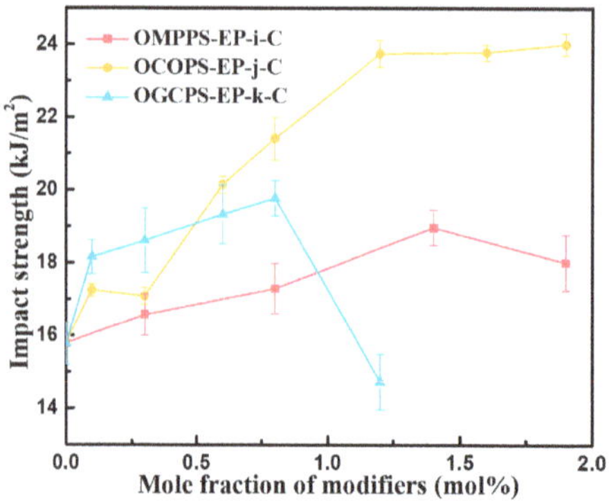

Figure 9. Impact strength of OMPPS-EP-i-C, OCOPS-EP-j-C, and OGCPS-EP-k-C.

In order to further analyze the relationship between the impact strength of the hybrid resins and the type and amount of modifier added, the micromorphology of the cured resins was investigated. The fracture surfaces of cured resins were observed by scanning electron microscopy (SEM), and the results are shown in Figure 10. As can be seen from Figure 10, EP-C presents a smooth fracture surface, indicating that the break of the resin is a brittle fracture. With the addition of OMPPS, OCOPS, and OGCPS, the fracture surfaces of the cured resins become rough, indicating that the fracture mode changes from brittleness to toughness [25,38,39]. For OMPPS-modified epoxy resins and OGCPS-modified epoxy resins, the roughness of the fracture surfaces first increases and then decreases with increases in POSS load amount. For OCOPS-modified epoxy resins, the roughness of the fracture surfaces increases with the increase in POSS load amount. The results demonstrate the change in impact strength.

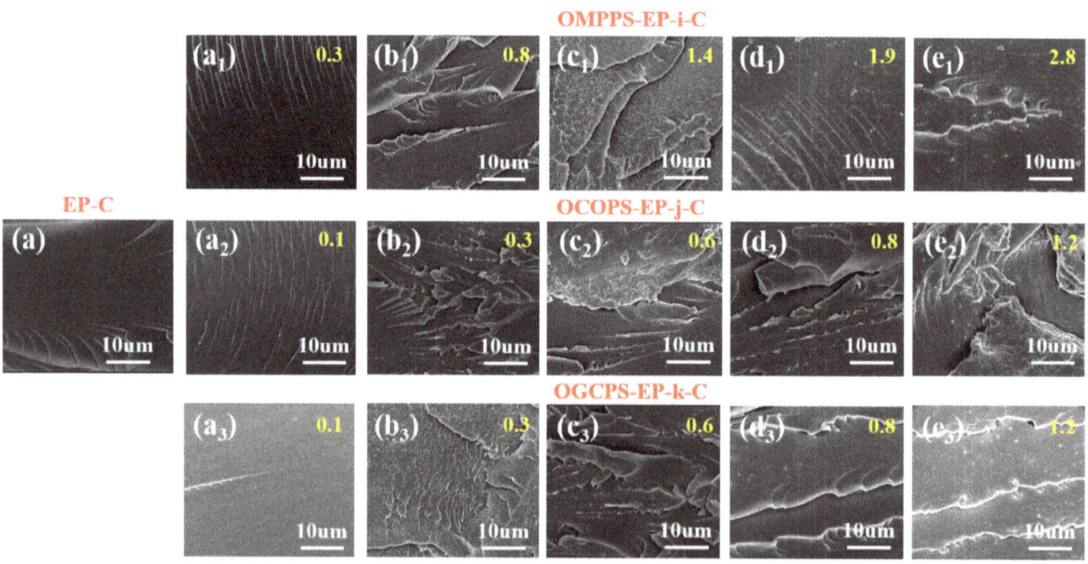

Figure 10. SEM images of the fracture surfaces of EP-C (**a**), OMPPS-EP-i-C (**a$_1$–e$_1$**), OCOPS-EP-j-C (**a$_2$–e$_2$**), and OGCPS-EP-k-C (**a$_3$–e$_3$**).

3.4.2. Fracture Toughness of Cured Hybrid Resins

Fracture toughness refers to the ability of a material to prevent the unstable propagation of macro cracks, which can generally be represented by the critical stress intensity factor (K_{IC}). The larger the K_{IC}, the better the fracture toughness of the material. Figure 11 shows the K_{IC} of OMPPS-EP-i-C, OCOPS-EP-j-C, and OGCPS-EP-k-C.

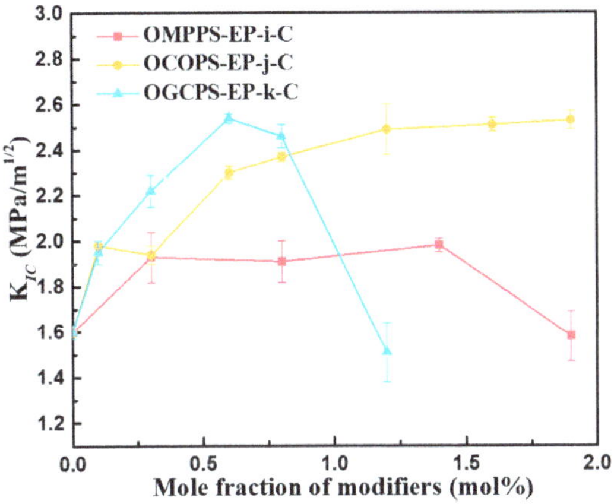

Figure 11. K_{IC} of OMPPS-EP-i-C, OCOPS-EP-j-C, and OGCPS-EP-k-C.

As shown in Figure 11, the K_{IC} of OMPPS-EP-i-C and OGCPS-EP-k-C increases first and then decreases with the increase in the amount of modifier added, but the toughening effect of OGCPS is much better than that of OMPPS. The K_{IC} of the pristine epoxy resin is 1.60 MPa·m$^{-1/2}$. When the addition amounts of OMPPS and OGCPS were 1.4 mol% and 0.6 mol%, respectively, the K_{IC} of the cured resin reached the maximum, which was 1.98 MPa·m$^{-1/2}$ and 2.54 MPa·m$^{-1/2}$. Compared with the K_{IC} of EP-C, it increased by 23.75% and 58.75%. The fracture toughness of OCOPS-modified epoxy resin increases with the increase in OCOPS addition. When the addition amount is 1.9 mol%, the K_{IC} of OCOPS-EP-1.9-C is 2.53 MPa·m$^{-1/2}$, which is increased by 58.13% and close to the toughness of the OGCPS-EP-0.6-C resin.

There are two factors that affect the fracture toughness of the hybrid resins. One is the load capacity of OMPPS, OCOPS, and OGCPC in the epoxy resin, and the other is the different organic chain segments and functional groups in OMPPS, OCOPS, and OGCPC. When OMPPS or OGCPS is added to the epoxy resin, the fracture toughness of the hybrid resins first increases and then decreases, and there is a maximum value. After reaching the maximum value, further addition of OMPPS or OGCPS leads to a decrease in fracture toughness. This probably results from the aggregation of modifiers at a high load amount, which form stress concentration points within the resin when the resin is stressed [29,40]. The aggregation of the modifiers was observed by SEM, and the results are shown in Figure S5. Unlike OMPPS and OGCPS, the fracture toughness of the OCOPS-modified epoxy resins increases with the increase in OCOPS load amount. This may probably result from the long and flexible aliphatic chains at the eight corners of OCOPS.

In addition to the load capacity of OMPPS, OCOPS, and OGCPC, the difference in the organic segments of OMPPS, OCOPS, and OGCPC also leads to different toughening effects on the epoxy resin. OCOPS and OGCPS exhibit superior performance in enhancing the toughness of epoxy resins compared to OMPPS. This is primarily attributed to the longer molecular chain segments of OCOPS and OGCPS than those of OMPPS. These long flexible chain segments allow them to easily move and absorb more energy when the resins are subjected to stress. In addition, OGCPS exhibits better toughening effects on the epoxy resins compared to OCOPS before OGCPS agglomerates in the resin. This is possibly due to the presence of benzene rings in the organic chain segments and the glycidyl groups of OGCPS [36].

3.5. Thermal Properties of Cured Hybrid Resins

Dynamic mechanical analysis (DMA) was used to analyze the thermal mechanical properties of the cured resins, and the results are shown in Figure 12. As can be seen from Figure 12, OMPPS-EP-i-C, OCOPS-EP-j-C, and OGCPS-EP-k-C have a lower storage modulus (E′) than EP-C. All three modifiers have a caged hollow structure and flexible organic chain segments on the outside. Introducing them into the epoxy resin will lead to an increase in the free volume of the hybrid resin, causing a decrease in the storage modulus. The crosslinking density of the cured resins can be calculated using the data from the DMA curves, and the results are shown in Table S2. The results show the crosslinking density of the cured hybrid resins is lower than that of the cured pristine epoxy resin, which indicates the toughness contribution. The presence of a single peak in all tanδ curves indicates good compatibility between OMPPS, OCOPS, OGCPC, and the epoxy resin. The temperature corresponding to the peak is the glass transition temperature (T_g) of the cured resin. The T_g values of all epoxy resins are listed in Table 2. The introduction of OMPPS and OGCPS had no significant effect on the T_g of the cured hybrid resin. This is because OMPPS and OGCPS both have rigid internal Si-O-Si structures and external flexible organic chain segments, and the effects of the two on the T_g offset each other. However, the T_g of OCOPS-EP-j-C had a significant decreasing trend with increase in OCOPS addition. This may be because the external organic chain segments of OCOPS are longer or more flexible than those of OMPPS and OGCPS. At a lower loading, the effect of the rigid Si-O-Si structure on the T_g and the effect of flexible chain segments on the T_g offset each other, so the T_g of OCOPS-EP-0.1-C is close to that of EP-C. However, at higher addition amounts, due to the longer molecular chain of OCOPS, OCOPS has a stronger plasticizing effect, which exceeds the effect of the rigid Si-O-Si structure on the T_g of the modified epoxy resin [41]. Therefore, the T_g of OCOPS-EP-j-C decreases.

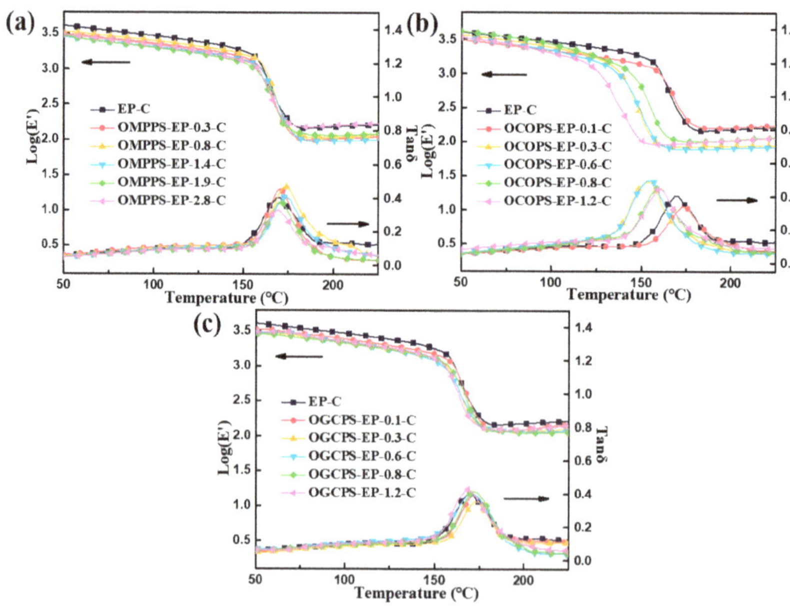

Figure 12. DMA curves of EP-C, OMPPS-EP-i-C (**a**), OCOPS-EP-j-C (**b**), and OGCPS-EP-k-C (**c**) in nitrogen.

Table 2. T_gs and T_{d5}s of EP-C, OMPPS-EP-i-C, OCOPS-EP-j-C, and OGCPS-EP-k-C.

Sample	T_g (°C)	T_{d5} (°C)
EP-C	169	317
OMPPS-EP-0.3-C	170	341
OMPPS-EP-0.8-C	173	347
OMPPS-EP-1.4-C	173	348
OMPPS-EP-1.9-C	171	349
OMPPS-EP-2.8-C	170	345
OCOPS-EP-0.1-C	174	356
OCOPS-EP-0.3-C	156	333
OCOPS-EP-0.6-C	155	204
OCOPS-EP-0.8-C	157	136
OCOPS-EP-1.2-C	158	125
OGCPS-EP-0.1-C	166	355
OGCPS-EP-0.3-C	168	356
OGCPS-EP-0.6-C	167	357
OGCPS-EP-0.8-C	167	356
OGCPS-EP-1.2-C	170	353

Thermogravimetric (TGA) analysis was used to determine the thermal stability of the resins, and the results are shown in Figure 13. The T_{d5} values of the cured hybrid resins are listed in Table 2. As shown in Table 2, we can see that the introduction of OMPPS and OGCPS can improve the thermal stability of the cured hybrid resins. Due to the short arms of OMPPS, most of which are composed of high energy Si-O and Si-C bonds [42,43], this is the key to the improvement of the thermal stability of OMPPS-EP-i-C. In contrast, OGCPS has a longer molecular chain segment than OMPPS, but the molecular chain of OGCPS contains a benzene ring structure, which also ensures that OGCPS-EP-k-C has high thermal stability [44]. However, for OCOPS-EP-j-C, the thermal stability is much lower than that of EP-C. There may be two reasons. Firstly, the presence of long flexible chains in the arms of OCOPS enhances the flexibility of the cross-linked structure when incorporated into the epoxy resin, subsequently compromising the thermal stability of the epoxy resin. Secondly, the reactive groups of OCOPS are carboxyl groups, and the ester groups formed during the curing reaction with epoxy groups may lead to a further reduction in the thermal stability of OCOPS-EP-j-C.

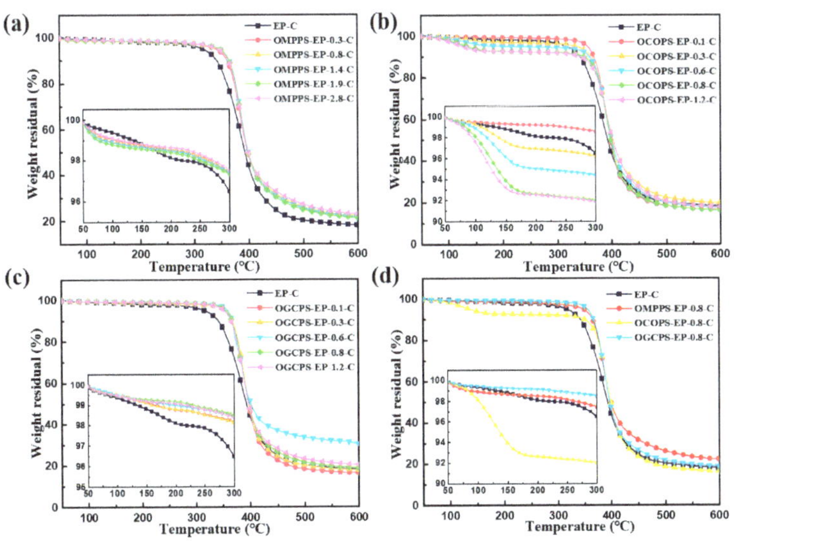

Figure 13. TGA curves of EP-C, OMPPS-EP-i-C (**a**), OCOPS-EP-j-C (**b**), and OGCPS-EP-k-C (**c**), and modifier-EP-0.8-C (**d**) in nitrogen.

4. Conclusions

Three reactive polyhedral oligomeric silsesquioxanes with a thiol group (OMPPS), a carboxyl group (OCOPS), and an epoxy group (OGCPS) are utilized to modify 3-(oxiran-2-ylmethoxy)-*N*,*N*-bis(oxiran-2-ylmethyl)aniline/4,4′-methylenebis(2-ethylaniline) (epoxy resin) and obtain hybrid resins with good thermal and mechanical properties. The hybrid resins display low viscosity in the temperature range of 45~85 °C and a low curing temperature (110 °C), which is beneficial for the liquid molding of the resin composites. The addition of OMPPS, OCOPS, and OGCPS greatly improves the toughness of the epoxy resin without sacrificing the mechanical and thermal properties of the epoxy resin. The critical stress intensity factor (K_{IC}), impact strength, and T_{d5} of OGCPS-EP-0.6-C reach 2.54 MPa·$m^{-1/2}$, 19.33 kJ/m^2, and 357 °C, showing increases of 58.75%, 22.48%, and 12.62%, respectively, compared to EP-C. The toughness of cured hybrid resins first increases and then decreases with the increase in the addition amount of OMPPS or OGCPS, reaching the maximum value when the addition amount is 1.4 mol% and 0.6 mol%, respectively. The toughness of OCOPS-EP-j-C increases with the increase in OCOPS addition. OCOPS, with long and flexible chain arms, exhibits a better toughening effect. This study provides a strategy for the design of polyhedral oligomeric silsesquioxane toughening agents and application in an epoxy resin, demonstrating an effective toughening method for the resin matrices of advanced composites.

Supplementary Materials: The following supporting information can be downloaded at: https://www.mdpi.com/article/10.3390/polym16131877/s1, Scheme S1: Synthetic route of HDS; Scheme S2: Synthetic route of DGHDS; Figure S1: (a) FT-IR spectra of HDS and DGHDS and (b) ^{1}H NMR spectra of HDS and DGHDS; Figure S2. MALDT-TOF-MS spectra of OMPPS, OCOPS, and OGCPS; Figure S3. The sample diagram of four mechanical properties tests; Figure S4: Linear fitting of T_i and T_p vs. heating rate for the pristine resin; Figure S5: SEM images of the fracture surface of OMPPS-EP-0.8-C (a_1), OMPPS-EP-1.9-C (a_2), OGCPS-EP-0.6-C (b_1), and OGCPS-EP-1.2 (b_2); Table S1: DSC analysis data of OMPPS-EP-i, OCOPS-EP-j, and OGCPS-EP-k hybrid resins; Table S2: The cross-linking density of cured epoxy resins. Reference [45] is cited in the supplementary materials.

Author Contributions: Conceptualization, F.H. and W.L. (Weibo Liu); methodology, C.W., W.L. (Weibo Liu) and F.H.; validation, F.H., C.W. and W.L. (Weibo Liu); formal analysis, J.Q., W.L. (Weiping Liu); investigation, L.W., Y.C. and Y.F.; writing—original draft preparation, W.L. (Weibo Liu); writing—review and editing, F.H.; supervision, Z.L.; project administration, J.Q. All authors have read and agreed to the published version of the manuscript.

Funding: This research was funded by the National KeyR&D Program of China (No. 2022YFB3709201) and the Fundamental Research Funds for the Central Universities (No. JKD01241701).

Institutional Review Board Statement: Not applicable.

Data Availability Statement: Data are contained within the article.

Conflicts of Interest: Zuozhen Liu and Jianhua Qian were employed by the company HuaChang Polymers Co., Ltd. Weiping Liu was employed by the company Manufacturing Center of Composite Materials for Commercial Aircraft, Shanghai Aircraft Manufacturing Co., Ltd. The remaining authors declare that the research was conducted in the absence of any commercial or financial relationships that could be construed as a potential conflict of interest.

References

1. Paolillo, S.; Bose, R.K.; Santana, M.H.; Grande, A.M. Intrinsic self-healing epoxies in polymer matrix composites (PMCs) for aerospace applications. *Polymers* **2021**, *13*, 2073–4360. [CrossRef]
2. Xu, N.; Zhu, T.; Yang, Z.; Han, M. Fabrication and optimization of La0.4Sr0.6Co0.2Fe0.7Nb0.1O3-δ electrode for symmetric solid oxide fuel cell with zirconia based electrolyte. *J. Mater. Sci. Technol.* **2017**, *33*, 1329–1333.
3. Zhao, Y.S.; He, Y.H.; Yang, K.R.; Wang, X.P.; Bai, J.H.; Du, B. Improving the surface insulating performance of epoxy resin/Al_2O_3 composite materials by extending chain of liquid epoxy resin with Me-THPA. *High Volt.* **2020**, *5*, 472–481. [CrossRef]
4. Zhao, Y.X.; Xu, R.; Xiao, Y.; Wang, H.L.; Zhang, W.; Zhang, G.Y. Mechanical performances of phenolic modified epoxy resins at room and high temperatures. *Coatings* **2022**, *12*, 643. [CrossRef]

5. Zhang, X.L.; Liu, M.; Chen, Y.; He, J.C.; Wang, X.L.; Xie, J.; Li, Z.W.; Chen, Z.M.; Fu, Y.H.; Xiong, C.X.; et al. Epoxy resin/hollow glass microspheres composite materials with low dielectric constant and excellent mechanical performance. *J. Appl. Polym. Sci.* **2022**, *139*, E52787. [CrossRef]
6. Qin, J.Y.; Zhao, H.P.; Qin, Z.L.; Zhang, W.C.; Yang, R.J. Effect of polyhedral oligomeric silsesquioxanes with different structures on dielectric and mechanical properties of epoxy resin. *Polym. Compos.* **2021**, *42*, 3445–3457. [CrossRef]
7. Neves, R.M.; Ornaghi, H.L., Jr.; Zattera, A.J.; Amico, S.C. Toughening epoxy resin with liquid rubber and its hybrid composites: A systematic review. *J. Polym. Res.* **2022**, *29*, 340. [CrossRef]
8. Thomas, R.; Yumei, D.; Yuelong, H.; Le, Y.; Moldenaers, P.; Weimin, Y.; Czigany, T.; Thomas, S. Miscibility, morphology, thermal, and mechanical properties of a DGEBA based epoxy resin toughened with a liquid rubber. *Polymer* **2008**, *49*, 278–294. [CrossRef]
9. Jones, A.R.; Watkins, C.A.; White, S.R.; Sottos, N.R. Self-healing thermoplastic-toughened epoxy. *Polymer* **2015**, *74*, 254–261. [CrossRef]
10. Wu, Z.J.; Yi, X.S.; Wilkinson, A. Interlaminar fracture toughness of carbon fibre/RTM6-2 composites toughened with thermoplastic-coated fabric reinforcement. *Compos. Part B* **2017**, *130*, 192–199. [CrossRef]
11. Bekeshev, A.; Mostovoy, A.; Shcherbakov, A.; Zhumabekova, A.; Serikbayeva, G.; Vikulova, M.; Svitkina, V. Effect of Phosphorus and Chlorine Containing Plasticizers on the Physicochemical and Mechanical Properties of Epoxy Composites. *J. Compos. Sci.* **2023**, *7*, 178. [CrossRef]
12. Eissa, M.M.; Samy, M.; Ramadan, A.M.; Amin, A. Amino-terminated hyperbranched polymer for toughness improvement of epoxy/clay nanocomposites. *Polym. Bull.* **2015**, *72*, 3147–3168. [CrossRef]
13. Shiravand, F.; Ascione, L.; Persico, P.; Carfagna, C.; Brocks, T.; Cioffi, M.O.H.; Puglisi, C.; Samperi, F.; Ambrogi, V. A novel hybrid linear-hyperbranched poly(butylene adipate) copolymer as an epoxy resin modifier with toughening effect. *Polym. Int.* **2016**, *65*, 308–319. [CrossRef]
14. Qi, Y.; Fan, Q.; Li, J.; Cao, Q.; Pan, X.; Pan, Y.; Jian, X.; Weng, Z. Toughened and reinforced the petroleum-based epoxy resin via thermotropic liquid crystal bio-based counterpart. *Compos. Commun.* **2023**, *44*, 101771. [CrossRef]
15. Wang, T.T.; Huang, P.; Li, Y.Q.; He, N.; Fu, S.Y. Epoxy nanocomposites significantly toughened by both poly(sulfone) and graphene oxide. *Compos. Commun.* **2019**, *14*, 55–60.
16. Xu, Z.G.; Song, P.G.; Zhang, J.; Guo, Q.P.; Mai, Y.W. Epoxy nanocomposites simultaneously strengthened and toughened by hybridization with graphene oxide and block ionomer. *Compos. Sci. Technol.* **2018**, *168*, 363–370. [CrossRef]
17. Park, Y.T.; Qian, Y.Q.; Chan, C.; Suh, T.; Nejhad, M.G.; Macosko, C.W.; Stein, A. Epoxy toughening with low graphene loading. *Adv. Funct. Mater.* **2015**, *25*, 575–585. [CrossRef]
18. Wang, F.Z.; Drzal, L.T.; Qin, Y.; Huang, Z.X. Enhancement of fracture toughness, mechanical and thermal properties of rubber/epoxy composites by incorporation of graphene nanoplatelets. *Compos. Part A* **2016**, *87*, 10–22. [CrossRef]
19. Wang, Y.T.; Wang, C.S.; Yin, H.Y.; Wang, L.L.; Xie, H.F.; Cheng, R.S. Carboxyl-terminated butadiene-acrylonitrile-toughened epoxy/carboxyl-modified carbon nanotube nanocomposites: Thermal and mechanical properties. *Express. Polym. Lett.* **2012**, *6*, 719–728. [CrossRef]
20. Rahman, M.M.; Hosur, M.; Ludwick, A.G.; Zainuddin, S.; Kumar, A.; Trovillion, J.; Jeelani, S. Thermo-mechanical behavior of epoxy composites modified with reactive polyol diluent and randomly-oriented amino-functionalized multi-walled carbon nanotubes. *Polym. Test.* **2012**, *31*, 777–784. [CrossRef]
21. Dharmavarapu, P.; Reddy, M. Mechanical, low velocity impact, fatigue and tribology behaviour of silane grafted aramid fibre and nano-silica toughened epoxy eomposite. *Silicon* **2021**, *13*, 1741–1750. [CrossRef]
22. Pham, T.D.; Vu, C.M.; Choi, H.J. Enhanced fracture toughness and mechanical properties of epoxy resin with rice husk-based nano-silica. *Polym. Sci. Ser. A* **2017**, *59*, 437–444. [CrossRef]
23. Sahoo, S.K.; Mohanty, S.; Nayak, S.K. Study of thermal stability and thermo-mechanical behavior of functionalized soybean oil modified toughened epoxy/organo clay nanocomposite. *Prog. Org. Coat.* **2015**, *88*, 263–271. [CrossRef]
24. Li, S.; Liu, Y.; Liu, Y.; Wang, Q. Synergistic effect of piperazine pyrophosphate and epoxy-octavinyl silsesquioxane on flame retardancy and mechanical properties of epoxy resin. *Compos. Part B Eng.* **2021**, *223*, 109115. [CrossRef]
25. Liu, H.; Liu, M.; Zhang, P.; Xue, K.; Yao, T.; Liu, L.; Huang, Y. POSS-polyurethane prepolymer strengthened and toughened CF/epoxy resin composites for room and simulated arctic ambient temperature. *Polymer* **2024**, *294*, 126692. [CrossRef]
26. Han, R.; Ma, X.; Cai, L.; Zhang, Z.; Fang, Y.; Wang, J. Low viscosity and low temperature curing reactive POSS/epoxy hybrid resin with enhanced toughness and comprehensive thermal performance. *RSC Adv.* **2024**, *14*, 7263–7275. [CrossRef] [PubMed]
27. Ni, C.; Ni, G.; Zhang, L.; Mi, J.; Yao, B.; Zhu, C. Syntheses of silsesquioxane (POSS)-based inorganic/organic hybrid and the application in reinforcement for an epoxy resin. *J. Colloid Interface Sci.* **2011**, *362*, 94–99. [CrossRef] [PubMed]
28. Mishra, K.; Pandey, G.; Singh, R.P. Enhancing the mechanical properties of an epoxy resin using polyhedral oligomeric silsesquioxane (POSS) as nano-reinforcement. *Polym. Test.* **2017**, *62*, 210–218. [CrossRef]
29. Zhang, C.X.; Li, T.X.; Song, H.; Han, Y.Q.; Su, H.B.; Wang, Y.M.; Wang, Q. Epoxy Resin/POSS Nanocomposites with Toughness and Thermal Stability. *J. Photopolym. Sci. Technol.* **2017**, *30*, 25–31. [CrossRef]
30. Cao, J.; Fan, H.; Li, B.-G.; Zhu, S. Synthesis and evaluation of Double-Decker Silsesquioxanes as modifying agent for epoxy resin. *Polymer* **2017**, *124*, 157–167. [CrossRef]
31. Konnola, R.; Parameswaranpillai, J.; Joseph, K. Mechanical, thermal, and viscoelastic response of novel in situ CTBN/POSS/epoxy hybrid composite system. *Polymer. Compos.* **2016**, *37*, 2109–2120. [CrossRef]

32. Zhang, C.; Li, T.; Song, H.; Han, Y.; Dong, Y.; Wang, Y.; Wang, Q. Improving the thermal conductivity and mechanical property of epoxy composites by introducing polyhedral oligomeric silsesquioxane-grafted graphene oxide. *Polymer. Compos.* **2018**, *39*, E1890–E1899. [CrossRef]
33. Yang, J.; Zhang, Y.; Hao, M.; Zhi, J.; Qian, X. Synergistically toughened epoxy resin based on modified-POSS triggered interpenetrating network. *Polymer* **2023**, *268*, 125719. [CrossRef]
34. Misasi, J.M.; Jin, Q.; Knauer, K.M.; Morgan, S.E.; Wiggins, J.S. Hybrid POSS-hyperbranched polymer additives for simultaneous reinforcement and toughness improvements in epoxy networks. *Polymer* **2017**, *117*, 54–63. [CrossRef]
35. Kopesky, E.T.; Haddad, T.S.; McKinley, G.H.; Cohen, R.E. Miscibility and viscoelastic properties of acrylic polyhedral oligomeric silsesquioxane–poly(methyl methacrylate) blends. *Polymer* **2005**, *46*, 4743–4752. [CrossRef]
36. Hao, N.; Böhning, M.; Schönhals, A. Dielectric properties of nanocomposites based on polystyrene and polyhedral oligomeric phenethyl-silsesquioxanes. *Macromolecules* **2007**, *40*, 9672–9679. [CrossRef]
37. Ke, F.; Zhang, C.; Guang, S.; Xu, H. POSS Core star-shape molecular hybrid materials: Effect of the chain length and POSS content on dielectric properties. *J. Appl. Polym. Sci.* **2013**, *127*, 2628–2634. [CrossRef]
38. Robertson, R.E.; Mindroiu, V.E.; Cheung, M.-F. Fracture in epoxy matrix resins. *Compos. Sci. Technol.* **1985**, *22*, 197–207. [CrossRef]
39. Sahagun, C.M.; Morgan, S.E. Thermal control of nanostructure and molecular network development in epoxy-amine thermosets. *ACS Appl. Mater. Inter.* **2012**, *4*, 564–572. [CrossRef]
40. Zhang, W.; Huang, J.; Guo, X.; Zhang, W.; Qin, Z.; Yang, R. Preparation and mechanism of toughening and flame retardance of epoxy resin using novel silsesquioxane molecules. *React. Funct. Polym.* **2023**, *190*, 105645. [CrossRef]
41. Turan, D.; Sirin, H.; Ozkoc, G. Effects of POSS particles on the mechanical, thermal, and morphological properties of PLA and plasticised PLA. *J. Appl. Polym. Sci.* **2011**, *121*, 1067–1075. [CrossRef]
42. Abbasi, A.; Salimi, A.; Bouhendi, H.; Karimi, M. A study of the thermal properties of alumina/glycidoxy propyl POSS/epoxy adhesives. *Int. J. Adhes. Adhes.* **2023**, *124*, 103368. [CrossRef]
43. Lian, Q.; Chen, H.; Luo, Y.; Li, Y.; Cheng, J.; Liu, Y. Toughening mechanism based on the physical entanglement of branched epoxy resin in the non-phase-separated inhomogeneous crosslinking network: An experimental and molecular dynamics simulation study. *Polymer* **2022**, *247*, 124754. [CrossRef]
44. Zeng, K.; Zheng, S. Nanostructures and surface dewettability of epoxy thermosets containing hepta(3,3,3-trifluoropropyl) polyhedral oligomeric silsesquioxane-capped poly(ethylene oxide). *J. Phys. Chem. B* **2007**, *111*, 13919–13928. [CrossRef]
45. Ohashi, S.; Kilbane, J.; Heyl, T.; Ishida, H. Synthesis and characterization of cyanate ester functional benzoxazine and its polymer. *Macromolecules* **2015**, *48*, 8412–8417. [CrossRef]

polymers

MDPI

Article

Investigation of Macroscopic Mechanical Behavior of Magnetorheological Elastomers under Shear Deformation Using Microscale Representative Volume Element Approach

Ilda Abdollahi * and Ramin Sedaghati *

Department of Mechanical, Industrial and Aerospace Engineering, Concordia University, Montreal, QC H3G 1M8, Canada
* Correspondence: i_abdoll@encs.concordia.ca (I.A.); ramin.sedaghati@concordia.ca (R.S.)

Abstract: Magnetorheological elastomers (MREs) are a class of smart materials with rubber-like qualities, demonstrating revertible magnetic field-dependent viscoelastic properties, which makes them an ideal candidate for development of the next generation of adaptive vibration absorbers. This research study aims at the development of a finite element model using microscale representative volume element (RVE) approach to predict the field-dependent shear behavior of MREs. MREs with different elastomeric matrices, including silicone rubber Ecoflex 30 and Ecoflex 50, and carbonyl iron particles (CIPs) have been considered as magnetic particles. The stress–strain characteristic of the pure silicon rubbers was evaluated experimentally to formulate the nonlinear Ogden strain energy function to describe hyper-elastic behavior of the rubbery matrix. The obtained mechanical and magnetic properties of the matrix and inclusions were integrated into COMSOL Multiphysics to develop the RVE for the MREs, in 2D and 3D configurations, with CIP volume fraction varying from 5% to 40%. Periodic boundary condition (PBC) was imposed on the RVE boundaries, while undergoing shear deformation subjected to magnetic flux densities of 0–0.4 T. Comparing the results from 2D and 3D modeling of isotropic MRE-RVE with the experimental results from the literature suggests that the 3D MRE-RVE can be effectively used to accurately predict the influence of varying factors including matrix type, volume fraction of magnetic particles, and applied magnetic field on the mechanical behavior of MREs.

Keywords: magnetorheological elastomers (MREs); representative volume element (RVE); periodic boundary condition (PBC); finite element method; hyper-elastic materials

1. Introduction

Magnetorheological (MR) materials are a class of smart materials with the unique ability to change their physical or mechanical characteristics rapidly in less than few milliseconds in response to an external magnetic field. These materials are fabricated by dispersing magnetic particles inside nonmagnetic host matrices. MR fluid is a commonly used MR material which comes with some drawbacks, such as magnetic particle sedimentation, sealing problems, and environmental contamination [1–5]. These challenges are not encountered by MR elastomers (MREs), because the magnetic particles are often bonded by a carrier matrix, like rubber. MREs stand as multi-functional materials, exhibiting the capability of dynamically altering their mechanical properties, including stiffness, and damping capacity, in response to an external magnetic field. This characteristic is measured by defining magnetorheological (MR) effect as the ratio of the change in the field-dependent physical or mechanical property to the value of the same property when no field is applied [6–8].

MREs comprise three fundamental components: magnetic particles, nonmagnetic elastic matrices, and additives [9,10]. Due to the application of a magnetic field during the curing process, it is possible to fabricate MREs with an anisotropic particle-formed microstructure, and when no field is applied during the curing process, prepared MREs have an isotropic particle-formed microstructure [11,12]. MREs show great potential to be incorporated into the design of intelligent devices in a variety of engineering disciplines [13–15].

Jolly et al. [16] has pioneered the investigation on the mechanical response of anisotropic elastomer composites with embedded CIP under the application of magnetic fields, revealing significant changes in shear modulus of MREs in response to the magnetic field. Davis [17] proposed a phenomenological model to predict the shear modulus of isotropic and anisotropic MREs, with and without the application of the external magnetic field. His study suggested that magnetic particles with 27% volume fraction is an optimal content with respect to MR effect. Berasategi et al. [18] studied silicone-based isotropic and anisotropic MREs containing CIP concentrations ranging from 5% to 30% volume content, indicating changes in storage modulus and loss modulus as CIP content increased. Vatandoost et al. [19] investigated pre-strain effects on compression mode dynamic characteristics of isotropic and anisotropic MREs. Their results revealed that pre-strain has a significantly nonlinear impact on the elastic and loss moduli. Syam et al. [20] conducted a finite element analysis on MREs' behavior at the microscale, using the COMSOL software (v. 5.4). However, they used linear material model for silicone rubber as the matrix material. Their results showed increased stiffness in both linear and torsional modes under the application of an external magnetic field. Asadi Khanouki et al. [21] examined isotropic and anisotropic silicone rubber based MREs with various contents of CIP, and proposed a microscale modeling that was validated with the experimental results. Dargahi et al. [22] fabricated different MRE samples in terms of rubber matrix and ferromagnetic particle contents and conducted static and dynamic shear tests on the samples. Their results showed a significant 1672% increase in storage modulus under 0.45 T magnetic flux density.

Sun et al. [23] conducted a finite element analysis on the shear deformation of isotropic MREs under the application of an external magnetic field, using the concept of representative volume element (RVE) in 2D configuration, in COMSOL. However, they approximated the nonlinear B-H properties of magnetic particles using a linear model and assumed the relative permeability of CIP to be 100. Inspired by their work, Xu et al. [24] conducted the 3D modeling of isotropic MREs in tensile mode under the application of a magnetic field in COMSOL. They used a linear material model to describe the nonlinear hyper-elastic behavior of the host elastomeric matrix, and the linear magnetic model was used to describe the magnetic properties of CIP, thus ignoring the saturation. Kiarie et al. [25] adopted a 2D RVE approach using COMSOL to predict the magnetic field-induced strain in MREs, and they employed the Mooney–Rivlin nonlinear material model for the host rubber; however, their problem did not include any mechanical load or displacement imposed on the RVE. Li et al. [26] studied the magnetic field-induced shear behavior of MREs using a 2D RVE approach in COMSOL for both isotropic and anisotropic MREs; however, they assumed the magnetic particles (hydroxy iron powder) to have a very high relative permeability of 5000, and the silicone rubber behavior to be like a linear elastic matrix.

Hence, due to the prohibitive and expensive experimental procedures to characterize the behavior of MREs under the influence of various factors, the development of a promising analytical model to predict the mechanical properties of MREs offers several benefits over experiments, including controlled simulations, broader scenario exploration, cost-effectiveness, and environmental sustainability. To the best of the authors' knowledge and by reviewing the literature, there is a gap in the literature concerning taking the nonlinear behavior of MRE components into account in FE modeling process, and studies so far have faced issues in taking this inherent nonlinearity into account. Therefore, in this research study, the representative volume element (RVE) approach as an FE modeling scheme was effectively utilized to model the shear deformation of MREs under the influence of an external magnetic field, considering material and magnetic nonlinearities into account. An appropriate RVE size was defined for modeling the MRE, and periodic boundary condition (PBC) was imposed on the RVE boundaries. Experiments were then carried out on host rubber samples (silicone rubber) to obtain the stress–strain data. The extracted data was used to formulate the Ogden strain energy to describe the nonlinear hyper-elastic behavior of the rubber material. The other nonlinearity in the MRE is attributed to the magnetic behavior of CIP, which was described through a B-H curve, considering the saturation, instead of using a high relative permeability as used in previous studies.

The RVE was then generated in COMSOL Multiphysics (v. 6.0) in 2D and 3D configurations, and pure shear deformation was incrementally applied on the RVE, while PBC was imposed on the RVE boundaries. Simultaneously, a homogeneous magnetic field was created in the surrounding air domain perpendicular to the shear direction, and the Maxwell stress tensor was defined on the CIP inclusion. The influence of different factors like CIP volume fraction, magnetic field intensity, and host rubber's mechanical behavior were investigated on the shear behavior of the MRE. The results revealed the significant credibility of the developed 3D model in predicting the MREs' shear behavior, exhibiting a substantial potential to be used for the prediction of MREs' mechanical behavior as a reliable alternative to costly experiments.

2. Materials and Methods

2.1. Representative Volume Element (RVE)

A fundamental goal in the field of heterogeneous materials physics is to determine the effective mechanical properties of these materials. Among the proposed techniques, representative volume element (RVE) homogenization stands out as a method that utilizes a statistically homogeneous representation of heterogeneous materials at a microscale to derive their effective properties on a macroscale. Many researchers have suggested different aspects to be considered in the evaluation of the RVE size; however, these definitions share a core idea that RVE is defined as the minimal volume element when compared to the macroscopic dimensions of the structure, yet large enough to include a substantial amount of information about the microstructure, that exhibits an equivalent target attribute or behavior to that of the whole material at a macroscopic scale [27–30].

2.2. The RVE Size for Models Containing Hard Particles

To obtain an accurate estimation of the effective properties, it is essential to establish a correlation between the size of the RVE and the various morphological, mechanical, and thermal factors associated with microstructures [31,32]. El Moumen et al. [30] adopted a combined numerical–statistical approach in order to investigate the variation in the RVE size with respect to different parameters in the microstructure. The geostatistical parameter, integral range "A" in the context of composite materials, was used to relate the size of the RVE with other microstructure parameters and was defined by previous researchers [33–35]. The integral range for random microstructures with a volume fraction ϕ may be expressed as follows [32]:

$$A = \frac{\phi}{V_{RVE}} \tag{1}$$

For a stationary random function z, the variance of z, $D_z^2(V)$, over the volume V is attainable as a function of the integral range A and the point variance S_z^2 [30] as follows:

$$D_z^2(V) = S_z^2 \frac{A}{V} \tag{2}$$

Considering a stochastic microstructure composed of two distinct phases, denoted as F_1 with its distinct real characteristics, namely, z_1 and F_2 with characteristics z_2, where phase F_1 occupies a volume percentage of ϕ, and phase F_2 occupies a volume fraction of $(1-\phi)$, the point variance S_z^2 of the random variable z in the context of a two-phase material may be provided as follows [30]:

$$S_z^2 = \phi\,(1-\phi)(z_1 - z_2)^2 \tag{3}$$

Given the current context, whereby the mechanical properties are represented as the random variable z, substituting Equation (3) in Equation (2), the volume variance, $D_z^2(V)$, can be evaluated as follows:

$$D_z^2(V) = \phi\,(1-\phi)(z_1 - z_2)^2\,\frac{A}{V} \tag{4}$$

where $D_z^2(V)$ is the variance of the volume V, and A is the integral range. To determine the RVE parameters, the number of realizations n and the absolute error ε_{abs} were used to express $D_z^2(V)$ as follows [26]:

$$4D_z^2(V) = {\varepsilon_{abs}}^2\,n \tag{5}$$

It should be noted that the determination of the size of the RVE involves defining the volume at which the number of realizations is equivalent to one [36]. Therefore, using Equations (1), (4), and (5) and considering $n(V = V_{RVE}) = 1$, we can write the following:

$$V_{RVE} = \frac{4\,\phi^2\,(1-\phi)(z_1 - z_2)^2}{{\varepsilon_{abs}}^2\,V_{RVE}} \tag{6}$$

Defining the contrast ratio (c) in mechanical characteristics as $c = \frac{z_2}{z_1}$, in which z_2 represents the matrix property (phase F_2) and z_1 corresponds to the property of phase F_1, the representativity of the estimated characteristics in random microstructures may be determined by considering the volume size, with respect to the desired relative error $\varepsilon_r = \frac{\varepsilon_{abs}}{z_2}$. Hence, using Equation (6), the final expression for RVE size would be the following:

$$V_{RVE}(c,\phi) = \frac{2|1-c|\phi\,\sqrt{(1-\phi)}}{\varepsilon_r} \tag{7}$$

Equation (7) provides a clear relationship between the representative volume element (RVE) of random microstructures and both the volume fraction and contrast ratio, while also accounting for a desirable and fixed relative error.

For MREs, the two phases include soft elastomeric phase (F_2) impregnated with hard solid spherical inclusions, phase (F_1), represented by the micron-sized carbonyl iron powder (CIP). Thus, for MREs, the contrast ratio c which here represents the ratio of the modulus of the rubbery matrix to that of solid inclusions is nearly negligible compared with unity. Therefore, using Equation (7) and assuming $c = 0$, the RVE size can be determined for different particle volume fractions and the desired relative error. Tuning the aforementioned parameters to a negligible relative error, the equation leads us to an RVE with one CIP inclusion in a cubic rubber matrix, which is the same RVE size presented by Davis [17].

2.3. Periodic Boundary Conditions

The commonly employed boundary conditions in micromechanics fall into two categories of uniform boundary conditions: Dirichlet, also known as Displacement Boundary Condition (DBC), and Neumann, also known as Traction Boundary Condition (TBC). However, TBC tends to overestimate the effective material properties, while DBC underestimates them. Furthermore, researchers have also developed periodic boundary conditions (PBC) which are typically applied to unit cells in cases where the heterogeneous material exhibits a periodic structure [37–39]. It has been reported that PBC yields more precise effective modulus estimates compared with other conventional boundary conditions and is less sensitive to the RVE size or inclusion position in the unit cell [40]. In the following, PBC in 2D and 3D configurations are briefly discussed.

2.3.1. Periodic Boundary Conditions in 2D

Considering the periodic structure of the macroscopic body and a square unit cell, for each boundary pair, compatibility is essential in line with the periodicity assumption. This implies that the deformation of each boundary pair is identical, and the stress vectors have opposite signs on each pair [37,38]. Smit et al. [41] derived the suitable displacement boundary conditions as follows:

$$\mathrm{u}_{12} - \mathrm{u}_{v_4} = \mathrm{u}_{11} - \mathrm{u}_{v_1}\mathrm{u}_{22} - \mathrm{u}_{v_2} = \mathrm{u}_{21} - \mathrm{u}_{v_2}\mathrm{u}_{v_3} - \mathrm{u}_{v_2} = \mathrm{u}_{v_4} - \mathrm{u}_{v_1} \tag{8}$$

where u_{ij} denotes the displacement vector associated with any material point situated on the corresponding boundary Γ_{ij}, while u_{v_i} represents the displacement vector attributed to each vertex v_i. To eliminate rigid body motions, it is necessary to impose $\mathrm{u}_{v_k} = 0$ for any k within the set $k \in \{1,2,4\}$.

The micro–macro relations for the total stress and strain tensors can be established as follows:

$$\sigma_{ij\ (macro)} = \langle \sigma_{ij} \rangle,\ \varepsilon_{ij\ (macro)} = \langle \varepsilon_{ij} \rangle \tag{9}$$

where $\sigma_{ij\ (macro)}$ and $\varepsilon_{ij\ (macro)}$ signify the macroscopic total stress and strain tensors, respectively, while $\langle \sigma_{ij} \rangle$ and $\langle \varepsilon_{ij} \rangle$ represent the corresponding microscopic averages over the surface of the unit cell (i.e., $\langle \sigma_{ij} \rangle = \frac{1}{S} \int_R \sigma_{ij}\, dy;\ S = \int_R dy$).

Utilizing the averaged elastic constitutive equations in Equation (9), the expressions for the affective elastic properties can be obtained as follows:

$$E_{11}^{eff} = \frac{\sigma_{11\ (macro)}}{\varepsilon_{11\ (macro)}},\ \nu_{12}^{eff} = -\frac{\varepsilon_{22\ (macro)}}{\varepsilon_{11\ (macro)}} \tag{10}$$

where E_{11}^{eff} is the effective Young's modulus, and ν_{12}^{eff} denotes the effective Poisson's ratio

2.3.2. Periodic Boundary Conditions in 3D

A 3D RVE featuring a periodic microstructure can be considered as a cubic structure which contains both fiber and matrix constituents. This cube is bounded by six surfaces, ensuring that any two parallel surfaces always maintain parallel alignment along either the x-, y-, or z-axes. Each two parallel surfaces are distinguished by assigning an index (*POS/NEG*) based on their location on the associated coordinate axis. For example, assuming a cubic RVE with the dimension denoted as *D*, the X_{POS} surface corresponds to the y–z plane situated at the maximum x-axis cubic dimension (i.e., x = D), while the X_{NEG} surface is located at the minimum x axis cubic dimension (i.e., x = 0).

Each of these surfaces consists of nodes, and nodes located on the X_{POS} surface are referred to as $X_{POSnodes}$. Similar terminology applies to the other five faces. The set of nodes for a specific surface is defined as S_{np}, where n represents the reference frames (X, Y, Z), and p encompasses either the positive or negative faces along a given axis. Consequently, similar to the concept of PBC in 2D configuration, the mathematical expressions governing the imposition of periodic deformation on all nodes in the three dimensions of the 3D RVE domain can be mathematically described as follows [42]:

$$\begin{gathered} U_i^{S_{XPOS}} - U_i^{S_{XNEG}} - U^{N_2} + U^{N_1} = 0 \\ U_i^{S_{YPOS}} - U_i^{S_{YNEG}} - U^{N_3} + U^{N_1} = 0 \\ U_i^{S_{ZPOS}} - U_i^{S_{ZNEG}} - U^{N_4} + U^{N_1} = 0 \end{gathered} \tag{11}$$

Enforcing periodic boundary conditions requires stress equilibrium across opposite surfaces within the RVE domain. For every surface S_{np} in the 3D RVE, a specific unit outward normal vector is defined as n_{np}. Assuming that the domain is experiencing stress, the condition for stress equilibrium across opposing pairs of surfaces is achieved in the following situations:

$$\sigma\, n_{XPOS}\,(y,z) = -\sigma\, n_{XNEG}\,(y,z)$$

$$\sigma\, n_{YPOS}\,(x,z) = -\sigma\, n_{YNEG}\,(x,z) \tag{12}$$

$$\sigma\, n_{ZPOS}\,(x,y) = -\sigma\, n_{ZNEG}\,(x,y)$$

where σ is the stress tensor.

Ultimately, the volume-averaged stress within the periodically deformed RVE domain can be expressed as follows:

$$\langle\sigma\rangle = \frac{1}{V}\int_V \sigma\, dV \tag{13}$$

where V is the RVE volume, and the volume-averaged stress is denoted as $\langle\sigma\rangle$. Under the assumption of global periodicity within the RVE domain, the overall macroscopic stress and the global strain are expressed as $\langle\sigma\rangle = \sigma_{macro}$ and ε_{macro} respectively. ε_{macro} is determined based on the displacements of the retained nodes, which are calculated as follows:

$$u_2 = \varepsilon_{macro}(X_2 - X_1),\;\; u_3 = \varepsilon_{macro}(X_3 - X_1),\quad u_4 = \varepsilon_{macro}(X_4 - X_1) \tag{14}$$

Here, u_i represents the displacement vector of retained node, i, in relation to its coordinate position, X_i, where $u_i = [u_{i,x}, u_{i,y}, u_{i,z}]$. It should be noted that u_1 is restricted to be zero to avoid rigid body motion.

The effective properties could be evaluated consequently in the same fashion as explained in 2D. For example, the following are the effective shear moduli obtained from simple shear deformation along XY, XZ, and YZ planes, respectively.

$$G_{12}^{eff} = \frac{\tau_{12\,(macro)}}{\gamma_{12\,(macro)}}\;,\; G_{13}^{eff} = \frac{\tau_{13\,(macro)}}{\gamma_{13\,(macro)}}\;,\; G_{23}^{eff} = \frac{\tau_{23\,(macro)}}{\gamma_{23\,(macro)}} \tag{15}$$

where τ and γ are shear stress and shear strain, respectively.

2.4. Maxwell's Stress Tensor

The force acting on a point charge q moving with velocity v in the presence of both an electric field E and a magnetic field B is described by the Lorentz force equation which is a fundamental concept in electromagnetism and is expressed as follows [43]:

$$F = q(E + v \times B) \tag{16}$$

Using the Maxwell equations in electromagnetism and the Lorentz force, we can introduce Maxwell Stress tensor as follows [43]:

$$T_{ij} = \epsilon_0\left(E_i\, E_j - \frac{1}{2}\delta_{ij}\, E^2\right) + \frac{1}{\mu_0}\left(B_i\, B_j - \frac{1}{2}\delta_{ij}\, B^2\right) \tag{17}$$

where μ_0 and ϵ_0 denote vacuum permeability and permittivity, respectively, and δ_{ij} is the Kronecker delta which is defined as follows:

$$\begin{cases} \delta_{ij} = 1 & ;i = j \\ \delta_{ij} = 0 & ;i \neq j \end{cases} \tag{18}$$

In this study, MREs are only exposed to uniform magnetic fields; thus, the first part of Equation (17), containing the electric field, is eliminated, yielding the Maxwell stress tensor as follows:

$$T_{ij} = \frac{1}{\mu_0}\left(B_i\, B_j - \frac{1}{2}\delta_{ij}\, B^2\right) \tag{19}$$

2.5. Governing Equations

Considering the basic balance principles of continuum mechanics, including the linear momentum and angular momentum balance principles, there exists the equation of mechanical equilibrium and is as follows [44]:

$$\nabla \cdot \sigma + \rho f = \rho \dot{v} \tag{20}$$

Here, σ represents the stress tensor, ρ is density, f represents the body forces, and v is the velocity. In the case of static or quasi static conditions ($\dot{v} = 0$), the force balance equation simplifies to the following:

$$\nabla \cdot \sigma + \rho f = 0 \tag{21}$$

When coupling magnetic and elastic behavior, different methods can be used to define body forces and stresses. The deformation of the material due to a magnetic field can be incorporated into the force balance equation in terms of the body force represented by the magnetic force per unit volume, denoted as f_m. Assuming that the body force due to the weight is negligible, Equation (21) can be described as follows:

$$\nabla \cdot \sigma + \rho \left(f_m \right) = 0 \tag{22}$$

Alternatively, Equation (22) can be expressed in terms of the total stress tensor $\boldsymbol{T}$ as follows:

$$\nabla \cdot \boldsymbol{T} = 0 \ ; \qquad \boldsymbol{T} = \sigma + T \tag{23}$$

Here, $\boldsymbol{T}$ contributes to the sum of mechanical and magnetic stress tensors as $\boldsymbol{T} = \sigma + T$, where T is defined in Equation (19).

2.6. Characterization of Elastomeric Matrix and Magnetic Inclusion in MREs

In order to generate the RVE for predicting the shear modulus of MRE, two datasets are required: the mechanical and magnetic properties of the pure rubber material (the matrix) and carbonyl iron particle (CIP) (the ferromagnetic inclusions). As for the rubber material, silicone rubber was chosen and produced in the laboratory using two different types of silicone rubber, represented as Ecoflex 30 and Ecoflex 50 (Smooth-On, Macungie, PA, USA). In order to fabricate identical samples in terms of shape and dimensions, two rectangular molds of 37 × 6 × 3 [mm] were fabricated using a 3D-printer (Original Prusa i3 MK3S+, Prusa Research, Prague, Czech Republic), as shown in Figure 1a.

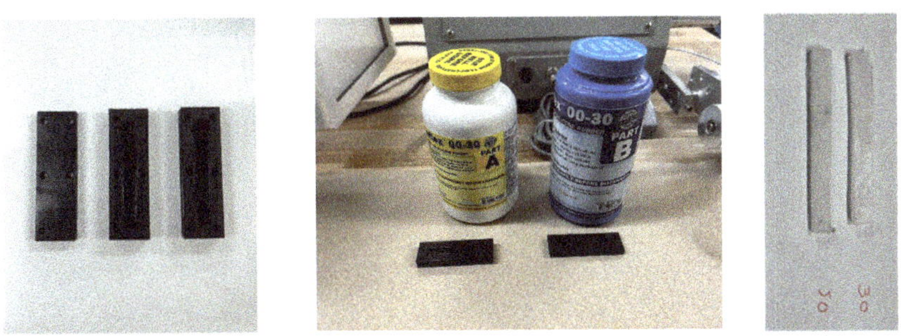

(a) (b) (c)

Figure 1. (**a**) The identical molds fabricated by the 3D printer, (**b**) parts A and B for fabricating silicone rubber Ecoflex 30 to be mixed and cured, and (**c**) the cured final samples (30 indicates Ecoflex 30, and 50 refers to Ecoflex 50).

To fabricate the silicone rubber samples, platinum-based silicone rubber from Smooth-On, Macungie, PA, USA was used, comprising two parts to be thoroughly mixed and cured. The two parts denoted as the rubber part (A) and the catalyst part (B), as shown in Figure 1b, were added, and stirred in a 50–50 weight fraction. The primary mixture was then placed in the conditioning vacuum mixer (THINKY: ARV-200, THINKY CORPORATION, Laguna Hills, CA, USA) for 40 s under 2000 rpm to be thoroughly mixed and degassed. The final mixture was then poured into the molds and cured at room temperature for 15 h. Finally, the vulcanized samples were removed from the molds and were ready for conducting the tensile test. Figure 1c shows the fabricated samples.

2.6.1. Characterization of the MRE's Elastomeric Matrix Using the Uniaxial Tensile Test

In pursuit of determining the viscoelastic properties of silicone rubber samples, the two identical rectangular samples with dimensions of 37 × 6 × 3 [mm] underwent pure tensile to failure test with 30 [mm/min] velocity, using an MTS machine (F1505-IM, Mark–10, Copiague, NY, USA), under identical conditions. Figures 2 and 3 illustrate the three significant steps of the conducted test for silicone rubber Ecoflex 30 and Ecoflex 50, respectively.

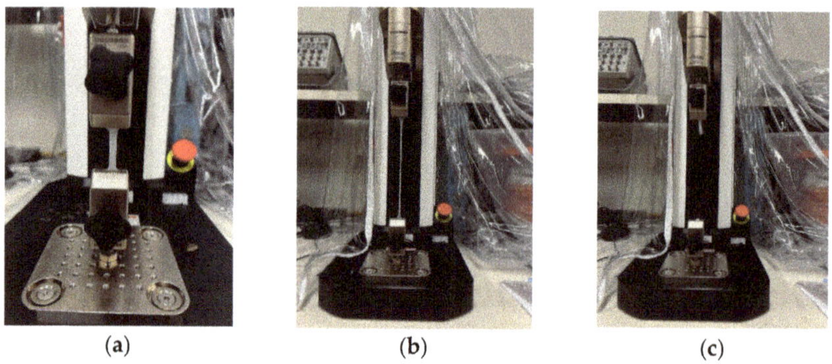

Figure 2. (**a**) Silicone rubber sample (Ecoflex 30) assembled on the MTS machine prior to the tensile test, (**b**) sample after the final steps of tensile test, and (**c**) the failed sample.

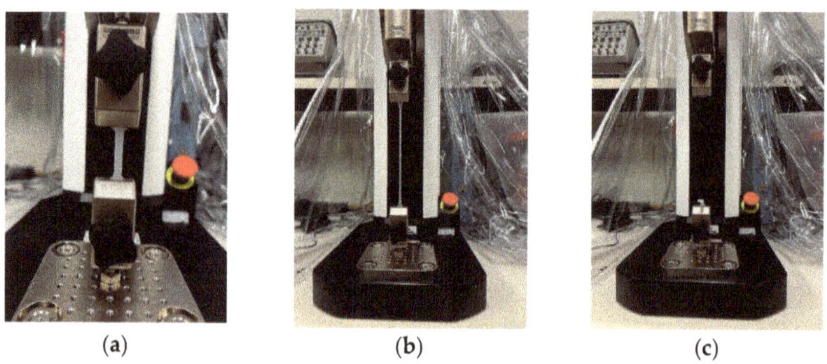

Figure 3. (**a**) Silicone rubber sample (Ecoflex 50) assembled on the MTS machine prior to the tensile test, (**b**) sample after the final steps of tensile test, and (**c**) the failed sample.

The extracted force–displacement experimental data for silicone rubber Ecoflex 30 and Ecoflex 50 are shown and compared in Figure 4. For the given force, silicone rubber Ecoflex 30 experiences larger displacement compared with Ecoflex 50 due to its lower stiffness. The force–displacement data were then used to obtain stress–stretch data that were subsequently utilized to identify the material parameters of hyper-elastic Ogden material model [45]. Compared with other hyper-elastic models such as Neo-Hookean and Mooney–Rivlin, the Ogden model was shown to provide the better prediction [45,46].

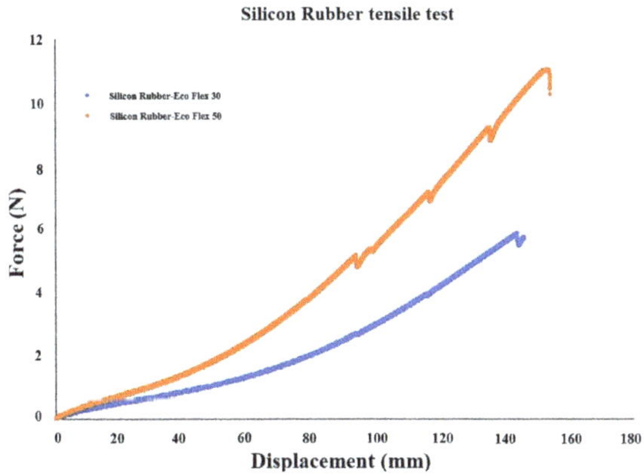

Figure 4. The extracted raw data of the conducted pure tensile to failure test for silicone rubber-Ecoflex 30 and Ecoflex 50.

To characterize the behavior of the rubber material, the strain energy function in the theory of hyper-elasticity, which represents the stored energy in the material during the deformation and is denoted as W, is employed. This energy function is dependent on the principal stretches (λ_1, λ_2, and λ_3) which are stretch ratios defined as deformed length divided by the original length for the unit fibers oriented along the principal directions [47,48]. Taking into consideration that rubber materials can be generally considered incompressible, we have $\lambda_1\lambda_2\,\lambda_3 = 1$. The principal Cauchy stresses, denoted as σ_i (i = 1, 2, 3), are intricately connected to the stretches through the derivative of the strain energy function, as expressed by the following equation [46]:

$$\sigma_i = \lambda_i \frac{\partial W}{\partial \lambda_i} - L \tag{24}$$

Here, the index i does not represent a dummy index, and there is no summation over it, and L serves as an unknown Lagrange multiplier, associated with the aforementioned incompressibility constraint. For the case of pure tension uniaxial tensile test, we have $\sigma_2 = \sigma_3 = 0$, and hence, by expressing the following equation and using Equation (24), we can effectively eliminate the unknown Lagrange multiplier L.

$$(\sigma_1 - \sigma_2) = (\sigma_1 - \sigma_3) = \sigma_1 \tag{25}$$

The Ogden strain energy function can accurately describe the nonlinear behavior of hyper-elastic materials, and for incompressible materials, it takes the following form [46]:

$$W = \sum_{p=1}^{N} \frac{\mu_p}{\alpha_p} \left(\lambda_1{}^{\alpha_p} + \lambda_2{}^{\alpha_p} + \lambda_3{}^{\alpha_p} - 3\right) \tag{26}$$

Here, each μ_p and α_p represents material characteristic parameters, to be determined using experimental data. For practical application, the summation in Equation (26) is confined to a finite number of terms. However, to maintain consistency with classical theory of incompressible isotropic elasticity, these constant parameters must adhere to the following condition:

$$\sum_{p=1}^{N} \mu_p\, \alpha_p = 2G \tag{27}$$

where N is a positive integer, and G is the shear modulus of the material in its undeformed stress-free (natural) configuration, which implies that $\sum_{p=1}^{N} \mu_p\, \alpha_p > 0$. In the present research study, the three-term Ogden model ($N = 3$) was adopted due to its better accuracy compared to one-term and two-term Ogden model [45]. Using the three-term Ogden strain energy function in Equation (26), and considering Equations (24) and (25), the principal value of the Cauchy stress can be obtained as follows:

$$\sigma = \sum_{p=1}^{3} \mu_p \left(\lambda_i^{\alpha_p - 1} - \lambda_i^{-\left(\frac{1}{2}\alpha_p + 1\right)} \right) \tag{28}$$

The extracted data from the pure tensile test was then employed to determine the material parameters in Ogden strain-energy function through least squares (LS) optimization technique. Let us consider a vector $\Lambda = [\Lambda_1, \Lambda_2, \ldots, \Lambda_m]^T$, representing a collection of experimental deformation values, and associated vector $S = [S_1, S_2, \ldots, S_m]^T$, corresponding to stress values, in which m represents the number of datasets.

For the given deformation vector Λ, using Ogden material model, the principal Cauchy stress in Equation (28) can be expressed as $\sigma\ (\mu_p, \alpha_p)$, in which the material parameters μ_p and α_p are unknown. It is noted that, for the three-term Ogden model, $p = 1$ to 3, and thus, the number of unknown material parameters are 6 (i.e., $\mu_1, \mu_2, \mu_3, \alpha_1, \alpha_2, \alpha_3$). A least square minimization problem has been subsequently formulated to identify the material parameters in order to minimize the error between experimental and model results. The error function may thus be defined as follows:

$$E_r = \sum_{j=1}^{m} [1 - \frac{\sigma(\mu_p\ , \alpha_p)}{S_j}\ ]^2 \tag{29}$$

Now, considering Equation (28), the optimization problem can be formulated as follows:
Find the design variables: $\{\mu_1, \mu_2, \mu_3, \alpha_1, \alpha_2, \alpha_3\}$
To minimize E_r;

$$E_r = \sum_{j=1}^{m} [1 - \frac{\sum_{p=1}^{3} \mu_p \left(\lambda_j^{\alpha_p - 1} - \lambda_j^{-\left(\frac{1}{2}\alpha_p + 1\right)} \right)}{S_j}]^2 \tag{30}$$

Subjected to: $\sum_{p=1}^{3} \left(-\mu_p \alpha_p\right) < 0$.

The optimization problem in Equation (30) has been solved using stochastic-based Genetic Algorithm (GA) and hybrid method based on the combination of GA and gradient-based Sequential Quadratic Programming (SQP) method. In the hybrid method, the optimal solution from GA has been fed into the SQP as the initial point in an attempt to accurately identify the global optimum solution. The identified optimal Ogden material parameters using GA and GA + SQP for both Ecoflex 30 and Ecoflex 50 silicone rubbers are provided in Table 1. The basic mechanical and magnetic properties of silicone rubber are also presented in Table 2.

Table 1. The optimized parameters gained through curve fitting the experimental data with the Ogden strain energy function.

Material	Optimization Method	μ_1 (Pa)	α_1	μ_2 (Pa)	α_2	μ_3 (Pa)	α_3	G (kPa)
Silicone Rubber Ecoflex 50	GA	0.0849×10^6	1.0000	0.0005×10^6	5.5354	-0.0001×10^6	−1.000	43.884
	GA + SQP	0.0001×10^6	1.0000	0.0088×10^6	3.4927	-0.0076×10^6	−5.3724	35.833
Silicone Rubber Ecoflex 30	GA	-6.9874×10^6	1.6760	5.2412×10^6	1.5601	1.8713×10^6	1.9162	25.850
	GA + SQP	4.3891×10^6	1.4354	-4.3380×10^6	1.4529	0.0163×10^6	3.1195	24.141

Table 2. Material properties of silicone rubber.

Material	Material Properties	Value
Silicone Rubber	Density ρ	920 $(\mathrm{Kg/m^3})$
	Poisson ratio ν	~0.5 (incompressible material)
	Magnetic Relative Permeability μ_r	2

Figures 5 and 6, respectively, show the comparison of stress–stretch response of silicone rubber Ecoflex 30 and Ecoflex 50 samples extracted from experiments with those obtained using the Ogden model based on optimal material parameters identified using GA and GA + SQP. Results clearly show that Ogden material model, with optimal material parameters identified through GA + SQP, provides reasonable agreement with the experimental data.

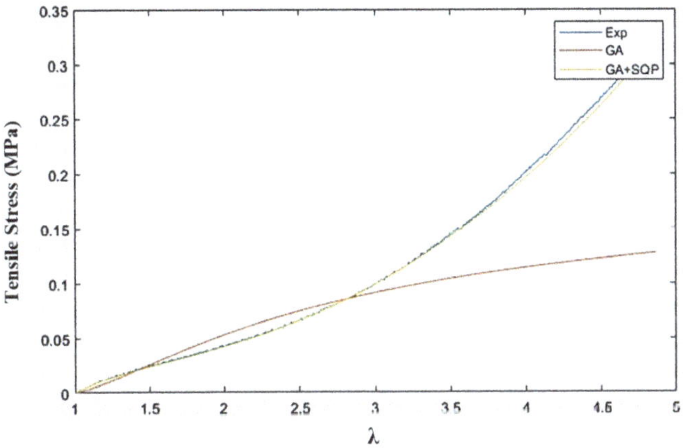

Figure 5. Curve-fitted plots for silicone rubber Ecoflex 30 using least-square method using GA and hybrid GA + SQP methods.

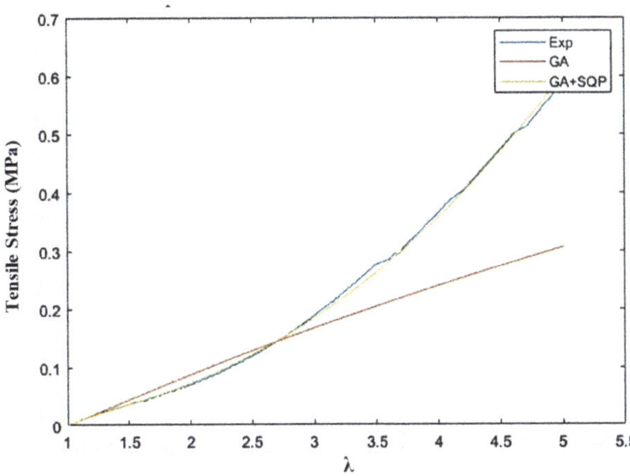

Figure 6. Curve-fitted plots for silicone rubber Ecoflex 50 using least-square method using GA and hybrid GA + SQP methods.

2.6.2. Magnetic Properties of Carbonyl Iron Particles

The magnetic properties of CIPs, in the form of hysteresis *B-H* curve, were provided by the manufacturer, BASF SE, Ludwigshafen, Germany, as depicted in Figure 7a. Using the experimental *B-H* data, the following equation can be effectively used to predict *B-H* response of CIPs up to saturation [49].

$$B(H) = B_s\left(1 - ae^{-bH}\right) + \mu_0 H \quad (31)$$

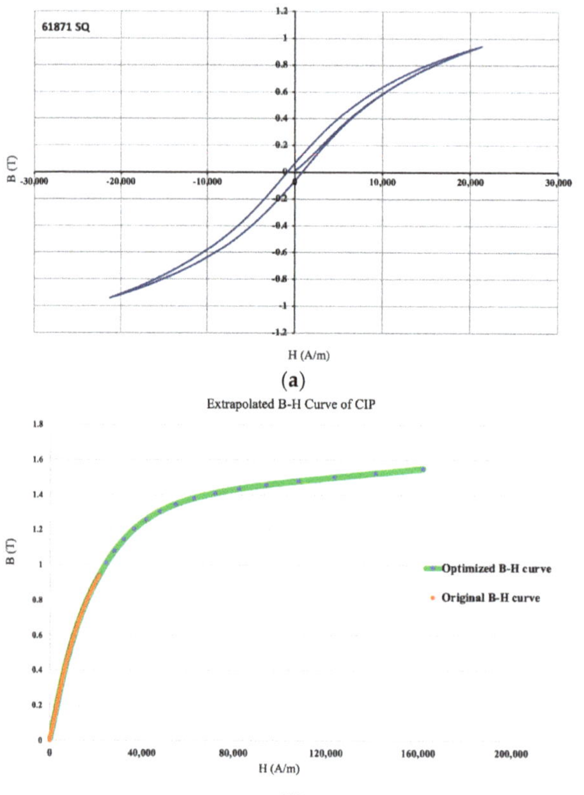

Figure 7. (**a**) B-H curve for CIP, provided by the manufacturer (BASF SE, Ludwigshafen, Germany), and (**b**) extrapolated B-H curve of CIP. Vertical axis represent magnetic flux density (B) in Tesla, and the horizontal axis is the field intensity in A/m.

In which, *B* and *H* are magnetic flux density and magnetic field intensity, respectively. B_s is the magnetic flux density at saturation, *a* and *b* are unknown magnetic parameters, and $\mu_0 = 4\pi \times 10^{-7}\left[\frac{N}{A^2}\right]$ is the vacuum permeability. Using Equation (31), the following *B-H* curve, shown in Figure 7b, was interpolated and extrapolated for CIP, using the provided COMSOL Multiphysics plug-in.

The identified material properties to be considered in the modeling, along with the optimized parameters for interpolation and extrapolation conducted on the *B-H* curve of CIP, using Equation (31) are provided in Table 3.

Table 3. Material properties of air and CIP.

Material	Material Properties	Value
Air	Relative Permeability μ_r	1
CIP	Young's Modulus E	210 GPa
	Poisson ratioν	0.33
	Density ρ	7870 Kg/m^3
Optimized Extrapolation Parameters for CIP B-H Curve	a	6.1746
	b	6.1943 m/A
	B_s	1.38 T

3. Results

3.1. Modeling the 2D Isotropic MRE-RVE in COMSOL

The 2D MRE-RVE was generated in COMSOL using a simple cube containing one CIP inclusion. The mechanical and magnetic data associated with each part (matrix, inclusion, and the surrounding air domain) were defined precisely, according to the previous section. In order to validate the model, the first modeling was conducted on silicone rubber Ecoflex 50 with 15% volume fraction of CIP to compare with the experimental results from the literature [21].

As for the meshing pattern, a user-defined mesh approach was employed to discretize the matrix, CIP, and the surrounding air. This methodology ensures precise control over meshing details, allowing for a finer mesh size in specific regions, such as boundaries, and coarsening where needed, especially within the air domain. A mesh sensitivity analysis was performed to determine the most efficient number of elements, balancing computational cost with the attainment of reasonable results. Results for mesh sensitivity for 2D MRE-RVE with 15% volume fraction of CIP and exposed to magnetic flux density of 0.2 T is provided for Ecoflex 50, in Figure 8 as an example. A relative error between the shear modulus obtained from the MRE-RVE modeling and the experimental results [21] is then defined as follows: $\frac{G_{MRE-RVE}-G_{Exp}}{G_{Exp}} \times 100$. It can be observed that the decrease in the relative error is negligible when the number of elements exceeds 1500.

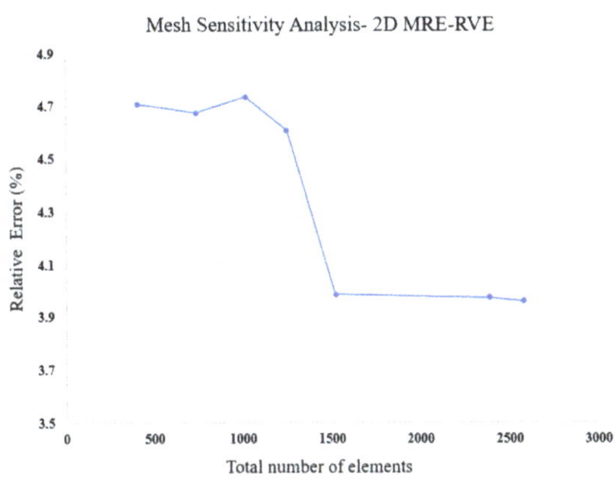

Figure 8. Mesh sensitivity analysis graph for 2D MRE-RVE model, Ecoflex 50 (ϕ = 15%, B = 0.2 T).

Figure 9 shows the FE model of the 2D MRE-RVE. As depicted in Figure 9a and as previously discussed, the mesh employed in the air domain gets coarser as it recedes the RVE boundaries. This is considered as the minimal mechanical or magnetic loading and displacements expected within this region. In contrast, the mesh is finely dispersed around the inclusion, using four boundary layers to ensure the necessary precision in that region. This is essential due to the concentrated interaction of magnetic and mechanical forces within this area. Figure 9b illustrates the boundary layers surrounding the inclusion. It is noteworthy that a total number of 3348 triangular elements were used to discretize the entire MRE-RVE including the surrounding air domain.

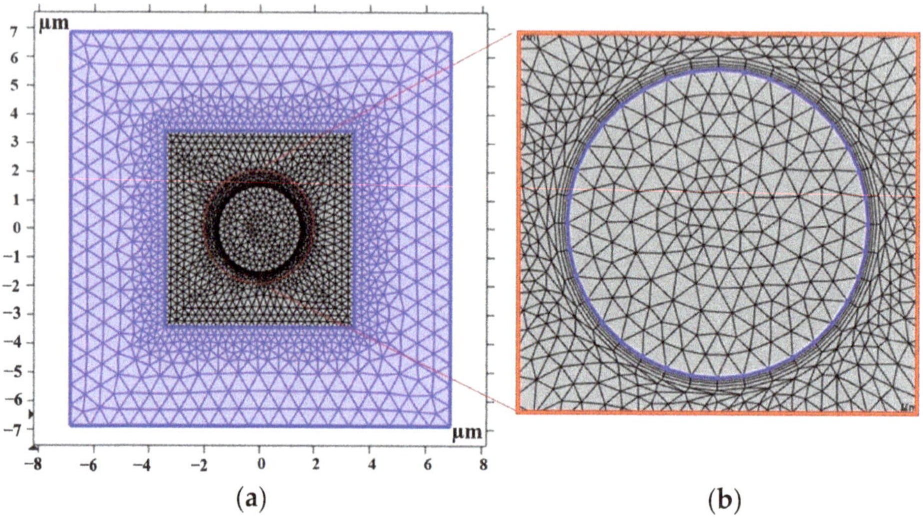

Figure 9. (**a**) The mesh pattern of MRE-RVE (the grey square containing the circle) surrounded by the air domain (the purple square) and (**b**) the boundary layers (the grey discretized circles) implemented to enhance precision around the inclusion (the blue circle). The cut-out section is magnified by the scale of 3.5.

Shear Deformation of Isotropic MRE-RVE

Once the MRE-RVE is constructed, a shear deformation is incrementally applied on the top face of RVE up to 30% shear strain, while the periodic boundary conditions are enforced on the edges. The shear deformation is conducted under magnetic field flux densities ranging from 0 to 0.7 T, applied perpendicular to the shear direction. Figure 10 provides an illustration of the applied magnetic field on the RVE and the distortion of the magnetic field around CIP inclusion, as being absorbed by the inclusion. The induced uniform magnetic flux density inside the inclusion is also obvious in this figure. Maxwell stress tensor is applied on the inclusion boundaries and in combination with the mechanical stress, the total shear stress generated in the RVE is calculated. Figure 11 presents the Maxwell stress distribution at the CIP boundaries, under a magnetic field of 0.4 T, when the RVE undergoes 30% shear strain. Figure 12 shows the shear deformation experienced by MRE-RVE under 30% shear strain, while the periodic boundary conditions are being applied on the RVE boundaries. It is noted that the results are provided for the silicone rubber Ecoflex 50 MRE-RVE, with 15% volume fraction of CIP under the application of 0.4 T magnetic flux density as an example. In Figures 10–12, the smaller square indicates the RVE boundaries, while the bigger one represents the air domain boundaries.

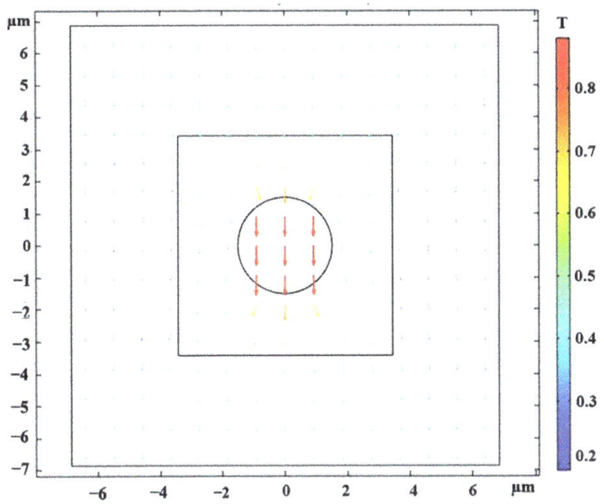

Figure 10. The magnetic field distortion around the CIP inclusion (the circle), while the small square indicates the RVE boundaries, and the big square is the air domain boundary. The color definition bar depicts the magnetic flux density in Tesla.

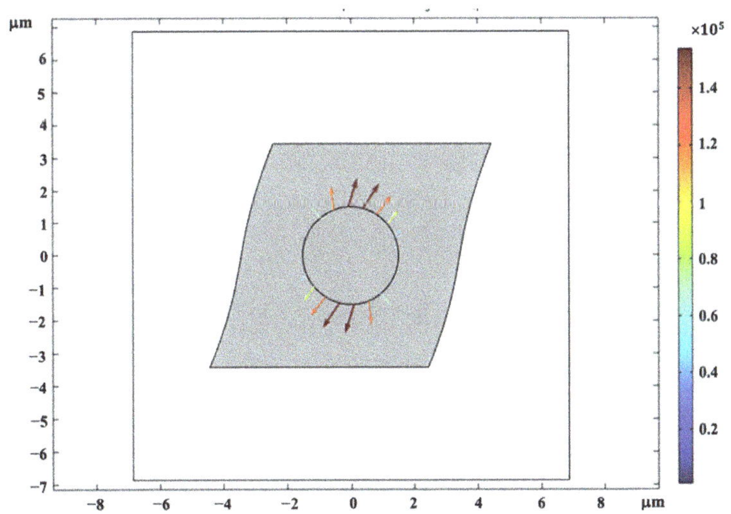

Figure 11. The Maxwell stress distribution generated at the CIP boundaries under magnetic flux density of 0.4 T, at 30% shear strain. The color definition bar depicts the Maxwell stress in Pa.

Finally, the pure shear analysis was conducted to obtain shear stress–shear strain response of MRE-RVE under the application of different magnetic flux densities. Figure 13 represents the homogenized shear stress versus shear strain behavior of the silicone rubber Ecoflex 50 MRE-RVE containing 15% CIP in volume fraction, under an external magnetic field ranging from 0 to 0.7 T. Examination of results in Figure 13 reveals that shear modulus, representing the slope of the shear stress–shear strain curves, substantially increases by increasing the magnetic field intensity. For instance, at nearly 30% shear strain, the generated shear stress increases almost over 60% from nearly 19 kPa to almost 31 kPa by increasing the magnetic flux density from 0 to 0.7 T, respectively. Results also show that the change in modulus decreases as the magnetic flux density increases, confirming the saturation phenomenon. The variation in the shear modulus with respect to the magnetic flux density obtained from the 2D isotropic MRE-RVE, and its comparison with the reported experimental results is shown in Figure 14.

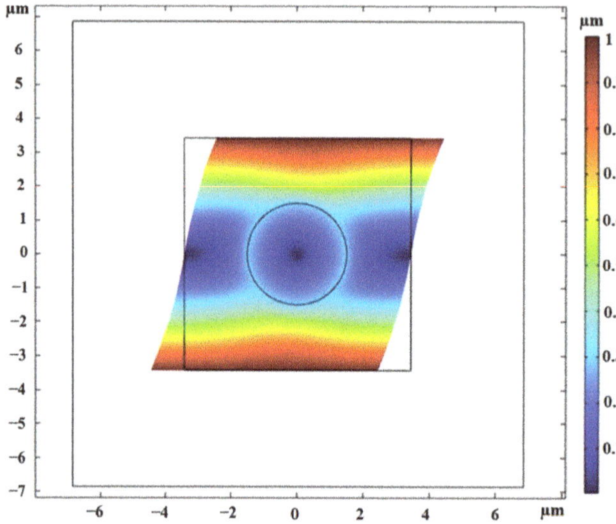

Figure 12. The shear deformation of MRE-RVE under 30% shear strain, while the periodic boundary conditions are applied on the RVE boundaries. The small square frame shows the RVE boundaries before deformation and the deformed RVE is colored. The color definition bar depicts the displacement magnitude in µm.

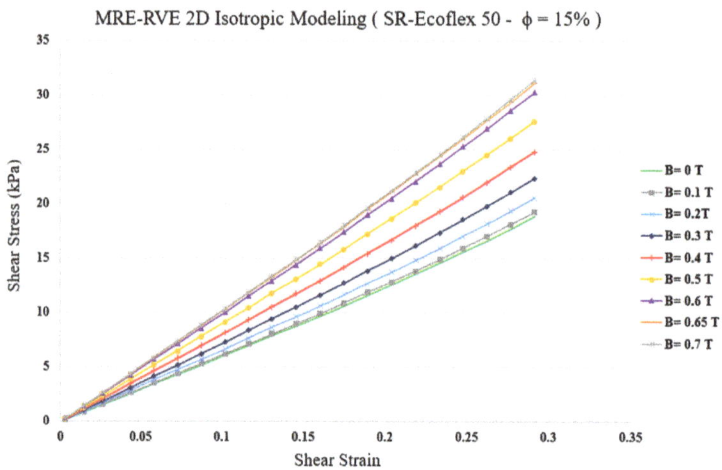

Figure 13. Shear stress–shear strain plot for 2D MRE-RVE under different applied magnetic fields.

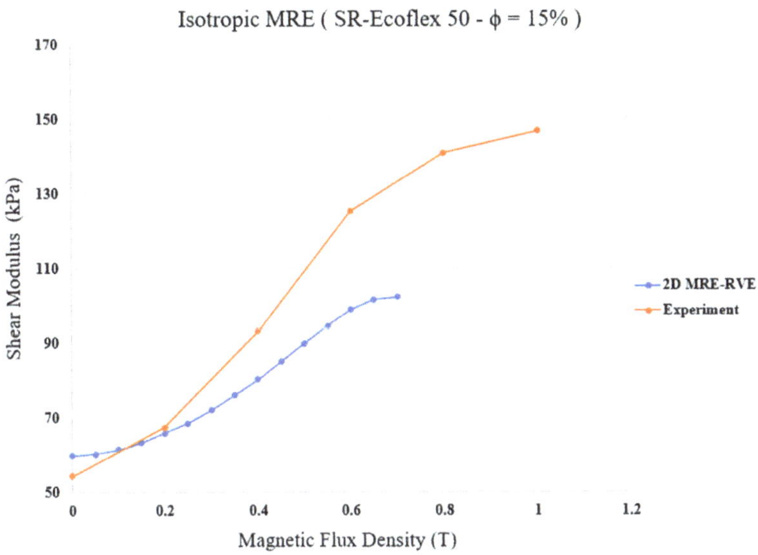

Figure 14. Comparison of shear modulus versus magnetic flux density obtained from COMSOL FE modeling of 2D isotropic MRE-RVE with the experimental results [21] for silicone rubber Ecoflex 50-MRE with 15% volume fraction of CIP.

Results show that the zero-field shear modulus of MRE-RVE obtained from COMSOL FE modeling is 59.9 kPa, which is 10% higher than the 54.43 kPa zero-field shear modulus of MRE given by experimental results. Results from 2D MRE-RVE model substantially deviates from experiential results as magnetic flux density increases beyond 0.2 T. Moreover, Figure 14 illustrates that the field-induced shear modulus of MRE-RVE reaches saturation at magnetic flux density of nearly 0.65 T as also evident from Figure 13, while that of the experiment keeps increasing up to 0.8 T of applied magnetic field. The differences between modeling and experimental results are quantified in Table 4.

Table 4. Comparing the results of 2D MRE-RVE and the experiments.

	Zero-Field Shear Modulus (kPa)	Maximum Shear Modulus (kPa)	Saturation Magnetic Flux Density (T)	MR-Effect ($\frac{\Delta G}{G_0}$)
MRE-RVE	59.9	102.27	0.65	70.73%
Experiments [21]	54.43	146.59	0.8	169%

Results show that while 2D isotropic MRE-RVE model may provide acceptable results for shear modulus results under lower magnetic field, it cannot capture the magneto-mechanical behavior of MREs under higher magnetic fields. For example, the differences between shear moduli from 2D FE modeling and experiment are 10% at 0 T, −11% at 0.4 T, and −20% at 0.7 T.

The developed 2D isotropic MRE-RVE FE model was subsequently used to qualitatively investigate the effect of CIP volume fraction on the shear modulus. Figure 15a–f show the results for the shear stress–shear strain response behavior of MREs under different magnetic flux densities for CIP volume fraction ranging from 5% to 40%. Results show that increasing the volume fraction of CIP yields higher field-induced shear modulus. For instance, when increasing the magnetic flux density from 0 to 0.7 T, and under shear strain of 30%, the shear stress increases from nearly 16 kPa to almost 23 kPa (44%) and from 35 kPa to 65 kPa (86%) for CIP volume fraction of 5% and 40%, respectively.

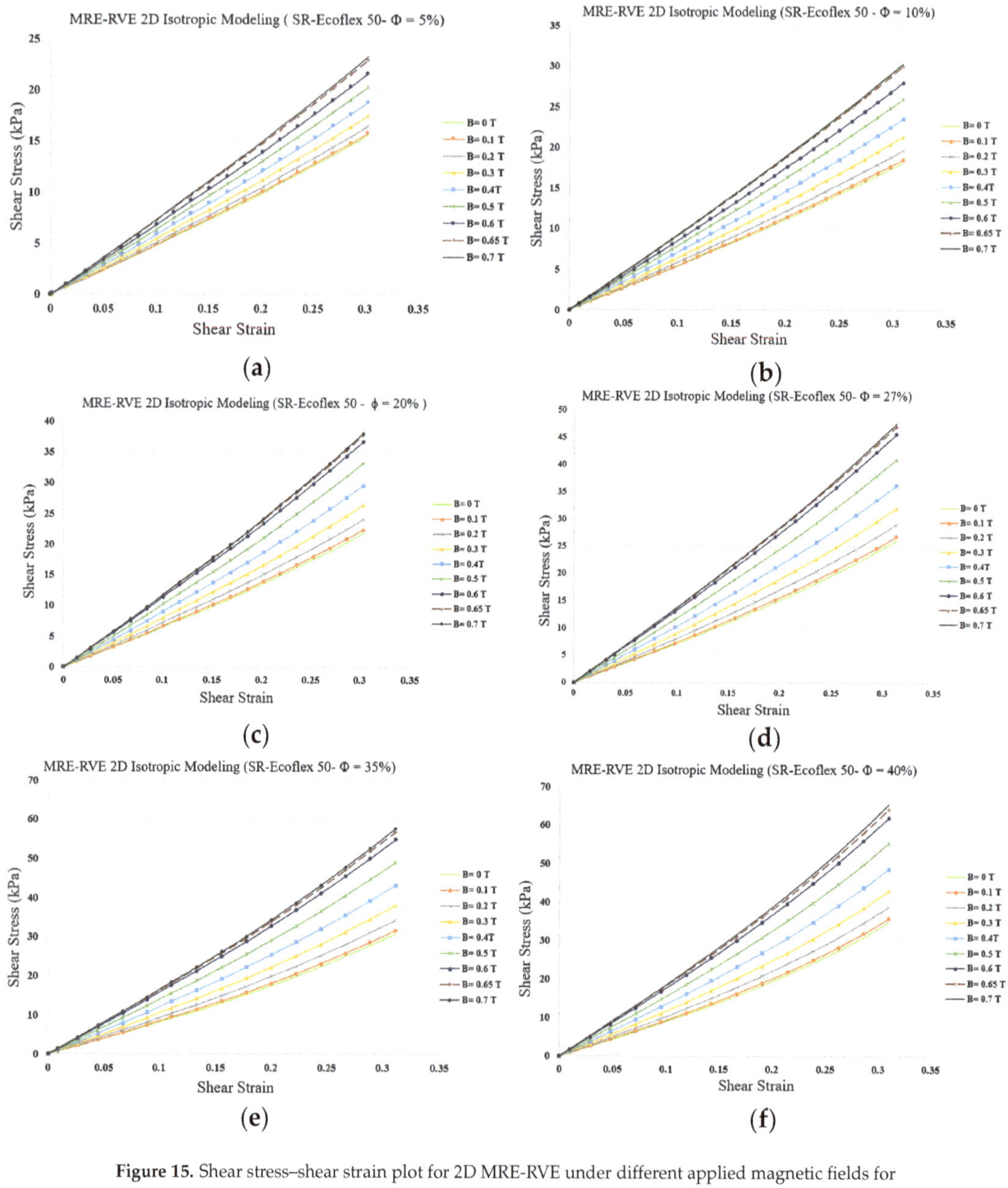

Figure 15. Shear stress–shear strain plot for 2D MRE-RVE under different applied magnetic fields for silicone rubber (SR) Ecoflex 50 with (**a**) 5%, (**b**) 10%, (**c**) 20%, (**d**) 27%, (**e**) 35%, and (**f**) 40% of CIP in volume fraction.

As the CIP volume fraction increases, the gap between two subsequent curves in each figure (Figure 15a–f) increases, implying that the influence of magnetic field on shear modulus, and consequently, the MR effect increase as the volume fraction increases. The variation in MR effect with respect to CIP volume fraction is shown in Figure 16. Results suggest that the MR effect increases by increasing the CIP volume fraction. Although the MR effect is supposed to reach a maximum at around $\phi = 27\%$ and then drop [17], the 2D modeling is thus not able to capture this behavior and the MR effect keeps increasing as the volume fraction goes up.

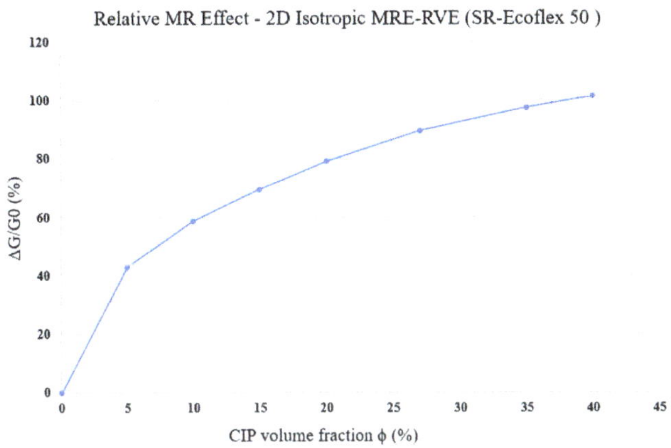

Figure 16. MR effect behavior in 2D modeling of silicone rubber (SR) Ecoflex 50 MRE-RVE with respect to CIP volume fraction.

The same procedure of FE modeling used for silicone rubber Ecoflex 50 MRE-RVE, was also conducted on silicone rubber Ecoflex 30 MRE-RVE. The influence of different magnetic flux densities, ranging from 0–0.7 T, was also studied on the shear stress–shear strain response of the silicone rubber Ecoflex 30 MRE-RVEs, containing various CIP content. Figure 17 shows the MR effect with respect to CIP volume fraction for 2D MRE-RVE with different matrix materials (Ecoflex 30 and Ecoflex 50). Results clearly show that MR effect in the MRE-RVE with the softer matrix (Ecoflex 30) reaches a maximum of 154%, while that in Ecoflex 50 MRE-RVE reaches a maximum of 100%, both containing 40% CIP in volume fraction. Although showing a higher relative MR effect in MREs with softer matrix is anticipated due to the experimental data in the literature [21,22], the relative MR effect is supposed to reach a peak at the optimum CIP volume fraction, and then decrease as the volume fraction goes up [17], and the 2D MRE-RVE modeling cannot capture this behavior.

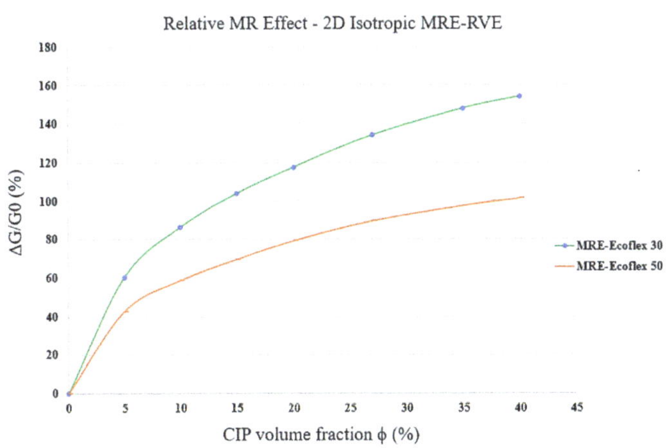

Figure 17. Comparison of the MR effect behavior in 2D isotropic MRE-RVEs with different matrix materials (Ecoflex 50 and Ecoflex 30) with respect to CIP volume fraction.

The difference in the results could be attributed to the incapability of 2D model to capture the whole physical phenomenon. It is noted that, in 2D RVE model, an extruded depth should be assigned to the plane geometry. Thus, the inclusion is in fact considered as a short cylindrical fiber which is different from the geometry of the nearly spherical inclusion in reality.

3.2. Modeling the 3D Isotropic MRE-RVE in COMSOL

As suggested in the previous section, the 2D MRE-RVE was not able to properly capture the coupled magneto-mechanical response of MREs in shear deformation. Hence, the modeling approach has been extended to 3D. Therefore, the 3D MRE-RVE was generated in COMSOL in the same fashion as that of 2D modeling. One CIP inclusion was generated and placed inside a simple cube of matrix material and surrounded by a larger cube of air. The mechanical and magnetic data associated with each part (matrix, magnetic particle inclusion, and the surrounding air domain) was also defined precisely, as explained before. To validate the model, we initiated the modeling process for silicone rubber Ecoflex 50 containing 15% volume fraction of CIP. Subsequently, we conducted a comparison with the experimental data reported in the literature [21].

In order to create a complex and customized structure, the RVE underwent a detailed meshing procedure. We decided to utilize the "user-defined mesh" method, similar to the approach used in the 2D configuration. A tetrahedron mesh type is used, as it provides more flexibility for meshing the curved boundaries, here, the spherical magnetic particle.

A methodical mesh sensitivity analysis was then systematically performed to reach the optimal mesh pattern, ensuring that computational resources were not needlessly burdened. Results of the relative error between the shear modulus obtained from the 3D MRE-RVE modeling and the experimental results [21], defined as: $\frac{G_{MRE-RVE}-G_{exp}}{G_{exp}} \times 100$, for different number of elements are provided in Figure 18. Just as described in the 2D modeling section, we explored various meshing configurations while creating the 3D model. Results in Figure 18 show that the relative error between the shear modulus obtained from 3D MRE-RVE and experiments decreases as the total number of tetrahedron elements increases, indicating the convergence of the shear modulus. Hence, based on this finding and by evaluating the computational cost, we opted for the mesh configuration consisting of 30,492 tetrahedral elements to balance between the computational cost and accuracy.

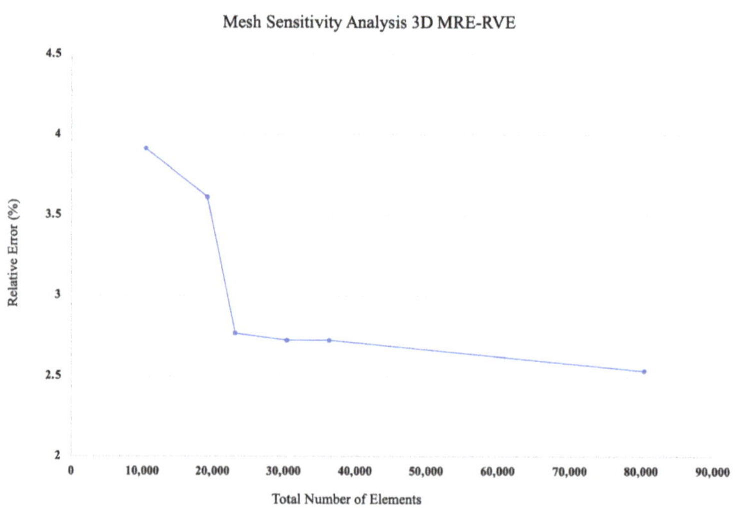

Figure 18. Mesh sensitivity analysis graph for 3D MRE-RVE model, (ϕ = 15%, B = 0.1 T) using different meshing schemes.

A visual representation of the mesh pattern applied in the 3D MRE-RVE modeling is provided in Figure 19a. As previously explained, the mesh density in the air domain progressively coarsens as it moves away from the RVE boundaries, for the anticipation of minimal mechanical or magnetic loading and displacements in this particular zone. Conversely, the mesh is finely adjusted in the vicinity of the inclusion, to ensure the required precision in that area due to the intensified interaction of magnetic and mechanical forces. Figure 19b further illustrates the mesh quality in all regions.

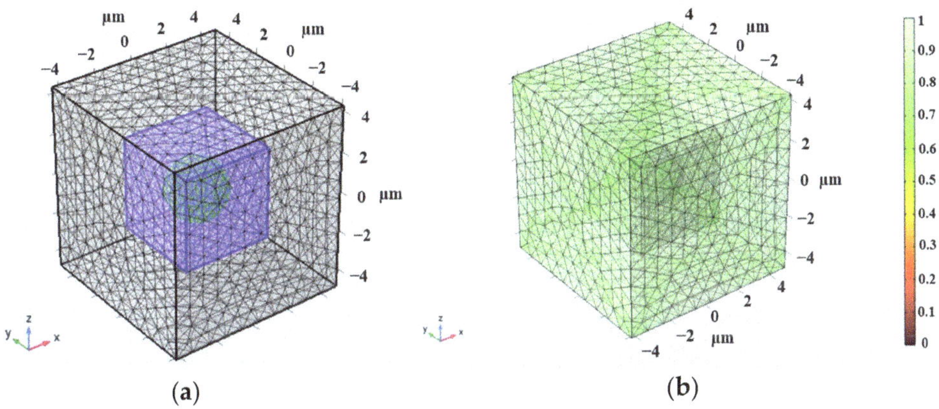

Figure 19. (**a**) The mesh pattern of MRE-RVE (the blue cubic matrix containing the green spherical inclusion) surrounded by the air domain (the grey cube) and (**b**) the mesh quality in all regions with the color bar representing the quality of mesh on the scale of 0 to 1.

As for shear analysis, the RVE was systematically subjected to incremental pure shear deformation, gradually reaching a shear strain of 30%. To maintain consistency, the periodic boundary conditions were imposed along all surface boundaries. Concurrently, a magnetic field was applied perpendicular to the shear direction, spanning a range of magnitudes from 0 to 0.4 T. The visual representation in Figure 20 clearly portrays the magnetic field's interaction with the RVE, particularly highlighting the distortion of the field as it encounters the CIP inclusion. The high induced magnetic flux density within the inclusion is clearly visible. In the same fashion, with the 2D modeling, the Maxwell stress tensor was applied to the boundaries of the CIP inclusion, calculated at each step of the analysis, and simultaneously interpreted as a mechanical load into the modeling in solid mechanics physics, interpreting the whole coupled problem as a solid mechanic problem with periodic boundary conditions applied accordingly. Integrating this stress with the mechanical stress developed by the shear deformation, the overall shear stress generated within the 3D MRE-RVE was determined. Figure 21a,b illustrate the Maxwell stress distribution at the CIP boundaries under a magnetic field of 0.1 T, when the RVE undergoes 30% shear strain.

Figure 22 illustrates a visualization of the shear deformation of the MRE-RVE under shear strain of 30% and the shear stress distribution throughout the entire MRE-RVE. The RVE boundaries are consistently depicted by the smaller cube, while the larger cube delineates the boundaries of the air domain. It is worth noting that Figures 20–22 feature the silicone rubber Ecoflex 50 MRE-RVE with a 15% volume fraction of CIP, subjected to a magnetic field of 0.1 T.

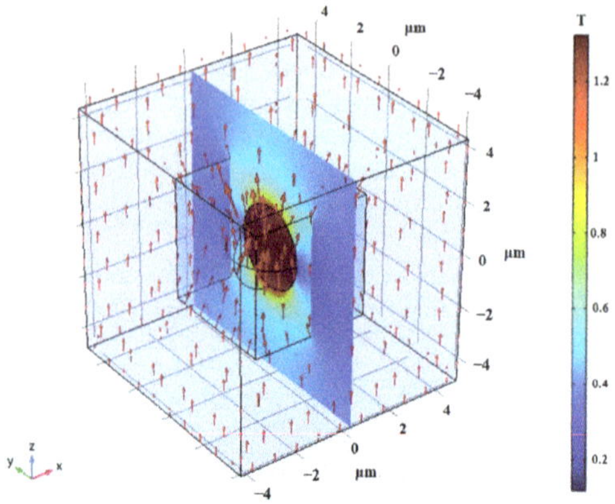

Figure 20. The magnetic field distortion around the CIP inclusion inside the 3D RVE, red arrows represent the magnetic field intensity and direction (0.1 T, upward), the color definition bar describes the magnetic flux density (T) in the air, and MRE-RVE domain, referring to the hypothetical cut out surface in the middle of the model.

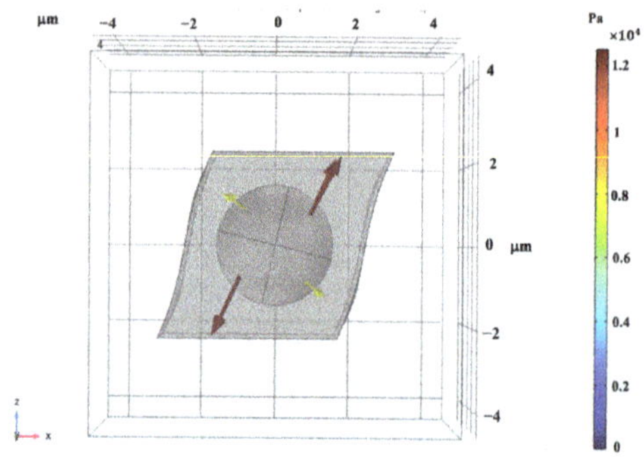

(**a**)

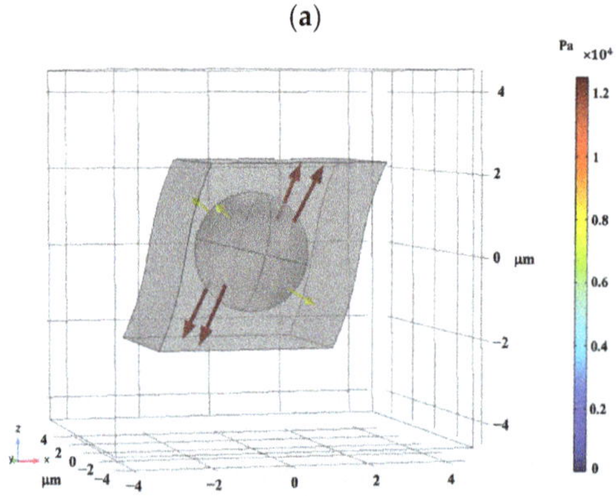

(**b**)

Figure 21. The Maxwell stress distribution generated at the CIP boundaries under magnetic flux density of 0.1 T, at 30% shear strain: (**a**) x–z plane view and (**b**) spatial 3D view. The arrows represent the Maxwell stress, and the color definition bar depicts the Maxwell stress in Pa.

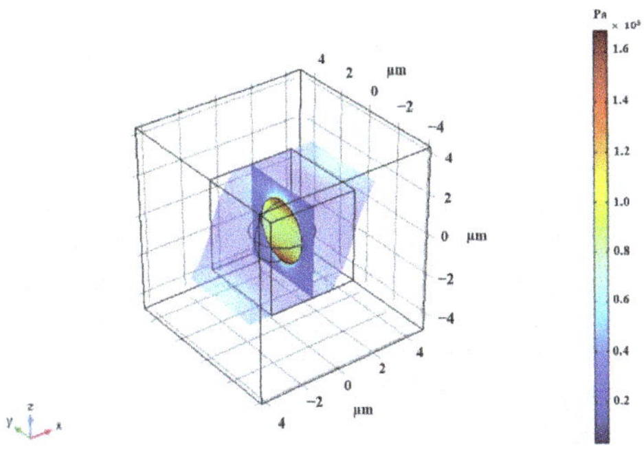

Figure 22. The shear deformation of MRE-RVE under 30% shear strain, while the periodic boundary conditions are being applied on the RVE boundaries, subjected to a magnetic field of 0.1 T. The smaller cubic frame shows the RVE boundaries before deformation and the deformed RVE is colored. The color definition bar depicts the Tresca stress in Pa.

3.2.1. MRE-RVE-Silicone Rubber Ecoflex 50

The pure shear analysis was subsequently conducted on MRE-RVE with the silicone rubber Ecoflex 50 as the matrix material under the application of magnetic flux densities ranging from 0 to 0.4 T. Results for the homogenized shear stress versus shear strain for MRE-RVE containing 15% CIP in volume fraction under the application of varied magnetic flux densities are shown in Figure 23.

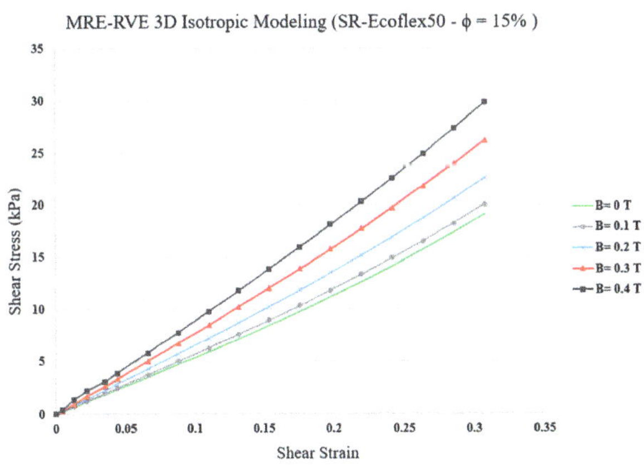

Figure 23. Shear stress–shear strain plot for 3D MRE-RVE (silicone rubber (SR) Ecoflex 50), containing 15% of CIP in volume fraction under different applied magnetic fields.

Examination of the results in Figure 23 reveals that, as expected, the shear modulus increases by increasing the magnetic field intensity. In other words, as the magnetic field increases, the MRE-RVE experiences higher shear stress while undergoing the same amount of shear strain. For example, when undergoing 30% shear strain, the MRE experiences roughly 18 kPa of shear stress when no magnetic field is applied; however, by increasing the magnetic field up to 0.4 T, the shear stress reaches almost 30 kPa, indicating a 66% increase. The variation in the predicted field-dependent shear modulus with respect to the applied magnetic flux density and its comparison with reported experimental results [21] are shown in Figure 24. It is observed that, unlike the 2D MRE-RVE model, the 3D MRE-RVE can accurately predict the field-dependent shear modulus of the MRE up to 0.4 T. For instance, the zero-field shear modulus of MRE-RVE obtained from COMSOL FE 3D modeling is 55.15 kPa, which is only 1.3% higher than the 54.43 kPa zero-field shear modulus of MRE obtained experimentally.

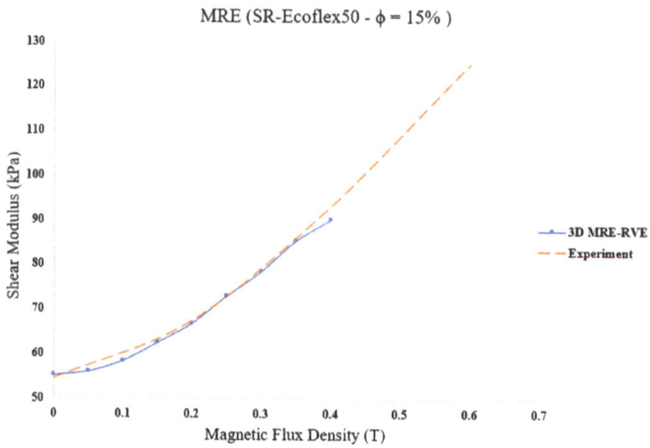

Figure 24. Shear modulus versus magnetic flux density for 3D isotropic MRE-RVE (silicone rubber (SR) Ecoflex 50).

In order to assure that the whole behavior of the MRE is captured accurately in the FE modeling, the coefficient of determination (R^2) is determined, which is defined as $R^2 = 1 - \frac{SS_{res}}{SS_{tot}}$. SS_{res} represents the sum of squared residuals (the differences between the predicted values and the actual values), and SS_{tot} is the total sum of squares, which measures the total variance of the predicted variable. R^2 close to one describes a perfect agreement between the predicted values and the actual values. R^2 was determined between the results from the modeling and the ones obtained from experiments [21], and is found to be 0.99.

We have attempted to evaluate the shear response behavior of the 3D-RVE modeling for higher magnetic flux densities, beyond 0.4 T. However, the model fails due to the complex interaction between the mechanical and magnetic loads. Analyzing the results, we realized that the issue is likely due to the abrupt change in the material properties between an extremely soft rubber and a rigid inclusion, along with the accumulated mechanical and magnetic nonlinearity associated with the stress and material behavior.

The 3D MRE-RVE model was then effectively utilized to investigate the influence of CIP volume fraction, magnetic field, and matrix stiffness on the shear deformation response behavior of MREs, under varying magnetic flux densities. Results for shear stress-shear strain response of MREs with matrix Ecoflex 50 concerning different CIP volume fraction, ranging from 5% to 40%, are illustrated in Figure 25a–f.

As suggested by the results in Figure 25a–f, the nonlinearity in the stress–strain curves increases by increasing the volume fraction of CIP and by increasing the applied magnetic field. Results also show that increasing the volume fraction of CIP yields a substantial increase in the magnetic field induced shear stress at the given shear strain. For instance, at 30% shear strain and under a magnetic flux density of 0.4 T, the MRE-RVE containing 5% CIP experiences a shear stress of roughly 17 kPa, while the shear stress reaches 110 kPa for MRE-RVE containing 40% CIP. From a broader perspective, as the CIP volume fraction increases, the gap between the stress–strain curves increase, indicating an enhancement in the relative MR effect. However, the most widened gaps are observed in MRE-RVE with 27% CIP, indicating that the maximum relative MR effect occurs at the volume fraction of 27%.

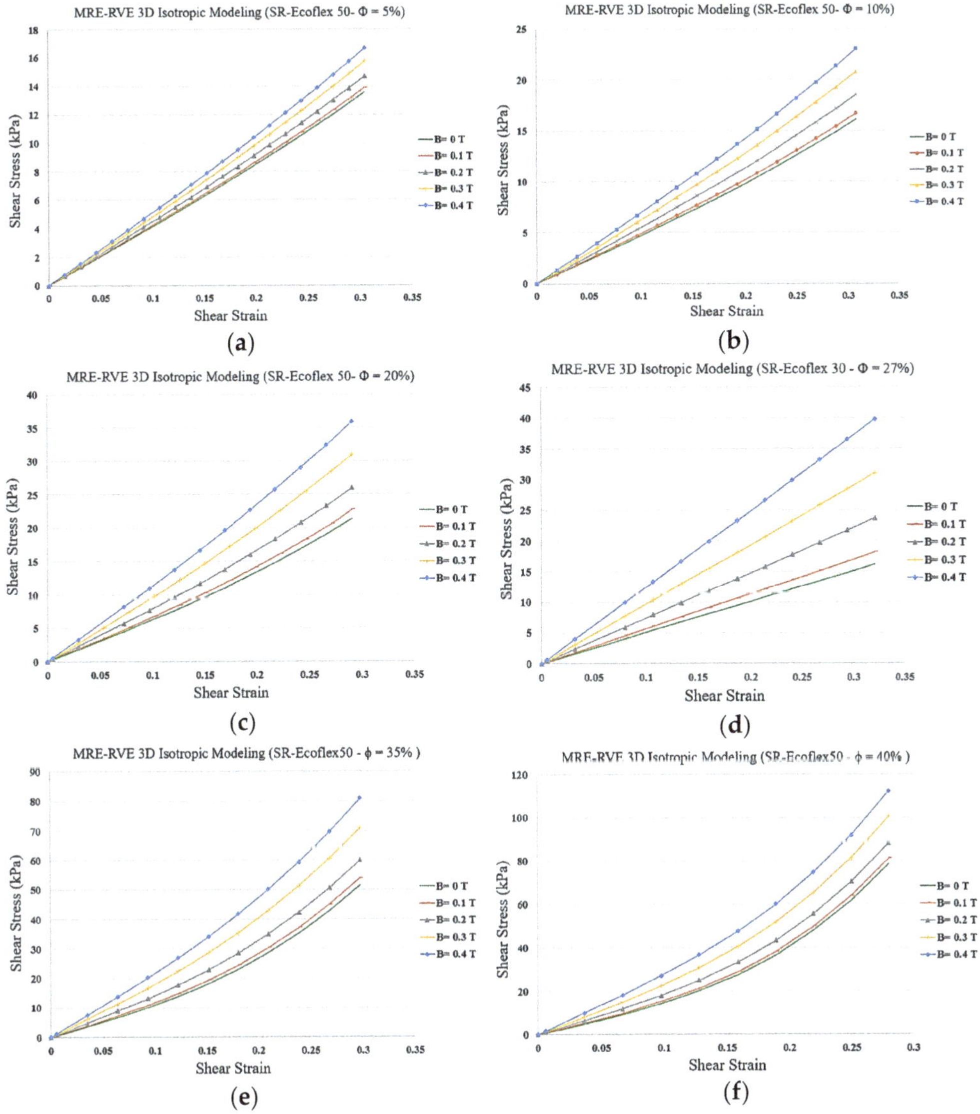

Figure 25. Shear stress–shear strain plot for 3D isotropic MRE-RVE under different applied magnetic fields for silicone rubber (SR) Ecoflex 50 with (**a**) 5%, (**b**) 10%, (**c**) 20%, (**d**) 27%, (**e**) 35%, and (**f**) 40% of CIP in volume fraction.

To have a better understanding of the influence of CIP volume fraction, the MR effect of the 3D MRE-RVEs with different volume fractions of CIP has been evaluated. It is noted that, in determining the MR effect, the maximum shear modulus was evaluated at magnetic flux density of 0.4 T. The results are shown in Figure 26, suggesting that the relative MR effect initially increases as the CIP content increases, reaching to a maximum level of 92% at 27% volume fraction as predicted before and then decreases with further increasing the volume fraction of CIP. This is in agreement with results reported by Davis [17].

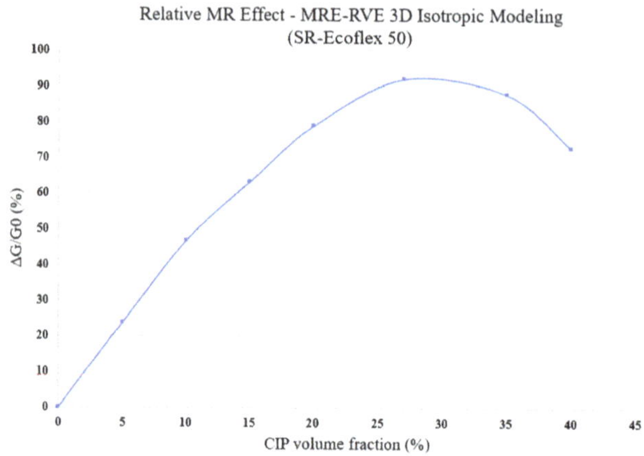

Figure 26. Relative MR effect for silicone rubber (SR) Ecoflex 50 MRE-RVE versus CIP volume fraction.

3.2.2. Silicone Rubber Ecoflex 30 MRE-RVE

The same procedure of FE modeling used for silicone rubber Ecoflex 50 MRE-RVE was conducted on MRE-RVE with silicone rubber Ecoflex 30 as the matrix material. The influence of different magnetic flux densities, ranging from 0 to 0.4 T, was also studied on the shear stress–shear strain response of the silicone rubber Ecoflex 30 MRE-RVEs containing various CIP content. Figure 27 shows the shear stress–shear strain behavior of Ecoflex 30 MRE-RVEs with 15% volume fraction of CIP, under varied magnetic flux densities ranging from 0 to 0.4 T. Results in Figure 27 clearly suggest that Ecoflex 30 MRE-RVE exhibits lower stiffness and less nonlinear behavior compared with Eoflex 50 MRE-RVE in Figure 23. Moreover, a closer look at the gaps between two subsequent curves in both Figures 23 and 27 reveals that, in Ecoflex 30 MRE-RVE, the stress–strain curves are more widened than those in Ecoflex 50 MRE-RVE, indicating a more pronounced relative MR effect obtained from the MRE with softer matrix material under identical conditions.

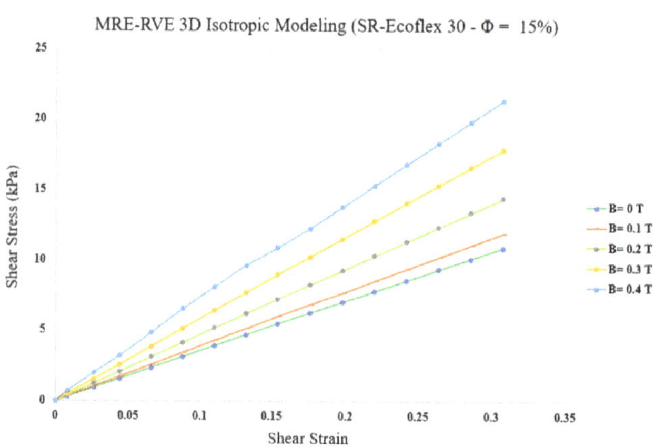

Figure 27. MRE-RVE (silicone rubber (SR) Ecoflex 30) shear stress–strain behavior under the application of different magnetic fields.

Similar to Ecoflex 50 MRE-RVE, the effect of volume fraction of CIP on the shear stress–shear strain response of Ecoflex 30 MRE-RVE has also been investigated, and results are shown in Figure 28a–f. Results suggest an escalation in the shear modulus and the relative MR effect when the CIP content goes up. The effect of volume fraction of CIP on the relative

MR effect can be better understood in Figure 29. Results show that the relative MR effect for Ecoflex 30 MRE-RVE reaches its peak at CIP volume fraction of nearly 35%, while the Ecoflex 50 MRE-RVE experienced its maximum relative MR effect at 27% CIP content (Figure 26). Figure 30 illustrates the comparison of the results for MR effect with respect to CIP content for MRE-RVE with Ecoflex 50 and Ecoflex 30 as the host matrices. Results show that the maximum MR effect obtained from the softer matrix (Ecoflex 30) was observed to be nearly 166% at 35% volume fraction of CIP compared with nearly 92% MR effect at 27% volume fraction CIP for silicone rubber Ecoflex 50. It can be observed that the maximum relative MR effect in MREs with the softer matrix is noticeably higher than that of MREs with silicone rubber Ecoflex 50, which has also been confirmed by other studies [21,22].

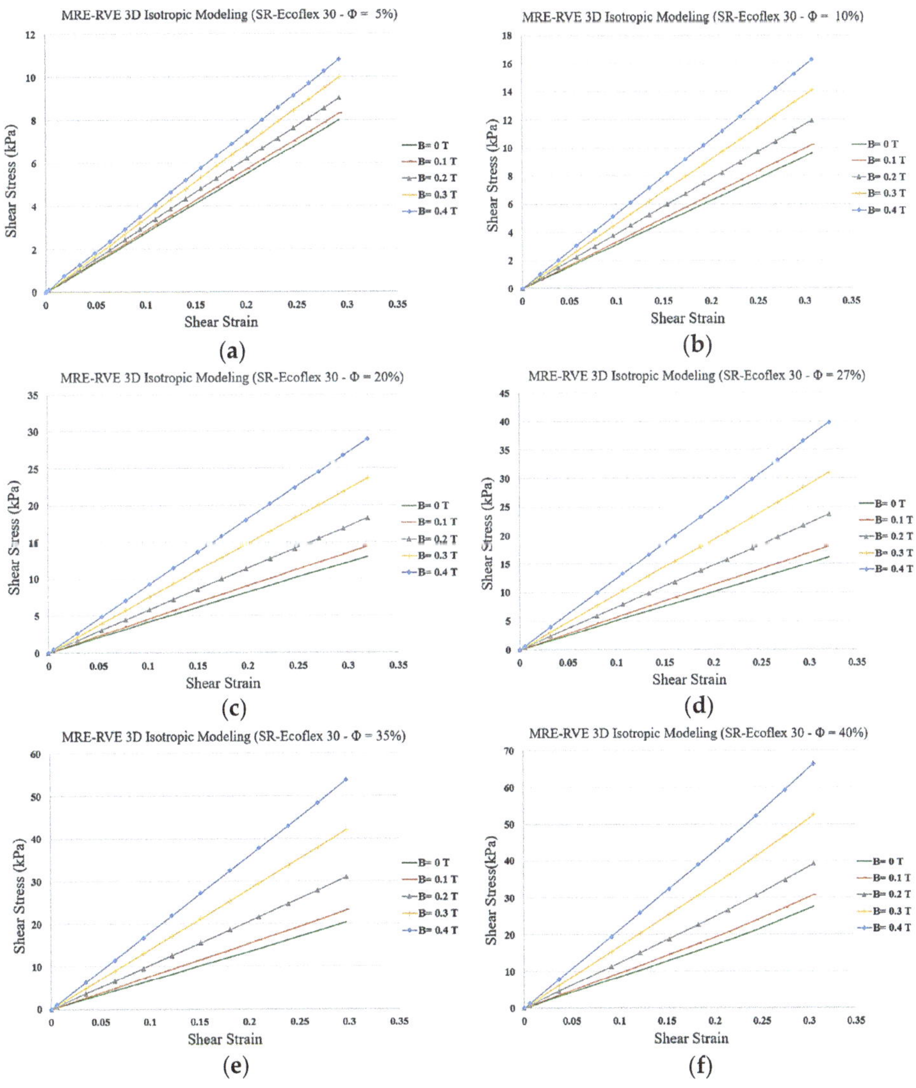

Figure 28. Shear stress–shear strain plot for 3D isotropic MRE-RVE under different applied magnetic fields for silicone rubber (SR) Ecoflex 30 with (**a**) 5%, (**b**) 10%, (**c**) 20%, (**d**) 27%, (**e**) 35%, and (**f**) 40% of CIP in volume fraction.

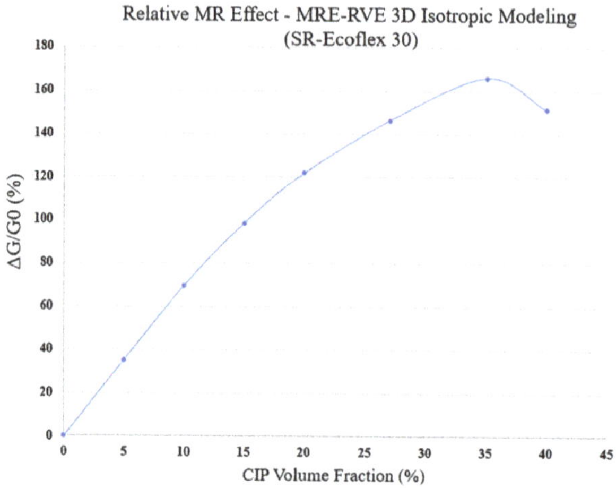

Figure 29. Relative MR effect versus CIP volume fraction obtained from 3D isotropic MRE-RVE for silicone rubber (SR) Ecoflex 30.

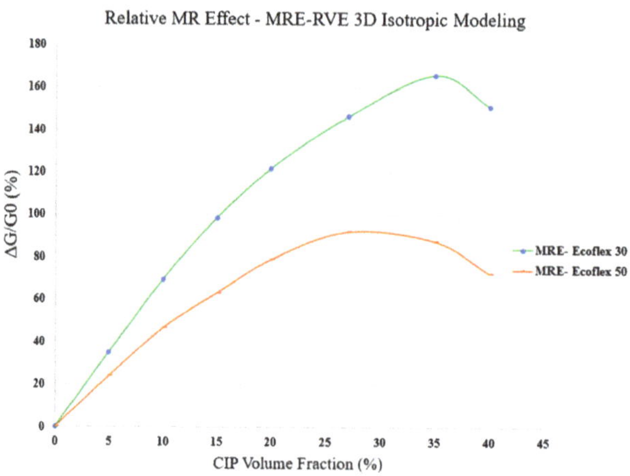

Figure 30. Comparing MR effect versus CIP volume fraction obtained from 3D isotropic MRE-RVE for silicone rubber Ecoflex 30 and silicone rubber Ecoflex 50.

4. Conclusions

In this study, an FE model based on the representative volume element (RVE) approach was proposed to model the shear deformation of MREs under the influence of an external magnetic field, while taking all the nonlinearities into account. Experiments were carried out on host rubber samples (silicone rubber) to formulate the highly nonlinear Ogden strain energy to describe hyper-elastic behavior of the rubbery matrix. The magnetic behavior of CIP was described through a nonlinear B-H curve. The appropriate RVE size for modeling the MREs was defined, and the MRE-RVE was generated in COMSOL Multiphysics in 2D and 3D configurations. The MRE-RVE underwent incremental pure shear deformation, while periodic boundary conditions (PBCs) were imposed on the RVE boundaries. Simultaneously, a homogeneous magnetic field was applied perpendicular to the shear direction, and the Maxwell stress tensor was defined on the CIP inclusion. The study focused on isotropic MREs, investigating the influence of varied magnetic flux densities, CIP content, and the host rubber's hyper-elastic behavior on the shear modulus of the MREs. The MR effect behavior of the MRE-RVEs was also studied and compared.

The analysis started with a comprehensive study on the 2D MRE-RVEs, generated as a simple square with one circular inclusion inside, surrounded by air. The MRE-RVE underwent incremental pure shear deformation up to 30%, and the impact of magnetic flux densities ranging from 0 to 0.7 T was investigated. The results showed that the shear modulus in the MRE-RVE keeps a positive correlation with the magnetic flux density and the CIP content. Comparing the 2D results for Ecoflex 50 with the experimental results in the literature [21] revealed that although the 2D modeling can predict the MRE's behavior within ±20% difference with the experimental results, and shows the saturation effect, it cannot accurately predict the variation in MR effect with respect to CIP volume fraction. The relative MR effect in the 2D MRE-RVE keeps increasing by increasing the CIP volume fraction without showing any peak MR effect. These differences between the results obtained from the modeling and the experiments could be attributed to the incapability of the 2D model to capture the physical shape of the inclusions.

Subsequently, 3D MRE-RVE was developed in the COMSOL environment. The modeling was conducted on a simple cubic MRE-RVE with one spherical inclusion inside, and the whole RVE was surrounded by a larger cube of air. The study was conducted on MRE-RVEs concerning varied magnetic flux densities (ranging 0–0.4 T), CIP content, and different host elastomers (silicone rubber Ecoflex 30 and Ecoflex 50). As expected, the positive correlation between the shear modulus and the magnetic flux density as well as the CIP content was suggested by the results, in both silicone rubber based MREs. Comparing the 3D results for Ecoflex 50 with the experimental results in the literature [21] revealed that the results hold a perfect agreement with the experiments, offering a coefficient of determination (R^2) of 0.99. Exploring the MR effect in the 3D MRE-RVE showed that the relative MR effect in the MRE-RVE with the softer matrix material (Ecoflex 30) is higher than that of Ecoflex 50 MRE-RVE. The 3D modeling was also able to predict the MR effect variation with respect to CIP volume fraction accurately. The results suggested that the relative MR effect in Ecoflex 30 MRE-RVE keeps increasing up to 35% volume fraction of CIP, peaking at 166%, while the MR effect in MRE-RVE with Ecoflex 50 reaches a maximum of 92% at 27% CIP content. Results suggest that using a softer matrix material leads to a higher relative MR effect with higher optimum volume fraction.

Author Contributions: Conceptualization, R.S.; methodology, I.A. and R.S.; software, I.A.; validation, I.A. and R.S.; formal analysis, I.A.; investigation, I.A.; resources, I.A. and R.S.; data curation, I.A.; writing—original draft preparation, I.A.; writing—review and editing, R.S.; supervision, R.S.; project administration, R.S.; funding acquisition, R.S. All authors have read and agreed to the published version of the manuscript.

Funding: This research was funded by the Natural Sciences and Engineering Research Council of Canada (NSERC) under grant No. RGPIN-2021-03482.

Institutional Review Board Statement: Not applicable.

Data Availability Statement: The datasets presented in this article are not readily available because the data are part of an ongoing study. Requests to access the datasets should be directed to the corresponding author.

Acknowledgments: The authors hereby acknowledge the help of Alireza Moezi in conducting the experimental tests.

Conflicts of Interest: The authors declare no conflicts of interest.

References

1. Carlson, J.D.; Jolly, M. MR Fluid, foam and elastomer devices. *Mechatronics* **2000**, *10*, 555–569. [CrossRef]
2. Bastola, A.K.; Paudel, M.; Li, L.; Li, W. Recent progress of magnetorheological elastomers: A review. *Smart Mater. Struct.* **2020**, *29*, 123002. [CrossRef]
3. Saber, A.; Sedaghati, R. The Modeling of Magnetorheological Elastomers: A State-of-the-Art Review. *Adv. Eng. Mater.* **2023**, *25*, 2300182. [CrossRef]
4. Li, Y.; Li, J.; Li, W.; Du, H. A state-of-the-art review on magnetorheological elastomer devices. *Smart Mater. Struct.* **2014**, *23*, 123001. [CrossRef]

5. Liu, Y. A review on magnetorheological elastomers. *Adv. Eng. Technol. Res.* **2023**, *4*, 503–514. [CrossRef]
6. Bastola, A.K.; Hossain, M. A review on magneto-mechanical characterizations of magnetorheological elastomers. *Compos. Part B Eng.* **2020**, *200*, 108348. [CrossRef]
7. Jaafar, M.F.; Mustapha, F.; Mustapha, M. Review of current research progress related to magnetorheological elastomer material. *J. Mater. Res. Technol.* **2021**, *15*, 5010–5045. [CrossRef]
8. Cantera, M.A.; Behrooz, M.; Gibson, R.F.; Gordaninejad, F. Modeling of magneto-mechanical response of magnetorheological elastomers (MRE) and MRE-based systems: A review. *Smart Mater. Struct.* **2017**, *26*, 023001. [CrossRef]
9. Díez, A.G.; Tubio, C.R.; Etxebarria, J.G.; Lanceros-Mendez, S. Magnetorheological Elastomer-Based Materials and Devices: State of the Art and Future Perspectives. *Adv. Eng. Mater.* **2021**, *23*, 2100240. [CrossRef]
10. Li, W.H.; Zhang, X.Z.; Du, H. Magnetorheological Elastomers and Their Applications. *Adv. Elastomers I* **2013**, *11*, 357–374.
11. Li, W.; Zhang, X. Research and Applications of MR Elastomers. *Recent Pat. Mech. Eng.* **2010**, *1*, 161–166. [CrossRef]
12. Ubaidillah; Sutrisno, J.; Purwanto, A.; Mazlan, S.A. Recent progress on magnetorheological solids: Materials, fabrication, testing, and applications. *Adv. Eng. Mater.* **2015**, *17*, 563–597. [CrossRef]
13. Sharif, U.; Sun, B.; Hussain, S.; Ibrahim, D.S.; Adewale, O.O.; Ashraf, S.; Bashir, F. Dynamic behavior of sandwich structures with magnetorheological elastomer: A review. *Materials* **2021**, *14*, 7025. [CrossRef]
14. Faizal Johari, M.A.; Mazlan, S.A.; Ubaidillah Harjana Abdul Aziz, S.A.; Nordin, N.A.; Johari, N.; Nazmi, N. An Overview of Durability Evaluations of Elastomer-Based Magnetorheological Materials. *IEEE Access* **2020**, *8*, 134536–134552. [CrossRef]
15. Ginder, J.M.; Nichols, M.E.; Elie, L.D.; Tardiff, J.L. Magnetorheological elastomers: Properties and applications. In *Smart Structures and Materials 1999: Smart Materials Technologies*; SPIE: Newport Beach, CA, USA, 1999; pp. 131–138.
16. Jolly, M.R.; Carlson, J.D.; Munoz, B.C.; Todd, A. The magnetoviscoelastic response of elastomer composites consisting of ferrus particles embedded in a polymer matrix. *J. Intell. Mater. Syst. Struct.* **1996**, *7*, 613–622. [CrossRef]
17. Davis, L.C. Model of magnetorheological elastomers. *J. Appl. Phys.* **1999**, *85*, 3348–3351. [CrossRef]
18. Berasategi, J.; Salazar, D.; Gomez, A.; Gutierrez, J.; Sebastián, M.S.; Bou-Ali, M.; Barandiaran, J.M. Anisotropic behaviour analysis of silicone/carbonyl iron particles magnetorheological elastomers. *Rheol. Acta* **2020**, *59*, 469–476. [CrossRef]
19. Vatandoost, H.; Sedaghati, R.; Rakheja, S.; Hemmatian, M. Effect of pre-strain on compression mode properties of magnetorheological elastomers. *Polym. Test.* **2021**, *93*, 106888. [CrossRef]
20. Syam, T.M.I.; Muthalif, A.G.A.; Salem, A.M.H.; Hejazi, A.A.A. 3D numerical modelling and analysis of a magnetorheological elastomer (MRE). *J. Vibroeng.* **2020**, *22*, 1251–1265. [CrossRef]
21. Asadi Khanouki, M.; Sedaghati, R.; Hemmatian, M. Experimental characterization and microscale modeling of isotropic and anisotropic magnetorheological elastomers. *Compos. Part B Eng.* **2019**, *176*, 107311. [CrossRef]
22. Dargahi, A.; Sedaghati, R.; Rakheja, S. On the properties of magnetorheological elastomers in shear mode: Design, fabrication and characterization. *Compos. Part B Eng.* **2019**, *159*, 269–283. [CrossRef]
23. Sun, S.; Peng, X.; Guo, Z. Study on macroscopic and microscopic mechanical behavior of magnetorheological elastomers by representative volume element approach. *Adv. Condens. Matter Phys.* **2014**, *2014*, 232510. [CrossRef]
24. Xu, H.; Ye, L. Characterization and Simulation of Magnetorheological Elastomer Filled with Carbonyl Iron and NdFeB Particles under Uniaxial Tension, Compression, and Pure Shear Modes. Master's Thesis, The University of Sydney, Sydney, Australia, 2019.
25. Kiarie, W.M.; Jiles, D.C. Modeling of Effect of Particle Size on Macroscopic Behavior of Magnetorheological Elastomers. In Proceedings of the COMSOL Conference 2020, Virtual, 7–8 October 2020.
26. Li, R.; Ma, S.; Jin, F.; Yin, S.; Nian, C.; Xin, N.; Yu, Q. Coupled multi-physics field simulation research of magneto-rheological elastomeric magneto-shear mechanical properties based on COMSOL. In Proceeding of the International Conference on Optical Technology, Semiconductor materials, and Devicesc (OTSMD 2022), Longyan, China, 14 February 2023.
27. Hill, R. Elastic properties of reinforced solids: Some theoretical principles. *J. Mech. Phys. Solids* **1963**, *11*, 357–372. [CrossRef]
28. Hashin, Z. Analysis of Composite Materials. *J. Appl. Mech.* **1983**, *50*, 481–505. [CrossRef]
29. Evesque, P. Fluctuations, Correlation and Representative Elementary Volume (REV) in Granular Materials. *Poudres Grains* **2005**, *11*, 6–17.
30. El Moumen, A.; Kanit, T.; Imad, A. Numerical evaluation of the representative volume element for random composites. *Eur. J. Mech. A/Solids* **2021**, *86*, 104181. [CrossRef]
31. Madi, K.; Forest, S.; Jeulin, D.; Boussuge, M. Estimating RVE sizes for 2D/3D viscoplastic composite materials. In *Matériaux 2006*; HAL Open Science: Dijon, France, 2006; Volume 12, p. hal-00144481.
32. El Moumen, A.; Imad, A.; Kanit, T.; Hilali, E.; El Minor, H. A multiscale approach and microstructure design of the elastic composite behavior reinforced with natural particles. *Compos. Part B Eng.* **2014**, *66*, 247–254. [CrossRef]
33. Moumen AEl Kanit, T.; Imad, A.; Minor, H.E.L. Effect of overlapping inclusions on effective elastic properties of composites. *Mech. Res. Commun.* **2013**, *53*, 24–30. [CrossRef]
34. Kanit, T.; Forest, S.; Galliet, I.; Mounoury, V.; Jeulin, D. Determination of the size of the representative volume element for random composites: Statistical and numerical approach. *Int. J. Solids Struct.* **2003**, *40*, 3647–3679. [CrossRef]
35. Jeulin, D. Morphology and effective properties of multi-scale random sets: A review. *Comptes Rendus Mec.* **2012**, *340*, 219–229. [CrossRef]

36. Willot, F.; Jeulin, D. Elastic behavior of composites containing Boolean random sets of inhomogeneities. *Int. J. Eng. Sci.* **2009**, *47*, 313–324. [CrossRef]
37. Suquet, P. Elements of homogenization for inelastic solid mechanics. *Lect. Notes Phys.* **1987**, *272*, 193–278.
38. Anthoine, A. Derivation of the in-plane elastic characteristics of masonry through homogenization theory. *Int. J. Solids Struct.* **1995**, *32*, 137–163. [CrossRef]
39. Terada, K.; Hori, M.; Kyoya, T.; Kikuchi, N. Simulation of the multi-scale convergence in computational homogenization approaches. *Int. J. Solids Struct.* **2000**, *37*, 2285–2311. [CrossRef]
40. Van Der Sluis, O.; Schreurs, P.J.G.; Brekelmans, W.A.M.; Meijer, H.E.H. Overall behaviour of heterogeneous elastoviscoplastic materials: Effect of microstructural modelling. *Mech. Mater.* **2000**, *32*, 449–462. [CrossRef]
41. Smit, R.J.M.; Brekelmans, W.A.M.; Meijer, H.E.H. Prediction of the large-strain mechanical response of heterogeneous polymer systems: Local and global deformation behaviour of a representative volume element of voided polycarbonate. *J. Mech. Phys. Solids* **1999**, *47*, 201–221. [CrossRef]
42. Okereke, M.I.; Akpoyomare, A.I. A virtual framework for prediction of full-field elastic response of unidirectional composites. *Comput Mater. Sci.* **2013**, *70*, 82–99. [CrossRef]
43. Davis, A.; Onoochin, V. The Maxwell Stress Tensor and Electromagnetic Momentum. *Prog. Electromagn. Res. Lett.* **2020**, *94*, 151–156. [CrossRef]
44. Pao, Y.-H. Electromagnetic forces in deformable continua. *Mech. Today* **1978**, *4*, 209–305.
45. Kim, B.; Lee, S.B.; Lee, J.; Cho, S.; Park, H.; Yeom, S.; Park, S.H. A comparison among Neo-Hookean model, Mooney-Rivlin model, and Ogden model for Chloroprene rubber. *Int. J. Precis. Eng. Manuf.* **2012**, *13*, 759–764. [CrossRef]
46. Ogden, R.; Saccomandi, G.; Sgura, I. Fitting hyperelastic models to experimental data. *Comput. Mech.* **2004**, *34*, 484–502. [CrossRef]
47. Bergström, J. Elasticity/Hyperelasticity. In *Mechanics of Solid Polymers*; Elsevier: Amsterdam, The Netherlands, 2015; pp. 209–307. [CrossRef]
48. Rackl, M. Curve Fitting for Ogden, Yeoh and Polynomial Models. In Proceeding of the Scilab TEC, 7th International Scilab Users Conference, Paris, France, 21–22 May 2015.
49. Rao, D.K.; Kuptsov, V. Effective Use of Magnetization Data in the Design of Electric Machines with Overfluxed Regions. *IEEE Trans. Magn.* **2015**, *51*, 6100709. [CrossRef]

polymers

MDPI

Article

Carbon Black Functionalized with Serinol Pyrrole to Replace Silica in Elastomeric Composites

Federica Magaletti [1], Martina Galbusera [1], Davide Gentile [1], Ulrich Giese [2], Vincenzina Barbera [1,*] and Maurizio Galimberti [1,*]

[1] Department of Chemistry, Materials and Chemical Engineering "G. Natta", Politecnico di Milano, Via Mancinelli 7, 20131 Milan, Italy; federica.magaletti@polimi.it (F.M.); martina.galbusera@mail.polimi.it (M.G.); davide.gentile@polimi.it (D.G.)

[2] Deutsches Institut für Kautschuktechnologie e. V., Eupener Straße 33, 30519 Hannover, Germany; ulrich.giese@dikautschuk.de

* Correspondence: vincenzina.barbera@polimi.it (V.B.); maurizio.galimberti@polimi.it (M.G.)

Abstract: Elastomer composites for dynamic mechanical applications with a low dissipation of energy are of great importance in view of their application in tire compounds. In this work, furnace carbon black functionalized with 2-2,5-dimethyl-1*H*-pyrrol-1-yl-1,3-propanediol (SP) was used in place of silica in an elastomer composite based on poly(styrene-co-butadiene) from solution anionic polymerization and poly(1,4-cis-isoprene) from *Hevea Brasiliensis*. The traditional coupling agent used for silica was also used for the CB/SP adduct: 3,3′-bis(triethoxysilylpropyl)tetrasulfide (TESPT). The composite with the CB/SP + TESPT system revealed a lower Payne effect, higher dynamic rigidity, and lower hysteresis, compared to the composite with CB + TESPT, although the latter composite had a higher crosslinking density. The properties of the silica and the CB/SP + TESPT-based composites appear similar, though in the presence of slightly higher hysteresis and lower ultimate properties for the CB/SP-based composite. The use of CB in place of silica allows us to prepare lighter compounds and paves the way for the preparation of tire compounds with lower environmental impacts.

Keywords: pyrrole methodology; functionalization; elastomer composites; Payne effect; hysteresis

1. Introduction

Tires are of fundamental importance for the mobility of people and goods, and they can play a key role in the development of sustainable transport, which is a pillar of climate action, which is the Sustainable Development Goal n. 13 in the 2030 Agenda for Sustainable Development [1]. The annual global freight and passenger traffic volumes were forecasted by the United Nations to grow by 50% and 70%, respectively, from 2015 to 2030. It is estimated that the cars on the road will be about 2.4 billion [2] and the tires will be 3.4 billion units [3]. Tires contribute around 20 percent of a vehicle's carbon footprint, and about 90% of this contribution comes from the use stage of a tire [4]. The carbon emission of a car tire during its use ranges from 550 kg CO_2 eq to 840 kg CO_2 eq [5] and is mainly due to the rolling resistance (RR), which signifies "the energy consumed per unit distance of travel as a tire rolls under load" [6].

Reinforcing fillers [7–9] are responsible for most of the hysteresis of a rubber compound for tires; hence, of the rolling resistance of a tire, the filler networking causes the dissipation of energy. Precipitated silica [10,11] is the preferred filler for a rubber compound with low hysteresis [12–14]. Indeed, silica is a nanostructured filler, with a high surface area and activity and with silanols on its surface, which allow a reaction with molecules able to favor compatibilization with the hydrocarbon matrix and, most of all, the reaction with elastomer chains. The covalent bond between the elastomer and silica, established via a coupling agent, reduces filler networking [7–9], hence the hysteresis of a compound. A sulfur-based silane, such as bis(triethoxysilylpropyl)tetrasulfide (TESPT), is the coupling agent most widely used [15,16], also at the industrial scale. Its chemical structure is shown in Figure 1.

(a) 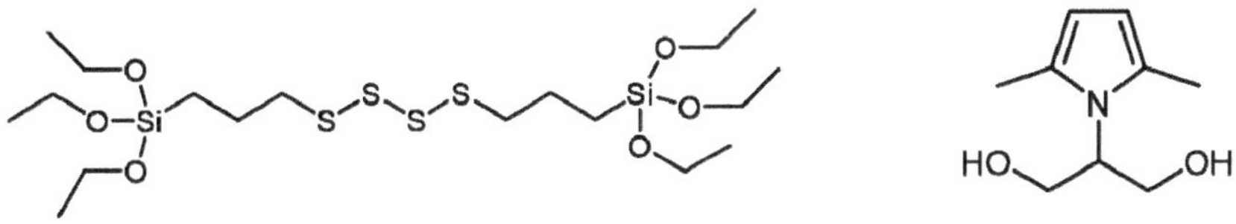 (b)

Figure 1. Chemical structures of bis(triethoxysilylpropyl)tetrasulfide (TESPT) (**a**) and 2-(2,5-dimethyl-1*H*-pyrrol-1-yl)-1,3-propanediol (serinol pyrrole, SP) (**b**).

Silica, however, presents important drawbacks, such as a high compound viscosity, which leads to worse processability and a shorter shelf life, and the need for higher energy and special equipment for mixing. Moreover, silica is corrosive and abrasive, and its adduct with silane is adhesive with the metal surfaces of mixers. Furthermore, the condensation of silanols with TESPT releases ethanol, which is then burnt to carbon dioxide. It has been estimated [17] that 8.44×10^4 tons of CO_2 are released worldwide per year because of the use of TESPT in a tire tread and that the emission value of volatile organic compounds (VOCs) due to the use of coupling agents in the tire industry is 130,000 m^3 [18]. Research for the reduction in VOC emissions is thus essential for the sustainable development of the rubber industry [19–21]. Finally, silica-based compounds have no electrical conductivity, and this is a major drawback for an application such as the one in tire compounds.

The best way to avoid the problems of silica would be to use carbon black [22], which is a nanostructured, high-surface-area reinforcing filler, easily compatible with the elastomer matrix, with lower supramolecular interactions compared to silica and which has been widely used in the rubber world since the beginning of the last century. Moreover, CB has a lower density than silica; hence, its use at the same volume fraction leads to lighter rubber compounds. However, CB cannot establish a chemical bond with elastomer chains, and this leads to remarkable compound hysteresis. The solution lies in the functionalization of CB. However, it is acknowledged [23,24] that the functionalization reactions of CB are troublesome and are often based on dangerous chemicals and harsh reaction conditions Some of the authors reported the pyrrole methodology [24–28] as a simple and versatile method, characterized by a high atom efficiency and a very low amount of waste, to covalently graft a variety of functional groups on a sp^2 carbon allotrope. In brief, a domino reaction occurs [27]: a pyrrole compound (PyC), obtained from the neat reaction of a primary amine with 2,5-hexanedione, undergoes carbocatalyzed oxidation in the benzylic position and the oxidized PyC gives rise to a cycloaddition reaction with the carbon substrate. Adducts of CB were prepared with pyrrole compounds containing sulfur-based functional groups [24], reactive with unsaturated elastomer chains, such as thiol -SH and disulfide -SS- groups. It was shown that CB adducts with sulphurated PyC, used in a hybrid filler system with CB in an elastomer composite based on unsaturated elastomers, led to more efficient crosslinking [29], with an increase in dynamic rigidity of about 20% and to a reduction in hysteresis of about 10% at 70 °C, with similar/better ultimate tensile properties. However, it was not reported if these adducts could replace silica. Adducts of CB were prepared as well with 2-(2,5-dimethyl-1*H*-pyrrol-1-yl)-1,3-propanediol (serinol pyrrole, SP) as PyC [26]. Its chemical structure is shown in Figure 1. The CB/SP adduct was used for the partial replacement of silica in a composite based on unsaturated elastomers, and TESPT was used as the coupling agent. More efficient crosslinking, a lower Payne effect, and lower hysteresis were obtained for the composites with CB/SP. The OH groups appeared to facilitate the interaction with silica and with silane, and hence with the polymer chains. Head space analysis suggested the occurrence of the reaction of silane with CB/SP. However, a complete substitution of silica was not performed. Hence, the objective of substituting silica with a CB/PyC adduct in an elastomeric composite suitable as a tire compound, obtaining the same dynamic–mechanical properties, has not yet been achieved.

The objective of this work was to use a CB/SP adduct as the reinforcing filler of an elastomer composite suitable for a tire-tread compound, at least with the aim of reproducing the properties of a silica-based composite, without using silica. The CB/SP adduct was prepared according to the reported procedure [26] and was characterized by means of thermogravimetric analysis, nitrogen adsorption measurements with the Brunauer, Emmet, and Teller (BET) technique, and with scanning electron microscopy (SEM). Poly(1,4-*cis*-isoprene) from *Hevea Brasiliensis* (Natural Rubber, NR) and poly(styrene-co-butadiene) from anionic solution polymerization were the elastomers used. The CB/SP-based composite was compared with composites containing either pristine silica or pristine CB by using the same volume fraction of the fillers and the same processing procedure. The CB grade was CBN234, and the silica was a precipitated silica with a high surface area (165 m^2/g). TESPT was used in all the composites, assuming it could act as a coupling agent for the silica and CB/SP. The aim of this study was to verify the ability of CB/SP to behave as silica, and the issue of ethanol release was neglected. The crosslinking density and the curing kinetics, the dynamic–mechanical properties in the shear, and in the axial mode and tensile properties were investigated. To evaluate the interaction between the CB and CB/SP fillers with the elastomer chains, the Kraus equation was used by applying the equilibrium swelling method [30,31].

The advantages of using CB in place of silica have been reported above. However, it could be argued that CB is not the ideal filler for the sustainable development of elastomer composites, as it is a material from a fossil fuel. Fortunately, other families of carbon materials, either biobased or circular, are appearing on the scene and could become useful reinforcing materials for elastomeric composites. The most important are the chars from the pyrolysis of end-of-life (EOL) tires [32] or from a biomass (biochar) [33]. It is worth analyzing if these chars can already fulfill the requirements of high-performance elastomer composites. In the scientific literature, rubber composites with either char from EOL tires [34–38] or from biochar [39–42] were reported. The former carbon materials were compared with furnace CB, such as CBN330 [35], N550 [36,37], and N660 [33], also by using combinations of them [34], in elastomers such as EPDM [35] and SBR [34,35,37]. A smaller size [36], a similar [35–37] or larger [34] surface area, and a lower structure [35,37] were observed. It has been said [38] that char from EOL tires is a significant challenge because of its impurities. A reduction in the tensile modulus at 300% [37] was reported. The tensile strength [34] and the tear strength [35,37] were higher for the furnace CB, and the detachment of rubber chains during breakage and lower reinforcement were observed in the presence of the char from tires [36]. In conclusion, it was stated that the char from tires can be used for rubber formulations [37], but CB was found to be a more active filler [34,35]. Biochar was obtained from different sources, such as lignin [41], wood-derived waste biomass [40], and agricultural waste from citrus trees trim [42], and was used in composites based on NR [40] and SBR [41,42], in comparison with CBN330 [40,41]. Biochar is said to have a higher ash content, lower carbon content, larger particle size, and more surface functional groups, [39,40], though the grade from lignin [41] could have a high specific surface area with a mesoporous structure. The elasticity, the strengthening impact at higher elongations, and the tensile strength of rubber samples filled with biochar were found to be lower [40]. Only 25% of CB could be replaced with biochar without losing its mechanical properties [43]. In a nutshell, the following comment was reported for biochar, which also appears to be valid for char from EOL tires: the successful use "relies on the ability to modify its properties" [39]. The pyrrole methodology could indeed modify the properties of these more sustainable carbon materials, making them suitable as fillers for elastomer composites. In light of this perspective, our research aimed at demonstrating the potential of the pyrrole methodology, working with traditional furnace carbon black. The industrial development of the pyrrole methodology and, in particular, of serinol pyrrole has been recently announced by a major player in the tire industry [42].

2. Materials and Methods

2.1. Materials

2.1.1. Chemicals

Acetone, 2,5-hexanedione, and 2-amino-1,3-propanediol were from Sigma-Aldrich. They were used without purification.

The following chemicals were used for the preparation of the elastomeric composites discussed in this paper: TESPT (3,3′-bis(triethoxysilylpropyl)tetrasulfide from Flexys, Ann Arbor, Michigan), TDAE (H&R Olwerke Schindler GmbH, Hamburg, Germany), RIOWAX BM01 (SER S.p.A., Torino, Italy), ZnO (Zincol Ossidi, Bellusco, MB, Italy), stearic acid (Sogis, Milan, Italy), 1,3-dimethyl butyl)-*N*′-Phenyl-p-phenylenediamine (6PPD from Eastman, Kingsport, TN, USA), sulfur (S from Solfotecnica, Cotignola, Italy), *N*-tert-butyl-2-benzothiazyl sulfenamide (TBBS from Lanxess Chemical, Shangai, China), and *N*-(Cyclohexylthio)phthalimide (PVI, Brenntag, S.p.A., Milan, Italy).

2.1.2. Elastomers

Poly(1,4-*cis*-isoprene) was from Hevea brasiliensis (NR) (EQR-E.Q. Rubber, BR-THAI, Eastern GR. Thailand–Chonburi), the trade name was SIR20, and Mooney viscosity (ML (1 + 4)100 °C) was equal to 73 Mooney units (MUs). Solution Styrene Butadiene, not functionalized (SPRINTAN™ SLR 4630 from Trinseo, Milano, Italy), was used. The chemical composition of the S-SBR was: Styrene 25%, Butadiene 75%, and TDAE 37.5 phr. The butadiene comonomer had a vinyl content of 63%. Other properties: Tg = −29 °C; Mooney viscosity (ML (1 + 4) 100 °C) = 55 MU.

2.1.3. Fillers

Carbon black N234 was provided by Birla Carbon (America, USA, Atlanta). The following data were reported in the technical data sheet: Oil absorption number (OAN): 125 mL/1000, nitrogen specific surface area (NSA): 119 m^2/g, and statistical thickness surface area (STSA): 112 m^2/g.

White micropearls of silica ZEOSIL 1165MP were from Solvay. The data in the technical data sheet were as follows: Specific surface area: 140–180 m^2/g, loss on drying (2 h @ 105 °C) $\leq$ 8.0%, and soluble salts (as Na_2SO_4) < 2.0%. In this work, the surface area was determined by using the BET method. Samples were evacuated at 200 °C for 2 h and N_2 adsorption isotherms were recorded at 77 K in a liquid nitrogen bath by using a MICROACTIVE TRISTAR ® II PLUS apparatus (Micromeritics Instrument Corporation, Norcross, GA, USA). The specific surface area (SSA) was 160 m^2/g.

2.2. Preparation of SP and CB/SP Adduct

2.2.1. Synthesis of 2-2,5-Dimethyl-1*H*-pyrrol-1-yl-1,3-propanediol (SP)

The key points in the synthesis of SP are: the absence of solvents, to reduce the waste; the absence of acidic substances, to prevent the oligomerization of the pyrrole compound [44]; and the removal of water, to shift the equilibrium to SP.

In a round-bottom flask, carefully washed and dried, and equipped with a condenser (used to avoid the loss of hexanedione) and a magnetic stirrer, 16.23 g of serinol (0.1781 mol) was suspended in 21 mL of 2,5-hexanedione (0.1780 mol, 1 eq); the temperature was raised to 150 °C and stirring was performed for 2 h. The condenser was then removed, to remove water from the reaction environment, and the reaction mixture was stirred for a further 30 min. The product was collected as a brown, viscous liquid without any further purification (27.73 g, 0.1639 mol). The yield was calculated by using the following expression: 100 × (weighed mass of SP)/(theoretical mass of SP). It was calculated to be about 92%.

^1H NMR ($CDCl_3$, 400 MHz); δ (ppm) = 2.27 (s, 6H), 3.99 (m, 4H); 4.42 (quintet, 1H); 5.79 (s, 2H). ^{13}C NMR (DMSO-d_6, 100 MHz); δ (ppm) = 127.7, 105.9, 43.72, 71.6, 61.2, 13.9.

2.2.2. Preparation of Adducts of CB N234 with SP

The objective was the preparation of a covalent adduct of CB with SP. As mentioned in the introduction, covalent bonds are formed though a domino reaction, which begins with the oxidation of the pyrrole compound. Hence, a key point of the preparation is the presence of oxygen, which is obtained by continuously fluxing air in the reaction flask.

A total of 15 g of SP and 200 mL of acetone was poured in a 250 mL round-bottom flask. The mixture was sonicated for 5 min to promote the complete solubilization of the pyrrole compound in the solvent. In a 1 L round-bottom flask, 150 g of Pristine CB N234 was weighted. The solution of SP and acetone was poured in the 1 L round-bottom flask and 300 mL of pure acetone was added. This was made to completely cover the powder of CB N234. The mixture was sonicated for 20 min. The solvent was then removed under reduced pressure at 40 °C and the resulting dry powder was poured into an aluminum backing tray and put in a preheated oven, at 160 °C for 2 h, to perform the functionalization reaction.

After the reaction, the CB-SP adduct was allowed to reach room temperature.

2.3. *Preparation of Elastomer Composites*

Elastomer Composite

Formulations are shown in Table 1. The quantities of ingredients are expressed in parts per hundred rubber (phr). The reference composite was a composite with silica as the only filler. Silica was completely substituted either with CBN234 or with CB/SP. The amount of CB was calculated to have the same volume fraction of silica. The amounts of the other ingredients were the same in all the composites. The processing procedure is shown in Figure 2.

Table 1. Formulation of S-SBR/NR-based composites with CB/SP.

Ingredient [a]	Silica	CB N234	CB/SP
S-SBR 4630	70	70	70
NR (SIR—20)	30	30	30
TESPT	5.2	5.2	5.2
Silica	70	0	0
CB N234	5.2	65.2	5.2
CB/SP	0	0	60

[a] Other ingredients (phr): TDAE 5, riowax BM01 1, stearic acid 2, ZnO 2.5, 6PPD 2, sulfur 2, TBBS 1.8, and N-cyclohexylthiophthalimide 0.5.

The mixing was performed in a Brabender®-type internal mixer with a chamber volume of 55 cm^3. The so-called non-productive mixing was performed as follows: The rubber, reduced into crumbs, was introduced into the mixer, the temperature was brought to 140 °C, and mastication was performed for 1 min, with rotors rotating at 60 rpm. The filler and half of the amount of ingredients, such as TDAE, stearic acid, wax, and TESPT, was added, performing the mixing for 1.5 min at 140 °C. The remaining amount of the same ingredients was then added, and mixing was carried out for a further 1.5 min, at the same temperature. Ingredients such as wax, ZnO, 6PPD, and CBN234 were then added, and mixing was performed at the same temperature for 2 min. The composite discharged from the internal mixer was passed 5 times through a two-roll mill operating at 50 °C, with the front roll rotating at 30 rpm and the back roll rotating at 38 rpm, with a 1 cm nip between the rolls. The so-called productive mixing was performed as follows: The masterbatch obtained from the non-productive mixing was fed to the internal mixer at 60 °C and mixed for 1 min. The vulcanization ingredients, sulfur and TBBS, were added and mixing was performed for a further 2 min. The composites were discharged and finally homogenized by passing them 5 times through a two-roll mill operating at nominal room temperature, with the front roll rotating at 30 rpm and the back roll rotating at 38 rpm, with a 1 cm nip between the rolls.

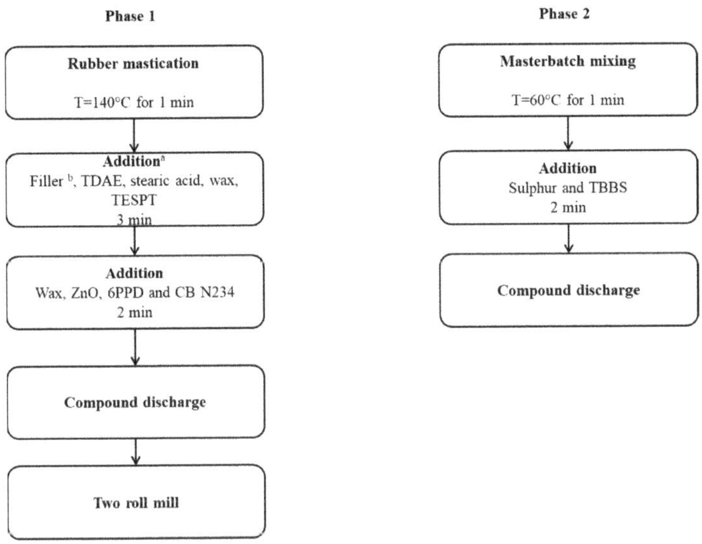

Figure 2. Block scheme of the processing procedure for the composites of Table 1. (a) for details see text. (b) Filler: Silica or CB N234 or CB/SP.

2.4. *Characterization Techniques*

2.4.1. Thermogravimetry Analysis (TGA) of CB/SP Adducts

Thermogravimetry analyses were performed with a TGA TA instrument Q500. The following method was adoperated: heating under air flux from 30 °C to 300 °C at 10 °C/min, isothermal step at 300 °C for 15 min, heating to 550 °C at 20 °C/min, isothermal step for 15 min, heating to 900 °C at 10 °C/min, isothermal step for 3 min, and final isothermal step for 30 min. In this study, the amount of SP in the CB/SP adduct is expressed in parts per hundred carbon (phc), which are estimated by using Equation (1):

$$\text{phc} = 100 \cdot \frac{x}{(100 - x)} \tag{1}$$

where x = (weight loss)$_{150\ °C–900\ °C}$.

The yield of the functionalization process, Y_f, from the weight of reactants to the estimation of the mass loss of the CB/SP adduct via TGA was calculated by means of Equation (2):

$$Y_f(\%) = 100 \cdot \frac{ml_w}{ml_{uw}} \tag{2}$$

The SP mass % in CB/SP adducts after washing is ml_w, while the SP mass % in CB/SP before washing is ml_{uw}.

2.4.2. Brunauer–Emmett–Teller (BET)

Specific surface area analyses were carried out via the Micromeritics TriStar 3000 instrument. Nitrogen was used as the gas and the method used followed ASTM 6556 [45].

Total surface area (NSA: nitrogen surface area) was determined in the $0.05 < p/p_0 < 0.2$ range, while the external surface area (STSA: specific thickness surface area) was determined in the p/p_0 range from 0.2 to 0.5.

2.4.3. Crosslinking

A rubber process analyzer (Monsanto R.P.A. 2000, Alpha Technologies Hudson, Hudson, OH, USA) was adoperated to perform crosslinking. A total of 5 g of rubber composite was weighed and put in the rheometer. Measurements were carried out at a frequency of 1.7 Hz and an oscillation angle of 6.98%. The sample, loaded at 50 °C, underwent a first strain sweep (0.2–25% strain) to cancel the thermo-mechanical history of the rubber composite; it was then maintained at 50 °C for 10 min and then underwent another strain sweep at 50 °C to measure the dynamic–mechanical properties at low deformations of the uncured sample. The crosslinking reaction was then carried out at 170 °C for 10 min. A torque–time curve was created. The minimum achievable torque (M_L), the maximum achievable torque (M_H), the time required to achieve a torque value equal to M_L + 1 (t_{S1}), and the time required to reach 90% of the maximum torque (t_{90}) were measured.

2.4.4. Structure of the Crosslinking Network

Following the procedure described in an article by the same authors [46], the samples were swollen in n-heptane for two days in a nitrogen atmosphere to allow the diffusion of the reagents. The n-heptane was then removed and the samples were washed with light petroleum and dried overnight at room temperature under reduced pressure. The samples were then immersed in toluene (200 mL) and placed in a glass tube under nitrogen. The tube was then sealed and left in the dark for 72 h, the time required for equilibrium swelling to be achieved. After this time, the samples were dried by blotting with filter paper. Finally, the samples were quickly sealed in a clean container and weighed. The samples were then dried under vacuum at 70 °C for 24 h to remove the solvent and weighed again in the dried state. This provides the amount of sorbed solvent and the weight of the dry network.

To calculate the crosslinking network, we used the Flory–Rehner equation (Equation (3)):

$$\nu_e = \frac{-\left[\ln(1 - V_r) + V_r + \chi_1 V_r^2\right]}{\frac{V_1\left(V_r^{1/3} - V_r\right)}{2}} \tag{3}$$

where ν_e is defined as the effective number of chains in a real network per unit volume; χ_1 is the polymer solvent interaction parameter; and V_1 is the molecular volume of the solvent. Vr is defined as the volume fraction of the polymer in a swollen network in equilibrium with pure solvent. The mathematical definition is reported in Equation (4):

$$V_r = \frac{\frac{m_{dr}}{\rho_{dr}}}{\frac{m_{dr}}{\rho_{dr}} + \frac{m_s}{\rho_s}} = \frac{V_{dr}}{V_{dr} + V_s} \tag{4}$$

where m_{dr} is the weight of dry rubber, ρ_{dr} the density of dry rubber, m_s *is* the weight of the absorbed solvent, and ρ_s is the density of the solvent. For the mono- and di-sulfidic crosslink measurements, the procedure followed was very similar.

A total of 100 mg of crosslinked composite was placed in a 20 mL beaker with 100 mL of heptane. The sample was left to stand for 24 h. After this time, 3.8 mL of propanethiol and 4 mL of piperidine were added, and the mixture was left at room temperature for 2 h. The mixture was then washed 3 times in 50 mL of heptane and then filtered. The solid was left in heptane for a further 24 h and then washed in light petroleum. The sample was filtered under vacuum and dried under reduced pressure for 24 h. The procedure was identical to that described above. The correction for the presence of the filler was taken into account.

2.4.5. Dynamic–Mechanical Analysis in the Shear Mode: The Strain Sweep Test

The shear dynamic–mechanical characteristics of rubber composites were evaluated by performing strain sweep tests in a rubber process analyzer (Monsanto R.P.A. 2000). As reported in Section 2.4.3, the crude sample was subjected to a first strain sweep, then held at 50 °C for ten minutes, followed by another strain sweep at 50 °C. Data from the second strain sweep were collected and are reported in the text below to discuss the behavior of uncured samples. The crosslinking was then carried out, as reported in Section 2.4.4, and the shear dynamic–mechanical properties of the cured samples were then assessed after 20 min at 50 °C using a 0.2–25% strain sweep at a frequency of 1 Hz. Shear storage and loss moduli (G′, G″), and subsequently Tan δ, were measured characteristics.

All the tests performed with the rubber process analyzer were kept under quality control, determining the repeatability and the reproducibility of the measurements [47]. Data reported in the manuscript were characterized by a variation coefficient in the range from 0.5% to 1%.

2.4.6. Dynamic–Mechanical Analysis in the Axial Mode

The elastomeric compound was rolled up to obtain a long cylinder. This cylinder was then cut into smaller cylinders and vulcanized (at 170 °C for 20 min) to produce cylindrical test pieces with dimensions of 25 mm in length and 12 mm in diameter. An Instron dynamic device in the traction–compression mode was employed to perform dynamic mechanical measurements, and was maintained at the predetermined temperatures (10, 23, and 70 °C) throughout the entire experiment. The cylinder was pre-loaded to a 25% longitudinal deformation with respect to the original length. The compression was subjected to a dynamic sinusoidal strain in compression with an amplitude of around 3.5% regarding the length under pre-load, at a frequency of 100 Hz. The values of the dynamic storage modulus (E′), loss modulus (E″) and loss factor (tan δ) were calculated as the ratio between E″ and E′.

All the tests performed with the rubber process analyzer were kept under quality control, determining the repeatability and the reproducibility of the measurements [47]. Data reported in the manuscript were characterized by a variation coefficient in the range from 0.5% to 1%.

2.4.7. Tensile Test

Standard dumbbells made from 10 cm by 10 cm by 1 mm vulcanized compound plates were used to perform tensile tests at room temperature with a Zwick Roell Z010 and an optical extensometer. Measurements were performed at 1 mm/min. Stresses at different elongations (σ_{50}, σ_{100}, and σ_{300}), stress at break (σ_B), elongation at break (ε_B), and the energy required to break were measured according to Standard ISO 37/UNI 6065 [48].

2.4.8. Digital Filler Dispersion Analysis through Transmission Electron Microscopy

The digital image analysis of filler dispersion through TEM was carried out following three steps: (1) the vulcanized compound was sliced with a diamond cryo-cutter under liquid nitrogen to obtain sections with thicknesses around 100 nm, (2) sections were observed in the transmission electron microscope and 10 images were acquired with a digital camera for each sample, and (3) the acquired images were visually analyzed to determine the amount of undispersed filler particles.

TEM characterizations were performed with a TEM, LIBRA® 120, Zeiss, Oberkochen (Baden-Württemberg, Germany), with an acceleration voltage of 120 kV.

Ultra-thin sections must be prepared from the sample material in order to analyze it using the TEM. For this purpose, a cryo-ultramicrotome (type: Leica EM FC 6, Leica Microsystem, Wetzlar, Germany) using a diamond knife (type: Diatome 35°) at a temperature of −80 °C was used to produce sections with a width approx. 100 µm and a thickness approx. 100 nm. The preparation was carried out using the wet-cutting method, in which the sections are floated on a mixture of dimethyl sulphoxide (DMSO) and water (50/50) and transferred to a copper mesh using a loop. To stabilise the sections in the transmission electron microscope, they were placed on a 400 mesh copper carrier coated with poly(vinyl formal) (Plano).

2.4.9. Scanning Electron Microscopy (SEM)

The SEM investigation was carried out using a Zeiss EVO 50 EP SEM (Zeis Vision Care, Castiglione Olana, Italy) coupled with an EDS spectrometer (Bruker Quantax 200 6/30, Bruker, Billerica, MA, USA). CB powders were applied on aluminum stubs with the aid of a conductive carbon bioadhesive. Although CB is already conductive, the samples were coated with gold using an Ewards S150B sputter coater (Perkin Elmer, Milan, Italy) and evaluated.

3. Results and Discussion

3.1. Preparation and Characterization of SP and CB/SP Adduct

Serinol pyrrole was prepared, as previously reported [26], by mixing 2-amino-1,3-propanediol (serinol) with 2,5-hexanedione and heating at 150 °C for 2 h 30 min, achieving a yield of 92%. Details are in the Materials and Methods Section.

The CB/SP adduct was prepared as summarized in the block diagram in Figure 3.

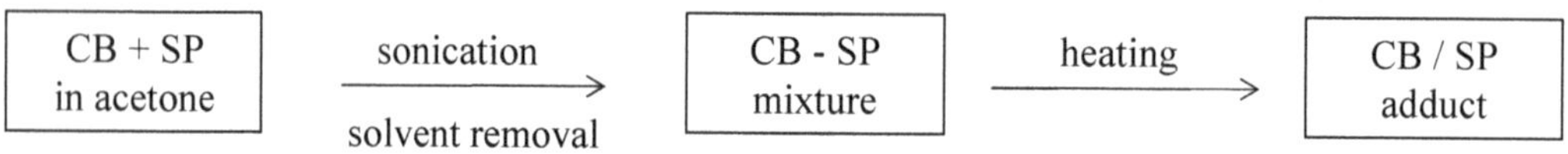

Figure 3. Block diagram for the preparation of the CB/SP adduct.

In brief, CB and SP were mixed in acetone, and a homogenous mixture was formed via sonication. This procedure is used at the lab scale, but both the organic solvent and sonication can be avoided at the pre-industrial scale [42], where a spray dryer [49] can be used. The physical mixture, obtained via the removal of the solvent at a reduced pressure, was heated at 150 °C for 2 h. As reported in the introduction, a domino reaction occurred [27], with the formation of a covalent CB/SP adduct, bearing OH functional groups. Research is in progress to reduce the temperature of the reaction, for example, by using a pre-oxidized pyrrole compound.

3.1.1. Thermogravimetric Analysis (TGA)

The amount of the organic modifier in the CB/SP adduct was estimated by means of TGA, performed on both pristine CB N234 and on the CB/SP adduct. The analyses were performed in a temperature range from 30 °C to 900 °C, with a heating rate of 10 °C/min under N_2. A TGA traces are reported in Figures S1 and S2 in the Supplementary Materials, and the data are shown in Table 2.

Table 2. Mass losses of pristine CB and CB/SP adduct from TGA.

Sample	Temperature Range					Amount of SP (phc)
	T < 150 °C	150 °C < T < 550 °C	550 < T < 900 °C	T > 900 °C	Residue	
SP	4.3	93.9	0.3	0.3	2	
CB N234	0.8	1.2	1.1	96.7	0.2	-
CB/—P—UW [a]	0.4	7.2	1.9	90.1	0.4	10
CB/—P—W [b]	0.5	5.3	1.7	90.6	1.9	6.8

[a] UW: unwashed. Before acetone extraction; [b] W: washed. After acetone extraction.

The mass loss peak of SP occurred at about 250 °C, as is visible in Figure S2 in the Supplementary Materials. The loss below 150 °C can be attributed to water, considering the hydrophilic character of SP (easily soluble in water). A three-step decomposition profile was observed for the carbon materials. The first mass loss below 150 °C is usually attributed to low molar mass substances, such as water or acetone used for washing the adduct [23,25,26]. The mass loss at 150 °C < T < 900 °C can be attributed to the alkenylic defects of the carbon substrate and to the organic modifier. The carbon substrate accounts for the mass loss at T > 900 °C. The residue is due to the inorganic components. For CB/SP, the peak of the largest mass loss below 900 °C was at about 450 °C (see Figure S2 in the Supplementary Materials). A small peak was visible in the TGA trace of the unwashed CB/SP sample at about 330 °C, whereas it was absent from the trace of the washed adduct. These findings indicate that the CB/SP adduct is based on a strong interaction between the pyrrole compound and the carbon substrate, the covalent bond arising from the cycloaddition, the last step of the domino functionalization reaction [27]. The CB/SP adduct was analyzed before and after acetone washing and the content of SP in the adduct was expressed in parts per hundred carbon (phc), calculated by applying Equation (1) (see the Materials and Methods Section). In the unwashed sample, the SP content was estimated to be 10, as expected, whereas 6.8 phc of SP were found in the washed sample. This value led us to evaluate a functionalization yield of 68% by applying Equation (2) (see the Materials and Methods Section). A very low amount of residue was observed. A nominal value of 10 phc of PyC in the adduct corresponds to 108 mmol OH/100 g of adduct. In the case of the silica grade used in this study, the amount of OH groups can be calculated [17] to be 176 mmol(OH)/100 g(silica), or 44.2 mmol(OH)/100 g(silica) by considering the most reactive isolated silanols. Hence, the functionalization of CB with SP led to the introduction of at least a comparable amount of OH groups.

3.1.2. Brunauer–Emmett–Teller (BET) Surface Area of CB Samples

The total surface area (NSA) and the external surface area (STSA) of pristine CBN234 and washed CB/SP adduct were determined by means of nitrogen adsorption measurements, performed with the BET technique. Data are shown in Table 3.

Table 3. BET total and external surface areas for pristine CB and CB/SP.

Sample	Total Surface Area (NSA) [m^2/g]	External Surface Area (STSA) [m^2/g]
CBN234	112	108.8
CB/SP	86.2	71.0

The functionalization with SP led to an appreciable reduction in both the total and the specific surface areas.

3.1.3. Scanning Electron Microscopy

The SEM investigation was carried out on gold-sputtered samples of CB and CB/SP, at different magnifications (250× and 2000×). Sixteen micrographs were taken for each sample and representative images were selected and are visible in Figure 4. Pristine CB images are shown in Figure 4a for the lower magnification and in Figure 4A for the higher magnification; CB/SP images are presented in Figure 4b for the lower magnification and in Figure 4B for the higher magnification.

The images reveal the higher distribution of aggregate sizes for CB/SP, which shows small but also large aggregates (up to more than 280 microns), compared to pristine CB (maximum size of about 80 microns). It can be said that the OH on the CB surface favors the aggregation of CB particles.

Figure 4. SEM images at lower (250×) and higher magnitudes (2000×). Pristine CB: (**a**) lower magnitude, (**A**) higher magnitude; CB/SP: (**b**) lower magnitude, (**B**) higher magnitude.

3.2. Elastomer Composites with the CB/SP Adduct

The unwashed CB/SP adduct was used for the preparation of the elastomer composite. The selection of the unwashed sample was made in view of possible industrial development; indeed, the adduct would not be washed in an industrial process. A traditional formula for a tire-tread compound was used, with S-SBR and NR as the elastomers and 70 phr of silica in the reference silica composite. CB and CB/SP adduct were used at the same volume fraction, and this allowed us to reduce the density of the compound from 1.21 g/cm^3 to 1.15 g/cm^3.

3.2.1. Crosslinking

Sulfur-based crosslinking was performed at 170 °C for 20 min. Values of minimum modulus, M_L, maximum modulus, M_H, induction crosslinking time, t_{s1}, optimum crosslinking time, t_{90}, and curing rate are visible in Table S1 in the Supplementary Materials. The crosslinking curves are shown in Figure 5.

A higher M_L value was obtained for the silica composite, as expected, considering that M_L is an index of the viscosity of the composite. In the presence of composites with the same elastomer matrix, the same filler volume fraction, and the same amount of so-called small ingredients, different M_L values should be ascribed to the extent of the filler network, which is reasonably higher in the presence of a filler with pronounced supramolecular interactions, such as silica. The M_H value is due to both the crosslinking and the filler networks. Indeed, the amplitude achieved in the shear test (see the Materials and Methods Section) is not sufficient to completely remove the filler network. The M_H and $(M_H - M_L)$ values were remarkably higher for the silica composite. It is worth commenting on the differences between the CB and the CB/SP composite. Faster curing was obtained with the CB composites, as expected. CB/SP produced the lowest values for induction and optimum time of vulcanization and the highest value for curing rate. These findings are in line with what has already been reported [26] and could be due to sp^3 nitrogen atoms on the CB surface.

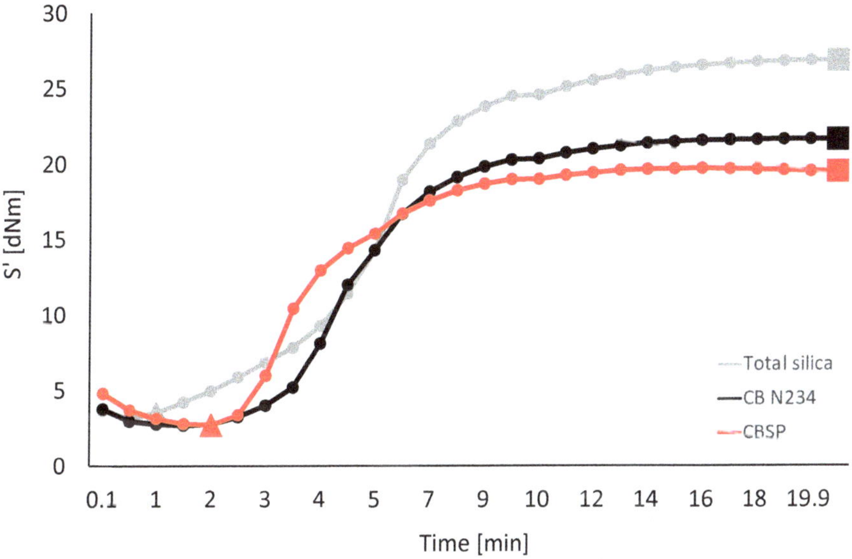

Figure 5. Torque vs. time measured during the crosslinking of the composites in Table 1. The triangles identify the experimental value of M_L; the squares identify the experimental value of M_H.

The structure of the crosslinking network was studied by performing swelling experiments, as explained in the Materials and Methods Section. Values of total crosslinks, mono- and di-sulphides, and poly-sulphides are presented in Table 4.

Table 4. Structure of the crosslinking network of composites in Table 1.

	Composites		
	Silica	**CB**	**CB/SP**
Total crosslinks (mol/g)	3.24	4.92	3.29
Mono- and di-sulphides (% mass)	47.90	46.10	47.90
Poly-sulphides (% mass)	52.10	53.90	52.10

The same values were obtained for the silica and the CB/SP composites, whereas a higher number of total crosslinks was achieved with the CB composite. In a previous study by some of the authors [46], it was shown that the total number of crosslinks increased with the surface area of the sp^2 carbon allotrope. Differences in the distribution of sulfur bridge lengths among the three composites cannot really be appreciated, though the higher amount of mono- and di-sulfides in the case of CB-SP could also be attributed to the presence of the basic nitrogen on the CB surface, which is expected to interact with the sulfur chains.

3.2.2. Shear Dynamic–Mechanical Properties

Dynamic–mechanical properties were determined in the shear mode by applying the procedure described in the Materials and Methods Section, exploring a strain amplitude ranging from 0.2% to 25% and measuring the storage (G′) modulus and the loss (G″) modulus. Values of $G'_{\gamma min}$, $G'_{\gamma max}$, ΔG′, $\Delta G'/G'_{\gamma min}$, G''_{max}, and $Tan\delta_{max}$ (Tan delta = G″/G′) are reported in Table 5, while G′ vs. strain and Tan delta vs. strain curves are shown in Figure S3 and Figure S4, respectively, in the Supplementary Materials.

Table 5. Dynamic–mechanical properties of S-SBR4630 compounds obtained through strain sweep experiments.

	Silica	CB	CB/SP
$G'_{\gamma min}$ [Mpa]	3.93	3.62	3.35
$G'_{\gamma max}$ [MPa]	1.66	1.38	1.27
ΔG′ [a] [MPa]	2.27	2.24	2.07
$\Delta G'/G'_{\gamma min}$	0.58	0.62	0.62
G''_{max} [MPa]	0.49	0.56	0.39
$tan\delta_{max}$	0.18	0.24	0.20

[a] $\Delta G' = G'_{\gamma m—n} - G'_{\gamma max}$.

Lower values of $G'_{\gamma min}$ and ΔG′ were obtained with the CB composites, in particular with the CB/SP composite. The values of $\Delta G'/G'_{\gamma min}$ appear to be similar for the three composites. ΔG′ and $\Delta G'/G'_{\gamma min}$ values provide indication of the Payne effect [47–49], which is essentially due to the filler networking phenomenon and causes the dissipation of energy. The higher values of $G'_{\gamma max}$ for the silica composite appear to be in line with the higher M_H value. The $tan\delta_{max}$ values are similar for the silica and the CB/SP composites, whereas they are is appreciably higher for the CB composite. These results suggest that the functionalization of CB with SP does play a role, favoring the reduction in the filler networking phenomenon, whose result is definitely not due to the presence of TESPT alone. The lower Tan delta values are also in line with the slightly better dispersion of CB/SP in the rubber matrix, documented by TEM analysis.

3.2.3. Dynamic–Mechanical Properties in the Axial Mode

Axial dynamic–mechanical properties were measured in compression, as described in the Material and Methods Section, determining E′, E″, and Tan delta (E″/E′) values at 10 °C, 23 °C, and 70 °C, respectively. The data are presented in Table 6, and the values at the different temperatures of E′ and Tan delta are plotted vs. the temperature in Figure 6 (Figures 6a and 6b, respectively).

Table 6. Axial dynamic–mechanical properties in the axial mode of composites in Table 1.

	T [°C]	Silica	CB	CB/SP
E′ [MPa]	10	15.30	14.89	16.52
	23	12.08	11.53	12.87
	70	8.17	7.99	8.24
E″ [MPa]	10	7.87	8.02	8.22
	23	4.66	4.85	4.96
	70	1.23	1.52	1.42
Tanδ	10	0.52	0.54	0.50
	23	0.39	0.42	0.39
	70	0.15	0.19	0.17

(a)

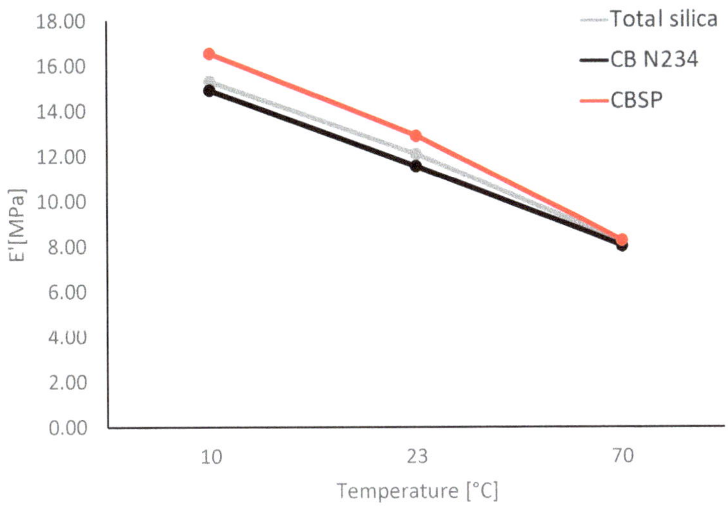

(b)

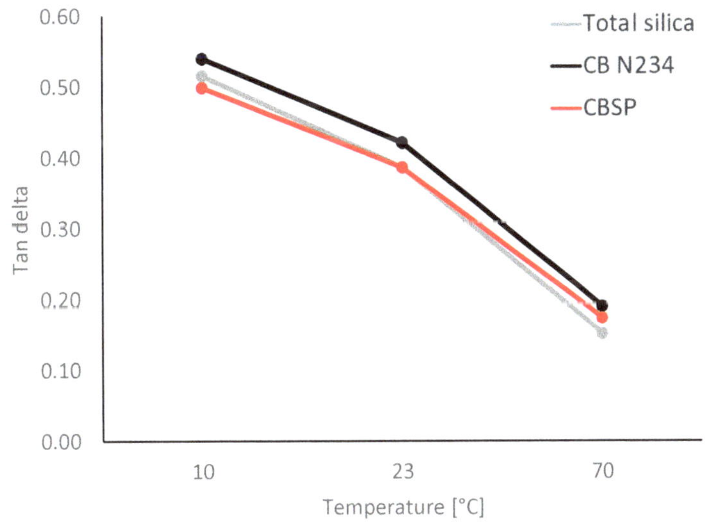

Figure 6. Storage modulus E′ vs. temperature (**a**) and Tan delta vs. temperature (**b**) for composites in Table 1.

The composite with CB/SP revealed the highest dynamic rigidity at all the temperatures. The CB composite has the lowest dynamic rigidity and the highest hysteresis. The procedure adopted for the measurements allows us to exclude the influence of the filler network. Indeed, a pre-strain was applied to remove the interactions involving filler aggregates. Moreover, the higher E′ values obtained with CB/SP compared with CB cannot be attributed to the total crosslinks, which are higher for the latter composite. Hence, the system formed by CB/SP + TESPT appears capable of effectively reinforcing the elastomer matrix, establishing a stable interaction between CB and the elastomer chains. The Tan delta values, which are an indication of the composite's hysteresis, are similar for the silica and the CB/SP composites. However, a crossover of the lines due to silica and to CB/SP can be observed in Figure 6. It is known that a silica-based composite, compared to a CB-based composite, has lower hysteresis at a high temperature (low frequency) and similar/higher hysteresis at a low temperature (high frequency). This is typical of silica-based composites and promotes road traction and low rolling resistance when the composites are used in tire treads. This behavior can be explained by the chemical bond between silica and the elastomer chains that arises from the reaction of silanols with silane. As reported above, the amount of OH on the CB surface was 108 mmoles/100 g of CB/SP adduct, an amount which is at least comparable to that of the OH groups in silica. The slightly different dependence of hysteresis on the temperature could be explained by the different reactivity of the OH groups from SP compared with silanols. It is worth mentioning that the highest dynamic rigidity would allow us to reduce the amount of filler in the CB/SP composite, thus obtaining a reduction in hysteresis. It is worth highlighting the difference between the CB and CB/SP composites. Also, these results indicate that CB/SP is able to interact with TESPT and then with elastomer chains.

3.2.4. Tensile Properties

Tensile properties were measured by means of quasi-static measurements, measuring the stresses $\sigma_{50\%}$ $\sigma_{100\%}$ $\sigma_{200\%}$ $\sigma_{300\%}$ σ_{break}, the elongation at break, ε_{break}, and the energy at break. The data are reported in Table 7 and the stress–strain curves are presented in Figure S5 in the Supplementary Materials.

Table 7. Tensile properties of S-SBR 4630 compounds obtained through stress–strain experiments.

	Composite		
	Silica	**CB**	**CB/SP**
$\sigma_{50\%}$ **[MPa]**	2.19	2.29	2.04
$\sigma_{100\%}$ **[MPa]**	3.83	4.74	3.61
$\sigma_{200\%}$ **[MPa]**	9.60	12.86	8.71
$\sigma_{300\%}$ **[MPa]**	17.08	-	14.50
$\sigma_{300\%}/\sigma_{100\%}$	4.45	-	4.01
σ_{break} **[MPa]**	18.53	15.37	15.13
ε_{break} **[%]**	320.94	229.79	313.30
Energy break [MJ/m^3]	25.71	15.23	21.66

The silica composite reveals the highest stresses at all the strains and the best ultimate properties. The tensile properties of the CB/SP composite are not that far off and are better than those of the CB composite. These findings confirm that the CB/SP + TESPT system can promote the reinforcement of an elastomer matrix. They also suggest that the formulation of the composite should be properly tuned, as this has been performed in recent decades for silica-based composites.

3.2.5. Transmission Electron Microscopy for the Distribution of Filler Analysis

Transmission electron microscopy (TEM) analysis was performed on the three vulcanized composites, with the aim to investigate the filler dispersion, 12 micrographs were taken for each composite, at low and high magnifications, and representative images were selected. TEM micrographs can be seen in Figure 7 for the silica composite (Figure 7a,A), the CB composite (Figure 7b,B), and the CB/SP composite (Figure 7c,C).

According to the formula of the three compounds in all images, the two phases of the polymer can be observed. The light phase (white to light gray) corresponds to NR, whereas the dark-gray phase is to be assigned to S-SBR. The black agglomerates of CB are distributed in the direction of the SBR phase, which can be explained by the chemical compatibility of CB with the aromatic systems in the SBR (Figure 7b,B,c,C). Furthermore, an even distribution of small aggregates, with a very low amount of agglomerates as well primary particles (size of appr. 25–35 nm) and absence of flaws can be observed for the silica and the CB composites, both at high and low magnifications. In addition, the micrographs of the CB/SP composite material show the presence of a small number of large particles (size of appr. 250 nm), which are not part of the filler; nevertheless, a clear identification cannot be made from the image itself (Figure 7c,C). The TEM images show only a small part of the sample in two dimensions (2D). But, overall, small particles with a size of approximately 30 nm are well dispersed in the SBR phase (dark gray). A lower $\Delta G'$ was observed for the CB/SP system in comparison with the others in the amplitude sweep experiments (Table 4 and Figure S3 in the Supplementary Materials). This finding is in line with the slightly better dispersion shown in the TEM image (Figure 7).

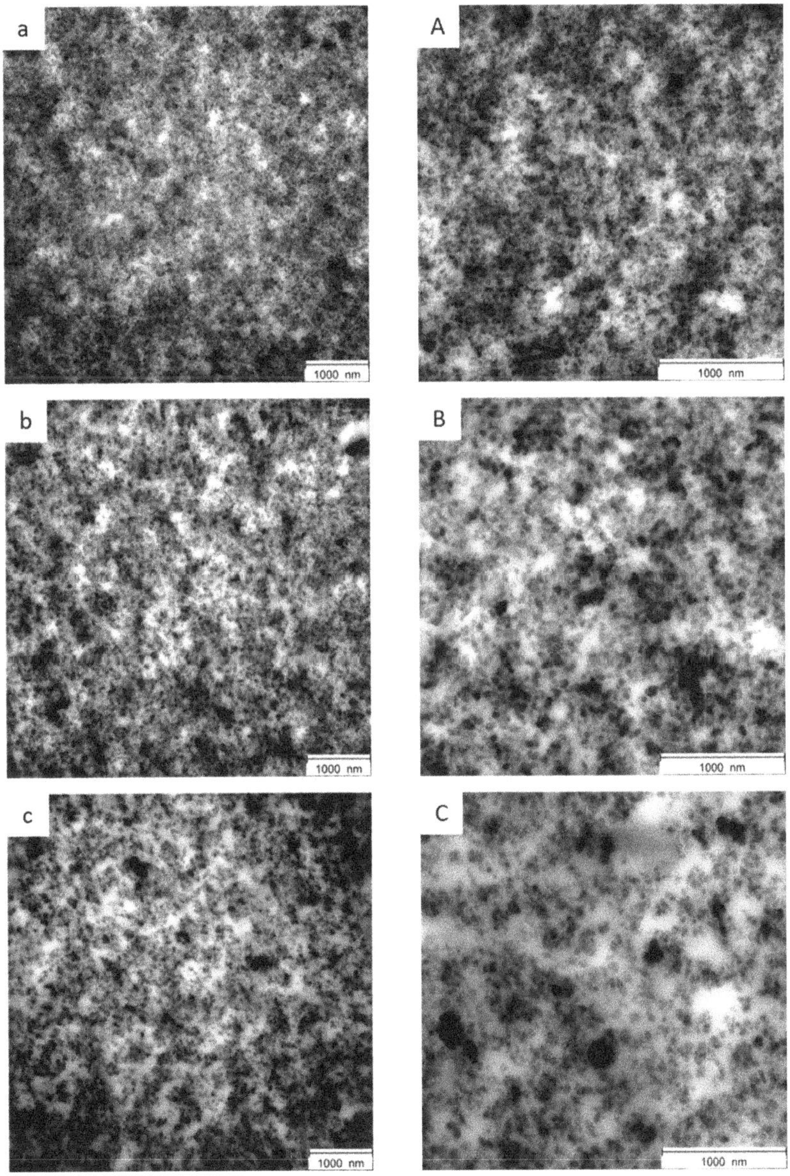

Figure 7. TEM images at lower and higher magnitudes. Silica (**a**) lower magnitude, (**A**) higher magnitude; CBN234: (**b**) lower magnitude, (**B**) higher magnitude; CB/SP: (**c**) lower magnitude, (**C**) higher magnitude.

3.2.6. Comparison of the Behavior of CB/PyC Adducts in Place of Silica

As reported in the introduction, CB/PyC adducts with PyC containing sulfur atoms were used as reinforcing fillers in elastomer composites based on S-SBR and NR [24,29]. The objective in common with the work reported in this manuscript is the preparation of composites with dynamic–mechanical properties similar to those of a silica-based compound, hence suitable for a tire tread. It is worthwhile to compare the results obtained with the two families of pyrrole compounds. Normalized data are shown in Table S2 in the Supplementary Materials: the composite based on pristine CB was taken as a reference and the relative values of dynamic–mechanical (in the shear and axial modes) and tensile properties were calculated.

Compared with the composite with pristine CB, the PyC grafted on CB led to more efficient curing (lower induction time and higher curing rate), to the reduction in $\Delta G'$, to higher dynamic rigidity (E' values at both 10 °C and 70 °C), and to lower hysteresis at 70 °C. The PyC with sulfur atoms appeared to be more efficient in promoting these effects. The CB/SP adduct led to lower stresses at different elongations and to similar stresses with a larger elongation at break. Similar tensile properties were obtained with SHP and SSP with somewhat lower values with the former PyC. These comparisons confirm that CB bearing functional groups reactive with the elastomer chain, either directly or through a coupling agent, create to composites with improved dynamic–mechanical reinforcement and lower hysteresis. The direct reactivity, promoted by the sulfur atoms, appears to be more efficient.

4. Conclusions

This work demonstrates that silica is not the only filler for reinforcing elastomer composites for dynamic mechanical applications with low energy dissipation. CBN234 functionalized with serinol pyrrole and thus bearing hydroxy functional groups was used in place of silica, at the same volume fraction, and the same coupling agent traditionally employed for silica, silane TESPT, was capable of promoting the interaction of CB/SP with the elastomer chains. The composite with the CB/SP + TESPT system shows better dynamic–mechanical properties than the composite with pristine CB + TESPT, despite the lower number of total crosslinks. The dynamic rigidity increased by 11%, the hysteresis was reduced by 11%, the stress at break decreased by 2%, and elongation at break increased by 11% and 36%. The properties of the silica and the CB/SP composites are comparable. Differences, such as the higher dynamic rigidity at a low temperature, and the poorer ultimate tensile properties could be ascribed to the worse filler dispersion. Indeed, CB/SP is a polar filler, and appropriate processing and composite environments, e.g., with suitable covering agents, should be developed. The use of CB in place of silica allows the preparation of lighter (5% in the present study), electrically conductive compounds.

This study demonstrates that the dynamic–mechanical reinforcement with low hysteresis of an elastomer composite is promoted by a high-surface-area nanostructured filler that can establish a chemical bond with elastomer chains. To date, this behavior is a monopoly of silica. The significance of the results reported here lies in demonstrating the role of filler features, and, in particular, of the chemical reactivity of the filer surface, regardless of the nature of the filler. This paves the way for the development of circular fillers, such as the chars coming from biosources and from end-of-life tires.

Supplementary Materials: The following supporting information can be downloaded at: https://www.mdpi.com/article/10.3390/polym16091214/s1, Figure S1: Thermograph from TGA analysis of CB/SP, Figure S2: Thermogravimetric curves for SP (a), CB (b), CB/SP UW (c); CB/SP W (d); Figure S3: Storage modulus curves for S-SBR 4630 compounds, Figure S4: Tan delta curves for SBR 4630 compounds, Figure S5. Tensile curves of S-SBR 4630 compounds obtained through stress strain experiments; Table S1: Data from the crosslinking of composites of Table 1; Table S2: Normalized data for curing, dynamic-mechanical (in the shear and axial modes) and tensile properties for composites with CB/PyC as the reinforcing filler [a].

Author Contributions: Conceptualization, F.M., D.G., V.B. and M.G. (Maurizio Galimberti); Methodology, F.M.; Validation, D.G.; Formal analysis, F.M., M.G. (Martina Galbusera), D.G. and U.G.; Investigation, F.M., M.G. (Martina Galbusera), D.G. and U.G.; Resources, V.B.; Data curation, F.M., M.G. (Martina Galbusera) and U.G.; Writing—original draft, F.M. and M.G. (Maurizio Galimberti); Writing—review and editing, F.M., V.B. and M.G. (Maurizio Galimberti); Visualization, D.G.; Supervision, V.B. and M.G. (Maurizio Galimberti); Project administration, M.G. (Maurizio Galimberti); Funding acquisition, M.G. (Maurizio Galimberti). All authors have read and agreed to the published version of the manuscript.

Funding: This research was funded by Made in Italy–Circular and Sustainable (MICS) Extended Partnership funded by the European Union Next-Generation EU (Piano Nazionale di Ripresa e Resilienza (PNRR)–Missione 4, Componente 2, Investimento 1.3–D.D. 1551.11-10-2022, PE00000004). Pirelli Tyre is gratefully acknowledged for financing the PhD activities of Federica Magaletti, and for the financial support for the research.

Institutional Review Board Statement: Not applicable.

Data Availability Statement: Data are contained within article and Supplementary Materials.

Conflicts of Interest: The authors declare no conflict of interest.

References

1. Available online: https://www.un.org/sustainabledevelopment/development-agenda/ (accessed on 2 January 2024).
2. Global Mobility Report 2017, Tracking Sector Performance. Sustainable Mobility for All. 2017. Available online: https://documents.worldbank.org/en/publication/documents-reports/documentdetail/920101508269072500/global-mobility-report-2017-tracking-sector-performance (accessed on 25 April 2024).
3. Available online: https://www.reportlinker.com/p05379599/?utm_source=GNW (accessed on 2 January 2024).
4. Pirelli Tyre. Global High Value—Company presentation—October 2019. Available online: https://corp-assets.pirelli.com/corporate/Pirelli_9M_2019_results_presentation.pdf (accessed on 25 April 2024).
5. Dong, Y.; Zhao, Y.; Hossain, M.U.; He, Y.; Liu, P. Life cycle assessment of vehicle tires: A systematic review. *Clean. Environ. Syst.* **2021**, *2*, 100033. [CrossRef]
6. Hall, D.E.; Moreland, J.C. Fundamentals of Rolling Resistance. *Rubber Chem. Technol.* **2001**, *74*, 525–539. [CrossRef]
7. Medalia, A.I.; Kraus, G. Reinforcement of Elastomers by Particulate Fillers. In *The Science and Technology of Rubber*, 2nd ed.; Mark, J.E., Erman, B., Eirich, F.R., Eds.; Elsevier Academic Press: Cambridge, MA, USA, 1994; Chapter 8; pp. 387–418.
8. Donnet, J.B.; Custodero, E. Reinforcement of Elastomers by Particulate Fillers. In *The Science and Technology of Rubber*, 3rd ed.; Mark, J.E., Erman, B., Eirich, F.R., Eds.; Academic Press: Cambridge, MA, USA, 2005; pp. 367–400.
9. Fröhlich, J.; Niedermeier, W.; Luginsland, H.-D. The effect of filler–filler and filler–elastomer interaction on rubber reinforcement. *Compos. Part A Appl. Sci. Manuf.* **2005**, *36*, 449–460. [CrossRef]
10. Legrand, A.P. On the silica edge. In *The Surface Properties of Silicas*; Legrand, A.P., Ed.; Wiley and Sons: New York, NY, USA, 1998; pp. 1–20.
11. Leblanc, J.L. Rubber–filler interactions and rheological properties in filled compounds. *Prog. Polym. Sci.* **2002**, *27*, 627–687. [CrossRef]
12. Chevalier, Y.; Morawski, J.C. Precipitated Silica with Morphological Properties, Process for Producing It and Its Application, Especially as a Filler. Legrand. EP0157703 B1, 6 April 1984.
13. Zou, H.; Wu, S.; Shen, J. Polymer/silica nanocomposites: Preparation, characterization, properties, and applications. *Chem. Rev.* **2008**, *108*, 3893–3957. [CrossRef] [PubMed]
14. Li, X.; Liu, J.; Zheng, Z. Recent progress of elastomer–silica nanocomposites toward green tires: Simulation and experiment. *Polym. Int.* **2023**, *72*, 764–782. [CrossRef]
15. Vilmin, F.; Bottero, I.; Travert, A.; Malicki, N.; Gaboriaud, F.; Trivella, A.; Thibault-Starzyk, F. Reactivity of bis [3-(triethoxysilyl) propyl] tetrasulfide (TESPT) silane coupling agent over hydrated silica: Operando IR spectroscopy and chemometrics study. *J. Phys. Chem. C* **2014**, *118*, 4056–4071. [CrossRef]
16. Reuvekamp, L.A.E.M.; Brinke, J.W.T.; van Swaaij, P.J.; Noordermeer, J.W.M. Effects of time and temperature on the reaction of TESPT silane coupling agent during mixing with silica filler and tire rubber. *Rubber Chem. Technol.* **2002**, *75*, 187–198. [CrossRef]
17. Locatelli, D.; Bernardi, A.; Rubino, L.R.; Gallo, S.; Vitale, A.; Bongiovanni, R.; Barbera, V.; Galimberti, M. Biosourced *Janus* Molecules as Silica Coupling Agents in Elastomer Composites for Tires with Lower Environmental Impact. *ACS Sustain. Chem. Eng.* **2023**, *11*, 2713–2726. [CrossRef]
18. Ye, N.; Zheng, J.; Ye, X.; Xue, J.; Han, D.; Xu, H.; Wang, Z.; Zhang, L. Performance enhancement of rubber composites using VOC-Free interfacial silica coupling agent. *Compos. Part B Eng.* **2020**, *202*, 108301. [CrossRef]
19. Liu, C.; Guo, M.; Zhai, X.; Ye, X.; Zhang, L. Using epoxidized solution polymerized styrene-butadiene rubbers (ESSBRs) as coupling agents to modify silica without volatile organic compounds. *Polymers* **2020**, *12*, 1257. [CrossRef]

20. Zhai, X.; Chen, Y.; Han, D.; Zheng, J.; Wu, X.; Wang, Z.; Li, X.; Ye, X.; Zhang, L. New designed coupling agents for silica used in green tires with low VOCs and low rolling resistance. *Appl. Surf. Sci.* **2021**, *558*, 149819. [CrossRef]
21. Zhai, X.; Han, D.; Zhang, G.; Chen, X.; Wu, X.; Li, X.; Zheng, J.; Ye, X.; Zhang, L. A novel VOC free bio-based coupling agent and its using strategy facing green tires with excellent dynamic performances. *Compos. Sci. Technol.* **2022**, *223*, 109413. [CrossRef]
22. Donnet, J. Carbon Black, Revised and Expanded. In *Science and Technology*, 2nd ed.; Donnet, J.B., Bansal, R.C., Wang, M.-J., Eds.; Marcel Dekker: New York, NU, USA, 1993.
23. Suner, S.S.; Kurt, S.B.; Demirci, S.; Sahiner, N. The advances in functionalized carbon nanomaterials for drug delivery. In *Functionalized Carbon Nanomaterials for Theranostic Applications*; Elsevier: Amsterdam, The Netherlands, 2023; pp. 197–241.
24. Prioglio, G.; Naddeo, S.; Giese, U.; Barbera, V.; Galimberti, M. Bio-Based Pyrrole Compounds Containing Sulfur Atoms as Coupling Agents of Carbon Black with Unsaturated Elastomers. *Nanomaterials* **2023**, *13*, 2761. [CrossRef] [PubMed]
25. Locatelli, D.; Barbera, V.; Brambilla, L.; Castiglioni, C.; Sironi, A.; Galimberti, M. Tuning the solubility parameters of carbon nanotubes by means of their adducts with Janus pyrrole compounds. *Nanomaterials* **2020**, *10*, 1176. [CrossRef] [PubMed]
26. Barbera, V.; Bernardi, A.; Palazzolo, A.; Rosengart, A.; Brambilla, L.; Galimberti, M. Facile and sustainable functionalization of graphene layers with pyrrole compounds. *Pure Appl. Chem.* **2018**, *90*, 253–270. [CrossRef]
27. Guerra, S.; Barbera, V.; Vitale, A.; Bongiovanni, R.; Serafini, A.; Conzatti, L.; Brambilla, L.; Galimberti, M. Edge functionalized graphene layers for (ultra) high exfoliation in carbon papers and aerogels in the presence of chitosan. *Materials* **2019**, *13*, 39. [CrossRef] [PubMed]
28. Barbera, V.; Brambilla, L.; Milani, A.; Palazzolo, A.; Castiglioni, C.; Vitale, A.; Bongiovanni, R.; Galimberti, M. Domino Reaction for the Sustainable Functionalization of Few-Layer Graphene. *Nanomaterials* **2019**, *9*, 44. [CrossRef]
29. Aprem, A.S.; Joseph, K.; Thomas, S. Recent developments in crosslinking of elastomers. *Rubber Chem. Technol.* **2005**, *78*, 458–488. [CrossRef]
30. Flory, P.J. Statistical mechanics of swelling of network structures. *J. Chem. Phys.* **1950**, *18*, 108–111. [CrossRef]
31. Kraus, G. Swelling of Filler-Reinforced Vulcanizates. *Rubber Chem. Technol.* **1964**, *37*, 6–13. [CrossRef]
32. Afash, H.; Ozarisoy, B.; Altan, H.; Budayan, C. Recycling of Tire Waste Using Pyrolysis: An Environmental Perspective. *Sustainability* **2023**, *15*, 14178. [CrossRef]
33. Danesh, P.; Niaparast, P.; Ghorbannezhad, P.; Ali, I. Biochar production: Recent developments, applications, and challenges. *Fuel* **2023**, *337*, 126889. [CrossRef]
34. Berki, P.; Göbl, R.; Karger-Kocsis, J. Structure and properties of styrene-butadiene rubber (SBR) with pyrolytic and industrial carbon black. *Polym. Test.* **2017**, *61*, 404–415. [CrossRef]
35. Sagar, M.; Nibedita, K.; Manohar, N.; Kumar, K.R.; Suchismita, S.; Pradnyesh, A.; Reddy, A.B.; Sadiku, E.R.; Gupta, U.; Lachit, P.; et al. A potential utilization of end-of-life tyres as recycled carbon black in EPDM rubber. *Waste Manag.* **2018**, *74*, 110–122. [CrossRef]
36. Martínez, J.D.; Cardona-Uribe, N.; Murillo, R.; García, T.; López, J.M. Carbon black recovery from waste tire pyrolysis by demineralization: Production and application in rubber compounding. *Waste Manag.* **2019**, *85*, 574–584. [CrossRef] [PubMed]
37. Gao, N.; Wang, F.; Quan, C.; Santamaria, L.; Lopez, G.; Williams, P.T. Tire pyrolysis char: Processes, properties, upgrading and applications. *Prog. Energy Combust. Sci.* **2022**, *93*, 101022. [CrossRef]
38. Greenough, S.; Dumont, M.-J.; Prasher, S. The physicochemical properties of biochar and its applicability as a filler in rubber composites: A review. *Mater. Today Commun.* **2021**, *29*, 102912. [CrossRef]
39. Zare, Y.; Rhee, K.Y. An innovative model for conductivity of graphene-based system by networked nano-sheets, interphase and tunneling zone. *Sci. Rep.* **2022**, *12*, 15179. [CrossRef]
40. Jiang, C.; Bo, J.; Xiao, X.; Zhang, S.; Wang, Z.; Yan, G.; Wu, Y.; Wong, C.; He, H. Converting waste lignin into nano-biochar as a renewable substitute of carbon black for reinforcing styrene-butadiene rubber. *Waste Manag.* **2020**, *102*, 732–742. [CrossRef]
41. Hasanin, M.S.; El-Aziz, M.E.A.; El-Nagar, I.; Hassan, Y.R.; Youssef, A.M. Green enhancement of wood plastic composite based on agriculture wastes compatibility via fungal enzymes. *Sci. Rep.* **2022**, *12*, 19197. [CrossRef] [PubMed]
42. Available online: https://corporate.pirelli.com/var/files2020/EN/PDF/PIRELLI_ANNUAL_REPORT_2020_ENG.pdf (accessed on 2 January 2024).
43. Thanh, L.T.; Okitsu, K.; Van Boi, L.; Maeda, Y. Catalytic technologies for biodiesel fuel production and utilization of glycerol: A review. *Catalysts* **2012**, *2*, 191–222. [CrossRef]
44. Broadbent, H.S.; Burnham, W.S.; Sheeley, R.M.; Olsen, R.K. Novel heterotricyclic systems: 2, 6-dioxa-and 2-oxa-6-thia-10-azatricyclo-[5.2. 1.04, 10] deeanes; 2, 6-dioxa-11-azatricyclo [5.3. 1.04, 11] undecane; and 9, 13-dioxa-14-azatetracyclo [6.5. 1.02, 7.011, 14] tetradeca-2, 4, 6-triene. *J. Heterocycl. Chem.* **1976**, *13*, 337–348. [CrossRef]
45. *ASTM 6556*; Standard Test Method for Carbon Black—Total and External Surface Area by Nitrogen Adsorption. ASTM International: West Conschohocken, PA, USA, 2019.
46. Musto, S.; Barbera, V.; Cipolletti, V.; Citterio, A.; Galimberti, M. Master curves for the sulphur assisted crosslinking reaction of natural rubber in the presence of nano- and nano-structured sp2 carbon allotropes. *Express Polym. Lett.* **2017**, *11*, 435–448. [CrossRef]
47. Zanobini, A.; Sereni, B.; Catelani, M.; Ciani, L. Repeatability and Reproducibility techniques for the analysis of measurement systems. *Measurement* **2016**, *86*, 125–132. [CrossRef]

48. Standard ISO 37/UNI 6065. Available online: https://www.eurolab.net/it/testler/malzeme-testleri/iso-37-kaucuk-vulkanize-veya-termoplastik-cekme-gerilimi-uzama-ozelliklerinin-tayini/ (accessed on 2 January 2024).
49. Flory, P.J.; Rehner, J. Statistical Mechanics of Cross-Linked Polymer Networks I. Rubberlike Elasticity. *J. Chem. Phys.* **1943**, *11*, 512–520. [CrossRef]

polymers MDPI

Article

Enhancing Rubber Vulcanization Cure Kinetics: Lowering Vulcanization Temperature by Addition of MgO as Co-Cure Activator in ZnO-Based Cure Activator Systems

Md Najib Alam [1], Vineet Kumar [1], Seok U Jeong [2] and Sang-Shin Park [1,*]

1 School of Mechanical Engineering, Yeungnam University, 280, Daehak-ro, Gyeongsan 38541, Republic of Korea; mdnajib.alam3@gmail.com (M.N.A.); vineetfri@gmail.com (V.K.)
2 Graduate School of Mechanical Engineering, Yeungnam University, 280, Daehak-ro, Gyeongsan 38541, Republic of Korea; fffu123@naver.com
* Correspondence: pss@ynu.ac.kr

Abstract: Vulcanization is a chemical modification of rubber that requires a considerable amount of thermal energy. To save thermal energy, the kinetics of rubber vulcanization should be improved. In this article, the curing properties of rubber vulcanization are thoroughly investigated using the moving die rheometer (MDR) technique. To enhance the kinetics in different stages of ZnO-based sulfur vulcanization systems, small amounts of MgO were added. The results revealed that the small amount of 1 to 2 phr (per hundred grams of rubber) of MgO in the controlled 5 phr ZnO-based curing systems can significantly improve the curing kinetics. For example, the optimum curing time of 1 phr MgO added to the 5 phr ZnO-containing semi-efficient vulcanization system at different temperatures was more than half that of the controlled 5 phr ZnO-only compound. While maintaining a similar rate of vulcanization, the vulcanization temperature can be reduced by up to 20 °C by using MgO as a co-cure activator, which exhibits similar or better rheometric mechanical properties compared to the controlled compounds. With the addition of MgO as a co-cure activator, the vulcanization reactions become very fast, enabling vulcanization to be completed, even at the boiling point of water (100 °C) with an affordable curing time (<1 h). By reducing the vulcanization temperature, the scorch safety time can be enhanced in the ZnO/MgO-based binary cure activator-containing vulcanizates. Overall, MgO could be a potential candidate as a co-cure activator with ZnO for the vulcanization of rubber, offering better economical and eco-friendly methods.

Keywords: rubber; cure activator; magnesium oxide; zinc oxide; vulcanization kinetics

1. Introduction

Among different polymeric materials, rubber has wide industrial applications [1]. However, raw rubber has limited applications because it is soft in hot weather and stiff in cold weather. To maintain uniformity in mechanical properties with variable environmental temperatures, raw rubber must be processed. The processing of rubber is performed via curing, which produces cross-linking between the rubber chains, transforming individual rubber macromolecules into a connected network that can sustain its mechanical properties over a wider temperature range and deformation [2]. Many rubber products, as we see in our daily lives, are processed rubber.

After the invention of the cross-linking process known as vulcanization of rubber by Charles Goodyear in 1839 through simple heating of rubber and sulfur, numerous curing ingredients have been developed to reduce the vulcanization time and enhance cure efficiency [3,4]. Currently, sulfur-based rubber vulcanization is the most common method, generally performed at 140–150 °C, with the required curing time depending on the type of rubber and the efficiency of the accelerator, which can vary from 10 to 30 min.

Recently, global warming has become a challenging issue, and the minimum use of thermal energy is recommended, along with reduced reliance on fossil fuels [5]. In the rubber industry, the use of thermal energy can be minimized by enhancing the curing rate without increasing vulcanization temperature and time. This is only achievable by selecting faster curing ingredients for rubber vulcanization. While higher temperatures can enhance the curing rate, they may also degrade important properties [6]. Additionally, at elevated temperatures, the rate of release of environmentally polluting gases such as H_2S, CS_2, and secondary amines from vulcanized rubber, as byproducts can increase [6,7]. Therefore, reducing vulcanization temperature is urgently needed from both economic and environmental perspectives.

In rubber vulcanization, accelerators and activators play a key role in the kinetics of the process. A very fast curing system can reduce the vulcanization time, but it may have drawbacks such as low scorch safety and an increased reversion rate [8]. These properties significantly affect the vulcanizate properties, leading to improper vulcanization throughout the rubber matrix. It has been found that lowering the vulcanization temperature can enhance scorch safety and other mechanical properties, although it takes a longer time to cure and may not be economically favorable. Currently, thiuram-type accelerators combined with ZnO-based cure activators serve as a conventionally faster curing system for rubber vulcanization. Recently, Alam et al. [9,10] observed that cure activities, especially the cure rate, can be further enhanced by using MgO as a single or co-cure activator with ZnO in rubber vulcanization. Although the optimum curing time, rate of cross-linking, and certain mechanical properties were enhanced, they had very low scorch safety and reduced cross-link density, possibly due to fast initiation of the cross-linking reaction and increased reversion [9,10]. Hence, new opportunities can arise by enhancing curing kinetics to reduce vulcanization temperature, achieving better scorch safety, reduced reversion rate, and improved cross-linking. Investigation shows that ZnO has many important roles in different stages of accelerated sulfur vulcanization [11]. Based on the previous reports, it was evident that either ZnO or zinc-based compounds are necessarily important to obtain vulcanizates with industrial value [12,13]. After ZnO, MgO has recently gained much attention from rubber scientists because of its inexpensive, abandoned, low toxicity, and high potential to enhance the kinetics of vulcanization in different rubber systems [14–21]. Most importantly, it was found that ZnO and MgO might undergo synergistic activities during accelerated sulfur vulcanization [10] and can enhance the kinetics as well as the mechanical properties.

Kinetic studies are necessarily important to understand the overall vulcanization reaction. A few important techniques, such as differential scanning calorimetry (DSC) and oscillating disc rheometer (ODR) tests, are employed to study the kinetics of vulcanization [22]. The main drawbacks of DSC studies include their limited ability to accurately predict the induction period, particularly for lower exothermic vulcanization kinetics [23]. However, ODR is considered best because, along with kinetic data, it also provides certain mechanical properties of the vulcanized rubber. The ODR curve is based on considering stiffness, which is directly proportional to the degree of vulcanization compound versus time of vulcanization. Based on the ODR curve, Coran proposed a model for accelerated sulfur vulcanization where four main kinetic steps can be involved [24,25]. In the first kinetic step, a sulfurating complex is formed. In the second step, cross-linking precursors are formed, followed by activation of the sulfurating complex. In the third step, polysulfide rubber networks are formed. In the fourth and final step, these polysulfides turn into mature cross-links [22,24–26]. The general schematics of rubber vulcanization, considering the kinetic model of accelerated sulfur vulcanization, can be found elsewhere [11,22].

Studies by Alam et al. [10], Zhao et al. [27], and Hadi et al. [28] have revealed the potential role of cure activators such as zinc oxide in the kinetics of rubber vulcanization. They investigated the influence of zinc oxide on the cure activity of natural rubber-based curing systems. Their findings suggest that the curing rate can decrease after reaching an optimum level (1–2 g per hundred grams of rubber) of zinc oxide in efficient vulcanization

(EV) and semi-efficient vulcanization (SEV) systems [10,27]. However, for conventional vulcanization systems, the curing rate may increase with higher amounts of zinc oxide (>5 phr) [28]. While the rate of vulcanization may decrease with certain vulcanization systems, higher amounts of zinc oxide can lead to the production of more cross-link precursors, thereby enhancing cross-link density and mechanical properties [27].

Previous studies by Alam et al. [9,10] concluded that the level of ZnO can be substantially reduced by utilizing zinc-based binary accelerator systems or employing MgO/ZnO binary cure activator systems. These approaches lead to improved vulcanization kinetics at static temperature, as well as enhancements in tensile strength, fracture toughness, and elongation at break properties. However, such reductions in ZnO levels come at the cost of compromised cross-link density and mechanical modulus values [9,10]. Currently, although the level of zinc oxide can be decreased in many rubber formulations [29–38], for certain applications like heavy-duty truck tire manufacturing, the amount cannot be compromised. These applications require a high degree of cross-linking to achieve elevated modulus values and low heat build-up properties [11]. From this perspective, reducing the vulcanization temperature while enhancing the vulcanization kinetics could prove to be an effective approach for reducing environmental pollution, production costs, and overall performances. While previous studies [9,10] explored the possibility of reducing ZnO levels using MgO as a co-cure activator, detailed kinetic studies necessary for potential reductions in vulcanization temperature remain unexplored. Lowering the vulcanization temperature with improved curing kinetics not only decreases production costs but also enhances the quality of rubber products, safeguarding them from issues like high-temperature scorching and reversion due to overcuring, particularly in thicker rubber products.

In this paper, our objective is to reduce the vulcanization temperature with an affordable time in accelerated sulfur vulcanization without the loss of vulcanizate properties, especially the level of rheometric mechanical properties. To improve the kinetics of rubber vulcanization, MgO was utilized as a secondary cure activator in the ZnO-based cure activator systems. To understand the mechanism of sulfur cross-linking in the presence of MgO and ZnO co-cure activators, different vulcanization systems, namely EV and SEV systems with high to moderate accelerator-to-sulfur ratios, are chosen. Because cure activators primarily form accelerator–activator complexes, a higher concentration of accelerator can facilitate a better understanding of the mechanism of accelerated sulfur vulcanization. It is believed that MgO plays a critical role in the in situ formation of zinc accelerator complexes, and their activity and amounts depend on accelerator-to-sulfur ratios. The curing studies were conducted followed by the ODR technique with different vulcanization temperatures ranging from 100 to 140 °C. The different kinetic parameters have been correlated with Arrhenius-type relationships to investigate the role of MgO in the activation energies.

2. Materials and Methods

2.1. Materials

Natural rubber (NR, RSS-3) and the curing additives such as zinc oxide, stearic acid, accelerator tetramethyl thiuram disulfide (TMTD), and sulfur were obtained from the Thai Rubber Research Institute, Bangkok, Thailand. Magnesium oxide light in fine powder form was purchased from AppliChem PanReac, Bangkok, Thailand.

2.2. Mixing of Curing Ingredients with Rubber

The mixing of curing ingredients was performed at room temperature in a laboratory-size open two-roll mill. The speeds of the front and rear rolls were fixed at 20 and 24 rpm, respectively, to maintain a 1.2 friction ratio. Initially, raw rubber was masticated for five minutes with a 2 mm nip gap between the two rolls. Then, cure activators (ZnO, MgO, and stearic acid) were added and milled for another five minutes, reducing the nip gap to 1 mm. After that, accelerator (TMTD) and sulfur were added together and milled for another five minutes. The mixing time was fixed at 15 min for all compounds. The compositions of rubber compounds are provided in Table 1. Different curing systems can be identified from the abbreviation in the mix composition and can be realized from their respective accelerator-to-sulfur ratios: 4.32 for EV and 0.96 for SEV systems, as presented in Table 1. For example, the EV/5-ZnO/1-MgO compound belongs to the EV system, where 5 phr ZnO along with 1 phr MgO were used as cure activators.

Table 1. Mixing composition of different ingredients in phr (per hundred gram of rubber).

Mix Compositions	Formulation						Vulcanization Systems (Accelerator/Sulfur)
	NR	ZnO	MgO	Stearic Acid	TMTD	Sulfur	
EV/5-ZnO (control-EV)	100	5	-	2	2.16	0.5	EV (4.32)
EV/5-ZnO/1-MgO	100	5	1	2	2.16	0.5	
EV/5-ZnO/2-MgO	100	5	2	2	2.16	0.5	
EV/5-ZnO/3-MgO	100	5	3	2	2.16	0.5	
EV/5-MgO	100	-	5	2	2.16	0.5	
SEV/5-ZnO (control-SEV)	100	5	-	2	1.44	1.5	SEV (0.96)
SEV/5-ZnO/1-MgO	100	5	1	2	1.44	1.5	
SEV/5-ZnO/2-MgO	100	5	2	2	1.44	1.5	
SEV/5-ZnO/3-MgO	100	5	3	2	1.44	1.5	
SEV/5-MgO	100	-	5	2	1.44	1.5	

2.3. Measurement of Curing Properties

Curing properties of the compounded rubber were investigated by a Moving Die Rheometer (MDR, Rheo Tech MD+, Model no. A022S, MonTech USA, Columbia City, IN, USA). The instrument provides a torque versus time plot at a constant vulcanization temperature, from which rheometric mechanical properties, such as minimum torque (M_L), highest torque (M_H), ΔM ($M_H - M_L$), torque after 300 s of overcuring (M_{300}), and percentage of reversion in torque at 300 s of overcuring (R_{300}), as well as kinetic parameters such as scorch time (t_2, considering 2% progress on Δ torque), optimum curing time (t_{90}, considering 90% progress on Δ torque), the actual rate of cross-linking {$R_v = (Mt_{90} - Mt_2)/(t_{90} - t_2)$}, and the cure rate index {$CRI = 100/(t_{90} - t_2)$}, were obtained. The different curing data are presented in the Supplementary Materials (Tables S1–S5).

The kinetic parameters, such as scorch safety time and the rate of cross-linking values, were fitted with the Arrhenius equation of temperature-dependent rate as follows:

$$k = Ae^{-\frac{E_a}{RT}}$$

where k denotes the rate of a process, A is the pre-exponential factor, E_a is the energy of activation for that process, R is the universal gas constant, and T is the temperature in absolute scale. Taking the logarithm, the above equation turns into:

$$\ln k = \ln A - \frac{E_a}{RT}$$

Replacing k as t_2 and R_v at different temperatures, different E_a values can be obtained. Since t_2 is represented as time, its inverse can be regarded as rate and follows the above equation very precisely. On the other hand, R_v is the true rate of vulcanization; hence, the E_a value obtained can be regarded as the activation energy of vulcanization.

3. Results and Discussion

Cure curves (rheographs) of EV systems with different types of cure activator at different temperatures are shown in Figure 1a–d. In the rheographs, three distinct zones, such as induction, cure, and overcure, can be clearly identified [22]. It can be seen that the induction and cure zones of ZnO-only compounds can be shortened by the addition of a small amount of MgO as a secondary cure activator. This suggests that MgO has a considerable effect on the kinetics of rubber vulcanization. With increasing vulcanization temperatures from 110 °C to 140 °C, the different time zones are further shortened, and the overcure zones show lower reversion resistances. Tables S1–S5 provide the reversion properties concerning M_{300} and R_{300}. It is evident from the tables that the M_{300} value decreases gradually with increasing vulcanization temperature, while the R_{300} values remain notably low for hybrid cure activator systems at lower temperatures. Therefore, implementing low-temperature vulcanization is desirable to achieve improved reversion resistance in rubber when utilizing binary cure activator systems. It can also be found from the rheographs that the ZnO-only compound shows a slight marching tendency in the overcure zone, whereas the MgO-only compound has a higher reversion tendency in the overcure zone. However, in the presence of both cure activators, ZnO and MgO, a perfect equilibrium cure curve can be obtained in the overcure zones. A better equilibrium curve might indicate better age-resistant rubber since it has no change in the degree of cross-linking with time and temperature after complete curing [39,40].

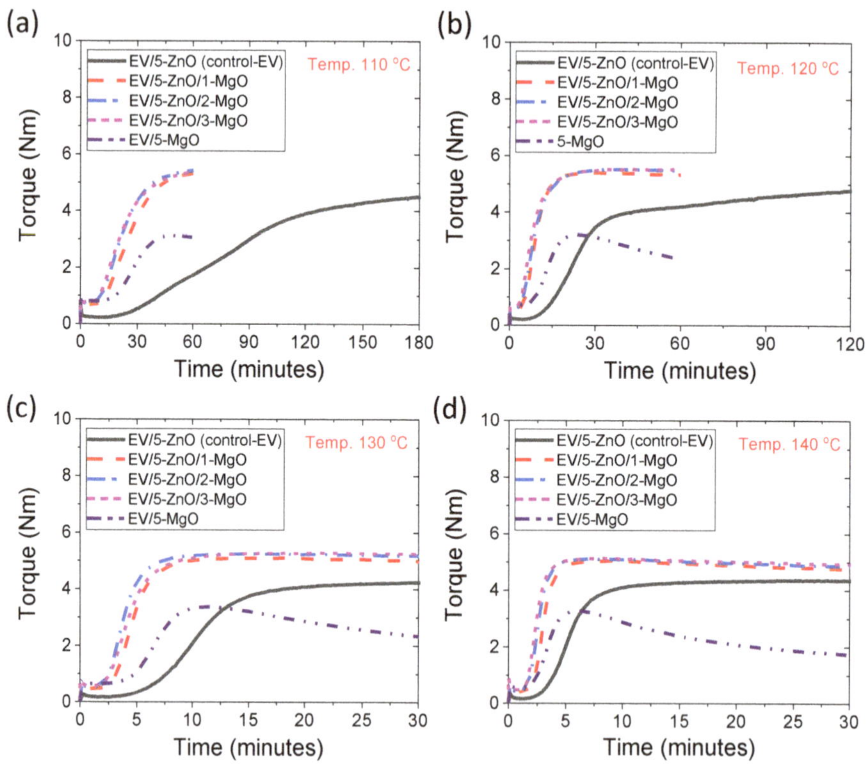

Figure 1. Rheographs for EV curing systems at different vulcanization temperatures; (**a**) 110 °C, (**b**) 120 °C, (**c**) 130 °C, and (**d**) 140 °C.

Rheographs for different SEV curing systems are provided in Figure 2a–e. Rheographs of SEV curing systems are similar to those for EV curing systems. However, higher values in the rheometric torque after curing have been found for SEV vulcanization systems compared to EV curing systems. Also, the tendency of reversion in the overcure zone is more pronounced in the SEV curing systems compared to EV curing systems, and the reversion rates are increased with increasing vulcanization temperature (Tables S1–S5). The elevated reversion rates observed in SEV curing systems might result from the formation of increased quantities of polysulfide linkages in the presence of a lower concentration of accelerator–activator complexes, as opposed to the higher concentration observed in EV curing systems. Polysulfide linkages have a tendency to form cyclic sulfides, which can break the linear sulfur bridges during prolonged heating, thereby impacting the reduction in rheometric torque values [8,39].

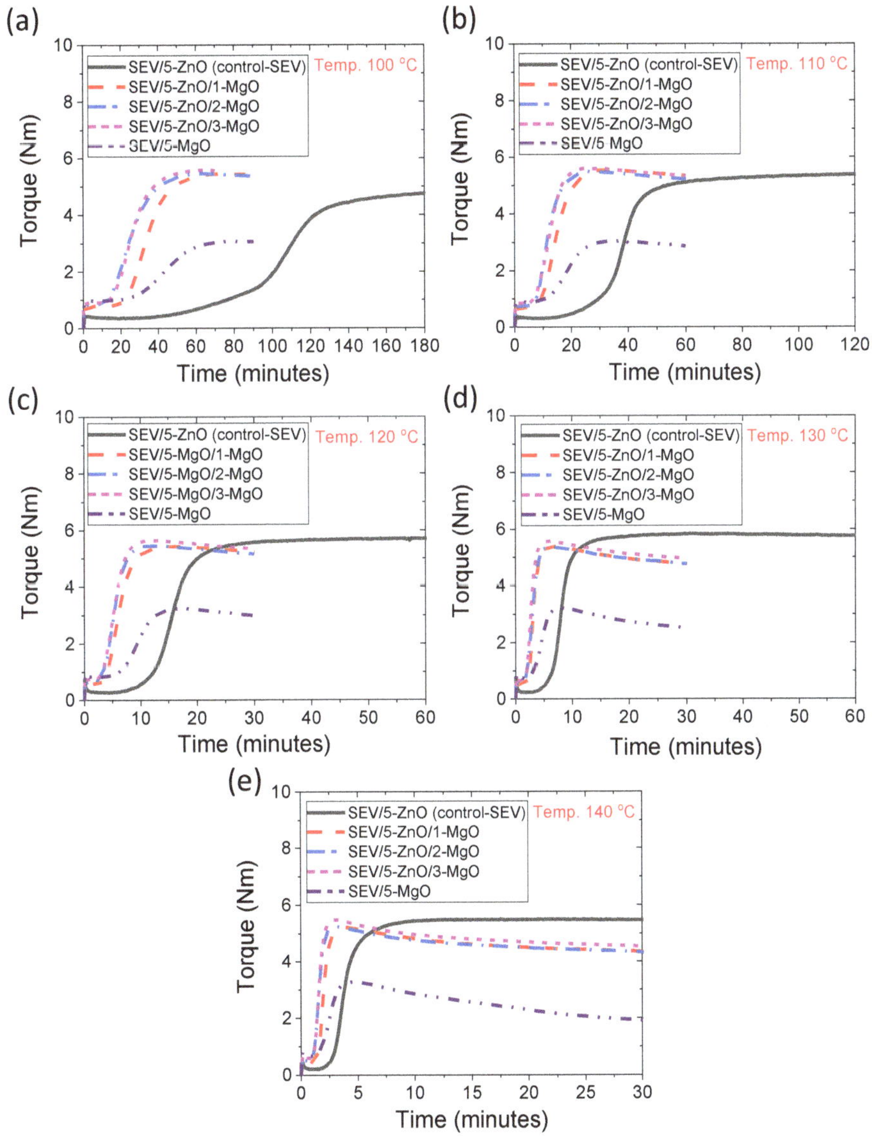

Figure 2. Rheographs for SEV curing systems at different vulcanization temperatures; (**a**) 100 °C, (**b**) 110 °C, (**c**) 120 °C, (**d**) 130 °C, and (**e**) 140 °C.

Rheometric mechanical properties are important for predicting final vulcanizate properties. For example, torque is proportional to stiffness, and ΔM is proportional to cross-link density [22]. The different mechanical properties of the EV curing systems are provided in

Figure 3a–c. From Figure 3c, the lowest torque (M_L) gradually decreases with increasing temperature for all EV curing systems. It is interesting to note that by increasing the amounts of MgO in the systems, M_L values can be enhanced. This might indicate that a higher amount of accelerator can be grafted to the rubber chains, resulting in higher viscosity in the compound incorporating MgO, which decomposes the accelerator and attaches to the rubber chains more effectively than ZnO [9,10]. From Figure 3b, it can be seen that the MgO-only compound shows the lowest values in the M_H, and binary cure activators containing compounds have higher M_H values. It is believed that when MgO interacts with the accelerator, it forms Mg-dithiocarbamate, which initially converts polysulfide cross-links into cyclic sulfides. This process reduces the number of cross-links, contrasting with the in situ formation of Zn-dithiocarbamate [9,10,39]. Consequently, MgO-only compounds exhibit the lowest M_H value compared to other activator systems. The ΔM values in Figure 3c follow a similar trend as M_H values. It can be seen from Figure 3b,c that M_H and ΔM can slightly increase with temperature and then decrease. This could be due to improved interactions between the curatives through sulfur diffusion [41] above the melting point of elemental sulfur near 110 °C; then the decrease might be due to reduced efficiency of the curing reagent at elevated temperatures. The improved ΔM in the binary activator systems could be advantageous in mechanical properties compared to ZnO as a single cure activator in the EV curing system.

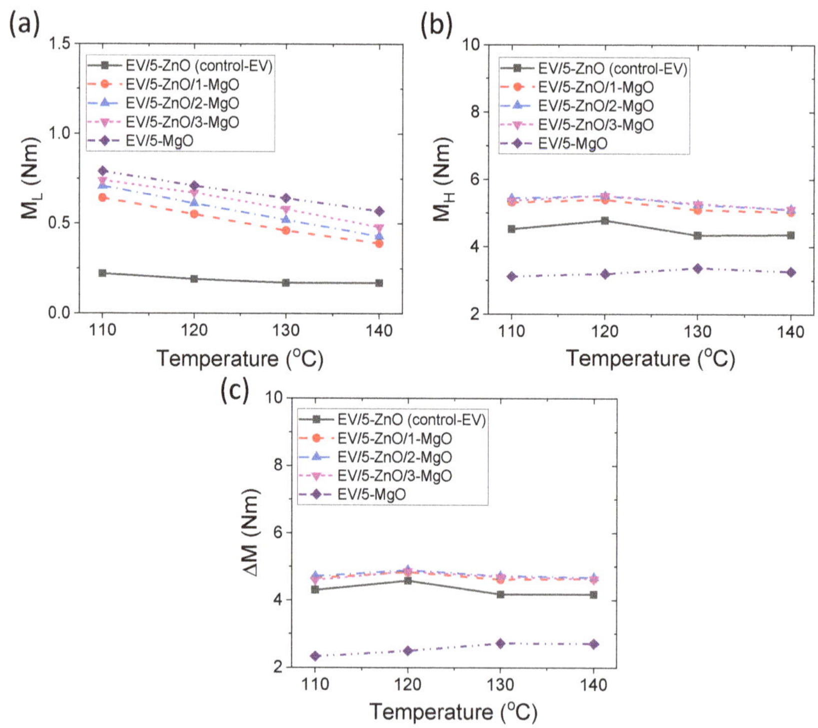

Figure 3. Rheometric mechanical properties for EV curing systems; (**a**) M_L, (**b**) M_H, (**c**) ΔM.

The different rheometric mechanical properties for SEV curing systems are provided in Figure 4a–c. The M_L values for both SEV and EV curing systems follow a similar behavior with vulcanization temperature (Figure 4a). The M_L values are slightly higher in SEV curing systems than in EV curing systems, which could be due to the higher molecular weight of cross-link precursors attached to rubber chains with an increased number of sulfur atoms in the former cases than in the latter cases. From Figure 4b, it is evident that M_H values are better for binary cure activator compounds at lower temperatures compared to ZnO-only compounds. The ZnO-only compound shows the best value at 130 °C of vulcanization temperature. From Figure 4c, it is evident that ΔM values remain unchanged with increasing vulcanization temperature for MgO-containing binary cure activator compounds. However, for the ZnO-only compound, ΔM values increase up to 130 °C and then decrease with further increasing temperature. From this observation, it can be concluded that in the presence of MgO as a co-cure activator, the achieved cross-link density may become independent of temperature increase, and vulcanization can be achieved at low temperatures, even below the melting point of sulfur. The constant ΔM for MgO-added vulcanizates at low temperatures may suggest that a better and constant amount of cross-link precursors may form at fixed ZnO contents and undergo final cross-links with more specific sulfur-ranked cross-links such as disulfide cross-links [17]. On the other hand, the ZnO-only compound forms an increased amount of cross-link precursor with increasing temperature and provides increased ΔM up to 130 °C. Hence, from a mechanical point of view, vulcanization can be possible at very low temperatures without loss in mechanical properties and the degree of cross-linking using MgO as a co-cure activator along with ZnO.

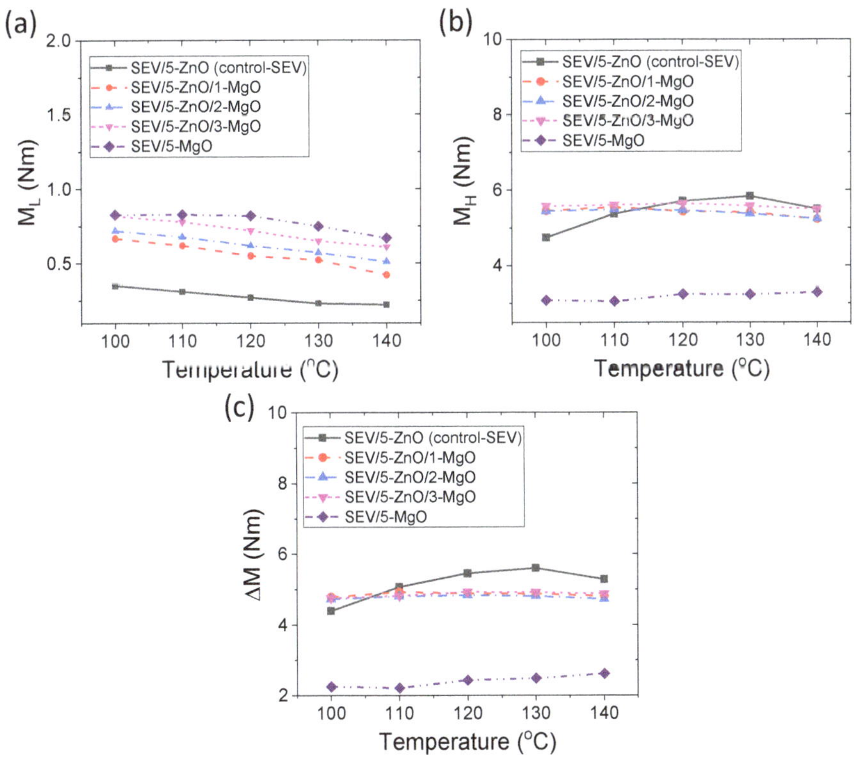

Figure 4. Rheometric mechanical properties for SEV curing systems; (**a**) M_L, (**b**) M_H, (**c**) ΔM.

The different curing parameters for EV curing systems are plotted in Figure 5a–d. From Figure 5a, MgO-containing compounds have lower t_2 compared to ZnO-only compounds. It is interesting to note that MgO and ZnO-based binary cure activator-containing compounds have much lower t_2 values than both MgO and ZnO as single cure activator compounds. This result may indicate that MgO and ZnO-based in-situ compounds may interact to form a more reactive sulfurating complex [9,10]. From this figure, it can also be evident that t_2 reduction occurs more rapidly in ZnO-only compounds than in MgO-containing compounds with increasing vulcanization temperature. Figure 5b represents the variation in the optimum curing times with different temperatures. From this figure, it can be found that the t_{90} value decreases more rapidly for the control-EV system than for all other MgO-containing compounds. It can also be seen that t_{90} is more than double compared to MgO-containing compounds. Hence, a catalytic amount of MgO could be effective in reducing the vulcanization time more economically. From Figure 5c, the rate of vulcanization is much lower for the control-EV system compared to MgO-added vulcanizates. With increasing vulcanization temperature, the rate increases more rapidly in ZnO- and MgO-based binary cure activator systems compared to single cure activator systems. It is evident that the addition of MgO to the binary cure activator-containing compounds can result in more than doubling of the vulcanization rate (R_v) compared to the control-EV system. Figure 5d suggests that CRI values similarly increase with rising temperature. It is important to note that a higher CRI value only signifies the completion of the first stage of vulcanization, not the actual rate of vulcanization. Therefore, while the R_v values of ZnO-only and MgO-only compounds may not differ significantly, their CRI values vary considerably. This is because the MgO-only compound achieves a lower ΔM value within a shorter period of time, resulting in a higher CRI value.

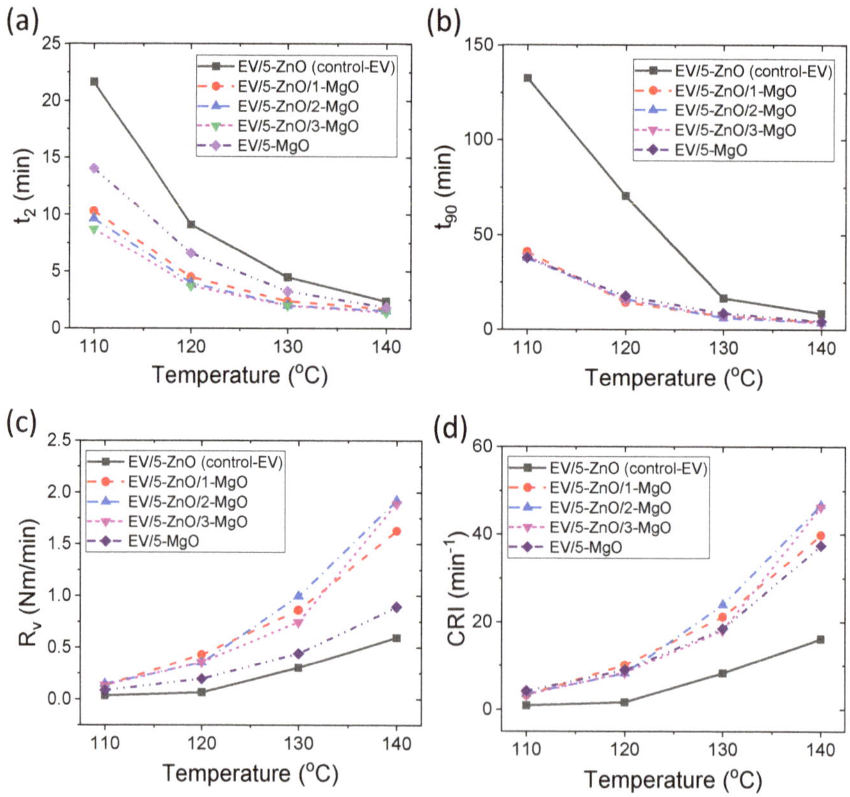

Figure 5. Curing properties for EV curing systems; (**a**) t_2, (**b**) t_{90}, (**c**) R_v, and (**d**) CRI.

Different curing parameters for SEV curing systems are provided in Figure 6a–d. From Figure 6a, t_2 of different SEV compounds shows a similar behavior with temperature as EV curing systems. However, there is an advantage in better t_2 values in SEV curing systems at lower temperatures compared to EV curing systems. From Figure 6b, it is evident that t_{90} values for different SEV curing systems also follow a similar behavior with increasing temperature as EV curing systems. It is interesting to note that MgO does not have a significant effect on t_{90} values after 1 phr in EV curing systems, but it can reduce the t_{90} value up to 2 phr in SEV curing systems. Hence, only 1 or 2 phr of MgO could be sufficient for extensive reduction in the curing time. From Figure 6c, R_v values of different SEV curing systems follow a similar behavior as EV curing systems with temperature but with higher R_v values in the former curing systems than in the latter curing systems. Figure 6d illustrates the CRI values in SEV curing systems. These values exhibit similar trends with temperature and are influenced by the nature of the cure activators. Interestingly, the rate of vulcanization in MgO-only compounds is lower compared to ZnO-only compounds at various temperatures in SEV curing systems, contrary to the behavior observed in EV curing systems. Therefore, it can be concluded that MgO-only acts as a less effective cure activator for achieving a faster vulcanization rate. However, it accelerates the intermolecular interactions, as evidenced by the CRI values, thereby reducing the vulcanization time. From Figures 5b,c and 6b,c, binary cure activator-containing compounds have similar t_{90} and R_v values at 120 °C of vulcanization temperature compared to the control compounds at the vulcanization temperature of 140 °C. Hence, at least 20 °C of vulcanization temperature can be reduced by the addition of just 1 or 2 phr of MgO to the binary cure activator-containing vulcanizates.

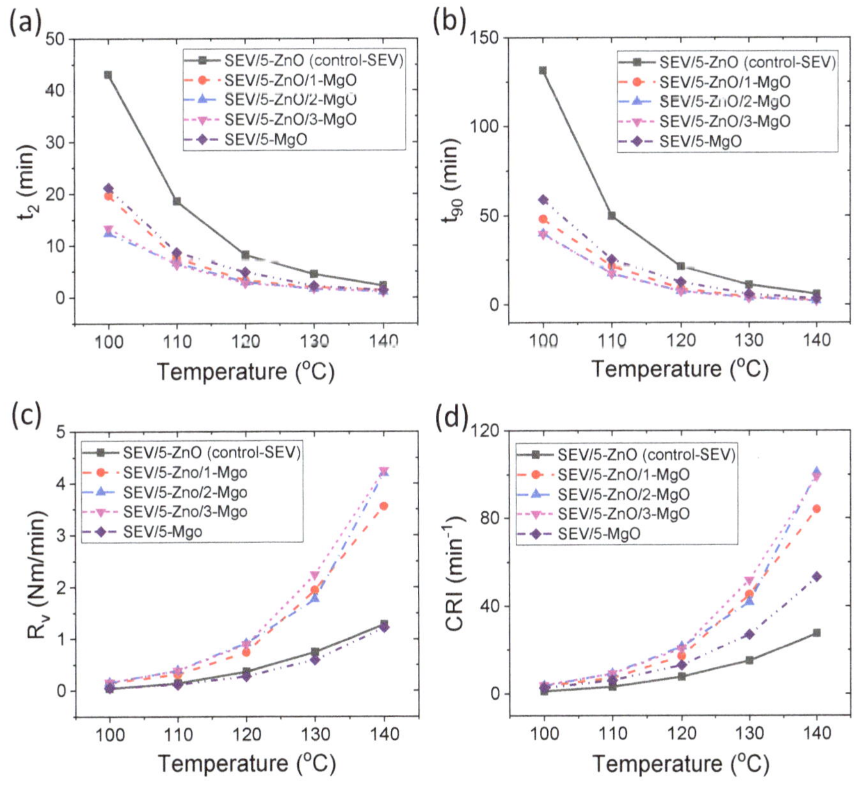

Figure 6. Curing properties for SEV curing systems; (**a**) t_2, (**b**) t_{90}, (**c**) R_v, and (**d**) CRI.

According to the Coran kinetic model [22–24], there is a rate involved in the formation of cross-link precursors from the active sulfurating complex. Since the formation of cross-link precursors does not provide increasing torque, the rate cannot be calculated directly [22]. However, the time of formation of the cross-linking precursors should be lower than the scorch safety time. Here we consider scorch safety time as just only 2%

achievement of the full torque (ΔM); hence, the scorch safety time, which is a specific point, can be roughly considered to correlate with the rate of formation of cross-link precursor. As the rate formation of cross-links by the activation of cross-link precursors may occur according to the Coran kinetic model [22–24], the rate of formation of cross-links can be used to determine the activation energy of the cross-linking reaction. Different plots on Arrhenius equations with respect to t_2 and R_v and their corresponding activation energies for EV curing systems are provided in Figure 7a–d. From Figure 7a, it can be seen that the plots are very linear and hence can be useful to predict the scorch safety time at other temperatures beyond the experimental temperatures. From the activation energy plots in Figure 7b, it can be concluded that the addition of MgO can reduce the activation energy of the formation of cross-link precursors. This could be due to faster decomposition of accelerator by MgO that can enhance the rate of formation of cross-link precursors [9,10]. In this perspective, ZnO has a slight delay in decomposing the accelerator. The different Arrhenius plots in Figure 7c also have shown excellent linearity. From the corresponding activation energy of vulcanization in Figure 7d, the addition of MgO can largely reduce the activation energy of vulcanization. It was found that ZnO or zinc ions can stabilize the cross-link precursors through coordination bonding [11,42], whereas MgO may not stabilize the cross-link precursors and have lower stability that can more easily decompose and form cross-link precursors than ZnO-only compounds. Because of the lowering activation energies of cross-link precursors and cross-link formation, ZnO- and MgO-based binary compounds show a much higher rate of vulcanization compared to single cure activator-containing compounds (Figure 5c).

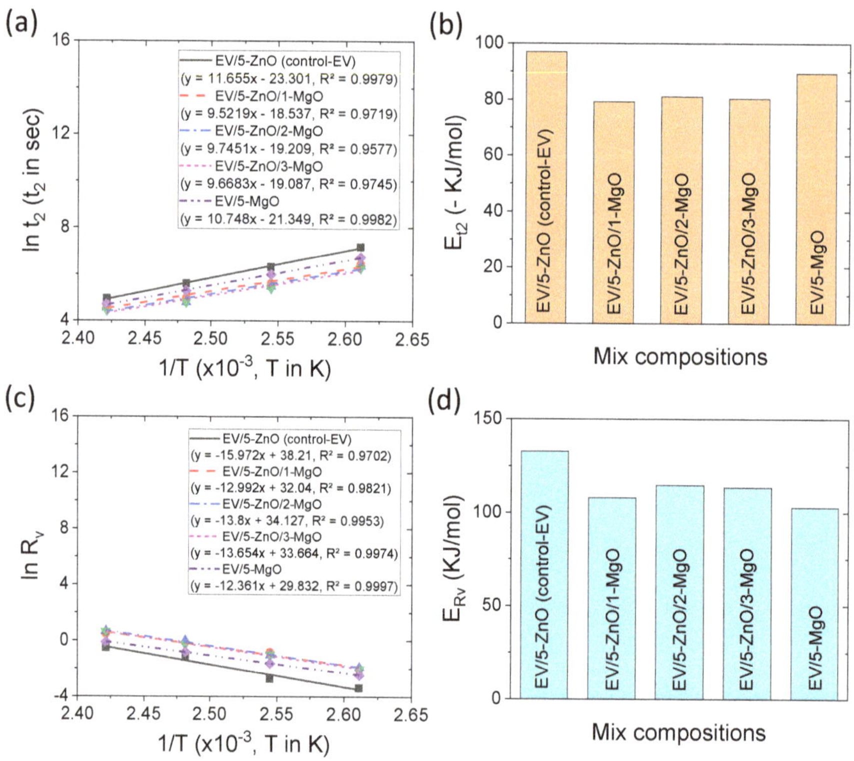

Figure 7. Arrhenius plots for activation energies of EV curing systems: (**a**) Arrhenius plot with respect to scorch safety time, (**b**) activation energy for cross-link precursor formation, (**c**) Arrhenius plot with respect to true cross-linking rate, and (**d**) activation energy of cross-linking formation.

Different Arrhenius plots and the corresponding activation energies for SEV curing systems are provided in Figure 8a–d. From Figure 8a, the Arrhenius plots with respect to t_2 are also very linear and can be useful to determine t_2 at any arbitrary temperatures beyond the experimental temperatures. The corresponding activation energy of cross-link precursors can be found in Figure 8b. From this figure, it can be seen that like EV curing systems, the addition of MgO to the SEV curing systems can reduce the energy of activation. The lower activation energies in the binary curing systems could be due to the synergistic effect of simultaneous decomposition of accelerator by MgO and stabilization of cross-link precursors through Zn-complexation [11,42]. The Arrhenius plots with respect to R_v for SEV curing systems have been provided in Figure 8c. The plots are very linear and follow the Arrhenius relationship. The corresponding activation energies are provided in Figure 8d. It has been found that there are no significant differences in the activation energies after the addition of MgO in the binary cure activator systems from the control-SEV system. If we compare the activation energies in Figures 7b,d and 8b,d, it can be found that a decreasing accelerator-to-sulfur ratio can reduce the activation energies of different vulcanization steps. It is believed that with increasing sulfur content in the accelerator-to-sulfur ratio, the chain length of sulfur bridges in the cross-link precursors increases and becomes easier to decompose with temperature.

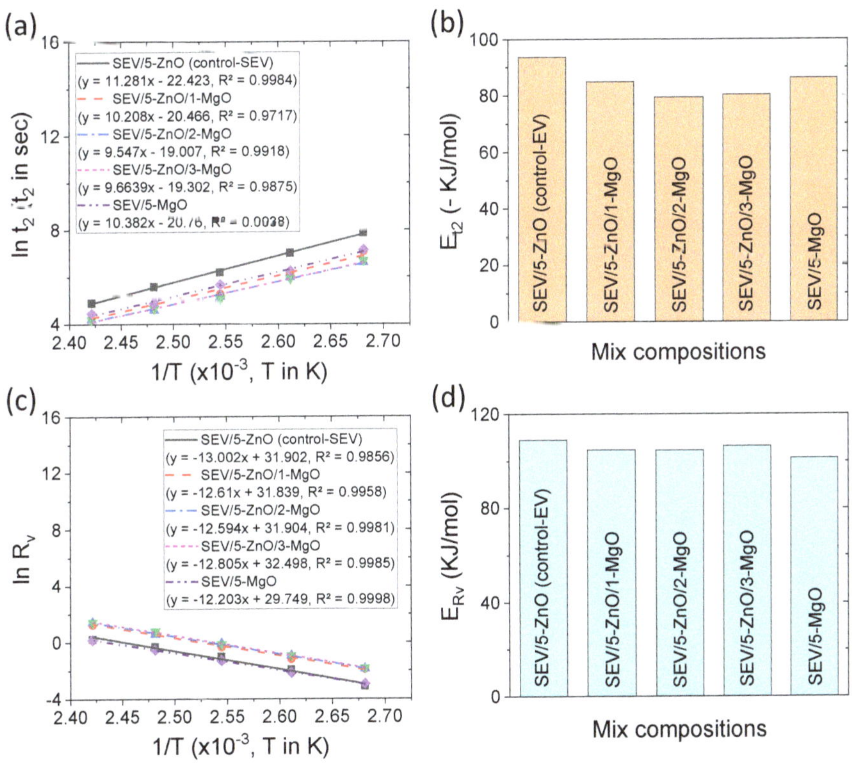

Figure 8. Arrhenius plots for activation energies of SEV curing systems: (**a**) Arrhenius plot with respect to scorch safety time, (**b**) activation energy for cross-link precursor formation, (**c**) Arrhenius plot with respect to true cross-linking rate, and (**d**) activation energy of cross-linking formation.

Joseph et al. [43] extensively discussed the detailed mechanism of accelerated sulfur vulcanization, focusing on the formation of zinc-complex compounds by accelerators and activators. Typically, stearic acid and amines can serve as ligands to these zinc complexes.

While the vulcanization mechanism is well understood, it involves complex pathways that can be challenging to represent concisely. However, Alam et al. [44] simplified the mechanism of accelerated sulfur vulcanization to facilitate a better understanding of the interactions between accelerators and activators. The accelerated sulfur vulcanization in the presence of ZnO- and MgO-based binary cure activators can be understood from the different paths presented in Figure 9 (paths 1–6). From path 1, the accelerator undergoes sulfur incorporation into its molecule to form a polysulfide form [44]. This compound reacts with H_2S from primary cross-links as shown by S-S radicals to the rubber chains, and the thiuram compound converts to its reduced state as thiocarbamic acid [44,45] through path 2. Some of the thiocarbamic acid decomposes to form free amine and CS_2 and reduces the amount of accelerator concentration from the vulcanization system followed by path 3. However, in the presence of cure activators such as MgO or ZnO, the thiocarbamic acid is trapped as a thiocarbamate compound followed by paths 4 and 5. These thiocarbamate compounds can undergo oxidation with added sulfur and again produce sulfur cross-links as shown in paths 1 and 2. According to previous reports, it was postulated that MgO can react faster with in situ generated thiocarbamic acid compared to ZnO and results in a higher rate of vulcanization [10]. However, MgO-activated compounds produced a lower cross-link density represented by lower ΔM value. In the presence of ZnO, the rate of vulcanization is low because it slowly converts thiocarbamic acid into a zinc-based active sulfurating complex. However, it can give a higher cross-link density to the rubber compound as evidenced by higher ΔM values, which might be due to the catalytic activity of zinc-based cross-linking precursors [42]. In the presence of both MgO and ZnO, the kinetically controlled magnesium-thiocarbamate may undergo mutual interactions with ZnO to produce a thermodynamically more stable zinc thiocarbamate-type active sulfurating complex as shown by path 6.

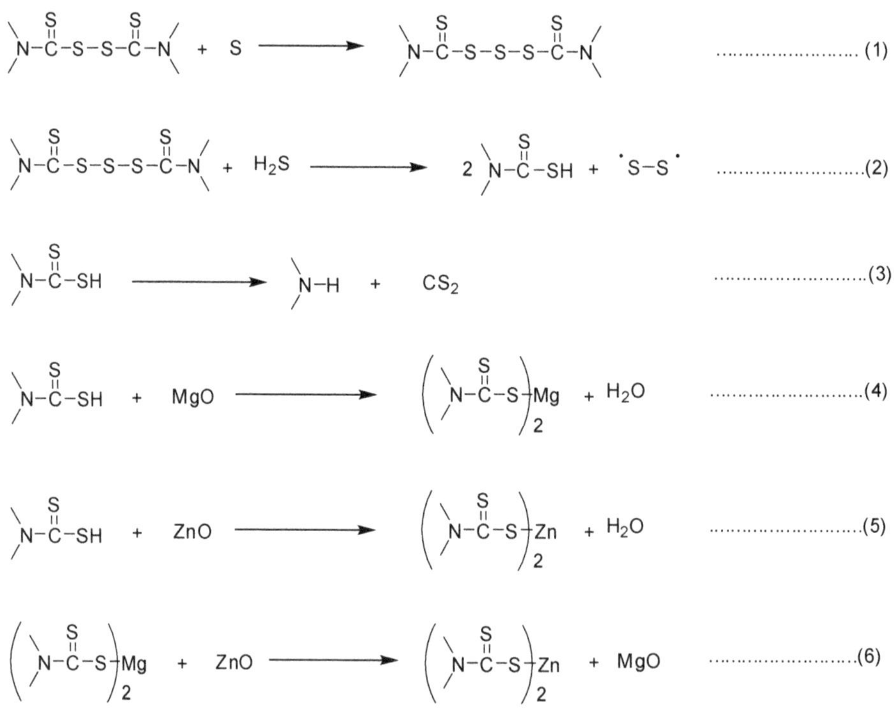

Figure 9. Possible interactions between accelerator, activators, and sulfur in ZnO- and MgO-based binary cure activator systems.

It has been found that with increasing sulfur content from EV to SEV curing systems, the ΔM values for MgO- and ZnO-containing binary cure activator systems have lower values compared to the control-SEV system at higher temperatures. It is believed that the concentration of thermodynamically controlled zinc dithiocarbamate can be increased with temperature to enhance the ΔM value in the control-SEV vulcanizate. Moreover, it was believed that zinc ions can better stabilize the cross-link precursors than magnesium ions. During the cross-linking process, some zinc-based precursors undergo cross-linking and some regenerate new cross-link precursors. Due to such type of catalytic activity, ZnO-based cure activator always produces a higher cross-link density. On the other hand, MgO may not stabilize the cross-link precursor, but it can faster decompose the accelerator or cross-link precursor to generate polysulfide cross-links as evidenced by lower ΔM values. In the presence of both cure activators, some catalytic and non-catalytic cross-linking precursors may form and show faster decomposition of cross-link precursors with an average value of cross-link density. Moreover, it was observed that a higher amount of stable disulfide cross-links was formed in Zn/Mg-hybrid oxide rather than ZnO-only compounds with a higher number of mono-sulfidic cross-links [17]. Since disulfide links contain double the sulfur atoms than mono-sulfidic cross-links, disulfide-rich vulcanizate is obviously expected to have lower cross-link density. Although ZnO- and MgO-containing binary cure activators introduce lower ΔM values in some cases, due to the presence of disulfide cross-links in the vulcanizate, it might have better fatigue resistance properties and could be important in the tire industry to produce low-fatigue tires [46]. Thus, in the presence of MgO and ZnO, the compounds not only include better rheometric mechanical properties but also provide faster curing kinetics.

4. Conclusions

This article represents the reduction in vulcanization temperature by the addition of MgO as a co-cure activator in different ZnO-based curing systems such as EV and SEV. Curing study results indicate that MgO as a co-cure activator can enhance the curing kinetics at different stages of rubber vulcanization. The improvements in the curing kinetics via the addition of MgO can result in a reduced optimum curing time with improved rate of vulcanization in both curing systems. With increasing sulfur content and reducing accelerator concentrations from EV to SEV curing systems, the rate of vulcanization can be further improved, but it results in lower delta torque in the ZnO- and MgO-based co-cure activators containing vulcanizates compared to the control-SEV curing system. Hence, to obtain better mechanical properties in the ZnO- and MgO-based co-cure activating vulcanization system, the accelerator-to-sulfur ratio should not be lower than in the SEV vulcanization system. Overall, a very small amount, approximately 1 to 2 phr of MgO, can substantially reduce the vulcanization temperature by about 20 °C in comparison to the controlled vulcanizates while retaining similar vulcanization kinetics with better rheometric mechanical properties. The reduced optimum curing time by using ZnO and MgO-based binary cure activators suggests that the vulcanization of rubber can be performed even below 100 °C with a reasonable vulcanization time and better curing kinetics. The substantial improvements in the curing kinetics via the addition of MgO- to the ZnO-based curing systems might be due to reduced activation energies in different vulcanization stages followed by mutual interactions between the two cure activators.

Supplementary Materials: The following supporting information can be downloaded at: https://www.mdpi.com/article/10.3390/polym16070876/s1, Table S1: Specific rheometric mechanical and kinetic parameters at 100 °C; Table S2: Specific rheometric mechanical and kinetic parameters at 110 °C; Table S3: Specific rheometric mechanical and kinetic parameters at 120 °C; Table S4: Specific rheometric mechanical and kinetic parameters at 130 °C; and Table S5: Specific rheometric mechanical and kinetic parameters at 140 °C.

Author Contributions: Conceptualization, M.N.A.; methodology, M.N.A.; validation, M.N.A., V.K., S.U.J. and S.-S.P.; formal analysis, M.N.A., V.K. and S.U.J.; investigation, M.N.A.; data curation, M.N.A.; writing—original draft preparation, M.N.A.; writing—review and editing, M.N.A., V.K., S.U.J. and S.-S.P.; visualization, M.N.A. and S.-S.P.; supervision, S.-S.P.; project administration, S.-S.P.; funding acquisition, S.-S.P. All authors have read and agreed to the published version of the manuscript.

Funding: This research received no external funding.

Institutional Review Board Statement: Not applicable.

Data Availability Statement: Data are contained within the article.

Conflicts of Interest: The authors declare no conflicts of interest.

References

1. Thomas, S.; Stephen, R. (Eds.) *Rubber Nanocomposites: Preparation, Properties, and Applications*; John Wiley & Sons: Hoboken, NJ, USA, 2010.
2. White, J.L. Rheological Behavior and Processing of Unvulcanized Rubber. In *Science and Technology of Rubber*, 3rd ed.; Mark, J.E., Erman, B., Eirich, F.R., Eds.; Academic Press: Cambridge, MA, USA, 2005; pp. 237–319.
3. Akiba, M.A.; Hashim, A.S. Vulcanization and crosslinking in elastomers. *Prog. Polym. Sci.* **1997**, *22*, 475–521. [CrossRef]
4. Ghosh, P.; Katare, S.; Patkar, P.; Caruthers, J.M.; Venkatasubramanian, V.; Walker, K.A. Sulfur vulcanization of natural rubber for benzothiazole accelerated formulations: From reaction mechanisms to a rational kinetic model. *Rubber Chem. Technol.* **2003**, *76*, 592–693. [CrossRef]
5. Bian, Q. Waste heat: The dominating root cause of current global warming. *Environ. Syst. Res.* **2020**, *9*, 8. [CrossRef]
6. Alam, M.N.; Mandal, S.K.; Debnath, S.C. Bis (N-benzyl piperazino) thiuram disulfide and dibenzothiazyl disulfide as synergistic safe accelerators in the vulcanization of natural rubber. *J. Appl. Polym. Sci.* **2012**, *126*, 1830–1836. [CrossRef]
7. Lee, N.; Lee, B.K.; Jeong, S.; Yi, G.Y.; Shin, J. Work environments and exposure to hazardous substances in Korean tire manufacturing. *Saf. Health Work.* **2012**, *3*, 130–139. [CrossRef] [PubMed]
8. Bornstein, D.; Pazur, R.J. The sulfur reversion process in natural rubber in terms of crosslink density and crosslink density distribution. *Polym. Test.* **2020**, *88*, 106524. [CrossRef]
9. Alam, M.N.; Kumar, V.; Potiyaraj, P.; Lee, D.J.; Choi, J. Synergistic activities of binary accelerators in presence of magnesium oxide as a cure activator in the vulcanization of natural rubber. *J. Elastomers Plast.* **2022**, *54*, 123–144. [CrossRef]
10. Alam, M.N.; Kumar, V.; Park, S.S. Advances in Rubber Compounds Using ZnO and MgO as Co-Cure Activators. *Polymers* **2022**, *14*, 5289. [CrossRef]
11. Heideman, G.; Datta, R.N.; Noordermeer, J.W.; Baarle, B.V. Influence of zinc oxide during different stages of sulfur vulcanization. Elucidated by model compound studies. *J. Appl. Polym. Sci.* **2005**, *95*, 1388–1404. [CrossRef]
12. Qin, X.; Xu, H.; Zhang, G.; Wang, J.; Wang, Z.; Zhao, Y.; Wang, Z.; Tan, T.; Bockstaller, M.R.; Zhang, L.; et al. Enhancing the performance of rubber with nano ZnO as activators. *ACS Appl. Mater. Interfaces* **2020**, *12*, 48007–48015. [CrossRef]
13. Mostoni, S.; Milana, P.; Di Credico, B.; D'Arienzo, M.; Scotti, R. Zinc-based curing activators: New trends for reducing zinc content in rubber vulcanization process. *Catalysts* **2019**, *9*, 664. [CrossRef]
14. Kar, S.; Bhowmick, A.K. Nanostructured magnesium oxide as cure activator for polychloroprene rubber. *J. Nanosci. Nanotechnol.* **2009**, *9*, 3144–3153. [CrossRef] [PubMed]
15. Guzmán, M.; Vega, B.; Agulló, N.; Giese, U.; Borrós, S. Zinc oxide versus magnesium oxide revisited. Part 1. *Rubber Chem. Technol.* **2012**, *85*, 38–55. [CrossRef]
16. Guzmán, M.; Vega, B.; Agulló, N.; Borrós, S. Zinc oxide versus magnesium oxide revisited. Part 2. *Rubber Chem. Technol.* **2012**, *85*, 56–67. [CrossRef]
17. Guzmán, M.; Reyes, G.; Agulló, N.; Borrós, S. Synthesis of Zn/Mg oxide nanoparticles and its influence on sulfur vulcanization. *J. Appl. Polym. Sci.* **2011**, *119*, 2048–2057. [CrossRef]
18. Roy, K.; Alam, M.N.; Mandal, S.K.; Debnath, S.C. Preparation of zinc-oxide-free natural rubber nanocomposites using nanostructured magnesium oxide as cure activator. *J. Appl. Polym. Sci.* **2015**, *132*. [CrossRef]
19. Siti, N.Q.M.; Kawahara, S. Evaluating performance of magnesium oxide at different sizes as activator for natural rubber vulcanization. *Int. J. Adv. Chem. Eng. Biol. Sci.* **2016**, *3*, 97–101.
20. Silva, A.A.; Rocha, E.B.; Furtado, C.R.; Sousa, A.M.F.; Carvalho, N.M. Magnesium Oxide Biosynthesized with Camellia sinensis Extract as Activator in Nitrile Rubber Vulcanization. *Polym. Bull.* **2021**, *78*, 6205–6219. [CrossRef]
21. Da Silva, A.A.; de Souza, M.A.; F de Sousa, A.M.; Carvalho, N.M.; Furtado, C.R. Magnesium oxide synthesized with Alpinia zerumbet leaf extracts as a sustainable alternative to zinc oxide in nitrile rubber compounds: A comparative vulcanization kinetics investigation. *J. Appl. Polym. Sci.* **2024**, *141*, e54945. [CrossRef]
22. Sirqueira, A.S.; Soares, B.G. The effect of functionalized ethylene propylene diene rubber (EPDM) on the kinetics of sulfur vulcanization of normal rubber/EPDM blends. *Macromol. Mater. Eng.* **2007**, *292*, 62–69. [CrossRef]

23. Arrillaga, A.; Zaldua, A.M.; Atxurra, R.M.; Farid, A.S. Techniques Used for Determining Cure Kinetics of Rubber Compounds. *Eur. Polym. J.* **2007**, *43*, 4783–4799. [CrossRef]
24. Coran, A.Y. Vulcanization. Part VI. A model and treatment for scorch delay kinetics. *Rubber Chem. Technol.* **1964**, *37*, 689–697. [CrossRef]
25. Coran, A.Y. Vulcanization. Part VII. Kinetics of sulfur vulcanization of natural rubber in presence of delayed-action accelerators. *Rubber Chem. Technol.* **1965**, *38*, 1–14. [CrossRef]
26. Ding, R.; Leonov, A.I. A kinetic model for sulfur accelerated vulcanization of a natural rubber compound. *J. Appl. Polym. Sci.* **1996**, *61*, 455–463. [CrossRef]
27. Zhao, F.; Wu, C.; Zhang, P. Influence of Zinc Oxide Amount on Cure Kinetics of Sulfur-Vulcanized Natural Rubber. *J. Macromol. Sci. Part B* **2011**, *50*, 1162–1170. [CrossRef]
28. Hadi, F.A.; Kadhim, R.G. A Study of the Effect of Nano Zinc Oxide on Cure Characteristics and Mechanical Properties of Rubber Composites. *J. Phys. Conf. Ser.* **2019**, *1234*, 012043. [CrossRef]
29. Heideman, G.; Datta, R.N.; Noordermeer, J.W.M.; van Baarle, B. Zinc loaded clay as activator in sulfur vulcanization: A new route for zinc oxide reduction in rubber compounds. *Rubber Chem. Technol.* **2004**, *77*, 336–355. [CrossRef]
30. Heideman, G.; Datta, R.N.; Noordermeer, J.W.M.; van Baarle, B. Effect of zinc complexes as activator for sulfur vulcanization in various rubbers. *Rubber Chem. Technol.* **2005**, *78*, 245–257. [CrossRef]
31. Basu, D.; Das, A.; Stöckelhuber, K.W.; Wagenknecht, U.; Heinrich, G. Advances in layered double hydroxide (LDH)-based elastomer composites. *Prog. Polym. Sci.* **2014**, *39*, 594–626. [CrossRef]
32. Przybyszewska, M.; Zaborski, M.; Jakubowski, B.; Zawadiak, J. Zinc chelates as new activators for sulphur vulcanization of acrylonitrile-butadiene elastomer. *Express Polym. Lett.* **2009**, *3*, 256–266. [CrossRef]
33. Sahoo, S.; Maiti, M.; Ganguly, A.; George, J.J.; Bhowmick, A.K. Effect of zinc oxide nanoparticles as cure activator on the properties of natural rubber and nitrile rubber. *J. Appl. Polym. Sci.* **2007**, *105*, 2407–2415. [CrossRef]
34. Wang, Z.; Lu, Y.; Liu, J.; Dang, Z.; Zhang, L.; Wang, W. Preparation of nano-zinc oxide/EPDM composites with both good thermal conductivity and mechanical properties. *J. Appl. Polym. Sci.* **2011**, *119*, 1144–1155. [CrossRef]
35. Maciejewska, M.; Sowińska, A.; Kucharska, J. Organic Zinc Salts as Pro-Ecological Activators for Sulfur Vulcanization of Styrene–Butadiene Rubber. *Polymers* **2019**, *11*, 1723. [CrossRef] [PubMed]
36. Yang, Z.; Huang, Y.; Xiong, Y. A functional modified graphene oxide/nanodiamond/nano zinc oxide composite for excellent vulcanization properties of natural rubber. *RSC Adv.* **2020**, *10*, 41857–41870. [CrossRef] [PubMed]
37. Wu, S.; Xiao, C.; Kong, S.; Li, B.; Yang, Z.; Tang, Z.; Liu, F.; Guo, B. Carbon nanodots as an eco-friendly activator of sulphur vulcanization in diene-rubber composites. *Compos. Commun.* **2021**, *25*, 100755. [CrossRef]
38. Wang, Z.; Hou, Z.; Liu, X.; Gu, Z.; Li, H.; Chen, Q. Preparation of Zinc Oxide with Core–Shell Structure and Its Application in Rubber Products. *Polymers* **2023**, *15*, 2353. [CrossRef] [PubMed]
39. Blackman, E.J.; McCall, E.B. Relationships between the structures of natural rubber vulcanizates and their thermal and oxidative aging. *Rubber Chem. Technol.* **1970**, *43*, 651–663. [CrossRef]
40. Alam, M.N.; Debnath, S.C.; Boondamnoen, O.; Kumar, K.N.; Kim, J.H.; Choi, J. Synergistic combination of 2-mercaptobenzothiazole (MBT) and nitrosoamine-safe thiuram disulfide as advanced rubber vulcanizing accelerators. *J. Elastomers Plast.* **2022**, *54*, 1061–1077. [CrossRef]
41. Kemp, A.R.; Malm, F.S.; Stiratelli, B. Solubility and Diffusion of Sulfur in Synthetic Elastomers. *Rubber Chem. Technol.* **1944**, *17*, 693–703. [CrossRef]
42. Coran, A.Y. Vulcanization. In *Science and Technology of Rubber*; Academic Press: Cambridge, MA, USA, 1994; pp. 339–385.
43. Joseph, A.M.; George, B.; Madhusoodanan, K.N.; Alex, R. Current status of sulphur vulcanization and devulcanization chemistry: Process of vulcanization. *Rubber Sci.* **2015**, *28*, 82–121.
44. Alam, M.N.; Mandal, S.K.; Debnath, S.C. Effect of zinc dithiocarbamates and thiazole-based accelerators on the vulcanization of natural rubber. *Rubber Chem. Technol.* **2012**, *85*, 120–131. [CrossRef]
45. Craig, D.; Juve, A.E.; Davidson, W.L. Tetramethylthiuram Disulfide Vulcanization of Extracted Rubber. I. Introduction and Compounding. *Rubber Chem. Technol.* **1951**, *24*, 254–262. [CrossRef]
46. Cox, W.L.; Parks, C.R. Effect of curing systems on fatigue of natural rubber vulcanizates. *Rubber Chem. Technol.* **1966**, *39*, 785–797. [CrossRef]

polymers

MDPI

Article

Electrically Conductive Natural Rubber Composite Films Reinforced with Graphite Platelets

Veerapat Kitsawat, Saranrat Siri and Muenduen Phisalaphong *

Bio-Circular-Green Economy Technology & Engineering Center, BCGeTEC, Department of Chemical Engineering, Faculty of Engineering, Chulalongkorn University, Bangkok 10330, Thailand; kit.veerapat@gmail.com (V.K.); saranrat13030@gmail.com (S.S.)
* Correspondence: muenduen.p@chula.ac.th; Tel.: +66-2-218-6875

Abstract: Green natural rubber (NR) composites reinforced with synthetic graphite platelets, using alginate as a thickening and dispersing agent, were successfully developed to improve mechanical properties, chemical resistance, and electrical conductivity. The fabrication was performed using a latex aqueous microdispersion process. The research demonstrated the effective incorporation of graphite platelets into the NR matrix up to 60 parts per hundred rubbers (phr) without causing agglomeration or phase separation. Graphite incorporation significantly improved the mechanical strength of the composite films. NR with 60 phr of graphite exhibited the highest Young's modulus of 12.3 MPa, roughly 100 times that of the neat NR film. The reinforcement also strongly improved the hydrophilicity of the composite films, resulting in a higher initial water absorption rate compared to the neat NR film. Moreover, the incorporation of graphite significantly improved the chemical resistance of the composite films against nonpolar solvents, such as toluene. The composite films exhibited biodegradability at about 21% to 30% after 90 days in soil. The electrical conductivity of the composite films was considerably enhanced up to 2.18×10^{-4} S/cm at a graphite loading of 60 phr. According to the improved properties, the developed composites have potential applications in electronic substrates.

Keywords: alginate; composites; electrical conductivity; graphite; latex; natural rubber

1. Introduction

The excessive accumulation of plastic waste, including the widespread distribution of microplastics in the oceans, has become a major global concern. This has generated interest in the development of biodegradable materials to replace non-biodegradable petroleum-based products [1]. Natural rubber (NR) is an incredibly versatile polymer that is widely used in various industries. It is a biopolymer of *cis*-1,4-polyisoprene, containing numerous double bonds in its structure [2,3]. Thailand has been known as the world's largest NR producer and exporter [4]. The production of NR primarily originates from *Hevea Brasiliensis*, commonly known as the Para rubber tree. It is extensively utilized in the production of various rubber products, including automotive tires, gloves, condoms, clothes, threads, adhesives, medical devices, and more. When NR is exported, it usually undergoes processing to create intermediate products, such as ribbed smoke sheets (RSS), un-smoked sheets (USS), block rubber (standard Thai rubber, STR), concentrated latex, compound rubber, and crepe rubber. However, Thailand sometimes is confronted by NR problems, such as fluctuating and unstable pricing caused by supply and demand imbalances [5].

Although NR exhibits many exceptional properties, including high flexibility, elasticity, excellent tear strength, toughness, and abrasion resistance, it also has limitations in terms of poor chemical resistance, limited temperature usage due to temperature sensitivity caused by environmental factors such as weather, sunlight, and ozone exposure [6–8]. The incorporation of fillers has been shown to be an effective strategy to improve the properties of NR. Fillers are substances that serve to enhance specific attributes of the material. Among them, carbonaceous fillers have received significant amounts of attention for their ability to enhance the electrical conductivity, thermal conductivity, fire resistance, and mechanical properties of polymers [9–12]. Common carbonaceous fillers include graphite (G), graphene (GP), carbon black (CB), fullerenes, carbon fibers (CF), and carbon nanotubes (CNT) [13].

Graphite is a microscale crystalline carbon allotrope commonly employed in the fabrication of polymer composites. It can be obtained naturally or synthetically. The structure of graphite is made up of parallel layers of graphene sheets with carbon atoms organized in a hexagonal pattern via strong covalent bonds. Van der Waals forces hold the parallel sheets together, giving the material anisotropic properties [14–16]. The π orbitals and free electrons are indeed evenly distributed throughout the graphene layers, which contributes to its exceptional electrical and thermal conductivity. Furthermore, graphite offers the advantages of being chemically and thermally stable, cost-effective, readily available, and versatile [17]. These factors make it appealing for various applications on commercial scales.

Both graphite and NR are hydrophobic or non-polar materials. However, due to graphite's high density, surface energy, and limited wettability, it is challenging to disperse it uniformly in the NR matrix [18–20]. In most cases, rubber composites are produced using the melt compounding technique, employing an internal mixer and a two-roll mill. However, when mixing small-sized fillers, there is a possibility of producing an inconsistent mixture or encountering agglomerated fillers. To address these challenges, this research employs the process of latex aqueous microdispersion to achieve more effective distribution and dispersion of graphite in the NR matrix.

Alginate is a biopolymer derived primarily from brown seaweed. It has captured researchers' attention due to its easy availability, non-toxic nature, biocompatibility, gel-forming property, and ability to degrade naturally [21,22]. When incorporated into NR, it offers a range of benefits, such as improving biodegradability, water-retention properties, processability, and both chemical and thermal stability [22–24]. Moreover, it acts as an efficient dispersing and thickening agent, increasing the viscosity of the rubber matrix. This elevated viscosity minimizes agglomeration and sedimentation, leading to improved filler dispersion and enhancing the uniformity and stability of the composites. Previous studies demonstrated that alginate could facilitate the mixing of various fillers in NR matrices [25–29]. This insight encourages our exploration into incorporating alginate to disperse graphite platelets, aiming to produce uniform composite films with high graphite loading. Alginate is anticipated to prevent the restacking and agglomeration of graphite platelets.

The objective of this study is to fabricate unvulcanized NR composite films incorporated with synthetic graphite platelets, which was facilitated by the addition of alginate as a thickening and dispersing agent. The effects of graphite as a reinforcing agent on the mechanical and thermal properties, hydrophilicity, chemical resistance, electrical conductivity, and biodegradability of NR composite films were investigated. The analysis involves comparing the properties of the composite films with different graphite loading as well as evaluating their performance against neat NR films.

2. Materials and Methods

2.1. Material

Natural rubber latex (NRL) with 60% dry rubber content (DRC) was purchased from the Rubber Research Institute of Thailand, Bangkok, Thailand. Synthetic graphite in the platelet form with a density of 2.26 g/cm^3 and a particle size of less than 20 μm was purchased from Sigma-Aldrich (Buchs, Switzerland). The sodium salt of alginic acid (sodium alginate, ALG) was purchased from Acros Organics (Thermo Fisher Scientific Co., Ltd., Geel, Belgium).

2.2. Fabrication of Natural Rubber, Graphite, and Alginate (NR-G-ALG) Composite Films

The amount of graphite used in this study varied at 0, 20, 30, 40, 50, and 60 phr, corresponding to weight percentages of the composite film (wt.% of 0, 15.4, 21.4, 26.7, 31.3, and 35.3, respectively. The preparation of composite films was carried out using the latex aqueous microdispersion process, as previously reported by our research group [25,26,30–32]. The alginate solution was prepared by dispersing ALG powder in deionized water. The mixture was mechanically stirred and heated at 40 °C for 2 h until complete dissolution of ALG powder, resulting in a 1% *w/v* ALG solution. Afterward, graphite was added into 30 mL ALG solution and thoroughly mixed the slurry under mechanical stirring at 500 rpm for 45 min. Subsequently, the slurry was exposed to sonication in ultrasonic cleaner at 500 watts for 10 min to further disperse the filler particles and ensure homogeneity of the mixture. Following this, 5 g of 60% DRC NRL was gradually added into the slurry while continuously mixing by mechanical stirring until the homogenous mixture was obtained. Thereafter, the mixture was poured into a glass Petri dish and dried at 40 °C for 20 h in an air-circulating oven to obtain NR composite films. Afterward, the composite films were washed with deionized water to remove any impurities, followed by drying in an oven at 40 °C to eliminate residual moisture. The composite films were stored in an airtight container prior to characterization. The composite films of NR combined with graphite and ALG were denoted as NR-Gx-ALG, where x represents the content of graphite. For instance, the label NR-G20-ALG indicates the incorporation of 20 phr of graphite in the composite film. The illustrative schematic diagram for the fabrication of NR-G-ALG composite films is shown in Figure 1.

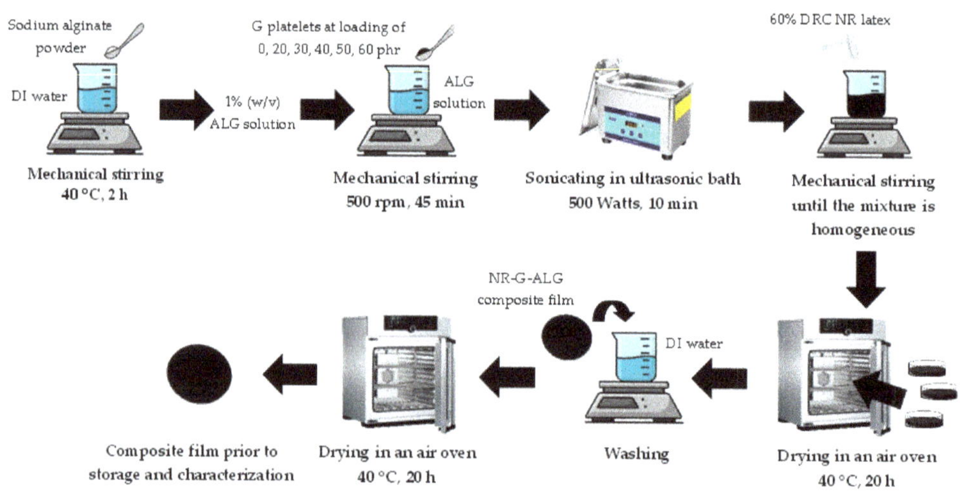

Figure 1. The illustrative schematic diagram for the fabrication of NR-G-ALG composite films.

2.3. Characterization

The functional groups and possible polymer–filler interaction in the NR and NR composite films were analyzed using a Fourier transform infrared spectrometer (FTIR) (Spectrum One, Perkin Elmer, Shelton, CT, USA) by examining wavenumbers from 4000 to 400 cm^{-1} with a resolution of 4 cm^{-1}. FTIR spectra were acquired in Transmission mode.

The morphology of G platelets, the surface and cross-section of the composite films, and the distribution of platelets in the composite films were examined using a field-emission scanning electron microscope (FE-SEM) (Quanta 250 FEG, Hillsboro, OR, USA) equipped with an energy-dispersive X-ray spectroscopy system. The dried sample films were coated with a thin layer of gold using a JEC-1100E sputter coater and imaged at an accelerating voltage of 15 kV. The average size of graphite platelets was determined using ImageJ 1.54d software.

The crystallinity and structural information of G platelets and the composite films were investigated using an X-ray diffractometer (XRD) (Bruker D8 Discover, Karlsruhe, Germany) with Cu-Kα radiation in the 2θ range of 5–60°. The operation conditions included an accelerating voltage of 40 kV and an electric current of 40 mA. The degree of crystallinity was calculated using the formula below.

$$\text{Degree of Crystallinity } (\%) = \frac{\text{Crystalline area} \times 100}{\text{Total area}} \quad (1)$$

The mechanical properties of the NR and NR composite films, including tensile strength, Young's modulus, and elongation at break, were evaluated using a universal testing machine (UTM H10 KM, Redhill, UK) in accordance with the ASTM D882 standard [33]. The rectangular specimens with dimensions of 6 × 0.5 cm^2 were placed between the upper and lower jaws of the instrument, and a controlled tensile force was gradually applied until the point of breakage. At least five specimens were tested for each composition to ensure statistical significance.

The thermal stability of the NR, G, and NR composite films, which refers to their ability to resist thermal degradation assessed by measuring weight loss, was tested via thermogravimetric analysis (TGA) (NETZSCH STA 449 F3 Jupiter, Selb, Germany). The test was conducted in a nitrogen atmosphere with a temperature range from 30 to 700 °C and a heating rate of 10 °C/min. The sample weights employed in the test ranged from 20 to 30 mg.

The thermal properties of the NR, G, and NR composite films, such as the melting temperature (T_m), degradation temperature (T_d), and glass transition temperature (T_g) of the polymer, were analyzed using differential scanning calorimetry (DSC) (NETZSCH DSC 204 F1 Phoenix, Selb, Germany). The samples were heated in a nitrogen atmosphere from −100 to 300 °C at a rate of 10 °C/min, and thermal transitions, whether endothermic or exothermic, were recorded. The sample weights employed in the test ranged from 5 to 15 mg.

Water absorption capacity (WAC) serves as an indicator of the composite films' ability to interact with water molecules, which corresponds to the water-retention capabilities of the composite films. To measure WAC, the sample films were cut into 2 × 2 cm^2 and immersed in deionized water at room temperature. After a certain period of time, the films were removed from the water, and excess water at the surface was wiped off using KIMWIPES™ Delicate Task Wipers (Kimberly-Clark Corporation, Irving, TX, USA). The weight of the wet films was then measured, and the procedure was repeated until no further weight change was observed. At least three specimens were tested for each composition. WAC was calculated by using the formula where W_w and W_d represent the weight of wet and dry specimen films, respectively, as follows:

$$\text{WAC } (\%) = \frac{W_w - W_d}{W_d} \times 100\% \quad (2)$$

Toluene uptake (TU) serves as an indicator of the composite films' capacity to absorb the non-polar solvent, toluene. This characteristic relates to the films' chemical resistance. To measure TU, the sample films were cut into 2 × 2 cm^2 and immersed in toluene at room temperature. After a certain period of time, the films were removed from toluene. The weight of the swollen films was measured, and the procedure was repeated until no further weight change was observed. At least three specimens were tested for each composition. The TU value was calculated using the formula where W_b and W_a represent the weight of the specimen films before and after immersion in toluene, respectively, as follows:

$$\text{TU } (\%) = \frac{W_a - W_b}{W_b} \times 100\% \quad (3)$$

The biodegradability test of the NR and NR composite films in soil was conducted for a duration of 3 months. Each composition of the composite films was cut into 5×5 cm^2 and buried in soil flowerpots at a depth of 10 cm below the surface under ambient weather and climate conditions. The soil used was loam soil, primarily composed of coconut coir and composted raintree leaves. The moisture content in the soil was maintained within the range of 30 to 60% during the experiments. The samples were meticulously excavated and weighed after 1, 2, and 3 months of burial. For weight loss calculation, the excavated samples were rinsed and washed with deionized water and then dried in an air-circulating oven at 40 °C for 24 h before being weighed. The biodegradability rate of each composition was evaluated by the following equation, where W_f is the weight of the films after being excavated from the soil, and W_i is the weight of the films before being buried in the soil.

$$\text{Biodegradability } (\%) = \frac{W_i - W_f}{W_i} \times 100\% \tag{4}$$

The electrical properties of the composite films were evaluated using Electrochemical Impedance Spectroscopy (EIS) and Cyclic Voltammetry (CV) techniques (Squidstat Plus, Tempe, AZ, USA) and analyzed via VersaStudio 2.60.6 software. The sample films were securely positioned between two electrode plates. EIS testing was conducted over a frequency range from 10^5 Hz to 1 Hz with an AC voltage amplitude of 250 mV_{rms} at room temperature, and the Nyquist plot and Bode plot were obtained. The CV test was conducted at a scan rate of 0.1 V/s, running for 3 cycles. Ensuring precise and reliable measurements, the samples were carefully prepared and handled to prevent any surface defects or contamination that could impact the electrical properties. At least three specimens were tested for each composition. The electrical conductivity was calculated using the following equation, where σ represents the electrical conductivity $(\Omega \cdot cm)^{-1}$ or (S/cm), L is the sample thickness (cm), $|Z|$ is the calculated impedance magnitude (Ω), A is the electrode cross-sectional area (cm^2), Z' is the real part of the complex impedance (Ω), and Z'' is the imaginary part of the complex impedance (Ω).

$$\text{Electrical Conductivity } (\sigma) = \frac{L}{|Z| \times A} \tag{5}$$

where $|Z|$ is determined by the following equation:

$$|Z| = \sqrt{|Z'|^2 + |Z''|^2} \tag{6}$$

3. Results and Discussion

3.1. Fourier Transform Infrared (FTIR) Spectroscopy

The FTIR spectrum for NR, G, and NR composite films is illustrated in Figure 2. The neat NR primarily consists of *cis*-1,4-polyisoprene, and its spectrum exhibits distinct peaks at specific wavenumbers. The NR spectrum exhibits characteristic peaks at various wavenumbers, each corresponding to specific chemical functional groups, as follows. The stretching vibrations of the -OH group is at 3292 cm^{-1}, the asymmetric stretching vibration of the CH_3 group is at 2960 cm^{-1}, the symmetric stretching vibration of the -CH_2 group is at 2916 cm^{-1}, the symmetric stretching of -CH_2 is at 2852 cm^{-1}, the asymmetrical -CH_3 overtone deformation is at 2725 cm^{-1}, and the C=C stretching is at 1660 cm^{-1} [26,34,35]. The N-H in-plane bending and C-N stretching are attributed to the peptide bonds in proteins, observable at approximately 1541 cm^{-1} [36]. The C-H deformation is at 1446 cm^{-1}, the peak at about 1375 cm^{-1} is ascribed to –CH_3 bending, the C-H out-of-plane bending is at 833 cm^{-1}, and the C-C-C in-plane bending is at 570 cm^{-1} [26,32,36]. In contrast, the spectrum of G lacks significant peaks and remains almost entirely flat due to its high-order structure and chemical inertness. The same observation has been previously reported [37–39].

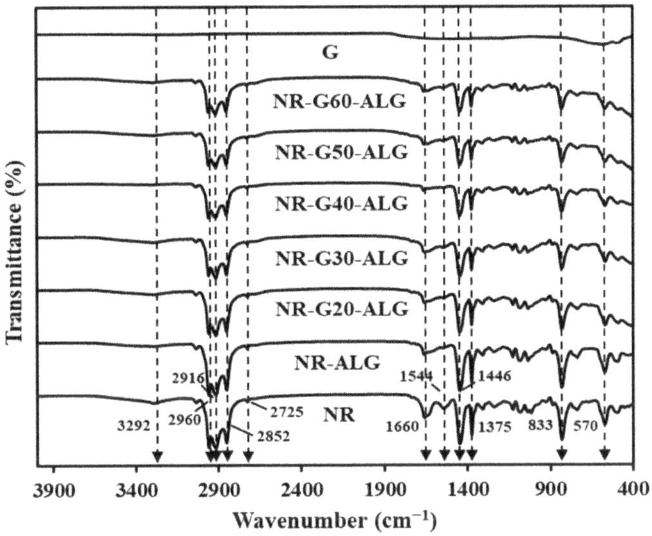

Figure 2. FTIR spectra of NR, G, NR-ALG, and NR-G-ALG composite films.

Across all the samples of composite films, the primary peaks assigned to the NR chemical structure were present. It was shown that no significant difference was observed in terms of peak numbers among the composites and NR samples. This consistency suggested the absence of active functional groups on the surface of G and indicated a lack of significant chemical interactions among NR, G, and ALG. Notably, the peaks that were indicative of NR behaviors exhibited a decrease in intensity with increasing G loading.

3.2. Field Emission-Scanning Electron Microscope (FE-SEM)

FE-SEM analysis was conducted on NR, G, and NR-G-ALG composite films. As shown in Figure 3, G platelets display a platelet-like and planar structure, with an average size of approximately 22.0 ± 7.1 µm (measured by ImageJ program). Figure 4 shows the uniform and smooth texture of the NR films. The NR-G-ALG films exhibited a comparatively rougher surface with a degree of roughness increasing with the higher amount of G loading. Nevertheless, these composite films demonstrated consistent uniformity, with the G platelets evenly dispersed and well-integrated into the NR matrix, as shown in both the surface and cross-sectional views. This indicated the effective dispersion of G platelets facilitated by ALG. The NR-G-ALG films showed the integration of G platelets in the NR matrix without phase separation and agglomeration, suggesting that ALG could serve as a suitable dispersing agent, enhancing the distribution and dispersion of G in the NR matrix. The low-concentration aqueous ALG solution exhibited high viscosity and contained numerous -OH and -COOH groups capable of intercalating into graphene layers [40]. This intercalation may prove beneficial in preventing the agglomeration of G platelets. During the film fabrication process, it was observed that, in the absence of ALG, G platelets faced challenges in achieving uniform dispersion within the NR matrix, and it resulted in the formation of clusters or agglomerates, leading to non-uniform composite films that might impact the overall properties of the composites.

Figure 3. FE-SEM images of graphite (G) platelets.

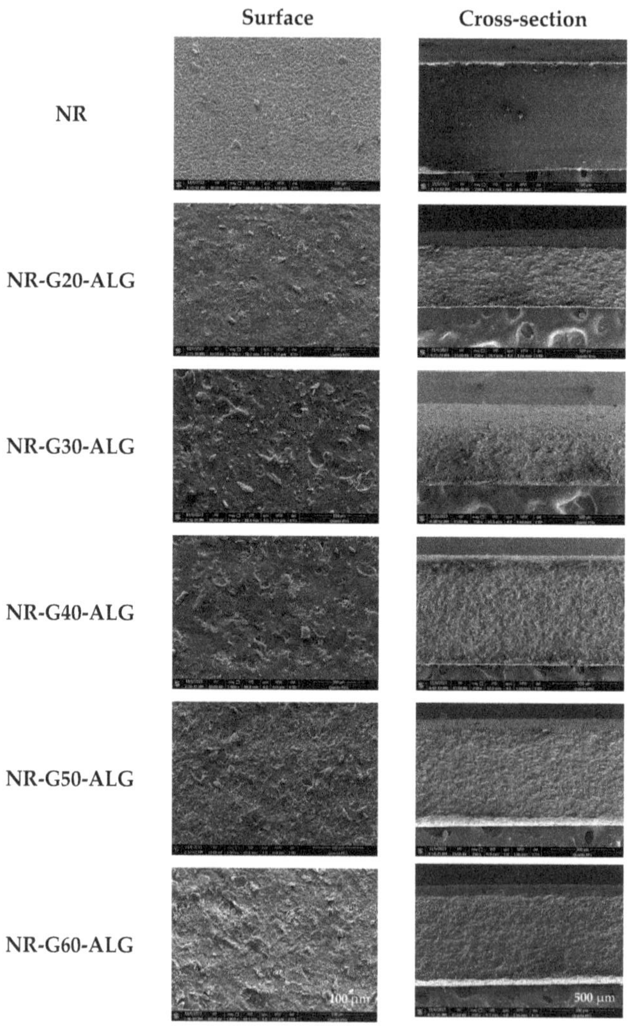

Figure 4. FE-SEM images of the surface morphologies and cross-sections of NR and NR-G-ALG at various loadings of 20 to 60 phr.

3.3. X-ray Diffraction (XRD)

In Figure 5, the XRD diffractograms of NR, G, and NR composite films are illustrated. The XRD diffractogram of G platelets exhibited a distinct, intense peak at 2θ = 26.4° and a minor peak at 54.5°, corresponding to the (002) and (004) crystallographic diffraction planes, confirming the highly ordered hexagonal crystalline lattice of G (JCPDS No. 00-041-1487) [41,42]. The neat NR film revealed a broad peak centered around at 2θ = 18°, indicating its amorphous polymer chain structure. The NR-G-ALG composite films displayed diffraction peaks corresponding to both NR and G characteristics. The intensity of the G peak increased with increasing G loading. The degree of crystallinity of the G platelets was determined to be 91.1%, where the degrees of crystallinity of NR composite films with G incorporations at 20, 30, 40, 50, and 60 phr were at 27.8%, 30.7%, 36.6%, 46.8%, and 48.3%, respectively. With the integration of G in the NR matrix, the XRD result showed an increase in the crystallinity of NR composite films related to G concentration.

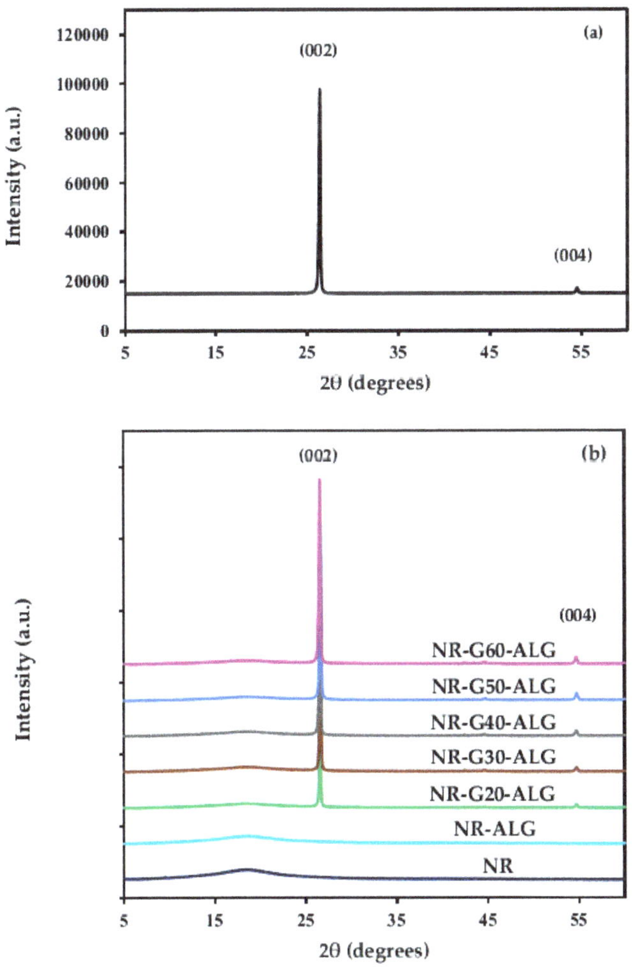

Figure 5. XRD diffractograms of (**a**) G (**b**) NR, NR-ALG, and NR-G-ALG composite films.

3.4. Mechanical Properties

The mechanical properties of the NR composite films, reinforced with various loadings of G and the neat NR film, were analyzed in Figure 6. The neat NR film exhibited tensile strength, Young's modulus, and elongation at break values of 5.33 MPa, 0.12 MPa, and 992.4%, respectively. The values of tensile strength and Young's modulus were significantly enhanced with the introduction of G platelets. For instance, the tensile strength of the NR film reinforced with 20 phr of G increased to 8.47 MPa, representing a substantial improvement of 1.6-fold over the neat NR film. However, with the increased incorporation of G beyond 20 phr, the tensile strength gradually decreased. This could be attributed to a reduction in the thickness of the NR matrix interlayers, which at higher G loadings might become insufficient to adequately coat all available filler surfaces [43]. Furthermore, because G possesses a high density and surface area [18], it is prone to agglomeration, especially at very high loading, which could result in reduced interfacial adhesion between the filler and matrix. Additionally, the structure of reinforcing fillers can have an effect on brittleness in fracture and reduced fracture toughness, thus may have a potential impact on the material's mechanical properties [44]. Under a study on polysulfone composites reinforced with different types of graphite, it was revealed that the initial addition of artificial graphite (AG) up to a certain amount could improve the tensile strength of the composite. However, when the AG loading exceeded 50% by weight, the tensile strength of the composite decreased, and further additions of AG continued to reduce the tensile strength [43].

Young's modulus quantifies a material's stiffness and its resistance to deformation under stress. Graphite, as a carbonaceous micro filler, has been known for its rigidity and stiffness [45]; therefore, the addition of G resulted in a sequential increase in Young's modulus of the composite films. The maximum Young's modulus value was achieved with NR composite films with G incorporation at 60 phr, reaching 12.33 MPa, or approximately 100 times greater than that of the neat NR film, which exhibited a low value of Young's modulus at only 0.12 MPa.

Due to the inherent elasticity and flexibility of NR, its elongation at break is exceptionally high. Therefore, the addition of a stiff and rigid filler resulted in a reduction in the elongation at the break of the NR composite. The decrease in elongation at break can be attributed to the heightened rigidity introduced by the presence of G, which restricts the material's capacity to stretch and deform before reaching the point of failure. It was shown that the elongation at break of the NR composite films with G incorporations varied between 480% and 377%, depending on the loading content of G, whereas the elongation of the neat NR film was 992%.

Our observation is aligned with prior investigations. The reinforcement of thermally reduced graphite oxide (RGOT) in the NR matrix showed increased tensile strength upon incorporating RGOT up to a certain amount [45]. However, the higher RGOT loading beyond a specific loading threshold (10 phr) led to a gradual decline in tensile strength and a reduction in the composite's elasticity [39]. Under the reinforcement of NR with carboxylated MWCNTs dispersed with sodium dodecyl sulfate (SDS), a 10-fold increase in Young's modulus and a 2-fold rise in tensile strength was obtained when compared to the neat NR [46]. The enhancement in mechanical strength could be related to the creation of a rigid network of reinforcing fillers within the NR latex structure, which might be facilitated by cross-linking via the functional groups found at the surface of the CNTs and the organic molecules present in the NR latex [46].

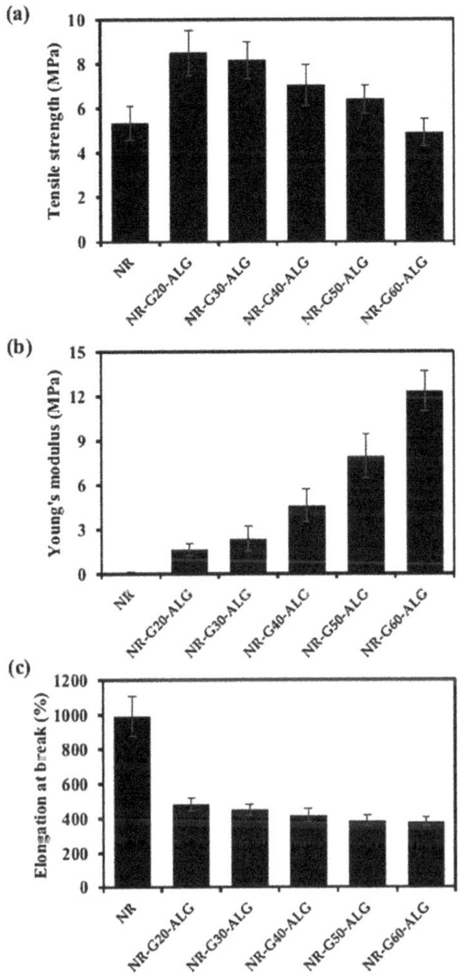

Figure 6. Mechanical properties of NR and NR-G-ALG composite films: (**a**) tensile strength, (**b**) Young's modulus, and (**c**) elongation at break.

3.5. Thermal Properties

Thermal stability is a crucial characteristic of electronic applications and materials, influencing both their processing and potential utility in high-temperature applications. The thermal properties of NR, G, and NR-G-ALG composite films were explored using TGA and DSC, as depicted in Figures 7 and 8, respectively. Table 1 presents thermal data for all samples, including decomposition temperatures at 5%, 10%, and 50%wt ($T_{5\%}$, $T_{10\%}$, and $T_{50\%}$) and residual mass at 700 °C.

The addition of ALG influences the decomposition temperature values, as its lower degradation temperature precedes that of NR. In the composite films, decomposition temperatures demonstrate an upward trend with increasing G content. Additionally, the incorporation of G contributes to an increased residual mass at 700 °C, highlighting its role in enhancing the thermal stability of the composite films. The crystalline nature of G platelets introduces a barrier effect, enhancing resistance to thermal degradation and impeding the release of decomposition products into the gas phase from the polymer [47].

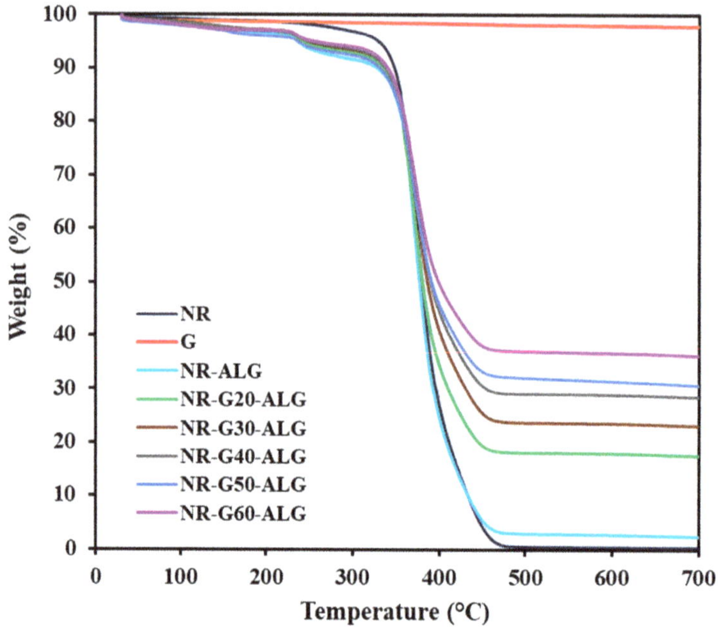

Figure 7. TGA curves of the NR, G, NR-ALG, and NR-G-ALG composite films.

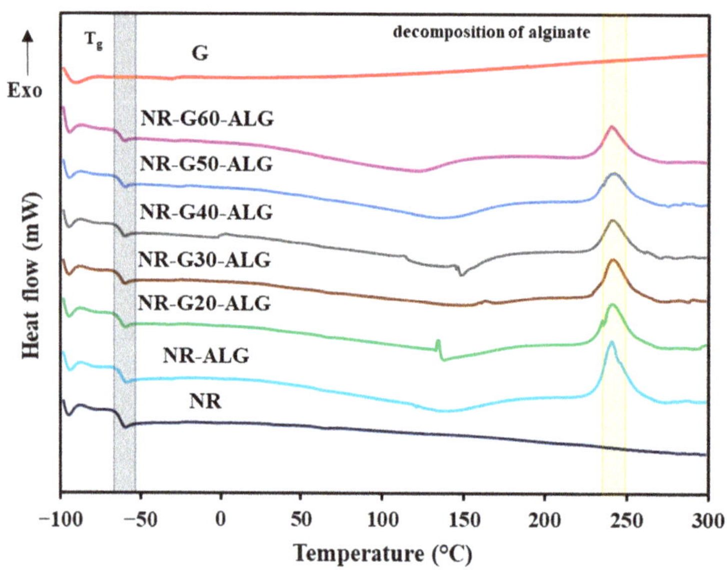

Figure 8. DSC curves of the NR, G, NR-ALG, and NR-G-ALG composite films.

Figure 7, which presents TGA, reveals that NR experienced an initial weight loss within the temperature range of 330–480 °C. The total weight loss of NR reached approximately 99.63%, indicating nearly complete degradation. This substantial weight loss can be attributed to NR's poor thermal resistance. In contrast, The TGA curve for G confirmed its exceptional thermal stability, with the weight remaining at approximately 97.98%, even when exposed to temperatures as high as 700 °C.

Table 1. Thermal data for NR, G, and NR-G-ALG composite films.

Sample	Graphite Loading (phr)	$T_{5\%}$ (°C)	$T_{10\%}$ (°C)	$T_{50\%}$ (°C)	Residual Mass at 700 °C (wt.%)
NR	-	330.1	347.8	379.3	0.37
NR-ALG	-	236.7	328.8	376.7	2.41
NR-G20-ALG	20	242.5	335.6	380.3	17.54
NR-G30-ALG	30	248.2	338.3	385.9	23.18
NR-G40-ALG	40	250.2	337.7	389.7	28.54
NR-G50-ALG	50	246.6	331.2	391	30.70
NR-G60-ALG	60	257.2	340.4	399.1	36.27
G	100	-	-	-	97.98

For NR-ALG and NR-G-ALG composite films, the thermal degradation of the composite exhibited a 2-step process. The first degradation occurred at T_d around 220 °C, likely due to the presence of ALG, and underwent a decomposition process involving the removal of moisture and surface-bound water [48]. The second weight loss occurred within the range of 330–480 °C, corresponding to NR degradation, as previously mentioned. The total remaining weight of NR-G-ALG ranged from 17.54 to 36.27%, related to G content in the composite films. Therefore, the incorporation of G led to higher residue levels, owing to its exceptional thermal resistance of G even at temperatures exceeding 700 °C.

The DSC thermograms presented in Figure 8 depict the thermal characteristics of NR, G, NR-ALG, and NR-G-ALG composite films. The thermogram corresponding to the NR film highlights its amorphous nature, with a glass transition temperature (T_g) of −63.5 °C. This transition temperature marks the point at which the material transforms from a rigid to a more flexible state, reflecting the NR's inherent properties. In the case of G, there are no distinct thermal peaks observed in its thermogram. The absence of significant peaks in the thermogram points to the exceptional thermal stability exhibited by G. Notably, upon analyzing the thermograms of all NR-G-ALG composite films, no significant change in T_g was observed. The broad endothermic peaks observed in the temperature range of 80–150 °C were assigned to the evaporation of water molecules associated with the ALG polymer [49]. Additionally, the exothermic peaks, centered at around 240 °C, were attributed to the decomposition of the alginate polymers, resulting in the formation of carbonized residue [50].

3.6. Water Absorption Capacity (WAC)

The water absorption capacity (WAC) of NR and NR composite films is presented in Figure 9. In the case of the neat NR film, a gradual increase in WAC is observed over a 20-day period. However, with the addition of ALG, a polar molecule known for its hydrophilic and water-retention properties, a remarkable change in the water affinity of the composite films is evident. Therefore, the initial water uptake rate of NR composite films in the first 2 days is much higher than that of neat NR. After 2 days of water immersion, the NR-G-ALG films exhibited nearly constant levels of water absorption.

The water absorption behavior in filled polymer composites depends on various factors, including temperature, filler properties (such as functionality, polarity, and specific surface area), filler loading and morphology, filler volume fraction, and porosity [51,52]. After 20 days of immersion, the WAC for NR composites reinforced with varying G loadings ranged from 38% to 52%. The NR composite films with G incorporations at 60 phr demonstrated the highest WAC. Many studies have investigated the water absorption of graphite or graphene-filled polymer composites and reported that the WAC tended to decrease with increasing graphite or graphene content. This effect might be attributed to the barrier created by the platelet structure of the fillers, which hinders water transportation, as well as the hydrophobic nature of the fillers [53–55].

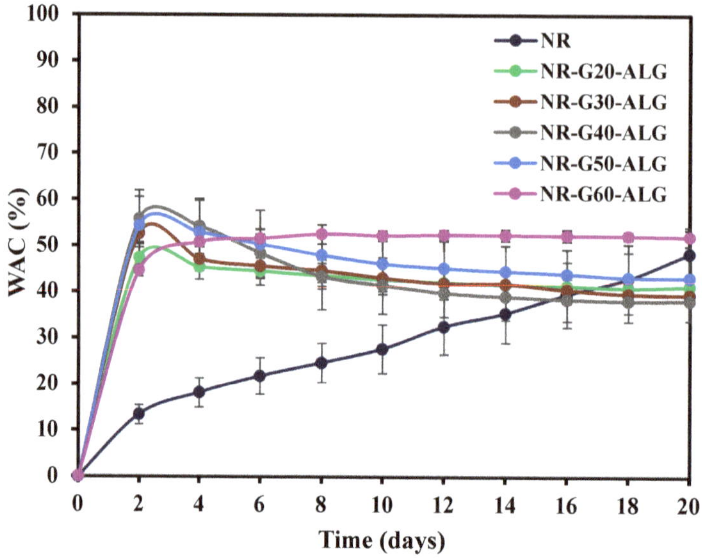

Figure 9. The water absorption capacity results of the NR and NR-G-ALG composite films.

In this study, after the equilibrium, the WAC of NR-G-ALG composite films loading with G platelets from 20 to 50 phr demonstrated slightly lower as compared to that of NR film (WAC ≈ 48%). However, with the incorporation of G at 60 phr, the WAC of the film increased to approximately 52%, which was slightly higher than that of NR film. Some previous studies also reported a similar result. For instance, Alo and Otunniyi studied graphite-filled high-density polyethylene (HDPE) and epoxy composites and found that the WAC increased with high graphite content, which could be attributed to poor interfacial adhesion [56]. Additionally, it was reported that an increase in WAC in graphite-filled polymers could be due to increased porosity and holes present [57,58]; this aligns with the voids observed from the FE-SEM image of composite films at high G loading of NR-G60-ALG with an approximate size of holes of 1.65 ± 0.66 μm (figure not shown). The high loading of G in the NR matrix appears to augment the presence of voids in the composite, potentially explaining the increase in WAC.

3.7. Toluene Uptake

The toluene uptakes of NR and NR composite films following an 8 h immersion in toluene are presented in Figure 10. The NR film demonstrated the highest toluene uptake, reaching 3848%, and this uptake increased further with a longer immersion time. Toluene and NR are both non-polar substances; therefore, NR films exhibit a strong affinity for this solvent. The high uptake rate and poor chemical resistance of NR have been previously reported by many studies.

In the case of NR-G-ALG films, the toluene uptake was significantly lower than that of the NR film. Although some slight swelling occurred, the composite films were able to maintain their structural integrity. Moreover, with longer immersion in toluene, NR-G-ALG films exhibited a consistently lower toluene uptake. As the G loading increased, the toluene uptake decreased remarkably. The composite film with the lowest toluene uptake was NR-G60-ALG, which exhibited an uptake of 714.9% after the 8 h toluene immersion. NR-G60-ALG showed a much lower toluene uptake at approximately 18.6% of that of NR film. The decrease in toluene uptake could be attributed to the chemical stability of graphite.

Additionally, the presence of the filler network reduced the mobility of the rubber chains, resulting in a slower swelling process within the rubber network [35]. This reduction was further facilitated by the formation of a framework-like network structure involving the fillers, effectively limiting the permeation of toluene into the rubber chains [59]. This phenomenon aligns with the findings of a similar study by Ismail and Khalaf, who explored graphite-filled styrene–butadiene rubber (SBR) composites. Their research revealed that graphite-filled SBR exhibited superior toluene resistance, and as the concentration of graphite increased, swelling decreased due to improved graphite dispersion and strong interactions between graphite and the polymer matrix [60]. Additionally, ALG is a polar molecule. The incorporation of polar components could indeed play a crucial role in reducing the toluene uptake rate and absorption capacity [26]. This observation also implies that the composite films achieve good dispersion and distribution of G within the NR matrix.

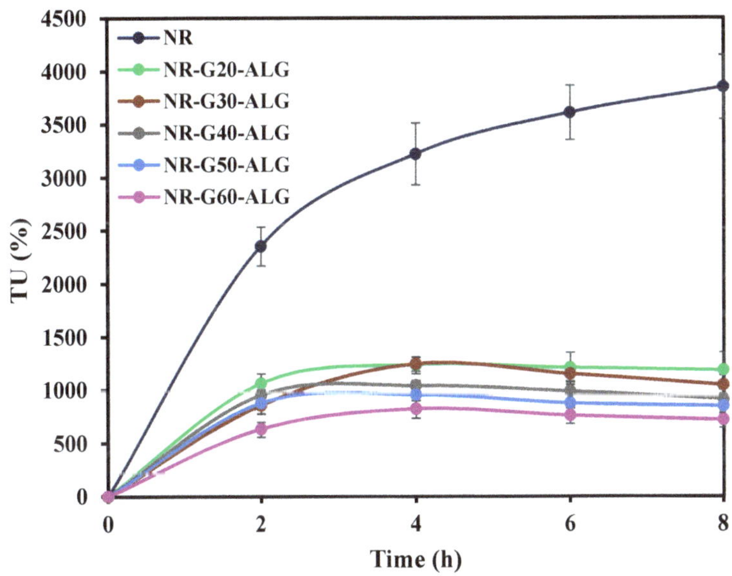

Figure 10. The toluene uptake results of the NR and NR-G-ALG composite films.

3.8. Biodegradation in Soil

Biodegradability is a crucial property of environmentally friendly materials. Biodegradation in soil is driven by microorganisms that break down complex organic compounds into simpler molecules. Ultimately, this process results in the production of water, carbon dioxide (under aerobic conditions), or methane (under anaerobic conditions) [61,62]. Figure 11 illustrates the relationship between biodegradability and the duration of burial. The NR and NR composite films underwent a 90-day burial period. Because of its intrinsic amorphous nature, NR readily accommodates microorganisms within its structure. NR film is a bio-based and biodegradable product, which could be broken down by numerous bacteria and fungi [63]. Following over 90 days of burial, the neat NR film exhibited a progressive increase in biodegradation, reaching a value of 38.8%.

With the incorporation of G platelets, which are a highly stable carbon allotrope known for its crystallinity and non-toxic nature in the environment, the weight loss of NR-G-ALG composite films was significantly lower compared to that of the neat NR film. However, the weight loss rate also increased over time. After 90 days of burial, NR-G60-ALG demonstrated the lowest weight loss at 21.0%, whereas NR-G20-ALG exhibited higher weight loss at 29.4%. The incorporation of G into the composite films is believed to contribute to an increase in their degree of crystallinity. In general, crystalline polymer tends to have denser molecular packing, thus slowing down its rate of breakdown [64].

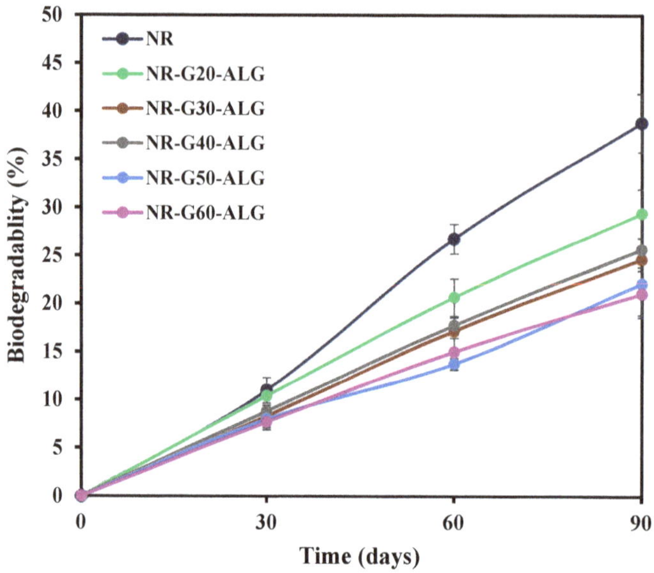

Figure 11. Biodegradation test results of the NR and NR-G-ALG composite films.

A visual analysis of NR and NR composite films is presented in Figure 12. It was observed that the NR films displayed black spots after being buried in the soil, indicating the likely influence of microbial activity and environmental factors during the burial. Over time, the composite film developed wrinkles, and certain sections of the film degraded. Some disintegration into small pieces was also observed, indicating microbial activity's role in degrading the film. This result underlines the biodegradation of NR composite films, which is favorable for environmentally friendly and sustainable applications.

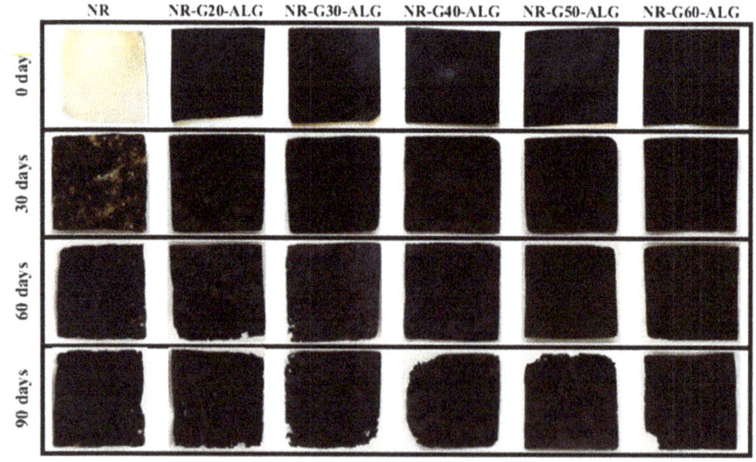

Figure 12. The visual analysis of biodegradation test of NR and NR-G-ALG composite films for the duration of 90 days.

3.9. Electrical Properties

Electrochemical impedance spectroscopy (EIS) assessed the electrical conductivity of NR and NR composite films across a frequency range from 10^5 Hz to 1 Hz. NR-G10-ALG films were intentionally fabricated by using G filler for improved electrically conductive properties. The Nyquist plots in Figure 13 depict the relationship between real and imaginary impedances of fabricated NR and NR composite films, exhibiting behavior dependent on G loading content. Smaller semicircle diameters indicate higher electrical conductivity due to lower impedance magnitude [65].

The incorporation of G platelets into NR composite films is expected to enhance their electrical conductivity owing to G's intrinsic electrical properties. This improvement is attributed to the even distribution of π orbitals and free electrons across the G layers, along with the influence of high crystallinity, which significantly contributes to the enhanced electrical and thermal conductivity of the composite films.

The Nyquist plots exhibited distinct behaviors corresponding to various G loading levels. In the NR plot, the largest semicircle, signifying its inherent electrical insulation properties, was observed. The calculated electrical conductivity was 1.43×10^{-11} S/cm, which was the lowest among all samples. With the increase in G loading, the semicircle in the composite films becomes smaller due to the introduction of a conductive filler. The calculated electrical conductivities were 1.14×10^{-9}, 5.59×10^{-9}, and 8.55×10^{-9} S/cm for NR composites with G loading of 10, 20, and 30 phr, respectively.

As the G loading continues to increase, reaching 40 and up to 60 phr, the signal shifts from a "semicircle" to a "linear-like" graph characteristic, where the intercept of the plot on the real axis representing the impedance, denoting the bulk resistance [66]. This transition suggests a significant movement toward conductivity within the composite, as revealed by the observed electrical behavior. The incorporation of G notably lowers the x-intercept of the plot, signifying an improvement in conductivity. The measured electrical values for NR composites with G loadings of 40, 50, and 60 phr were 2.77×10^{-5}, 5.26×10^{-5}, and 2.18×10^{-4} S/cm, respectively. Remarkably, a significant increase in conductivity is observed between G loadings of 30 and 40 phr. This observation suggests that the percolation threshold of G within the NR matrix falls approximately within this range. It indicates a conductivity increase in approximately four orders of magnitude compared to the 30 phr loading and roughly six orders of magnitude compared to the neat NR. The enhanced conductivity falls within the range observed for semiconducting materials [67].

In Figure 14, the Bode plot exhibits a linear relationship between impedance magnitude and frequency. With an increase in the incorporation of G loading, there is a consecutive decrease in impedance magnitude. Among the samples, NR-G60-ALG demonstrated the lowest impedance magnitude across all frequency ranges. In samples with G loading higher than 30 phr, the linear behavior of impedance no longer follows the previous crescent trend observed in those with lower G loadings and the neat NR. In fact, it tends toward a plateau, appearing as a horizontal line with no slope. This characteristic indicates that the impedance is no longer influenced by frequency in lower frequencies, suggesting the achievement of electrical percolation [65].

Figure 15 displays the graph illustrating the relationship between electrical conductivity and G loading, showcasing a distinctive S-shape percolation transition [45]. The conductivity rises proportionally with increasing G loading until a certain threshold, beyond which only moderate increases are evident with additional G incorporation. This behavior indicates the development of a continuous conductive network after surpassing the percolation threshold, facilitating efficient electron transport throughout the composite. These conductive networks form as a result of the integrated conductive fillers. At the critical point of filler concentration, the material transitions from an insulating to a conducting state. The initial development of conduction pathways leads to a drastic increase in electrical conductivity.

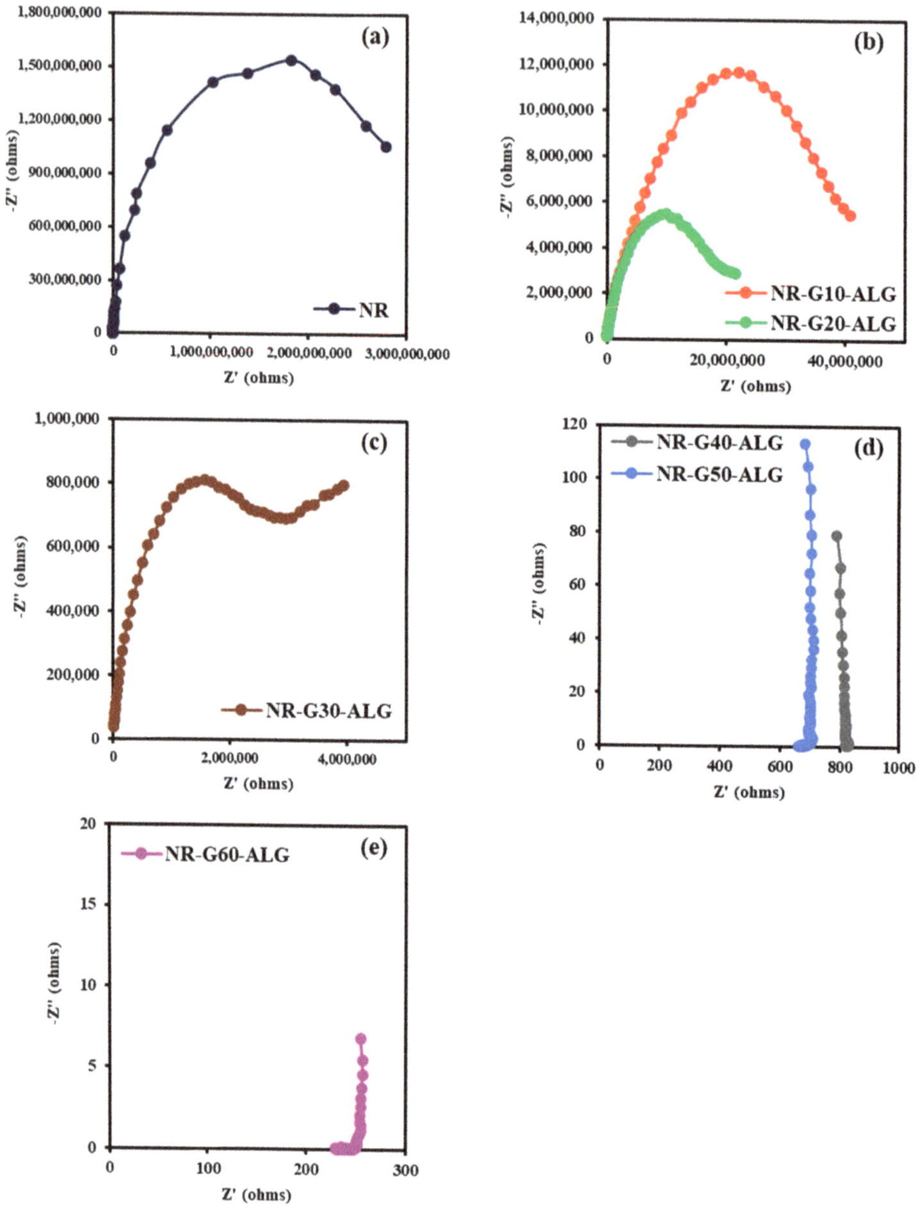

Figure 13. Nyquist plots of NR and NR-G-ALG across a frequency range of 10^5 Hz to 1 Hz, exhibited within various impedance ranges: (**a**) NR, (**b**) NR-G10-ALG and NR-G-20-ALG, (**c**) NR-G30-ALG, and (**d**) NR-G40-ALG, NR-G50-ALG, and (**e**) NR-G60-ALG.

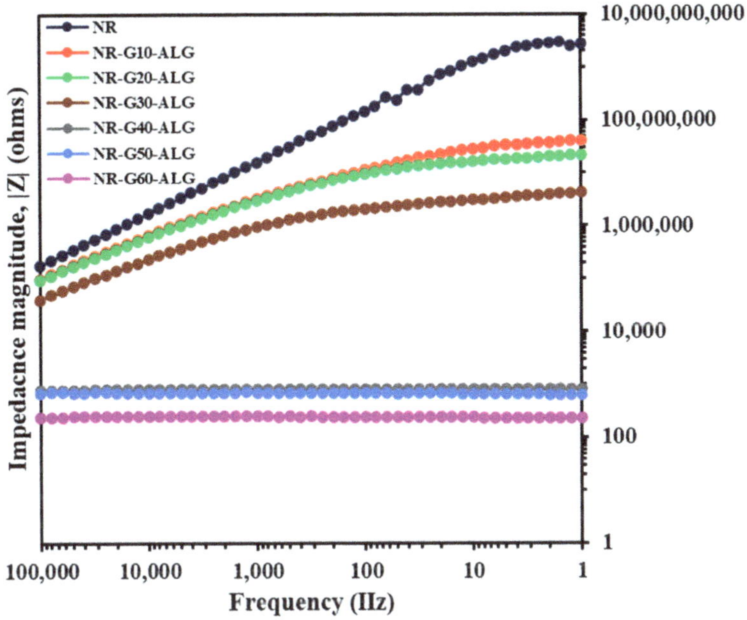

Figure 14. The Bode plot depicts the impedance magnitude of NR and NR-G-ALG with G loading ranging from 10 to 60 phr over a frequency range of 10^5 Hz to 1 Hz.

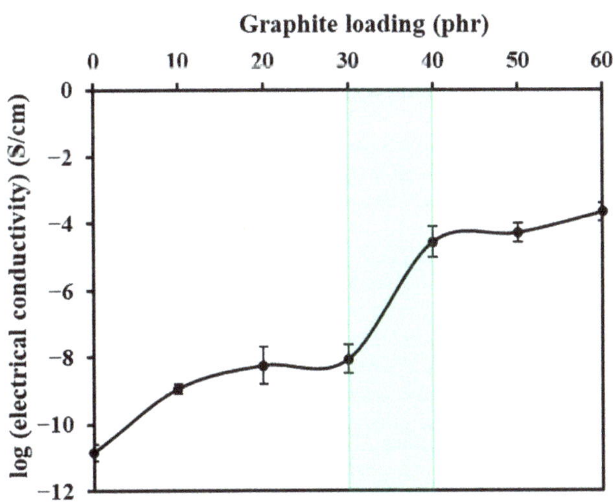

Figure 15. The relationship between log (electrical conductivity) and G loading, highlighting the percolation threshold within the graphite loading range of 30 to 40 phr.

For CV measurements (Figure 16), the NR and NR composite films were assessed within a potential range of −1.0 V to 1.0 V. In NR and NR-G-ALG composites with low G loading (up to 30 phr), the closed-loop rectangular shape in the CV graphs suggests non-linear electrical behavior, resembling capacitance behavior. This behavior may be linked to incomplete conductive pathways, resulting in a non-uniform electrical response. However, as the G loading increases, there is a noticeable increase in the slope of the graphs. Beyond the percolation threshold, the observed linear-like behavior in the CV graphs may indicate the establishment of a continuous conductive network within the composite. This network could facilitate efficient electron transport, leading to more linear and predictable electrical responses.

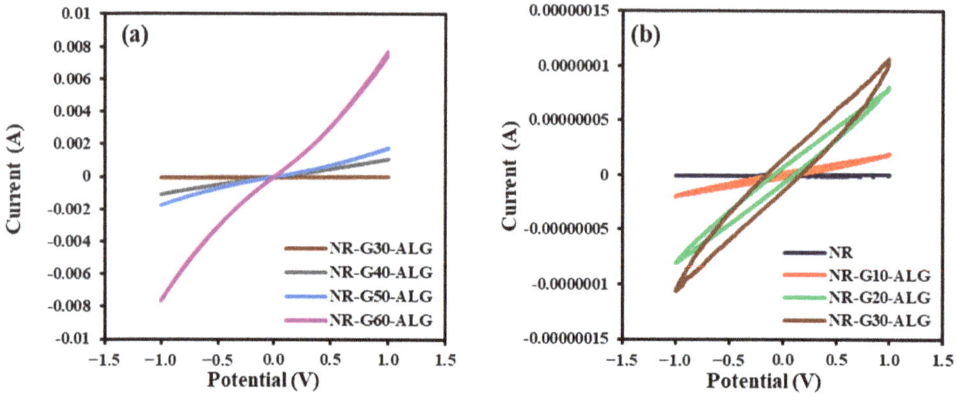

Figure 16. The cyclic voltammetry (CV) measurement of NR and NR-G-ALG, taken at a scan rate of 0.1 V/s, displays the graph representing the 2nd cycle (**a**) at G loading above percolation threshold and (**b**) at G loading below percolation threshold.

4. Conclusions

Unvulcanized NR composites reinforced with synthetic graphite platelets (G) at different loadings from 20 to 60 phr were successfully fabricated using sodium alginate as a dispersing and thickening agent. FTIR analysis indicated no chemical interaction between NR, G, and ALG. FE-SEM images depict the effective dispersion of G without any phase separation in the NR matrix, evident from both surface and cross-sectional perspectives. The reinforcement with G resulted in an increased degree of crystallinity compared to the inherent amorphous nature of NR. NR-G-ALG films displayed remarkably improved tensile strength and Young's modulus, whereas elongation at break decreased. The maximum tensile strength of 8.5 MPa was attained when incorporated with G at 20 phr, while the highest Young's modulus of 12.3 MPa was obtained when incorporated with G at 60 phr. The NR composite films reinforced with G exhibited significantly enhanced chemical resistance, as evidenced by a toluene uptake that was only 18.6% of that observed in the neat NR film. The biodegradation test in soil over a three-month period demonstrated that the composite films achieved a biodegradation value ranging from 21.0% to 29.4%. Additionally, the electrical conductivity of the NR composite films exhibited a remarkable enhancement, reaching an approximate seven-order of magnitude increase compared to that of the neat NR film. The maximum conductivity at 2.18×10^{-4} S/cm was obtained from the NR composite film with G loading at 60 phr. The percolation threshold was identified to be between 30 and 40 phr of G loading. This research underscores the noteworthy enhancements in the degree of crystallinity, mechanical properties, chemical resistance, and electrical conductivity of NR composites reinforced with G platelets, indicating their promising potential as a conductive substrate for electronic applications.

Author Contributions: Conceptualization, M.P.; methodology, V.K., S.S. and M.P; software, V.K.; validation, M.P.; formal analysis, V.K.; investigation, V.K. and M.P.; resources, M.P.; data curation, V.K.; writing—original draft preparation, V.K.; writing—review and editing, V.K. and M.P.; visualization, V.K.; supervision, M.P.; project administration, M.P.; funding acquisition, M.P. All authors have read and agreed to the published version of the manuscript.

Funding: This research is funded by Thailand Science Research and Innovation Fund, Chulalongkorn University (No. 6641/2566).

Institutional Review Board Statement: Not applicable.

Data Availability Statement: The supporting data for the results of this study can be obtained upon request from the corresponding author.

Acknowledgments: We also would like to thank Soorathep Kheawhom from the Center of Excellence on Advanced Materials for Energy Storage, Chulalongkorn University, for his kind assistance in the analysis of electrical properties.

Conflicts of Interest: The authors declare no conflicts of interest.

References

1. Arif, Z.U.; Khalid, M.Y.; Sheikh, M.F.; Zolfagharian, A.; Bodaghi, M. Biopolymeric sustainable materials and their emerging applications. *J. Environ. Chem. Eng.* **2022**, *10*, 108159. [CrossRef]
2. Herculano, R.; Tzu, L.; Silva, C.; Brunello, C.; De Queiroz, A.; Kinoshita, A.; Graeff, C. Nitric Oxide Release Using Natural Rubber Latex as Matrix. *Mater. Res.-Ibero-Am. J. Mater.* **2011**, *14*, 355–359. [CrossRef]
3. Tran, T.N.; Pasetto, P.; Pichon, C.; Bruant, D.; Brotons, G.; Nourry, A. Natural rubber based films integrating Zosteric acid analogues as bioactive monomers. *React. Funct. Polym.* **2019**, *144*, 104343. [CrossRef]
4. Sowcharoensuk, C. Industry Outlook 2022–2024: Natural Rubber. Available online: https://www.krungsri.com/en/research/industry/industry-outlook/agriculture/rubber/io/rubber-2022 (accessed on 23 November 2023).
5. Doner, R. The Risks of Unbalanced Rubber Sector Development. Available online: https://www.bangkokpost.com/opinion/opinion/369082/the-risks-of-unbalanced-rubber-sector-development (accessed on 11 September 2023).
6. Reowdecha, M.; Dittanet, P.; Sae-oui, P.; Loykulnant, S.; Prapainainar, P. Film and latex forms of silica-reinforced natural rubber composite vulcanized using electron beam irradiation. *Heliyon* **2021**, *7*, e07176. [CrossRef] [PubMed]
7. Wei, Y.-C.; Zhu, D.; Xie, W.-Y.; Xia, J.-H.; He, M.-F.; Liao, S. In-situ observation of spatial organization of natural rubber latex particles and exploring the relationship between particle size and mechanical properties of natural rubber. *Ind. Crops Prod.* **2022**, *180*, 114737. [CrossRef]
8. Phetarporn, V.; Loykulnant, S.; Kongkaew, C.; Seubsai, A.; Prapainainar, P. Composite properties of graphene-based materials/natural rubber vulcanized using electron beam irradiation. *Mater. Today Commun.* **2019**, *19*, 413–424. [CrossRef]
9. Mohammad, H.; Stepashkin, A.A.; Tcherdyntsev, V.V. Effect of Graphite Filler Type on the Thermal Conductivity and Mechanical Behavior of Polysulfone-Based Composites. *Polymers* **2022**, *14*, 399. [CrossRef]
10. Khan, A.; Savi, P.; Quaranta, S.; Rovere, M.; Giorcelli, M.; Tagliaferro, A.; Rosso, C.; Jia, C.Q. Low-Cost Carbon Fillers to Improve Mechanical Properties and Conductivity of Epoxy Composites. *Polymers* **2017**, *9*, 642. [CrossRef]
11. Qin, M.; Xu, Y.; Cao, R.; Feng, W.; Chen, L. Efficiently Controlling the 3D Thermal Conductivity of a Polymer Nanocomposite via a Hyperelastic Double-Continuous Network of Graphene and Sponge. *Adv. Funct. Mater.* **2018**, *28*, 1805053. [CrossRef]
12. Ray, M.; George, J.J.; Chakraborty, A.; Bhowmick, A.K. An Investigation of the Electromagnetic Shielding Effectiveness of Ethylene Vinyl Acetate Elastomer Reinforced with Carbon Nanofillers. *Polym. Polym. Compos.* **2010**, *18*, 59–65. [CrossRef]
13. Kasgoz, A.; Akın, D.; Ayten, A.İ.; Durmus, A. Effect of different types of carbon fillers on mechanical and rheological properties of cyclic olefin copolymer (COC) composites. *Compos. Part B Eng.* **2014**, *66*, 126–135. [CrossRef]
14. Charlier, J.C.; Gonze, X.; Michenaud, J.P. Graphite Interplanar Bonding: Electronic Delocalization and van der Waals Interaction. *Europhys. Lett.* **1994**, *28*, 403. [CrossRef]
15. Alofi, A.; Srivastava, G.P. Thermal conductivity of graphene and graphite. *Phys. Rev. B* **2013**, *87*, 115421. [CrossRef]
16. Jin, J.; Leesirisan, S.; Song, M. Electrical conductivity of ion-doped graphite/polyethersulphone composites. *Compos. Sci. Technol.* **2010**, *70*, 1544–1549. [CrossRef]
17. Sadasivuni, K.K.; Ponnamma, D.; Thomas, S.; Grohens, Y. Evolution from graphite to graphene elastomer composites. *Prog. Polym. Sci.* **2014**, *39*, 749–780. [CrossRef]
18. Li, Y.; Liu, G.; Wang, L.; Zhang, J.; Xu, M.; Shi, S.Q. Multifunctional conductive graphite/cellulosic microfiber-natural rubber composite sponge with ultrasensitive collision-warning and fire-waring. *Chem. Eng. J.* **2022**, *431*, 134046. [CrossRef]
19. Han, Y.; Lai, K.C.; Lii-Rosales, A.; Tringides, M.C.; Evans, J.W.; Thiel, P.A. Surface energies, adhesion energies, and exfoliation energies relevant to copper-graphene and copper-graphite systems. *Surf. Sci.* **2019**, *685*, 48–58. [CrossRef]
20. Li, Z.; Wang, Y.; Kozbial, A.; Shenoy, G.; Zhou, F.; McGinley, R.; Ireland, P.; Morganstein, B.; Kunkel, A.; Surwade, S.P.; et al. Effect of airborne contaminants on the wettability of supported graphene and graphite. *Nat. Mater.* **2013**, *12*, 925–931. [CrossRef]
21. Martínez-Cano, B.; Mendoza-Meneses, C.J.; García-Trejo, J.F.; Macías-Bobadilla, G.; Aguirre-Becerra, H.; Soto-Zarazúa, G.M.; Feregrino-Pérez, A.A. Review and Perspectives of the Use of Alginate as a Polymer Matrix for Microorganisms Applied in Agro-Industry. *Molecules* **2022**, *27*, 4248. [CrossRef]
22. Reddy, S.G. Alginates—A Seaweed Product: Its Properties and Applications. In *Properties and Applications of Alginates*; Irem, D., Esra, I., Tugba, K.-G., Eds.; IntechOpen: Rijeka, Croatia, 2021; Chapter 2.
23. Jiang, Y.; Pang, X.; Deng, Y.; Sun, X.; Zhao, X.; Xu, P.; Shao, P.; Zhang, L.; Li, Q.; Li, Z. An Alginate Hybrid Sponge with High Thermal Stability: Its Flame Retardant Properties and Mechanism. *Polymers* **2019**, *11*, 1973. [CrossRef]

24. Shen, Q.; Wu, M.; Xu, C.; Wang, Y.; Wang, Q.; Liu, W. Sodium alginate crosslinked oxidized natural rubber supramolecular network with rapid self-healing at room temperature and improved mechanical properties. *Compos. Part A Appl. Sci. Manuf.* **2021**, *150*, 106601. [CrossRef]
25. Panitchakarn, P.; Wikranvanich, J.; Phisalaphong, M. Synthesis and characterization of natural rubber/coal fly ash composites via latex aqueous microdispersion. *J. Mater. Cycles Waste Manag.* **2018**, *21*, 134–144. [CrossRef]
26. Sintharm, P.; Phisalaphong, M. Green Natural Rubber Composites Reinforced with Black/White Rice Husk Ashes: Effects of Reinforcing Agent on Film's Mechanical and Dielectric Properties. *Polymers* **2021**, *13*, 882. [CrossRef] [PubMed]
27. Supanakorn, G.; Varatkowpairote, N.; Taokaew, S.; Phisalaphong, M. Alginate as Dispersing Agent for Compounding Natural Rubber with High Loading Microfibrillated Cellulose. *Polymers* **2021**, *13*, 468. [CrossRef] [PubMed]
28. Krainoi, A.; Boonkerd, K. Mechanical/electrical properties and strain sensibility of epoxidized natural rubber nanocomposite filled with carbon nanotube: Effect of sodium alginate as a surfactant on latex technology process. *Express Polym. Lett.* **2023**, *17*, 850–866. [CrossRef]
29. Suppanucroa, N.; Nimpaiboon, A.; Boonchuay, K.; Khamkeaw, A.; Phisalaphong, M. Green composite sponge of natural rubber reinforced with cellulose filer using alginate as a dispersing agent. *J. Mater. Res. Technol.* **2023**, *27*, 3119–3130. [CrossRef]
30. Phomrak, S.; Phisalaphong, M. Reinforcement of Natural Rubber with Bacterial Cellulose via a Latex Aqueous Microdispersion Process. *J. Nanomater.* **2017**, *2017*, 4739793. [CrossRef]
31. Phomrak, S.; Phisalaphong, M. Lactic Acid Modified Natural Rubber–Bacterial Cellulose Composites. *Appl. Sci.* **2020**, *10*, 3583. [CrossRef]
32. Supanakorn, G.; Taokaew, S.; Phisalaphong, M. Ternary composite films of natural rubber, cellulose microfiber, and carboxymethyl cellulose for excellent mechanical properties, biodegradability and chemical resistance. *Cellulose* **2021**, *28*, 8553–8566. [CrossRef]
33. *ASTM D882*; Standard Test Method for Tensile Properties of Thin Plastic Sheeting. ASTM International: West Conshohocken, PA, USA, 2018.
34. Sintharm, P.; Nimpaiboon, A.; Liao, Y.-C.; Phisalaphong, M. Bacterial cellulose reinforced with skim/fresh natural rubber latex for improved mechanical, chemical and dielectric properties. *Cellulose* **2022**, *29*, 1739–1758. [CrossRef]
35. Krainoi, A.; Boonkerd, K. Novel hybrid natural rubber nanocomposites with carbon nanotube and cellulose nanofiber for strain-sensitive sensor. *Ind. Crops Prod.* **2022**, *187*, 115455. [CrossRef]
36. Rolere, S.; Liengprayoon, S.; Vaysse, L.; Sainte-Beuve, J.; Bonfils, F. Investigating natural rubber composition with Fourier Transform Infrared (FT-IR) spectroscopy: A rapid and non-destructive method to determine both protein and lipid contents simultaneously. *Polym. Test.* **2015**, *43*, 83–93. [CrossRef]
37. Kundi, A.H.W.; Waseem, I. Synthesis of Graphene Nano Sheets by the Rapid Reduction of Electrochemically Exfoliated Graphene Oxide Induced by Microwaves. *J. Chem. Soc. Pak.* **2016**, *38*, 11.
38. Bera, M.; Yadav, C.; Gupta, P.; Maji, P. Facile One-Pot Synthesis of Graphene Oxide by Sonication Assisted Mechanochemical Approach and Its Surface Chemistry. *J. Nanosci. Nanotechnol.* **2018**, *18*, 902–912. [CrossRef]
39. Wipatkrut, P.; Poompradub, S. Exfoliation approach for preparing high conductive reduced graphite oxide and its application in natural rubber composites. *Mater. Sci. Eng. B* **2017**, *218*, 74–83. [CrossRef]
40. Hu, M.; Zhou, Y.; Nie, W.; Chen, P. Functionalized Graphene Nanosheets with Fewer Defects Prepared via Sodium Alginate Assisted Direct Exfoliation of Graphite in Aqueous Media for Lithium-Ion Batteries. *ACS Appl. Nano Mater.* **2018**, *1*, 1985–1994. [CrossRef]
41. Al-Hartomy, O.A.; Gupta, R.K.; Al-Ghamdi, A.A.; Yakuphanoglu, F. High performance organic-on-inorganic hybrid photodiodes based on organic semiconductor-graphene oxide blends. *Synth. Met.* **2014**, *195*, 217–221. [CrossRef]
42. Chimupala, Y.; Junploy, P.; Hardcastle, T.; Westwood, A.; Scott, A.; Johnson, B.; Brydson, R. Universal synthesis method for mixed phase TiO_2(B)/anatase TiO_2 thin films on substrates via a modified low pressure chemical vapour deposition (LPCVD) route. *J. Mater. Chem. A* **2016**, *4*, 5685–5699. [CrossRef]
43. Mohammad, H.; Stepashkin, A.A.; Laptev, A.I.; Tcherdyntsev, V.V. Mechanical and Conductive Behavior of Graphite Filled Polysulfone-Based Composites. *Appl. Sci.* **2023**, *13*, 542. [CrossRef]
44. Alshammari, B.A.; Wilkinson, A.N.; AlOtaibi, B.M.; Alotibi, M.F. Influence of Carbon Micro- and Nano-Fillers on the Viscoelastic Properties of Polyethylene Terephthalate. *Polymers* **2022**, *14*, 2440. [CrossRef]
45. Alshammari, B.A.; Al-Mubaddel, F.S.; Karim, M.R.; Hossain, M.; Al-Mutairi, A.S.; Wilkinson, A.N. Addition of Graphite Filler to Enhance Electrical, Morphological, Thermal, and Mechanical Properties in Poly (Ethylene Terephthalate): Experimental Characterization and Material Modeling. *Polymers* **2019**, *11*, 1411. [CrossRef] [PubMed]
46. Bhattacharyya, S.; Sinturel, C.; Bahloul, O.; Saboungi, M.-L.; Thomas, S.; Salvetat, J.-P. Improving reinforcement of natural rubber by networking of activated carbon nanotubes. *Carbon* **2008**, *46*, 1037–1045. [CrossRef]
47. Feng, C.; Ni, H.; Chen, J.; Yang, W. Facile Method to Fabricate Highly Thermally Conductive Graphite/PP Composite with Network Structures. *ACS Appl. Mater. Interfaces* **2016**, *8*, 19732–19738. [CrossRef] [PubMed]
48. Huq, T.; Salmieri, S.; Khan, A.; Khan, R.A.; Le Tien, C.; Riedl, B.; Fraschini, C.; Bouchard, J.; Uribe-Calderon, J.; Kamal, M.R.; et al. Nanocrystalline cellulose (NCC) reinforced alginate based biodegradable nanocomposite film. *Carbohydr. Polym.* **2012**, *90*, 1757–1763. [CrossRef] [PubMed]
49. Abulateefeh, S.; Taha, M. Enhanced drug encapsulation and extended release profiles of calcium–alginate nanoparticles by using tannic acid as a bridging cross-linking agent. *J. Microencapsul.* **2014**, *32*, 96–105. [CrossRef] [PubMed]

50. Soares, J.; Santos, J.; Chierice, G.; Cavalheiro, É. Thermal behavior of alginic acid and its sodium salt. *Eclet. Quim.* **2004**, *29*, 57–63. [CrossRef]
51. Deng, H.; Reynolds, C.T.; Cabrera, N.O.; Barkoula, N.M.; Alcock, B.; Peijs, T. The water absorption behaviour of all-polypropylene composites and its effect on mechanical properties. *Compos. Part B Eng.* **2010**, *41*, 268–275. [CrossRef]
52. Turmanova, S.; Genieva, S.; Vlaev, L. Obtaining Some Polymer Composites Filled with Rice Husks Ash-A Review. *Int. J. Chem.* **2012**, *4*, 62–89. [CrossRef]
53. Phuangngamphan, M.; Okhawilai, M.; Hiziroglu, S.; Rimdusit, S. Development of highly conductive graphite-/graphene-filled polybenzoxazine composites for bipolar plates in fuel cells. *J. Appl. Polym. Sci.* **2019**, *136*, 47183. [CrossRef]
54. Sheshmani, S.; Ashori, A.; Arab Fashapoyeh, M. Wood plastic composite using graphene nanoplatelets. *Int. J. Biol. Macromol.* **2013**, *58*, 1–6. [CrossRef]
55. Wang, J.; Wang, X.; Xu, C.; Zhang, M.; Shang, X. Preparation of graphene/poly(vinyl alcohol) nanocomposites with enhanced mechanical properties and water resistance. *Polym. Int.* **2011**, *60*, 816–822. [CrossRef]
56. Alo, O.A.; Otunniyi, I.O. Comparative study of flexural and physical properties of graphite-filled immiscible polypropylene/epoxy and high-density polyethylene/epoxy blends. *Polym. Polym. Compos.* **2021**, *29*, S1103–S1112. [CrossRef]
57. Prolongo, S.G.; Moriche, R.; Jiménez-Suárez, A.; Sánchez, M.; Ureña, A. Advantages and disadvantages of the addition of graphene nanoplatelets to epoxy resins. *Eur. Polym. J.* **2014**, *61*, 206–214. [CrossRef]
58. Gautam, R.K.; Kar, K.K. Synergistic Effects of Carbon Fillers of Phenolic Resin Based Composite Bipolar Plates on the Performance of PEM Fuel Cell. *Fuel Cells* **2016**, *16*, 179–192. [CrossRef]
59. Nie, J.; Huang, J.; Fan, J.; Cao, L.; Xu, C. Strengthened, Self-Healing, and Conductive ENR-Based Composites Based on Multiple Hydrogen Bonding Interactions. *ACS Sustain. Chem. Eng.* **2020**, *8*, 13724–13733. [CrossRef]
60. Ismail, M.N.; Khalaf, A.I. Styrene–butadiene rubber/graphite powder composites: Rheometrical, physicomechanical, and morphological properties. *J. Appl. Polym. Sci.* **2011**, *120*, 298–304. [CrossRef]
61. Nezha Tahri, J.; Wifak, B.; Hanane, S.; Naïma El, G. Biodegradation: Involved Microorganisms and Genetically Engineered Microorganisms. In *Biodegradation*; Rolando, C., Francisca, R., Eds.; IntechOpen: Rijeka, Croatia, 2013; Chapter 11.
62. Manikandan, S.; Vickram, S.; Sirohi, R.; Subbaiya, R.; Krishnan, R.Y.; Karmegam, N.; Sumathijones, C.; Rajagopal, R.; Chang, S.W.; Ravindran, B.; et al. Critical review of biochemical pathways to transformation of waste and biomass into bioenergy. *Bioresour. Technol.* **2023**, *372*, 128679. [CrossRef]
63. Bosco, F.; Mollea, C. Biodegradation of Natural Rubber: Microcosm Study. *Water Air Soil Pollut.* **2021**, *232*, 227. [CrossRef]
64. Pantani, R.; Sorrentino, A. Influence of crystallinity on the biodegradation rate of injection-moulded poly(lactic acid) samples in controlled composting conditions. *Polym. Degrad. Stab.* **2013**, *98*, 1089–1096. [CrossRef]
65. Barbosa, R.; Gonçalves, R.; Blanco, G.E.d.O.; Saccardo, M.C.; Tozzi, K.A.; Zuquello, A.G.; Scuracchio, C.H. Multi-sensing properties of hybrid filled natural rubber nanocomposites using impedance spectroscopy. *Electrochim. Acta* **2022**, *435*, 141341. [CrossRef]
66. Chai, K.L.; Aung, M.M.; Noor, I.M.; Lim, H.N.; Abdullah, L.C. Observation of ionic conductivity on PUA-TBAI-I2 gel polymer electrolyte. *Sci. Rep.* **2022**, *12*, 121. [CrossRef] [PubMed]
67. Kaur, G.; Adhikari, R.; Cass, P.; Bown, M.; Gunatillake, P. Electrically conductive polymers and composites for biomedical applications. *RSC Adv.* **2015**, *5*, 37553–37567. [CrossRef]

polymers

MDPI

Article

Thermodynamic Parameters of Crosslinked Elastomers (BR, SBR and NBR) and Their Blends

César Leyva-Porras *, Iván A. Estrada-Moreno *, Claudia I. Piñón-Balderrama, Sergio G. Flores-Gallardo and Alfredo Márquez-Lucero

Advanced Materials Research Center (CIMAV), Complejo Industrial Chihuahua, Miguel de Cervantes No. 120, Chihuahua 31136, Mexico; claudia.pinon@cimav.edu.mx (C.I.P.-B.); sergio.flores@cimav.edu.mx (S.G.F.-G.); alfredo.marquez@cimav.edu.mx (A.M.-L.)

* Correspondence: cesar.leyva@cimav.edu.mx (C.L.-P.); ivan.estrada@cimav.edu.mx (I.A.E.-M.); Tel.: +52-61-4439-1100 (ext. 2011) (C.L.-P.); +52-61-4439-1100 (ext. 2404) (I.A.E.-M.)

Abstract: Herein, a methodology is employed based on the Flory–Rehner equation for estimating the Flory–Huggins interaction parameter (χ_{12}*) of crosslinked elastomer blends. For this purpose, binary elastomer blends containing polybutadiene rubber (BR), styrene–butadiene rubber (SBR) and nitrile–butadiene rubber (NBR), were prepared in a mixing chamber at a temperature below the activation of the crosslinking agent. Swelling tests with benzene were employed to determine the crosslinked fraction, finding that after 20 min of thermal annealing, the BR and NBR were almost completely crosslinked, while the SBR only reached 60%. Additionally, the BR-SBR blend increased by 2–3 times its volume than its pure components; this could be explained based on the crosslink density. From the mechanical tests, a negative deviation from the rule of mixtures was observed, which suggested that the crosslinking was preferably carried out in the phases and not at the interface. Furthermore, tensile tests and swelling fraction (ϕ_{sw}) results were employed to determine the average molecular weight between two crosslinking points (M_c), and subsequently χ_{12}*. Calculated χ_{12}* values were slightly higher than those reported in the literature. The calculated thermodynamic parameters for the blends showed positive ΔG_{mix} values and endothermic behavior, suggesting their immiscible nature.

Keywords: thermodynamic parameters; elastomer blends; Flory–Huggins interaction parameter; swelling of crosslinked rubbers

1. Introduction

In recent years, elastomers have been used as an integral part in the development of chemical sensors and actuators, such as artificial noses, distributed sensors, and super-absorbent materials [1–3]. In this type of application, the ability of these elastomers to absorb low-polarity substances is of great importance and one of their great advantages. Currently, there is little literature about methods to increase the absorption capacity of these substances by elastomers. However, there is a large amount of literature on super-absorbent hydrogels for aqueous solutions [4,5]. There are many ways to obtain new polymers that present certain superior physicochemical properties, but commonly, these processes involve complicated and expensive polymerization reactions. On the other hand, sometimes it is only necessary to mix the polymer with some other material that may impart new properties [6]. Elastomer blends are important since they present better performance in their final properties, which are not achievable by the individual elastomers [7]. For example, among the well-known properties of hydrogenated acrylonitrile-butadiene rubber (HNBR) are chemical and thermal degradation, while the use of chloroprene rubber (CR) provides the flame resistance property. The blending of these rubbers when crosslinked with silver oxide (Ag_2O) improved the homogeneity and miscibility, resulting in a material with significant resistance to thermo-oxidative factors [8]. Within the field of polymer blends, several works have reported the behavior of miscible and immiscible blends, variations in polarity and unsaturation, differences in molecular weights and branching of polymers [9,10].

Rubbers such as polybutadiene (BR), poly(butadiene-*co*-styrene) (SBR) and poly(butadiene-*co*-acrylonitrile) (NBR) are of great importance in the production of tires, engineering resins, adhesives, seals, hoses and gaskets, and in materials that are oil resistant. Globally, its demand requires a consumption of approximately 58% of the total production of butadiene, which is used as monomer for their production [11–13]. The study of mixing any of these three elastomers with other polymers has been studied extensively [14–21]. However, works reporting PB-NBR and SBR-NBR blends are scarce, particularly PB-SBR. For example, blends of SBR and NBR have been used to promote the uptake of chlorinated hydrocarbons, where the polarity of the hydrocarbon and the SBR fraction affect the degree of absorption [22]. Laser scanning confocal microscopy (LSCM) was used to demonstrate phase separation in PB-SBR blends. It was found that the SBR domains presented a fine structure of the order of sub-micrometers, while the continuous phase of PB presented domains separated by 10 μm [23]. The incorporation of *trans*-polyoctylene rubber (TOR) improved the mechanical properties and crosslinking density of SBR-recycled NBR (rNBR) blends. This is because the compatibilizer agent improves the adhesion between the SBR and rNBR phases [24]. In this sense, the addition of mesoporous silica particles as a compatibilizer agent improves the mechanical behavior of SBR-NBR blends, since the silica surface promotes the dispersion of the NBR phase [25]. Because of the high surface area and the presence of polar groups on the surface of the nanofillers, the physical interactions and interfacial adhesion between blends of BR-SBR-carbon black containing isocyanate modified graphite nanoplatelets (i-MG) were enhanced, showing an increase in mechanical and thermal properties, hardness and abrasion resistance [26]. In the case of carboxylated NBR and SBR blends reinforced with graphene oxide (GO), the increase in the GO content causes an increase in the crosslinking density and consequently a decrease in the swelling ratio, which makes the XNBR-SBR blend more resistant to solvents [27]. In 70/30 SBR/BR blends with silica as filler, the increase in the content of free fatty acids (FFA) in palm oil (PO) increased the crosslinking density, which was reflected in the increase in mechanical properties. such as modulus, tensile strength and abrasion resistance [28]. In SBR-NBR blends where the SBR was modified with dichlorocarbene, the degree of swelling in the presence of *n* heptane decreased with the NBR content, from 75.3 to 59.9 and 24.3%, for the 70/30, 50/50 and 30/70 of SBR/NBR blends, respectively. This was caused by the polar character of NBR, which restricts swelling [29].

One of the most important reasons for the use of compatible blends is the ease with which they are prepared, since only involve the use of a mixing instrument and raw materials. For example, sensors have been developed to detect hydrocarbon leaks, the principle of which is based on the swelling of the polymer upon contact with the hydrocarbon [30]. This swelling must not only be rapid, but must also have certain properties such as mechanical resistance. Because the glass transition temperature of elastomers is below room temperature, they are flexible, while at temperatures above 80 °C they are viscoplastically deformable. One way to impart rigidity to the structure is through chemical reactions in the solid state, known as a crosslinking reaction, which results in a three-dimensional network that prevents the movement of the polymer chains. The crosslinking reaction is largely determined by the type of crosslinking agent, process, temperature and time [31]. The number of crosslinks formed is known as the crosslink density and has a large influence on the final mechanical properties of the elastomer. Blends are commonly used to improve the processability of rubbers. This improvement may consist of either lowering the viscosity or producing a material less prone to fracture during melt flow. Side effects such as swelling of an extrudate after leaving a forming die can also be affected by mixing.

The importance of knowing the thermodynamic parameters of elastomer blends lies in the possibility of predicting their miscibility. Phase homogeneity in rubber mixtures is promoted when both individual components share properties such as viscosity, solubility parameters and polar groups [32]. The equilibrium miscibility criterion for a polymer blend is given by the Gibbs free energy of mixing (ΔG_{mix}), which must be negative; while for the stability of the mixture, its second derivative with respect to the concentration must be positive [33]. For a polymer–solvent system, the thermodynamic properties are calculated from the Flory–Huggins model, using solubility values to estimate the interaction parameter (χ). However, although this equation is widely used due to its simplicity, it does not allow the interaction parameter for polymer blends exposed to a solvent to be calculated. Therefore, it is necessary to find a method that allows the average interaction parameter for polymer blends subjected to the absorption of solvents to be estimated.

Thus, in the present work, a phenomenological methodology based on the Flory–Rehner equation is presented to determine the interaction parameter of crosslinked binary elastomer blends and the calculation of the thermodynamic parameters. This approach is based on swelling experiments with an organic solvent and the measurement of the mechanical properties from specimens of the blends. The contribution of this work relies on the calculation of the Flory–Huggins interaction parameter and the thermodynamic parameters of elastomer blends of immiscible nature.

2. Physicochemical Models

The Flory–Huggins model is based on the division of ΔG_{mix} into enthalpic and entropic terms, and the evaluation of these two terms separately according to Equation (1):

$$\Delta G_{mix} = \Delta H_{mix} - T\Delta S_{mix} \tag{1}$$

The entropy of mixing (ΔS_{mix}) is evaluated from the possible number of arrangements of the molecules in the network. The mixing enthalpy (ΔH_{mix}) is calculated as the change in the interaction energies between the molecular surfaces during the mixing process. These terms are defined as follows:

$$\Delta S_{mix} = -R(n_1 \ln\phi_1 + n_2 \ln\phi_2) \tag{2}$$

$$\Delta H_{mix} = RT\, n_1 \phi_2 \chi \tag{3}$$

where the volume fractions ϕ_1 and ϕ_2 represent the total number of lattice points occupied by solvent molecules and polymer segments, respectively, n_1 is the number of moles of solvent, n_2 is the number of moles of polymer, and χ is the Flory interaction parameter [34].

The polymer–solvent binary system is based on the network model, where the effect of the difference in size between the polymer and solvent molecules on the mixing entropy is pointed out. The quantitative calculation of the mixing enthalpy allows the introduction of a dimensionless value called the Flory–Huggins interaction parameter, for the thermodynamic description of polymer solutions. This parameter considers the specific interactions between polymer segments and solvent molecules within a particular system. Then, after substituting the entropy and the enthalpy of mixing in Equation (1), the well-known Flory–Huggins relationship is obtained (Equation (4)) [35]:

$$\Delta G_{mix} = RT\,(n_1 \ln\phi_1 + n_2 \ln\phi_2 + n_1 \phi_2 \chi) \tag{4}$$

One of the simplest ways to calculate the interaction parameter is employing the solubility parameters of the polymer and the solvent (Equation (5)), where V_1 is the molar volume of the solvent, and δ_1 and δ_2 are the solubility parameters of the polymer and the solvent, respectively [36]:

$$X = 0.34 + V_1/RT(\delta_1 - \delta_2)^2 \tag{5}$$

According to this equation, the concept of the solubility parameter implies that the interaction parameter must always be positive, while the lowest possible value is 0.34, applicable to mixtures of components with exactly the same value of the solubility parameters. However, this equation, although it is a widely used approximation due to its simplicity, does not allow the interaction parameter for polymer blends exposed to a solvent to be calculated. For describing the elasticity phenomenon of crosslinked rubbers subjected to swelling due to the absorption of a solvent, the Flory–Rehner Equation (6) is used [36]:

$$\ln(1 - \phi_{sw}) + \phi_{sw} + \chi_{12}\phi_{sw}^{2} + (\rho V_1/M_c)(\phi_{sw}^{1/3} - \phi_{sw}/2) = 0 \quad (6)$$

where ϕ_{sw} is the inverse of the swelling ratio of V_0 and V, the volumes of the dry and swollen gel, respectively. ρ is the density of the polymer, V_1 is the molar volume of the solvent, χ_{12} is the Flory–Huggins interaction parameter of the blend and M_c is the average chain size between two crosslinks.

Additionally, from the stress–strain curves, it is possible to determine the effect of the nature and distribution of the crosslinks in the rubber, since changes in the module, elongation and stress are influenced by alterations in the structure of the polymeric chains [37]. Now, the shear modulus (G) of a rubber is a direct measure of the number of elemental chains in a unit of volume (N_v), as

$$G = kTN_v \quad (7)$$

with k being the constant of Boltzmann and T the temperature. Frequently, the number of chains is expressed through the average molecular weight of the chain between two crosslinks (M_c), rubber density (ρ) and the universal gas constant (R). The result is:

$$G = RT\rho/M_c \quad (8)$$

Then, by determining the Young's modulus (E) and taking into account that the relationship of the shear modulus with the Young's modulus is G = E/3, the molecular weight between two crosslinking units can be estimated from tensile tests [35] as

$$M_c = 3RT\rho/E \quad (9)$$

Then, from the swelling experiments, the swelling ratio can be determined, while from the mechanical tests, the average molecular weight between two crosslinks is determined. Knowing both parameters, the Flory–Huggins interaction parameter of the blend is calculated, solving the Flory–Rehner equation.

3. Materials and Methods

3.1. Materials

Polybutadiene rubber (BR) (Solprene 200, Dynasol, Altamira, Mexico), poly(butadiene-*co*-styrene) (SBR) (Emulprene 1778, Industrias Negromex, Altamira, Mexico) and poly(butadiene-*co*-acrylonitrile) (NBR) (Paracril BJLT-M50, Quimipol, Mexico city, Mexico) were used as raw materials for preparing the rubber blends. Dicumyl Peroxide (DCP) (99.02%, Kaucho Quimico, Mexico city, Mexico) was employed as the crosslinking agent. Benzene (99.9%, Merck, Naucalpan de Juarez, Mexico) was employed as the solvent for the Soxhlet extractions and swelling tests.

3.2. Rubber Blends Preparation

The mixing of the rubbers with the crosslinking agent was carried out in an internal mixing chamber (Brabender, South Hackensack, NJ, USA). A chamber type VI was used with a gear ratio of 3:2, and a volume capacity of 60 cm^3 when using Cam-type blades. Heating and cooling were carried out through an electrical resistance and circulating air, respectively. Mixing was carried out at a temperature of 90 °C, and 40 RPM, for 15 min. The mass of the rubber (60 g) was added and mixed for 2 min. At this time, 1 phr (parts per 100 of resin) of the crosslinking agent was added into the chamber. At the end of the mixing time, the sample was removed from the chamber and identified according to Table 1:

Table 1. Sample name and content of the blends.

Sample	BR (g)	SBR (g)	NBR (g)	DCP (g)
P	60	0	0	0.6
S	0	60	0	0.6
N	0	0	60	0.6
PS	30	30	0	0.6
PN	30	0	30	0.6
SN	0	30	30	0.6

3.3. Crosslinking

The crosslinking of the rubber samples with DCP was carried out in a forced air heating oven (LindBerg Blue, Thermo Fisher Scientific Inc., Mexico city, Mexico), with a precision temperature controller of ±0.1 °C. The crosslinking was carried out at 175 °C, and different times of 2, 4, 6, 10 and 20 min. This crosslinking temperature was selected to ensure at least a 50% crosslinking efficiency. These experiments were carried out by duplicate.

3.4. Crosslinked Fraction in the Blend and Swelling Tests

Once the samples were crosslinked, the gel fraction or the crosslinked fraction of the blends was calculated. For this, a solid–liquid Soxhlet extraction was carried out, with benzene as solvent. The extraction process was repeated for 24 h, to ensure that all the non-crosslinked fraction was dissolved. The sample was then dried at 90 °C for 24 h, to evaporate the solvent. Subsequently, dried sample was weighed and the crosslinked fraction (Θ) was determined according to Equation (10):

$$\Theta = [(W_4 - W_1)/W_2] \times 100 \tag{10}$$

where W_1 is the initial weight of the extraction capsule, W_2 is the weight of the rubber sample and W_4 is the dried weight of the remaining rubber sample in the capsule after extraction process.

Once the Soxhlet extraction was completed and the solvent was cooled to room temperature, the excess of solvent was removed using absorbent paper, then the capsule containing the swollen elastomer was weighed. The weight was recorded as W_3. The swelling ratio or the volume of solvent absorbed per unit volume of crosslinked polymer ($1/\Phi_{sw} = V/V_0$) was determined according to Equation (11):

$$1/\Phi_{sw} = V/V_0 = 1 + \frac{W_3 - W_4}{W_4}\left(\frac{\rho_2}{\rho_1}\right) \tag{11}$$

where W_3 is the swollen mass of the rubber and solvent, ρ_1 and ρ_2 are the densities of the solvent (benzene = 880 kg/m^3) and the elastomer sample (values reported in Table S1), respectively [38]. The value of $1/\Phi_{sw}$ indicates the times the initial volume of the rubber increases when it is immersed in the solvent. These experiments were carried out in duplicate.

3.5. Tensile Tests

Tensile tests were carried out on an Instron Universal Machine, model 4468 (Norwood, MA, USA) with a 500 kg load cell. The specimens and the tension test were carried out in accordance with the ASTM D-638 standard [39], with dimensions of the specimens of 12 cm × 2 cm × 0.3 cm, deformation rate of 500 mm/min and temperature of 25 °C. Young's modulus was determined from the ratio of stress and deformation ($E = \sigma/\varepsilon$).

3.6. Density of Rubber Blends

To determine the density of the samples, an Ultrapycnometer 1000 (Quantachrome, Boynton Beach, FL, USA) was used. The instrument determines the density of a solid from the displaced volume of Helium gas.

4. Results and Discussion

4.1. Crosslinked Fraction and Swelling Ratio

Table S1 shows the values of the masses of the elastomers and their blends during the Soxhlet extraction process, the calculated crosslinked fraction and density values. Clearly, there is no significant effect of the initiator on the density of the samples, because the volume of the sample remains constant during crosslinking, and only some chemical bonds change position when the hydrogen extraction is carried out by the radical peroxide. Furthermore, NBR showed the highest density value (980 Kg/m^3), while SBR and BR showed the lower values of 930 and 890 (Kg/m^3), respectively. On the other hand, when calculating the density for the different blends, intermediate values of the components are obtained, suggesting that they follow a mixture rule behavior.

Figure 1 shows the crosslinked fraction versus time for the elastomers and their blends. At 2 min, the SBR and NBR samples crosslink rapidly, with values close to 30% and 70%, respectively. At this crosslinking time, the BR presents only about 1% of crosslinking. This behavior may be caused by the difference in the reaction rate of DCP in each of the elastomers, suggesting that this rate is in the following order: NBR > SBR > BR. However, at 4 min of crosslinking, BR is now the elastomer with the highest crosslinked fraction, while SBR shows the lowest percentage of crosslinking. On the other hand, the NBR that initially crosslinked very quickly, after 4 min, has increased its crosslinked fraction by only 5%. At the end of the crosslinking time (20 min), it can be observed that the BR and NBR were almost completely crosslinked ($\Theta = 1$), while the SBR only reached 60% of crosslinked fraction. All this can be explained by the number of active sites or double bonds in the chains of the different elastomers. For example, when observing the chemical structure of each of the elastomers, for BR there is one double bond for every four carbons, while for NBR there is one double bond for every seven carbons, and for SBR, one for every twelve carbons.

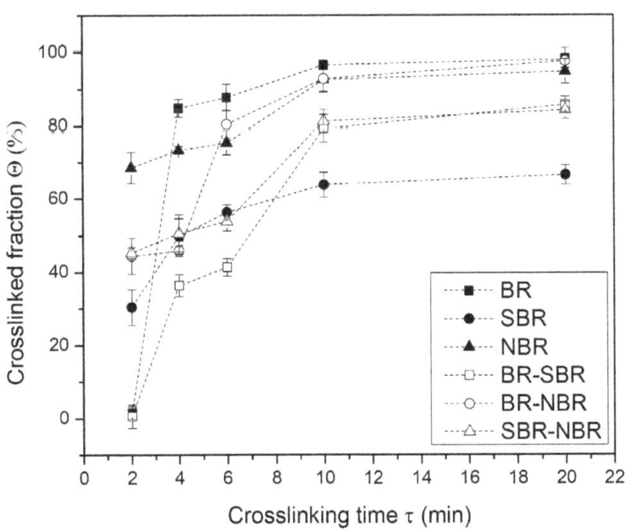

Figure 1. Crosslinked fraction (Θ) versus crosslinking (τ) time for the elastomers and their blends.

Table S1 includes the results of the calculated swelling ratio. In general, all elastomers and their blends presented a similar behavior, where the swelling ratio decreased with crosslinking time. Consequently, the highest swelling values occurred at the shortest crosslinking times. Among the elastomers, BR presented the highest swelling values, while NBR was the least swollen. Figure 2 shows the effect of blending the elastomers on the swelling ratio. The mixing effect showed interesting results, since for some of the blends, a positive deviation from the rule of mixtures was observed. That is, there was an increase in swelling for certain crosslinking times. For example, the BR-SBR blend crosslinked for 2 min presented a swelling value of 412 times, while the individual BR and SBR elastomers presented swelling values of 165 and 32, respectively. This blend presented the same behavior at crosslinking times of 4 and 6 min, but the swelling values were not as high. The same performance was presented by the BR-NBR and SBR-NBR blends at 4 min of crosslinking time. According to Marzzoca et al. [40], while two rubbers may be virtually insoluble, the blend prepared by intense mechanical means can be macroscopically homogeneous, and once the blend has been crosslinked, it can perform as a single-phase system. On the other hand, according to Wang [41], the PS domains within the SBS are more continuous for the sample processed in an internal mixing chamber than for the sample obtained by casting. This was observed by transmission electron microscopy (TEM) as a hexagonal structure for the PS domains in the sample prepared by casting (SBS_I), while in the samples prepared by compression (SBS_{II}) and in the internal mixing chamber (SBS_{III}), the microstructure was more distorted, and almost no hexagonal domains of PS were observed. These observations suggested a phase separation in the samples SBS_{II} and SBS_{III}. Although phase separation may exist at the microscopic level, the mixture can behave as a single phase at the macroscopic level due to the effect of crosslinking. On the other hand, hemp fibers have been used to mechanically reinforce ethylene–propylene–diene–terpolymer (EPDM), which hindered the absorption of solvent within the blend, decreasing the degree of swelling and increasing the crosslink density [3].

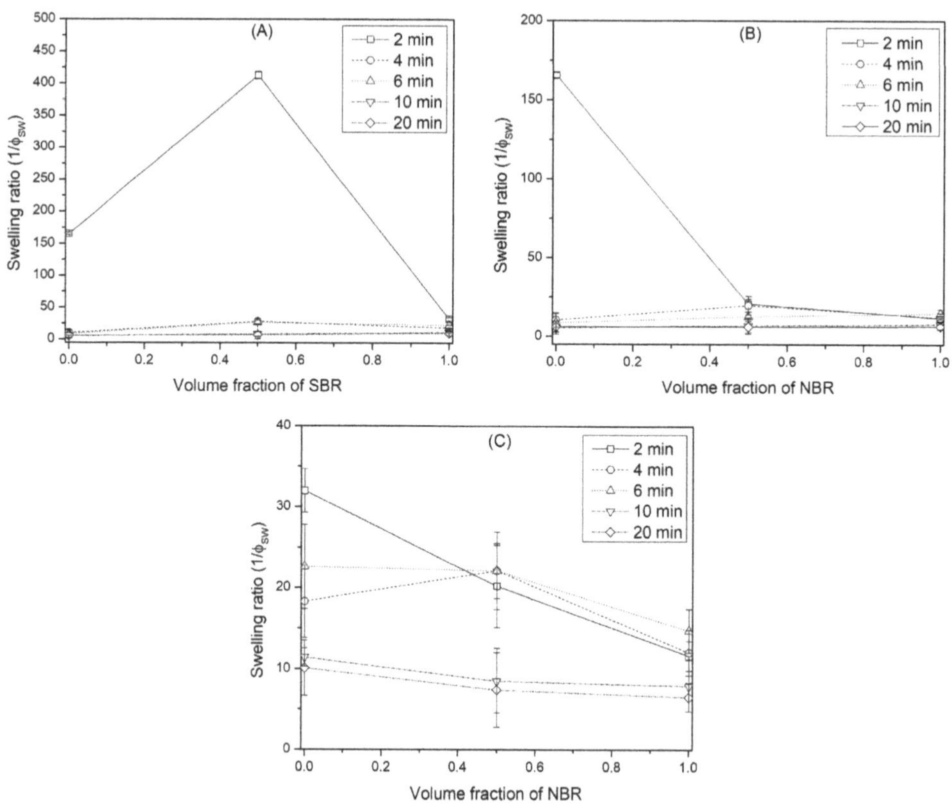

Figure 2. Effect of mixing the elastomers on the swelling ratio for each of the blends: (**A**) BR-SBR, (**B**) BR-NBR and (**C**) SBR-NBR.

4.2. Tensile Tests of the Elastomers and Their Blends

Figure S1 shows the stress–strain curves for the elastomers and their blends. For pure elastomers, as the stress increases, the deformation increases, until reaching a maximum where the specimen breaks. The NBR specimen presented the maximum deformation, being approximately 500%, while the BR presented the minimum with a value of 4%. The mixtures behave in the same way, where the BR-NBR specimen reached up to 300% deformation. On the other hand, it is evident that, at low strains, the stress value is higher for samples that contain a greater number of crosslinks. Thus, samples with longer crosslinking time become more rigid. In this sense, BR specimens were the most rigid, since they presented the lower deformation, while the most elastic were the NBR specimens.

Young's modulus was determined from the linear region of the stress–strain curve for each of the specimens. Young's modulus values are presented in Table S1. Figure 3 shows Young's modulus values versus the crosslinking time for the elastomers and their blends. For all samples, by increasing the crosslinking time, the Young's modulus increases. With the exception of the BR-SBR samples, the other elastomers and blends presented very similar values and slope. However, the BR-SBR sample presented the highest values of Young's modulus and a more pronounced slope. Evidently, this is caused by the differences in the crosslinking fraction previously described. Similar results were reported by Smejda-Krzewicka et al. for blends of chloroprene (CR) and BR, crosslinked with zinc oxide [31]. They found that the blend with the lowest swelling degree presented both the highest crosslinking density and tensile strength.

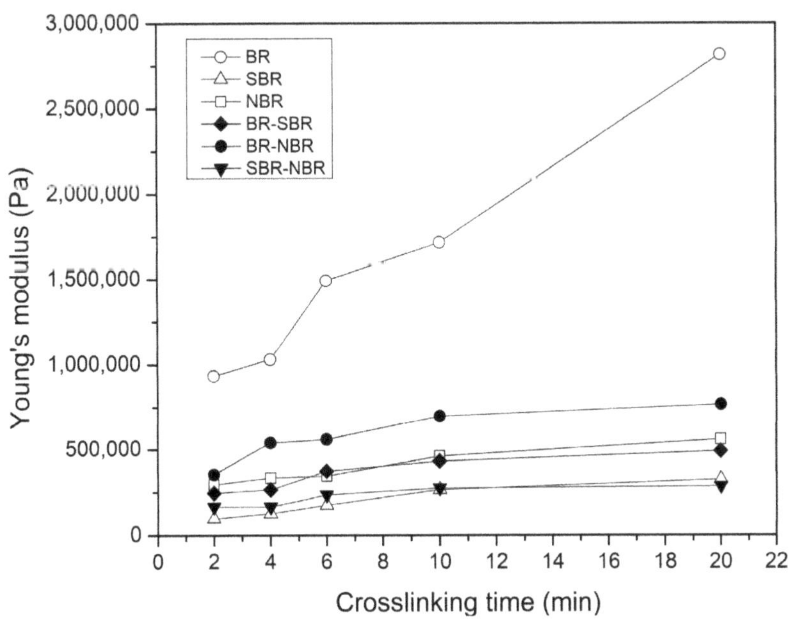

Figure 3. Young's modulus values versus the crosslinking time for the elastomers and their blends.

For the blends, the effect of adding another elastomer reduces the Young's modulus, which indicates a negative deviation from the rule of mixtures. This effect is more evident at higher crosslinking times, while it is attenuated as the crosslinking time decreases. This behavior can be explained in terms of the difference in diffusivity of the DCP in each of the elastomers during the mixing and crosslinking stages. Nah et al. found that the velocity of the initiator and an accelerator in NBR is lower than that for natural rubber (NR) [42]. Furthermore, if during mixing, the initiator diffuses into both phases, and during crosslinking, the DCP preferentially migrates to one of the phases, then crosslinking will be favored within one of the phases. Gardiner et al. employed an optical analysis to investigate the diffusion of crosslinking substances in elastomer blends. They found that the migration of these substances is related to diffusion during crosslinking and not to transfer during mixing [43]. The addition of SiO_2 nanoparticles in mixtures of polypropylene (PP) and polyolefin elastomer (POE) promotes phase refining, which significantly increases the Young's modulus [44].

4.3. Average Molecular Weight between Crosslinks

It is well known that crosslinked polymers absorb when immersed in organic solvents; the volume and absorption rate will depend on the temperature, molecular weight of the solvent, crosslinking density of the polymer and the polymer–solvent interactions, in addition to the added additives [45]. Hence, the determination of the average molecular weight between crosslinks (M_c) is important to predict other properties. Then, based on the knowledge of the Young's modulus and Equation (9), the M_c value was calculated for each of the elastomers and their blends. These values are reported in Table S1. Figure 4 shows the M_c values plot against the Young's modulus. An increase in Young's modulus is observed as the length of the M_c chain decreases. The shape of the curve is due to the inverse relationship between M_c and Young's modulus. The decrease in M_c indicates the occurrence of more crosslinking events of shorter average length, caused by the increase in the crosslinking time. The decrease in M_c results in an increase in Young's modulus, causing the material to become more rigid. In this sense, BR sample showed the lowest M_c values, while the SBR sample showed the highest values. The M_c values for NBR and the blends remained in between. In mixtures of NBR with polyvinyl chloride (PVC), an inverse relationship between Mc and Young's modulus was also found. This behavior was attributed to the formation of weak Van der Waals bonds at a concentration of 40 phr of PVC [46].

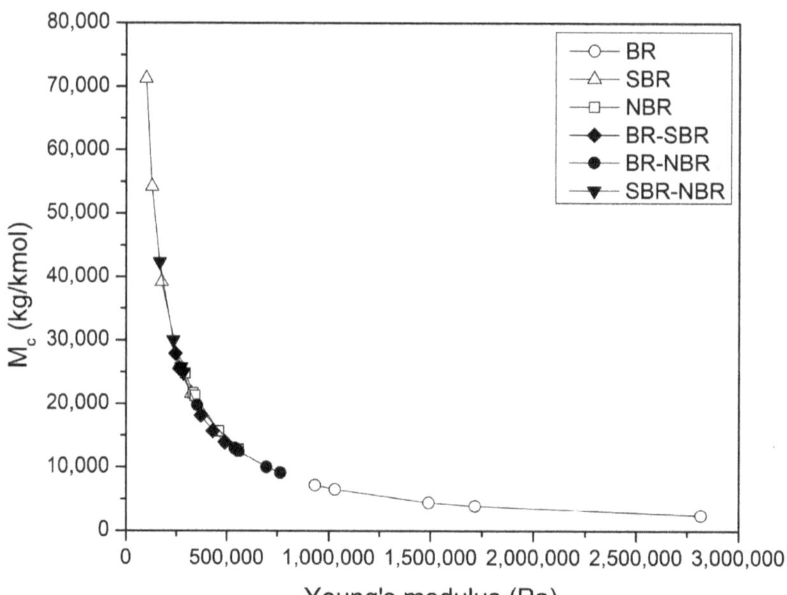

Figure 4. M_c values plot against the Young's modulus for the elastomers and their blends.

4.4. Determination of the Flory–Huggins Interaction Parameter

Once the values of M_c and $1/\phi_{sw}$ have been calculated, it is possible to solve the Flory–Rehner equation for X_{12}. In this work, the interaction parameter (χ_{12}*) will be defined as the resultant parameter of solving Equations (6), (9) and (11). This dimensionless parameter takes into account the specific interactions between the polymer chains of the blend and the solvent molecules for a given polymer–solvent system [47]. Likewise, the Flory–Huggins criterion for complete miscibility of the polymer and the solvent throughout the composition range is $\chi_{12} < 0.5$.

Figure 5 shows the interaction parameter (χ_{12}*) against the chain length M_c, for the elastomers and their blends. Overall, χ_{12}* tends to decrease as M_c increases, which is presented at short crosslinking times, or low crosslinking fractions. Most of the χ_{12}* values are slightly higher than the miscibility criterion, indicating the low interaction between the solvent and the elastomers and blends. The only two points below the miscibility criterion correspond to the BR and BR-SBR samples, both with a crosslinking time of 2 min. The spread of the points on the graph suggests an inverse dependence between χ_{12}* and M_c, indicating that for crosslinked elastomers and their blends, there is a narrow range of interaction parameters, rather than a single value. Table 2 shows the range of χ_{12}* values obtained at different values of crosslinking fractions. Likewise, for comparative purposes, values of χ calculated from the solubility parameters expressed in Equation (5) are included. All the values calculated herein are slightly higher than those calculated from the solubility parameters. Equation (5) only allows to introduce a single polymer for the calculation of the interaction parameter, and consequently, the resulting value is valid only for the polymer–solvent system employed. Additionally, strictly speaking, whenever specific interactions such as those with hydrogen bonds and donor–acceptor interactions are involved, the solubility parameter concept must fail [35]. Therefore, the values of χ calculated from Equation (5) and reported in Table 2 should be considered as reference values for comparative purposes only and not as absolute values. On the other hand, the Flory–Rehner equation provides the change in free energy upon swelling of a polymer gel. The derived interaction parameter (χ_{12}*) is a quantitative measure of the degree of interaction between the crosslinked polymer and solvent molecules [48,49]. A lower value implies higher polymer–solvent interaction and thus indicates higher solubility.

Several works have also reported different values of the interaction parameter for the same material. Martter et al. evaluated the interaction parameter for a series of branched and star-type polybutadienes, concluding that the interaction parameter (i) does not vary monotonically with the number of arms in the star, (ii) decreases with increasing concentration of the star and (iii) is almost invariable with increasing the length of the linear PB chain [50]. Schwahn et al. presented the calculation of the interaction parameter from experiments in a low-angle neutron diffractometer (SANS), for a series of mixtures of polybutadiene with polybutadiene of different vinyl content and different molar volume. Among their results, they reported that the interaction parameter has a strong inverse dependence on the molar volume [51]. Alberda et at., through DSC studies, experimentally determined the phase separation of blends of polystyrene with poly(styrene-*co*-4-vinylpyridine) and poly(styrene-*co*-2-vinylpyridine), and calculated the interval of the interaction parameter as $0.30 < \chi_{S\text{-}4VP} < 0.35$ and $0.09 < \chi_{S\text{-}2VP} < 0.11$, respectively [52]. Okay et al. prepared GDP with sulfur monochloride at low crosslinking concentrations, and reported a three-term equation to estimate the interaction parameter as a function of the volume fraction of the polymer [53]. Carbognani et al. studied asphaltenes swollen with polar and non-polar hydrocarbons, from which they obtained the solubility parameters of asphaltenes [54]. They proposed the use of an iterative method to simultaneously determine M_c, χ_{12} and δ_2, which is based on the assumption that each asphaltene has a single pair of M_c and δ_2 values. The χ_{12} values obtained in this way were lower than those obtained experimentally and from the literature. Habeeb Rahiman and Unnikrishnan studied the SBR-NBR system at different compositions and with three different crosslinking agents (sulfur, DCP and mixed). They determined that in the presence of chlorinated hydrocarbons, the value of χ is always higher for the blends crosslinked with DCP, and lower when sulfur was used. Likewise, the value of χ can vary, depending on the polarity of the solvent and the SBR content in the blend. The reported that values of χ were found in the range of 0.6–0.64

and 0.36–0.42, for carbon tetrachloride (CCl_4) and chloroform ($CHCl_3$), respectively [22]. Paul and Ebra-Lima reported the interaction parameter for crosslinked pure gum rubber membranes in the presence of 12 organic solvents. Among these solvents, the reported values of χ for benzene and CCl_4 were 0.474 and 0.353, respectively [55].

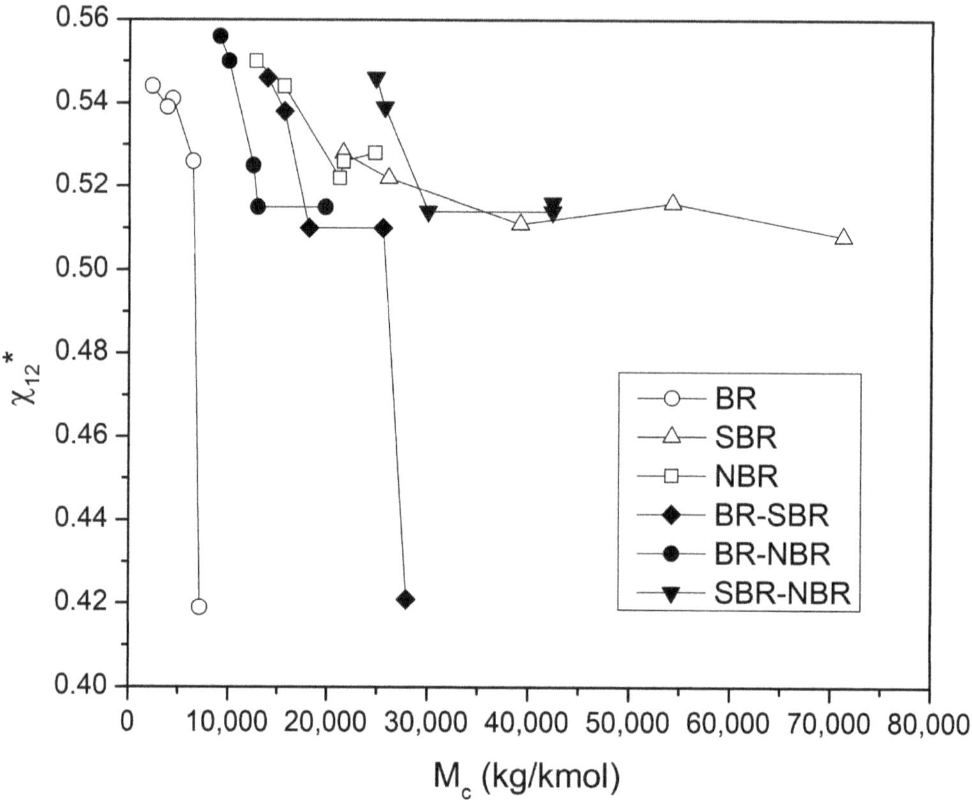

Figure 5. Interaction parameter (χ_{12}*) against the chain length M_c, for the elastomers and their blends.

Table 2. Range of values of χ_{12}* at different crosslinking fractions for the elastomers and their blends.

Sample	Range of χ_{12}*	Crosslinking Fraction (Θ)	χ Calculated from Equation (5)
BR	0.447–0.569	0.01–0.98	0.4369
SBR	0.510–0.535	0.3–0.66	0.4039
NBR	0.524–0.557	0.68–0.94	0.3449
BR-SBR	0.434–0.553	0.006–0.85	0.4204
BR-NBR	0.516–0.561	0.44–0.97	0.3909
SBR-NBR	0.515–0.550	0.45–0.84	0.3744

4.5. Evaluation of the Thermodynamic Parameters

After the interaction parameter for the elastomers and their blends has been estimated, it is possible to determine the values of the thermodynamic parameters (ΔG_{mix}, ΔH_{mix}, and ΔS_{mix}) from Equations (1)–(3) or by solving the Flory–Huggins relationship from Equation (4). Table S2 shows the calculated values of the thermodynamic parameters, while Figure 6 shows the ΔG_{mix} for the elastomer blends. The ΔG_{mix} values are positive, which means that the swelling phenomenon does not take place spontaneously. However,

ΔG_{mix} tends to decrease at low crosslinking times, suggesting more spontaneous swelling. Positive ΔG_{mix} values indicate moderate swelling, where $0 < 1/\phi_{sw} < 1$. For blends, the ΔG_{mix} is lower than for individual elastomers. This effect is more prominent at low crossing times (2, 4 and 6 min), while at times greater than 10 min, the behavior is the opposite. For example, the blends of BR-SBR at 2, 4 and 6 min of crosslinking, BR-NBR at 4 min, and SBR-NBR at 4 and 6 min presented this behavior.

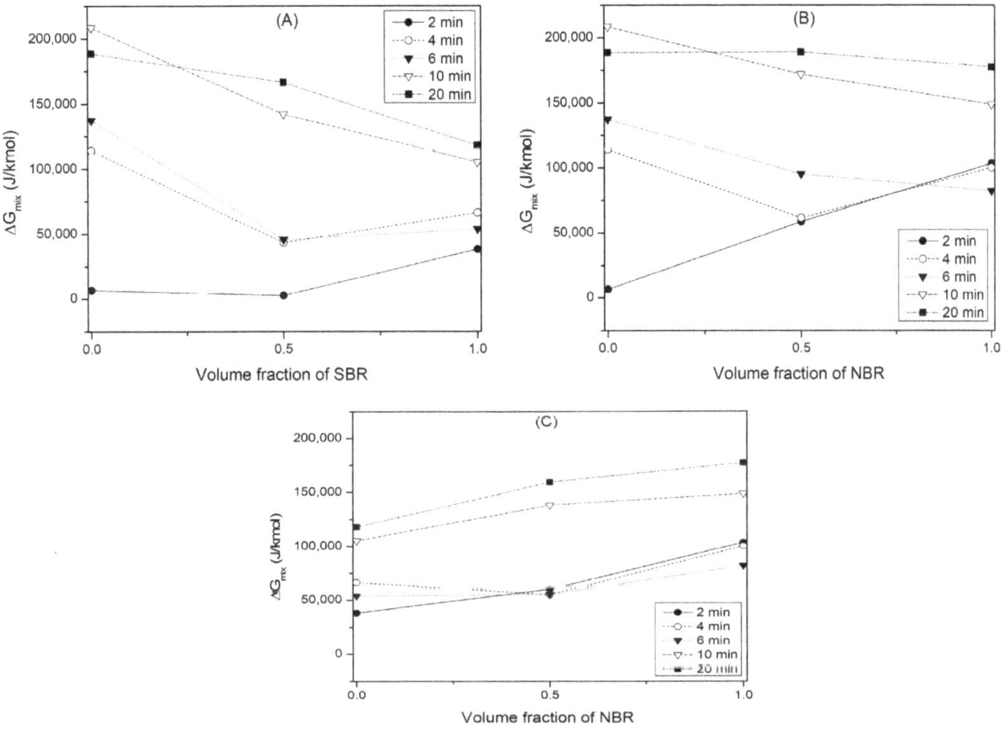

Figure 6. Effect of blending on the ΔG_{mix} of the elastomer blends. (**A**) BR-SBR, (**B**) BR-NBR and (**C**) SBR-NBR.

The enthalpy values are positive and tend to increase with crosslinking time, indicating that the solvent absorption phenomenon is an endothermic process. In an endothermic process, the formation of a site for the subsequent absorption of the solvent is necessary first [56]. Thus, the swelling can be improved by increasing the temperature of the system.

The ΔS_{mix} is always positive and tends to increase with crosslinking time. However, the values are relatively low, suggesting few configuration modes of the polymer chains between the two elastomers in the blend. On the other hand, when comparing the values of ΔS_{mix} against those of ΔH_{mix}, a low contribution of ΔS_{mix} on the ΔG_{mix} was observed which results in positive values of this thermodynamic parameter.

5. Conclusions

The thermodynamic parameters (ΔG_{mix}, ΔH_{mix} and ΔS_{mix}) of crosslinked elastomers of immiscible nature and their blends were reported herein. The study was based on a phenomenological approach to determine the Flory–Huggins interaction parameter (χ_{12}*), from swelling experiments and tensile tests. The samples prepared at shorter crosslinking times presented a greater swelling ratio and lower values in the Young's modulus. The effect of mixing on the swelling ratio showed a positive deviation from the rule of mixtures for crosslinking times lower than 6 min. The calculated values of the average molecular weight between crosslinks (M_c) decrease with the Young's modulus, indicating that the longer the crosslinking time, there is a greater occurrence of crosslink bonds but of shorter average length. The χ_{12}* showed values slightly above the miscibility criterion ($\chi_{12} < 0.5$), an inverse relationship with M_c, as well as a range of values instead of a single value for the same elastomer–elastomer solvent system. The determination of the ΔG_{mix} indicated that the swelling phenomenon of elastomers and their blends is not a spontaneous process, while the ΔH_{mix} showed the endothermic nature of the absorption process. The mixing of the elastomers showed a positive effect at low values of crosslinking time, where the blends presented lower ΔG_{mix} values than those of the individual components.

Supplementary Materials: The following supporting information can be downloaded at: https://www.mdpi.com/article/10.3390/polym16030351/s1, Figure S1: Stress–strain curves for the elastomers (BR, SBR and NBR) and their blends (BR-SBR, BR-NBR and SBR-NBR); Table S1: Sample identification and crosslinking time (τ). Values determined experimentally: Weights from the Soxhlet extraction process, crosslinking density (Θ), swelling ratio (Φ_{sw}), density (ρ) and Young's modulus (YM). Calculated values for parameters M_c and χ_{12}*; Table S2: Calculated values of the thermodynamic parameters.

Author Contributions: Conceptualization, C.L.-P. and A.M.-L.; data curation, I.A.E.-M. and C.I.P.-B.; formal analysis, C.L.-P. and S.G.F.-G.; funding acquisition, I.A.E.-M., C.I.P.-B. and S.G.F.-G.; investigation, C.L.-P. and C.I.P.-B.; methodology, C.L.-P.; project administration, A.M.-L.; resources, C.L.-P.; software, I.A.E.-M.; supervision, C.L.-P.; validation, C.L.-P., S.G.F.-G. and A.M.-L.; visualization, I.A.E.-M.; writing—original draft, C.L.-P.; writing—review and editing, C.L.-P. All authors have read and agreed to the published version of the manuscript.

Funding: This research received no external funding.

Institutional Review Board Statement: Not applicable.

Data Availability Statement: Some of the results and the discussion reported herein are portions of the master's thesis in materials science for C.L.-P. presented in 2004, which can be consulted in the institutional repository at the link: "https://cimav.repositorioinstitucional.mx/jspui/".

Acknowledgments: The authors thank the Advanced Materials Research Center (CIMAV) for the support in carrying out this research. C.L.-P. thanks CONAHCyT for the support granted through scholarship No. 206410.

Conflicts of Interest: The authors declare no conflict of interest.

References

1. Pearce, T.C.; Schiffman, S.S.; Nagle, H.T.; Gardner, J.W. *Handbook of Machine Olfaction: Electronic Nose Technology*; John Wiley & Sons: Hoboken, NJ, USA, 2006.
2. Ludeelerd, P.; Niamlang, S.; Kunaruksapong, R.; Sirivat, A. Effect of elastomer matrix type on electromechanical response of conductive polypyrrole/elastomer blends. *J. Phys. Chem. Solids* **2010**, *71*, 1243–1250. [CrossRef]
3. Stelescu, M.D.; Airinei, A.; Bargan, A.; Fifere, N.; Georgescu, M.; Sonmez, M.; Nituica, M.; Alexandrescu, L.; Stefan, A. Mechanical Properties and Equilibrium Swelling Characteristics of Some Polymer Composites Based on Ethylene Propylene Diene Terpolymer (EPDM) Reinforced with Hemp Fibers. *Materials* **2022**, *15*, 6838. [CrossRef] [PubMed]
4. Christova, D.; Velichkova, R.; Goethals, E.J.; Du Prez, F.E. Amphiphilic segmented polymer networks based on poly(2-alkyl-2-oxazoline) and poly(methyl methacrylate). *Polymer* **2002**, *43*, 4585–4590. [CrossRef]
5. Bajpai, A.K.; Shrivastava, M. Swelling kinetics of a hydrogel of poly(ethylene glycol) and poly(acrylamide-co-styrene). *J. Appl. Polym. Sci.* **2002**, *85*, 1419–1428. [CrossRef]
6. Chungprempree, J.; Charoenpongpool, S.; Preechawong, J.; Atthi, N.; Nithitanakul, M. Simple preparation of polydimethylsiloxane and polyurethane blend film for marine antibiofouling application. *Polymers* **2021**, *13*, 2242. [CrossRef]
7. Datta, S. Elastomer blends. In *Science and Technology of Rubber*; Elsevier: Amsterdam, The Netherlands, 2005; pp. 529–554.
8. Smejda-Krzewicka, A.; Mrozowski, K.; Strzelec, K. Interelastomer Reactions Occurring during the Cross-Linking of Hydrogenated Acrylonitrile-Butadiene (HNBR) and Chloroprene (CR) Rubbers Blends in the Presence of Silver (I) Oxide (Ag2O) and Mechanical Properties of Cured Products. *Materials* **2023**, *16*, 4573. [CrossRef]
9. Liu, S.; Qiu, J.; Han, L.; Luan, J.; Ma, X.; Chen, W. Effect of SEBS Molecular Structure and Formula Composition on the Performance of SEBS/PP TPE for Automotive Interior Skin. *Polymers* **2023**, *15*, 2753. [CrossRef]
10. Zhang, P.; Wang, X.; Yang, J.; Zhang, Y. Effect of Acetylated SEBS/PP for Potential HVAC Cable Insulation. *Materials* **2021**, *14*, 1811. [CrossRef] [PubMed]

11. Brydson, J.A. Styrene—Butadiene Rubber. In *Developments in Rubber Technology—2: Synthetic Rubbers*; Springer: Berlin/Heidelberg, Germany, 1981; pp. 21–49.
12. White, W.C. Butadiene production process overview. *Chem. Biol. Interact.* **2007**, *166*, 10–14. [CrossRef]
13. Sone, T. Industrial synthetic method of the rubbers. 1. Butadiene rubber. *Int. Polym. Sci. Technol.* **2016**, *43*, 49–54. [CrossRef]
14. Fujimoto, K.; Yoshimiya, N. Blends of cis-1, 4-polybutadiene with natural or styrene butadiene rubber. *Rubber Chem. Technol.* **1968**, *41*, 669–677. [CrossRef]
15. Varkey, J.T.; Thomas, S.; Rao, S.S. Rheological behavior of blends of natural rubber and styrene–butadiene rubber latices. *J. Appl. Polym. Sci.* **1995**, *56*, 451–460. [CrossRef]
16. Ismail, H.; Suzaimah, S. Styrene butadiene rubber/epoxidized natural rubber blends: Dynamic properties, curing characteristics and swelling studies. *Polym. Test.* **2000**, *19*, 879–888. [CrossRef]
17. Nair, T.M.; Kumaran, M.G.; Unnikrishnan, G. Mechanical and aging properties of cross-linked ethylene propylene diene rubber/styrene butadiene rubber blends. *J. Appl. Polym. Sci.* **2004**, *93*, 2606–2621. [CrossRef]
18. Higazy, A.A.; Afifi, H.; Khafagy, A.H.; El-Shahawy, M.A.; Mansour, A.M. Ultrasonic studies on polystyrene/styrene butadiene rubber polymer blends filled with glass fiber and talc. *Ultrasonics* **2006**, *44*, e1439–e1445. [CrossRef]
19. Salgueiro, W.; Somoza, A.; Marzocca, A.J.; Torriani, I.; Mansilla, M.A. A SAXS and swelling study of cured natural rubber/styrene–butadiene rubber blends. *J. Polym. Sci. Part B Polym. Phys.* **2009**, *47*, 2320–2327. [CrossRef]
20. Alimardani, M.; Abbassi-Sourki, F. New and emerging applications of carboxylated styrene butadiene rubber latex in polymer composites and blends: Review from structure to future prospective. *J. Compos. Mater.* **2015**, *49*, 1267–1282. [CrossRef]
21. Jovanović, S.; Samaržija-Jovanović, S.; Marković, G.; Jovanović, V.; Adamović, T.; Marinović-Cincović, M. Mechanical properties and thermal aging behaviour of polyisoprene/polybutadiene/styrene-butadiene rubber ternary blend reinforced with carbon black. *Compos. Part B Eng.* **2016**, *98*, 126–133. [CrossRef]
22. Habeeb Rahiman, K.; Unnikrishnan, G. The behaviour of styrene butadiene rubber/acrylonitrile butadiene rubber blends in the presence of chlorinated hydrocarbons. *J. Polym. Res.* **2006**, *13*, 297–314. [CrossRef]
23. Jinnai, H.; Yoshida, H.; Kimishima, K.; Funaki, Y.; Hirokawa, Y.; Ribbe, A.E.; Hashimoto, T. Observation of fine structure in bicontinuous phase-separated domains of a polymer blend by laser scanning confocal microscopy. *Macromolecules* **2001**, *34*, 5186–5191. [CrossRef]
24. Noriman, N.Z.; Ismail, H.; Rashid, A.A. Properties of styrene butadiene rubber/recycled acrylonitrile-butadiene rubber (SBR/NBRr) blends: Effect of the addition of trans-polyoctylene rubber. *J. Appl. Polym. Sci.* **2012**, *126*, E56–E63. [CrossRef]
25. Perez, L.D.; Sierra, L.; Lopez, B.L. Effect of the filler characteristics on the miscibility of styrene-butadiene rubber and nitrile-butadiene rubber blends. *Polym. Eng. Sci.* **2008**, *48*, 1986–1993. [CrossRef]
26. Malas, A.; Pal, P.; Das, C.K. Effect of expanded graphite and modified graphite flakes on the physical and thermo mechanical properties of styrene butadiene rubber/polybutadiene rubber (SBR/BR) blends. *Mater. Des.* **2014**, *55*, 664–673. [CrossRef]
27. Zhang, X.; Xue, X.; Yin, Q.; Jia, H.; Wang, J.; Ji, Q.; Xu, Z. Enhanced compatibility and mechanical properties of carboxylated acrylonitrile butadiene rubber/styrene butadiene rubber by using graphene oxide as reinforcing filler. *Compos. Part B Eng.* **2017**, *111*, 243–250. [CrossRef]
28. Boonrasri, S.; Thipchai, P.; Sae-Oui, P.; Thanakkasaranee, S.; Jantanasakulwong, K.; Rachtanapun, P. Property Improvements of Silica-Filled Styrene Butadiene Rubber/Butadiene Rubber Blend Incorporated with Fatty-Acid-Containing Palm Oil. *Polymers* **2023**, *15*, 3429. [CrossRef] [PubMed]
29. Ramesan, M.T.; Kuriakose, B.; Pradeep, P.; Alex, R.; Varghese, S. Compatibilization of SBR/NBR Blends Using Chemically Modified Styrene Butadiene Rubber: Part I: Effect of Chlorine Content of Compatibilizer. *Int. Polym. Process.* **2001**, *16*, 183–191. [CrossRef]
30. Mendoza, M.; Carrillo, A.; Márquez, A. New distributed optical sensor for detection and localization of liquid hydrocarbons: Part II: Optimization of the elastomer performance. *Sens. Actuators A Phys.* **2004**, *111*, 154–165. [CrossRef]
31. Smejda-Krzewicka, A.; Rybiński, P.; Bradło, D.; Żukowski, W. The Morphology, Mechanical and Dynamic Properties, Fire Hazard and Toxicity of Chloroprene and Butadiene Rubber Composites Cross-Linked with Zinc. *Materials* **2023**, *16*, 1240. [CrossRef]
32. Hess, W.M.; Herd, C.R.; Vegvari, P.C. Characterization of immiscible elastomer blends. *Rubber Chem. Technol.* **1993**, *66*, 329–375. [CrossRef]
33. Mangaraj, D. Elastomer blends. *Rubber Chem. Technol.* **2002**, *75*, 365–427. [CrossRef]
34. Flory, P.J. *Principles of Polymer Chemistry*; Cornell University Press: Ithaca, NY, USA, 1953.
35. Munk, P.; Aminabhavi, T.M. *Introduction to Macromolecular Science*; Wiley Interscience: New York, NY, USA, 1989.
36. Sperling, L.H. *Introduction to Physical Polymer Science*; John Wiley & Sons: Hoboken, NJ, USA, 2005.
37. George, S.C.; Thomas, S. Effect of nature and extent of crosslinking on swelling and mechanical behavior of styrene–butadiene rubber membranes. *J. Membr. Sci.* **1999**, *163*, 1–17. [CrossRef]
38. Hedden, R.C.; Saxena, H.; Cohen, C. Mechanical properties and swelling behavior of end-linked poly(diethylsiloxane) networks. *Macromolecules* **2000**, *33*, 8676–8684. [CrossRef]
39. *ASTM Standard D638*; Standard Test Method for Tensile Properties of Plastics. ASTM International: West Conshohocken, PA, USA, 2014. [CrossRef]
40. Marzocca, A.J.; Cerveny, S.; Méndez, J.M. Some considerations concerning the dynamic mechanical properties of cured styrene–butadiene rubber/polybutadiene blends. *Polym. Int.* **2000**, *49*, 216–222. [CrossRef]

41. Wang, C. Tear strength of styrene–butadiene–styrene block copolymers. *Macromolecules* **2001**, *34*, 9006–9014. [CrossRef]
42. Nah, C.; Han, S.; Jo, B.W.; Kim, W.D.; Chang, Y. Influences of trans-polyoctylene rubber on the physical properties and phase morphology of natural rubber/acrylonitrile–butadiene rubber blends. *J. Appl. Polym. Sci.* **2002**, *86*, 125–134. [CrossRef]
43. Gardiner, J.B. Curative diffusion between dissimilar elastomers and its influence on adhesion. *Rubber Chem. Technol.* **1968**, *41*, 1312–1328. [CrossRef]
44. Lee, S.H.; Kontopoulou, M.; Park, C.B. Effect of nanosilica on the co-continuous morphology of polypropylene/polyolefin elastomer blends. *Polymer* **2010**, *51*, 1147–1155. [CrossRef]
45. Gopalan Nair, K.; Dufresne, A. Crab shell chitin whisker reinforced natural rubber nanocomposites. 1. Processing and swelling behavior. *Biomacromolecules* **2003**, *4*, 657–665. [CrossRef] [PubMed]
46. Abu-Abdeen, M.; Elamer, I. Mechanical and swelling properties of thermoplastic elastomer blends. *Mater. Des.* **2010**, *31*, 808–815. [CrossRef]
47. Burrell, H.; Immergut, B. *Polymer Handbook*; Wiley Interscience: New York, NY, USA, 1975.
48. Flory, P.J.; Rehner, J., Jr. Statistical mechanics of cross-linked polymer networks II. Swelling. *J. Chem. Phys.* **1943**, *11*, 521–526. [CrossRef]
49. Paulin, J.A.; Lopez-Aguilar, J.E.; Fouconnier, B.; Vargas, R.O.; Lopez-Serrano, F. Revisiting the Flory–Rehner equation: Taking a closer look at the Flory–Huggins interaction parameter and its functionality with temperature and concentration with NIPA as a case example. *Polym. Bull.* **2021**, *79*, 6709–6732. [CrossRef]
50. Martter, T.D.; Foster, M.D.; Yoo, T.; Xu, S.; Lizzaraga, G.; Quirk, R.P.; Butler, P.D. Nonuniversal behavior of the thermodynamic interaction parameter in blends of star and linear polybutadiene. *Macromolecules* **2002**, *35*, 9763–9772. [CrossRef]
51. Schwahn, D.; Willner, L. Phase Behavior and Flory–Huggins Interaction Parameter of Binary Polybutadiene Copolymer Mixtures with Different Vinyl Content and Molar Volume. *Macromolecules* **2002**, *35*, 239–247. [CrossRef]
52. Alberda van Ekenstein, G.; Meyboom, R.; Ten Brinke, G.; Ikkala, O. Determination of the Flory– Huggins interaction parameter of styrene and 4-vinylpyridine using copolymer blends of poly(styrene-co-4-vinylpyridine) and polystyrene. *Macromolecules* **2000**, *33*, 3752–3756. [CrossRef]
53. Okay, O.; Durmaz, S.; Erman, B. Solution cross-linked poly(isobutylene) gels: Synthesis and swelling behavior. *Macromolecules* **2000**, *33*, 4822–4827. [CrossRef]
54. Carbognani, L.; Rogel, E. Solvent swelling of petroleum asphaltenes. *Energy Fuels* **2002**, *16*, 1348–1358. [CrossRef]
55. Paul, D.R.; Ebra-Lima, O.M. Pressure-induced diffusion of organic liquids through highly swollen polymer membranes. *J. Appl. Polym. Sci.* **1970**, *14*, 2201–2224. [CrossRef]
56. Mathai, A.E.; Singh, R.P.; Thomas, S. Transport of substituted benzenes through nitrile rubber/natural rubber blend membranes. *J. Membr. Sci.* **2002**, *202*, 35–54. [CrossRef]

polymers

MDPI

Article

Improved Heat Dissipation of NR/SBR-Based Tire Tread Compounds via Hybrid Fillers of Multi-Walled Carbon Nanotube and Carbon Black

Mehmet Kodal [1,2,3,*], Nazlı Yazıcı Çakır [1], Rumeysa Yıldırım [2], Nursel Karakaya [4] and Güralp Özkoç [3,4,5]

1 Department of Chemical Engineering, Kocaeli University, 41001 Kocaeli, Türkiye; nazliyazici93@gmail.com
2 Polymer Science and Technology Graduate Program, Kocaeli University, 41001 Kocaeli, Türkiye; yildirimmrumeysa@gmail.com
3 Sabancı University Nanotechnology Research and Application Center, 34956 Istanbul, Türkiye; guralp.ozkoc@istinye.edu.tr
4 Xplore Instruments B.V., 6135 KT Sittard, The Netherlands; nursel.karakaya@xplore-together.com
5 Department of Chemistry, Istinye University, 34396 Istanbul, Türkiye
* Correspondence: mehmet.kodal@kocaeli.edu.tr; Tel.: +90-262-303-3540

Abstract: The development of thermally conductive rubber nanocomposites for heat management poses a formidable challenge in numerous applications, notably within the realm of tire technology. Notably, rubber materials are characterized by their inherently low thermal conductivity. Consequently, it becomes imperative to incorporate diverse conductive fillers to mitigate the propensity for heat build-up. Multi-walled carbon nanotubes (MWCNTs), as reinforcement agents within the tire tread compounds, have gained considerable attention owing to their extraordinary attributes. The attainment of high-performance rubber nanocomposites hinges significantly on the uniform distribution of MWCNT. This study presents the influence of MWCNTs on the performance of carbon black (CB)-reinforced natural rubber (NR)/styrene butadiene rubber (SBR) tire compounds prepared via high shear melt mixing. Morphological analysis showed a good distribution of MWCNTs in the NR/SBR/CB compound. The vulcanization parameters, such as the maximum and minimum torque, cross-linking density, hardness, abrasion resistance, tensile strength, and Young modulus, exhibited a progressive improvement with the addition of MWCNT. Remarkably, adding MWCNT into CB improved the heat conductivity of the NR/SBR/CB compounds, hence decreasing the heat build-up. A percolation mode was also proposed for the hybrid carbon fillers based on the data obtained.

Keywords: NR; SBR; hybrid fillers; distribution; mechanical properties

1. Introduction

Natural rubber (NR) stands as an exceptionally sustainable material, distinguishing itself as the sole agricultural product within the spectrum of rubber varieties. NR exhibits a myriad of distinctive physical properties, encompassing remarkable attributes like high elasticity, minimal heat generation, exceptional resistance to fatigue crack propagation, and more. These inherent properties render its applicability across a diverse array of engineering domains, including but not limited to its utilization in the construction of tires and shock absorbers [1–3]. NR can be effectively compounded with synthetic rubbers to enhance its mechanical characteristics, including tensile strength, resilience, tear strength, fatigue resistance, and fracture toughness. It has been documented that the incorporation of NR into styrene butadiene rubber (SBR) blends can lead to enhancements in oxidative stability [4]. Rubber blends find their most significant application within the tire manufacturing industry. SBR exhibits superior attributes in terms of crack resistance, wet traction, and resistance to adverse weather conditions when compared to natural rubber. Conversely, NR surpasses SBR in terms of strength, heat dissipation, and low-temperature performance. Therefore, the synergistic qualities of NR and SBR have led to their extensive use in the production of tires [5]. It is worth noting that tire compositions typically incorporate substantial quantities of carbon black (CB) to enhance reinforcement and provide desirable properties related to fatigue resistance and abrasion [6]. Nevertheless, the utilization of these fillers poses certain challenges. Their relatively high density contributes to an increase in the specific gravity of rubber compounds. Furthermore, their efficacy in reinforcing rubber materials is most pronounced when employed at substantial loadings [7–9]. It is essential to note that, in the context of NR vulcanizates, an increase in CB ratios is accompanied by considerable heat build-up, which, in turn, adversely affects various performance attributes of vehicle tires. Enhancing the performance of vehicle tires entails addressing issues such as the mitigation of heat accumulation, the augmentation of wet handling characteristics, and the reduction in rolling resistance [10,11]. In pursuit of these objectives, researchers have explored the application of hybrid nanofillers, encompassing the use of CB or combinations of two or three distinct fillers. This approach has garnered attention for its potential in the development of advanced, high-performance rubber composites [6,12–14]. Researchers have undertaken investigations into the partial substitution of CB with carbon nanotubes (CNTs). Their studies have unveiled a synergistic interaction between CNT and CB, resulting in a significant enhancement in crack propagation resistance within NR composites. Moreover, this concurrent usage of CNT and CB has demonstrated notable efficiency in mitigating heat build-up in NR vulcanizates [2,15].

CNTs are cylindrical molecular structures composed of carbon, characterized by nanometer-scale diameters and micrometer-scale lengths. Their extraordinary mechanical properties, such as a remarkably high aspect ratio, expansive surface areas, exceptional tensile strength of 150–180 GPa, moduli ranging from 640 GPa to 1 TPa, and remarkable elasticity, endow them with substantial reinforcing capabilities [7,9,16–18]. The enhancement of properties in polymer/CNT composites primarily hinges on factors including the uniform dispersion of CNTs, interactions between the filler and the matrix, and the inherent attributes of the CNTs themselves. Due to relatively weak intermolecular forces such as van der Waals and π-π interactions that govern the assembly of CNTs, these nanotubes exhibit a propensity to intertwine, forming sizeable aggregations referred to as "bundles" [19–21]. The attainment of high-performance NR nanocomposites characterized by homogeneous dispersion of nanofillers necessitates the resolution of two prominent challenges: aggregation and inadequate matrix–filler interactions. The establishment of robust interactions between NR and nanofillers, along with the effective alleviation of filler–filler agglomerations, is of paramount importance in realizing property enhancements within NR-based nanocomposites. Several well-regarded processing techniques for the fabrication of NR nanocomposites include methods such as melt/mechanical mixing, solution mixing, and latex casting [22].

Within the solution mixing method, the polymer solution is intricately combined with the nanofiller dispersion through means such as ultrasonication, stirring, or shear mixing. Subsequently, the resultant polymer–nanofiller dispersion is cast into a mold and allowed to undergo solvent evaporation. This approach offers the advantage of achieving a more refined nanofiller dispersion. Nevertheless, the removal of solvents upon completion of this procedure presents a notable challenge. Moreover, the method is afflicted by drawbacks such as elevated costs, substantial solvent consumption, and environmental concerns related to solvent disposal [22,23]. Fakhrúl-Razi et al. [24] undertook the preparation of nanocomposites comprising NR and multi-walled carbon nanotubes (MWCNTs) through the solution casting method. Toluene was used as the solvent for dispersing the MWCNTs, and NR was similarly dispersed in toluene. The two solutions were subsequently amalgamated, resulting in a final solution containing a well-balanced blend of MWCNTs within the NR matrix, which underwent further processing. This methodology facilitated the attainment of a uniformly dispersed MWCNT structure within the NR, thereby yielding enhancements in mechanical, chemical, and physical properties. In another study, the fabrication of NR/MWCNT nanocomposites was executed through the incorporation of MWCNTs at various loading levels, employing two distinct methodologies: mechanical mixing and solution mixing by Hanafi Ismail et al. [25] The process entailed the initial swelling of natural rubber in toluene under sustained agitation. Concurrently, multi-walled carbon nanotubes were dispersed in toluene and subjected to magnetic stirring for a duration of 20 min. Following this interval, the MWCNT dispersion was amalgamated with the rubber–toluene solution and agitated vigorously for a period of 2 h. Subsequently, the solvents within the resultant dispersions were subjected to evaporation within a fume chamber until the weight of the rubber-MWCNT composite sample reached a constant value. Once this criterion was met, the sample was removed from the chamber and prepared for further mixing with additional rubber additives, accomplished on a laboratory-scale two-roll mill, resulting in an improved MWCNT dispersion in the rubber matrix [25]. In an effort to enhance the dispersion of MWCNTs within the NR matrix, a solution-mixing method was undertaken by Huang et al. [26]. Initially, MWCNTs were exfoliated and dispersed in water through ultrasonication, with the aid of tannic acid (TA) serving as a dispersing agent. Subsequently, this MWCNT dispersion was combined with interspersed SiO_2 microspheres (m-SiO_2). It is noteworthy that a notable attribute of this process is the ability to uniformly disperse both m-SiO_2 and MWCNTs within the NR matrix through direct mixing of the dry hybrid filler and NR within a double roller mill. This method capitalizes on the distinctive and synergistic dispersing effects between m-SiO_2 and MWCNTs, in addition to the dispersing efficacy conferred by TA. The ultimate research findings underscore the excellent mechanical, electrical, and thermal properties [26].

The latex mixing method involves the utilization of rubber in its latex form. The dispersion of nanofillers is integrated with the polymer latex and subsequently undergoes a casting and drying procedure. Alternatively, latex coagulation can be employed to foster substantial interactions between the nanofiller and the polymer matrix. In this particular context, the coagulation of the latex and the nanofiller dispersion is achieved through the addition of an acidic agent, followed by the incorporation of additional curatives into the polymer nanocomposite utilizing mechanical techniques. It is noteworthy that this method obviates the necessity for highly toxic and expensive organic solvents, thereby establishing itself as an ecologically sustainable alternative to solution mixing, as elucidated [22,27]. Nonetheless, the substantial volume of deionized water employed for the dispersion of CNT results in the dilution of the latex, inducing a propensity for CNTs to settle within the diluted latex medium and, consequently, giving rise to a secondary agglomeration phenomenon [28,29]. The preparation of nanocomposites based on NR has been accomplished through a latex stage mixing technique featuring the incorporation of single-walled carbon nanotubes (SWNTs) by Anoop Anand K. SWNTs were initially dispersed in water using ultrasonication and subsequently stabilized through the introduction of a surfactant. This stabilized aqueous dispersion of SWNTs was then blended with NR latex to investigate the impact of the nanofiller on the rheological properties of the NR latex. Subsequent to the compounding of the latex and the curing process, high-quality composite films were obtained, characterized by a pronounced improvement in their mechanical properties. Remarkably, the composite film, including 2 phr SWNTs, exhibited a substantial increase of 56% in tensile strength and a 63% increase in modulus in comparison to pure NR films [30]. In another study, Laiyun Wei et al. stated that the utilization of the latex mixing method was pivotal in enhancing the dispersion of graphene oxide (GO)/CNT hybrid fillers and CB within the NR matrix. This enhanced dispersion mechanism contributed to the development of a more robust and efficient filler network, consequently leading to significant enhancements in the modulus and energy dissipation properties of NR composites. The introduction of GO/CNT hybrid fillers and CB into the NR matrix resulted in a substantial improvement in resistance to fatigue crack propagation and a reduction in heat generation, thereby enhancing the overall performance of NR composite materials [2]. In another work, polyvinylpyrrolidone (PVP) was employed as a dispersant for MWCNTs, and the effectiveness of various dispersion techniques, including high-speed shear, water bath ultrasonication, and tip ultrasonication, was systematically compared to assess their exfoliation capabilities. Following this, the appropriateness of employing nonionic PVP for the purpose of achieving high-speed shear exfoliation and dispersion of MWCNTs was thoroughly investigated and fine-tuned, while classical anionic sodium dodecyl benzene sulphonate (SDBS) and cationic cetyltrimethylammonium bromide (CTAB) were utilized as control agents for comparative assessment. Ultimately, an aqueous dispersion of MWCNTs, which had been noncovalently modified by PVP, hereinafter denoted as PVP-m-MWCNTs, was amalgamated with natural latex to produce a composite material comprising NR and MWCNTs, designated as W-NMC. A comparative assessment, conducted in relation to the conventional dry-mixing procedure (D-NMNCs), elucidated that the wet mixing methodology implemented in this investigation yielded W-NMCs distinguished by their superior tensile properties, reduced compression temperature rise, enhanced resistance to DIN abrasion, increased electrical and thermal conductivity, augmented resilience against ice and wet-slip, and diminished rolling resistance. These improvements can be attributed to the enhanced dispersion of MWCNTs and the reinforcement of the interfacial interaction between NR-MWCNTs within the W-NMNCs, aligning with the research findings of Wei Xiao [31]. An investigation conducted by Song et al. focused on the development of high-performance nanocomposites comprising MWCNTs and SBR through a latex mixing process, with the utilization of lower MWCNT quantities. To augment MWCNT dispersion in the aqueous phase for latex mixing, a diverse array of surfactants was systematically employed to modify the MWCNTs. In a targeted approach, a suspension of MWCNTs was incorporated into the SBR latex to optimize the dispersion of MWCNTs within the elastomeric matrix, consequently bolstering interfacial strength. The successful realization of uniformly dispersed MWCNTs throughout the SBR matrix yielded composite materials distinguished by their exceptional mechanical, thermal, and electrical properties, even at low MWCNT loadings [32].

In the melt/mechanical mixing method, the production of nanocomposites is achieved through the utilization of equipment such as an internal mixer, two-roll mill, ball mill, or similar apparatus. This method finds substantial favor within industrial applications owing to its environmentally conscientious attributes, cost effectiveness, elevated efficiency, and expeditious production capabilities. Notably, the solvent-free characteristic represents a pivotal advantage, rendering it a highly regarded alternative to solution mixing [22,23]. Hanafi Ismail et al. prepared NR/MWCNT samples by using a laboratory-sized two-roll mill in accordance with ASTM D3184 guidelines [33]. The authors found a more favorable dispersion of MWCNTs within the NR matrix when employing the solution mixing method, in contrast to the mechanical mixing method. In terms of mechanical properties, the tensile modulus exhibited a direct proportionality with the MWCNT loading, while conversely, the tensile strength and elongation at break displayed an inverse relationship. Importantly, NR/MWCNT nanocomposites prepared via the solution mixing method exhibited superior tensile characteristics in comparison to composites produced through mechanical mixing techniques [25].

Laboratory-scale formulation development traditionally relies on small-scale internal mixers and two-roll mills, which are encumbered with several significant drawbacks. These conventional compounding methods entail labor-intensive procedures, requiring at least two personnel for the preparation of a single batch. Furthermore, these methods pose problems in terms of contamination, especially when dealing with different types of samples, necessitating separate Banbury machines for black and nonblack samples. Moreover, there is an extended turnaround time involved in the process. In addition to these operational challenges, safety concerns are associated with the open mixing area of two-roll mills. This process typically demands a substantial material volume, thus exacerbating the cost implications, especially when dealing with expensive additives. Moreover, the spatial requirements of both open mills and internal mixers are substantial, consuming approximately 6 square meters of laboratory space, particularly due to the dusting associated with carbon black additives. However, the adoption of high-torque laboratory twin-screw micro-compounders, which have served the plastic industry for over three decades, presents a more efficient alternative for the formulation of new rubber compounds. This method streamlines sample preparation, contributing to the overall cost effectiveness of research and development efforts. Micro-compounders offer the advantage of being operated by single personnel, characterized by ease of operation, reduced labor intensity, and a remarkably expedited compound preparation time, typically as short as 5 min. These devices require only minimal quantities of materials, which is particularly advantageous when working with costly additives. Additionally, micro-compounders enable the formulation of more precise compound compositions and occupy significantly less laboratory space compared to their conventional counterparts. Their ease of cleaning and enhanced safety features further underscore the advantages of this modern approach to rubber compounding at the laboratory scale.

Hence, the primary aim of this research is to investigate the influence of a hybrid microstructure developed within a tire tread compound composed of an NR/SBR blend and augmented by the inclusion of MWCNT and CB. This study seeks to elucidate the impact of this hybrid filler system on various properties, encompassing the mechanical, thermal, rheological, morphological, and electrical and thermal conductivity aspects of the composites. The distribution of these hybrid fillers within the NR/SBR matrix was accomplished through the use of a high-torque lab-scale twin-screw micro-compounder, a method until now unreported in the literature. Within this context, MWCNTs were introduced into the NR/SBR composite, in conjunction with the reinforcing CB, within the formulation of a tire tread composition. This approach aims to provide deeper insights into the intricate dynamics of this composite system, particularly with regard to the structure-property relationships arising from the distinctive synergism inherent in a macro-nano dual filler system.

2. Materials and Methods

2.1. Materials

NR and SBR were purchased from Baria Rubber, Ho Chi Minh City, Vietnam. N330-type CB was purchased from Elkim, Istanbul, Türkiye. Insoluble sulfur, zinc oxide, stearic acid, N-(cyclohexylthio)phthalimide (PVI), 2-mercaptobenzothiazole (MBT), and N cyclohexyl-2-benzothiazolesulfenamide (CBS) were used for the vulcanization system. 2,2,4-trimethyl-1,2-dihydroquinoline (TMQ) and N-Isopropyl-N′-phenyl-1,4-phenylenediamine (IPPD) were used as antiaging agents. Naphthenic oil was obtained from Petro Yağ, Kocaeli, Türkiye. MWCNT, with an average diameter of 8–10 nm, a length of 1.5 μm, a purity of 92%, and an electrical conductivity of 98 S/m, was purchased from Nanografi, Ankara, Türkiye (Table 1). According to the manufacturer of MWCNT used in this study, it was obtained via the chemical vapor deposition method.

Table 1. Materials used in the study.

Materials	Suppliers
NR	Baria Rubber Co., Ho Chi Minh City, Vietnam
SBR	Baria Rubber Co., Ho Chi Minh City, Vietnam
CB	Elkim/N330, Istanbul, Türkiye
MWCNT	Nanografi, Ankara, Türkiye
Naphthenic Oil	Petro Yağ, Kocaeli, Türkiye
Zinc Oxide	RubberChem, Istanbul, Türkiye
Stearic Acid	RubberChem, Istanbul, Türkiye
IPPD	RubberChem, Istanbul, Türkiye
TMQ	RubberChem, Istanbul, Türkiye
MBT	RubberChem, Istanbul, Türkiye
CBS	RubberChem, Istanbul, Türkiye
PVI	RubberChem, Istanbul, Türkiye
Sulfur	RubberChem, Istanbul, Türkiye

2.2. Processing

A laboratory-scale high-torque twin-screw micro-compounder, namely, an Xplore Instruments MC 15 HT (from Sittard, The Netherlands), was employed for the preparation of the formulations, namely, the NR/SBR-based compounds. Initially, natural rubber (NR) was subjected to mastication within the MC 15 HT for a duration of 30 s, with the addition of peptizer as a processing aid. Subsequently, the premasticated NR was combined with styrene butadiene rubber (SBR), carbon black (CB), multi-walled carbon nanotubes (MWCNT), and other constituents, excluding sulfur and accelerators. This compound was prepared within the MC 15 HT micro-compounder, operating at a rotational speed of 100 rpm. The temperature of the machine's barrel was meticulously regulated via water-cooling jackets to maintain it at a constant temperature of 60 °C. Following an elapsed time of 2 min, sulfur and accelerator agents were introduced into the compound, and a mixing phase of 1 min was ensured. The resulting rubber strands were subsequently extricated from the compounder's die by adjusting the position of the valve. An illustration of the compounding equipment can be observed in Figure 1. The vulcanization process for the NR/SBR composites was executed within a hot press, sustaining a temperature of 160 °C. The optimal curing temperature and duration were ascertained through analyses employing a differential scanning calorimeter (DSC) and a moving die rheometer (MDR), respectively. Recipe 1, Recipe 2, and Recipe 3 correspond to the MWCNTs contents of 0, 3, and 7 phr. The total amount of filler is 30 phr for each sample, as seen in Table 2.

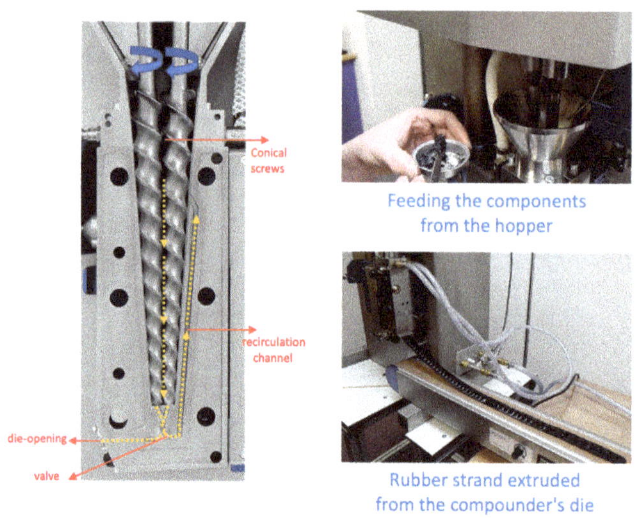

Figure 1. MC 15 HT twin-screw micro-compounder (Xplore Instruments, The Netherlands).

2.3. Characterization

The assessment of CB and MWCNT distribution within cured NR/SBR composites was conducted through the utilization of scanning electron microscopy (SEM) and DisperGrader analysis. In the SEM examination, the composite surfaces underwent a gold coating process to prevent sample charging artifacts. Additionally, the assessment of CB and MWCNT distribution within the NR/SBR matrix, as well as the overall distribution quality of the cured NR/SBR samples, was carried out in accordance with the ISO 11345 Method C standard [34], employing DisperGrader analyses. Five tests on different parts of each test piece were carried out for DisperGrader analyses.

Table 2. Rubber compound recipes (phr).

Components	Recipe 1	Recipe 2	Recipe 3
NR	70	70	70
SBR	30	30	30
CB	30	27	23
MWCNT	0	3	7
Oil	10	10	10
ZnO	3	3	3
Stearic Acid	2	2	2
IPPD	2	2	2
TMQ	1	1	1
MBTS	0.4	0.4	0.4
CBS	1	1	1
PVI	0.2	0.2	0.2
Sulfur	2.0	2.0	2.0

The investigation of the thermal properties of NR/SBR compounds was undertaken using a differential scanning calorimeter (DSC), specifically a Mettler Toledo DSC-1 Star instrument. DSC analyses were executed within a controlled N_2 atmosphere, employing a gradual heating rate of 10 °C per minute over a temperature range between 25 °C to 300 °C.

Significant rheological parameters such as optimum cure time, scorch time, and cure rate index (CRI) were derived through the analysis of rheometer curves. A moving die rheometer (MDR) from Alpha Technologies was employed for these purposes, following the guidelines outlined in ASTM D-5289 [35]. The rheometer curves, specifically collected at an elevated temperature of 160 °C, were employed to characterize the pertinent rheological parameters. CRI, a measure of the speed of vulcanization reaction, was calculated using the following Equation (1):

$$CRI = \frac{100}{t_{90} - t_{s2}} \quad (1)$$

where t_{s2} is the scorch time and t_{90} is the optimum cure time.

The cross-link density (CLD) of the vulcanizates was subsequently calculated through the application of the Flory–Rehner equation. This computation involved the utilization of key factors such as swelling ratios in toluene, the density of the rubber matrix, volume fractions pertaining to both the polymer and solvent, as well as the critical polymer–solvent interaction parameter, χ. The densities of Recipes from 1 to 3 were measured as 1.052, 1.012, and 0.997 g/cm^3, respectively. In addition, solvent–polymer interaction value of all Recipes is 0.350.

The determination of cross-link densities in rubber materials can also be achieved through the examination of stress relaxation behaviors in accordance with the classical rubber elasticity theory based on the neo-Hookean law. According to ASTM D8363-20 [36], crosslink density can be deduced from TSSR-measurements by considering the initial slope of the stress-temperature curve. This method is based on the neo-Hookean material model, which applies well to ideal rubber. The initial slope of the stress–temperature curve, the absolute value of initial stress, σ_0, was used to calculate the crosslink density using Equation (2) given below:

$$\nu = \frac{\sigma_0}{RT[\lambda - \lambda^{-2}]} \quad (2)$$

The thermoelastic behavior of vulcanizates is examined at constant elongation in a temperature range from 30 °C to 120 °C. When a stretched elastomer sample is heated, there is an increase in stress, which, in filled elastomers, is superimposed by physical and chemical relaxation processes and is partially compensated for [37]. The proportion of the polymer layer adsorbed on the filler surface is also considered. It is assumed here that the polymer segments are immobilized within the adsorbed polymer layer at room temperature, which leads to an apparent increase in the filler content. For the experimental execution, a constant tensile strain of 50% was applied to a dumbbell-shaped test specimen (Type 5A, ISO 527 [38]). Subsequently, the specimen underwent a preconditioning phase, allowing for isothermal relaxation at 30 °C over a duration of 2 h without any additional heating. Following this initial stage, the specimen was subjected to a controlled heating process, commencing at 30 °C and concluding at 120 °C, with a heating rate of 3 °C per minute. This process continued until full stress relaxation was achieved. The relaxation spectrum H(T) is calculated from Equation (3), the negative derivative of the non-isothermal relaxation modulus, $E_{non-iso} = \sigma/\varepsilon_0$, which, at a constant heating rate, $\beta = \Delta T/t$, follows from the tensile stress ascertained, $\sigma(T)$.

$$H(T) = -\Delta T. \left[\frac{dE_{non-iso}}{d\Delta T}\right]_\beta \quad (3)$$

The tensile properties of the cured composites composed of NR and SBR were determined through the utilization of an Instron Universal Testing Machine (Model 3345), with strict adherence to the standards set forth by ASTM D412 [39]. The testing was conducted at a crosshead speed of 500 mm/min. To gauge the hardness of the samples, measurements were performed using a Zwick Shore A-type durometer, adhering to the specifications outlined in ASTM D2240 [40]. To characterize the abrasion resistance of composites, a DIN Abrasion test was carried out. This assessment followed the guidelines established by ISO 4649 [41] and employed a DIN abrasion resistance tester. The methodology is based on evaluating the volume changes induced by the frictional forces arising from the abrasive action.

Thermal stability of the composites was determined using thermal gravimetric analysis (TGA, Mettler Toledo), under a N_2 atmosphere and at a heating rate of 10 °C/min from ambient temperature to 600 °C.

The thermal conductivity of the disk-shaped CB and CB/MWCNT-reinforced NR/SBR composites was carried out according to the ASTM E1530 standard [42]. Multiple trials, exceeding three in number, were undertaken for each sample. Surface resistivity measurements were conducted on compression-molded specimens with a 15 V potential difference, and a current of up to 1.5 mA was applied between parallel conductive probes. Measurements were performed at standard room-temperature conditions.

3. Results and Discussions

3.1. Morphological Analyses of Samples via Scanning Electron Microscopy (SEM) and DisperGrader

To achieve the desirable properties of CNT-reinforced polymer composites, it is important to establish a homogeneous distribution of CNTs in the polymer matrix. When the distribution of CNTs fails to meet the desired criteria, it impedes the complete realization of their reinforcing potential. As seen in Figure 2, for both CB and CB/MWCNT-reinforced NR/SBR composites, a smooth surface morphology was obtained. In a broader context, the distribution of carbon black within the polymer matrix is intricately linked to the diminishment of carbon black aggregate dimensions. It is evident that the distribution of carbon black is uniform across all compound compositions. It is widely acknowledged that achieving uniform distribution of CNTs within a polymer matrix poses a substantial challenge when employing conventional polymer processing techniques, such as two-roll mixing, Banbury mixing, extrusion, and injection molding. This challenge arises from the intricate entanglement and robust π-π stacking interactions prevalent among CNTs [26,43]. As seen in Figure 2, the SEM images reveal a homogeneous distribution of MWCNTs within the NR/SBR matrix, displaying an absence of discernible aggregation. Furthermore, an examination of the cryogenically fractured surfaces of the NR/SBR/CB/MWCNT composites demonstrates smoother morphology. Moreover, the concentration of MWCNT particles increased as the MWCNT content increased as expected. Furthermore, it is noteworthy that all rubber compounds displayed a nearly identical average particle size. Moreover, the distribution of carbon black was quantitatively assessed via DisperGrader analyses, yielding an 80% distribution, which attests to the commendable homogeneity observed in all composite materials (Figure 3).

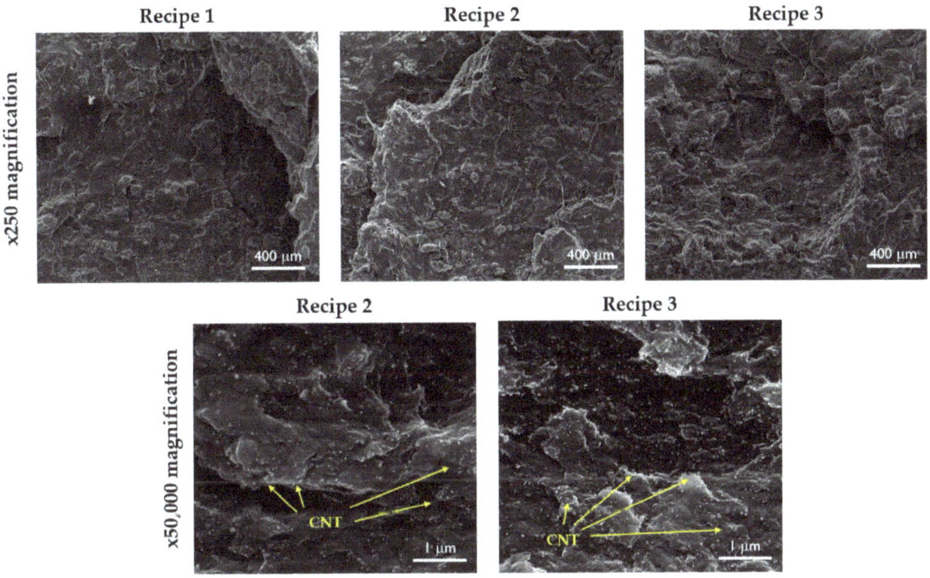

Figure 2. Cryogenically fractured surface morphologies of NR/SBR/CB and NR/SBR/CB/MWCNT composites.

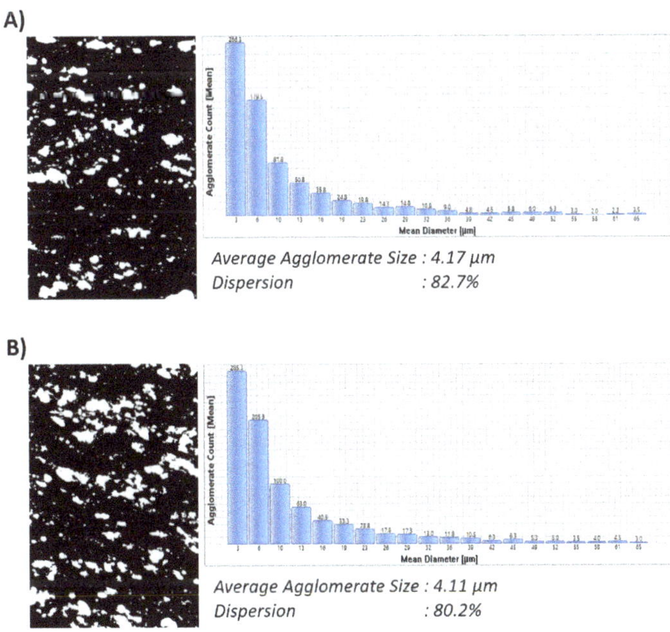

Figure 3. *Cont.*

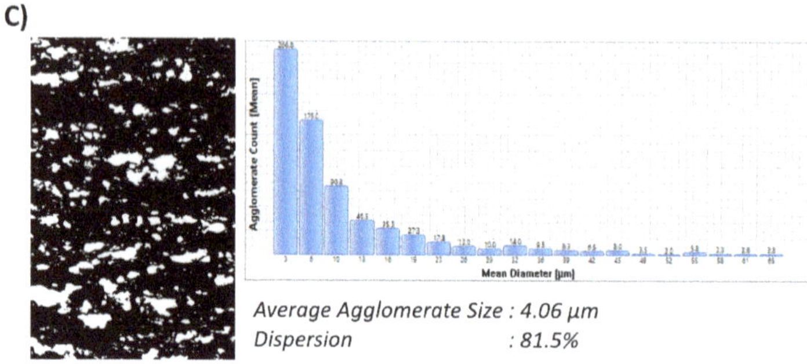

Figure 3. Carbon black dispersion in NR/SBR and NR/SBR/MWCNT composites; (**A**) Recipe 1, (**B**) Recipe 2, (**C**) Recipe 3, (Magnification 100×).

The phase structure of the particulate-filled composite materials is subject to influence from several pivotal factors, including surface characteristics, viscosity, the blend ratio of individual components, and the intricacies of the compounding process [44]. The flow characteristics within an intermeshing corotating twin-screw extruder primarily manifest as a state of simple shear in the radial direction, coupled with the presence of elongational flow in the axial direction [45,46]. In contrast to a Banbury mixer, it is noteworthy that the entire composition within an intermeshing twin-screw corotating extruder undergoes continuous exposure to shear stresses throughout the processing cycle. This resulted in a homogeneous distribution of MWCNT and CB within the polymer matrix. Moreover, the synergistic intercalation among string-like CNTs and CB particles serves to effectively mitigate their reagglomeration tendencies, culminating in the establishment of a well-developed network architecture. This phenomenon consequently lends credibility to the assertion that the distribution of carbon black fillers in the NR matrix has been significantly enhanced [2,10,47].

3.2. Curing and Rheological Properties of Compounds

In order to investigate the influence of CB and CB/MWCNT hybrid fillers on the thermal properties of NR/SBR tire tread compounds, an examination was conducted using differential scanning calorimetry (DSC) on uncured NR/SBR compounds. The DSC curves are graphically presented in Figure 4, and the results are tabulated in Table 3. Evidently, all examined samples demonstrated exothermic peaks attributed to the cross-linking of the materials. Notably, the reaction enthalpy exhibited a proportional increase corresponding to the escalating carbon nanotube content within the compounds. Furthermore, a discernible shift toward higher values was observed in the onset temperature of crosslinking ($T_{x\text{-link,onset}}$). This phenomenon can be ascribed to the elevated viscosity and barrier effect induced by CNT nanoparticles, consequently delaying the initiation of the vulcanization reaction [48]. Moreover, several studies have suggested that the incorporation of CNTs may result in the absorption of fundamental accelerator species, leading to a retardation of the vulcanization process [49,50]. In summary, it can be deduced that augmenting the quantity of MWCNTs and achieving a uniform distribution of MWCNTs within the matrix increases the adsorption of accelerator compounds, thereby postponing the vulcanization process.

Rheological assessments were conducted at a selected curing temperature of 160 °C, yielding several key parameters, including the scorch time (t_{s2}), optimum cure time (t_{90}), cure extent, maximum and minimum torque values (MH and ML), as well as the cure rate index (CRI). The term "cure extent" signifies the cross-link density (CLD) of a vulcanizate and is calculated as the disparity between the MH and ML values. A summary of these significant findings is provided in Table 4. The graphical representation in Figure 5 illustrates the rheometer curves consisting of CB and CB/MWCNT hybrid fillers reinforced

NR/SBR composites exhibited the ability to attain a plateau region immediately following the vulcanization process.

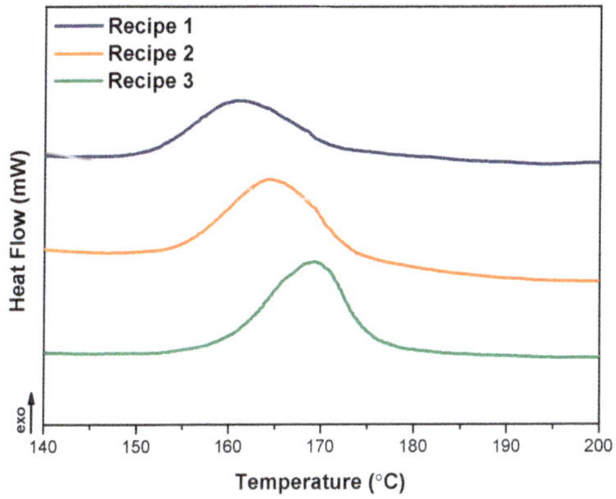

Figure 4. DSC thermograms of compounds.

Table 3. DSC test results of compounds.

Compounds	$T_{x\text{-}link,onset}$ (°C)	$T_{x\text{-}link,peak}$ (°C)
Recipe 1	148.6	161.3
Recipe 2	151.3	164.5
Recipe 3	154.1	169.5

Table 4. Rheological properties of compounds.

Compounds	ML (dNm)	MH (dNm)	t_{s2} (min)	t_{90} (min)	Cure Extent (dNm)	Cure Rate Index (CRI)
Recipe 1	1.25	7.41	1.27	2.78	6.16	66.22
Recipe 2	1.43	7.88	1.34	3.00	6.45	60.24
Recipe 3	1.63	8.72	1.35	3.19	7.09	54.34

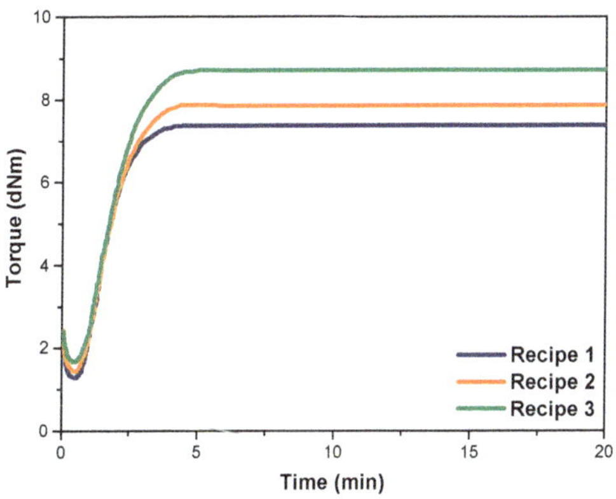

Figure 5. Rheometer curves of compounds.

The scorch time, denoted as t_{s2}, represents the premature vulcanization of rubber samples. As seen in Table 4, the scorch time of NR/SBR vulcanizates exhibited an increase when MWCNT nanoparticles were incorporated into NR/SBR. Notably, this effect is more pronounced with the addition of a higher content of MWCNT. Therefore, it can be concluded that there is a discernible delay in the scorch time, providing a more generous processing window in the presence of MWCNTs. This delay in scorch time can be attributed to the strong adsorption capacity of the CNT surface for vulcanization agents and accelerants, which, in effect, postpones the onset of vulcanization processes [51,52]. Similarly, the optimum cure time prolongs from 2.78 min to 3.00 min and 3.19 min upon the addition of 3 phr and 7 phr of MWCNTs, respectively. This indicates that the addition of MWCNTs prolonged the duration required for the complete curing reaction, which is consistent with lower cure rate index values. This situation can be attributed to the fact that the introduction of well-distributed MWCNT nanoparticles in the NR/SBR matrix permits an adequate timeframe for the development of chemical interactions between MWCNTs and the rubber matrix. This, in turn, facilitates the completion of the vulcanization reaction, leading to the establishment of a robust cross-link network, thereby enabling efficient load transfer between the filler and the rubber matrix [51].

The minimum torque value aligns with the initial viscosity characteristics of the blends comprising unvulcanized rubber and filler. As the concentration of MWCNTs in the blend rises, the minimum torque value (ML) exhibits a discernible increment (Table 4). Throughout the process of curing, the maximum torque (MH) exhibited a progressive rise in the NR/SBR compounds, including MWCNTs. Notably, the time required to attain maximum torque in the nanocomposite containing MWCNTs exceeded that of the NR/SBR/CB. This observation suggests a comparatively slower vulcanization rate for NR/SBR/MWCNT nanocomposites, a trend corroborated by findings from cure rate index values. Concurrently, it is worth noting that the NR/SBR nanocomposites with 7 phr MWCNT nanoparticles demonstrated the highest maximum torque among the materials under consideration. This phenomenon can be attributed to the exceptionally high modulus and higher surface area of MWCNTs, which effectively constrained alterations in the molecular configuration of the polymer and, consequently, increased the modulus of the rubber composites [49,53]. Furthermore, subjecting the composite to elevated torque during the compounding process has enabled the homogeneous distribution of MWCNTs within the rubber matrix, which leads to an increase in the interactions between the rubber and filler components. This, in turn, culminated in an improvement in the reinforcement of the NR/SBR matrix. It is evident that the addition of MWCNTs in NR/SBR/CB compounds promoted the formation of a network structure. This observation aligns with the variations in minimum and maximum torque (cure extent values), which are typically indicative of cross-link density, as shown in Table 4. This was due to the enhanced distribution of MWCNTs in the rubber matrix. As mentioned in the literature, a higher value for MH–ML difference serves as an indicator of superior MWCNT distribution. This is substantiated by the fact that effective cross-linking within the entrapped rubber clusters within MWCNT aggregations can be significantly impeded [31,54–58].

Recipe 3 (MWCNT loading of 7 phr) reveals the highest ML (viscosity index), MH (crosslinking density index), and, consequently, difference in torque (strength index) compared to those with a lower content of MWCNT. This observation has confirmed the possible diffusion of curatives from the SBR phase to the NR phase with higher crosslinking density.

3.3. *Cross-Link Density Measurements and Stress Relaxation Behaviors of Samples*

The temperature-dependent relaxation modulus of NR/SBR/CB composites with varying MWCNT loadings at 3 and 7 phr can be seen in the graphical representation provided in Figure 6. The composite material comprising NR/SBR filled with hybrid fillers consisting of carbon nanotubes exhibited slightly higher initial modulus at lower temperatures and elevated modulus at specific temperatures when compared to NR/SBR/CB composites. As reported by Vennemann et al. [57,59,60], the marginal escalation in stress within the temperature span of up to 40 °C is primarily ascribed to the entropic elastic forces exerted by the rubber polymer chains. Above 45 °C, the stress decreases linearly with temperature before a strong stress decay that occurs at temperatures above 110 °C. The chemical coupling of rubber to the filler surface, which contrasts with the Payne effect, can also be detected when used with low filler content. The intermediate decrease in stress in the temperature range reflects physically induced stress relaxation caused by polymer filler interactions. The subsequent thermal processes occurring in the temperature range spanning 40 °C to 105 °C are predominantly linked to the phenomenon of debonding, where the interfacial connections between the rubber and the filler material undergo disruption. However, the scission of rubber chains emerges as the dominant mechanism responsible for the substantial reduction in stress observed within the temperature range surpassing 105 °C. In the context of the present study, it becomes evident that the stress within all composites diminishes notably between 45 °C and 120 °C. However, for NR/SBR composites with higher MWCNT content, it happens at an elevated stress relaxation modulus within this temperature range. Moreover, as seen from Table 5, NR/SBR compounds with 7 phr MWCNT nanoparticles exhibited the highest initial stress σ0 and T10 values, which is associated with the thermoelastic inversion point, which can be attributed to higher cross-link density, consistent with cure extent values. All these observations may be attributed to the higher amount of crosslink structures, as well as intensified filler–filler and filler–rubber interactions (as depicted in Figure 2). These combined factors contribute to the augmentation of NR/SBR/CB/MWCNT composites when contrasted with NR/SBR composites that lack the incorporation of MWCNTs. Notably, this phenomenon aligns with the discernibly greater torque difference (as outlined in Table 4), the elevated 100% modulus (as indicated in Figure 7), and the enhanced cross-link density (as reported in Table 5) [61]. Moreover, relaxation spectra (H(T)) of NR/SBR compounds, including single and hybrid filler, in the temperature range from 30 °C to 120 °C can be seen in Figure 7. As seen in Figure 7, the relaxation spectra of the NR/SBR compound with 7 phr MWCNT showed an upward trend in comparison with the other compounds. This can be attributed to the improved distribution of MWCNT in the rubber matrix, an enhanced interfacial area between MWCNT and NR/SBR resulted in improved initial relaxation modulus and increased physical desorption [62,63].

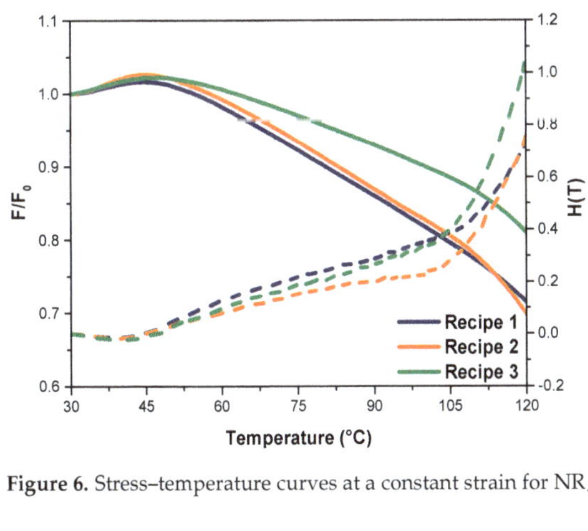

Figure 6. Stress–temperature curves at a constant strain for NR/SBR compounds.

Table 5. TSSR results (σ_0 and T10) and the cross-link density (CLD) of compounds from Flory–Rehner swelling measurements and TSSR method.

Compounds	T10 (°C) *	Initial Stress, σ_0 (MPa)	TSSR (mol/m^3)	Flory–Rehner (mol/m^3)
Recipe 1	80.5	0.58	97.8	231.8
Recipe 2	83.0	0.40	117.9	304.3
Recipe 3	100.2	0.79	132.9	383.7

* T10 stands for the temperature at which the force (F) decreased by 10% from its initial value (F_0).

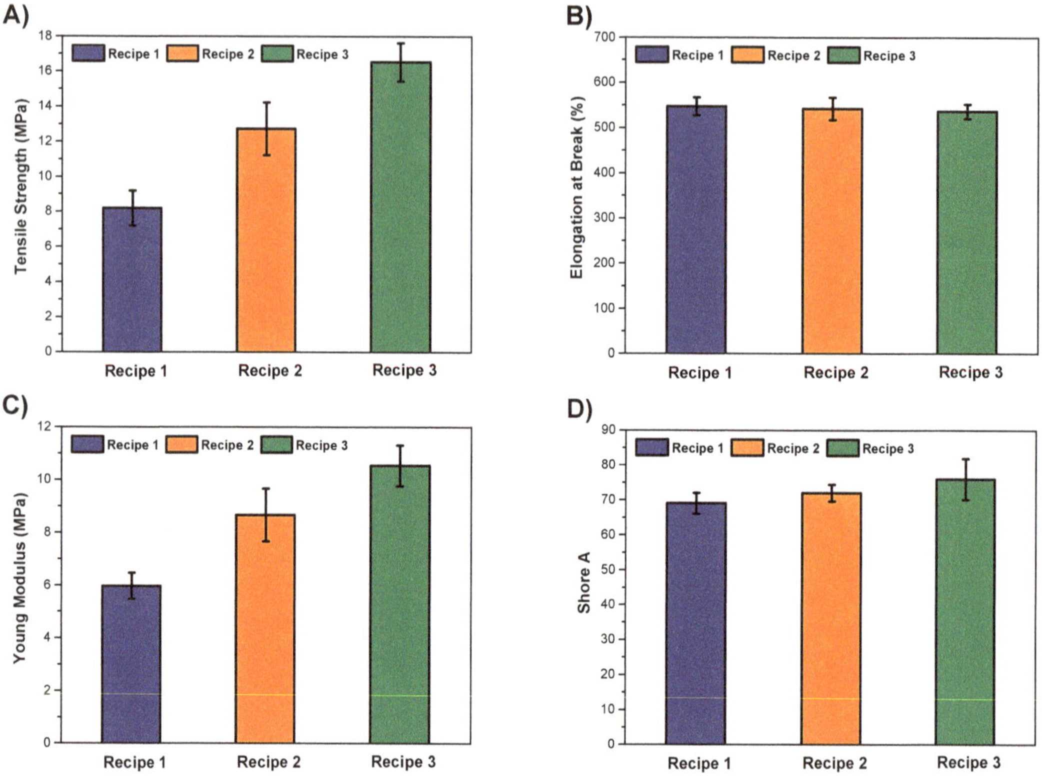

Figure 7. (**A**) Tensile strength, (**B**) elongation at break, (**C**) Young modulus, (**D**) Shore A values of composites.

At a constant strain, the slope of the stress versus temperature plot can be used to determine the cross-link density (CLD) of a material [64,65]. As shown in Table 5, the CLD of MWCNT-reinforced tire tread NR/SBR compounds is higher than that of CB-reinforced compounds. Moreover, it increased as the MWCNT content increased, which is consistent with cure extent values. In addition, the determination of crosslink densities in the vulcanizates was also conducted through the utilization of the equilibrium swelling method. As presented in Table 5, there is a discernible upward trend observed in the cross-link density of the vulcanizates as the MWCNT content is increased. This can be primarily attributed to the establishment of physical cross-links between the CNTs and the polymer matrix, thereby immobilizing the polymer chains and impeding solvent transport, as reported by previous studies [66–68]. Moreover, as mentioned before, a homogeneous distribution of MWCNTs in NR/SBR/CB matrix using a high-torque lab-scale micro-compounder resulted in an improvement in the CLD of composite materials.

3.4. Mechanical Properties

To evaluate the effect of hybrid filler on the mechanical properties of NR/SBR tire tread compounds, tensile, hardness, and abrasion tests were carried out. The changes in mechanical properties are shown in Figure 7 and tabulated in Table 6.

Table 6. Tensile test, Shore D, and abrasion test results of compounds.

Compounds	Tensile Strength (MPa)	Elongation at Break (%)	Young Modulus (MPa)	Shore D	Abrasion Loss (%)	Abrasion Loss (mm^3)
Recipe 1	8.2 ± 1.1	548 ± 20	6.0 ± 0.5	69 ± 3	19.8	191.1
Recipe 2	12.7 ± 1.5	543 ± 26	8.7 ± 1.5	72 ± 2	19.4	188.5
Recipe 3	16.5 ± 1.1	536 ± 16	10.5 ± 0.8	76 ± 6	10.4	131.3

The reinforcing effect of fillers in rubber matrices is the result of several molecular mechanisms involved in this complex phenomenon: the rubber network, the hydrodynamic effect, the filler–filler interactions, and the filler–rubber interactions [19,69–71]. In contrast to NR/SBR/MWCNT composites, the incorporation of carbon nanotubes has yielded a striking increase in the material properties of tensile strength within the rubber matrix. At an MWCNT concentration of 3 phr, a notable enhancement of 55% in the tensile strength was observed in comparison to the NR/SBR/CB. Upon further increasing the MWCNT content to 7 phr, a remarkable upsurge of 102% was realized in the tensile strength. The absence of conspicuous MWCNT aggregates within the rubber matrix can be easily seen in the SEM images. Instead, a good distribution of MWCNT nanoparticles indicates the establishment of robust interactions between the MWCNTs and the polymer matrix. Consequently, these interactions endow the MWCNTs with the capacity to effectively transmit stress throughout the rubber matrix, thereby substantiating their pivotal role in enhancing the tensile strength of the resultant nanocomposites [24,49].

Figure 7C,D illustrate the variation in the Young modulus and hardness values of NR/SBR/CB at various MWCNT loading ratios. As seen, both the modulus and hardness values of composites are significantly enhanced with the incorporation of MWCNT nanoparticles. This trend underscores the notable impact of MWCNTs in augmenting the rigidity of the macromolecular chains in NR/SBR to a greater extent than carbon black [25]. Moreover, the good distribution of MWCNTs in the rubber matrix created a larger interfacial area between hybrid fillers and NR/SBR matrix that significantly improved the stiffness of the resultant composites. Additionally, in line with findings by Shanmugaraj et al. [50], nanoparticles with substantial aspect ratios contribute to the formation of supplementary entanglements and physical cross-links within the natural rubber matrix. Consequently, NR/SBR/CB/MWCNT hybrid nanocomposites exhibited enhanced stiffness. However, the elongation at break values slightly decreased with the addition of MWCNT nanoparticles to the rubber matrix, as expected. The reduction in elongation at break can be ascribed to the diminished deformability arising from the stiffened interface between MWCNTs and the rubber matrix [72]. The inclusion of fillers serves to fortify the rubber composite, thereby engendering elevated abrasion resistance and reduced volume loss. As shown in the SEM images, it is evident that the MWCNT particles exhibited homogeneous morphologies, thereby expanding the surface area available for interaction between the rubber matrix and the filler. This characteristic augmentation in the contact surface area contributes to a notable reduction in the volume loss of the NR/SBR/CB composite, thereby indicating an improvement in the abrasion resistance, as shown in Table 6 [73].

3.5. TGA Results

The thermal stability of samples was evaluated via TGA analysis. Table 7 and Figure 8 show the thermal degradation data and TGA curves of single- and hybrid-filler-filled NR/SBR compounds, respectively. The temperatures at 5% weight loss (T_{d5}), 10% weight loss (T_{d10}), and maximum weight loss (T_{dmax}) of NR/SBR, including reinforcements, can be seen in Table 7. Figure 8 reveals a uniform single-step degradation pattern observed across all composite materials within the temperature range of 300–450 °C under a nitrogen atmosphere. Specifically, in the NR/SBR/CB without MWCNTs, the initiations of degradation (T_{d5} and T_{d10}) occurred at approximately 301.0 and 339.1 °C. However, in the presence of MWCNTs, the initial decomposition temperature started at lower temperatures, indicating the rate of degradation increased with the incorporation of MWCNTs into the NR/SBR/CB. This was attributed to the high thermal conductivity and specific surface areas of MWCNTs, which can accelerate the transfer of heat, as mentioned in the following section [48].

Table 7. TGA test results of compounds.

Compounds	T_{d5} (°C)	T_{d10} (°C)	T_{dmax} (°C)	Char Yield (%)
Recipe 1	301.0	339.1	376.0	19.3
Recipe 2	282.9	328.5	377.3	20.3
Recipe 3	286.4	329.0	377.0	23.2

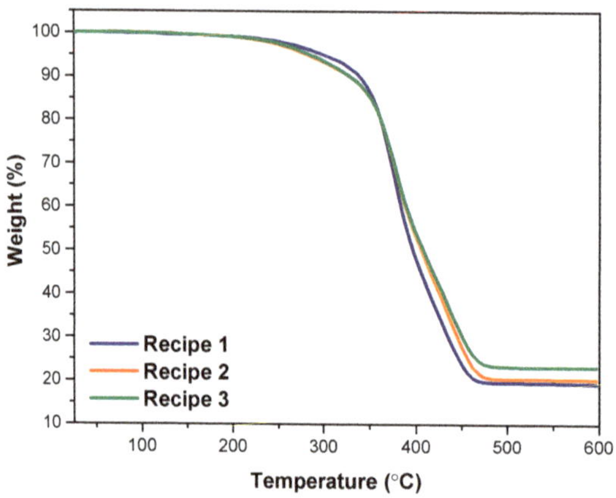

Figure 8. TGA curves of compounds.

3.6. Electrical and Thermal Conductivity Test Results

The thermal conductivity test results of CB- and CB/MWCNT-filled NR/SBR composites are given in Table 8. The thermal dissipation and heat buildup (HBU) are closely related to the thermal transport properties of tire rubber. The enhanced thermal conductivity is beneficial to decrease the HBU, which, in turn, improves the service life of tire rubber [11]. As shown in Table 8, the thermal conductivity of NR/SBR/CB composite is the lowest, which is 0.56 W/mK. The thermal conductivity of MWCNT-reinforced tire tread composites is significantly increased. Among them, the thermal conductivity of the NR/SBR/CB composite, including 7 phr MWCNT, is the highest, reaching 0.91 W/mK, increased by 62.5% in comparison with the NR/SBR/CB composite. As mentioned in the literature, enhancing the distribution state of fillers holds paramount importance in the restoration of the desired thermal attributes in composites. The utilization of "polydispersity" or a "hybrid filler" approach serves as a viable method to advance the thermal properties of these composites. This approach entails the incorporation of smaller fillers at the interface, preventing agglomeration and ensuring uniform distribution alongside larger fillers. A hybrid filler system, encompassing thermally conductive fillers of varied sizes, shapes, or chemical properties, has been demonstrated to optimize distribution within the matrix and foster the creation of efficient thermal conduction pathways. For instance, in contrast to three-dimensional (3D) filler structures, one-dimensional (such as carbon nanotubes) and two-dimensional (2D) fillers with a high aspect ratio possess the capability to forge continuous, thermally conductive pathways within the polymer matrix, even when present in relatively low volume fractions [74–80]. Therefore, it can be concluded that improved distribution of MWCNTs in rubber matrix through high-torque micro-compounder facilitated a phenomenon known as the "dilution effect" arising from the reduction in interstitial gaps between adjacent filler particles as the filler content is increased. Consequently, the hindrance encountered by phonon-mediated thermal transport through the filler network is diminished [81].

Table 8. Electrical and thermal conductivities of compounds.

Compounds	Surface Resistivity (Ohm/cm^2)	Thermal Conductivity (W/m·K)
Recipe 1	4.5×10^7	0.56
Recipe 2	6.5×10^6	0.58
Recipe 3	1.4×10^6	0.91

Table 8 summarizes the electrical conductivity of NR/SBR compounds as a function of both CB and MWCNT content. Numerous critical factors, including the specific nanotube variant, the nature of the polymer matrix, the interactions between fillers and the matrix, as well as the orientation of the fillers, play pivotal roles in shaping the electrical characteristics of the material. Carbon nanotubes exhibit the capacity to confer sufficient electrical conductivity at substantially lower filler content levels, owing to their elevated aspect ratios, thus preserving the intended mechanical characteristics of the resultant material. An alternative strategy for enhancing electrical conductivity has been elucidated in studies involving the concurrent integration of hybrid fillers comprising carbon nanotubes and carbon black. It has been empirically validated that the introduction of carbon nanotubes into nanocomposites already containing carbon blacks engenders an improved overall distribution, primarily due to the synergistic interactions between these disparate fillers. This synergy results in the formation of interconnected structures, which serve to bridge the interstitial gaps between previously unconnected particles [82–84].

As depicted in Table 8, the addition of MWCNTs, even at low loadings, significantly increased the electrical conductivity of the NR/SBR/CB composites. A sudden decrease in the surface resistivity of samples was obtained as the MWCNT content increased to 3 phr. At a specific concentration of conductive particles, which is commonly referred to as the percolation threshold, a contiguous network of fillers is established throughout the matrix, thereby facilitating a marked transition from an insulating to a conducting state [19]. This critical threshold marks the stage at which the filler particles coalesce to form an interconnected network within the matrix. The spatial proximity of filler particles at this threshold is such that it enables the dielectric breakdown of the matrix material. Consequently, at this juncture, electrons can effectively traverse the filled rubber matrix, facilitated by mechanisms such as hopping or quantum tunneling, thereby precipitating a marked and abrupt shift in electrical resistivity [81]. As the electrical conductivity of the filler and the matrix are different by many orders of magnitude, the percolation of a filler must achieve electrical conductivity in the otherwise insulating matrix. The volume loading corresponding to a sharp increase in electrical conductivity is known as the percolation threshold. The incorporation of MWCNT as a second filler at 3 phr results in a decrease in the electrical resistivity of the NR/SBR/CB composite by an order of magnitude. The results indicated that the percolation threshold of NR/SBR was achieved at a 3 phr MWCNT loading level. In this situation, the CB aggregates were found to be uniformly distributed throughout the composite material, forming robust connections with the MWCNTs (Figure 9). This configuration established novel electron conduits within the interconnected MWCNT-CB-MWCNT networks that traverse the NR/SBR matrix [85]. In a general conclusion, as a consequence of enhanced distribution techniques, there is a discernible shift in the percolation threshold toward a lower nanotube content. Consequently, the assessments of electrical resistivity assume significance as an indirect means to appraise the quality of distribution. As highlighted in many parts of this manuscript, the state of distribution of MWCNTs plays a pivotal role in regulating performance, notably the electrical characteristics of composites reinforced with MWCNTs.

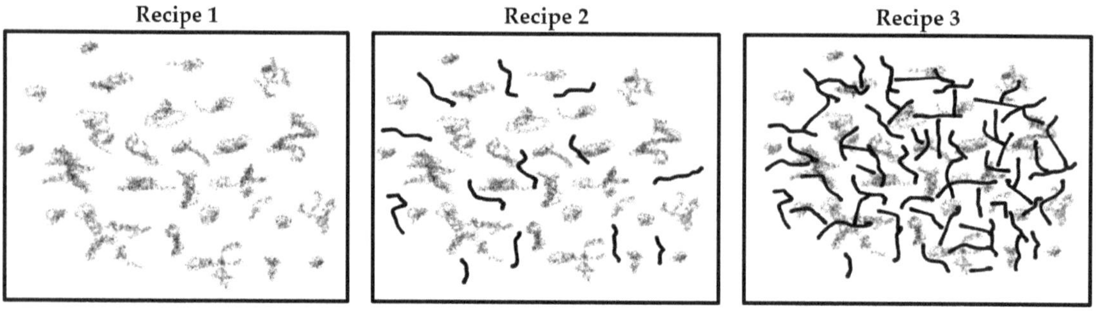

Figure 9. The distribution mechanism of CB and CB/MWCNT in NR/SBR matrix.

4. Conclusions

The intrinsic low thermal conductivity inherent in tires contributes to the phenomenon of heat build-up and concurrent material degradation. In light of this, there is a recognized necessity to enhance the thermal conductivity of tires, a goal that can be effectively realized through the incorporation of conductive fillers. Accordingly, carbon-based fillers emerge as particularly auspicious additives for augmenting the thermal characteristics of polymer composites, owing to their exceptional thermal properties. Therefore, in this study, it was aimed to decrease the heat accumulation properties as well as to improve the mechanical, rheological, and morphological properties of NR/SBR-based tire tread compounds using thermally conductive carbon nanotube nanoparticles. For this purpose, different tire thread compound recipes, including different loading levels of MWCNT, were prepared. The improved distribution of MWCNT/CB hybrid fillers was achieved by using a laboratory-scale high-torque micro-compounder, serving as an economically viable and expeditious research facility, presenting opportunities for the manipulation of small quantities of materials.

The incorporation of MWCNT nanoparticles in the NR/SBR compounds significantly developed the thermal conductivity of the samples, resulting in heat removal from the tire tread compound, which can improve the service life of a tire. Moreover, the enhancement in MWCNT distribution within the NR/SBR compound had a discernible impact on mechanical, rheological, and stress relaxation characteristics. Specifically, with an increase in the distribution of MWCNT, several key vulcanization parameters, such as scorch time and cross-link density, were notably amended. Moreover, the homogeneous distribution of MWCNT created more interfacial interaction and interfacial area between the fillers and rubber matrix; therefore, mechanical properties such as tensile strength and Young modulus were substantially developed. In addition, the findings of the study revealed that the percolation threshold for the NR/SBR composite occurred at a loading level of 3 phr of MWCNTs. It was observed that CB aggregates were uniformly distributed throughout the tire tread compound, effectively establishing robust interconnections with the MWCNTs.

Author Contributions: M.K.: conceptualization, design, material preparation, analyses, original draft preparation, writing review, and editing; N.Y.Ç.: conceptualization, design, material preparation, analyses, original draft preparation, writing review, and editing; R.Y.: conceptualization, design, material preparation, analyses, original draft preparation, writing review, and editing; N.K.: conceptualization, design, material preparation, analyses, original draft preparation, writing review, and editing; G.Ö.: conceptualization, design, material preparation, analyses, original draft preparation, writing review, and editing. All authors have read and agreed to the published version of the manuscript.

Funding: This research received no external funding.

Institutional Review Board Statement: Not applicable.

Data Availability Statement: The data presented in this study are available upon request from the corresponding author.

Conflicts of Interest: Author Nursel Karakaya was employed by the company Xplore Instruments B.V. The remaining authors declare that the research was conducted in the absence of any commercial or financial re-lationships that could be construed as a potential conflict of interest.

References

1. Roberts, A.D. *Natural Rubber Science and Technology*; Oxford University Press: Oxford, UK, 1988; ISBN 0198552254.
2. Wei, L.; Fu, X.; Luo, M.; Xie, Z.; Huang, C.; Zhou, J.; Zhu, Y.; Huang, G.; Wu, J. Synergistic Effect of CB and GO/CNT Hybrid Fillers on the Mechanical Properties and Fatigue Behavior of NR Composites. *RSC Adv.* **2018**, *8*, 10573–10581. [CrossRef]
3. Trabelsi, S.; Albouy, P.A.; Rault, J. Stress-Induced Crystallization around a Crack Tip in Natural Rubber. *Macromolecules* **2002**, *35*, 10054–10061. [CrossRef]
4. Ismail, M.N.; El-Sabbagh, S.H.; Yehia, A.A. Fatigue and Mechanical Properties of NR/SBR and NR/NBR Blend Vulcanizates. *J. Elastomers Plast.* **1999**, *31*, 255–270. [CrossRef]
5. Joseph, R.; George, K.E.; Francis, D.J. Studies on the Cure Characteristics and Vulcanizate Properties of 50/50 NR/SBR Blend. *J. Appl. Polym. Sci.* **1988**, *35*, 1003–1017. [CrossRef]
6. Parameswaran, S.K.; Bhattacharya, S.; Mukhopadhyay, R.; Naskar, K.; Bhowmick, A.K. Excavating the Unique Synergism of Nanofibers and Carbon Black in Natural Rubber Based Tire Tread Composition. *J. Appl. Polym. Sci.* **2021**, *138*, 49682. [CrossRef]
7. Gumede, J.I.; Carson, J.; Hlangothi, S.P.; Bolo, L.L. Effect of Single-Walled Carbon Nanotubes on the Cure and Mechanical Properties of Reclaimed Rubber/Natural Rubber Blends. *Mater. Today Commun.* **2020**, *23*, 100852. [CrossRef]
8. Farahani, T.D.; Bakhshandeh, G.R.; Abtahi, M. Mechanical and Viscoelastic Properties of Natural Rubber/ Reclaimed Rubber Blends. *Polym. Bull.* **2006**, *56*, 495–505. [CrossRef]
9. Boonmahitthisud, A.; Chuayjuljit, S. NR/XSBR Nanocomposites with Carbon Black and Carbon Nanotube Prepared by Latex Compounding. *J. Met. Mater. Miner.* **2012**, *22*, 77–85.
10. Dong, B.; Liu, C.; Lu, Y.; Wu, Y. Synergistic Effects of Carbon Nanotubes and Carbon Black on the Fracture and Fatigue Resistance of Natural Rubber Composites. *J. Appl. Polym. Sci.* **2015**, *132*, 42075. [CrossRef]
11. Chen, L.; Yu, G.; Zhuang, T.; Pang, X. Reinforcing Effects of Carbon Nanotubes on Graphene/Trans-1,4-Polyisoprene/Natural Rubber Composites. *J. Polym. Res.* **2022**, *29*, 387. [CrossRef]
12. Sattayanurak, S.; Noordermeer, J.W.M.; Sahakaro, K.; Kaewsakul, W.; Dierkes, W.K.; Blume, A. Silica-Reinforced Natural Rubber: Synergistic Effects by Addition of Small Amounts of Secondary Fillers to Silica-Reinforced Natural Rubber Tire Tread Compounds. *Adv. Mater. Sci. Eng.* **2019**, *2019*, 5891051. [CrossRef]
13. Cao, L.; Sinha, T.K.; Tao, L.; Li, H.; Zong, C.; Kim, J.K. Synergistic Reinforcement of Silanized Silica-Graphene Oxide Hybrid in Natural Rubber for Tire-Tread Fabrication: A Latex Based Facile Approach. *Compos. Part B Eng.* **2019**, *161*, 667–676. [CrossRef]
14. Kong, L.; Li, F.; Wang, F.; Miao, Y.; Huang, X.; Zhu, H.; Lu, Y. High-Performing Multi-Walled Carbon Nanotubes/Silica Nanocomposites for Elastomer Application. *Compos. Sci. Technol.* **2018**, *162*, 23–32. [CrossRef]
15. Junkong, P.; Kueseng, P.; Wirasate, S.; Huynh, C.; Rattanasom, N. Cut Growth and Abrasion Behaviour, and Morphology of Natural Rubber Filled with MWCNT and MWCNT/Carbon Black. *Polym. Test.* **2015**, *41*, 172–183. [CrossRef]
16. Kaushik, B.K.; Majumder, M.K. Carbon Nanotube: Properties and Applications. In *Carbon Nanotube Based VLSI Interconnects: Analysis and Design*; Kaushik, B.K., Majumder, M.K., Eds.; Springer: New Delhi, India, 2015; pp. 17–37, ISBN 9788132220466.
17. Jogi, B.F.; Sawant, M.; Kulkarni, M.; Brahmankar, P.K. Dispersion and Performance Properties of Carbon Nanotubes (CNTs) Based Polymer Composites: A Review. *J. Encapsulation Adsorpt. Sci.* **2012**, *2012*, 69–78. [CrossRef]
18. Kummerlowe, C.; Vennemann, N.; Pieper, S.; Siebert, A.; Nakaramontri, Y. Preparation and Properties of Carbon-Nanotube Composites with Natural Rubber and Epoxidized Natural Rubber. *Polimery* **2014**, *59*, 811–818. [CrossRef]
19. Bernal-Ortega, P.; Bernal, M.M.; Blume, A.; González-Jiménez, A.; Posadas, P.; Navarro, R.; Valentín, J.L. Sulfur-Modified Carbon Nanotubes for the Development of Advanced Elastomeric Materials. *Polymers* **2021**, *13*, 821. [CrossRef]
20. Bokobza, L. Natural Rubber Nanocomposites: A Review. *Nanomaterials* **2018**, *9*, 12. [CrossRef]
21. Rafiee, R. *Carbon Nanotube-Reinforced Polymers: From Nanoscale to Macroscale*; Elsevier: Amsterdam, The Netherlands, 2018; ISBN 9780323482219.
22. Sethulekshmi, A.S.; Saritha, A.; Joseph, K. A Comprehensive Review on the Recent Advancements in Natural Rubber Nanocomposites. *Int. J. Biol. Macromol.* **2022**, *194*, 819–842. [CrossRef]
23. Fawaz, J.; Mittal, V. Synthesis of Polymer Nanocomposites: Review of Various Techniques. In *Synthesis Techniques for Polymer Nanocomposites*; Mittal, V., Ed.; John Wiley & Sons: Hoboken, NJ, USA, 2014; pp. 1–30, ISBN 9783527670307.
24. Fakhru'l-Razi, A.; Atieh, M.A.; Girun, N.; Chuah, T.G.; El-Sadig, M.; Biak, D.R.A. Effect of Multi-Wall Carbon Nanotubes on the Mechanical Properties of Natural Rubber. *Compos. Struct.* **2006**, *75*, 496–500. [CrossRef]
25. Ismail, H.; Ramly, F.; Othman, N. Multiwall Carbon Nanotube-Filled Natural Rubber: The Effects of Filler Loading and Mixing Method. *Polym. Plast. Technol. Eng.* **2010**, *49*, 260–266. [CrossRef]

26. Huang, K.; Jiang, L.; Du, Y.; Wang, J.; Zheng, X.; Jiang, X.; Sui, Y.; Che, M.; Xu, Q.; Li, L.; et al. Multifunctional NR/MWCNTs Nanocomposites Constructed via Combining Volume Exclusion of SiO2 Microspheres with Interface Reinforcement of Tannic Acid. *Eur. Polym. J.* **2021**, *151*, 110424. [CrossRef]
27. Malas, A. Rubber Nanocomposites with Graphene as the Nanofiller. In *Progress in Rubber Nanocomposites*; Thomas, S., Maria, H.J., Eds.; Woodhead Publishing: Cambridge, UK, 2017; pp. 179–229, ISBN 9780081004289.
28. Guo, X.; Guo, S.; Liu, G.; Bai, L.; Liu, H.; Xu, Y.; Zhao, J.; Chai, H.; Jian, X.; Guo, L.; et al. Improving Dispersion of Carbon Nanotubes in Natural Rubber by Using Waterjet-Produced Rubber Powder as a Carrier. *Polymers* **2023**, *15*, 477. [CrossRef] [PubMed]
29. Bian, H.; Xue, J.; Hao, G.; Hao, Y.; Xie, M.; Wang, C.; Wang, Z.; Zhu, L.; Xiao, Y. High Thermal Conductivity Graphene Oxide/Carbon Nanotubes/Butyl Rubber Composites Prepared by a Dry Ice Expansion Pre-Dispersion Flocculation Method. *J. Appl. Polym. Sci.* **2022**, *139*, 51897. [CrossRef]
30. Anand, K.A.; Jose, T.S.; Alex, R.; Joseph, R. Natural Rubber-Carbon Nanotube Composites through Latex Compounding. *Int. J. Polym. Mater.* **2009**, *59*, 33–44. [CrossRef]
31. Xiao, W.; Yu, S.; Cao, X.; Su, K.; Qu, Q.; Tan, Y.; Zhao, F.; Zhao, S.; Zhang, G.; Gao, A.; et al. High-Speed Shear Dispersion of MWCNTs Assisted by PVP in Water and Its Effective Combination with Wet-Mixing Technology for NR/MWCNTs Nanocomposites. *Polym. Compos.* **2022**, *43*, 3858–3870. [CrossRef]
32. Song, S.H. Effect of Surfactants on MWNT Filled Elastomer Composites by Latex Mixing. *Polym. Eng. Sci.* **2018**, *58*, 1843–1848. [CrossRef]
33. *ASTM D3184*; Standard Practice for Rubber—Evaluation of NR (Natural Rubber). ASTM International: West Conshohocken, PA, USA, 2022.
34. *ISO 11345*; Assessment of Carbon Black and Carbon Black/Silica Dispersion. Rapid Comparative Methods. ISO: Geneva, Switzerland, 2006.
35. *ASTM D-5289*; Standard Test Method for Rubber Property—Vulcanization Using Rotorless Cure Meters. ASTM International: West Conshohocken, PA, USA, 2019.
36. *ASTM D8363-20*; Standard Test Method for Rubber—Stress Relaxation in Tension Under Non-Isothermal Conditions. ASTM International: West Conshohocken, PA, USA, 2020.
37. Vennemann, N.; Schwarze, C.; Kummerlöwe, C. Determination of Crosslink Density and Network Structure of NR Vulcanizates by Means of TSSR. *Adv. Mater. Res.* **2014**, *844*, 482–485. [CrossRef]
38. *ISO 527*; Plastics—Determination of tensile properties. ISO: Geneva, Switzerland, 2012.
39. *ASTM D412*; Standard Test Methods for Vulcanized Rubber and Thermoplastic Elastomers—Tension. ASTM International: West Conshohocken, PA, USA, 2021.
40. *ASTM D2240*; Standard Test Method for Rubber Property—Durometer Hardness. ASTM International: West Conshohocken, PA, USA, 2021.
41. *ISO 4649*; Rubber, Vulcanized or Thermoplastic—Determination of Abrasion Resistance Using a Rotating Cylindrical Drum Device. ISO: Geneva, Switzerland, 2017.
42. *ASTM E1530*; Standard Test Method for Evaluating the Resistance to Thermal Transmission by the Guarded Heat Flow Meter Technique. ASTM International: West Conshohocken, PA, USA, 2019.
43. Lu, Y.; Liu, J.; Hou, G.; Ma, J.; Wang, W.; Wei, F.; Zhang, L. From Nano to Giant? Designing Carbon Nanotubes for Rubber Reinforcement and Their Applications for High Performance Tires. *Compos. Sci. Technol.* **2016**, *137*, 94–101. [CrossRef]
44. Abdelsalam, A.A.; Araby, S.; El-Sabbagh, S.H.; Abdelmoneim, A.; Hassan, M.A. A Comparative Study on Mechanical and Rheological Properties of Ternary Rubber Blends. *Polym. Polym. Compos.* **2021**, *29*, 15–28. [CrossRef]
45. Yang, H.-H.; Manas-Zloczower, I. Flow Field Analysis of the Kneading Disc Region in a Co-Rotating Twin Screw Extruder. *Polym. Eng. Sci.* **1992**, *32*, 1411–1417. [CrossRef]
46. Yazıcı, N.; Kodal, M.; Özkoç, G. Lab-Scale Twin-Screw Micro-Compounders as a New Rubber-Mixing Tool: 'A Comparison on EPDM/Carbon Black and EPDM/Silica Composites'. *Polymers* **2021**, *13*, 4391. [CrossRef] [PubMed]
47. Yang, G.; Liao, Z.; Yang, Z.; Tang, Z.; Guo, B. Effects of Substitution for Carbon Black with Graphene Oxide or Graphene on the Morphology and Performance of Natural Rubber/Carbon Black Composites. *J. Appl. Polym. Sci.* **2015**, *132*, 41832. [CrossRef]
48. Strommer, B.; Schulze, D.; Schartel, B.; Böhning, M. Networking Skills: The Effect of Graphene on the Crosslinking of Natural Rubber Nanocomposites with Sulfur and Peroxide Systems. *Polymers* **2022**, *14*, 4363. [CrossRef]
49. Sui, G.; Zhong, W.H.; Yang, X.P.; Yu, Y.H.; Zhao, S.H. Preparation and Properties of Natural Rubber Composites Reinforced with Pretreated Carbon Nanotubes. *Polym. Adv. Technol.* **2008**, *19*, 1543–1549. [CrossRef]
50. Shanmugharaj, A.M.; Bae, J.H.; Lee, K.Y.; Noh, W.H.; Lee, S.H.; Ryu, S.H. Physical and Chemical Characteristics of Multiwalled Carbon Nanotubes Functionalized with Aminosilane and Its Influence on the Properties of Natural Rubber Composites. *Compos. Sci. Technol.* **2007**, *67*, 1813–1822. [CrossRef]
51. Yan, N.; Wu, J.K.; Zhan, Y.H.; Xia, H.S. Carbon Nanotubes/Carbon Black Synergistic Reinforced Natural Rubber Composites. *Plast. Rubber Compos.* **2009**, *38*, 290–296. [CrossRef]
52. Kueseng, K.; Jacob, K.I. Natural Rubber Nanocomposites with SiC Nanoparticles and Carbon Nanotubes. *Eur. Polym. J.* **2006**, *42*, 220–227. [CrossRef]

53. Shao, H.Q.; Wei, H.; He, J.H. DYNAMIC PROPERTIES AND TIRE PERFORMANCES OF COMPOSITES FILLED WITH CARBON NANOTUBES. *Rubber Chem. Technol.* **2018**, *91*, 609–620. [CrossRef]
54. Valentini, L.; Bittolo Bon, S.; Hernández, M.; Lopez-Manchado, M.A.; Pugno, N.M. Nitrile Butadiene Rubber Composites Reinforced with Reduced Graphene Oxide and Carbon Nanotubes Show Superior Mechanical, Electrical and Icephobic Properties. *Compos. Sci. Technol.* **2018**, *166*, 109–114. [CrossRef]
55. Zhang, H.; Zheng, S.; Zheng, L.; Wang, M.J. Silica Masterbatches Produced with Liquid Phase Mixing Part IV. Volume Variation and SEM Observations upon Stretching. *Polym. Eng. Sci.* **2021**, *61*, 1368–1378. [CrossRef]
56. Zhang, X.; Wang, J.; Jia, H.; You, S.; Xiong, X.; Ding, L.; Xu, Z. Multifunctional Nanocomposites between Natural Rubber and Polyvinyl Pyrrolidone Modified Graphene. *Compos. Part B Eng.* **2016**, *84*, 121–129. [CrossRef]
57. Le, H.H.; Pham, T.; Henning, S.; Klehm, J.; Wießner, S.; Stöckelhuber, K.W.; Das, A.; Hoang, X.T.; Do, Q.K.; Wu, M.; et al. Formation and Stability of Carbon Nanotube Network in Natural Rubber: Effect of Non-Rubber Components. *Polymer* **2015**, *73*, 111–121. [CrossRef]
58. Le, H.H.; Abhijeet, S.; Ilisch, S.; Klehm, J.; Henning, S.; Beiner, M.; Sarkawi, S.S.; Dierkes, W.; Das, A.; Fischer, D.; et al. The Role of Linked Phospholipids in the Rubber-Filler Interaction in Carbon Nanotube (CNT) Filled Natural Rubber (NR) Composites. *Polymer* **2014**, *55*, 4738–4747. [CrossRef]
59. Vennemann, N.; Wu, M.; Heinz, M. Experimentelle Untersuchungen Und Entwicklung Eines Modells Der Thermoelastischen Eigenschaften Russgefüllter SBR-Vulkanisate. *KGK* **2011**, *64*, 40–46.
60. Vennemann, N.; Wu, M.; Heinz, M. Thermoelastic Properties and Relaxation Behavior of S-SBR/Silica Vulcanizates. *Rubber World* **2012**, *246*, 18–23.
61. Damampai, K.; Pichaiyut, S.; Das, A.; Nakason, C. Internal Polymerization of Epoxy Group of Epoxidized Natural Rubber by Ferric Chloride Filled with Carbon Nanotubes: Mechanical, Morphological, Thermal and Electrical Properties of Rubber Vulcanizates. *Express Polym. Lett.* **2022**, *16*, 812–826. [CrossRef]
62. Kitisavetjit, W.; Nakaramontri, Y.; Pichaiyut, S.; Wisunthorn, S.; Nakason, C.; Kiatkamjornwong, S. Influences of Carbon Nanotubes and Graphite Hybrid Filler on Properties of Natural Rubber Nanocomposites. *Polym. Test.* **2021**, *93*, 106981. [CrossRef]
63. Nakaramontri, Y.; Kummerlöwe, C.; Vennemann, N.; Wisunthorn, S.; Pichaiyut, S. Effect of Bis(Triethoxysilylpropyl) Tetrasulfide (TESPT) on Properties of Carbon Nanotubes and Conductive Carbon Black Hybrid Filler Filled Natural Rubber Nanocomposites. *Express Polym. Lett.* **2018**, *12*, 867–884. [CrossRef]
64. Yazıcı, N.; Opar, E.; Kodal, M.; Tanören, B.; Sezen, M.; Özkoç, G. A Novel Practical Approach for Monitoring the Crosslink Density of an Ethylene Propylene Diene Monomer Compound: Complementary Scanning Acoustic Microscopy and FIB-SEM-EDS Analyses. *Polym. Polym. Compos.* **2022**, *30*, doi. [CrossRef]
65. Karaağaç, B.; Cengiz, S.C.; Bayram, T.; Şen, M. Identification of Temperature Scanning Stress Relaxation Behaviors of New Grade Ethylene Propylene Diene Elastomers. *Adv. Polym. Technol.* **2018**, *37*, 3027–3037. [CrossRef]
66. Zhan, Y.H.; Liu, G.Q.; Xia, H.S.; Yan, N. Natural Rubber/Carbon Black/Carbon Nanotubes Composites Prepared through Ultrasonic Assisted Latex Mixing Process. *Plast. Rubber Compos.* **2011**, *40*, 32–39. [CrossRef]
67. Pojanavaraphan, T.; Magaraphan, R. Prevulcanized Natural Rubber Latex/Clay Aerogel Nanocomposites. *Eur. Polym. J.* **2008**, *44*, 1968–1977. [CrossRef]
68. Kader, M.A.; Kim, K.; Lee, Y.S.; Nah, C. Preparation and Properties of Nitrile Rubber/Montmorillonite Nanocomposites via Latex Blending. *J. Mater. Sci.* **2006**, *41*, 7341–7352. [CrossRef]
69. Payne, A.R. The Dynamic Properties of Carbon Black Loaded Natural Rubber Vulcanizates. Part II. *J. Appl. Polym. Sci.* **1962**, *6*, 368–372. [CrossRef]
70. Payne, A.R. The Dynamic Properties of Carbon Black-Loaded Natural Rubber Vulcanizates. Part I. *J. Appl. Polym. Sci.* **1962**, *6*, 57–63. [CrossRef]
71. Payne, A.R.; Whittaker, R.E. Low Strain Dynamic Properties of Filled Rubbers. *Rubber Chem. Technol.* **1971**, *44*, 440–478. [CrossRef]
72. Shanmugharaj, A.M.; Hun Ryu, S. Influence of Aminosilane-Functionalized Carbon Nanotubes on the Rheometric, Mechanical, Electrical and Thermal Degradation Properties of Epoxidized Natural Rubber Nanocomposites. *Polym. Int.* **2013**, *62*, 1433–1441. [CrossRef]
73. Zainal Abidin, Z.; Mamauod, S.N.L.; Romli, A.Z.; Sarkawi, S.S.; Zainal, N.H. Synergistic Effect of Partial Replacement of Carbon Black by Palm Kernel Shell Biochar in Carboxylated Nitrile Butadiene Rubber Composites. *Polymers* **2023**, *15*, 943. [CrossRef]
74. Shahamatifard, F.; Rodrigue, D.; Mighri, F. Thermal and Mechanical Properties of Carbon-Based Rubber Nanocomposites: A Review. *Plast. Rubber Compos.* **2023**, *52*, 483–505. [CrossRef]
75. Shahamatifard, F.; Rodrigue, D.; Park, K.W.; Frikha, S.; Mighri, F. Natural Rubber Nanocomposites: Effect of Carbon Black/Multi-Walled Carbon Nanotubes Hybrid Fillers on the Mechanical Properties and Thermal Conductivity. *Polym. Technol. Mater.* **2021**, *60*, 1686–1696. [CrossRef]
76. Wang, H.; Li, L.; Chen, Y.; Li, M.; Fu, H.; Hou, X.; Wu, X.; Lin, C.T.; Jiang, N.; Yu, J. Efficient Thermal Transport Highway Construction within Epoxy Matrix via Hybrid Carbon Fibers and Alumina Particles. *ACS Omega* **2020**, *5*, 1170–1177. [CrossRef]
77. Song, J.; Li, X.; Tian, K.; Ma, L.; Li, W.; Yao, S. Thermal Conductivity of Natural Rubber Nanocomposites with Hybrid Fillers. *Chin. J. Chem. Eng.* **2019**, *27*, 928–934. [CrossRef]
78. Ismail, H.; Ramly, A.F.; Othman, N. The Effect of Carbon Black/Multiwall Carbon Nanotube Hybrid Fillers on the Properties of Natural Rubber Nanocomposites. *Polym. Plast. Technol. Eng.* **2011**, *50*, 660–666. [CrossRef]

79. Yuan, W.; Xiao, Q.; Li, L.; Xu, T. Thermal Conductivity of Epoxy Adhesive Enhanced by Hybrid Graphene Oxide/AlN Particles. *Appl. Therm. Eng.* **2016**, *106*, 1067–1074. [CrossRef]
80. Rattanasom, N.; Saowapark, T.; Deeprasertkul, C. Reinforcement of Natural Rubber with Silica/Carbon Black Hybrid Filler. *Polym. Test.* **2007**, *26*, 369–377. [CrossRef]
81. Sae-Oui, P.; Thepsuwan, U.; Thaptong, P.; Sirisinha, C. Comparison of Reinforcing Efficiency of Carbon Black, Conductive Carbon Black, and Carbon Nanotube in Natural Rubber. *Adv. Polym. Technol.* **2014**, *33*, 21422. [CrossRef]
82. Bokobza, L.; El Bounia, N.E. Reinforcement Effects of Multiwall Carbon Nanotubes in Elastomeric Matrices: Comparison with Other Types of Fillers. *Compos. Interfaces* **2008**, *15*, 9–17. [CrossRef]
83. Bokobza, L.; Rahmani, M.; Belin, C.; Bruneel, J.L.; El Bounia, N.E. Blends of Carbon Blacks and Multiwall Carbon Nanotubes as Reinforcing Fillers for Hydrocarbon Rubbers. *J. Polym. Sci. Part B Polym. Phys.* **2008**, *46*, 1939–1951. [CrossRef]
84. Gojny, F.H.; Schulte, K. Functionalisation Effect on the Thermo-Mechanical Behaviour of Multi-Wall Carbon Nanotube/Epoxy-Composites. *Compos. Sci. Technol.* **2004**, *64*, 2303–2308. [CrossRef]
85. Nakaramontri, Y.; Nakason, C.; Kummerlöwe, C.; Vennemann, N. Enhancement of Electrical Conductivity and Other Related Properties of Epoxidized Natural Rubber/Carbon Nanotube Composites by Optimizing Concentration of 3-Aminopropyltriethoxy Silane. *Polym. Eng. Sci.* **2017**, *57*, 381–391. [CrossRef]

polymers

MDPI

Article

Enhanced Low-Density Silicone Foams Blown by Water–Hydroxyl Blends

Ingrid Rebane [1,*], Karl Jakob Levin [1], Uno Mäeorg [2], Urmas Johanson [1], Peeter Piirimägi [3], Tauri Tätte [1] and Tarmo Tamm [1,*]

1 Institute of Technology, University of Tartu, Nooruse 1, 50411 Tartu, Estonia; karl.jakob.levin@ut.ee (K.J.L.); urmas.johanson@ut.ee (U.J.)
2 Institute of Chemistry, University of Tartu, Ravila 14a, 50411 Tartu, Estonia; uno.maeorg@ut.ee
3 Estelaxe Ltd., Kiviliuuru 2, 65605 Võru, Estonia
* Correspondence: ingrid.rebane@ut.ee (I.R.); tarmo.tamm@ut.ee (T.T.)

Abstract: Water, alcohols, diols, and glycerol are low-cost blowing agents that can be used to create the desired silicone foam structures. Although their combined use can be beneficial, it remains unclear how it affects the physical properties of the resulting materials. We conducted a comparative study of these hydroxyl-bearing blowing agents in fumed silica- and mica-filled polymer composite systems for simultaneous blowing and crosslinking to obtain a low-density, uniform porosity and superior mechanical properties. The foams were optimized for a uniform open-pore structure with densities ranging from 75 to 150 $kg \cdot m^{-3}$. Varying the diol chain length (C_n) from one to seven carbons can alter the foam density and structure, thereby enhancing the foam tensile strength while maintaining a low density. Replacing 10 mol% of water with 1,4-butanediol decreased the density by 26%, while increasing the specific strength by 5%. By combining glycerol and water blowing, the resulting foams exhibited a 30% lower apparent density than their water-blown analogs. The results further showed that $C_n > 4$ alkane chain diols had an odd–even effect on the apparent density and cell wall thickness. All foamable compositions had viscosities of approximately 7000 cSt and curing times below 2 min, allowing for quick dispensing and sufficient time for the foam to cure in semi-industrial volumes.

Keywords: open-cell cellular structures; lightweight materials; blowing agents; silicone foams; porous PDMS; elastomeric foams; injection molding

1. Introduction

Polysiloxane-based materials have become one of the most preferred materials in applications in which chemical inertness, inherent fire retardancy, and excellent physicomechanical properties are desired [1]. Their elastic ternary structures with tunable density and mechanical properties are especially attractive for medical and wearable electronics applications, as well as in the automotive and construction industries.

Extensive research has been conducted on the preparation methods for polysiloxane-based foams, often based on two-component premixture systems (e.g., Sylgard®, Rhodorsil RTFoam, or Elastosil® LR series) in which the exact composition is proprietary or a trade secret [2–7]. Unfortunately, to achieve a low-density rubber foam with the desired physicomechanical properties, the ready-made compositions must be significantly altered, which often have not been optimized for elastomeric foams [8]. Nevertheless, the synthesis of a silicone network, which constitutes the ternary polymer scaffold, is a relatively simple process that requires combining a selection of functionalized prepolymers, suitable crosslinking agent(s), and, depending on the reaction mechanism, a suitable catalyst [9]. By incorporating reinforcing fillers and additives, it is possible to significantly modify the strength and other mechanical properties of pristine polysiloxane elastomers, limited by the nonpolar nature of the Si-O-Si backbone and helical polymer chain [6,8].

Although silicone foams (SIFs) are commercially available, mainly in the form of sheets, it is important to further study the possibilities of enhancing their properties and improving the procedures used for manufacturing expandable elastomeric materials. By altering the chemical composition and, thus, tuning the reaction rates, it is possible to change the foam density, cell wall thickness, cell size, and structural homogeneity, all of which eventually directly affect the mechanical properties of the foam [10–12].

As low-density foams are often desirable, numerous physical and chemical blowing methods have been applied, of which saturation with supercritical CO_2 [13] and sacrificial templating (citric acid [14], NaCl [5], sugar [15–17], and water [18]) have been reported to achieve porous end results. In addition, by generating oxygen as a by-product, simultaneous foaming during crosslinking has been reported by Yan et al. [19]. Often, these methods restrict the production of foams in large quantities and yield inhomogeneous morphologies. Using either chemical or physical blowing or their combination, the timing of crosslinking to capture the evolving gas is another aspect that needs to be considered [20].

In addition to inherent blowing by dehydrocondensation between silanol-functionalized (OH-PDMS) and hydrogen-functionalized (PMHS) silicones to propagate expansion, blowing agents are typically used [11,21,22]. The high Si-H bond reactivity allows thermodynamically highly favored nucleophilic substitution by water, alcohols, or alkoxysilanes [23].

Water is not an uncommon additive in silicone foams; it has been applied as a liquefier to minimize the split and tear during foam expansion [24], as a pore-former in water–silicone emulsions [25,26], and as an additional or primary blowing agent [2,3,27,28]. Recent studies have applied water as a blowing agent in commercially available prepolymer mixtures and described its softening effect on foams [3]. Monoalcohols (e.g., ethanol and isopropyl alcohol) are standard blowing agents in the literature and have been described as effective additives [2,29]. Although monoalcohols, which have only one reaction function, do not form crosslinks, each alcohol molecule contributes to foam rise by donating a hydrogen atom to the evolving H_2 (see Figure 1).

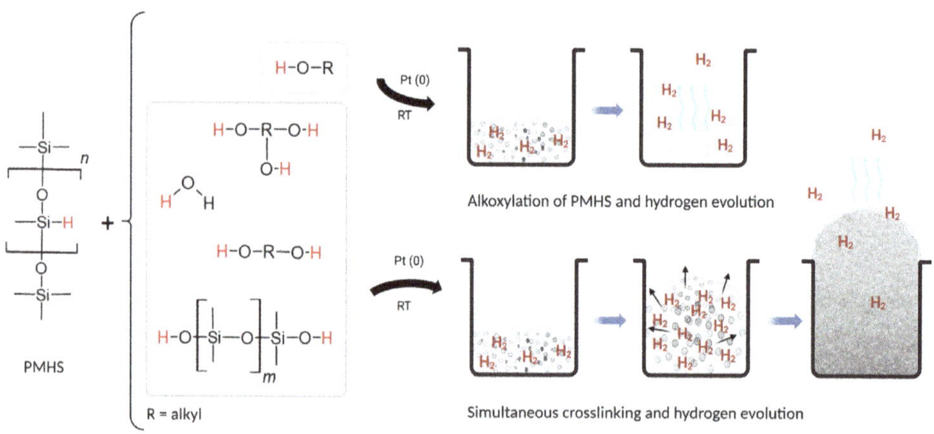

Figure 1. Alkanols, as blowing agents, simultaneously raise and crosslink the polymer mixture. Monoalcohols, as an exception, contribute to foam rise but do not act as crosslinkers. The silanol-terminated PDMS reacts with PMHS similarly, although as a high molecular weight polymer has less impact on the crosslinking.

A recent report showed the use of isopropyl alcohol and water on a silicone foam's microstructure based on a commercial two-part premix, although the foam had closed cells and the densities were not discussed [2]. Conversely, dialcohols (diols) can form crosslinks, the rigidity of which can potentially be reduced by increasing the length of the alkane chain between the two reaction sites. The additional hydroxyl groups in glycerol molecules (propane-1,2,3-triol) can contribute significantly to the blowing, but their performance depends strongly on the steric accessibility to the hydroxyl groups [30]. Previously, a combination of ethanol in glycerol has been shown to produce medium-density foams, whereby increasing the ethanol content in glycerol (0.5–1.1%) and an increase in the glycerol/ethanol mixture content, in general, lowers the relative density of the foam by approximately 15–20%, remaining near 300 kg·m^{-3} [31,32]. There have only been a handful of reports on the use of water–alkanol, especially water–glycerol mixtures, but their application in morphology enhancement is gaining more interest in continuous processes using injection molding. A comparison of silicone foams and their physicomechanical properties found in recent studies is listed in Table A1 in Appendix A.

This study aimed to design a curable silicone polymer base to withstand the expansion of the foam during synthesis and injection molding, allowing for the achievement of low-density polysiloxane foams (SIFs), preferably ≤100 kg·m^{-3}. The tunability of the desired physicomechanical properties is essential for such foams to suit various applications. While water as a blowing and crosslinking agent (Figure 2) has generally shown promising results in terms of the apparent density and homogeneity of foams [27], it is somewhat lacking in creating optimal mechanical properties. Therefore, one option is to partially replace water with mono-, di-, or polyalcohols, allowing for the tuning of the rates of both addition and dehydrocondensation reactions.

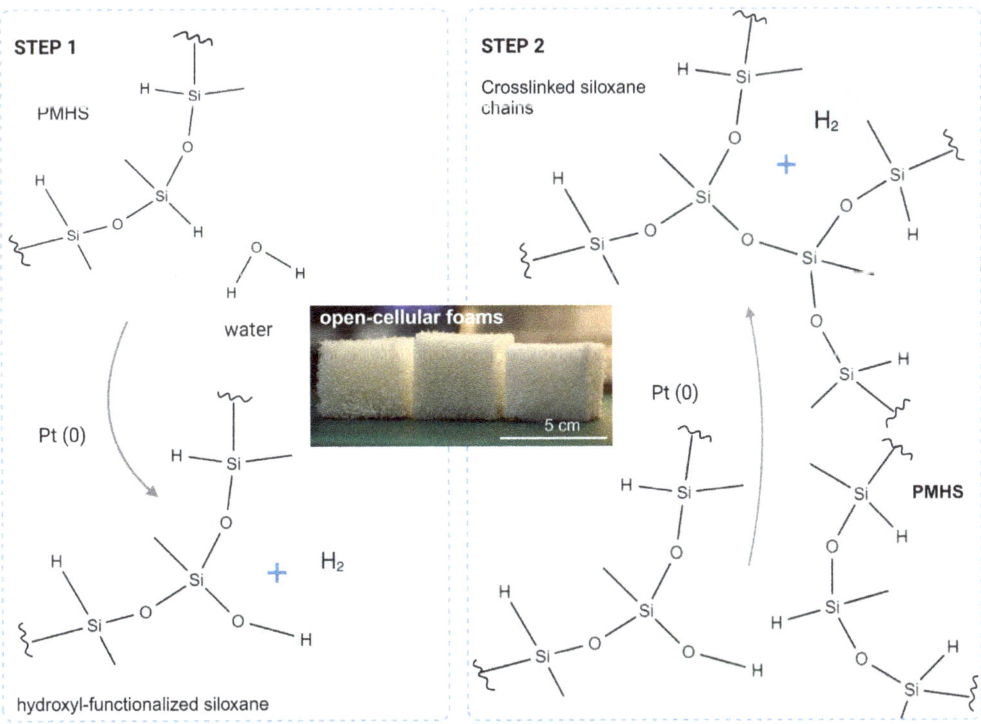

Figure 2. The catalyzed dehydrocondensation between water and poly(methylhydrosiloxane) (PMHS) is a two-step process, resulting in crosslinks between the PMHS chains and a porous structure that is enhanced by the evolution of hydrogen during both steps (step 1 and step 2).

In our study, SIFs were obtained from a combination of polysiloxane prepolymers, crosslinked, and expanded at room temperature in the presence of a platinum catalyst (Pt(0)) and injection molded in a continuous process. The elastomer network of SIFs was strengthened by incorporating fumed silica (FS) and muscovite mica into the premixtures, both of which have been used in silicone elastomers and have been shown to enhance their physicomechanical properties and thermal stability [31]. In this designed silicone base, we varied the mole percentages (mol%) of the mono- and dialcohol solutions in water, which were applied as additional blowing additives. Several combinations showed a significant strengthening effect on the foam, resulting from alterations in the crosslinking and dehydrocondensation reaction kinetics.

2. Materials and Methods

2.1. Preparation of Polysiloxane Elastomer Foams

All foams used in the physicomechanical analysis were prepared by combining vinyl- and hydroxyl-functionalized poly(dimethylsiloxane) (5k cSt), poly(methyl hydro)siloxane (100% H, 25–35 cSt), platinum(0)-1,3-divinyl-1,1,3,3-tetramethyldisiloxane complex (Karstedt's catalyst) obtained from Hubei Chem, moderator (2,4,6,8-tetramethyl-2,4,6,8-tetravinylcyclotetrasiloxane), and general strengthening fillers: hexamethyldisilazane (HMDS)-treated fumed silica from Gelest Inc. (Morrisville, PA, USA) and muscovite mica ($KAl_2(Si_3Al)O_{10}(OH)_2$) from OMYA, Elnesvågen, Norway.

Methanol (Sigma-Aldrich, St. Louis, MO, USA), ethane-1,2-diol (Sigma-Aldrich, 99%), and propane-1,2-diol (Fluka Honeywell, distributor HNK Analüüsitehnika OÜ, Estonia, pur.), propane-1,3-diol (Ferak, Berlin, Germany, 96%), propane-1,2,3-triol (Lah-ner, Cambridge, MA, USA), butane-1,4-diol (Acros Organics, Morris Plains, NJ, USA), pentane-1,5-diol (Fluka, puriss., GC grade), and hexane-1,6-diol (Fluka, pur.), and heptane-1,7-diol (Fluka, pur.) were dried on molecular sieves before use.

For the prepolymer mixtures, we used a standalone mixer with a PTFE-covered rotary blade. In this study, the SIF matrix comprised OH-PDMS (100 parts), V-PDMS (50 parts), PMHS (20 parts), fumed silica (20 parts), and mica (5 parts). Karstedt's catalyst Pt(0) (25 ppm) and four equivalents of moderator were added. The additional blowing blend, either water or a water–hydroxyl blend, was combined with the prepolymer during mixing. Next, the premixtures were combined and injected using an in-house injection molding device, dispensing a volume of 500 mL of premixture per foam sample.

Changes in the composition were made by adjusting the mole percentage (shown as mol%) based on the reactive hydroxyl groups (-OH) in alkanols and water (H-OH). We considered the water molecule to consume two equivalents of silicon hydride, reacting stepwise: initially, a dehydrocondensation reaction involving Si-H and HO(R-)OH groups, followed by a possible dehydrocondensation reaction with the formed silanol (Si-OH) and another available hydride (Si-H) in the vicinity (Figure 2).

2.2. Determining the Physicomechanical Properties of the Prepared Foams

We used the pycnometry technique to measure apparent density ($kg \cdot m^{-3}$). The expanded measurement uncertainty was U(density, molded) = 3 $kg \cdot m^{-3}$ with a confidence interval of 95% (k = 1.9), which showed the excellent reproducibility of the foam using the injection molding technique.

We measured the elongation and tensile strength of the foam samples using a motorized test stand (model AEL-A-1000, 3–1000 N range, and 300 $mm \cdot min^{-1}$ test speed) following the procedures in the ISO 1798:2008 [33] and ASTM D 3574-17 [34] standards. The test specimens were die-cut from a flat sheet of material with the foam rising in the thickness direction and were free of ragged edges. The foam specimens were secured using a screw-type jagged plate grip that exerted uniform pressure across the gripping surface. Before the mechanical measurements, the samples were thoroughly post-cured in an oven at 80 °C for at least 2 h. All measurements were performed in triplicate, and the expanded measurement uncertainty for each foam sample was calculated.

The morphology of the foam was determined using SEM (TM-3000, Hitachi High-Technologies Corporation, Tokyo, Japan) at 15 kV, for which the cubical samples were sputter-coated with a 10 nm thick gold layer (Leica EM ACE600 Sputter Coater, Wetzlar, Germany). We measured the average cell diameter and cell wall thickness of each foam from the cross-section of the torn test samples using the Fiji application in ImageJ 1.54f [35]. The results are expressed as the average of 100 measurements for each sample.

Infrared spectra were collected using FTIR (Bruker, Billerica, MA, USA, Platinum-ATR, diamond crystal) to characterize the structure of the silicone foams, uncured mixture, and PMHS. The spectra can be found in the Supplementary Files.

3. Results and Discussion

3.1. Reaction Mechanisms of SIF Expansion

The foam synthesis experiments revealed that varying the mole percentages of the water and alkanol as blowing and crosslinking agents allowed for distinct changes in the microstructure and mechanical properties of the foam. Some of these effects are described and evaluated in this section to outline a clear path for the preparation of silicone foams with the desired morphology and mechanical properties. The use of water as the blowing agent in our composition resulted in low-density foams with a uniform structure. Despite this considerable advantage, the mechanical properties of such foams depend on the relatively short and stiff crosslinks, which may result in a mechanically weak and brittle polymer scaffold. Although water is not an alkanol, it contributes to dehydrogenation with both H-O bonds in a two-step process, converting the Si-H on the PMHS chain to a Si-OH and releasing hydrogen as a side product (step 1, Figure 2). The formed Si-OH groups readily react with the closest unreacted Si-H bond in the vicinal PMHS chain in the presence of a catalyst, producing additional hydrogen and forming a crosslink between siloxane chains (step 2, Figure 2) [36].

As the length of the formed crosslinks and the crosslink density affect the mechanical properties of the polymer network, dialcohols (diols), which are the shortest alkanol analogs of water, can be considered as candidates for tuning the physicomechanical properties of the foam. Similarly, via the dehydrocondensation reaction, diols contribute to strengthening the polymer material but also influence the apparent density of the resulting foam.

3.2. Tuning Structure and Mechanical Properties of Foams with Diols

The use of water as the reactive foam blowing additive produced low-density foams with uniform structures. To increase the tensile strength at low densities, water was partially replaced with diol. We were interested in industrially feasible dialcohols ranging from ethane-1,2-diol (ethylene glycol) to heptane-1,7-diol. We tested diol contents ranging from 5 mol% to 100 mol% in the primary blowing composition. However, increasing the diol/water ratio increases the diol volume fraction in the premixture, causing a significant decrease in viscosity. In addition, because the accompanying increase in the exothermic effect of the dehydrocondensation reaction is undesirable, we chose to use 10 mol% in this comparison. Depending on the blowing blend used, the viscosities of the polymer mixture components remained at approximately 7000 $\pm$ 200 cSt. By partially replacing water with a diol and considering the reaction stoichiometry and molarity, a change in the foam structure was immediately observed (Figure 3).

The alkane chain length C_n from n = 2 to n = 7 does not correlate linearly with the apparent density (Figure 4A,B), which was similar for ethane- and both propane diols; we achieved the minimum density value with 10 mol% butane-1,4-diol. An increase in the density in the order $C_4 < C_6 < C_5 < C_7$ could also be explained by the odd–even effect seen in alkanes [37]. The odd-numbered alkanes (n = 5, 7, and 9) exhibit up to 30 times slower dynamics than the even-numbered alkanes near their melting points. Because the molecular structure of a diol consists of an alkane chain, such an analogy could be applied to diols. This could explain the resulting higher densities of foams with propane-1,3-diol, pentane-1,5-diol, and heptane-1,7-diol and the lower densities for those with an even number of carbons in the alkane chain (n = 4 and 6) (Figure 4).

Assuming that slower dynamics affect crosslinking in the polymer mixture, the evolving gas gathers and is more likely to escape, allowing the gas bubbles to coalesce with the thickening of the pore walls. It has previously been reported that rheological properties, such as the viscosity of the mixture, affect cell growth and control bubble coalescence during crosslinking [9,38].

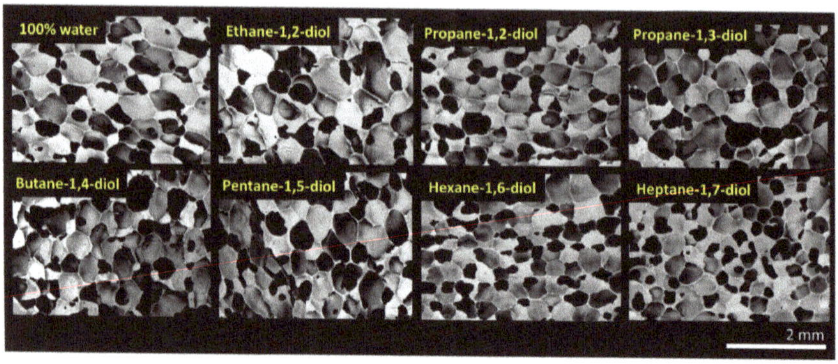

Figure 3. Comparison of the morphologies of foams blown with 10 mol% diols in water and 100 mol% water. Compared to water-only blowing, the 10 mol% replacement of water with diols caused slight structural changes in the foam morphology.

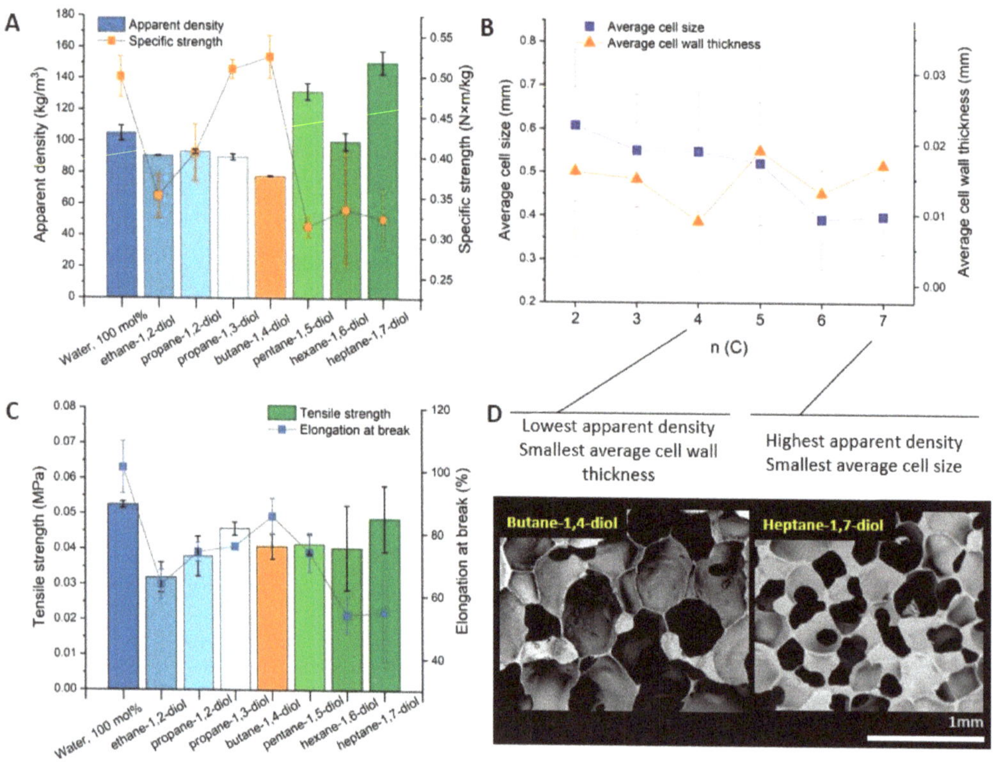

Figure 4. Changes in the foam density and structure for 100 mol% water-only blown foams compared to foams blown with 10 mol% diol solutions in water (**A**). The respective changes in elongation at break, tensile strength, and specific strength dependent on the apparent density are depicted in images (**A**,**C**). A general trend in cell size decrease is observed along with an increase in the chain length of the diol (**B**,**D**).

In addition, the experimental data based on the foam morphology (Table A2 in the Appendix A) suggest that the increase in apparent density has a strong correlation with the average wall thickness. The average cell size for foams prepared with 10 mol% diol substitution in water showed a clear decreasing trend with the increasing chain length (C_n). Interestingly, the cell wall thickness correlated with the apparent density of the foams, supporting the odd–even effect proposed earlier.

As seen from the tensile strength measurements (Figure 4C), the foams with 10 mol% propane-1,3-diol and butane-1,4-diol exhibited the highest specific strengths ($N{\cdot}m{\cdot}kg^{-1}$) at the lowest apparent densities (Figure 4A). Interestingly, with 10 mol% butane-1,4-diol, the resulting cell wall was thinner than that of the foams blown with 10 mol% propane-1,3-diol, although the average cell sizes were similar. In conclusion, butane-1,4-diol was the most effective crosslinking blowing agent in the 10 mol% solutions for producing lightweight, firm foams with the highest specific strength among the diols, surpassing the water-only foams. The tensile strength of the water-only foams remained the highest, next to the other 10 mol% diol–water blown foams, and can be beneficial in applications in which this property is more important than the apparent density.

3.3. *Tuning of Foam Morphology with Methanol*

Although using water or water–diol combinations was successful in terms of the apparent density and pore uniformity, the obtained material exhibited a memory effect and a lack of stiffness under load, which is assumed to be a consequence of the small pore sizes and an increased ratio of the crosslinks. One option is to tune the rigidity of the foam by increasing its pore size and pore wall thickness [39]. This could be achieved by using monoalcohols, neither significantly altering the composition of the prepolymer mixture nor the amount of evolved hydrogen. We intended to suppress both the crosslinking and hydrogen evolution, allowing for the formation of larger voids. However, methanol has shown rapid Si-H conversion in dehydrocondensation reactions with PMHS in previous research [40]. By varying the mole percentage (mol%) of the methanol OH groups in the alcohol–water mixtures, we observed changes in the morphology, apparent density, and tensile strength. In general, an increase in the ratio of methanol (and the respective OH groups) was accompanied by a significant increase in the pore size (Figure 5A,C,E).

Methanol does not form crosslinks between siloxane chains; therefore, a lower crosslinking degree supports a slower polymer mixture solidification. We suggest that the steric hindrance of the increasing number of -OCH_3 functionalized chains retards the overall crosslinking process by restricting the access to the -H sites on the PMHS chain. Our previous research found that the dehydrogenative coupling reaction of PMHS with methanol is rapid during the initial stage but gradually slows down, allowing for the slower evolution and convolution of hydrogen [40]. The resulting crosslinking degree decreased with an increase in the mole percentage of methanol in water while maintaining a constant number of -OH groups. This way, smaller gas-filled voids have time to coalesce, allowing for the formation of larger voids in the solidifying polymer mixture [39]. As depicted in Figure 5B,C, the increase in the methanol/water ratio caused an increase in both the cell wall thickness and cell diameter, accompanied by an increase in the apparent density of the foam. Similarly, Tan et al. showed that increasing the content of ethanol as a single blowing agent has a similar effect on foam morphology [29]. In our experiments, we kept the -OH/-H stoichiometric ratio in methanol/water mixtures constant, so changes in viscosity and crosslink density, and not the change in hydrogen volume, were the key aspects that affected the resulting structure.

Based on these measurements, the tensile strength did not change significantly in the 10–25 mol% range of methanol in water mixtures (see Figure 5D). This range corresponds to apparent densities of 108–115 $kg{\cdot}m^{-3}$. A further increase in the methanol content caused a significant change in the structure of the foam. The increase in the average wall thickness and pore diameter accompanied lower elongation at break values, which may result from a higher occurrence of defects, leading to premature breakage of the test sample. For foams with 50 mol% methanol, the ratio of the pore diameter and the test piece cross-sectional area did not allow for consistent tensile test measurements (see Figure 5C and cell wall and diameter data in Table A3 in Appendix A).

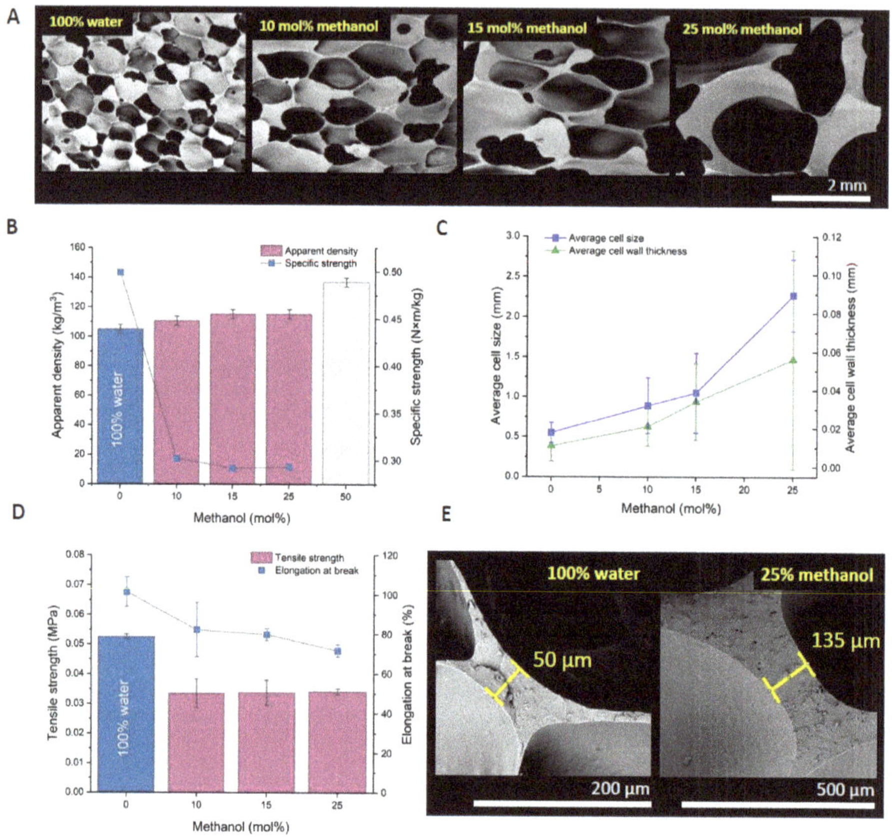

Figure 5. (**A**) Effect of increasing the methanol mol% in water on the morphology of the cured foam. (**B**) A higher methanol/water ratio resulted in higher densities, partially due to the decreased number of crosslinks formed and (**C**,**E**) increase in cell size and wall thickness. (**D**) Both the tensile strength and elongation at break remained relatively constant despite structural changes (**A**,**C**).

Consequently, replacing 10 mol% of water with methanol was sufficient to maintain a low density (110 kg·m^{-3}, +5% compared to 100% water-blown foam) with an altered morphology. Nevertheless, the tensile strength and elongation at break values were significantly lower (decreases of nearly 20% and 40 %, respectively). Loss in mechanical properties is not the expected advancement, although such foams could offer increased compression strength from a morphological aspect.

3.4. Optimizing Tensile Strength and Density with Diols

When increasing the mol% of a specific diol in water, for example, butane-1,4-diol from 5 to 50 mol%, we can see a general increase in the apparent density and tensile strength and a decrease in elongation at break (Figure 6A,B).

The lowest densities were reached for 5 and 10 mol% butane-1,4-diol, but concentrations above 50 mol% had clearly undesirable density-increasing effects (>150 kg·m^{-3}). Although there is no significant difference in the elongation at break from 5 to 20 mol%, the general trend seems to be the loss in elongation together with the density increase. The average pore diameter did not significantly change when the butane-1,4-diol content was increased from 5 mol% to 35 mol%; the dominant effect for density increase was the increase in cell wall thickness near and above 50 mol% butane-1,4-diol solutions (Figure 6C).

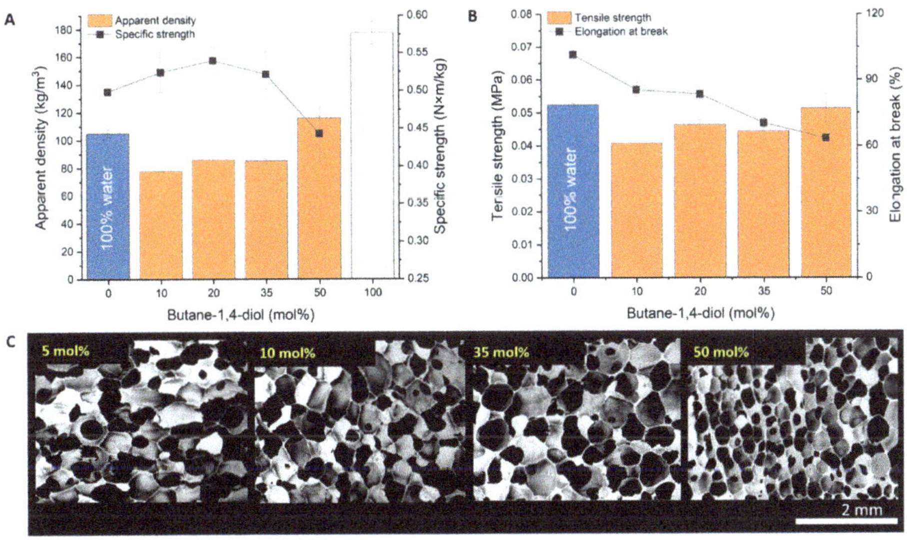

Figure 6. Comparison of the apparent density (**A**) and tensile strength (**B**) of foams synthesized with water and butane-1,4-diol/water solutions. The specific strength near the apparent density of 80 kg·m^{-3} is relatively high, particularly for the foam with 10 mol% butane-1,4-diol. Increasing the butane-1,4-diol above 35mol% in water changes the cell diameter and cell wall thicknesses drastically (**C**).

3.5. Comparison of Water and Water–Diol Blown Foams

Considering the density of the foam, the specific strength is relatively high near the apparent density of 80 kg·m^{-3}, especially for foams with 10 mol% of butane-1,4-diol (0.52 N·m·kg^{-1}, combined uncertainty U_c = 0.07), being comparable with 100% water-blown foam (density = 100 kg·m^{-3} and specific strength 0.55 (U_c = 0.02) N·m·kg^{-1}) (Figure 6A). Relatively good results were also obtained with 20 mol% butane-1,4-diol blown foam, which still had a lower apparent density than 100% water-blown foams; in addition, the tensile strength was the highest among foams with a density of <90 kg·m^{-3}. Hence, it had a considerably high specific strength value of 0.54 (U_c = 0.05) N·m·kg^{-1}.

We did not consider a further increase in the diol content for the reasons mentioned above, and to increase the viscosity during the reaction, we would have needed to alter the main composition of the premix and adjust the catalyst concentration, making the results incomparable. Maintaining a reasonable viscosity is crucial for trapping the evolving gas during foam expansion and elastomer curing processes. We also tested the 100 mol% butane-1,4-diol blown foam but could not perform tensile strength measurements because of its brittle and weak structure. In addition, the resulting foam density is undesirable (>120 kg·m^{-3}).

3.6. Effect of Functional Group Position

When the location of the hydroxyl group in the alkane chain is varied, the access to reactive sites also changes. For propane-1,3-diol, there is less steric hindrance in the linear chain end-capped with hydroxyl groups compared to the vicinal hydroxyl groups in propane-1,2-diol, which may encounter hindered reactivity following the dehydrocondensation reaction with the first available reactive site [41]. For glycerol (propane-1,2,3-triol), the steric hindrance at carbons C_1 and C_3 is lower than that at the second carbon, C_2. Depending on the polymer chain entanglement and vicinity of the PMHS chain and catalyst, the -OH attached to C_2 is also prone to react, resulting in an additional bond that may cause the rigidity of the crosslinked material, thus altering the elasticity of the foam (lower elongation at break and lower tensile strength) (Figure 7).

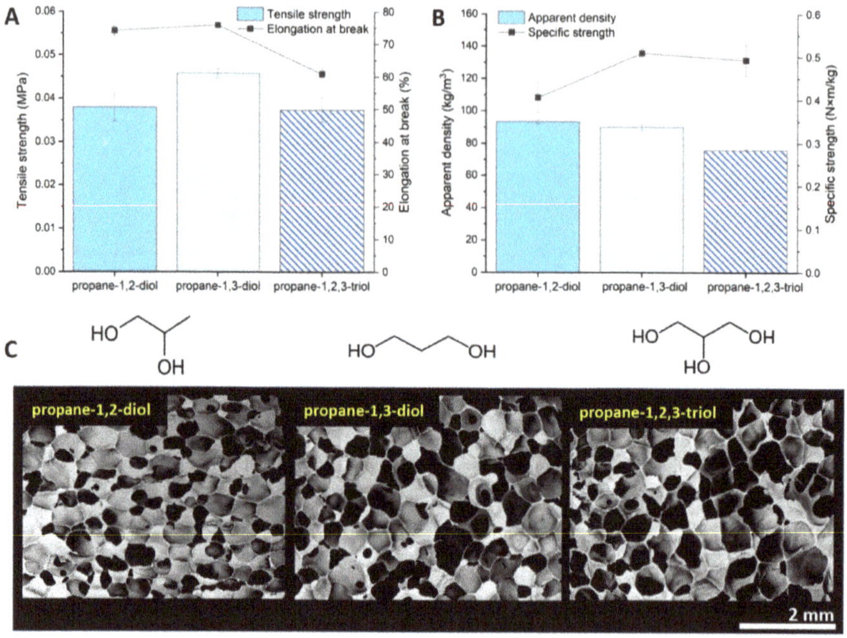

Figure 7. (**A**) Potentially reactive hydroxyl groups in propanols with distinct steric hindrance (accessibility of the reaction site in the respective molecule) and their effect on the tensile strength and elongation at break (**A**), specific strength (**B**), and apparent density (**C**). The importance of the OH groups in di- or triols lies in their ability to react with the PMHS chain and form a crosslink, adding or decreasing the flexibility of the crosslinked molecular structure.

Nevertheless, foams blown and simultaneously crosslinked with 10 mol% glycerol in water have an apparent density of 76 $kg \cdot m^{-3}$, which is approximately 30% lower than that of the water-only blown foam. Perhaps the lower density is unsurprising since there are initially three hydroxyl groups in each glycerol molecule that form crosslinks with the Si-H bond. With each crosslink formation, the viscosity of the reaction mixture increased, thus preventing the evolution of gas from escaping. In addition, Nakagawa et al. reported that the vicinal OH groups in triols, compared to monoalcohols attached to polymer chains, show significant retardation in chain dynamics [30]. It is important to stress that the R(-OH)/PM(-H)S ratio remained constant for all experiments, so the resulting moles of hydrogen would remain constant.

4. Conclusions

This study used water and its blends with methanol, glycerol, and dialcohols as additional blowing agents to expand fumed silica- and mica-reinforced polysiloxane elastomer networks during crosslinking. Owing to the dehydrogenative coupling mechanism, the water–hydroxyl blends offer additional crosslinking options for polysiloxane chains. Varying the molar ratios, structure of the diols, and alkane chain length allows for the physicomechanical properties and morphology of the foam to be easily tuned.

When water was partially substituted with a diol (10 mol%), the apparent density and tensile strength both gradually decreased from C_2 to C_4. From C_4 and above, the odd–even effect for alkane chains seems to describe the correlation between the diol chain length and the alternating values of the apparent density and cell wall thickness. Compared to the 100% water-blown foam, replacing 10 mol% with butane-1,4-diol resulted in the lowest density among all the diols, whereas the specific strength values remained comparable to that of the water-only blown foam. Further increasing the diol content caused the apparent density and tensile strength to increase. The dominant effect on the density increase is the increase in cell wall thickness near and above 50 mol% butane-1,4-diol solutions.

Compared to triols, diols allow for the fabrication of foams with higher elongation at break values, and dispensing and molding are more effective owing to the reasonable viscosity of the curing and expanding mixture. A similar conclusion can be drawn for the water-only blown foams. We find this aspect to be especially important for industrial production, where small cavities need to be filled. For triols, higher ratios may also result in decreased elongation at break values because of the increased number of relatively short and closely positioned crosslinks.

Methanol, as a monoalcohol, was introduced to the blowing blend to decrease the crosslinking degree while maintaining the amount of evolving gas. The resulting changes in the foam structure demonstrate the tunability of the foam's physicomechanical properties, including bypassing the memory effect in water-only blown foams, and they can be implemented in various cushioning applications, in which the density increase arising from the wall thickness and cell size increase is not an issue.

Chemical blowing in the form of inherent gas-generating reactions yields the most consistent, controllable, and repeatable results regarding the apparent density, pore structure, and mechanical properties. Finding a balanced formula in this multireaction system for a desired application is an ongoing process; however, our results show that the synthesis of moldable low-density elastomeric silicone foams with desired properties is achievable.

Supplementary Materials: The following supporting information can be downloaded at: https://www.mdpi.com/article/10.3390/polym15224425/s1, Figure S1: IR spectrum of a water-blown foam; Figure S2: FTIR spectra of uncured polymer blend with water; Figure S3: IR spectrum of a PMHS polymer; Figure S4: IR spectra comparison of PMHS, uncured blend, and cured foam; Figure S5: IR spectra comparison of SIFs.

Author Contributions: Conceptualization, I.R., T.T. (Tarmo Tamm) and U.M.; methodology, I.R., T.T. (Tauri Tätte) and K.J.L.; validation, I.R., K.J.L., U.J. and T.T. (Tarmo Tamm); formal analysis, I.R.; investigation, I.R. and K.J.L.; resources, T.T. (Tauri Tätte) and P.P.; data curation, I.R. and K.J.L.; writing—original draft preparation, I.R.; writing—review and editing, I.R. and T.T. (Tarmo Tamm); visualization, I.R.; supervision, T.T. (Tarmo Tamm), U.M. and U.J.; project administration, T.T. (Tarmo Tamm); funding acquisition, T.T. (Tarmo Tamm). All authors have read and agreed to the published version of the manuscript.

Funding: This research was part of a development project supported by SA Archimedes, grant no. 2014-2020.4.02.19-0155, 8 May 2019–17 July 2023. SA Archimedes funded APC.

Institutional Review Board Statement: Not applicable.

Data Availability Statement: The data that support the findings of this study are available in the Supplementary Materials of this article and from the corresponding author upon reasonable request.

Acknowledgments: I acknowledge my co-workers' and institutional partners' long-standing financial support and hard work. I would like to express my sincere appreciation to them for their efforts.

Conflicts of Interest: The authors declare no conflict of interest. The funders had no role in the study design; collection, analyses, or interpretation of the data; writing of the manuscript; or decision to publish the results. Author P. Piirimägi was employed by Estelaxe, Ltd. The remaining authors declare that the research was conducted in the absence of any commercial or financial relationships that could be construed as potential conflicts of interest.

Appendix A

Table A1. Comparative data on the physicomechanical properties of silicone-based foams produced by applying additional blowing agents or techniques.

Polymer System	Blowing Agent	Density, ($kg \cdot m^{-3}$)	Elongation at Break (%), E_b	Tensile Strength, (kPa)	Average Cell Size, (mm)	Average Wall Thickness, (mm)	Comments	Curing	Reference
				Water					
Vi-PDMS/OH-PDMS/PMHS	water	220	-	-	0.200–0.720	0.030–0.060	surfactant for W/O emulsion	RT (several minutes)	[26]
Vi-PDMS/OH-PDMS/PMHS	water (0.2–1.8 wt%)	214	-	-	0.510	-	Inhibitor used	15 min at RT + 2 h at 100 °C	[28]
Vi-PDMS/OH-PDMS/PMHS	water (1 wt%)	105	101	52.5	0.6, SD = 0.2	0.011, SD = 0.008	-	3 min at RT	This work
PDMS/CFPS/D-66	water (0.07 mL, 1.33%)	250	42	35	>1	-	isocyanate + water reaction	RT (3–5+ min)	[27]
PDMS/CFPS/D-17	water (0.07 mL, 1.33%)	240	-	-	>1	-	isocyanate + water reaction	RT (3–5+ min)	[27]
PDMS/PMDI	water (0.07 mL, 1.12%), CO_2	270	-	-	>1, 0.4–1.0	-	isocyanate + water reaction	RT (3–5+ min)	[27]
Elastosil LR 3003/50	water (1–3 phr)	–52%	-	-	0.005 (mm^2)	n/a	water mixed with silica (8:1)	140 °C/180 °C + 4 h/200 °C	[3]
				Alcohols					
Vi-PDMS/H-PDMS/PMHS	ethanol (0 wt%)	450	73	325.6	0.55	0.95		3 h at 80 °C	[29]
Vi-PDMS/H-PDMS/PMHS	ethanol (1.5 wt%)	200	31	52.8	1.6	1.4		3 h at 80 °C	[29]
Vi-PDMS/OH-PDMS/PMHS	1:9 methanol/water (1 wt%)	111	84	41.8	0.9, SD = 0.4	0.021, SD = 0.010		3 min at RT	This work
Vi-PDMS/OH-PDMS/PMHS	1:9 butane-1,4-diol/water (1 wt%)	78	86	40.8	0.6, SD = 0.2	0.009, SD = 0.007		3 min at RT	This work
				Other					
OH-PDMS/PMHS (Rhodorsil RTFoam 3240)	no additional blowing agent, H_2 from reaction	197	-	-	0.486 +/− 0.006	-		RT	[4]
VMQ/DBPMH/MHS	$scCO_2$	109–548	-	-	0.073–0.291	0.019	compounded rubber sheets + $scCO_2$	50 °C to 80 °C	[42]
				Combinative					
Sylgard 184	glycerol (137 phr)/ethanol (1 wt%)	282	-	-	1.37 [a]	-		120 °C	[32]
Sylgard 184	glycerol (110 phr)/ethanol (1.1 wt%)	300	-	-	1.63 [a]	-		120 °C	[32]
Vi-PDMS/OH-PDMS/PMHS	10 mol% glycerol/water (1 wt%)	76	61	37.3	0.632, SD = 0.141	0.016, SD = 0.010		3 min at RT	This work

[a] Values derived from foam cell size (mm^2) data for comparison (Mazurek et al 2019).

Table A2. Average cell diameter, cell wall thickness, and apparent density of foams with different mol% of diols in water with varying alkane chain lengths and structures (i.e., propane-1,2-diol). SD is the standard deviation of the mean of 100 measurements.

R-(OH)$_n$	Diol, (mol%)	Water, (mol%)	Average Wall Thickness, (mm)	Average Cell Size, (mm)	Apparent Density, (kg·m^{-3}) (U = 3 kg·m^{-3})
Water	0	100	0.011, SD = 0.008	0.6, SD = 0.2	105
Ethane-1,2-diol	10	90	0.016, SD = 0.012	0.6, SD = 0.2	91
Propane-1,2-diol	10	90	0.013, SD = 0.008	0.6, SD = 0.2	94
Propane-1,3-diol	10	90	0.015, SD = 0.008	0.6, SD = 0.2	90
Butane-1,4-diol	10	90	0.009, SD = 0.007	0.6, SD = 0.2	78
Pentane-1,5-diol	10	90	0.019, SD = 0.015	0.5, SD = 0.2	132
Hexane-1,6-diol	10	90	0.013, SD = 0.007	0.4, SD = 0.2	100
Heptane-1,7-diol	10	90	0.017, SD = 0.009	0.4, SD = 0.1	151

Table A3. Structural parameters for foams with increasing methanol content; SD is the standard deviation of 100 measurements.

R-(OH)n	Alkanol, (mol%)	Water, (mol%)	Average Wall Thickness, (mm)	Average Cell Size, (mm)	Apparent Density, (kg·m−3) (U = 3 kg·m−3)
Methanol	0	100	0.011, SD = 0.008	0.6, SD = 0.2	105
	10	90	0.021, SD = 0.010	0.9, SD = 0.4	111
	15	85	0.034, SD = 0.020	1.1, SD = 0.5	115
	25	75	0.056, SD = 0.056	2.3, SD = 0.5	115

References

1. Hamdani, S.; Longuet, C.; Perrin, D.; Lopez-cuesta, J.-M.; Ganachaud, F. Flame retardancy of silicone-based materials. *Polym. Degrad. Stab.* **2009**, *94*, 465–495. [CrossRef]
2. Kwak, Y.; Kang, Y.; Park, W.; Jo, E.; Kim, J. Fabrication of fine-pored polydimethylsiloxane using an isopropyl alcohol and water mixture for adjustable mechanical, optical, and thermal properties. *RSC Adv.* **2021**, *11*, 18061–18067. [CrossRef]
3. Marl, S.; Rüppel, A.; Hartung, M.; Klier, K.; Giesen, R.-U.; Heim, H. P. Liquid Silicone Rubber Foams Made with Water as Blowing Agent. *Adv. Eng. Mater.* **2022**, *24*, 2100382. [CrossRef]
4. Verdejo, R.; Saiz-Arroyo, C.; Carretero-Gonzalez, J.; Barroso-Bujans, F.; Rodriguez-Perez, M.A.; Lopez-Manchado, M.A. Physical properties of silicone foams filled with carbon nanotubes and functionalized graphene sheets. *Eur. Polym. J.* **2008**, *44*, 2790–2797. [CrossRef]
5. Chen, B.; Qian, Z.; Song, G.; Zhuang, Z.; Sun, X.; Ma, S.; Liang, Y.; Ren, L.; Ren, L. Gas-permeable and stretchable on-skin electronics based on a gradient porous elastomer and self-assembled silver nanowires. *Chem. Eng. J.* **2023**, *463*, 142350. [CrossRef]
6. Liu, J.; Yao, Y.; Li, X.; Zhang, Z. Fabrication of advanced polydimethylsiloxane-based functional materials: Bulk modifications and surface functionalizations. *Chem. Eng. J.* **2021**, *408*, 127262. [CrossRef]
7. Timusk, M.; Nigol, I.A.; Vlassov, S.; Oras, S.; Kangur, T.; Linarts, A.; Šutka, A. Low-density PDMS foams by controlled destabilization of thixotropic emulsions. *J. Colloid Interface Sci.* **2022**, *626*, 265–275. [CrossRef]
8. Mazurek, P.; Vudayagiri, S.; Skov, A.L. How to tailor flexible silicone elastomers with mechanical integrity: A tutorial review. *Chem. Soc. Rev.* **2019**, *48*, 1448–1464. [CrossRef] [PubMed]
9. Adam, M.; Bäumer, M.; Schowalter, M.; Birkenstock, J.; Wilhelm, M.; Grathwohl, G. Generation of Pt- and Pt/Zn-containing ceramers and their structuring as macro/microporous foams. *Chem. Eng. J.* **2014**, *247*, 205–215. [CrossRef]
10. Obi, B.E. Foaming Processes. In *Polymeric Foams Structure-Property-Performance*; Elsevier: Amsterdam, The Netherlands, 2018; pp. 131–188. [CrossRef]
11. Zhang, C.; Qu, L.; Wang, Y.; Xu, T.; Zhang, C. Thermal insulation and stability of polysiloxane foams containing hydroxyl-terminated polydimethylsiloxanes. *RSC Adv.* **2018**, *8*, 9901–9909. [CrossRef]
12. Dezaki, M.L.; Bodaghi, M. Soft Magneto-Responsive Shape Memory Foam Composite Actuators. *Macro Mater. Amp. Eng.* **2022**, *307*, 2200490. [CrossRef]
13. Yang, J.; Liao, X.; Wang, G.; Chen, J.; Guo, F.; Tang, W.; Wang, W.; Yan, Z.; Li, G. Gradient structure design of lightweight and flexible silicone rubber nanocomposite foam for efficient electromagnetic interference shielding. *Chem. Eng. J.* **2020**, *390*, 124589. [CrossRef]

14. Ghahramani, P.; Moradi-Dastjerdi, R.; Behdinan, K.; Naguib, H.E. Mechanical characterization of multifunctional highly porous carbon nanotube-reinforced foams. *Polym. Compos.* **2023**, *44*, 2093–2101. [CrossRef]
15. Gonzalez, J.; Iglio, R.; Barillaro, G.; Duce, C.; Tiné, M.R. Structural and Thermoanalytical Characterization of 3D Porous PDMS Foam Materials: The Effect of Impurities Derived from a Sugar Templating Process. *Polymers* **2018**, *10*, 616. [CrossRef] [PubMed]
16. Hwang, J.; Kim, Y.; Yang, H.; Oh, J.H. Fabrication of hierarchically porous structured PDMS composites and their application as a flexible capacitive pressure sensor. *Compos. Part B Eng.* **2021**, *211*, 108607. [CrossRef]
17. Wang, X.; Tang, Y.; Cheng, S.; Liu, C.; Gao, Q.; Lian, W.; Yuan, Y.; Li, A.; Li, C.; Guan, S. Polydimethylsiloxane Composite Sponge Decorated with Graphene/Carbon Nanotube via Polydopamine for Multifunctional Applications. *ACS Appl. Polym. Mater.* **2023**, *5*, 6022–6033. [CrossRef]
18. Abshirini, M.; Saha, M.C.; Cummings, L.; Robison, T. Synthesis and characterization of porous polydimethylsiloxane structures with adjustable porosity and pore morphology using emulsion templating technique. *Polym. Eng. Sci.* **2021**, *61*, 1943–1955. [CrossRef]
19. Yan, J.; Cao, J.; Xue, L.; Feng, S.; Zhang, H.; Wang, D. Thiol Oxidative Coupling Synthesis of Silicone Foams for Oil/Water Separation. *ACS Appl. Polym. Mater.* **2020**, *2*, 1634–1643. [CrossRef]
20. Grande, J.B.; Fawcett, A.S.; McLaughlin, A.J.; Gonzaga, F.; Bender, T.P.; Brook, M.A. Anhydrous formation of foamed silicone elastomers using the Piers–Rubinsztajn reaction. *Polymer* **2012**, *53*, 3135–3142. [CrossRef]
21. Mabe, A.; Perez, L.P.; Wu, A.; Wilson, T. *Effects of Varying Composition and Kinetics on the Microstructural and Mechanical Properties of Polysiloxane Foams*; LLNL-TR-748401; Lawrence Livermore National Laboratory (LLNL): Livermore, CA, USA, 2018. [CrossRef]
22. Coste, G.; Negrell, C.; Caillol, S. From gas release to foam synthesis, the second breath of blowing agents. *Eur. Polym. J.* **2020**, *140*, 110029. [CrossRef]
23. Brook, M.A. New Control over Silicone Synthesis using SiH Chemistry: The Piers-Rubinsztajn Reaction. *Chem. Eur. J.* **2018**, *24*, 8458–8469. [CrossRef] [PubMed]
24. Modic, F.J.; Striker, R.A. Low Viscosity Silicone Foam Compositions. EP0355380A2, 28 February 1990.
25. Riesco, R.; Boyer, L.; Blosse, S.; Lefebvre, P.M.; Assemat, P.; Leichle, T.; Accardo, A.; Malaquin, L. Water-in-PDMS Emulsion Templating of Highly Interconnected Porous Architectures for 3D Cell Culture. *ACS Appl. Mater. Interfaces* **2019**, *11*, 28631–28640. [CrossRef]
26. Guo, B.F.; Wang, P.H.; Cao, C.F.; Qu, Z.H.; Lv, L.Y.; Zhang, G.D.; Gong, L.-X.; Song, P.; Gao, J.-F.; Mai, Y.W.; et al. Restricted assembly of ultralow loading of graphene oxide for lightweight, mechanically flexible and flame retardant polydimethylsiloxane foam composites. *Compos. Part B* **2022**, *247*, 110290. [CrossRef]
27. Chruściel, J.J.; Leśniak, E. Preparation of flexible, self-extinguishing silicone foams. *J. Appl. Polym. Sci.* **2011**, *119*, 1696–1703. [CrossRef]
28. Cao, C.F.; Wang, P.H.; Zhang, J.W.; Guo, K.Y.; Li, Y.; Xia, Q.Q.; Zhang, G.D.; Zhao, L.; Chen, H.; Wang, L.; et al. One-step and green synthesis of lightweight, mechanically flexible and flame-retardant polydimethylsiloxane foam nanocomposites via surface-assembling ultralow content of graphene derivative. *Chem. Eng. J.* **2020**, *393*, 124724. [CrossRef]
29. Tan, Y.; Yao, J.; Zhu, H.-P. Effects of ethanol content on the properties of silicone rubber foam. *J. Polym. Eng.* **2020**, *40*, 543–550. [CrossRef]
30. Nakagawa, S.; Xia, J.; Yoshie, N. Quantifying the effects of cooperative hydrogen bonds between vicinal diols on polymer dynamics. *Soft Matter* **2022**, *18*, 1275–1286. [CrossRef]
31. Mazurek, P.; Hvilsted, S.; Skov, A.L. Green silicone elastomer obtained from a counterintuitively stable mixture of glycerol and PDMS. *Polymer* **2016**, *87*, 1–7. [CrossRef]
32. Mazurek, P.; Ekbrant, B.E.F.; Madsen, F.B.; Yu, L.; Skov, A.L. Glycerol-silicone foams—Tunable 3-phase elastomeric porous materials. *Eur. Polym. J.* **2019**, *113*, 107–114. [CrossRef]
33. *ISO 1798:2008*; Flexible Cellular Polymeric Materials—Determination of Tensile Strength and Elongation at Break. ISO: Geneva, Switzerland, 2008.
34. *ASTM D 3574-17*; Standard Test Methods for Flexible Cellular Materials—Slab, Bonded, and Molded Urethane Foams. ASTM: West Conshohocken, PA, USA, 2017. [CrossRef]
35. Schindelin, J.; Arganda-Carreras, I.; Frise, E.; Kaynig, V.; Longair, M.; Pietzsch, T.; Preibisch, S.; Rueden, C.; Saalfeld, S.; Schmid, B.; et al. Fiji: An open-source platform for biological-image analysis. *Nat. Methods* **2012**, *9*, 676–682. [CrossRef]
36. Liao, M.; Schneider, A.F.; Laengert, S.E.; Gale, C.B.; Chen, Y.; Brook, M.A. Living synthesis of silicone polymers controlled by humidity. *Eur. Polym. J.* **2018**, *107*, 287–293. [CrossRef]
37. Yang, K.; Cai, Z.; Jaiswal, A.; Tyagi, M.; Moore, J.S.; Zhang, Y. Dynamic Odd-Even Effect in Liquid *n*-Alkanes near Their Melting Points. *Angew. Chem. Int. Ed.* **2016**, *55*, 14090–14095. [CrossRef] [PubMed]
38. Jawhar, M.-C.D.; Blanc, D.; Chaumont, P.; Cassagnau, P. Study of the Coalescence Mechanisms during Silicone Foaming: Study of the Coalescence Mechanisms During Silicone Foaming. *Macromol. Mater. Eng.* **2014**, *299*, 336–343. [CrossRef]
39. Pang, Y.; Cao, Y.; Zheng, W.; Park, C.B. A comprehensive review of cell structure variation and general rules for polymer microcellular foams. *Chem. Eng. J.* **2022**, *430*, 132662. [CrossRef]
40. Rebane, I.; Mäeorg, U.; Johanson, U.; Ilisson, M.; Piirimägi, P.; Tamm, T. Kinetics of catalyzed dehydrocondensation of hydrogen functionalized siloxane. *J. Appl. Polym. Sci* **2022**, *139*, 52304. [CrossRef]

41. Sato, T.; Buchner, R. The cooperative dynamics of the H-bond system in 2-propanol/water mixtures: Steric hindrance effects of nonpolar head group. *J. Chem. Phys.* **2003**, *119*, 10789–10800. [CrossRef]
42. Song, P.; Zhang, Y.; Luo, Y.; Liao, X.; Tang, W.; Yang, J.; Tian, C.; Li, G. Design of lightweight silicone rubber foam for outstanding deformation recoverability based on supercritical CO_2 foaming technology. *J. Mater. Sci.* **2022**, *57*, 2292–2304. [CrossRef]

MDPI AG
Grosspeteranlage 5
4052 Basel
Switzerland
Tel.: +41 61 683 77 34

Polymers Editorial Office
E-mail: polymers@mdpi.com
www.mdpi.com/journal/polymers

www.ingramcontent.com/pod-product-compliance
Lightning Source LLC
Chambersburg PA
CBHW061506220926
51850CB00004B/180
9783725853274